The Study of Trace Fossils

ERRATA SHEET FOR

The Study of Trace Fossils

(pp. 16, 110–111, 114–115, 472, and 516)

mental regimes. These assemblages are re-current through time, whenever the requisite environmental conditions are repeated.

In marine environments, many of the factors that control the distribution of trace-making organisms—such things as temperature, food supply, and the intensity of wave or current agitation—tend to change progressively with water depth. Therefore, Seilacher's assemblage concept may also be

TABLE 2.1 Recurring Aquatic Trace Fossil Assemblages and their Environmental Implications.*

Name of Assemblage	Characteristic Lebensspuren [see Häntzschel (1962) for examples of genera]	Typical Benthonic Environment (generally related directly or indirectly to bathymetry)†
Scoyenia	Vertebrate tracks, trails, and burrows, mainly of aquatic or semi-aquatic species but also of terrestrial species coming to water; abundant insect and other arthropod traces; certain forms of Planolites; scattered snail and clam trails and shallow burrows. Local diversity and abundance generally less than in marine environments.	Nonmarine clastics, especially continental "red beds," floodplain deposits, etc., as the assemblage is presently defined. [But this "nonmarine" concept actually involves a diversity and zonation of habitats and traces (Chapter 19) that parallels the marine categories outlined below; the concept thus needs additional study and redefinition.]
Glossifungites	Vertical cylindrical, U-shaped, or sparsely ramified dwelling burrows; protrusive spreiten in some, developed mainly through growth of animals. Many species leave the burrows to feed (e.g., crabs); others are mostly suspension feeders (e.g., polychaetes, pholads); Diversity low, although burrows of given types may be abundant.	Marine littoral zone and sublittoral omission surfaces; stable, coherent substrates either in protected, low-energy settings (e.g., salt marshes; muddy quiet-water bars, flats, and shoals) or in areas of slightly higher energy where semiconsolidated substrates offer resistance to erosion (e.g., Frey and Howard, 1969, Pl. 4, fig. 3). [On rocky coasts and hardgrounds, Glossifungites and Skolithos assemblages are replaced by borers and scrapers, and bioerosion itself becomes a major feature of the environment (Chapters 11 and 18).]
Skolithos	Vertical cylindrical or U-shaped dwelling burrows; protrusive and retrusive spreiten, developed mostly in response to substrate aggradation or degradation (e.g., Diplocraterion, Corophioides); forms of Ophiomorpha consisting predominantly of vertical or steeply inclined burrow components. Animals mainly suspension feeders. Diversity low, although given kinds of burrows may be abundant.	Littoral and very shallow sublittoral zones; relatively high energy conditions; well-sorted, shifting sediments; abrupt erosion or deposition (e.g., beaches, inlet bars and shoals, tidal deltas). [Higher energy increases physical reworking and obliterates biogenic sedimentary structures, leaving preserved record of physical stratification (Chapter 8); for this reason, the very shallow sublittoral (shoreface) is typically barren of traces.]

the radiating worm burrow *Oldhamia* as a Cambrian index fossil (Kinahan, 1878). It was used to date the otherwise almost barren "Bray Series" of eastern Ireland. Recent investigations confirmed the validity of this conclusion for the *Oldhamia*-bearing localities (Crimes and Crossley, 1968; Rast and Crimes, 1969). Similarly, Mägdefrau (1934) recognized that the complex feeding burrow *Phycodes circinatum* is restricted to the Lower Ordovician and, by virtue of its widespread occurrence, could be a useful guide fossil. Little further work was attempted on the use of trace fossils in stratigraphy until Seilacher (1960, 1970) and Crimes (1968, 1969, 1970c) showed that

these fossils could date otherwise unfossiliferous lower Paleozoic successions.

To produce traces of stratigraphical value, an animal should ideally (1) be widely distributed, (2) leave a record of its activities on or within the sediment, (3) belong to a rapidly evolving group, and (4) be facies independent. Few benthic animals are truly facies independent; indeed, the facies control shown by trace fossils is one of their most characteristic features.

In Paleozoic strata, trace fossils occur most abundantly in shallow-water sandstone–shale sequences, many of which are devoid of body fossils. In Mesozoic and

TABLE 7.1   Known Ranges of Selected Trace Fossils.*

	Precambrian	Cambrian	Ordovician	Silurian	Devonian	Carboniferous	Permian	Triassic	Jurassic	Cretaceous	Paleocene	Eocene	Oligocene
Acanthorhaphe										▬	▬	▬	
Asterosoma		▬	▬	▬	▬	▬	▬	▬	▬	▬			
Astropolithon		▪											
Atollites										▬	▬	▬	
Belorhaphe										▬	▬	▬	▬
Crossopodia			▬	▬	▬	▬	▬	▬	▬	▬			
Cruziana		▬	▬	▬	▬	▪ ▪							
Desmograpton										▬	▬	▬	▬
Dictyodora		▬	▬	▬	▬								
Dimorphichnus		▬											
Diplichnites		▬	▬	▬	▬	▬							
Favreina										▬	▬	▬	▬
Glockeria										▬	▬	▬	▬
Helicolithus										▬	▬	▬	
Helicorhaphe										▬	▬	▬	▬

Tertiary rocks the greatest variety occurs in otherwise relatively fossil-poor, deep-water, distal turbidites. The common occurrence of trace fossils within otherwise unfossiliferous successions means that they have great potential for broad stratigraphical correlations. They might provide the exploration geologist with the first reliable indication of the age of strata. Many trace fossil genera seem to be restricted to one or several systems (Crimes, 1974), so an ichnofauna can often be assigned readily to the correct system on generic identification alone (Table 7.1). Where short-ranging genera are involved, assigning rocks to part of a system may be possible. For example, the occurrence of *Astropolithon*, *Plagiogmus*, and *Syringomorpha* may be taken to indicate Lower Cambrian age. Precise correlations, however, usually require identification to species level. Fortunately, the number of stratigraphically useful species so far described is sufficiently limited that correct identification may often be made in the field by a mapping geologist.

To date, the most detailed work has been attempted on trilobite traces from lower Paleozoic successions. Many trilobites were short ranging and also active, benthic

TABLE 7.1—Continued.

	Precambrian	Cambrian	Ordovician	Silurian	Devonian	Carboniferous	Permian	Triassic	Jurassic	Cretaceous	Paleocene	Eocene	Oligocene
Helminthoida (s.l.)			▬	▬	▬					▬	▬	▬	▬
Megagrapton										▬	▬	▬	▬
Oldhamia		▬	▬										
Ophiomorpha						▬	▬	▬	▬	▬	▬	▬	▬
Plagiogmus		▬											
Rusophycus		▬	▬	▬	▬	▬							
Spirophycus			▬	▬	▬	▬	▬	▬	▬	▬			
Spirorhaphe										▬	▬	▬	▬
Spongeliomorpha							▬	▬	▬	▬	▬	▬	▬
Squamodictyon						▬	▬	▬	▬	▬	▬	▬	▬
Subphyllochorda										▬	▬	▬	
Syringomorpha		▬											
Thalassinoides						▬	▬	▬	▬	▬	▬	▬	▬
Tomaculum				▬									
Urohelminthoida										▬	▬	▬	▬

* Compiled from various sources.

Vertical burrows were claimed in the Buckingham Sandstone and Areyonga Formation in Australia, both reputed to be about 800 million years old (Glaessner, 1969; Ranford et al., 1965). These scattered occurrences indicate that, although more traces of such antiquity will undoubtedly be found, they must be regarded as rare in rocks more than 750 million years old (see also Banks, 1970). More extensive collections were recorded by Webby (1970) from the Torrawangee Group in New South Wales (Australia) and by Glaessner (1969) from the well-known Ediacara Beds in South Australia. Both units are suggested to be about 600 million years old.

Banks (1970) reported that within a continuous succession spanning the Precambrian–Cambrian boundary in Finnmark (Norway), the abundance and diversity of traces increases abruptly in all facies in latest Precambrian and Early Cambrian times (see also Chapter 6). The earliest trilobite traces appear not far above the late Precambrian tillite but before the first preserved trilobites in both Finnmark (Banks, 1970) and Greenland (Cowie and Spencer, 1970). In Sweden, trilobite resting traces (*Rusophycus*) occur in the Hardeberga Sandstone, which is separated by the overlying *Diplocraterion* Sandstone from beds containing the oldest recorded trilo-

TABLE 7.2 Known Ranges of *Cruziana* Species.

	Cambrian			Tremadoc	Ordovician					Silurian		Devonian		
	Lower	Middle	Upper	Tremadoc	Arenig	Llanvirn	Llandeilo	Caradoc	Ashgill	Lower	Upper	Lower	Middle	Upper
C. cantabrica	■													
C. fasciculata	■													
C. dispar	■													
C. carinata	■													
C. barbata		■												
C. arizonensis		■												
C. semiplicata			■											
C. polonica			■											
C. jenningsi	■	■	■											
C. omanica				■	■	■	■	■	■					
C. imbricata					■									
C. furcifera				■	■									
C. rugosa					■	■								
C. goldfussi					■									
C. grenvillensis								■						

bites (Bergström, 1970). Near the Barrios de Luna in the Cantabrian Mountains, north Spain, I also found *Rusophycus* and *Cruziana* within the Herreria Formation, immediately above the Precambrian–Cambrian unconformity and about 1,000 m beneath the earliest preserved trilobites. According to Lotze (1961), the trilobites are Early Cambrian and the oldest in the Iberian Peninsula. In Australia, trilobite traces, including *Diplichnites* and *Cruziana*, occur below the first preserved trilobites in the Arumbera Formation (Glaessner, 1969), and Young (1972) also recorded trilobite traces below the earliest recorded trilobites in a late Precambrian–Cambrian succession in Canada. The persistent occurrence, at widespread localities, of trilobite traces stratigraphically below the earliest recorded trilobites is unlikely to be purely a matter of chance.

The sudden appearance of an animal as complex as a trilobite at the beginning of Cambrian times has been considered something of an enigma. A popular idea is that similar creatures existed for a long time during the Precambrian but did not develop hard skeletons until Early Cambrian times. Soft-bodied trilobites would be expected to leave characteristic traces but no body fossils. The common occurrence of trilobite traces below the first recorded tribolite body fossils might therefore tend to confirm the existence of an early, soft-bodied form.

One objection to the development of a trilobite from a soft-bodied form rests in their complex and powerful musculature.

TABLE 7.2—Continued.

| | Cambrian | | | Tremadoc | Ordovician | | | | | Silurian | | Devonian | | |
	Lower	Middle	Upper		Arenig	Llanvirn	Llandeilo	Caradoc	Ashgill	Lower	Upper	Lower	Middle	Upper
C. petraea								■						
C. flammosa								■						
C. perucca								■						
C. lineata								■						
C. carleyi					■	■	■	■						
C. pudica								■	■					
C. ancora										■	■			
C. acacensis											■			
C. pedroana											■			
C. quadrata											■			
C. dilatata								■	■	■	■			
C. uniloba											■			
C. lobosa													■	
C. rhenana												■	■	■

TABLE 20.1. Zonation, Terminology, and Environmental Characteristics in a Generalized Beach-Offshore Profile.

Boundaries between environmental zones	Mean high-water line / Mean low-water line / Wave base				
General (littoral) classification	Supralittoral	Eulittoral	Upper sublittoral	Middle sublittoral	Lower sublittoral
Beach-offshore classification	Dunes and backshore	Foreshore	Shoreface	Upper offshore	Lower offshore
Energy sources and sedimentary processes	Eolian reworking; currents and wave action during spring tides and storms	Bidiurnal submersion; mainly wave action; periodically high reworking rate; low rate of bioturbation	Strong wave action; breakers, surf; high reworking rate; low rate of bioturbation	Low wave influence to sea bottom; sedimentation prevails over reworking; high rate of bioturbation	Exceptional storm wave influence to sea bottom; currents are main energy source; scarce or no sedimentation or reworking; considerable bioturbation
Sediments	Fine sand	Fine sand with medium sand	Fine sand with medium sand	Silty fine sand	Silt to clay, or coarse relict sediments

1953a) and Ziegelmeier (1952, 1969). Röder (1971) investigated the ecology, behavior, and burrow patterns of the polychaete *Paraonis fulgens*. Schäfer (1939) described burrows and burrowing behavior of talitrid amphipods in beaches. Sedimentological and ichnological research in the deeper part of the German Bight was initiated by Reineck et al. (1967).

The beach, shoreface, and upper offshore areas of Norderney (East Frisian Islands) were investigated in 1970 and 1971. Results of this study are not yet published, but a brief report is sketched as follows.

In the backshore area, no lebensspuren of marine animals have been found. The foreshore area is a broad (150 m) zone, reflecting the tidal range of 2.4 m. Several beach runnels and ridges occur. The main lebensspuren found within this area are

burrows of the polychaete *Scolecolepis squamata*. They were found, in slightly varying abundance, in nearly all can cores investigated. Röder (1971) suggested that *S. squamata* prefers the upper foreshore and the beach ridges, whereas the lower foreshore and the beach runnels are populated by the polychaete *Paraonis fulgens*. Burrows of *P. fulgens*, exhibiting characteristic spiral and meandering parts, are the second most important lebensspuren of the foreshore.

In the submerged area, the boundary between shoreface and offshore (*sensu stricto*) is not yet determined. Presumably, the tidal currents represent an energy level below wave base similar to that of waves in the lower part of the shoreface (H.–E. Reineck, 1973, oral communication). Thus, a similar rate of reworking occurs in both areas. Zoologically, the nearshore area is

occurring in two phases. The first phase involves foot probing and anchorage, followed by movement raising the shell to a vertical position after it has been lying on the substrate surface. This is followed by the second phase, in which cycles of digging movements lead to burial (Fig. 22.1). Each cycle consists of six stages: (1) foot probing, accompanied by slight lifting of the shell, which acts as a penetration anchor; (2) siphon closure and continued probing; (3) adduction, producing increased internal pressure that leads to dilation and terminal anchoring of the foot, and water ejection from the mantle cavity; (4) pedal retractor contraction, causing the shell to be pulled into the sand, sometimes accompanied by a rocking motion caused by sequential contraction of the anterior and posterior retractors; (5) adductor relaxation, followed by valve gaping and loss of pedal anchorage; and (6) a static period, during which foot probing recommences, while the shell gapes to act as an anchor.

These two phases of burrowing are

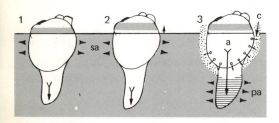

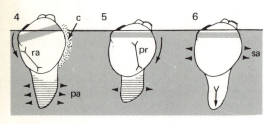

Fig. 22.1 Positions of a bivalve during six stages in the burrowing cycle. Emphasized are the terminal pedal (pa) and shell (sa) anchors, adduction (a), water ejection (o→), sand loosened around the shell (c), foot probing (>→) and contraction of the anterior (ra) and posterior (pr) retractor muscles. (Modified from Trueman, 1968b.)

generally characteristic of all bivalve mollusks. However, some notable exceptions are known in terms of the details of digging cycles. In *Lyonsia norvegica* (Ansell, 1967), movement is slow because stage 3, where terminal pedal anchorage and water ejection normally occur, is poorly developed; in fact, no water is ejected through the pedal gape. In *Glycymeris glycymeris* (Ansell and Trueman, 1967a), the mantle cavity extends so far dorsally that the visceral mass is not in contact with the valves. For this reason, adduction only indirectly effects the hemocoele, and transverse muscles (*G. glycymeris* has more of these than most other bivalves) assume the role of maintaining pressure in the foot during digging. Swelling of the foot therefore occurs in stage 1 of the digging cycle, rather than stage 3, and anchorage is increased in stage 4 by spreading of the two pedal flaps.

Mercenaria mercenaria (Ansell and Trueman, 1967b) differs slightly from most bivalves in that siphonal closure, characteristic of stage 2, may precede final foot probing (stage 1). Also, later digging cycles exhibit a secondary phase of siphonal movement, between stages 5 and 6, in which the siphons close and withdraw for about five seconds; reopening of the siphons follows their slow reextension to the surface. Movements of the siphons are accompanied by similar movements of the foot. These secondary movements may increase pressure in the hemocoele and mantle cavity, to aid in relaxation of the adductors and in opening the valves. Such aid is necessary because of the presence in *Mercenaria* of a weak hinge ligament. A secondary opening movement is also found in *Margaritifera margaritifera* (Trueman, 1968a) and in *Petricola pholadiformis* (Ansell, 1970). In all other respects, however, the burrowing of the above bivalves conforms to the six-stage cycle of other bivalves.

As a result of the repetition of this burrowing cycle, bivalves move into the

The Study of
TRACE FOSSILS

A Synthesis of Principles, Problems
and Procedures in Ichnology

Edited by ROBERT W. FREY

S P R I N G E R – V E R L A G

Berlin   Heidelberg   New York   1975

Robert W. Frey
Department of Geology
University of Georgia
Athens, Georgia 30602

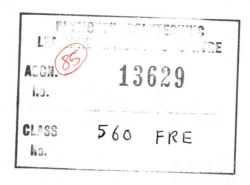
Library of Congress Cataloging in Publication Data

Frey, Robert W.
  The study of trace fossils.

  1. Trace fossils.   I. Title.
QE720.5.F73   560   74–30164

ISBN 3-540-06870-8 Springer-Verlag Berlin Heidelberg New York

ISBN 0-387-06870-8 Springer-Verlag New York Heidelberg Berlin

WALTER HÄNTZSCHEL (1904-1972) AND THE FOUNDATION OF MODERN INVERTEBRATE ICHNOLOGY

ADOLF SEILACHER

Geologisches-Paläontologisches Institut der Universität Tübingen
Tübingen, West Germany

Walter Häntzschel in 1969.
(*Photo courtesy of Marianne Häntzschel.*)

Walter Häntzschel, whose name is known to every ichnologist, died on May 10th, 1972, following an operation. In appreciation of his fundamental contributions, this book—which summarizes the present status of ichnology—is dedicated to Professor Häntzschel's memory.

Although a more detailed eulogy and a complete list of his publications have already been published (Lehmann, 1972), a few facts must be repeated here because they are essential to appreciate Walter Häntzschel's personality and scientific accomplishments. His life was molded by two world wars and their economic consequences: having grown up in Dresden as the only son of a schoolteacher and as a devoted amateur geologist, he entered the university of his home town at a time of inflation and unemployment. Becoming a scientist at that time would have meant starvation. Therefore, he studied toward an examination for high school teachers. But in order to satisfy his strong scientific interests, he devoted all of his free time as a volunteer at the local Zwinger Museum, curating paleontological treasures from the Cretaceous of Saxony. He even managed to complete a doctoral dissertation while already engaged in teaching at a local high school. Long before this time he had been in contact with Professor Rudolf Richter (Frankfurt), who by then had opened the new field of "actuopaleontology"—in which fossil phenomena are systematically compared with possible analogs from modern marine environments.

When Richter asked him to take over the new station ("Senckenberg am Meer") for marine geology and paleontology in Wilhelmshaven, Walter Häntzschel cheerfully accepted. From 1934 to 1938 he spent what was, perhaps, the happiest time of his life, exploring the Wadden Sea tidal flats for sedimentary structures, lebensspuren, and biostratinomic features that could possibly be applied to the interpretation of ancient sediments (Schäfer, 1964). The results were published in Senckenbergiana or Natur und Volk and were discussed with students and scientists who visited the station during the summer months in order to see geology in the making. At the same time, however, his duties included consulting for the local port authorities, to whom sedimentation meant not just a scientific problem but also a major threat for an important naval base.

During his Wilhelmshaven time, Walter Häntzschel was well aware that wet–sediment studies should occupy only a limited part of a geologist's life and that the personal experience must sooner or later be referred back to the outcrop. The opportunity came in 1938 with his appointment as a curator at the Dresden Museum, together with the hope for permanence for him and his newly founded family. But the Second World War soon destroyed this hope: he was drafted in 1942 and returned six years later, his health and hopes corroded by three years in a Russian prison camp. He found the city of Dresden and its museum destroyed, and the country divided by the iron curtain. He searched almost a year before he found a new place for the future: the Geological Institute of Hamburg University, where he stayed until his retirement in 1969.

In Hamburg, his major task was to rebuild the departmental library and collections, both of which were destroyed during the war, in addition to continuing his teaching and administration. Still he found time to publish a considerable number of papers, mostly about trace fossils and other sedimentary structures, and to critically review the complete trace fossil literature for Zentralblatt für Geologie und Paläontologie. But there was still another important service to the scientific community that does not appear in official records—Professor Häntzschel's personal messages that he sent out spontaneously whenever he came across some hidden paper that he thought would be of interest to a particular student or colleague, within or outside the country.

This life, without the glamor of an ambitious academic career or of economic success, was the background for the great contribution that will always be connected with Walter Häntzschel's name: the trace fossil volume of the Treatise on Invertebrate Paleontology (1962). To appreciate fully the significance of this book, a short review of the history of trace fossil research is necessary. (See also Chapter 1.)

The study of trace fossils had a first culmination in the last century, when most paleontologists were still convinced that they dealt with fossil seaweeds. Accordingly, they described, classified, and named trace fossils along with other fossils. An interesting thought is that Hall's concept of geosynclinal basins, in which sedimentation kept compensating the subsidence, was probably influenced by the "seaweed" theory.

When it gradually became clear that most "fucoids" are caused either by sedimentary processes or by burrowing and crawling creatures, the popularity of these "algal" fossils suddenly decreased. One reason for this indifference was that the sedimentary features could no more be considered as reliable guide fossils or photic-zone indicators. But the other reason was that they now fell into a taxonomic "no-man's-land." They were either omitted from textbooks, or listed under "incertae sedis" or "problematica"—a view that has left its mark even in the *Treatise* arrangement.

Only a few people continued their interest in this field. These were the detectives, fascinated by what they viewed not as novel collector's items but as the record of individual biologic events. Comparison with the tracks and burrows of recent organisms was the obvious starting point, and taxonomic identification of the producers was the primary goal. Modern tidal flats, particularly those of tropical seas, seemed to hold all the clues. But it soon became clear that ancient bedding planes received most of their biogenic markings after the beds were already covered by at least a veneer of subsequent sedimentation. This discovery meant that modern mud surfaces are not the best analogy and that perfectly preserved trace fossils are not an indication of the intertidal zone; they may have formed at any depth.

During this reorientation period, in which trace fossils emerged as a major tool for paleoecology and environmental sedimentology, workers were reluctant to prematurely immobilize this fluctuating field by superimposing a narrow grid of parataxonomy and nomenclature. Walter Häntzschel himself coined very few trace names. But at the same time he felt the need to systematically review all existing knowledge of trace fossil morphology. While others indulged in worldwide field studies and in the discovery of new applications, he chose

the tedious task of accumulating data from a literature scattered widely in history, countries, languages, and disciplines. Only a man of his diligence and modesty could fulfill this task. Tracing dubious names back to their original source and meaning usually requires much more patience and energy than coining new ones. Also required is a basic respect for other scientists' work, regardless of quality and interpretation.

Walter Häntzschel's *Treatise* volume, together with the more extensive bibliography in *Fossilium Catalogus* (1965), was a tremendous milestone in the history of trace fossil research because it provided the first comprehensive reference that was also free enough from interpretation to be accepted by the majority of specialists in the field. In a way, it provided the paradigm—in Kuhn's sense—on which all future research in the field can be based.

The new boom in trace fossil research would have been impossible without Walter Häntzschel's contribution. But with his endowment goes an obligation: in ichnology more than in any other field of paleontology, taxonomic decisions depend on the behavioral and preservational character of the material and on the particular author's interpretation. Also, errors in assigning lower categories cannot be smoothed out on a higher taxonomic level, because hardly any higher category is generally accepted. The alphabetic order of genera still remains best; every new trace fossil name claims equal status, and puts the full load of responsibility on its author.

Out of consideration for Walter Häntzschel, who devoted himself to clearing the nomenclatural jungle, we should not let the weeds grow again!

REFERENCES

Häntzschel, W. 1962. Trace fossils and problematica. *In* R. C. Moore (ed.), Treatise on invertebrate paleontology, Pt. W, Miscellanea. Geol. Soc. America and Univ. Kansas Press, p. W177–W245. (Until precluded by ili health shortly before his death, Prof.

Häntzschel continued his work on a forthcoming revision of the *Treatise* volume.)

————. 1965. Vestigia Invertebratorum et Problematica. Fossilium Catalogus I: Animalia, pars 108, 142 p.

Lehmann, U. 1972. Ein Nachruf. Mitt. Geol.-Paläont. Inst. Univ. Hamburg, Heft, 41: 15–128.

Schäfer, W. 1964. "Senckenberg am Meer" in Hinblick auf einen Geburtstag. Natur u. Museum, 94:444–446.

PROLOGUE

In a sense, the field of ichnology is both old and new. Its basic guiding principles were known to a few workers many years ago, and these principles are now being rediscovered by scores of current workers (paleontologists, stratigraphers, sedimentologists, paleoecologists, biologists, and others), who are adding their own bustle, momentum, and refinements to the subdiscipline. As is true in the development of any science, ichnologists have indeed gotten some occasional pebbles mixed in with their snowball; but they have also exposed many misconceptions and have made numerous positive gains.

Ichnology today is rapidly approaching that plateau at which the subdiscipline will settle comfortably into the ever-growing accumulation of "standard" but highly useful methods or procedures in geology. And that fact is perhaps the single most important message of this book: ichnology is not a new "magic wand," to render sister subdisciplines obsolete; but neither can it be glibly ignored by anyone seriously interested in ancient life or environmental reconstructions.

WHAT IS ICHNOLOGY?
HOW EFFECTIVE IS IT?

Numerous authors, myself included, are fond of introducing papers in ichnology with a statement to the effect that "trace fossils are valuable in paleoecology and facies analysis." No matter how firm our conviction, however, I suspect that we have often been too complacent in explaining it, or at times even in testing it.

Just how useful are trace fossils in paleoecology and environmental reconstructions, or in such "traditional" fields as paleon-

tology, stratigraphy, and sedimentology? More specifically, how do we undertake the study of trace fossils? What kinds of things do we look for? How do we know what we are looking at? What kinds of problems may we expect? Which things are or are not unique about trace fossils? What is a trace fossil?

In effect, this book represents our combined attempt to grapple with such questions. Of course, we did not always agree on the best answer to a given question. But at least we have tried to avoid "fond statements" and "smug answers." And this leads to what is perhaps the second most important message of the book: ichnology is not a mystical science that provides ready-made, unique answers on instant demand; research in this field requires the same thoroughness and devotion to detail necessary in any other discipline of geology, and involves the same kinds of subjective judgments and logical conclusions. In fact, ichnology represents a mingling of (and draws expertise from) numerous different disciplines — ethology, petrology, geochemistry, oceanography, etc., in addition to the fields already mentioned.

The foregoing also shows that trace fossils, like ripple marks or foraminifers, should not be studied outside of their overall geologic context. Therefore, the third most important message of the book is that we cannot reasonably study trace fossils without paying appropriate regard to other chemical, physical, and biological features contained in the same substrates.

You, the reader and user of this book, will ultimately decide just how effective we have been in demonstrating these three principles and in answering the above questions.

A NOTE ON ORGANIZATION
AND CONTENT OF THE BOOK

Broadly speaking, Parts I and II of the book are concerned with the conceptual and material "core" of ichnology: how the science developed historically, and the present "state of the art"; how trace fossils are formed, preserved, sampled, identified, classified, interpreted, and used in other fields of geology; and also how they grade into, or may be confused with, other phenomena. In practice, however, the main emphasis in these two parts is upon biogenic sedimentary structures made by marine invertebrates, the kinds of traces that traditionally have received the most attention from the most people. This traditional but disproportionate emphasis is compensated in the book by Part III, which stresses borings, plant and vertebrate traces, and other occurrences that merit equal study and utilization.

Part IV focuses upon the recent as a potential key to the past. Aquatic environments are stressed simply because most of the preserved record of ancient traces originated in such environments. Differences in the scope and content of these chapters stem largely from differences in the amount of specific information available from the respective environments.

The objectives of Part V overlap somewhat with those of Part IV, although the main emphasis in Part V is upon methodology and the kinds of results that can be obtained. Borings and plant and vertebrate traces are slighted among the discussions in Parts IV and V, but most of this information may be gleaned from appropriate chapters in Part III.

All in all, we have tried to cover the length, breadth, and depth of ichnology—at least in terms of the information currently available—and we hope that you will find this summary useful.

ROBERT W. FREY

PREFACE

In 1971 I published a review of ichnology (*Houston AAPG: SEPM Trace Fossil Field Trip Guidebook*) that I thought could be expanded rather easily into a worthwhile book on the subject. I probed that possibility for a while, thinking that I would write the book myself. As I began to outline the chapters in more detail, however, it soon became apparent that my personal knowledge of too many facets of ichnology scraped bottom all too soon. I quickly decided that a better book could be produced by soliciting specific contributions from other workers who, collectively, had first-hand experience with virtually every aspect of the field. That became the actual plan, the result of which is this book.

Now, looking over these contributions, I wonder why I ever thought that I *could* write such a book myself. In my humble opinion the contributors have done a commendable job, and I am deeply grateful to them. Significantly, the authors include biologists as well as geologists—a viable combination.

The original outline for the book was essentially my own, a copy sent to each contributor. But the individual authors responded twofold, expanding and refining their parts even more than I had dared hope for. The final product is truly "our" book and not "my" book.

Certain chapters do overlap slightly, as is apparent even from the table of contents; but the intended effect is to enhance continuity. Coherence through the book is especially desirable where the same basic topic is approached from two or more different viewpoints—as by paleontologists and sedimentologists, or by one worker concentrating on diverse traces found in a particular environmental setting and an-

other concentrating only on traces made by a certain group of organisms, regardless of their setting. Nevertheless, needless redundancy has hopefully been eliminated.

Some of the chapters are more specialized than others (because of the nature of particular topics); hence, these may be somewhat less familiar or "comprehensible" than others—depending upon the reader's own interests and background. Other differences in the scope and content of various chapters stem from the simple fact that a considerably greater backlog of previous work is available in certain facets of ichnology than in others. But we hope that all of the chapters will prove to be useful to anyone wishing to delve into them.

The only parts missing now from the original plan are a chapter on coprolites and one on invertebrate trace fossils in non-marine rocks. Unfortunately but unavoidably, these had to be abandoned during the project. Some of the information, however, has been recouped in other chapters.

Our overall objective has been to produce a comprehensive "textbook" of ichnology, a book that, despite its diverse topics and numerous contributors, is not only thorough in coverage but is also well organized and coherent—not "just another compendium" on the book market. Accordingly, I took considerable liberty in editing the original typescripts, trying to establish a more-or-less uniform style throughout the book and inserting cross-references and other bits of information wherever they seemed to be appropriate. I thank the authors for bearing with me in these alterations, and I hope we attained our objective.

In addition to uniform style and continuity, we tried to arrive at a basic standard in our conventions, classifications, and terminologies. Considering our diverse personal and scientific backgrounds, I believe that we were largely successful in this effort, especially in matters pertaining strictly to ichnology. One glaring exception, however, is our lack of agreement on a standard terminology for marine bathymetric zones; yet the authors have tried to make their respective meanings clear, and perhaps we may be excused for sidestepping a problem that belongs more to oceanography than to ichnology.

Each chapter was reviewed critically by at least two persons in addition to myself. In most cases the contributors reviewed each other's work, with an admirable display of cooperation. But "outside" reviewers also participated, and I sincerely appreciate their time and interest. Outside reviewers who kindly responded to my requests for the critical reading of various chapters include: D. V. Ager, *University of Wales*; Donald Baird, *Princeton University*; Barry Cameron, *Boston University*; M. R. Carriker, *Woods Hole, Massachusetts*; R. E. Carver, *University of Georgia*; K. E. Caster, *University of Cincinnati*; E. H. Colbert, *Museum of Northern Arizona*; B. R. Erickson, *Science Museum of Minnesota*; J. W. Evans, *Memorial University of Newfoundland*; Laing Ferguson, *Mount Allison University*; D. G. Frey, *Indiana University*; E. I. Friedman, *Florida State University*; Roland Goldring, *University of Reading*; D. E. Hattin, *Indiana University*; H. J. Hofmann, *Université de Montréal*; A. S. Horowitz, *Indiana University*; Wann Lang-

ston, *University of Texas*; E. D. McKee, *U. S. Geological Survey*; Anders Martinsson, *Uppsala Universitet*; N. D. Newell, *American Museum of Natural History*; P. R. Pinet, *University of Georgia*; H.-E. Reineck, *Senckenberg Institut*; A. S. Romer, *Harvard University*; B. K. Sen Gupta, *University of Georgia*; E. A. Stanley, *University of Georgia*; Curt Teichert, *University of Kansas*; E. R. Trueman, *University of Manchester*; and E. L. Yochelson, *U. S. Geological Survey*.

I am also very grateful to the contributors who relinquished all rights to royalties in order to lower the sales price of this book. The savings to the consumer were substantial.

For the dust-cover design, I am indebted to R. G. Bromley (he will be happy to answer any questions about it).

I must also acknowledge Vedia Vinluan, who skillfully retyped innumerable manuscript pages after I had scribbled all over the originals, and my wife, Sharon, for her considerable patience and understanding during the time that she endured this "book widowhood."

Finally, I can hardly end this preface without mentioning Walter Häntzschel, to whom the book is dedicated. He was one of the original collaborators on this project but was never able to finish the work. He will be missed by all of us.

ROBERT W. FREY
Athens, Georgia

CONTENTS

PART I
Introduction to Ichnology

CHAPTER 1

THE HISTORY OF INVERTEBRATE ICHNOLOGY

RICHARD G. OSGOOD, JR.

Department of Geology, College of Wooster
Wooster, Ohio, U.S.A.

SYNOPSIS

Ichnology developed slowly, and only during the past two decades has it attained worldwide status as a scientific discipline. With a few notable exceptions, most trace fossils were originally interpreted as fossil algae or "fucoids." Although Nathorst demonstrated the animal nature of "fucoids" in 1881, systematic studies of trace fossils were not initiated until Rudolf Richter's work in the 1920s. Even then ichnological studies lagged until the 1950s, when Seilacher (1953) provided both methodology and a satisfactory working classification. The pace of research was accelerated by Häntzschel's (1962, 1965) contributions in cataloging trace fossil genera and providing extensive bibliographic data. The developments of the last decade, mentioned only briefly herein, have been concerned mainly with continuous refinement of the discipline.

INTRODUCTION

This chapter provides a brief history of the development of ichnology, from its inception to the present day. This history will hopefully give an added dimension to the chapters that follow, although I have confined myself to traces of invertebrate origin. As Frey (1971) pointed out, vertebrate ichnology developed as a separate field; the reasons for this division will become apparent below (see Chapter 14 for a discussion of the historical aspects of that field).

A complete history of invertebrate ichnology has never been written, and regrettably, death has taken the man best qualified for the task—Walter Häntzschel. Short summaries were presented by Caster (1957), Häntzschel (1962), Osgood (1970), and Frey (1971). In my 1970 paper I sub-divided the history of ichnological studies into three sections. Although acknowledging that history perhaps cannot realistically be forced into artificial bundles, I follow the same approach here. The first period extends to the year 1881 and is entitled The Age of Fucoids. During this time many trace fossils were described under the guise of "fucoids," fossil marine algae. The years 1881 to 1920 mark The Period of Controversy, a time when the vegetable origin of "fucoids" was seriously questioned. Finally, we have The Development of a Modern Approach, initiated by the work of Rudolf Richter and extending to the present.

THE AGE OF FUCOIDS

One might expect that the study of ichnology started slowly, with the description and interpretation of a few forms, and then gradually gained momentum and sophistication, as the general field of paleontology developed. However, this is not true. Much of the early history of ichnology may be learned by studying Häntzschel's (1965) volume of *Fossilium Catalogus*, which contains data on trace fossils as well as problematic fossils of unknown affinities. I searched the volume for genera established prior to 1900, ignoring the problematic forms such as "Eozoon" and Walcott's numerous medusae, concentrating instead on taxa that are today clearly acknowledged as trace fossils. This search revealed 258 genera erected before 1900 for which Häntzschel recorded the original author's thoughts on the biological affinities of the organism. Of these, 115 were initially called trace fossils, 23 were considered body fossils, and 120 were assigned to the "fucoids." Indeed, anyone working with trace fossils is familiar with the "-*phycus*" suffix of many genera, e.g., *Rusophycus, Palaeophycus*.

Nevertheless, one unfamiliar with trace fossils might well inquire why nearly half of all genera of trace fossils named before the twentieth century were originally assigned to the plant kingdom. There are several reasons for this situation. First, in gross morphology many trace fossils, especially the branching forms, do resemble plants quite closely (Fig. 1.1). A second factor is that paleontology during the nineteenth century was primarily taxonomic and descriptive in tone. A wholesale rush to document the fossil record was underway, and in some cases descriptions were brief and illustrations nonexistent. In contrast, some of the works on "fucoids" are beautifully illustrated and are accompanied by long descriptions; several examples could be cited, among them Brongniart (1823, 1828), Lebesconte (1883), and Saporta (1884). Many of these authors, regardless of their views on Lamarckian or Darwinian evolution, undoubtedly reasoned that the fossil record of marine algae should rival that of the invertebrates. The workers

Fig. 1.1 *Chondrites bollensis.* Originally designated a "fucoid," this form is now interpreted as the feeding burrow of a sediment-ingesting animal. It fed by repeatedly mining out new branches, to produce this burrow system. The branches do not interpenetrate, a fact that Rudolf Richter used to demonstrate the non-algal origin of the genus.

were merely overzealous in their search. Moreover, the lack of knowledge concerning processes of fossilization and the formation of primary sedimentary structures in largely unstudied marine environments almost ensured that mistakes would be made. If the reader places himself in the mid-nineteenth century and considers the spirit and knowledge of the times, the large number of "fucoids" in the literature readily becomes understandable. The final result of this half-century of labor probably is best seen in work by Schimper (in Zittel, 1879–1890); the "*Algae incertae sedis*" cover 25 pages and are subdivided into 16 different groups. By 1880 certain genera, such as *Fucoides* and *Cruziana*, had accumulated so many species that they rivaled the brachiopod *Spirifer* and the nautiloid *Orthoceras* in their nomenclatural complexity.

Yet trace fossils as such were not ignored entirely by early paleontologists. Probably the most notable work was published by Edward Hitchcock in 1858. Some 200 pages long and entitled *Ichnology of New England*, it described the famous Triassic dinosaur-track beds of the Connecticut River Valley. This volume, and subsequent publications by Hitchcock, did not deal solely with vertebrates; Häntzschel (1965) listed more than two dozen genera of trace fossils erected by Hitchcock and attributed to invertebrates. Also, some of the most famous and enigmatic trace fossils were described prior to 1900. *Climactichnites*, which resembles the impression of a large tractor tread, was described by Logan in 1860 from the Cambrian Potsdam Sandstone of Canada. S. A. Miller (1880) established numerous genera in the Upper Ordovician of Ohio. Unfortunately, works such as these were a minority. Most paleontologists did not believe that trace fossil studies were as important as those dealing with "fucoids" or invertebrate body fossils. This attitude is manifest in the works of Hall (1852), who named and figured several "fucoid" genera from the Silurian rocks of New York State; trace fossils were figured and briefly discussed by Hall (1852, Pls. 11–16) but were not formally named. Regrettably, we have too many works on hand that refer simply to "tracks of Gasteropoda, Crustacea or other Marine Animals" (Hall, 1852, p. 26).

THE PERIOD OF CONTROVERSY

The unrestrained establishment of "fucoid" genera could not go unchallenged indefinitely. As early as 1864, Dawson questioned the vegetable origin of *Rusophycus*. He believed it to be a trilobite burrow. Similarly, a gifted amateur geologist, J. F. James (1884, 1885), determined that the many "fucoids" of Miller and Dyer (1878 a–b) from the Cincinnati area were either trace fossils or of inorganic origin. Generally these isolated attacks fell on deaf ears. Except for one or two instances, James was

not able to alter Miller and Dyer's original views.

The first concerted effort to disprove the "fucoid" notion was a monographic study by the Swedish paleobotanist Alfred Nathorst. This work was first published in 1873 in his native language; a slightly abridged version was translated into French in 1881. (Although James published his study three years after the translation of Nathorst's work, James apparently was unaware of the latter's work.) Nathorst's arguments against "fucoids" were well organized and were supported by evidence gained from studies of shallow marine environments along the Swedish coast. Most of his lithographs show plaster casts made of recent trails and primary sedimentary structures. Nathorst's attack was not confined to a few genera; instead, he proposed to dismiss nearly all the "fucoids." Many, he demonstrated, were traces of invertebrate activity; others were inorganic, e.g., structures that Hall and others had called "stems" were, in reality, groove casts. Such a large-scale work written by a paleobotanist could not go unanswered, and several paleontologists made attempts at rebuttal. Lebesconte (1883) endeavored to reestablish *Cruziana* as an alga, and the Marquis de Saporta (1884) wrote a monograph more than 100 pages long in which he attempted to refute Nathorst. In 1886 Lebesconte reenforced his arguments on *Cruziana*, as did Delgado (1910),[1] describing well-preserved cruzianae from the Silurian of Portugal. In the same year, 1886, Nathorst answered all three authors.

To discuss each work independently is not necessary in understanding the nature of the controversy. Instead, the major arguments are summarized below. Much of that discussion focused on the short, bilobed *Rusophycus* (Fig. 1.2) and the longer, bilobed bands of *Cruziana* (Fig. 1.3). Nathorst maintained that both genera were the result of trilobite burrowing activity.

[1] Published posthumously.

Fig. 1.2 *Rusophycus bilobatum*, convex hyporelief. A "fucoid" that proved to be a trilobite trace. The "rugae" resulted from digging by the legs. Silurian; Oneida County, New York.

The arguments used by the various authors also apply to many other "fucoids."

Nathorst (1881, 1886) pointed out a seemingly peculiar pattern in the preservation of "fucoids." Many are found as convex bodies on the undersides of beds. Moreover, the rock containing the "fucoid" is invariably underlain by finer grained material. He questioned whether an object as small as *Rusophycus* could impress itself 5 to 10 cm down into the mud. He stated that this ". . . would necessitate a weight and consistency . . . which is found in no present-day plants" (Nathorst 1881, p. 62: *transl.*).

Both Saporta (1881, 1882, 1884) and Lebesconte (1883, 1886) attempted to counter this argument. Saporta suggested a process he termed "fossilization en demi-relief." Briefly, the theory states that the

Fig. 1.3 *Cruziana*, convex hyporelief. First interpreted as a branching "fucoid," this *Cruziana* is now known to be the crawling trace of a trilobite moving horizontally, half-buried in sediment. Branching effect is given by a later trail crossing over one formed previously.

plant falls to the muddy bottom and is driven deeper into the mud when it is covered by sand. Decay is initiated at the plant–sand interface and proceeds down into the plant; sand fills the resulting void. The last part of the plant to decay is the surface in contact with the mud. Nathorst (1886) posed several questions with regard to Saporta's explanation. For example, why should decomposition be initiated at the sand–plant interface? More reasonably, he assumed that the first organs to decay would be those most sensitive to decay and most accessible to bacteria. Nathorst again brought up the question of pressure: how was the plant forced down into the mud? Moreover, he noted that a pressure differential must have existed because the points of greatest convexity (i.e., the point of maximum relief from the sole of the bed) is usually in the center of both *Cruziana* and *Rusophycus*. They taper at both ends and merge imperceptibly with the host rock. In a beautiful piece of logic, he went even further and asked: if "fossilization en demi-relief" was operative in soft plants, why could it not apply also to soft-bodied animals? Should not the annelids and the medusae have a larger representation in the fossil record? Neither Saporta nor later authors were able to refute this line of reasoning, and "fossilization en demirelief" never gained an appreciable degree of acceptance. Lebesconte (1883) did endeavor to negate Nathorst's reasoning by stating that he had in his collections specimens that did continue up into the host rock, i.e., they exhibited full relief instead of demi-relief. Nathorst (1886) replied that this easily could be explained by a trilobite burrowing down through the sand, moving along the sand–mud interface for a distance, and then moving back up into the sand.

Naturally the presence or absence of organic residues, such as carbon and (or) plant microstructure, became a major point of discussion. Nathorst maintained that he could find traces of neither in his Swedish material, whereas both Saporta (1884) and Lebesconte (1883, 1886) claimed that the striate markings on the lobes of *Rusophycus* and *Cruziana* represented fibrous networks. Lebesconte (1886) was so impressed by this idea that he altered his original "fucoid" interpretation and considered both genera to be sponges! One has difficulty generalizing on this facet of the controversy, but no evidence of plant microstructure has been found in either genus by more recent workers.

Several arguments revolved around the gross morphology of *Rusophycus* and *Cruziana*. Lebesconte (1883) argued that he had collected bilobed cruzianae that gradually converted laterally into unilobed forms. Nathorst (1886) countered that this change could be explained by a tilting of the trilobite's body, where the appendages on one side of the body lost contact with the substrate.

Many of the long *Cruziana* specimens appear to branch. Obviously, those in favor of a plant origin used this trait to their advantage. Nathorst argued that one trail simply truncated an earlier one. A study of Delgado's (1910) excellent illustrations demonstrates that Nathorst was correct.

In some instances the "algal adherents" took the offensive and attempted to place Nathorst at a disadvantage. Most of Nathorst's *Cruziana* material was collected from the Cambrian beds at Lugnås, Sweden; here *Cruziana* is found throughout the sequence, whereas trilobite body fossils are restricted to the upper part. Lebesconte (1886) raised this point: if the trilobite traces were present, why not the trilobites themselves? Nathorst's reply seems at first to be weak: he declared that the lithology of the lower part of the section did not provide optimum conditions for the preservation of body fossils. However, subsequent work has substantiated this view. As Seilacher (1964) and many other writers have demonstrated, one of the great advantages of trace fossils is that they occur in clastic rocks where body fossils may be rare. Indeed, the problem is one of preservation.

Another point frequently raised by the "fucoid" workers concerned the means of preservation of trace fossils. The opponents were hesitant to accept the widespread existence of trace fossils because they could not perceive how delicate markings, made on a mud bottom by such lightweight organisms as trilobites or annelids, could ever escape destruction by wave and current activity. This is a valid point (see Chapter 4). Although attempts at explanations have been made, this subject requires more detailed analysis by sedimentologists because it applies not only to organic traces but to flute markings and groove casts as well.

One must admire Nathorst's courage in attacking the "fucoid" problem. During the crucial years 1881–1886 he had virtually no supporters, and at times the arguments became quite heated. Gaudry (1883) went so far as to imply that Nathorst's primary motive in attempting to refute the "fucoids" was that he did not believe in Darwinism (this was not true). Nevertheless, Nathorst's work was successful. Maillard (1887) agreed with Nathorst's conclusions, and in 1895 Fuchs published a quarto monograph nearly 100 pages long in which he discussed many "fucoids" as trace fossils. It was the first major work in German on the subject. By the turn of the century the arguments had abated, and the number of "fucoids" widely accepted as algae were few in number. One of the latter was *Chondrites* (Fig. 1.1), the ramifying network of an unknown sediment-feeding animal. Although Nathorst believed the genus to be a trace fossil, the matter really was not settled until Richter's work in 1927. Nathorst's work is now nearly a century old, yet one occasionally still finds "fucoids" described in the literature (see also Fig. 2.12). Andrews (1970), in an index of the generic names of fossil plants, lists many "fucoid" genera (e.g., *Rusophycus*, *Chondrites*) as "?alga."

THE DEVELOPMENT OF A
MODERN APPROACH

Nathorst's work left a vacuum. "Fucoids" had been disproved, but no great rush to study trace fossils ensued. Undoubtedly there were several reasons for this apathy. A great deal of descriptive paleontology remained to be done on body fossils, and the nomenclatural problems within many "fucoid" genera were enough to confuse and discourage most prospective workers.

The first organized study of ichnology was initiated by Rudolf Richter in the 1920s. Richter's contributions to the development of ichnology cannot be overemphasized. Nathorst had shown in 1881 that uniformitarian principles were the key to understanding trace fossils. Richter applied these principles on an extensive scale. With the establishment of the marine institute Senckenberg am Meer at Wilhelmshaven in 1925, Richter could explore the vast mudflats of the North Sea. He studied recent traces and applied this knowledge to the fossil record, publishing many articles over a 25-year period. He correctly interpreted U-tubes, and was the first to understand the mechanics of formation of the branching burrow *Chondrites*. He demonstrated that the tight meanders of *Helminthoida* (see Fig. 6.5) were made by vermiform organisms grazing on thin, detritus-rich, organic layers of sediment.[2]

[2] See Chapter 6 for more detailed discussion of Richter's contributions.

Probably his greatest contribution to the application of trace fossils to geologic problems was the discovery of burrows in the Devonian Hunsrück Shale of the Rhineland. This deposit was thought to represent a classic euxinic environment. However, the presence of extensive burrows in some beds demonstrated that oxygen, in quantities sufficient to sustain bottom life, was present at least part of the time during Hunsrück deposition.

The progress in ichnology at that time was synthesized in 1935 by Othenio Abel. His remarkable book, *Vorzeitliche Lebensspuren*, more than 600 pages long, covers both vertebrate and invertebrate traces as well as coprolites and examples of osteological pathology in the fossil record. It was the standard reference work for more than 20 years.

With the durable work of Richter as a foundation and Abel's text as a reference, one wonders why ichnological studies did not progress more rapidly than they did from 1930 to 1950. A study of the literature does disclose several papers published during this period, most of them written by German authors, e.g., Hundt (1931), Mägdefrau (1934), Dahmer (1937), Häntzschel (1939), and Linck (1942). But, although ichnological research had achieved status in Germany, the dearth of papers published in French, English, and Russian shows that many scientists still regarded the field as an avocation. There are some exceptions, of course. Caster's (1938) detailed study of *Paramphibius* demonstrated the presence of limuloid trails in the Devonian of Pennsylvania. Howell (1943, 1946) published several papers, which dealt mostly with the vertical tubes *Skolithos* and *Monocraterion* in the Paleozoic of the Appalachians. During the same period, Brady (1939, 1947) described arthropod trails from the Permian of Arizona.

In addition to the limitations placed on scientific research by wartime conditions, I believe that at least three circumstances explain why invertebrate ichnology did not enjoy widespread acceptance until the

1950s—and indeed in the United States until the 1960s. The first reason involves taxonomy. As of 1950, the nomenclature of trace fossils remained in an incredible snarl. Although Abel had discussed many genera, his major goal was not an encyclopedic list. Thus, a prospective author still was faced with major problems regarding synonymies and previous work. Moreover, the rulings of the International Commission on Zoological Nomenclature were vague on the validity of trace fossil names, which were not included in listings of zoologic genera.

The second factor was that no one had yet made a clear, concise statement of the value of trace fossil studies in paleontology, zoology, and sedimentology. To be sure, with the advantage of hindsight this information may be obtained by reading Abel's book or the many works of Richter; but this assignment is an undertaking of some magnitude.

The third reason is more general and involves the development of paleoecology. As sedimentologists and paleontologists became interested in this field, some of them naturally turned to trace fossils for evidence of the nature of past environments. Thus, increased popularity in the general field of paleoecology enhanced the development of ichnology.

The potential contributions of trace fossils to diverse fields of geology were clearly elucidated by Seilacher in 1953. At the same time, he also provided both a sedimentological (i.e., preservational) and an ethological (behavioral) method of classification. Previous attempts at suprageneric classification had been made, e.g., Krejci-Graf (1932), but the value of Seilacher's pie-shaped diagram (see Fig. 3.2) is its simplicity. One can determine whether a given trace fossil was a feeding structure or a dwelling burrow without being able to identify the zoological affinities of the trace-maker. Undoubtedly, Seilacher has done more to promote trace fossil research in the past 20 years than any other person.

Seilacher's behavioral classification led to another important contribution. In 1955

he employed suites of trace fossils in facies analyses to illustrate that certain assemblages are diagnostic of shallow water whereas others are more indicative of deeper water. Since publishing his original work, he has refined the concept to include several additional suites and has tested it on rocks of various geologic ages. This facet of Seilacher's work has been extremely important to the development of methodology in trace fossil research. In many situations an inordinate amount of time may be spent attempting to determine the taxonomic affinities of the organism that produced the trace. Seilacher's studies show that this effort is not generally necessary. In facies analysis the critical part is mainly to establish whether a given form is a resting, feeding, or grazing trace, etc. (see Chapters 3 and 9).

Although Seilacher provided a clear concept of goals and methodology, the taxonomic problems at the generic level remained. Although this difficulty was partly alleviated by Lessertisseur's *Traces fossiles d'activite animale et leur signification paléobiologique*, published in 1955, the thankless task of cataloging trace fossil genera and determining synonymies was accomplished by Walter Häntzschel. In 1962 he contributed *Part W, Miscellanea* to *The Treatise on Invertebrate Paleontology*. For the first time, workers in the field had convenient access to concise descriptions and illustrations of genera that earlier were scattered through the literature. Equally valuable is Häntzschel's (1965) volume of *Fossilium Catalogus*, mentioned previously. It is more comprehensive than the treatise volume and contains a more extensive bibliography.

During the past decade the general pace of research has greatly intensified, and ichnological studies have been applied to many fields of geology. The advances can be mentioned only briefly here. The reader is referred to the paper by Frey (1971) for a more complete discussion. Also, Vialov (1966) published a comprehensive study of the value of trace fossil research, in Russian.

For the study of recent traces, the tidal flats of the North Sea still provide a fertile field [e.g., Schäfer's (1972) *Ecology and Palaeoecology of Marine Environments*].

In North America, the University of Georgia Marine Institute at Sapelo Island, Georgia, has become a focal point for research. Studies are underway there that will document in detail the ichnological assemblages and their relationships to the sedimentary facies of the area. In the field of palichnology, Seilacher's "ichnofacies" concept (see Table 2.1) has been applied in detailed analyses of rocks of various ages and geographic locations by many workers (e.g., Ager and Wallace, 1970; Osgood, 1970; Chamberlain, 1971). (See Chapters 7–9.) In sedimentology, important advances have been made in many areas. One notable example is the effect that boring organisms have upon indurated sediments (see Chapters 10–12). Trace fossils have also proved themselves useful as indices of rates of sedimentation and subaqueous erosion (for several examples, see Frey, 1971, p. 113–115; Fig. 8.5). Trace fossils continue to contribute to our knowledge of the ethology and morphology of ancient organisms. Trace fossils are also providing insight into one of the most vexing problems in paleontology: the apparent sudden appearance of metazoa in late Precambrian time. This subject is treated in greater depth in Chapters 6 and 7. Major advances have been achieved not only in application but in techniques of study as well. The development of new coring devices, epoxy resins for making casts of burrows, and x-radiography analysis has added greatly to our knowledge (see Chapter 23).

As Frey (1971) pointed out, the status and popularity of ichnology may be gauged by different methods. The most obvious index is the increasingly large number of papers published annually, in many languages. The worldwide status that ichnological work has achieved may also be demonstrated by the following events: a symposium on "Marine Burrowing Animals" sponsored by the Paleontological Society and the Society of Systematic Zoology at the convention of the American Association for the Advancement of Science in 1968; a technical session devoted entirely to ichnology at the Geological Society of American convention in 1970; an international conference on trace fossils held at Liverpool in 1970, and subsequent publication of the papers in the volume *Trace Fossils* by Crimes and Harper (eds.); a field trip to trace fossil localities sponsored by the Society of Economic Paleontologists and Mineralogists at the American Association of Petroleum Geologists convention in Houston in 1971; and a session on trace fossils under the auspices of the International Paleontological Union at the International Geological Congress in Montreal in 1972.

Like most developing fields, ichnology has its teething problems. Because the study of traces attracts zoologists, paleontologists, and sedimentologists, the literature remains scattered through several journals. Workers are kept up to date by the informal *Ichnology Newsletter*, which originated at the University of Georgia Marine Institute in 1968. Also, at present no universal agreement exists on terminology employed in trace fossil work. This problem is treated in Chapter 3.

ACKNOWLEDGMENTS

I wish to thank Charles Moke, Kenneth Caster, Curt Teichert, and Anders Martinsson for reviewing the manuscript and offering helpful suggestions. The figures were drawn by Krista Roche of Fredericksburg, Ohio.

REFERENCES

Abel, O. 1935. Vorzeitliche Lebensspuren. Jena, Gustav Fischer, 644 p.

Ager, D. V. and P. Wallace. 1970. The distribution and significance of trace fossils in the uppermost Jurassic rocks of the Boulonnais, northern France. In T. P. Crimes and J. C. Harper (eds.), Trace fossils. Geol. Jour., Spec. Issue 3:1–18.

Andrews, H. N. 1970. Index of generic names of fossil plants, 1820–1965. U.S. Geol. Surv., Bull. 1300, 354 p.

Brady, L. F. 1939. Tracks in the Coconino Sandstone compared with those of small, living arthropods. Plateau, 12:32–34.

————. 1947. Invertebrate tracks from the Coconino Sandstone of northern Arizona. Jour. Paleont., 21:466–472.

Brongniart, A. 1823. Observations sur les fucoides. Soc. Hist. Natur. Paris, Mém., 1:301–320.

————. 1828. Histoire des végétaux fossiles. Paris, 1:1–488.

Caster, K. E. 1938. A restudy of the tracks of Paramphibius. Jour. Paleont., 12:3–60.

————. 1957. Problematica. In H. S. Ladd (ed.), Treatise on Marine Ecology and Paleoecology, v. 2, Paleoecology. Geol. Soc. America, Mem. 67:1025–1032.

Chamberlain, C. K. 1971. Morphology and ethology of trace fossils from the Ouachita Mountains, southeast Oklahoma. Jour. Paleont., 45:212–246.

Crimes, T. P. and J. C. Harper (eds.). 1970. Trace fossils. Geol. Jour., Spec. Issue 3, 547 p.

Dahmer, G. 1937. Lebensspuren aus dem Taunusquarzit und den Siegener Schichten (Unterdevon). Jahrb. Preuss. Geol. Landesanst. 1936, 57:523–539.

Dawson, J. W. 1864. On the fossils of the genus Rusophycus. Canadian Natural., N. Ser., 1:363–367.

Delgado. J. F. N. 1910. (Ouvrage posthume) Terrains Paléozoiques du Portugal. Étude sur les fossiles des schistes à Néréites de San Domingos et des schistes à Néréites et à Graptolites de Barrancos. Serv. Geol. Portugal, 56:1–68.

Frey, R. W. 1971. Ichnology—the study of fossil and recent lebensspuren. In B. F. Perkins (ed.), Trace fossils, a field guide. Louisiana State Univ., School Geosci., Misc. Publ. 71–1:91–125.

Fuchs, T. 1895. Studien über Fucoiden und Hieroglyphen. Kaiserl. Akad. Wiss. Wien, math.-naturw. Kl., Denkschr. 62:369–448.

Gaudry, A. 1883. (Note on "A propos des Algues fossiles" by Saporta). Soc. Géol. France, Bull., Ser. 3:451–452.

Hall, J. 1852. Natural history of New York. Palaeontology of New York, v. 2. Albany, C. van Benthuysen, 362 p.

Häntzschel, W. 1939. Die Lebensspuren von Corophium volutator (Pallas) und ihre paläontologische Bedeutung. Senckenbergiana, 21:215–227.

————. 1962. Trace fossils and problematica. In R. C. Moore (ed.), Treatise on invertebrate paleontology, Pt. W, Miscellanea. Lawrence, Kan., Geol. Soc. America and Univ. Kansas Press, p. W177–W245.

————. 1965. Vestigia invertebratorum et Problematica. Fossilium Catalogus, 1: Animalia, pars 108, 142 p.

Hitchcock, E. 1858. Ichnology of New England. A report of the sandstone of the Connecticut Valley especially its footprints. Boston, W. White, 220 p.

Howell, B. F. 1943. Burrows of Skolithos and Planolites in the Cambrian Hardyston Sandstone at Reading, Pennsylvania. Publ. Wagner Free Inst. Sci., 3:3–33.

————. 1946. Silurian Monocraterion Clintonense burrows showing the aperture. Wagner Free Inst. Sci., Bull., 21:29–37.

Hundt, R. 1931. Eine Monographie des Lebensspuren des Unteren Mitteldevons Thüringens. Leipzig, 68 p.

James, J. F. 1884. The fucoids of the Cincinnati Group, Pt. 1. Cincinnati Soc. Nat. Hist., Jour., 7:124–132.

————. 1885. The fucoids of the Cincinnati Group, Pt. 2. Cincinnati Soc. Nat. Hist., Jour., 7:151–166.

Krejci-Graf, K. 1932. Definition der Begriffe Marken, Spuren, Fährten, Bauten, Hieroglyphen und Fucoiden. Senckenbergiana, 14:19–39.

Lebesconte, P. 1883. Présentation à la société des oeuvres posthumes de Marie Rouault, suivies d'une note sur les Cruziana et Rhysophycus. Soc. Géol. France, Bull., Ser. 3, 11:466–472.

————. 1886. Constitution générale du massif breton comparée à celle du Finisterre. Soc. Géol. France, Bull., Ser. 3, 14:776–820.

Lessertisseur, J. 1955. Traces fossiles d'activité animale et leur signification paléobiologique. Soc. Géol. France, Mém. 74, 150 p.

Linck, O. 1942. Die Spur *Isopodichnus*. Senckenbergiana, 25:232–255.

Logan, W. E. 1860. Remarks on the fauna of the Quebec group of rocks, and the primordial zone of Canada. Canadian Nat., 5:472–477.

Mägdefrau, K. 1934. Über *Phycodes circinatum* Reinh. Richter aus dem thüringischen Ordovicium. Neues Jahrb. Mineral., Geol., Paläont., Beil., 72:259–282.

Maillard, G. 1887. Considérations sur les fossiles décrits comme Algues. Soc. Paléont. Suisse, Mém. 14:1–40.

Miller, S. A. 1880. Silurian ichnolites, with definitions of new genera and species. Note on the habit of some fossil annelids. Cincinnati Soc. Nat. Hist., Jour., 2:217–229.

———— and C. B. Dyer. 1878a. Contributions to Palaeontology, No. 1. Cincinnati Soc. Nat. Hist., Jour., 1:24–39.

———— and C. B. Dyer. 1878b. Contributions to Palaeontology, No. 2. Cincinnati, private printing, 11 p.

Nathorst, A. G. 1873. Om några förmodade växtfossilier. Öfversigt af Kgl. Vetensk. Akad. Förhandl. 1873, 9:25–52 (1874).

————. 1881. Om spår af nagra evertebrerade djur M.M. och deras paleontologiska betydelse. (Mémoire sur quelques traces d'animaux sans vertebrés etc. et de leur portée paléontologique.) Kgl. Svenska Vetensk. Akad. Handl., 18, 104 p.

————. 1886. Nouvelles observations sur les traces d'Animaux et autres phénomènes d'origine purement mécanique décrits comme "Algues fossiles." Kgl. Svenska Vetensk. Akad. Handl., 21, 58 p.

Osgood, R. G. 1970. Trace fossils of the Cincinnati area. Palaeontographica Amer., 6 (41): 281–444.

Richter, R. 1927. Die fossilien Fährten und Bauten der Würmer, ein Überblick über ihre biologischen Grundformen und deren geologische Bedeutung. Paläont. Zeitschr., 9:193–240.

————. 1931. Tierwelt und Umwelt im Hunsrückschiefer zur Entstehung eins schwarzen Schlammsteins. Senckenbergiana, 13:299–324.

————. 1941. Marken und Spuren im Hunsrückschiefer. 3. Fahrten als Zeugnisse des Lebens auf dem Meeresgrunde. Senckenbergiana, 23:218–260.

Saporta, G. de. 1881. L'évolution du règne végétal, Les Cryptogames. Paris, Masson, 238 p.

————. 1882. A propos des algues fossiles. Paris, Masson, 82 p.

————. 1884. Les organismes problématiques des anciennes mers. Paris, Masson, 100 p.

Schäfer, W. 1972. Ecology and palaeoecology of marine environments. Edinburgh and Chicago, Oliver & Boyd and Univ. Chicago Press, 568 p.

Seilacher, A. 1953. Über die Methoden der Palichnologie. 1. Studien zur Palichnologie. Neues Jahrb. Geol. Paläont., Abh., 96:421–452.

————. 1955. Spuren und Lebensweise der Trilobiten; Spuren und Fazies im Unterkambrium. In O. H. Schindewolf and A. Seilacher, Beiträge zur Kenntnis des Kambriums in der Salt Range (Pakistan). Akad. Wiss. u. Lit. Mainz, math.-naturw. Kl., Abh. 1955, 10:342–399.

————. 1964. Biogenic sedimentary structures. In J. Imbrie and N. D. Newell (eds.), Approaches to paleoecology. New York, John Wiley, p. 296–316.

Vialov, O. S. 1966. The traces of vital activity of organisms and their paleontological significance. Acad. Sci. Ukraine S.S.R., 1966:1–164. (in Russian)

Zittel, K. A. 1879–1890. Handbuch der Paläontologie. Munich and Leipzig, 958 p.

CHAPTER 2

THE REALM OF ICHNOLOGY, ITS STRENGTHS AND LIMITATIONS

ROBERT W. FREY

Department of Geology, University of Georgia
Athens, Georgia, U.S.A.

SYNOPSIS

Ichnology is the study of all manner of gouges, scrapes, and traces made by living or ancient organisms. At first glance, these oddities might seem to offer little encouragement for serious study. But many looks later, a surprisingly sophisticated body of information begins to emerge, most of it unavailable from any other source. Fossils that once were dismissed simply as "indirect evidence of ancient life" or "secondary sedimentary structures" now are proving to be invaluable in interpreting many forms of ancient life and the associated sedimentological and environmental conditions.

Trace fossils are preserved in numerous places where body fossils are not, and they document several behavioral, ecological, and sedimentological traits that body fossils cannot. Even where both are present in representative quantities and are associated with physical sedimentary structures, trace fossils can yield information that is basic and valuable in its own right, broadening a picture that otherwise is needlessly (and at times misleadingly) narrow.

By the same token, trace fossils are inherently less useful than body fossils and physical sedimentary structures in many circumstances, and may yield obscure or ambiguous information in others. The question, then, is not simply which line of evidence to use in a given study, but rather, how all the evidence available can be brought to bear in that study. The actual situation is that, more often than not, trace fossils have been relegated to a minor role in paleoecology, sedimentology, and facies analysis where, if utilized, they could have made a very substantial contribution.

THE WORLD OF ICHNOLOGY

What is ichnology? We commonly say that it is the study of fossil and recent biogenic structures, or the study of tracks, trails, burrows, borings, and other traces made by recent and ancient organisms. But we really mean that we are interested in the goodness of fit between recent and ancient traces and tracemakers: the kinds of creatures that made the traces; the conditions under which these organisms lived; how, where, and when the traces were made and preserved; what influences these processes had upon other organisms and the chemical and physical environment; and how all this information can best be used to enrich our practical knowledge of geology and biology. The "fit" in many instances is strikingly good (Fig. 2.1).

Fig. 2.1 Comparison between recent and ancient lebensspuren. A, lower Miocene burrows in parallel stratified lenses intercalated with cross-stratified sand; bedding style similar to bars in modern Platte River. B, recent burrow (plastic cast) from transverse bar sands of Platte River. Both burrows presumably made by tiger beetles (Family Cicindelidae—Chapter 19) for hibernation; both observed near Valley, Nebraska. (Photos courtesy J. A. Fagerstrom.)

Ichnology embraces a remarkable array of organisms and their habits and habitats. Representatives from most major lineages of plants and animals are capable of making preservable traces, and they do so in virtually all substrates that can sustain life. The examples are as varied as echinoids that bite away schist and gneiss, dinosaur footprints on coral and rudist reefs, birds that burrow, lichens that etch rocks under the sea, and conodonts that were bored while the animal lived. In fact, traces may be made even where life cannot long be sustained; the tracks and grotesque death throes recorded in Pompeii ash and cinders are striking albeit morbid examples. Whatever the occurrence of traces, each has a story to tell.

In short, a tremendous reservoir of information is available here, much of it still untapped, some of it difficult to evaluate, but most of it very pertinent in one way or another to our studies of the present day and of earth and life history. The wonder is not that these resources are so diverse or ubiquitous, but that they were so complacently overlooked for so long.

The character and significance of our "ichnological heritage" is of course the subject of the entire book. Here I simply hope to lay a broad foundation for the chapters that follow, stressing some of the major strengths and weaknesses of our discipline.

COMPARISONS BETWEEN TRACE FOSSILS AND BODY FOSSILS

Trace fossils exhibit several characteristics that set them apart from body fossils. The main distinction is that trace fossils represent <u>behavior or activity</u> by organisms rather than actual body parts, or casts and molds of body parts, and thus require different kinds of analyses (see Chapter 3). Other practical differences are also noteworthy.

Seilacher (1964) outlined some of the major properties of biogenic sedimentary structures, including their (1) <u>long time</u>

range, (2) narrow facies range, (3) lack of secondary displacement, and (4) frequent occurrence in otherwise unfossiliferous rocks. This scheme is a convenient basis for comparing the potential for preservation and the temporal and facies relationships of trace fossils and body fossils generally.

Guide Fossils Versus Facies Fossils

Paleontologists and others who work with fossils usually make a careful distinction between facies fossils and guide (or index) fossils. Ideally, guide fossils represent small intervals of geologic time but are widespread geographically; the organisms evolved rapidly and were either broadly adapted to different environmental conditions, crossing many facies boundaries, or occupied an environment that was itself very broadly distributed, as was true of ammonites in Mesozoic seas. These fossils are very useful in biostratigraphy but are much less so in detailed paleoecology. Facies fossils, in contrast, are long-ranging forms having more restricted geographic distributions; the organisms evolved slowly and were narrowly adapted to particular environmental conditions. Such fossils are of little use in refined biostratigraphy but are very useful in paleoecology and environmental reconstruction. These concepts may be applied to trace fossils as well as to body fossils, at least in a general sense.

As a rule, trace fossils have much longer geologic ranges than do body fossils. The reason is not that trace-making organisms evolve more slowly than other kinds of organisms, but simply that speciation is reflected much more accurately by body fossils than by trace fossils. The mobile and hemisessile benthos seem to have designed themselves for a comparatively small number of substrate adaptations (represented by lebensspuren) that are repeated again and again in the rock record. Small-scale variations on a given behavioral plan may be discernible (e.g., Seilacher, 1967b), but only rarely can these differences be placed in a definite phylogenetic context. Indeed, the basic kinds of traces observed tend to be related more closely to behavioral patterns and environmental conditions than to phylogeny. (See Chapter 22.) Thus, the scanty contribution of trace fossils to biostratigraphy is compensated by their special contributions to paleoecology and facies analysis. For example, the trace fossil *Helminthoida* occurs in Phanerozoic rocks of many different ages, yet its environmental implications are essentially the same everywhere: the muddy deep-sea floor (Seilacher, 1967a; Macsotay, 1967; Crimes, 1973; Chapter 21).

Of course, certain kinds of trace fossils are more useful in biostratigraphy than others. Their relative usefulness depends mostly upon the correspondence between the morphology of traces and that of the tracemakers: the closer the resemblance, the more nearly trace fossils mirror biological speciation, the more reliable the traces in temporal zonation. Few invertebrates make such distinctive traces, although some trilobites (Chapters 6 and 7) and certain borers (Chapters 11 and 12) are exceptions. Vertebrate tracks are generally more useful (Chapter 14), but even here the behavioral and preservational aspects of track morphology must be weighed carefully against anatomical aspects.

Facies Relations of Trace Fossils

In practice, the long time ranges typical of trace fossils are an advantage not shared by most body fossils: such time ranges greatly facilitate ecological and environmental comparisons of rocks that differ widely in age. The most popularly known expression of this principle is Seilacher's (1964, 1967a) concept of "universal trace fossil facies," applicable mainly to marine invertebrates. The basic idea is that trace fossils generally reflect behavioral adaptations to specific environmental conditions; thus, particular assemblages of trace fossils tend to be characteristic of given environmental regimes. These assemblages are re-current through time, whenever the requisite environmental conditions are repeated.

In marine environments, may of the factors that control the distribution of trace-making organisms—such things as temperature, food supply, and the intensity of wave or current agitation—tend to change progressively with water depth. Therefore, Seilacher's assemblage concept may also be

TABLE 2.1   Recurrings Aquatic Trace Fossil Assemblage and their Environmental Implications.*

Name of Assemblage	Characteristic Lebensspuren [see Häntzschel (1962) for examples of genera]	Typical Benthonic Environment (generally related directly or indirectly to bathymetry)†
Scoyenia	Vertebrate tracks, trails, and burrows, mainly of aquatic or semi-aquatic species but also of terrestrial species coming to water; abundant insect and other arthropod traces; certain forms of *Planolites*; scattered snail and clam trails and shallow burrows. Local diversity and abundance generally less than in marine environments.	Nonmarine clastics, especially continental "red beds," floodplain deposits, etc., as the assemblage is presently defined. [But this "nonmarine" concept actually involves a diversity and zonation of habitats and traces (Chapter 19) that parallels the marine categories outlined below; the concept thus needs additional study and redefinition.]
Glossifungites	Vertical cylindrical, U-shaped, or sparsely ramified dwelling burrows; protrusive spreiten in some, developed mainly through growth of animals. Many species leave the burrows to feed (e.g., crabs); others are mostly suspension feeders (e.g., polychaetes, pholads); Diversity low, although burrows of given types may be abundant.	Marine littoral zone and sublittoral omission surfaces; stable, coherent substrates either in protected, low-energy settings (e.g., salt marshes; muddy quiet-water bars, flats, and shoals) or in areas of slightly higher energy where semiconsolidated substrates offer resistance to erosion (e.g., Frey and Howard, 1969, Pl. 4, fig. 3). [On rocky coasts and hardgrounds, *Glossifungites* and *Skolithos* assemblages are replaced by borers and scrapers, and bioerosion itself becomes a major feature of the environment (Chapters 11 and 18).]
Skolithos	Vertical cylindrical or U-shaped dwelling burrows; protrusive and retrusive spreiten, developed mostly in response to substrate aggradation or degradation (e.g., *Diplocraterion*, *Corophioides*); forms of *Ophiomorpha* consisting predominantly of vertical or steeply inclined burrow components. Animals mainly suspension feeders. Diversity low, although given kinds of burrows may be abundant.	Littoral and very shallow sublittoral zones; relatively high energy conditions; well-sorted, shifting sediments; abrupt erosion or deposition (e.g., beaches, inlet bars and shoals, tidal deltas). [Higher energy increases physical reworking and obliterates biogenic sedimentary structures, leaving preserved record of physical stratification (Chapter 8); for this reason, the very shallow sublittoral (shoreface) is typically barren of traces.]

viewed as the basis for a relative bathymetric scale (Table 2.1).

The assemblages are named for representative trace fossil genera, yet these genera

TABLE 2.1—Continued.

Name of Assemblage	Characteristic Lebensspuren [see Häntzschel (1962) for examples of genera]	Typical Benthonic Environment (generally related directly or indirectly to bathymetry)†
Cruziana	Abundant crawling traces, both epi- and intra-stratal; inclined U-shaped burrows having mainly protrusive spreiten (Rhizocorallium); forms of Ophiomorpha and Thalassinoides consisting of irregularly inclined to horizontal burrow components; scattered vertical cylindrical burrows. Animals mostly carnivores and suspension feeders, although some are deposit feeders. Diversity and abundance generally high.	Shallow sublittoral; below daily wave base but not storm wave base, to slightly quieter offshore-type conditions; moderate to relatively low energy; well-sorted silts and sands, to interbedded muddy and clean sands; appreciable but not necessarily rapid sedimentation. A very common type of depositional environment, overlapping with that of the Zoophycos assemblage.
Zoophycos	Relatively simple to complex, efficiently executed grazing traces and shallow feeding structures; spreiten typically gently inclined, distributed in delicate sheets, ribbons, or spirals (e.g., certain forms of Dictyodora, "flattened" forms of Zoophycos). Animals mostly deposit feeders. Diversity and abundance generally low.	Sublittoral to bathyal; quiet-water, offshore-type conditions; impure silts and sands; below storm wave base to upper continental slope or equivalent, in areas free of turbidity flows or relict sediments (where deposited food is scarce). A broad gradational "zone" intermediate between, and in many places indistinguishable from, the Cruziana and Nereites "zones," respectively.
Nereites	Complex grazing traces reflecting highly organized, efficient feeding behavior (e.g., Paleodictyon); spreiten structures typically nearly planar, on substrate surface (e.g., Phycosiphon); numerous crawling-grazing traces and sinuous fecal castings (e.g., Neonereites, Helminthoida, Cosmorhaphe; Chapter 21). Animals mostly deposit feeders and "scavengers." Local diversity and abundance generally low, but somewhat greater than in Zoophycos assemblage; net density of lebensspuren increased by virtue of very slow deposition.	Bathyal to abyssal; mostly very quiet waters, interrupted by turbidity flows; pelagic muds, typically bounded above and below by turbidite deposits. In terms of area occupied on the modern sea floor, this is the most important of the five marine "zones"; in terms of representation in the rock record, it is probably second in importance to the Cruziana "zone."

* Modified from Seilacher (1967a); compare with Fig. 7.2 and Fig. 9.1.
† The bathymetric zones listed here are equivalent to those outlined by Ager (1963, Fig. 2.3).

need not actually be present in a given occurrence of that assemblage, or may in fact occur in other assemblages; for example, the *Glossifungites* assemblage on the present Georgia coast consists almost exclusively of pelecypod borings (Frey and Howard, 1969, Pl. 4, fig. 3), and *Cruziana* has been reported from the *Scoyenia* assemblage (Bromley and Asgaard, 1972). *Zoophycos*, which occurs commonly in shallow- and deep-water deposits, is even more problematical when viewed individually (see Chapters 4, 6, and 17). The main point, however, is that the character of the overall assemblage is considerably more important in environmental interpretation than is any single member of that assemblage.

An equally important point is that, ultimately, Seilacher's scheme is not intended to be an unequivocal bathymetric scale. Here, as in all other "facies relations" of trace fossils (or body fossils), the fundamental consideration is not such inanimate things as water depth, distance from shore, mineral composition of sediments, or particular orogenic or physiographic settings, but rather the innate, dynamic environmental factors that actually governed the abundance and distribution of the original trace-making organisms. After we have reconstructed the set of physical, chemical, and biological conditions prevailing at the time the traces were made, we can begin to concern ourselves with the kind of place or deposit (beach, delta, crevasse-splay, or bathyal slope) in which the traces may have been formed. (See Crimes, 1973.)

Reconstructing the salient parameters of an ancient environment is not an easy task, of course. Modern ecological studies, involving such things as biomass, respiration and metabolism, and energy and nutrient cycling, show that the controlling factors can be remarkably subtle and complex (e.g., Odum, 1971; Valentine, 1973). Paleontologists and ichnologists cannot hope to document this entire spectrum of factors and their inner workings, yet ich-

nologists at least have the advantage of a preserved record of several behavioral responses to these conditions.

On a level at which environmental parameters can be resolved in the rock record, critical ecological factors especially include the kind and abundance of food, substrate consistency or coherence, and energy levels. Much has been written about these parameters, including various passages in this book.

Ecological factors may also be viewed in terms of "stimulus physiology" relationships. Seilacher (1953) and numerous others have noted that organisms display specific responses to external stimuli; the resulting directional adjustment by the organisms may be away from (negative), toward (positive), or transverse (neutral) to the source of the stimulus. Some of the more important stimuli that evoke animal reactions include touch (thigmotaxis), light (phototaxis), heat (thermotaxis), gravity (geotaxis), chemicals (chemotaxis), and currents (rheotaxis). Numerous other stimuli and reactions could be listed, including some that are uniquely ichnologic (e.g., phobotaxis, the ability of certain sediment-ingesting animals to mine an area thoroughly, without recrossing previous feeding swaths; see Chapter 6).

In most cases trace fossils record only the responses of organisms to, not the location or origin of, external stimuli; yet even the recorded responses can be very instructive in facies analyses. The ethological categories explained in Chapter 3 are more useful in classifying behavioral responses, but the "stimulus physiology" relationships should be weighed with these in our ultimate behavioral and environmental interpretations.

Potential for Preservation

Relation to Substrate

Unlike tests, shells, and other body parts, most lebensspuren cannot be transported away from their place of origin. Trace fossils, particularly biogenic sedimentary structures, thus retain a much closer relationship to original depositional conditions than do most body fossils, which is a decided advantage in paleoecology and facies analysis.

Biogenic sedimentary structures, whether produced by invertebrates, vertebrates, or plants, are ordinarily destroyed or greatly modified by the processes that rework or transport body fossils. But when these structures are preserved, they typically remain in place. The only notable exceptions are burrows that have thick, durable linings (dwelling tubes—see Fig. 2.5), which are sometimes washed out of the substrate and redeposited more or less intact; these reworked dwelling tubes are usually easily recognized as such in the rock record[1], and thus do not diminish the paleoecological advantage of in situ preservations generally.

Once the host sediments have been lithified, the contained trace fossils may be weathered out and reworked, of course, but in most cases these, too, are distinctive as such (e.g., Howard, 1966; Schloz, 1972). The specimens are apt to display irregular truncations or breakage, imbrications, and other characteristics of inter- and intraformational conglomerates.

Fecal pellets are perhaps the most typical example of transportable "sedimentary structures." They may accumulate in appreciable quantities locally, behaving more as sand than as clay during transport (Oertel, 1973). With compaction, they commonly are transformed into flasers, stringers, or thin layers of mud (accounting for the high clay content of many otherwise clean sands); but by this time they usually have been altered beyond recognition as individual biogenic structures.

Bioerosion structures have a different prospect for preservation. Borings in rock are ordinarily preserved in the same way as the rock. If it is destroyed by coastal or submarine erosion, or breaks up into blocks and clasts that are then scattered over the seabed, or is preserved in place, then so also are the contained borings. Boring organisms may in fact be largely responsible for the destruction of the substrate (Chapters 11 and 12). The same is generally true for borings or gnawings in shell, bone, or wood, except that these substrates are more easily reworked. Nevertheless, preserved bioerosion structures do retain their original spatial relationship to the substrate; the semantic difference between these and biogenic sedimentary structures is that here the host substrate itself is transported, not the lebensspuren individually.

In hardgrounds, an intriguing interplay exists between burrows and borings. Their formation and preservation are recounted in Chapters 17 and 18. The processes of formation and preservation of biogenic sedimentary structures are discussed more fully in Chapters 3 and 4.

Before leaving this topic, however, we should note a potential anomaly: the in situ preservation of traces does not invariably mean that tracemaking occurred contemporaneously or penecontemporaneously with sedimentation or, therefore, that conditions during deposition of host sediments were the same as those at the time the traces were made. In a prograding bar, spit, or tidal delta, the high-angle stratification would imply high-energy conditions and rapidly shifting sediments, a combination that excludes most kinds of tracemakers. Abundant trace fossils in such sediments would likely mean that local energy had diminished somewhat and that the substrate had become correspondingly more stable before its occupancy by mobile animals. A clue would be the cross-cutting relationships, of course: do the traces penetrate the high-angle laminae, or do the laminae truncate or otherwise modify or become incorporated within some of the traces? (e.g.,

[1] The name *Siphonites* has been applied to fossil examples (Häntzschel, 1962); yet trace fossil names should be based upon original morphology and configuration, not depositional history.

do very many of the traces grade into escape structures?).

Similarly, Kennedy and Sellwood (1970) and Asgaard and Bromley (1974) reported marine *Ophiomorpha* within "nonmarine" rocks (beneath an omission surface). The host sediments evidently accumulated initially under nonmarine conditions; the burrows were made much later, when marine waters stood over the same, unconsolidated sediments. From the standpoint of the tracemaker, the environment was marine; but the sediments alone would suggest a nonmarine environment. Such palimpsest substrates are found in many guises, and must be reckoned with in our interpretations of depositional history and paleoecology (see Chapter 5).

Diagenesis

Another advantage that trace fossils have over body fossils is that traces tend to be enhanced by the processes of solution, leaching, and mineralization that typically destroy or greatly modify tests, shells, bone, and wood (Fig. 2.2). Indeed, distinct traces may remain even in low-grade metamorphic rocks (e.g., Aceñolaza and Durand, 1973). These generalities apply more consistently to biogenic sedimentary structures than to bioerosion structures. In many situations, borings, scrapings, and gnawings are likely to suffer the same diagenetic or post-diagenetic fate as the host shell or bone; in others, however, leaching of shells may produce superior casts of the borings within them, and thereby greatly improve the ultimate state of the borings.

Most biogenic sedimentary structures are delimited by conspicuous breaks in the fabric of the substrate containing them, or the sediments later burying them. In many instances this break is purely a physical disruption, as when a crab or an elephant

Fig. 2.2 Comparison between preservation of burrows and shells in permeable siliceous sediments, same locality. A, *Ophiomorpha nodosa*, having thick knobby walls and smooth interior, enhanced by diagenesis. B, wispy arcuate films (dark)—all that remains of small pelecypod shells (probably *Mulinia* and *Donax*); even these are much better preserved than most shells at this locality. Friable Pleistocene sandstone, Florida. Quarter-dollar coin for scale.

steps in soft sediment. Among aquatic invertebrates, however, the physical trace is often lined in some manner by the organism. Gastropods and polychaetes commonly lay thin slimes to lubricate their path and facilitate locomotion. Dwelling burrows are typically lined with mucoid compounds, biogenically cemented sand grains, or horny or chitinophosphatic tubes that act as structural reinforcements. These linings perform a dual role: initially they offer resistance to mechanical collapse or erosion of traces during burial, and later they act as foci for biogeochemical reactions that further accentuate the traces.

Even simple physical disruptions can be enhanced by lithification and subsequent diagenesis, causing pronounced textural discontinuities between lebensspuren and their sediment fills or casts. But linings ordinarily give rise to proportionately more striking discontinuities, because chemicals tend to diffuse along these breaks in texture (Hanor and Marshall, 1971) and react with the organic compounds; the thicker linings may in fact be preserved themselves. These boundaries thus tend to be very conspicuous in petrographic thin sections and x-radiographs (see Chapter 23), and because of them, trace fossils commonly weather in relief at the outcrop.

Nevertheless, these processes can be carried to such an extreme that they modify or obliterate trace fossils just as surely as other processes do body fossils. Burrows in certain lithologies may have acted as foci for the formation of pyritic nodules or calcareous concretions, for example. The mineralization was evidently aided by decay of, and reactions with, the organic secretions and perhaps even with dead animals inside the burrows. Development of these diagenetic features ranges from simple burrow fills (as with many of the "burrow flints" in English chalk), through thick envelopes surrounding the burrows (as in the trace fossil *Tisoa*[2]), to extreme overgrowths and distortions in which the mineralization is no longer confined to the original configuration or trend of the burrow. Such nodules and concretions are potentially misleading to ichnologists, as discussed further in Chapters 4 and 5 and 16 to 18. The diagenesis and mineralization of lebensspuren remain an "open field" for geochemical and biochemical research.

The more "routine" processes that destroy tests and shells but accentuate traces are especially effective in permeable siliceous sediments (Fig. 2.2). In siltstones and sandstones, trace fossils may thus be the only indication of ancient life. Striking examples are the dinosaur tracks of the Connecticut Valley (Chapter 14) and trilobite traces in many of the "unfossiliferous" lower Paleozoic formations (Chapter 7). This situation has also given rise to the popular misconception that trace fossils are inherently more abundant in siltstones and sandstones than in other lithologies, particularly carbonates. The real difference, however, is simply that trace fossils in carbonates and other lithologies are less apt to exhibit such distinct textural and diagenetic breaks or to weather in relief (see Chapter 17); carbonate cementation and recrystallization tend to deemphasize traces. Furthermore, the relatively inconspicuous trace fossils in these rocks are apt to be overlooked by paleoecologists because of the attraction of more conspicuous, better preserved body fossils.

Species Diversity

In the sense of Lawrence (1968), body fossils constitute the primary record of ancient life, and trace fossils are a secondary or "redundant" expression of that life. This rationale is perfectly acceptable philosophi-

[2] In *Tisoa*, the enveloping concretion has in fact been given taxonomic significance; without it, the burrow undoubtedly would be referred to another genus, such as *Arenicolites* (Frey and Cowles, 1969). As with *Siphonites*, this practice is inconsistent with taxonomic principles (see Chapter 3).

cally, yet such semantic distinctions have considerably less immediate value in reconstructing ancient communities. In many marine assemblages the dominant body fossils consist of sessile species that would not have made lebensspuren. Numerous soft-bodied organisms, in contrast, are ordinarily represented in the fossil record only by their lebensspuren. In certain lithologies, even the hard-bodied tracemakers stand less chance of being preserved than do their traces. From a practical standpoint, therefore, trace fossils can hardly be considered "redundant" where they are the predominant, or indeed the only, fossils present.

Where trace fossils and body fossils occur together, the possibility remains, of course, that the same organism may be represented by a "primary" shell and a "secondary" trace. Distinguishing these "redundancies" is complicated by the fact that two or more species of tracemakers can sometimes make surprisingly similar traces, and a single kind of organism can sometimes make an equally surprising variety of traces; certain traces even reflect sexual dimorphism (Fig. 2.3) or ontogenetic changes in tracemakers (Forbes, 1973; Chapter 19). Nevertheless, careful analyses of trace fossil assemblages, from the viewpoints of animal behavior and the "functional morphology" of traces, can often give us a reasonably accurate indication of the diversity of trace-making species present.

As a hypothetical example of determining species diversity among ancient trace-making organisms, consider a marine assemblage containing the trace fossils *Skolithos, Chondrites, Zoophycos, Asteriacites, Rusophycus,* and *Cruziana* (Fig. 2.4). The behavioral and morphological differences represented by these traces are so striking that only the last two are at all likely to have been constructed by the same species of animal. The resting trace *Asteriacites* is totally unlike the resting trace *Rusophycus,* and the feeding burrow *Chondrites* is equally unlike the grazing trace *Zoophycos.* All these structures are very different from the dwelling burrow *Skolithos.* The crawling trace *Cruziana* might or might not be an extension of the resting trace *Rusophycus;* detailed morphological and preservational studies would hopefully resolve this ambiguity (cf. Crimes, 1970). At any rate, our analysis indicates that at least five, and possibly six, different kinds of trace-making animals lived in the ancient environment.

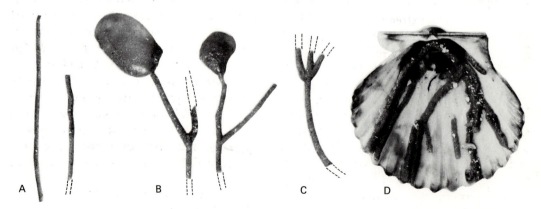

Fig. 2.3 Morphological (and functional) variations in dwelling tubes of recent polychaete *Clymenella mucosa.* A, unbranched tubes occupied by males. B, forked tubes occupied by spawning females (egg sacs attached, at level of substrate surface). C, "spurious" branching, seen occasionally. D, dwelling tubes attached to interior of bay scallop shell, *Argopecten irradians.* Tubes 3–5 mm in diameter. Sand flat; Beaufort, North Carolina. (From Frey, 1971.)

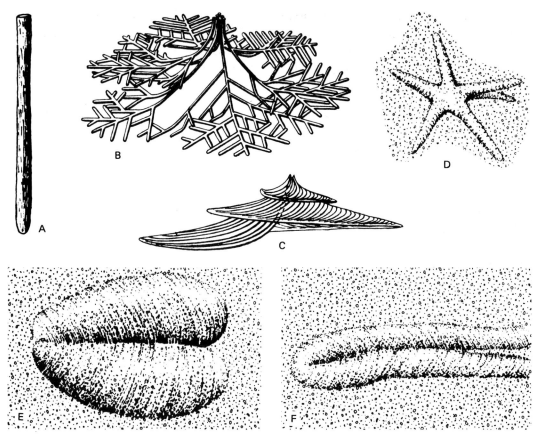

Fig. 2.4 Illustration of animal species diversity represented by trace fossils. A, dwelling structure *Skolithos*. B, feeding structure *Chondrites*. C, grazing trace *Zoophycos*. D, resting trace *Asteriacites*. E, resting trace *Rusophycus*. F, crawling trace *Cruziana*. Not to scale. (Adapted from various sources.)

In order to eliminate possible "redundancies" and thus help determine the ultimate species diversity in this ancient community, we would of course search through the associated body fossils to see whether they include a sea star capable of having produced the *Asteriacites*, or trilobites that might have made the *Cruziana* and *Rusophycus*; but we almost certainly would not find body fossil counterparts of *Skolithos*, *Chondrites*, and *Zoophycos*, which evidently are made by soft-bodied, unpreservable animals. Thus, whether we know the taxonomic identity of each organism or not, we have arrived at some actual numerical data that can be fed into many of the mathematical indices of diversity. By integrating this kind of information with that derived from the other body fossils and

trace fossils present—such things as coelenterates, brachiopods, bryozoans, and echinoderms, and the borings possibly contained in each of them—our reconstruction of the total community is enriched considerably. (See also Chapters 6, 9, and 20.)

Trace fossils usually are not so directly amenable to quantitative analyses, of course (e.g., the ambiguities noted above); but they do enhance our resolution of species diversity, and therefore cannot reasonably be ignored in community reconstructions.

RECONSTRUCTING ANCIENT TRACEMAKERS

Identity of Tracemakers

Ichnologists naturally would like to relate specific trace fossils to specific organisms in the original environment, although this is rarely possible. Roots are occasionally preserved within root-mottles, and clams and a few other kinds of animals are sometimes found in body-size burrows or borings obviously made by them. But the only other unequivocal examples of such relationships are the remarkable "accidents in preservation," many of them involving a catastrophe of some sort, in which the remains of a crawling animal are found in situ within a trail. Quarries in the famous Solnhofen Limestone of Bavaria have yielded numerous specimens of limulids quite literally at the end of their last trail (Goldring and Seilacher, 1971); the quarrymen reportedly search for the limulids simply by finding a trail in the rock and following it to its end!

Needless to say, the ichnologist's usual task is much more complicated, at least at the species and genus level (at higher taxonomic levels, comparisons between traces and their probable makers generally may be made with greater confidence; see Chapter 6). As mentioned previously, the closer the resemblance between a trace and its maker, or at least the part of the organism that made the trace, the easier our job in relating the two. If the tracemaker is not preserved, or if the trace bears little resemblance to the morphology of the tracemaker, then we must search for other kinds of clues. For example, shells carried by hermit crabs frequently leave tell-tale scrapes and gouges along the crab's trackway (Frey, 1971, Fig. 7); a tracemaker's own body parts (e.g., various carapace spines) might also leave gouges along a trackway, but those made by the hermit crab's shell are more distinctive because of their haphazard form and distribution.

The claws and other body parts of callianassid and glypheoid shrimp have been found within such burrows as *Thalassinoides* and *Ophiomorpha* (Sellwood, 1971; Bromley and Asgaard, 1972; and references cited therein), and skeletons of the beaver *Paleocastor* have been found within the

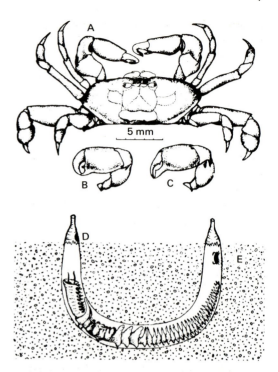

Fig. 2.5 Dwelling tube of recent polychaete *Chaetopterus variopedatus* (bottom) with commensal crab *Pinnixa chaetopterina* (A–E). A, male, dorsal view. Chela, in frontal view, of male (B) and female (C). Crab ordinarily resides in posterior end of dwelling tube (E) but can slash escape exits through tube walls (D) and can bury itself in adjacent sediment. [Adapted from Williams (1965).]

large spiraled burrow *Daimonelix* (Chapter 15); these associations are very credible. Nevertheless, most such evidence is only circumstantial; it does not in itself rule out the possibility that the body fossils are the remains of commensals or "nestlers," or even prey, preserved within the domiciles of other kinds of animals. Evans (1967) reported 30 species of nestlers residing within pholad borings, for example, and the shrimp *Callianassa* and the polychaete *Chaetopterus* commonly share their dwelling structures with various commensal crabs [e.g., Gray (1961), Howard and Dörjes (1972); see Fig. 2.5]. The small, thin-shell crabs are not especially well suited for preservation; but they are immanently more so than the polychaete, and in the rock record might be misinterpreted as the tracemaker.

Analysis of the functional morphology of body fossils preserved within, or in close association with, such burrows is a natural starting point in documenting possible tracemakers. The claws of *Callianassa*, *Upogebia*, *Glyphea*, and similar shrimp are obviously capable of manipulating sediment in a way necessary to construct *Thalassinoides* or *Ophiomorpha*, whereas those of the small commensal crabs are not (Fig. 2.5). Even this approach is not without pitfalls, however. Rice and Chapman (1971) showed that recent *Nephrops* (a lobster) and *Goneplax* (a crab), although possessing no obvious morphological adaptations for burrowing, nevertheless excavate prominent *Thalassinoides*-like structures (cf. Fig. 2.6A).

Comparisons between the diameter of the burrow interior and the size of more or less intact skeletal remains is also a starting point. The carapace of the commensal crab *Pinnixa* is dwarfed by the diameter of *Chaetopterus* dwelling tubes, for example, and this would be additional evidence

against the crabs as tracemakers. Yet even among undoubted tracemakers, we do not always see a close correspondence between the diameter of the animal and the diameter of its dwelling structure (Fig. 2.6). Furthermore, dimensions within a single burrow system may vary considerably (Bromley and Frey, 1974). In this, as in all other such interpretations, our conclusions must be tempered by consideration of the numerous different possibilities.

For many kinds of trace fossils, we cannot identify the tracemaker even at the phylum level. Köhnlein and Bergt (1971, Figs. 9 and 10) pictured structures made by plant roots and earthworms that are very similar. The recent marine eel *Gorgasia sillneri* makes burrows that resemble those of certain invertebrates (Clark, 1972). Scores of soft-bodied animals, including several phyla of "worms," make equally similar traces, and we have little hope of ever sorting them out taxonomically. In addition, the traces of one kind of organism may be reused or otherwise modified by cohabitants or secondary tracemakers (e.g., Shinn, 1972; Howard and Frey, 1973, p. 1182; Karplus et al., 1974; Chapter 18). For these kinds of ambiguities, we can more profitably address ourselves to analyses of the behavior represented by the traces, and their environmental implications, than to the identity of the tracemakers themselves. This idea can also be applied to studies of animal diversity, as mentioned previously.

Finally, as the traces become more and more specious or "nondescript," the possibility arises that the structures may in fact be of physical rather than biological origin (Colton, 1967; Voigt, 1972; Chapters 5 and 16).

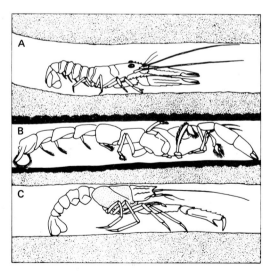

Fig. 2.6 Variations in burrow diameter relative to body size among selected decapods. A, recent lobster *Nephrops norvegicus*. B, female *Callianassa major*. C, fossil *Glyphea rosenkrantzi*. (From Bromley and Asgaard, 1972; cf. Fig. 22.10.)

Behavior of Tracemakers

The key to interpreting the biological, ecological, and environmental significance of trace fossils is to understand something of the behavior represented by those traces. Such activities as resting, crawling, grazing, feeding, dwelling, and escape have received considerable study (Fig. 2.4; see Chapter 3) that need not be reiterated here. Perhaps the most eloquent testimony to the practical value of these behavioral categories is the fact that this simple classification has well stood 20 years of testing and use since Seilacher (1953) defined it, and is the primary basis for the "recurring assemblage" concept (Table 2.1).

But the key to understanding the real scope of such behavioral studies and classifications is the word "practical" in the above sentence. As mentioned previously, the ecological factors that control the diversity and distribution of organisms can be remarkably subtle and complex, and so is the finite behavior of these organisms (e.g., MacGinitie and MacGinitie, 1968; Schäfer, 1972; Chapters 13 and 22). Seilacher wisely restricted his classification to a small number of general, highly useful categories, although an infinite number of more specific, individual behavioral patterns are discernible in nature.

Even a "simple" animal's manner of feeding may be much more subtle than we ordinarily suspect. Jacobsen (1967) showed that the well-known lugworm *Arenicola*— a deposit feeder that stopes sediment at the anterior end of its burrow—pumps water through the burrow not only for aeration but also to trap suspended food and incorporate it within the sediment about to be ingested. Semantically *Arenicola* is thus part deposit feeder and part "suspension" feeder; but we could not deduce such behavior from fossil burrows. Therefore, the "practical" aspect of this behavior is "intrastratal sediment ingestion," an activity that we could interpret: a "feeding structure."

In spite of severe limitations imposed by fossilization and "information loss," however, behavioral studies can disclose very sophisticated traits, reaching far beyond our merely characterizing a given trace

fossil as a feeding structure, etc. (e.g., Seilacher, 1967a, 1967b; Raup and Seilacher, 1969).

Such studies can also yield novel results. Müller and Nogami (1972) reported various kinds of borings within conodonts, for example, some of the damage being repaired subsequently by the conodont animal. We thus know more about the behavioral interrelationships among certain organisms than we know about the organisms themselves.

Whatever the nature of the trace being analyzed, its behavioral implications warrant our close attention. Many kinds of traces offer a remarkable challenge in interpretation, some even reflecting "spurious" behavior (Fig. 2.7; see also Chapter 16); and we need a great many more field and experimental observations on recent organisms (e.g., Myers, 1972; Chapter 22), with which to compare our ancient examples. But this broad topic is the real heart of ichnology.

TRACEMAKERS AS SEDIMENTOLOGIC AGENTS

Trace fossils may be considered both as paleontologic entities and as sedimentary structures (stretching the last category a bit to include borings, etc.). In this chapter I have mostly stressed the role of tracemaking organisms in paleobiology, yet the activities and characteristics of tracemakers are equally important in sedimentology. When the distribution and genesis of all sedimentary structures present are included in a given study, for example, trace fossils and their interrelationships with physical structures add a powerful dimension to our repertoire of facies indicators (Chapters 7–9).

In spite of the present and past ubiquity of boring, burrowing, crawling, or rooting organisms in various sedimentary environments, however, their role as sedimentologic agents has been vastly underemphasized in most of the otherwise pertinent literature. This aspect of sedimentology is treated more

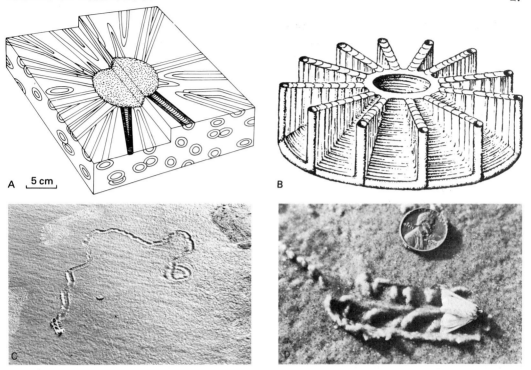

Fig. 2.7 Some lebensspuren reflecting "unusual" behavior. A, B, trace fossils representing complex foraging patterns. A, *Phoebichnus trochoides*—burrows radiating from central large shaft; mica orientation indicated for shaft and two radials; Triassic, Greenland. B, *Heliochone*—vertical and lateral concentric repetition; Devonian, Germany. C, D, struggles of a moth on its back result in snail-like "crawling trace" terminating in "arthropod resting trace." This particular occurrence stands very little chance of being preserved (a South Carolina beach, at low tide); but such "accidents" undoubtedly do get recorded in the rocks, and might be very misleading to ichnologists. [A, from Bromley and Asgaard, 1972; B, after Seilacher and Hemleben (1966); C, D, courtesy J. H. Howard III.]

fully in Chapter 8, but some salient points may be mentioned here.

A prevalent misconception, at least as the terms are generally understood, is "primary depositional fabric" and "primary sedimentary structure." Two facets of this concept need clarification. First is the semantics of "primary" itself. Many workers tend to equate "primary" strictly with physical processes of sedimentation and "secondary" strictly with biological processes of modification or reworking. But in nature, both processes often are gradational and contemporaneous; indeed, by the above definitions we can easily envision a "secondary" structure being modified by a "primary" one, as when biogenically graded sediments are rippled by gentle wave oscilla-

tions. More importantly, biogenic processes may in fact cause primary deposition, e.g., biogenic sediment trapping (Prokopovich, 1969). Realistically, therefore, "primary" and "secondary" (or tertiary and quaternary) should refer not simply to physiogenic versus biogenic but strictly to chronologic order of development—the original intent of the terms (cf. Pettijohn and Potter, 1964)—or informally, perhaps, to the magnitude of their importance in a given situation.

Second, many workers underestimate the reality of widespread biogenic reworking of sediments, and allude that they are examining a primary fabric when it in fact may be a secondary one (see Chapter 8). Before going very far in studies of sedimentary petrography, remnant magnetism in deep-sea cores, or even microfaunal distributions, for example, we need first to understand the fabric itself. (See Watkins, 1968; Simpson, 1969; Hanor and Marshall, 1971; Sarnthein, 1972.) Trace fossils and bioturbate textures are clear clues to sediment modification, of course, but these traces and biogenic textures may not be conspicuous in a given lithology, no matter how extensively the sediments have been reworked (see Chapter 23). Differences in color between burrow fills and the host sediments may be the only discernible evidence for bioturbation, and development of such colors depends largely upon the diagenesis and state of weathering of the sample (see Chapters 3, 4, and 17). With only a cursory look at such samples, important fabrics and textures easily can be overlooked.

Trace fossils and bioturbate textures have their limitations in sedimentology, just as they do in stratigraphy and paleontology. (A mud crack is a better indicator of substrate desiccation than is any biogenic sedimentary structure.) But they can provide a great deal of unambiguous, basic information (e.g., Goldring and Seilacher, 1971), and they can no more reasonably be ignored in sedimentology than in paleontology. (See Chapters 8 and 9.)

ICHNOLOGY AND UNIFORMITARIANISM

Uniformitarian principles are the ultimate basis for interpretations and reconstructions in ichnology (e.g., Fig. 2.1; see Stanley and Fagerstrom, 1974), just as they are in most other areas of sedimentary geology. This axiom is a recurrent theme throughout the book and hardly needs to be outlined in detail here.

One facet of uniformitarianism that perhaps does need to be stressed here, however, is that we should not be blindly obsessed with finding "unique," one-to-one analogs between certain kinds of recent and

ancient lebensspuren, or between such structures and the specific kinds of environments in which they occur. Variations in animal behavior or trace morphology, or their broad or overlapping environmental ranges, are equally as important as the "unique" indicators in our overall characterizations of recent and ancient assemblages —the search for the "ultimate truth" (see Crimes, 1973, p. 119).

Worthy of note in this regard are several kinds of traces described by Reineck (1973) and Reineck and Singh (1973) from cores taken on the continental shelf, slope, and abyss off east Africa. On the basis of shapes and sizes alone, we would have considerable difficulty relating these traces to our "standard" scale of bathymetry (Table 2.1). In addition to intense bioturbations by various animals and such recognizable feeding burrows as *Zoophycos* (see Chapters 6, 17, and 21), several types of burrows having distinct walls were documented. One of these (Fig. 2.8A) is a branched structure strongly resembling the pattern of bifurcation seen in burrow systems of shallow-water thalassinideans and similar shrimp. The walls are not thickly lined, but the character of the sediment fill suggests an abandoned dwelling burrow. Equally significant is a type of dwelling burrow having walls as thick as 3 mm (Fig. 2.8B); the overall burrow is vermiform, about 10 mm in diameter, and oriented predominantly horizontally. Deep-sea benthos are mostly deposit feeders and therefore do not ordinarily expend much energy in maintaining permanent domiciles; the above burrows represent obvious and important exceptions, however, reflecting behavioral responses to environmental conditions other than simply the depth of water or distance from shore.

Another burrow type observed in the African cores that is found commonly in shallow-water carbonates is the tiny, vertical, pyrite-filled structure *Trichichnus* (Fig. 2.8C). The pyrite evidently represents decay and sulfide enrichment of organic

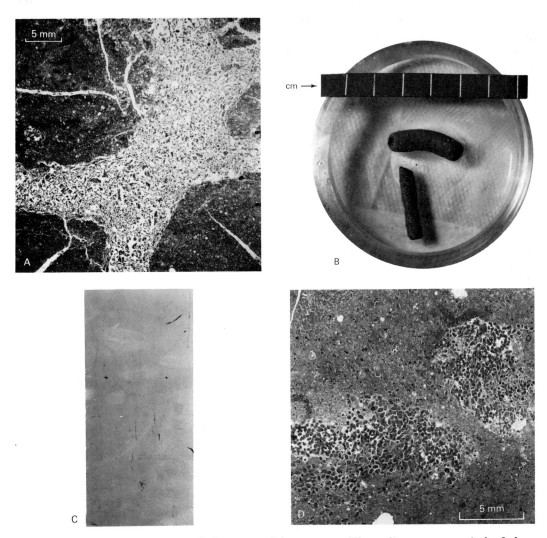

Fig. 2.8 Characteristics of selected deep-water lebensspuren. The sediments are typical of deep water, but such traces occur commonly in shallow-water deposits. A, thin section of ?thalassinidean burrow filled with foraminiferal sand; water depth, 930 m. B, fragments of thick-walled burrows washed free of matrix; burrow lining consists of agglutinated *Globigerina* ooze; 4,355 m. C, x-radiograph of peel of bioturbated *Globigerina* ooze containing numerous very small burrows party filled with pyrite; ca. 3,500 m. D, thin section of deep-sea clay showing concentrations of fecal pellets; 4,690 m. Off east Africa. (Photos courtesy H.-E. Reineck.)

slimes or burrow linings (Chapters 4 and 17). Off Africa, these structures occur mostly at water depths between 3,500 and 3,800 m, more than 1 m within the substrate (*Globigerina* ooze).

Distinct fecal pellets were also observed off Africa. In general, elongate fecal strings or simple, sinuous castings are more typical of deep-sea benthos (Chapter 21), but coherent pellets such as those found com-

monly in shallow-water environments do occur (Fig. 2.8D). The fecal pellets were found much more abundantly in deep-sea clays than in globigerinid oozes.

Reverse Uniformitarianism

In uniformitarianism we say that "the present is the key to the past." However, the reverse may also be true: in ichnology, the past is often more informative than the present, or at least insofar as we have been able to study the present. As Seilacher (1964) put it, the present has shown us "how the locks work" but has seldom provided exact "keys to the past."

The reasons for this disparity are varied, but the main one is simply a matter of logistics and technology. In Cretaceous chalk we know of pencil-thin marine burrows that evidently extended as much as 9 m into the original substrate (Chapter 16); even if an animal having such behavior is alive today, how can we hope to recover it and its entire burrow in a core? or set up an aquarium in order to watch its activities? This example is not as extreme as one might think. Polyester plastics are well suited for casting burrows such as those of the mantis shrimp *Squilla* (Frey and Howard, 1969, Pl. 4, fig. 2); the plastic works equally well for casting burrow systems of the ghost shrimp *Callianassa major* (Weimer and Hoyt, 1964), yet the resulting cast is so deep (3–4 m) and so extensively criss-crossed through the substrate that it simply cannot be recovered intact.

Even among the more "typical" occurrences of lebensspuren, intrastratal structures are much easier to study at the outcrop than in loose recent sediments. Most recent biogenic structures, especially aquatic ones, require special techniques; some spectacular advances have been made in our technology (Chapters 12, 22, and 23), but the methods available still are not always the answer to a given need. How do we recover a broad biogenic structure that consists of both open and filled parts? The burrow system of *Callianassa major* commonly contains sediment-filled cutoffs (Weimer and Hoyt, 1964, Pl. 123, fig. 5), and this kind of sampling problem is a general one. In *C. major* burrows, the cutoffs are obvious in cores and trenched substrates because of the thick burrow walls; but in many kinds of thinly lined burrows, we probably fail to see the cutoffs simply because they are so inconspicuous (Seilacher,

1957). Unless the cutoffs are accentuated by differences in color or sediment fabric, we generally notice only the last-formed part of the structure (Fig. 2.9). With the enhancement typically wrought by lithification and post-diagenetic processes, however, the entire burrow or burrow system is apt to be conspicuous in the eventual rock.

X-radiography of recent sediments is a partial answer to this problem (Chapters 22 and 23), yet the effective depth of penetration of x-rays is a severe limitation on the study of large, three-dimensional structures. One can increase the field of view in aquaria somewhat by using cryolite as a substrate (its refractive index is nearly the same as that of water), but then we tend to see only the animal and not its lebensspuren. (See Josephson and Flessa, 1972.)

Another problem in aquatic ichnology is that our data on recent traces are biased heavily in favor of ones found in the more accessible environments, even though these traces and environments may be poorly represented in the rock record. The surfaces of beaches, bars, and tidal flats have probably been examined in more detail by more people than all other marine environments combined, yet the traces found on these surfaces are the least likely of all to be preserved. Even shallow burrows in such places stand little chance of preservation. A paradox in this case—or an example of the importance of "information loss" (taphonomy) in ichnology—is that shallow burrowers can even be the dominant animals of an assemblage and yet be only poorly or not at all represented in the final fossil record. (See Chapter 20.) An example from the Georgia coast is the coquina clam *Donax variabilis*; its burrows are exceedingly abundant locally (Fig. 2.10), yet they are obliterated periodically even by slight increases in surf action, and are virtually unknown in equivalent Pleistocene sediments where they must have occurred in comparable numbers.

Shallow offshore areas and their lebensspuren are much better represented in the

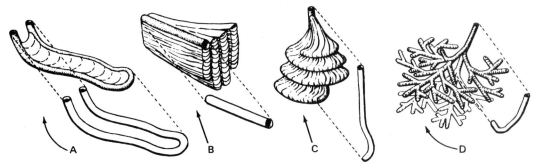

Fig. 2.9 Hypothetical differences between lebensspuren as observed in rock (total structure) and in recent substrates (extended parts of structure). A, *Rhizocorallium*. B, *Teichichnus*. C, *Daedalus*. D, *Chondrites*. (After Seilacher, 1957.)

rock record, and fortunately we now have the means of examining these areas in considerable detail (e.g., Chapter 20).

Deep-water benthos and their traces offer the greatest challenge in sampling, and correspondingly less is known about them. The organisms ordinarily cannot be recovered alive for aquarial studies, and large cores are difficult, expensive, and very time-consuming to take. Most of our information thus comes simply from sea-floor photographs (Chapter 21). These photos are very valuable because they show us many surficial traces that do have close analogs in the rock record (e.g., Seilacher, 1967a; Chapter 7); yet they also reveal the presence of various kinds of burrow openings, and we know very little about most of these intrastratal structures or their inhabitants (cf. Fig. 2.8A, B).

The Human Element

Many of the things discussed above might be labeled "problems in ichnology." Another facet of our science might be labeled "problems with ichnologists." The human element is an undeniable, highly subjective component of earth science, and ichnology is no exception. We see examples of this subjectivity at all levels of complexity, from trying to decide, by definition, just what is or is not a "trace fossil," to our artificial compartmentalization of continuous gradations in plant and animal behavior (e.g., Chapters 3 and 10). In itself, the human element is neither good nor bad, but simply the way we see nature. The problem comes in reconciling the different ways that we perceive the same "natural relationships."

From an immediate or practical standpoint, however, the resulting problems can be much more mundane than philosophical. For example, difficulties in comparing recent and ancient structures may be as simple (albeit sometimes subtle) as the particular way in which we orient the specimens. Many trace fossils are preserved better at lithologic interfaces than within beds (see Chapter 4); but like physical sedimentary structures, casts on the sole of the overlying bed may be considerably more distinctive and conspicuous than the original structure on the top of the host bed (Pettijohn and Potter, 1964). Our first tendency, especially if the specimen was collected as float, is to turn these sole casts upside down so that they appear to be on the "top" of the bed (Fig. 2.11); our second tendency may be to compare the inverted structures directly with recent traces observed in original configuration, e.g., as photographed on beaches, tidal flats, or the ocean floor. For simple horizontal burrows, the inversion may make little ultimate difference; but for U-shaped or branched vertical burrows and crawling or resting traces, the goodness of fit may range from "frustratingly poor" to "deceitfully good."

(See Heinberg, 1970.) Even when aware of the problem, however, we cannot always avoid it; numerous authors fail to specify which way is "up" in their published illustrations. Of course, many trace fossils are themselves good "way up" indicators (see Chapter 7).

A more important problem in ichnology, as in all other branches of sedimentary geology, is that some of the best or most striking analogs tend to be overused or overextended. In the literature on carbonate rocks, we see many instances in which oolitic limestones are compared indiscriminately with Bahamian oolite shoals, for example, and in ichnological literature we see equally uncritical comparisons between the trace fossil *Ophiomorpha* and burrows of the recent shrimp *Callianassa*, especially *C. major* (Weimer and Hoyt, 1964). The paper by Weimer and Hoyt certainly deserves its place as a classic, but subsequent literature shows that many writers lack an awareness of other possibilities. First, not all species of *Callianassa* construct thick, knobby burrow linings [e.g., *C. californiensis* does not (Warme, 1967)], and *Callianassa* species are not the only animals that do construct such burrows [e.g., *Upogebia pugettensis* does (Thompson, 1972)]. Certain species of *Callianassa* may be nestlers or even "borers" (Chapter 11; Bromley and Frey, 1974).

Second, even along the Georgia coast where Weimer and Hoyt did their work, the significance of *Callianassa major* is not unique; it is characteristic of "high energy beaches," but it also inhabits shoals, tidal flats, point bars in tidal streams, and other places (Frey and Howard, 1969; Howard and Frey, 1973). Furthermore, the environmental range of *C. major* overlaps that of the smaller species *C. biformis*, which makes equally distinctive *Ophiomorpha*-type burrows and which is found both intertidally and subtidally (Hertweck, 1972; Howard and Dörjes, 1972; Chapter 20). In a statistical sense these two burrow forms may represent intertidal to very shallow, inshore waters and somewhat deeper, offshore waters, respectively. But in the fossil record we would not wish to base our interpretation exclusively upon a few scattered burrows, and certainly not upon mere identification of the trace fossils as *Ophiomorpha*. In addition to the environments mentioned above, such burrows evidently can occur in the *Zoophycos* assemblage (Macsotay, 1967; see Table 2.1), or even at tremendous water depths (Kern and Warme, 1974).

Third, even the concept "high energy," as many workers interpret Weimer and Hoyt's work, needs additional qualification. The beaches inhabited by *Callianassa major* do represent some of the highest energy levels recorded along the Georgia coast, but the overall energy level of this coast is much lower than at practically any other place along the Atlantic seaboard (Tanner, 1960, Fig. 2). Thus, although these beaches are probably analogous to those of shallow epeiric seas, they are hardly our image of high-energy beaches along most open-ocean coasts.

As a final, somewhat related example, we frequently encounter the misconception that thalassinidean shrimp are also responsible for all burrows of the general *Thalassinoides* type. In practice, several other animals are capable of making such structures (e.g., Figs. 2.6, 19.2A; Frey and Howard, 1969, Pl. 4, fig. 2), and considerable additional study of them is needed. (See also Bromley and Frey, 1974.)

The foregoing examples help illustrate the thoroughness and thoughtfulness that should attend our comparisons of the past

◄ **Fig. 2.10** Profuse dwelling burrows of recent pelecypod *Donax variabilis*. A, substrate surface pocked with burrow openings. B, view of entire burrow, including inhalent and exhalent water canals. C, x-radiograph of sediment peel from can core, showing abundance and distribution of clams. Tidal flat; Cabretta Island, Georgia.

Fig. 2.11 Casts of physical and biogenic sedimentary structures, on sole of sandstone bed. Although the structures appear realistic in form, the ripple "crests" actually represent troughs, and vice-versa, and the tiny burrow casts are upside down. Silurian; Georgia. (From Frey and Chowns, 1972.)

and present. Equally important are our commonsense interpretations of given specimens or situations. A novel but laudable example is the use of dinosaur footprints as marine paleobathometers in the Glen Rose Limestone of Texas (see Perkins and Stewart, 1971). Some of the tracks seem to have been made on mudflats (they are associated with mudcracks), others in extremely shallow water (they are deeply impressed in the nondesiccated substrate), and still others in water just deep enough to buoy the animal's body slightly (the tracks are slurred and lightly impressed). Many of the tracks were impressed into living rudist colonies; afterward, rudists established themselves within the footprints. After submarine lithification, many of the Glen Rose footprints were occupied by various invertebrate tracemakers (see Fig. 18.5A).

In practice, of course, even the "best" of

our uniformitarian (or "actualistic") principles and commonsense observations cannot guarantee correct interpretations. A classic example of specious or ill-founded comparisons is the "fucoid" concept (see Chapters 1 and 6). Here, various early workers mistakenly identified scores of trace fossils as marine algae or "seaweeds." These workers commonly figured recent thallophytes in order to strengthen their comparisons with the ancient "fucoids" [in come cases the drawings of the fossils were even modified so that they more nearly resembled algae! (Häntzschel, 1962, p. W179)]. The true origin of these "seaweeds" is now generally well known (e.g., Fry and Banks, 1955, p. 37), yet the "fucoid" idea lingers on. For instance, natural history museums in several nations still feature dioramas of Paleozoic sea life in which seaweeds loom prominent (Fig. 2.12), although

Fig. 2.12   Reconstruction of Ordovician sea life. The seaweeds shown here are diagrammatic but bear striking resemblance to the trace fossil *Chondrites* (see Fig. 1.1), which actually may have been the basis for their "reconstruction." (Photo courtesy Field Museum of Natural History, Chicago— with my apologies for having used it as a specific example; numerous other dioramas could have been chosen.)

little real evidence is available for them. True fucoidal algae probably did inhabit such sea floors, thus the dioramas are plausible; but the actual "evidence" for the weeds consists mostly of *trace fossils*, not plant fossils. The "near accuracy" of the dioramas is almost a matter of serendipity.[3] To be technically correct, however, the

[3] In spite of the scanty fossil record of noncalcareous algae, paleobotanists with whom I have discussed the problem (e.g., G. F. Elliott and H. P. Banks, 1973, separate personal communications) feel that brown algae were probably diverse and widespread during the early Paleozoic. Algae in general have an extremely long geologic range, the major lineages having been established early in time; and occasional "glimpses" of probable brown algae are seen in the rock record (e.g., Fry and Banks, 1955; Phillips et al., 1972). Precambrian and lower Paleozoic black cherts have yielded good algal fossils. Circumstantial evidence includes accumulations of minute fossil gastropods of a kind that today live on such weeds, although no vestiges of seaweeds were found in association with the fossil snails.

Aside from the paucity of good fossil specimens of "soft" algae, the main problem at present is the much-needed but "thankless" task of completely sorting out the animal traces from the algal fossils (cf. Andrews, 1970), a task requiring expertise not only in paleobotany and ichnology but also in sedimentary petrography.

"seaweeds" in most dioramas should be shown not as plants extending up into the water but as burrows extending down into the substrate! Another consequence of this technicality is that, without better evidence for the abundance of true, photosynthetic algae, we cannot infer from such reconstructions alone that the associated trilobites, nautiloids, and brachiopods (Fig. 2.12) necessarily lived in "shallow, clear" waters.

We hope not to make such mistakes ourselves; but like our predecessors, we can only interpret the evidence as best we can, in light of current knowledge. We can, of course, improve our chances of success by gathering data from as many different sources as possible—physical, chemical, and biological—before making our ultimate interpretations.

CONCLUSIONS

The study of trace fossils involves several concepts and approaches that are somewhat unusual among other branches of geology and biology, perhaps imparting a "mystical" or esoteric quality to the research. But mysticism is merely a synonym of "unfamiliar," and in this and the following chapters we hope to dispel a great deal of the unfamiliarity and to show how ichnology is rapidly becoming a mature, well-founded subdiscipline, an integral part of the geological and biological sciences. To varying degrees of sophistication, ichnology can make substantial contributions in numerous areas of geology:

Paleontology and paleoecology: (1) fossil record of soft-bodied animals, (2) evolution of the metazoa, (3) evidence of activity by organisms, (4) evolution of behavior, (5) diversity in fossil assemblages, and (6) trophic levels in fossil assemblages.
Stratigraphy and structural geology: (1) biostratigraphy of "unfossiliferous" rocks, (2) correlation by marker beds, (3) structural attitude of beds, and (4) structural deformation of sediments.
Sedimentology: (1) production of sediment by boring organisms, (2) trapping of sediment by organisms, (3) alteration of grains by

sediment-ingesting animals, and (4) sediment reworking, including (a) destruction of initial fabrics and sedimentary structures and (b) creation of new fabrics and sedimentary structures.
Depositional environments: (1) bathymetry, (2) temperature and salinity, (3) depositional history, including (a) rates of deposition and (b) amounts of sediment deposited or eroded, (4) aeration of water and sediments, and (5) substrate coherence and stability.
Consolidation of sediments: (1) initial history of lithification and (2) measures of compaction.

ACKNOWLEDGMENTS

For their critical review of this chapter, I thank R. G. Bromley, Københavns Universitet; J. D. Howard, Skidaway Institute of Oceanography; and B. K. Sen Gupta, University of Georgia. I am also grateful to the several persons who supplied illustrations, and to the National Science Foundation for support in much of my original work on fossil and recent traces (grants GA-719, GA-10888, GA-22710, GA-30565, and GA-39999X).

REFERENCES

Aceñolaza, F. G. and F. Durand. 1973. Trazas fosiles del basamento cristalino del noroeste Argentino. Asoc. Geol. Cordoba, Bol., 2(1–2):45–56.

Ager, D. V. 1963. Principles of paleoecology. New York, McGraw-Hill, 371 p.

Andrews, H. N. 1970. Index of generic names of fossil plants, 1820–1965. U.S. Geol. Survey, Bull. 1300, 354 p.

Asgaard, U. and R. G. Bromley. 1974. Sporfossiler fra den mellem miocæne transgression i Søby-Fasterholt området. Dansk geol. Foren., Årsskrift 1973: 11–19. (in Danish, with English summary)

Bromley, R. G. and U. Asgaard. 1972. Notes on Greenland trace fossils. (Pts. 1–III). Geol. Survey Greenland, Rept. 49:1–30.

———— and R. W. Frey. 1974. Redescription of the trace fossil *Gyrolithes*, and taxonomic evaluation of *Thalassinoides*, *Ophiomorpha* and *Spongeliomorpha*. Geol. Soc. Denmark, Bull., 23:311–335.

Clark, E. 1972. The Red Sea's gardens of eels. Natl. Geogr., 142:724–735.

Colton, G. W. 1967. Late Devonian current directions in western New York with special reference to *Fucoides graphica*. Jour. Geol., 75:11–22.

Crimes, T. P. 1970. The significance of trace fossils in sedimentology, stratigraphy and palaeoecology, with examples from lower Palaeozoic strata. In T. P. Crimes and J. C. Harper (eds.), Trace fossils. Geol. Jour., Spec. Issue 3:101–126.

————. 1973. From limestones to distal turbidites: a facies and trace fossil analysis in the Zumaya flysch (Paleocene–Eocene), north Spain. Sedimentology, 20:105–131.

Evans, J. W. 1967. Relationship between *Penitella penita* (Conrad, 1837) and other organisms of the rocky shore. Veliger, 10:148–151.

Forbes, A. T. 1973. An unusual abbreviated larval life in the estuarine prawn *Callianassa kraussi* (Crustacea:Decapoda:Thalassinidea). Marine Biol., 22:361–365.

Frey, R. W. 1971. Ichnology—the study of fossil and recent lebensspuren. In B. F. Perkins (ed.), Trace fossils, a field guide. Louisiana State Univ., School Geosci., Misc. Publ. 71-1:91–125.

———— and T. M. Chowns. 1972. Trace fossils from the Ringgold road cut (Ordovician and Silurian), Georgia. In T. M. Chowns (comp.), Sedimentary environments in the Paleozoic rocks of northwest Georgia. Georgia Geol. Survey, Guidebook 11:25–55.

———— and J. Cowles. 1969. New observations on *Tisoa*, a trace fossil from the Lincoln Creek Formation (mid-Tertiary) of Washington. Compass, 47:10–22.

———— and J. D. Howard. 1969. A profile of biogenic sedimentary structures in a Holocene barrier island–salt marsh complex, Georgia. Gulf Coast Assoc. Geol. Socs., Trans., 19:427–444.

Fry, W. L. and H. P. Banks. 1955. Three new genera of algae from the Upper Devonian of New York. Jour. Paleont., 29:37–44.

Goldring, R. and A. Seilacher. 1971. Limulid undertracks and their sedimentological implications. Neues Jahrb. Geol. Paläont., Abh., 137:422–442.

Gray, I. E. 1961. Changes in abundance of the commensal crabs of *Chaetopterus*. Biol. Bull., 120:353–359.

Hanor, J. S. and N. F. Marshall. 1971. Mixing of sediment by organisms. In B. F. Perkins (ed.), Trace fossils, a field guide. Louisiana State Univ., School Geosci., Misc. Publ. 71-1:127–135.

Häntzschel, W. 1962. Trace fossils and problematica. In R. C. Moore (ed.), Treatise on invertebrate paleontology, Pt. W, Miscellanea. Lawrence, Kan., Geol. Soc. America and Univ. Kansas Press, p. W177–W245.

Heinberg, C. 1970. Some Jurassic trace fossils from Jameson Land (East Greenland). In T. P. Crimes and J. C. Harper (eds.), Trace fossils. Geol. Jour., Spec. Issue 3:227–234.

Hertweck, G. 1972. Georgia coastal region, Sapelo Island, U.S.A.: sedimentology and biology. V. Distribution and environmental significance of lebensspuren and in-situ skeletal remains. Senckenbergiana Marit., 4:125–167.

Howard, J. D. 1966. Sedimentation of the Panther Sandstone Tongue. Utah Geol. Mineral. Survey, Bull. 80:23–33.

———— and J. Dörjes. 1972. Animal–sediment relationships in two beach-related tidal flats; Sapelo Island, Georgia. Jour. Sed. Petrol., 42:608–623.

———— and R. W. Frey. 1973. Characteristic physical and biogenic sedimentary structures in Georgia estuaries. Amer. Assoc. Petrol. Geol., Bull., 57:1169–1184.

Jacobsen, V. H. 1967. The feeding of the lugworm, *Arenicola marina* (L.). Quantitative studies. Ophelia, 4:91–109.

Josephson, R. K. and K. W. Flessa. 1972. Cryolite: a medium for the study of burrowing aquatic organisms. Limnol. Oceanogr., 17:134–135.

Karplus, I. et al. 1974. The burrows of alpheid shrimp associated with gobiid fish in the northern Red Sea. Marine Biol., 24:259–268.

Kennedy, W. J. and B. W. Sellwood. 1970. *Ophiomorpha nodosa* Lundgren, a marine indicator from the Sparnacian of south-east England. Geologists' Assoc., Proc., 81:99–110.

Kern, J. P. and J. E. Warme. 1974. Trace fossils and bathymetry of the Upper Cretaceous Point Loma Formation, San Diego, California. Geol. Soc. America, Bull., 85:893–900.

Köhnlein, J. and K. Bergt. 1971. Untersuchungen zur Entstehung der biogenen Durchporung im Unterboden eingedeichter Marschen. Zeitschr. Acker- u. Pflanzenbau, 133:261–298.

Lawrence, D. R. 1968. Taphonomy and information losses in fossil communities. Geol. Soc. America, Bull., 79:1315–1330.

MacGinitie, G. E. and N. MacGinitie. 1968. Natural history of marine animals (2nd ed.). New York, McGraw-Hill, 523 p.

Macsotay, O. 1967. Huellas problematicas y su valor paleoecologico en Venezuela. Geos, 16:7–79.

Müller, K. J. and Y. Nogami. 1972. Entöken und Bohrspuren bei den Conodontophorida. Paläont. Zeitschr., 46:68–86.

Myers, A. C. 1972. Tube-worm-sediment relationships of *Diopatra cuprea* (Polychaeta: Onuphidae). Marine Biol., 17:350–356.

Odum, E. P. 1971. Fundamentals of ecology. W. B. Saunders, 574 p.

Oertel, G. F. 1973. Examination of textures and structures of mud in layered sediments at the entrance of a Georgia tidal inlet. Jour. Sed. Petrol., 43:33–41.

Perkins, B. F. and C. L. Stewart. 1971. Stop 7: Dinosaur Valley State Park. In B. F. Perkins (ed.), Trace fossils, a field guide. Louisiana State Univ., School Geosci., Misc. Publ. 71-1:56–59.

Pettijohn, F. J. and P. E. Potter. 1964. Atlas and glossary of primary sedimentary structures. New York, Springer-Verlag, 386 p.

Phillips, T. L. et al. 1972. Morphology and vertical distribution of Protosalvinia (Foerstia) from the New Albany Shale (Upper Devonian). Rev. Paleobot. Palynol., 14:171–196.

Prokopovich, N. P. 1969. Deposition of clastic sediments by clams. Jour. Sed. Petrol., 39:891–901.

Raup, D. M. and A. Seilacher. 1969. Fossil foraging behavior: computer simulation. Science, 166:994–995.

Reineck, H.-E. 1973. Schichtung und Wühlgefüge in Grundproben vor der ostafrikanischen Küste. Meteor Forschungsergebnisse, 16:67–81.

————— and I. B. Singh. 1973. Depositional sedimentary environments. New York, Springer-Verlag, 439 p.

Rice, A. L. and C. J. Chapman. 1971. Observations on the burrows and burrowing behaviour of two mud-dwelling decapod crustaceans, Nephrops norvegicus and Goneplax rhomboides. Marine Biol., 10:330–342.

Sarnthein, M. 1972. Stratigraphic contamination by vertical bioturbation in Holocene shelf sediments. 24th Internat. Geol. Congr., Montreal, Proc. 6:432–436.

Schäfer, W. 1972. Ecology and palaeoecology of marine environments. Edinburgh and Chicago, Oliver & Boyd and Univ. Chicago Press, 568 p.

Schloz, W. 1972. Zur Bildungsgeschichte der Oolithenbank (Hettangium) in Baden-Württemberg. Inst. Geol. Paläont. Univ. Stuttgart, Arb., 67:101–212.

Seilacher, A. 1953. Studien zur Palichnologie. I. Über die Methoden der Palichnologie. Neues Jahrb. Geol. Paläont., Abh., 96:421–452.

—————. 1957. An-aktualistisches Wattenmeer? Paläont. Zeitschr., 31:198–206.

—————. 1964. Biogenic sedimentary structures. In J. Imbrie and N. D. Newell (eds.), Approaches to paleoecology. New York, John Wiley, p. 296–316.

—————. 1967a. Bathymetry of trace fossils. Marine Geol., 5:413–428.

—————. 1967b. Fossil behavior. Scientific Amer., 217:72–80.

————— and C. Hemleben. 1966. Beiträge zur Sedimentation und Fossilführung des Hunsrückschiefers. 14. Spurenfauna und Bildungstiefe der Hunsrückschiefer (Unterdevon). Notizbl. hess. L.-Amt Bodenforsch., 94:40–53.

Sellwood, B. W. 1971. A Thalassinoides burrow containing the crustacean Glyphaea udressieri (Meyer) from the Bathonian of Oxfordshire. Palaeontology, 14:589–591.

Shinn, E. A. 1972. Worm and algal-built columnar stromatolites in the Persian Gulf. Jour. Sed. Petrol., 42:837–840.

Simpson, F. 1969. Interfacial assemblages of Foraminifera in the Carpathian flysch. Soc. Géol. Pologne, Ann., 39:471–486.

Stanley, K. O. and J. A. Fagerstrom. 1974. Miocene invertebrate trace fossils from a braided river environment, western Nebraska, U.S.A. Palaeogeogr., Palaeoclimatol., Palaeoecol., 15:63–82.

Tanner, W. F. 1960. Florida coastal classification. Gulf Coast Assoc. Geol. Socs., Trans., 10:259–266.

Thompson, R. K. 1972. Functional morphology of the hind-gut gland of Upogebia pugettensis (Crustacea, Thalassinidea) and its role in burrow construction. Unpubl. Ph.D. Dissert., Univ. California, Berkeley, 202 p.

Valentine, J. W. 1973. Evolutionary paleoecology of the marine biosphere. Englewood Cliffs, N.J., Prentice-Hall, 511 p.

Voigt, E. 1972. Tonrollen als potentielle Pseudofossilien. Natur u. Museum, 102:401–410.

Warme, J. E. 1967. Graded bedding in the recent sediments of Mugu Lagoon, California. Jour. Sed. Petrol., 37:540–547.

Watkins, N. D. 1968. Short period geomagnetic polarity events in deep-sea sedimentary cores. Earth Planet. Sci. Letters, 4:341–349.

Weimer, R. J. and J. H. Hoyt. 1964. Burrows of Callianassa major Say, geologic indicators of littoral and shallow neritic environments. Jour. Paleont., 38:761–767.

Williams, A. B. 1965. Marine decapod crustaceans of the Carolinas. Bureau Commercial Fish., Fish. Bull., 65(1):1–298.

CHAPTER 3

CLASSIFICATION
OF TRACE FOSSILS

SCOTT SIMPSON

Department of Geology, University of Exeter

Exeter, England

SYNOPSIS

Classification is one of the tools of scientific discovery, no less so in the study of trace fossils than in any other science. Many different kinds of classification are possible. Simple descriptive classifications, although better than nothing, are of very limited value. Trace fossils have three distinct and significant aspects, to each of which a unique classification attaches: (1) the preservational (stratinomic), which treats of the origin of the fossil in the rocks, (2) the behavioral (ethological), which treats of the biological function represented in the fossil, and (3) the phylogenetic (taxonomic), which is concerned with the identity of the organism that produced the fossil.

Relationships established in the stratinomic classification are limited in number and are in-formative mainly about the characteristics of sedimentation when and where fossilization took place. The ethological classification is generally the most appropriate, because it provides insight into the environment at the time of formation. The taxonomic classification is of very limited applicability; phyletic identifications are possible only in a few cases, mainly certain tracks or trails.

Agreed nomenclatural procedures, involving the acceptance of a code for a binomial system of names related to types, are beset with difficulties. But if any nomenclature is devised for international adoption, it should be based on a classification of trace fossils as the product of organic behavioral patterns.

INTRODUCTION

In this chapter we are concerned with the classification of trace fossils as natural objects; the discussion deals particularly with the methods of classification, but also to some extent with the procedures of nomenclature. The approach is philosophic, and no attempt was made to provide an exhaustive treatment of rigorously defined terms and concepts. This convention is followed for two reasons: (1) the whole subject of trace fossils has been approached so often via a historical, classificatory route, and (2) a critical review of specific terms and concepts—representing a consensus among 33 ichnologists from 12 countries—has recently been published (Frey, 1973); most of this information need not be repeated here.

PRINCIPLES OF ICHNOLOGICAL CLASSIFICATION

Classification consists in the orderly arrangement of data—the grouping together of like things and the separation of unlike things. It is one of the essential activities of science, whether for its own sake or as a tool in the discovery of new or unexpected relations.

Because natural objects are generally complex and possess many different, vari-

39

able characters, many different kinds of classification are usually possible. Thus, human beings may be classified according to sex, religion, language, kinship, blood group, and many other characteristics. No one classification is best—only a most suitable one for a particular purpose. However, some characters are much more profound or significant than others, generally because they were imposed at the moment of origin. These characters are the basis for the genetic classifications. More superficial types of classification are based on minor features and are generally called "descriptive." Descriptive schemes may be highly detailed but are of very limited applicability; they are the "working" or field classifications, which have been developed for particular purposes and tend to be of an ephemeral nature.

A unique genetic classification is the natural classification of biological systematics, which is based on organic descent (phylogenetic). It is genetic in a far more profound way than other genetic classifications.

A classification of trace fossils implies a definition of the concept "trace fossil." The concept as it is understood here includes borings (bioerosion structures) made in a hard substrate as well as the bioturbation structures made in an unconsolidated particulate substrate. These concepts of the differentiation of biogenic structures were rigorously defined and explained by Frey (1973) in a table reproduced here as Table 3.1.

Worth noting is that animals interacting with sediment may produce lamination and bedding and that algae produce stromatolites. These features were considered by Frey (1973) to form a separate group of biogenic sedimentary structures, the biostratification structures, besides the bioturbation structures and other entities that constitute the true trace fossils.

Central to the concept of a trace fossil is a shape defined by sedimentary boundaries, which can be explained in terms of an organism moving its parts within, or on, incoherent sediment or hard rock.

Examples agreeing with the above description will be acceptable to all geologists. But because of geologists' diverse lines of approach, they have widely differing ideas about what should be excluded or included at the periphery (cf. Chapters 10 and 13). Workers to whom the behavioral aspect of a trace fossil is of chief interest consider any practice unnatural that excludes borings from the various sorts of causative behavior (as is done by some authors), but perfectly natural to exclude stromatolites and the other, as yet unnamed, biostratification structures. In contrast, workers whose concern is mainly with the stratinomic aspects are quite happy to exclude borings. One sort of thing that is here excluded from the concept of trace fossil is exoskeletons that could be treated as body fossils, such as the "test" of the polychaete *Pectinaria*, and the permanent body-size tubes of other polychaete worms. In contrast, *Lanice* and *Sabellaria*, which —although also polychaetes—move about only within their permanently sessile dwelling tubes (unlike *Pectinaria*, which humps its house as it moves about), are appropriately classed as trace fossils (potentially!). Conversely, another sort of thing here treated as trace fossils, but not acceptable as such by all workers, are feces, fecal pellets, coprolites, castings, etc.

Our hope is that the consensus we represent will persuade other authors, by the logic of our arguments, to accept and use the concepts we recommend.

An essential feature of trace fossils is that they are, at one and the same time, (1) sedimentary structures (excepting borings), (2) traces of organic activity, and (3) the product of particular sorts of organisms. Three separate classifications apply to these three aspects. The first, which is concerned with trace fossils as sedimentary structures, is essentially morphological and descriptive.

TABLE 3.1   Basic Concepts in the Study of Biogenic Structures.*

Differentiation of Biogenic Structures

1. *Biogenic structure*—in ichnology, tangible evidence of activity by an organism, fossil or recent, other than the production of body parts. Embraces the entire spectrum of substrate traces or structures that reflect a behavioral function: biogenic sedimenatry structures, bioerosion structures, and other miscellaneous features representing activity. Excludes molds of body fossils that result from passive contact between body parts and the host substrate, but not imprints made by the body parts of active organisms.

   A. *Biogenic sedimentary structure*—biogenic structure produced by the activity of an organism upon or within an unconsolidated particulate substrate: bioturbation structures and biostratification structures.

      (1) *Bioturbation structure*—biogenic sedimentary structure that reflects the disruption of biogenic and physical stratification features or sediment fabrics by the activity of an organism: tracks, trails, burrows, and similar structures.

      (2) *Biostratification structure*—biogenic sedimentary structure consisting of stratification features imparted by the activity of an organism: biogenic graded bedding, byssal mats, certain stromatolites, and others.

   B. *Bioerosion structure*—biogenic structure excavated mechanically or biochemically by an organism into a rigid substrate: borings, gnawings, scrapings, bitings, and related traces.

Disciplines and Components

2. *Ichnology*—overall study of traces made by organisms, including their description, classification, and interpretation. Divisions include *palichnology* (= *paleoichnology*) for fossil traces, and *neoichnology* for recent ones.

   A. *Trace*—in ichnology, an individually distinctive biogenic structure, especially one that is related more or less directly to the morphology of the organism that made it: tracks, trails, burrows, borings, coprolites, fecal castings, and similar features, fossil or recent (= *lebensspur*). Excludes biostratification structures and other traces lacking diagnostic anatomical features.

   B. *Ichnocoenose*—assemblage of traces. Components include the *ichnofauna*, or animal traces, and the *ichnoflora*, or plant traces (such as algal borings).

   C. *Trace fossil*—fossil trace (= *ichnofossil*).

3. *Ethology*—in ichnology, the study or interpretation of the behavior of organisms as reflected by their traces.

* From Frey, 1973.

It is the preservational classification, which includes the stratinomical and toponomical facets. The second, which is concerned with trace fossils as the result of activity by organisms, is the ecological or ethological or behavioral classification; it is a genetic scheme. The third is the natural classification of biological systematics itself; this classification is applicable only to a very minor part of all trace fossils, because the identity of the animals concerned is usually not known.[1]

The prime stimulus to the recent growth of palichnology was the discovery that many of its phenomena are significant for the interpretation of problems in sedimentology, facies analysis, and diagenesis. The preservational and behavioral classifications mentioned above facilitate the discussion and understanding of the abundant palichnological data from sedimentary rocks. A system for naming the constituent units of the classifications obviously facilitates discussion. Nomenclatorial conventions are being worked out; a tendency at the present time is to try getting away from a rigid binomial system, as less emphasis is laid on the fossil as the product of a particular animal (taxonomy), and more on its preservational and behavioral aspects.

But before this recent growth, the subject was dominated by the traditional paleontological approach, together with all its preconceptions (see Chapter 1). The striking superficial similarity between some body fossils and trace fossils obscured the profound difference between these two categories of objects. The main concern of early workers was to discover the stratigraphic "age" or range of the fossils so that they might be used as "medals of creation." Consequently, a Linnean binomial nomenclature was unquestioningly applied.

Although much progress is being made with the behavioral approach and nomenclature, many well-defined and common trace fossils remain to be interpreted in terms of animal behavior. In order to take account of them, the use of a crude, descriptive morphological classification is often helpful. In practice, different workers have

used their own classifications for their own material, and few attempts have been made to erect a consistent and comprehensive framework on a purely morphological basis. Frey (1971, Table 2) and Ewing and Davis (1967) reproduced examples, however, and to set out another version seems worthwhile, in order to show by comparison just how arbitrary such classifications must be. Nevertheless, the example does give a bird's-eye view of the subject; it also provides a key for preliminary identification. (See Fig. 10.7.)

A CRUDE MORPHOLOGICAL CLASSIFICATION

Table 3.2 exemplifies a purely morphological, descriptive classification. No attempt was made to be comprehensive; no mention is made of many distinctive groups of forms. However, the forms chosen for inclusion are taken from the commoner ones and also those best displaying the range of variation. The examples in column 5 are illustrated in the *Treatise on Invertebrate Paleontology, Part W* (Häntzschel, 1962).

The horizontal rows of the table each describe a separate and distinctive form, having the various component characters described in the four vertical columns. Each column represents a rank of a hierarchical system, rising from rank I on the right to rank IV on the left. Thus, the units of rank II comprise one or more units of rank I, the units of rank III one or more from those of rank II, and so on.

By analogy with biological taxonomy, the characters of rank I may be said to define a "family" of trace fossils [e.g., Richter (1927) wrote of the pseudo-family Rhizocorallidae]; those of rank II, a "superfamily"; those of rank III, an "order"; and those of rank IV, a "class." In this sense, the names listed under the heading *Examples* are "genera."

The weaknesses contained in Table 3.2 are quite obvious, however. The characters

[1] No attempt is made in this book to discuss the extensive, but eclectic, classificatory schemes of Vialov (1972), which confuse the three distinct classifications of trace fossils just outlined. His schemes are united to bring together not only trace fossils but also, for instance, eggs, which are body fossils, and physiological processes such as egg laying, or extraneous events like wounds or fractures, which are not fossils at all. (See Chapter 14.)

TABLE 3.2 Purely Morphological Classification of Some Common Invertebrate Trace Fossils.

Rank IV	Rank III	Rank II	Rank I	Examples
A. Track-like trace on bedding plane		a. "Prods" or "scratches"; all alike	(1) Clustered "scratches"	*Paleohelcura*
			(2) Rows of "prods"	*Tasmanadia*
		b. "Prods" or "scratches" of different kinds	Rows of "prods"	*Kouphichnium*
B. Trail-like trace on bedding plane	1. Freely winding	a. Simple trail	(1) No ornament	*Gordia*
			(2) Transverse ornament	*Climactichnus*
		b. Bilobed trail	Transverse ornament	*Cruziana*
		c. Trilobed trail	Transverse ornament	*Scolicia*
	2. Windings in contact with one another; pattern on bedding plane	a. Simple trail	No ornament	*Helminthoidea*
		b. Bilobed trail	Transverse ornament	*Nereites*
C. Radially symmetrical in a horizontal plane	1. Without axial vertical structure	a. Five-rayed	Rays are grooved	*Asteriacites*
		b. Multirayed	Club-shaped rays	*Asterosoma*
	2. With vertical axial structure	a. Circular outline	Conical depression	*Histioderma*
		b. Multirayed	Radial branches	*Lennea*
D. Tunnels and shafts	1. Of uniform diameter	a. Vertical	(1) Isolated	*Tigillites*
			(2) *En masse*	*Skolithos*
		b. Horizontal	Winding	*Planolites*
		c. U-shaped		*Arenicolites*
		d. Regularly branching		*Chondrites*
	2. Variable diameter	Irregular network		*Thalassinoides*
E. Forms having a spreite		a. U-shaped	(1) Vertical plane	*Diplocraterion*
			(2) Horizontal plane	*Rhizocorallium*
		b. Spiral	Inclined plane	*Zoophycos*
		c. Branched	Vertical	*Phycodes*
F. Pouch shaped		a. Smooth surface		*Pelecypodichnus*
		b. Transverse ornament		*Rusophycus*
G. Miscellaneous		Net pattern		*Palaeodictyon*

actually accounted for are only a few of those that might have been used. The choice of characters is very arbitrary, and no two people—each putting together such a system—would produce similar results. The purpose of producing the classification here is to show that material of problematical origin may yet be dealt with in a logical manner, to yield results for biostratigraphy, facies analysis, and paleoecology. Lessertisseur (1955) produced a comprehensive classification of trace fossils that, although being very largely based on their behavioral origins, fills in all the gaps with a morpho-

logical scheme that has been largely followed in Table 3.2.

Worth noting is that the first two categories of rank IV in Table 3.2 seem, because of the use of the words "track" and "trail," to be behavioral and not purely morpho-

TABLE 3.3   Useful Descriptive-Genetic Terms.*

Tracks and Trails

Track—impression left in underlying sediment by an individual foot or podium.

Trackway—succession of tracks reflecting directed locomotion.

Trail—a continuous groove produced during locomotion by an animal having part of its body in contact with the substrate surface, or a continuous subsurface trace made by an animal traveling from one point to another.

Burrows

Burrow—excavation made within unconsolidated sediment. Excludes intrastratal trails.

Burrow system—highly ramified and (or) interconnected burrows.

Shaft—dominantly vertical burrow, or a dominantly vertical component of a burrow system having prominent vertical and horizontal parts.

Tunnel—dominantly horizontal burrow, or a dominantly horizontal component of a burrow system having prominent vertical and horizontal parts (= *gallery*).

Burrow lining—thickened burrow wall constructed by organisms as a structural reinforcement. May consist of (1) host sediments retained essentially by mucus impregnation, (2) peletoidal aggregates of sediment shoved into the wall, like mud-daubed chimneys, (3) detrital particles selected and cemented like masonry, or (4) leathery or felted tubes consisting mostly of chitinophosphatic secretions by organisms. Burrow linings of types 3 and 4 are commonly called *dwelling tubes.*

Burrow cast—sediments infilling a burrow (= *burrow fill*). Sediment fill may be either *active,* if done by animals, or *passive,* if done by gravity. Active fill is termed *back fill* wherever U-in-U laminae, etc., show that the animal packed sediment behind itself as it moved through the substrate.

Miscellaneous

Configuration—in ichnology, the spatial relationships of lebensspuren, including the disposition of component parts and their orientation with respect to bedding and (or) azimuth.

Spreite—blade-like to sinuous, U-shaped, or spiraled structure consisting of sets or cosets of closely juxtaposed, repetitious parallel or concentric feeding or dwelling burrows or grazing traces; individual burrows or grooves comprising the spreite commonly anastomose into a single trunk or stem (as in *Daedalus*) or are strung between peripheral "support" stems (as in *Rhizocorallium*). *Retrusive* spreiten are extended upward, or proximal to the initial point of entry by the animal, and *protrusive* spreiten are extended downward, or distal to the point of entry.

* From Frey, 1973.

TABLE 3.4   Classification of Preservational Processes.

1. *Stratinomical processes*—those processes leading to the formation of sedimentary structures through the interaction of sediment and organism during sedimentary accumulation.
    A. *Toponomical processes*—those processes responsible for trace fossils being preserved as relief features, typically at boundaries between different kinds of sediment.
    B. *Excretion and burial of fecal material* of deposit feeders.
2. Processes involving the *working over of sediment* by living organisms. Includes diverse kinds of bioturbation, among which are:
    A. *Sorting and selection of grains* leading to the formation of structures of a special lithology.
    B. *Plastering or lining of burrow walls.*
3. *Diagenetic processes*
4. *Boring and embedding*   as explained in the text.
5. *Deposit structuring*

logical. The point is that the words are available for use for material which, at a particular stage of growth of the subject, is still problematical in respect of its origin as a track or a trail. "Track-like" means no more than it says, and although a particular object may in fact be a track, if it cannot at present be referred confidently to any definite place in the genetic, behavioral classification, it has a place here. Similarly, "prods" and "scratches" mean prod-like and scratch-like.[2]

Although such classifications are not used very widely in ichnology, the terminologies devised for them can be very helpful in discussions or description. Some commonly used terms, descriptive and genetic, are defined in Table 3.3.

CLASSIFICATION ACCORDING TO MODE OF PRESERVATION

When the enormous variety of trace fossils is analyzed from a sedimentological point of view, we realize that trace morphology can be understood as resulting from a rela-

[2] The word "spreite," at E in the first column of the table, is of German coinage and refers to a web or septum-like structure that, like the web of a duck's foot spanning the space between its digits, connects the tubular limbs of certain forms. (See Table 3.3.)

tively small number of different processes active during sedimentation and diagenesis. These processes cause the preservation of a record of animal movement in most cases. The diversity of forms is seen to arise from the numerous different organisms that, by their various shapes or behavior, determine the actual detail that results from the same basic process. These structures are revealed by outcrop weathering or by the work of the preparator. (See Chapter 23.) This is the preservational approach, and here we are concerned with classification of the processes leading to preservation (Table 3.4).

Stratinomy

Etymologically, the word "stratinomy" is the same as "stratigraphy"; it was probably used first by Weigelt (1928), in the restricted sense of interaction between sedimentary accumulation and the burial and preservation of organisms. The term has failed to gain general support. It is retained here as a name for one of the five strongly contrasting, constituent parts of the "preservational processes" classification, which otherwise lacks a designation (see Table 3.4).

The major part of stratinomy is concerned with structures that may originate through organic activity during normal sedimentary accumulation, i.e., the bioturbation structures reiterated by Frey (1973). The term "toponomy" seems to be establishing itself for this subject.

The subject lends itself to organization into logical special classifications, of which several have been proposed (e.g., Table 3.5). In the first one, Seilacher (1953) distinguished full reliefs, semireliefs, and cleavage reliefs; exogene and endogene forms; and top (epigene) and bottom (hypogene) forms. Martinsson's (1965, 1970) scheme differs in approach but is equally valid; its greater simplicity is offset by its greater rigidity and more limited applicability. The main categories of preserved structures constituting the units of these classifications are illustrated in Figure 3.1 and in Figure 4.4.

Simpson (1957) showed that exhumation and subsequent reburial is a factor that plays a part in the toponomic process, and Seilacher's classification was extended by himself (1964a, 1964b) and by Webby (1969) to take account of this aspect (see Fig. 4.2). Seilacher always envisaged the possibility of certain full reliefs being the product of an active process of back-filling of evacuated tunnels, etc., by a deposit feeder. Such a process is very difficult to prove, but insofar as it may occur, it should be distinguished from a normal inorganic toponomic process, involving current or other hydraulic transport and sedimentation. Rather, it should be classified with the processes by which, in preserving the record of the organism's

activity, some alteration of the sediment (as by grain sorting) is recorded by the organism. These processes are treated below.

A stratinomic process that cannot well be classified under the heading of toponomy is the preservation of shaped fecal material of deposit feeders. The process gives rise to morphologically striking fossils (see Häntzschel et al., 1968), which are preserved by burial in the course of normal accumulation, as though they were body fossils. But if a body fossil must be part of an actual organism, coprolites are clearly not body fossils. The familiar Ordovician form *Tomaculum, Favreina* (Kennedy, et al., 1969), and *Chomatichnus* (Donaldson and Simpson, 1962) are interesting examples. Although coprolites are not very rare, no proper comparative study has yet been attempted. The most useful contribution on the subject is the annotated bibliography of coprolites by Häntzschel et al. (1968). (See Chapters 13 and 14.)

Alteration of Sediment by Organisms

Under this heading, a number of distinct processes may be grouped, e.g., the "stuffing" of part of a tunnel system to accommodate waste as the nutrient-bearing sediment is "mined" in another part. This process, referred to already, is deduced from the abundant fecal pellets filling some burrow systems. The organism voids its waste, in the form of fecal pellets of clay, after the nutrients have been consumed. We may often have difficulty determining whether the pellets occupying a tunnel have been formed in situ or have moved as a constituent of a pellet-bearing mud from the tunnel entrance under suction (see Ferguson, 1965), or by gravity (Osgood, 1970, p. 338–339).

The use of pellets by callianassid shrimps to form a wall or lining supporting their burrows is well known. In burrow forms such as *Ophiomorpha*, the pellets are cemented by "mucus" and pushed into the wall so as to line the surface completely.

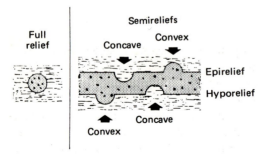

Fig. 3.1 Trace fossils as relief features. (Adapted from Seilacher, 1964a, 1964b.)

TABLE 3.5   Toponomic Classification of Bioturbation Structures.*

General Terms

Toponomy—primarily, the description and classification of lebensspuren with respect to their mode of preservation and occurrence (position on or within a stratum, or relative to the casting medium), and secondarily, the interpretation of the mechanical origin of traces.

Bioturbation—reworking of sediments by an organism.

Bioturbate texture—gross texture imparted to sediment by extensive bioturbation; typically consists of dense, contorted, or interpenetrating burrows or other traces, few of which are distinct morphologically. Where burrows are somewhat less crowded and are thus more distinct individually, the sediment is said to be *burrow mottled*.

Classification by Seilacher (1953, 1964a, 1964b)

1. *Full relief*—bioturbation structure preserved within a stratum.
2. *Semirelief*—bioturbation structure preserved at a lithological interface: boundary reliefs and cleavage reliefs.
   A. *Boundary relief*—semirelief not involving cleavage preservation: hyporeliefs and epireliefs.
      (1) *Hyporelief*—boundary relief occurring on the sole of a stratum; relief may be *concave* or *convex*.
      (2) *Epirelief*—boundary relief occurring at the top of a stratum; relief may be *concave* or *convex*.
   B. *Cleavage relief*—semirelief in which subsurface laminae are deformed during production of the surficial trace; parting of these laminae (as by weathering fissility) reveals vertical repetition of a given lebensspur, any isolated specimen of which resembles a single boundary relief.

Classification by Martinsson (1965, 1970)

Endichnion—bioturbation structure preserved within the main body of the casting medium.

Exichnion—bioturbation structure preserved outside the main body of the casting medium.

Epichnion—bioturbation structure preserved at the upper surface of the main body of the casting medium; may appear as a *ridge* or *groove*.

Hypichnion—bioturbation structure preserved at the lower surface of the main body of the casting medium; may appear as a *ridge* or *groove*.

* From Frey, 1973.

Another totally distinct process is exemplified by *Zoophycos*. Here a deposit-feeding animal, mining the sediment, sorted the larger particles of the deposit and plastered them against one side of the tunnel, and in so doing, displaced the tunnel to the other side. Thus, the animal made its own special sediment and, by repeating the process, preserved the form of the tunnel system as a spreite composed of the special sediment (Simpson, 1970).

Diagenetic Processes

Toponomic processes, as noted above, yield structures through the disturbance of surfaces between sediments of varying composition. No evidence ordinarily survives where the organic activity encountered no sedimentary boundaries. Sometimes, however, solutions moving preferentially in sediments that were more permeable have caused chemical changes that show the trace fossil structure by differences in color or texture. Thus, as a photographic film is developed to reveal the latent image in the negative, diagenetic solutions have "developed" the sedimentary surface. A good example is provided by the preservation of *Chondrites* from the Alpine flysch. The substance of the fossil is a dark-colored infilling of the tunnels, which are entirely enveloped in light-colored limestones. The dark color is due not to carbonaceous matter but to pyrite (Tauber, 1949). The pyritization probably took place during diagenesis.

Boring and Embedding

Boring by organisms, as distinct from burrowing, is the process of excavation of hard materials, such as shell, rock, wood, or vertebrate skeleton. The process is used for a variety of purposes, which are dealt with below in the behavioral classification. (See also Chapters 10–13, 17, and 18.) A distinction must be made between boring and embedding (Bromley, 1970); the latter is the lodging (dwelling) of an organism within the growing skeleton of another organism, which invests it. For example, a variety of organisms dwell within the skeletal aragonite of reef corals by allowing the coral to grow around and enclose them. The embedded organism may, in later growth, extend its lodging space by true boring.

The cavities may be preserved unaltered, or as molds. In the latter case the process is similar to the preservation of body fossils as internal and external molds, except that here a cavity is preserved, i.e., a sort of negative body fossil. The starting point is the cavity, resulting from boring or embedding, which may be formed in a calcium carbonate skeleton. After death of the organism responsible, sediment occupies the cavity. Subsequently, the skeleton is dissolved, and the mold of the cavity remains as the trace fossil. Alternatively, instead of the boring being made into skeletal material, it may be excavated in rocks or

some form of hardgrounds, such as those of chalk (see Chapter 18). In that case, the preserved structure falls into a normal toponomic category; it is a full-relief form and, according to Martinsson's (1970) scheme, an endichnial boring.

Deposit Structuring

A common mode of occurrence of *Chondrites* is the all-pervasive penetration of a bed of sediment, which gives the bed a characteristic mottling (Simpson, 1957). Several other burrow forms produce similar effects (e.g., Fig. 17.1). An allied effect is obtained when the ubiquitous organisms break down preexisting structure to produce a virtually homogenized bed, which can be recognized by a characteristic bioturbate texture. The pervasive character of the trace fossil may express the density of the population of the organism concerned. Or, it may reflect a slow rate of sediment deposition and extended accumulation together with a rather sparse population.

Only an arbitrary boundary can be drawn between special bed structures of this kind and the biostratification structures defined by Frey (1973), such as the graded bedding described by Rhodes and Stanley (1965), and the binding of sediment on mussel banks (McMaster, 1958).

CLASSIFICATION ACCORDING TO BEHAVIOR

Of the vast variety of different animals that inhabit environments in which sedimentary accumulation may lead to the preservation of a record of their activity, only a few are actually effective in this way. Certain types of activity, however, readily result in some sort of record and do appear very commonly in different places and different times. These common trace fossils have been grouped into five categories of an ethological classification, devised by Seilacher (1953), which has been found most useful for workers in the fields of facies analysis and sedimentation. Nothing very significant

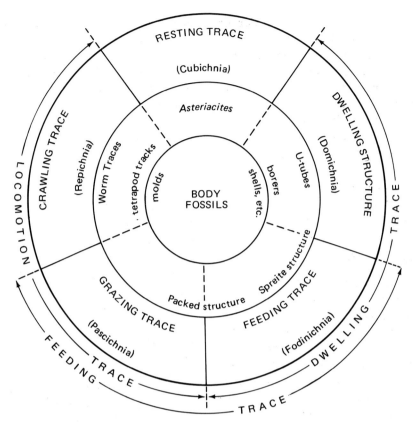

Fig. 3.2 Ethological classification. Diagram shows five main behavioral categories and their relations to one another and to body fossils. Escape structures (Fugichnia), which overlap with several other categories, are not shown. (From Osgood, 1970, after Seilacher, 1953.)

rests in the choice of precisely five categories. To devise a number of additional ones is easy; as many as ten have been suggested.[3] Most palichnologists are agreed on the basic five of Seilacher, and these are defined here. (In addition to the five, however, a sixth is here justified and defined.) Seilacher (1953) devised a clever circular diagram that shows the five categories of trace fossils and their relationship to one another (i.e., mainly the overlap of the different activities among individuals) and to body fossils. A translated version of this diagram is presented as Figure 3.2. Examples of most of the behavioral categories were given in Chapter 2 (Fig. 2.4).

[3] Not including the many compartments in Vialov's (1972) classification.

Resting Traces *(Cubichnia)*

Essentially, resting traces are impressions caused by an animal that temporarily interrupts its locomotion in search of rest or refuge. Well-defined relief features may reflect the morphology of the undersurface of the animal, the outline of which may be clearly defined. More typically, the relief results from movements of the animal made in taking refuge by burying itself just under the sediment surface or in the act of leaving. Some forms are transitional to the Repichnia and Domichnia. Examples: *Asteriacites, Lockeia (= Pelecypodichnus), Rusophycus* (see Häntzschel, 1962).

Crawling Traces *(Repichnia)*

Crawling traces include tracks produced by locomotion, with the aid of walking bristles or other appendages, and trails due to locomotion affected by muscular movements of the body. Normally the traces originate as impressions on a sediment surface and are most commonly preserved in low-lying areas and in aquatic environments. Several different movement patterns may be made by the same animal (e.g., walking, cantering, galloping). Arthropods may walk on the tips of walking appendages (*Diplichnites*) or may crawl by sweeping movements of the endopodites (*Cruziana*). Many tracks and trails are made just below the surface of the sediment, e.g., *Scolicia, Olivellites, Gyrochorte, Cruziana, Aulichnites*.

Grazing Traces (*Pascichnia*)

Grazing traces are recognized by the way in which, following an animal's distinctive behavior pattern, the surface of the sediment is systematically patterned by the physical structure imparted to it. The effect clearly results from biologic or genetic control—by *thigmotaxis* (touch stimulus)—resulting in the economical exploitation of nutritious sediment by deposit feeders. Two familiar examples are *Helminthoida* and *Nereites*. A close relationship exists with the Fodinichnia, which includes a group of deposit feeders exploiting sediment at depth, according to a well-defined pattern of behavior. At the same time, a close relationship exists with the Repichnia, because many grazing traces are in part locomotion traces.

Feeding Structures (*Fodinichnia*)

Feeding structures consist of temporary burrows or other traces of deposit feeders, excavated while in search of food within the sediment or at the sediment surface. They may follow a complex behavior pattern, making use of phobotactic sensitivity— avoidance of previously mined sediment— to achieve economical exploitation. Radial patterns predominate, but Us occur; e.g.,

Chondrites, Phycodes, certain *Diplocraterion, Rhizocorallium, Zoophycos*. A close relationship exists not only with the Pascichnia but also with the Domichnia.

Borings made through the shells of living bivalves by predators can be conveniently included in this category. (See Chapter 13.)

Dwelling Structures (*Domichnia*)

Dwelling structures include burrows and borings, more or less permanently inhabited, of suspension feeders. The structures are predominantly cyclindrical, having agglutinated or otherwise strengthened walls. Familiar forms include various U burrows and borings lacking a spreite: *Ophiomorpha, Skolithos, Arenicolites, Entobia*.

Escape Structures (*Fugichnia*)

Escape structures constitute a very distinctive group, only recently understood; a useful short account was given by Frey (1973). Most fossils are roughly cylindrical and were vertical at the time of their origin; they are made predominantly by bivalves and other suspension feeders, lacking a lining to the burrow. The overall structure is made by an organism uncovering itself after having been buried by an influx of sediment, or by burrowing deeper into the sediment to offset erosion at the substrate surface. Obvious transitions into the Cubichnia are known, as in the case of vertically repetitive *Asteriacites*, from which they differ in being produced by normally semisessile, not vagile organisms. A transition to feeding structures is also quite evident. For instance, *Diplocraterion* is a feeding burrow and has generally been so classified; but it is, at the same time, a typical escape structure and belongs more definitely to this category. Good illustrations of fossil examples were published by Goldring (1962), Hardy (1970), and De Raaf et al. (1965), and recent examples by Schäfer (1972). (See also Fig. 8.5 and Fig. 9.6.)

CLASSIFICATION ACCORDING TO PHYLOGENY

The third aspect of trace fossil classification (after preservation and behavior) concerns their being the product of particular sorts of animals, which are variously related to one another by natural descent (phylogeny). Seilacher (1953) referred to this as the "taxonomic aspect." When we can attribute a particular trace fossil to a taxon recognized by zoological nomenclature, we must realize that the relationship between it and the zoological taxon is of a kind totally different from that which subsists between it and a behavioral taxon. The zoological taxonomy and the ichnological or palichnological taxonomy are totally distinct.

For the great majority of trace fossils, the zoological identity of the originator of the trace is not known—in part probably because the animal in question lacked hard parts and therefore is not known as a body fossil. But even if hard parts exist, the likelihood that they can be related to a particular trace fossil is slight. (See Chapter 2.)

The exceptions to this generality are provided mainly by the Repichnia. In the case of the footprints produced by numerous vertebrates, identifying the genus and species is often possible. This facility is because the trace fossil is essentially an impression (mold) reproducing a part of the morphology of the animal in question; it is thus not different from the mold of a body fossil representing a fragment of skeleton. (See Chapter 14.)

The same relation holds for tracks made by various classes of arthropods. For the Trilobita, Seilacher (1970) showed that a great amount of morphological information, relating to the shape and dimension of the animal concerned, can be obtained from their burrowing traces; numerous species may be defined or identified. These species can be correlated with, but are not identical with, zoological species. As a group, trilobite traces are thus exceptional: they may be used for stratigraphical correlation because of their brief time ranges (see Chapters 6 and 7). They have such brief ranges precisely because they correspond so closely with zoological species.

NOMENCLATURE

Surprisingly, the two fundamentally different procedures of classification and nomenclature are often confused with one another. In fact, some authors fail to distinguish the two, using them as if they were synonyms. However, they differ in nature and purpose.

A classification is a systematic ordering of things or concepts, and its purpose is to facilitate the understanding of the relationships between them. A nomenclature is a system of rules and conventions for labeling the elements of a classification, and its purpose is to facilitate communication.

We may note in passing that the word "classification" has a very wide usage as noun and verb. It is applied to all sorts of natural or artificial objects or to concepts. The classifications designed for organisms are usually referred to as "systematics," e.g., brachiopod systematics. "Taxonomy" is a synonym.

Thus, we can devise as many different nomenclatures as we can classifications. In practice, however, ordinary language is normally sufficient in making reference to the constituent parts of a classification. In many other cases, where special words must be invented for the elements of a classification, we may use a special terminology. An example is the terminology of Martinsson's toponomic classification, mentioned previously. The word "nomenclature" itself has generally been reserved for cases where the classification includes very large numbers of elements of comparable status, and in which new elements are continually being added.

International Codes

Historically, trace fossils were first named on the Linnean binominal system because they were believed to be body fossils (see Chapter 1). Since their true nature has been discovered, a strong tendency to continue using a binomial scheme persists, although many words of caution have been spoken by various workers. Until recently palichnologists, although using such a scheme, have not always sought to follow the existing formal procedures either of botanical or zoological nomenclature. Instead, they have followed the general procedures informally and without compulsion or excessive consistency. Looking at trace fossils in their behavioral aspect, these workers prepared diagnoses for ichnogenera and ichnospecies and then treated them as if they were plant or animal genera or species.

Now, however, as a result probably of the rapid growth of the subject and the multiplication of names, the feeling is changing and the consensus is in favor of a formal binomial nomenclature. This systemization can be done in two different ways: either (1) as the extension of the existing international code of zoological nomenclature, or (2) as an independent code administered by a new international commission for trace fossil nomenclature.

The first procedure was selected by Häntzschel and Kraus (1972), who made a submission to the ICZN for the inclusion of trace fossil names among those to which its rules provide protection. The second procedure was considered by Sarjeant and Kennedy (1973) in that they devised an independent set of rules for trace fossil nomenclature; it is analogous to the Linnean systems of zoology and botany but does not define "ichnogenus" or "ichnospecies." They advocated the type procedure whereby the holotype is the ultimate repository of the name. They are aware of the difficulty that the understanding of a trace fossil often depends on understanding its field relations. But they realize that the acceptance of a photograph for a holotype is not a complete or satisfactory solution to the problem.

While dealing with international conventions or regulations, we may note that, in printing the names of trace fossils, the ordinary procedure is to use italics. This habit follows the usage for zoological and botanical nomenclature, as laid down in the respective codes. The reasons for italicizing trace fossil names are the same as those that apply to the italicizing of plant and animal names; it distinguishes names that are "official," i.e., which come under the protection of the international codes. The practice has various advantages in helping the reader to find his way through a paper and in the use of faunal lists. Thus, the recommendation against italicizing the names of ichnotaxa, which is contained in the submission to the ICZN by Häntzschel and Kraus (1972), is not supported by many palichnologists.[4]

Ichnogenus and Ichnospecies

In zoological systematics, the genus is defined in terms of the included species. Thus, the trivial name of an animal species applies to the most fundamental element in the hierarchy of classificatory units. In the general usage of palichnologists, the tendency is to make the ichnogenus the more fundamental of the two categories, ichnospecies and ichnogenus.

The tendency in ichnology has been to constitute the trace genus on the basis of morphotypes reflecting a characteristic behavior pattern and to name as species the less striking morphological variants, which may not correspond to any significant differ-

[4] An alternative suggestion by some paleontologists is the use of expanded Roman type for trace fossil names, to set them apart from body fossil genera and species. But, again, italicized names are needed in ichnology for the same reasons as in conventional paleontology; and use of expanded Roman type would probably have the undesirable effect of promoting rather than curtailing the informality of trace fossil nomenclature.

ence of behavior. Thus, Seilacher (1970) recognized species groups based not on difference of behavior but on the systematic morphology of the originators; some of his trace taxa (species groups) are based on the number and shape of the trilobite's endopodal claws, whereas subordinate (species) and superior (genus) taxa are based on behavior.

Another nomenclatural inconsistency shows the impossibility of a simple translation of body fossil nomenclature to trace fossil nomenclature. The trilobite trace *Cruziana* (genus) has the species *barbata* (Seilacher, 1970). But *barbata* is also a species of *Rusophycus*. Thus, in contrast to zoological nomenclature, the *same* trace species, defined on the *same* base, can belong to two different genera.

A POSTSCRIPT ON KINDS OF CLASSIFICATION

Semantic matters are not unimportant for accuracy in thought and expression. In discussions about the classification and nomenclature of trace fossils, two matters result in frequent confusion: (1) morphological classification and (2) the distinction between classifications of objects and classifications of the processes that give rise to them.

A morphological classification is concerned with objects, according to their form, and nothing else. Examples were considered in Table 3.1. Ideally, terms such as "burrow" have no place in the diagnosis of particular genera or species, because the terms involve interpretation and thus become elements of an ethological classification. Only geometrical characters, such as "cylinders," can be used in defining trace taxa.

The point is *not*, in final analysis, that the entities which one actually observes and classifies are purely morphological. If that were the point, one might maintain with equal reason that the phylogenetic classification of animals is "in fact" purely morphological. No! the point is that, in defining and identifying the units of the classification, some interpretation of the morphological features takes place. This interpretation means an explanation of the morphology in terms of the origin (genesis) of some selected aspect of the objects classified, such as the ethological or preservational aspects.

The other source of confusion comes from imagining that, when we think of a behavioral classification, we are dealing with behavior alone. A classification of behavior would then be dealing with processes (such as burrowing) rather than with the products (burrows), which are the actual objects being classified.

ACKNOWLEDGMENTS

This chapter owes a great deal to R. W. Frey, who made numerous suggestions of detail that have greatly improved it in content, clarity, and accuracy. But in addition, he made constructive criticisms that I have been glad to accept and which I acknowledge with gratitude.

I am also indebted to R. G. Bromley, L. Ferguson, and R. G. Osgood for helpful comments and criticisms that have enabled me to improve this chapter.

REFERENCES

Bromley, R. G. 1970. Borings as trace fossils and *Entobia cretacea* Portlock, as an example. In T. P. Crimes and J. C. Harper (eds.), Trace fossils. Geol. Jour., Spec. Issue 3:49–90.

De Raaf, J. F. M. et al. 1965. Cyclic sedimentation in the lower Westphalian of North Devon, England. Sedimentology, 4:1–52.

Donaldson, D. and S. Simpson. 1962. *Chomatichnus*, a new ichnogenus, and other trace fossils of Wegber Quarry. Liverpool and Manchr. Geol. Jour., 3:73–81.

Ewing, M. and R. A. Davis. 1967. Lebensspuren photographed on the ocean floor. Johns Hopkins Oceanogr. Stud., 3:259–294.

Ferguson, L. 1965. A note on the emplacement

of sediment in the trace fossil *Chondrites*. Geol. Soc. London, Proc., 1622:79–82.

Frey, R. W. 1971. Ichnology—the study of fossil and recent lebensspuren. In B. F. Perkins (ed.), Trace fossils, a field guide. Louisiana State Univ., School Geosci., Misc. Publ. 71-1:91–125.

————. 1973. Concepts in the study of biogenic sedimentary structures. Jour. Sed. Petrol., 43:6–19.

Goldring, R. 1962. The trace fossils of the Baggy Beds (Upper Devonian) of North Devon, England. Paläont. Zeitschr., 36:232–251.

Häntzschel, W. 1962. Trace fossils and problematica. In R. C. Moore (ed.), Treatise on invertebrate paleontology, Pt. W, Miscellanea. Lawrence, Kan., Geol. Soc. America and Univ. Kansas Press, p. W177–W245.

———— and O. Kraus. 1972. Names based on trace fossils (ichnotaxa): request for a recommendation. Bull. Zool. Nomencl., 29: 137–141.

———— et al. 1968. Coprolites, an annotated bibliography. Geol. Soc. America, Mem. 108:1–132.

Hardy, P. G. 1970. Aspects of paleoecology in the arenaceous sediments of Upper Carboniferous age in the area around Manchester. Unpubl. Ph.D. Dissert., Victoria Univ. Manchester, 211 p.

Kennedy, W. J. et al. 1969. A *Favreina-Thalassinoides* association from the Great Oolite of Oxfordshire. Palaeontology, 12:549–554.

Lessertisseur, J. 1955. Traces fossiles d'activité animale et leur signification paléobiologique. Soc. Géol. France, Mém. (N. Ser.) 74:1–150.

McMaster, R. L. 1958. Modification of underwater surface sediment layers by sea mussels (*Mytilus edulis*). Jour. Sed. Petrol., 28:515–516.

Martinsson, A. 1965. Aspects of a Middle Cambrian thanatotope on Öland. Geol. Fören. I Stockholm Förhand., 87:171–230.

————. 1970. Toponomy of trace fossils. In T. P. Crimes and J. C. Harper (eds.), Trace fossils. Geol. Jour., Spec. Issue 3:323–330.

Osgood, R. G., Jr. 1970. Trace fossils of the Cincinnati area. Palaeontographica Amer., 6(41):281–444.

Rhodes, D. C. and D. J. Stanley. 1965. Biogenic graded bedding. Jour. Sed. Petrol., 36:1144–1149.

Richter, R. 1927. Die fossilien Fährten und Bauten der Würmer. Paläont. Zeitschr., 9: 193–240.

Sarjeant, W. A. S. and W. J. Kennedy. 1973. Proposal of a code for the nomenclature of trace fossils. Canadian Jour. Earth Sci., 10: 460–475.

Schäfer, W. 1972. Ecology and palaeoecology of marine environments. Edinburgh and Chicago, Oliver & Boyd and Univ. of Chicago Press, 568 p.

Seilacher, A. 1953. Studien zur Palichnologie. I. Über die Methoden der Palichnologie. Neues Jahrb. Geol. Paläont., Abh., 98:87–124.

————. 1964a. Biogenic sedimentary structures. In J. Imbrie and N. D. Newell (eds.), Approaches to paleoecology. New York, John Wiley, p. 296–316.

————. 1964b. Sedimentological classification and nomenclature of trace fossils. Sedimentology, 3:253–256.

————. 1970. *Cruziana* stratigraphy of "nonfossiliferous" Palaeozoic sandstones. In T. P. Crimes and J. C. Harper (eds.), Trace fossils. Geol. Jour., Spec. Issue 3:447–476.

Simpson, S. 1957. On the trace fossil *Chondrites*. Geol. Soc. London, Quart. Jour., 112:475–499.

————. 1970. Notes on *Zoophycos* and *Spirophyton*. In T. P. Crimes and J. C. Harper (eds.), Trace fossils. Geol. Jour., Spec. Issue 3:505–514.

Tauber, A. F. 1949. Paläobiologische Analyse von *Chondrites furcatus* Sternberg. Jahrb. Geol. Bundesamt., 92:141–154.

Vialov, O. S. 1972. The classification of the fossil traces of life. 24th Internat. Geol. Congr., Montreal, Sect. 7:639–644.

Webby, B. D. 1969. Trace fossils (Pascichnia) from the Silurian of New South Wales, Australia. Paläont. Zeitschr., 43:81–94.

Weigelt, J. 1928. Rezente Wirbeltierleichen und Ihre Bedeutung für die Paläontologie. Paläont. Zeitschr., 9:327–328.

CHAPTER 4

PRESERVATION OF TRACE FOSSILS

A. HALLAM
Department of Geology and Mineralogy, University of Oxford
Oxford, England

SYNOPSIS

Trace fossils are normally preserved at or near interfaces, as a result of burrowing or crawling activity along the junction of successive beds of different lithology. The traces may be enhanced by diagenetic concentration of such minerals as calcite, chert, and pyrite, and hence may appear as concretionary nodules. The scour-fill activity of turbidity currents on the seabed may also play a preservational role in flysch sequences, and in some instances, a difference in orientation of grains within beds of uniform lithology may be sufficient to reveal trace fossils.

A variety of formal terms has been proposed to describe trace fossil preservation. Among the more useful are full reliefs and semi-reliefs, the latter being subdivisible into epireliefs and hyporeliefs, depending on their relationship to the main casting medium. Alternative terms are epichnia, endichnia, hypichnia, and exichnia.

Characteristics of the sedimentary matrix control the style of preservation to a considerable extent; for instance, compact clay retains finer impressions than does unconsolidated sand. Variations in sedimentation rate can affect the morphology of burrows, and the character of whole assemblages of trace fossils—as for instance in turbidite sequences—may be strongly influenced by the sedimentary bottom conditions.

INTRODUCTION

A thorough knowledge of the manner in which crawling and burrowing organisms produce preservable structures in soft sediment is obviously vital to a proper understanding of trace fossils; it can also cast light on reconstructing conditions of sedimentation. Furthermore, preservational factors can influence the composition of trace fossil assemblages in particular types of strata, so that failure to appreciate their significance may lead to errors in interpretation of the environments of deposition.

The classification and interpretation of trace fossil preservation is termed "toponomy." (See Chapter 3.) Several attempts at classifying different types of trace fossil preservation have been made in recent years; one result has been the introduction of a few simple terms and concepts that can be easily learned and understood.

TYPES OF PRESERVATION

An early, although frequently overlooked, attempt to classify preservational types was made by Simpson (1957), who incorporated a fourfold division: bed junction preservation, concealed bed junction preservation, diagenetic preservation, and burial preservation. (See also Fig. 8.5.)

Bed junction preservation is the most usual means by which trace fossils are preserved and is also the most obvious and straightforward. Trace fossils are most frequently visible where animal activity occurred at the interface between beds of different lithology, usually sandstone and shale or mudstone; the fossils appear in relief at bed junctions as casts of burrows or trails, e.g., the "hypichnia" and "exichnia" of Figure 4.4.

Concealed bed junction preservation (Fig. 4.1) signifies the case where individual masses of sediment are isolated within a different lithology—for instance, sand-filled burrows in shale. No connection with the overlying sandstone bed exists because the latter was removed by an episode of erosion. (See Chapter 18.) This type of preservation must be carefully distinguished from normal bed-junction preservation, in which sediments that fill downward-piping burrow structures are seen in horizontal or vertical section as isolated circles or ellipses but in three dimensions are seen to continue into the overlying bed (e.g., most specimens of *Chondrites*). In my experience, concealed bed junction preservation is an exceptional occurrence.

Diagenetic preservation is much more widespread than is concealed bed junction preservation. Simpson cited the example of *Chondrites* preserved as protuberances on the surface of concretionary limestone nodules, clearly indicating that segregation of $CaCO_3$ during early diagenesis was the agent responsible for preservation. In the English Jurassic, such trace fossils as *Chondrites, Rhizocorallium,* and *Thalassinoides* are commonly preserved as precompactional nodules or nodule protuberances in both calcite and siderite; the argillaceous sequence is otherwise rather uniform. Presumably, the migration of pore fluids was controlled by the reduced porosity of the infilled burrows compared with that of the matrix. This type of preservation is frequently observed at bed junctions in sequences of alternating argillaceous limestones and shales, but is readily distinguishable from normal bed junction preservation; the latter is characterized by primary alterations of sediment types during deposition (Hallam, 1964).

Small-diameter "burrows" are frequently preserved in pyrite, and appear in weathered rocks as ginger or reddish-brown structures because of oxidation to goethite (cf. Frey, 1970, p. 24–26). This form of preservation probably relates to the decay of organic compounds after death of the organism. Decomposition produces hydrogen sulfide, which reacts with iron in the sediment to form hydrotroilite, and this eventually crystallizes to pyrite. Phosphatic margins of the burrow *Kulindrichnus*, which may also relate to organic decay, were described by Hallam (1960a). Bromley (1967) gave an account of crustacean burrows in the English Chalk preserved as flint (or chert) concretions. Collophane was recorded as the possible cementing agent in certain Holocene burrows of the shrimp *Callianassa* (Weimer and Hoyt, 1964), although this has not been established conclusively.

Attention should also be drawn to the importance of burrow linings, which may be simple secretions of mucus, particulate walls agglutinated by mucoid compounds, or even thick chitinophosphatic structures, as in the dwelling tubes of the worm *Chaetopterus* (Frey, 1971, Fig. 6). K. L. Smith (in Frey, 1971, p. 102) has undertaken chemical analyses of burrow walls made by the recent shrimps *Callianassa* and *Upogebia*, and also Pleistocene forms of the burrow *Ophiomorpha*. All contained significant quantities of calcium, magnesium,

Fig. 4.1 Concealed bed junction preservation. (After Simpson, 1957.)

and sodium; trace amounts of copper and iron were detected among the recent burrows. Linings of opaline silica were described for *Ophiomorpha* systems from the Eocene of Mississippi (Hester and Pryor, 1972). The authors believe that this lining is most probably the result of activity by the organisms responsible for the burrows. The animals are presumed to have ingested clay-size, silica-rich material and then secreted it as a colloidal organic compound binding the quartz-grain pellets in the walls of *Ophiomorpha*.

Within highly reducing environments in fine-grained sediments, the respiratory or water-pumping activity of burrowing animals may give rise to oxygenated haloes around the burrows. This aeration is conspicuous in fresh sediment cores in Georgia estuaries, where light-colored, occupied burrows contrast sharply with the black matrix of otherwise oxygen-deficient muds (Howard and Frey, 1973). I have seen similar structures surrounding *Diplocraterion* walls in the Upper Triassic White Lias of southern England (Hallam, 1960b); they are enveloped by a pyritic rim.

Also worthy of note in this regard is Frey's (1971, p. 108) contention that biochemical alteration of sediments by animals may account for most of the differences in color commonly observed between burrow casts and host sediment in certain uniform lithologies.

Burial preservation is another rare case, invoked by Simpson (1957) specifically for infilled *Chondrites* burrows in the English Devonian; these had been excavated by currents winnowing away the soft matrix during a subsequent erosional phase. The mucus-bound burrow linings thus lay as sediment-filled "gloves" and were preserved if covered quickly enough by later sediments. The term "burial preservation" may also be used in a sense other than that of Simpson, involving the preservation of trace fossils in flysch sequences; here turbidity currents may scour burrows made in the pelagic muds and a short while afterward cast the scoured burrows with sand.

These four categories seem to cover the majority of trace fossils, but Simpson himself (1970) described an example that must be viewed as an exception. He considered the structure of *Zoophycos* to be revealed by the contrast between the composition of the enclosing sediment and that of material derived from it by a special sorting activity of the organism. (See Chapter 3.) Perhaps the difference in orientation of grains within a uniform lithology is more important in the preservation of trace fossils than has previously been recognized. It may be a common mode of preservation of spreiten structures, for instance.

A different approach to preservational classification was adopted by Seilacher (1953a, 1964) and Martinsson (1970), who proposed a series of terms to describe the position of the trace fossil with respect to the main casting medium, which is commonly sandstone or siltstone.

In Seilacher's widely adopted scheme, semireliefs—which are sculptures at, for instance, sand–clay interfaces—are distinguished from full reliefs—which are discrete bodies occurring within beds or at bed junctions (Fig. 4.2). Semireliefs are subdivided into epireliefs if they are situated on the upper surface, and hyporeliefs if on the undersurface, of the main casting medium. The genetic terms exogene, endogene, and pseudendogene refer to the way in which the trace fossils are interpreted to have formed. This terminology describes the assumed relation of the trace fossil to the contemporary sediment surface rather than to the trace producer. Active fill of the burrow by its producer is distinguished from passive fill by processes of sedimentation.

Certain slight modifications to Seilacher's scheme have been suggested. Osgood (1970) proposed that the producing organism be distinguished from the produced trace fossil by using the well-established ecological terms "epifaunal" and

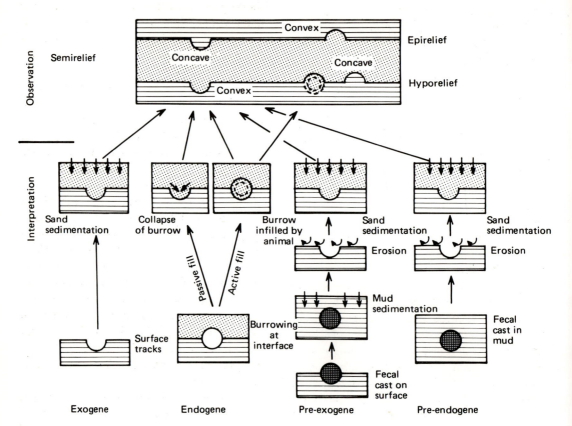

Fig. 4.2 Seilacher's classification of trace fossil preservation types, and their interpretation. (Adapted from Webby, 1969.)

"infaunal." Chamberlain (1971) suggested that the term "epigenic" is more precise than "exogenic," and added "intergenic" for endogenous structures produced at bed junctions. (See Fig. 4.3.)

A disadvantage of the above schemes is that they require the use of parallel descriptive and genetic terms, given combinations

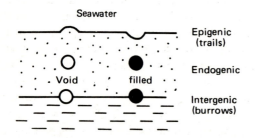

Fig. 4.3 Diagram to illustrate the toponomic terms proposed by Chamberlain (1971). These terms are genetic rather than descriptive.

of which may be awkward. The descriptive terms alone are sufficient for most purposes. (See Frey, 1973.)

A simpler scheme of classification, having terms that are better founded etymologically, is that of Martinsson (1970); he attached importance to trace fossils as interface phenomena of different magnitude, rather than as bed-junction phenomena alone. He divided trace fossils into epichnia, endichnia, hypichnia, and exichnia, depending on their relationship to the main casting medium (Fig. 4.4). These four categories, particularly the epichnia and hypichnia, may be further described by using commonplace terms, such as "ridges" or "grooves." Hence the "relief" and genetic terms of Seilacher become largely superfluous. Genetic aspects may be included, as in "hypichnial burrow cast," for instance.

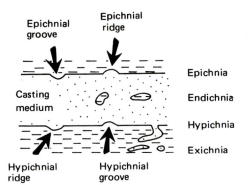

Fig. 4.4 Diagram to illustrate the toponomic terms proposed by Martinsson (1970).

This scheme is clear and unambiguous, and the few terms are easily mastered and comprehended.

One other aspect of trace fossil preservation that deserves at least passing mention is that of deformation due to compaction or tectonic disturbance. Plessmann (1966), for example, used deformed burrows to evaluate lateral and vertical compaction of Mesozoic and Tertiary sediments in Germany and Italy. (See also Chapter 7.)

RELATIONS BETWEEN TRACE FOSSILS AND THE MATRIX

The character of the host sediment has an important influence on the preservation of trace fossils. Of the animals burrowing in sand, for example, only a few leave lasting galleries or other distinct traces in the sand itself, but all can leave sharp impressions on underlying clay layers (Fig. 4.5). These impressions are cast by the overlying sand and therefore are more easily preserved than trails at the sediment–water interface. Such "cleavage reliefs" among trace fossils also result from feet or podia that penetrated subsurface laminae, thus producing "under tracks" that become progressively less distinct with depth (Osgood, 1970, Fig. 19).

Certain burrowing organisms, for example such crustaceans as *Callianassa* and *Upogebia*, apparently stabilize their galleries in sand from collapse by applying mucus liberally to the burrow walls; galleries made in mud possibly require less mucus. Much sediment may fall into the galleries from the substrate surface. If the passages are still occupied, the material may be either removed by the burrowers or plastered against the walls, producing in this way a series of concentric layers (Schäfer, 1972).

Simpson's (1957) evidence for burial preservation seems to suggest the existence of some sort of mucus envelope in the burrow *Chondrites*. The way in which such an intricate set of burrows as *Chondrites* gets plugged by sediment remains something of a mystery; both active and passive fill have been suggested (cf. Ferguson, 1965; Osgood, 1970). Chisholm (1970), describing *Teichichnus* in the Scottish Carboniferous, inferred that some of the sand laminae were formed while the animal remained in the burrow. This situation must be generally true of trace fossils exhibiting "septa" or spreiten structures, other than those made by epibenthos.

Sometimes one must infer that burrows were dug in compacted mud. This coherence is necessary for the preservation of delicate traces, such as the narrow subparallel ridges and grooves on the surface of many *Rhizocorallium*, which are usually interpreted as crustacean scratch marks (e.g., Chisholm, 1970).

Shinn (1968) discovered that open burrows of crustaceans exist beneath considerable thicknesses of unconsolidated lime mud in parts of the Florida–Bahamas region. For instance, open *Callianassa* burrows are found as deep as 2.5 m below the surface; living *Callianassa* do not penetrate deeper than 1 m, thus Shinn estimated that the burrows 2.5 m deep may have persisted unfilled for more than 1,000 years. Such structures evidently could survive lithification, which is pertinent to Bromley's (1967) account of *Thalassinoides* systems preserved in English Chalk hardgrounds. Exceptionally, the burrows remain empty; but more typically a soft sediment fill, which contrasts clearly with the hard

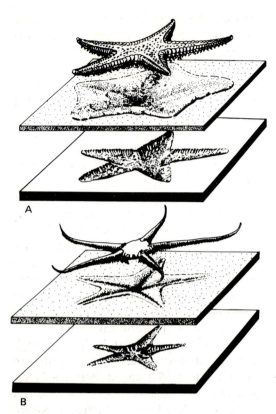

Fig. 4.5 Diagram to illustrate different impressions made by the same animal on sand and underlying mud layers. A, asteroid. B, ophiuroid. (After Seilacher, 1953b.)

matrix, is present. Partial reexcavation of these burrows and several periods of sedimentation may be detected. I have seen similar unfilled or clay-filled *Thalassinoides* systems in Bathonian limestone hardgrounds in southern England.

INFLUENCE OF SEDIMENTARY CONDITIONS ON TRACE FOSSIL ASSEMBLAGES

The actual pattern or process of sedimentation can strongly influence the character of the trace fossil assemblage that is ultimately preserved, thereby producing a physically induced bias. In environmental interpretation one must have a thorough appreciation of this factor, which will be illustrated by several examples.

In discussing shallow-water Devonian deposits in southwest England, Goldring (1964) recognized two trace fossil associations. The first, consisting of surface or near-surface traces, was associated with more or less continuous clastic sedimentation. The second type consisted only of traces made a certain distance below the sediment surface. This distribution signifies discontinuous sedimentation; only a fraction of the originally deposited sediment is preserved. Goldring (1962) interpreted protrusive *Diplocraterion* (the last-formed burrow underlying all previously formed ones) as reflecting adjustment of the burrowing animal to erosion of the topmost few centimeters of sediment. Conversely, retrusive *Diplocraterion* represent an adjustment to increased sedimentation (see Fig. 8.5).

The latter type of adjustment, where the burrowing animal moves upward through the sediment, is commonly signified by "nested cone" structures having the apex pointed downward (Fig. 4.6). These "escape structures" are readily produced in laboratory experiments with marine burrowers (Schäfer, 1956; Shinn, 1968; Fig. 8.5). The difference between *Skolithos* and *Monocraterion* horizons in the Lower Cambrian Pipe Rock of northwest Scotland is probably due to changes in rate of sedimentation, the latter structures being produced by upward migration of the *Skolithos* animal (Hallam and Swett, 1965).

A distinction was drawn by Chisholm (1970), in his study of *Teichichnus* associa-

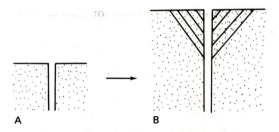

Fig. 4.6 Modification of simple vertical burrows with increased sedimentation. A, *Skolithos*. B, *Monocraterion*. (After Hallam and Swett, 1965.)

tions, between (1) unplugged, distorted teichichnians occurring with *Planolites* and thought to be formed not long after the mud was deposited, and (2) undistorted, plugged varieties suggesting excavation into partly compacted mud. The two groups are thought probably to represent Seilacher's (1967) *Cruziana* and *Glossifungites* "facies," respectively. (See Chapters 7 and 9.)

Whether delicate structures such as *Gyrochorte* are preserved in shallow-water sediments probably depends at least to some extent on the degree of bioturbation by other burrowers. For example, the abundance of delicate traces in marginal marine deposits of the English Bathonian correlates with the absence in the same beds of normally ubiquitous forms such as *Diplocraterion*, *Rhizocorallium*, and *Thalassinoides* (Hallam, 1970).

With deep-water associations (Seilacher's *Nereites* "facies"), the trace fossils may give useful information on sedimentary conditions. Seilacher (1962) argued, for instance, that the ichnology of Maastrichtian and Eocene flysch deposits in northern Spain supported the popular sedimentological interpretation of more or less instantaneous deposition from turbidity currents. Endobenthos had different depth ranges if the flysch sandstones were deposited instantaneously; shallow burrowers could only reach the base of thinner beds. Post-depositional trace fossils indeed showed the predicted relationship. *Neonereites* did not occur in beds thicker than 6 cm. The corresponding thicknesses for beds containing *Scolicia* and *Phycosiphon* are 8 and 14 cm, respectively. Most of the trace fossils, however, are pre-depositional endogenous forms; the traces were exposed by the stripping away of overlying mud, evidently very uniformly over a wide area.

Osgood (1970) questioned Seilacher's view that many trace fossils in turbidite sequences are endogenous forms, necessarily implying the widespread removal of a thin mantle of surface sediment. Just how these delicate traces can survive extensive scouring by fast-moving turbidity currents remains something of a problem, unless they are protected by a traction carpet or by some other means. Osgood cited evidence tentatively favoring good trace fossil preservation through deposition of silt by traction currents that result in the accumulation of cross-laminated beds. (See also Crimes, 1973.)

Trace fossil types in the lower Paleozoic flysch of Wales seem to be at least partly controlled by preservational factors (Crimes, 1970). In proximal environments, most surface or near-surface traces were removed by rapid turbidity currents; only deep burrows are preserved. The preservational potential of shallow traces increases with successively more distal environments.

A final general point is that the well-known bathymetric associations proposed by Seilacher (1967) are at least partly dependent on preservational factors. In particular, *Zoophycos*—although accorded a separate "facies" of moderate depth intermediate between the *Nereites* "facies" and the *Skolithos*, *Glossifungites*, or *Cruziana* "facies"—does range into shallow-water deposits; examples are Carboniferous limestones in northern England and Scotland and certain Jurassic limestones in Spain, where *Zoophycos* occurs in association with *Rhizocorallium*. Similarly, Osgood (1970) and Osgood and Szmuc (1972) found evidence of shallow-water *Zoophycos* in the Ordovician and lower Carboniferous of Ohio. Evidently *Zoophycos*, like *Chondrites*, has a considerable depth range and may occur wherever the surface sediment was largely undisturbed by waves, currents, or profound bioturbation.

ACKNOWLEDGMENTS

I would like to thank the editor of this volume for many constructive suggestions and for bringing to my attention a number of relevant references.

REFERENCES

Bromley, R. G. 1967. Some observations on burrows of thalassinidean Crustacea in chalk hardgrounds. Geol. Soc. London, Quart. Jour., 123:157–182.

Chamberlain, C. K. 1971. Morphology and ethology of trace fossils from the Ouachita Mountains, southeast Oklahoma. Jour. Paleont., 45:212–246.

Chisholm, J. I. 1970. *Teichichnus* and related trace fossils in the Lower Carboniferous at St. Monance, Scotland. Geol. Surv. Great Britain, Bull. 32:21–51.

Crimes, T. P. 1970. The significance of trace fossils in sedimentology, stratigraphy and palaeoecology with examples from lower Palaeozoic strata. In T. P. Crimes and J. C. Harper (eds.), Trace fossils. Geol. Jour., Spec. Issue 3:101–126.

————. 1973. From limestones to distal turbidites: a facies and trace fossil analysis in the Zumaya flysch (Paleocene-Eocene), north Spain. Sedimentology, 20:105–131.

Ferguson, L. 1965. A note on the emplacement of sediment in the trace-fossil *Chondrites*. Geol. Soc. London, Proc., 1622:79–82.

Frey, R. W. 1970. Trace fossils of Fort Hays Limestone Member of Niobrara Chalk (Upper Cretaceous), west-central Kansas. Univ. Kansas Paleont. Contr., Art. 53, 41 p.

————. 1971. Ichnology—the study of fossil and recent lebensspuren. In B. F. Perkins (ed.), Trace fossils, a field guide. Louisiana State Univ., School Geosci., Misc. Publ. 71-1:91–125.

————. 1973. Concepts in the study of biogenic sedimentary structures. Jour. Sed. Petrol., 43:6–19.

Goldring, R. 1962. The trace fossils of the Baggy Beds (Upper Devonian) of North Devon, England. Paläont. Zeitschr., 36:232–257.

————. 1964. Trace-fossils and the sedimentary surface in shallow-water marine sediments. In L. M. J. U. van Straaten (ed.), Deltaic and shallow marine deposits. Developments in Sedimentology, 1:136–143.

Hallam, A. 1960a. *Kulindrichnus langi*, a new trace-fossil from the Lias. Palaeontology, 3:64–68.

————. 1960b. The White Lias of the Devon coast. Geol. Assoc. London, Proc., 71:47–60.

————. 1964. Origin of the limestone–shale rhythm in the Blue Lias of England: a composite theory. Jour. Geol., 72:157–169.

————. 1970. *Gyrochorte* and other trace fossils in the Forest Marble (Bathonian) of Dorset, England. In T. P. Crimes and J. C. Harper (eds.), Trace fossils. Geol. Jour., Spec. Issue 3:189–200.

———— and K. Swett. 1965. Trace fossils from the Lower Cambrian Pipe Rock of the north-west Highlands. Scottish Jour. Geol., 2:101–106.

Hester, N. C. and W. A. Pryor. 1972. Blade-shaped crustacean burrows of Eocene age: a composite form of *Ophiomorpha*. Geol. Soc. America, Bull., 83:677–688.

Howard, J. D. and R. W. Frey. 1973. Characteristic physical and biogenic sedimentary ·structures in Georgia estuaries. Amer. Assoc. Petrol. Geol., Bull., 57:1169–1184.

Martinsson, A. 1970. Toponomy of trace fossils. In T. P. Crimes and J. C. Harper (eds.), Trace fossils. Geol. Jour., Spec. Issue 3:323–330.

Osgood, R. G., Jr. 1970. Trace fossils of the Cincinnati area. Palaeontographica Amer., 6(41):281–244.

———— and E. J. Szmuc. 1972. The trace fossil *Zoophycos* as an indicator of water depth. Bulls. Amer. Paleont., 62(271):5–22.

Plessmann, W. 1966. Diagenetische und kompressive Verfermung in der Oberkreide des Harz-Nordrandes sowie im Flysch von San Remo. Neues Jahrb. Geol. Paläont., Monat., 8:480–493.

Schäfer, W. 1956. Wirkungen der Benthos-Organismen auf den jungen Schichtverband. Senckenbergiana Leth., 37:183–263.

————. 1972. Ecology and palaeoecology of marine environments. Edinburgh and Chicago, Oliver & Boyd and Univ. Chicago Press, 568 p.

Seilacher, A. 1953a. Studien zur Palichnologie. I. Über die Methoden der Palichnologie. Neues Jahrb. Geol. Paläont., Abh., 96:421–452.

————. 1953b. Studien zur Palichnologie. II. Die Fossilen Ruhespuren (Cubichnia). Neues Jahrb. Geol. Paläont., Abh., 98:87–124.

————. 1962. Paleontological studies on turbidite sedimentation and erosion. Jour. Geol., 70:227–234.

————. 1964. Sedimentological classification and nomenclature of trace fossils. Sedimentology, 3:253–256.

————. 1967. Bathymetry of trace fossils. Marine Geol., 5:413–428.

Shinn, E. A. 1968. Burrowing in recent lime sediments of Florida and the Bahamas. Jour. Paleont., 42:879–894.

Simpson, S. 1957. On the trace-fossil *Chondrites*. Geol. Soc. London, Quart. Jour., 112:475–495.

————. 1970. Notes on *Zoophycos* and *Spirophyton*. In T. P. Crimes and J. C. Harper (eds.), Trace fossils. Geol. Jour., Spec. Issue 3:505–514.

Webby, B. D. 1969. Trace fossils (Pascichnia) from the Silurian of New South Wales, Australia. Paläont. Zeitschr., 43:81–94.

Weimer, R. J. and J. H. Hoyt. 1964. Burrows of *Callianassa major* Say, geological indicators of littoral and shallow neritic environments. Jour. Paleont., 38:761–767.

CHAPTER 5

FALSE OR MISLEADING TRACES

DONALD W. BOYD

Department of Geology, University of Wyoming
Laramie, Wyoming, U.S.A.

SYNOPSIS

Various features of sedimentary rocks may be mistaken for evidence of behavioral activity by organisms. Physical processes producing spurious traces include current action, movement of fluids through sediment, lightning, displacement of ambient mineral grains, deformation of unlithified strata, and shrinkage of sediment. Chemical processes yielding potentially deceptive products in this context include differential

precipitation or dissolution, silicification, dolomitization, and crystallization. False traces of organic origin include some molds of organisms and certain primary shell features.

The chronologic significance of bioerosion traces may be misinterpreted in cases where origin of substrate and time of penetration differ markedly in age.

INTRODUCTION

they greet with illustrious fervor
each least, lost gesture of life
Thomas John Carlisle

The fervor of a typical ichnologist certainly equals, if not surpasses, that of the archeologists who inspired this introductory couplet. Current enthusiasm for discovery, description, and interpretation of trace fossils will probably stimulate the imagining of "gestures" where no life existed. If so, a review of the possible sources of deception is justified.

The assignment is not without hazards. Those inorganic structures that first come to mind as examples of false traces are mere instances of superficial resemblance, easily recognized by most geologists who work with sedimentary rocks. It follows that the more sedimentologically sophisticated the reader, the greater the percentage of this chapter that will be superfluous. By contrast, the more nearly perfect the

deception, the greater the likelihood that the present treatment fails to analyze the phenomenon adequately. In fact, a few of the more enigmatic examples treated here eventually may be shown to have at least some genetic relation to organic behavior.

The dividing line between valid and spurious traces can be very subtle. As an example, consider the distinction between the discontinuous marks left by an active trilobite drifting downcurrent (e.g., Martinsson, 1965, Fig. 14) and those left by an empty shell bouncing across a similar substrate. The former pattern is a valid trace fossil whereas the latter is not, yet ultimate distinction between the two phenomena assumes a correct answer to the near-theological question of when the spirit actually left the body.

In this chapter, emphasis is upon features that might be mistaken for evidence of behavioral activity by an organism. Most of these features are direct results of physical or chemical processes, but some result from chemical modification of biological materials, and a few are wholly of biologic origin. The discussion also includes markedly anachronistic imprints and bioerosion structures, which might be mistaken by an observer for traces essentially contemporaneous in origin with the rock in which they are found. Not discussed are cases of deception wherein a biogenic structure has been interpreted erroneously in regard to type of behavior represented. Although justifiable in terms of the chapter's title, this topic is too subjective and limitless to warrant inclusion.

PHYSICAL ORIGIN

Current Action at the Depositional Surface

Tool Marks

A soft substrate registers the signatures of transient objects that contact it. These objects may be organisms moving under their own power, or they may be current-moved tools of various origins. Even such ephemeral things as gas bubbles are capable of producing rectilinear and curvilinear marks, both simple and compound, as they drift singly or in clusters across the sediment surface (Cloud, 1960). More commonly, current-transported shells function as tools, which leave distinctive and repetitive marks as they bound or roll across a soft substrate (Pettijohn and Potter, 1964, Pl. 68). Seilacher (1963) gave a fascinating analysis of patterns produced by rolling cephalopod shells. The character of these marks reflects not only original shell shape and ornamentation, but also the type of damage a given shell had sustained prior to and during transport. These aspects of form also influenced the type of motion (e.g., wheel-like rolling; wobbling; bounding) to which a shell was susceptible. The resulting diverse roll marks had been misinterpreted as lebensspuren of squids, tortoises, and fish.

A row of crescentic traces produced by

fin strokes of *Agonus cataphractus* on a modern sea bottom was illustrated by Schäfer (1972, Fig. 153). In distinguishing between such traces and skip marks, one might reason that successive imprints of a shell bounding end over end would likely differ in pattern. However, this criterion will not distinguish between all types of tool marks and their biogenic analogs. Buoyant tools, such as cephalopod shells and tree branches, can leave prod marks on the bottom as they bob in waves while drifting in shallow water. Repetition of pattern can be expected if the same morphologic extremity touches bottom repeatedly.

The Triassic sole marks shown in Figure 5.1 resemble certain resting traces, such as *Rusophycus* (Häntzschel, 1962, Fig. 131,5), as well as certain vertebrate trackways. However, I interpret them as casts of prod marks made by floating driftwood. This interpretation is supported by the fact that several of the grooves (now sole ridges) had overhanging roofs at one end, and these are consistently oriented. Furthermore, some of the grooves have an abrupt change in direction at the distal end (as defined by the previously mentioned overhangs). This aspect might have been produced as a temporarily stranded object rotated in the current and subsequently pulled free. However, a similar angularity characterizes tracks of certain clawed vertebrates. Permian tracks made by a small vertebrate having claws set almost at right angles to the long axis of the digit were illustrated by Sarjeant (1971, Pl. 6). (See Chapter 14.) Casts of the tracks have hooklike terminations reminiscent of some of the Triassic sole marks of Fig. 5.1, but they are more consistent in form, and they lack the fine longitudinal grooves that characterize the Triassic sole marks. Second-order sculpture is not characteristic of vertebrate tracks, whereas prod marks commonly include minor ridges and grooves parallel to their long axes, as a result of irregularities on the edge of the tool. Second-order sculpture on elongate resting or crawling traces

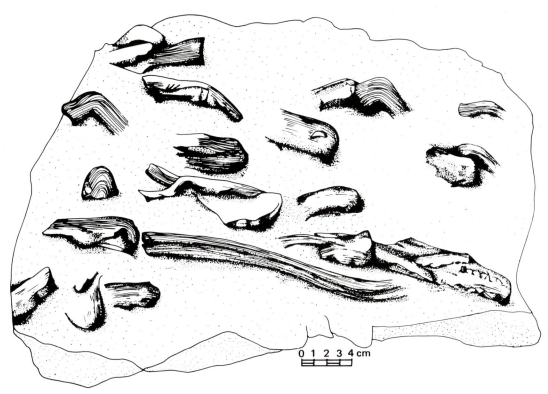

0 1 2 3 4 cm

Fig. 5.1 Casts of prod marks on sandstone sole. Abrupt change in direction of fine striations apparent on several casts. Downcurrent to right, as interpreted from overhanging ends of prod marks. Triassic, Wyoming.

commonly forms through action of appendages or by contraction and expansion of a muscular foot or body wall. Minor ridges and grooves thus produced tend to form high angles with respect to the long axis of the trace.

Grooves formed by drifting plants can resemble grazing or crawling traces. For example, kelp uprooted by storm waves may drag attached rocks over the bottom. Fronds of floating algae, such as *Sargassum*, may scrape the substrate while moving to and fro as the plant is stranded by an incoming tide. Concentric grooves in sand around coastal and desert plants are scribed by the windblown branches or blades (Barthel, 1966, Fig. 4). Similar sweep marks on Ordovician bedding surfaces originated when anchored crinoid stems were rotated by ocean currents (Osgood, 1970, Pl. 80, fig. 3). The difficulty in determining whether the more enigmatic Ordovician specimens are feeding traces or sweep marks was discussed by Osgood (1970, p. 396).

Priels and Current Crescents

Narrow erosion channels ("priels" of Martinsson, 1965, p. 194) typify some mud surfaces also marked by organic activity. Both phenomena are preserved as ridges on siltstone soles (Frey and Chowns, 1972, Pl. 4A), and small priel casts may be mistaken for hypichnial trace fossils.

Current crescents are horseshoe-shaped depressions eroded on the up-current side of an obstacle (Pettijohn and Potter, 1964, Pls. 91B, 92B, and 93). "*Blastophycus*" specimens from the Ordovician of Ohio are apparently not trace fossils but casts of current crescents associated with enrolled trilobites; Osgood (1970, p. 390) duplicated the form experimentally.

Turbidity Current Marks

As inferred from field relationships and demonstrated in several laboratories (e.g., Dżułyński and Simpson, 1966), turbidity currents can erode and deform bottom sediments over which they pass. Sculpture thus formed may be mistaken for trace fossils. For example, Hofmann (1971, p. 18) noted the convincing resemblance between *Taonichnites*, originally described as drag marks of a tentacle-bearing animal, and bedding surface features produced by investigators experimenting with flowing suspensions (e.g., Dżułyński and Walton, 1963).

Interference Ripple Marks

Tadpoles excavate depressions in soft sediment. Crowding results in encroachment of adjacent hollows, and a honeycomb-like arrangement of relatively uniform-size depressions, having polygonal bounding ridges, is sometimes produced. The closely packed depressions characterizing these dimpled surfaces have diameters of 10 to 45 mm and depths as much as 10 mm. Cameron and Estes (1971), impressed by the ephemeral nature of modern tadpole nests they observed, inferred that such structures are not likely to be preserved in the rock record. They reviewed the several published accounts of "fossil" tadpole nests and concluded that all the occurrences are probably interference ripple marks. Other authors (e.g., Ford and Breed, 1970) maintained that patches of small circular depressions in the midst of stream deposits probably are the result of tadpole activity, and that such features have been misidentified as interference ripples! As criteria for distinguishing between the two phenomena, Boekschoten (1964) suggested that interference ripples may cover more extensive surfaces than do most tadpole nests and commonly involve depressions of larger dimensions. The ripple pattern

has two dominant axes, unlike the polygonal pattern of the biogenic structure, and involves ridges characterized by convex slopes, in contrast to the concave ones bounding tadpole excavations.

Movement of Fluids Through Sediment

During movement of a fluid through sediment, a gathering tendency is commonly exhibited whereby migration is localized along certain routes. These pathways are predetermined to great extent by sediment inhomogeneities of various types. In any case, the concentrated activity of the fluid can not only deform and destroy lamination but can also erode sediment in its path. Subsequent deposition of this material, where the current reaches the sediment surface, can result in distinctive constructive features. Many of the internal and surficial phenomena resulting from localized gathering and migration of fluids through sediment have striking analogs among biogenic structures. Effects of the inorganic and organic agents probably are mistaken for one another in many geologic investigations. Air and water are the fluids of primary importance in this discussion, although hydrogen sulfide and other gasses generated by bacterial action within the sediment are also significant. (See also Schumm, 1970.)

Gas

Gas can produce distinctive sedimentary structures both within the sediment and at the sediment surface (Cloud, 1960). In cohesive sediment, such as aragonitic mud from the Bahama Banks, internal pockets are formed as gas accretion sites, and elongate passageways are produced by gas migration. As seen along the inner surface of a plastic core liner (Cloud's Fig. 1), cavities of this type are commonly about 1 cm in maximum dimension and about one-fourth that in minimum dimension. In his example, the pockets were produced in a sealed core of aragonitic mud during the first month after collecting. The gas was presumably generated through bacterial action. Similar fenestrae in fine-grained limestone would probably be identified as burrows by many geologists.

Probable gas trackways in Devonian limestone were discussed and illustrated by Cloud (1960). Several characteristics cited as evidence against burrow origin include marked irregularity in longitudinal profile, gradational margins, abrupt changes in course (from horizontal to vertical) and specific relations between pyrite, sparry dolomite, and calcitic matrix. Cloud concluded that the cavities, many of which are now filled with coarse-grained carbonate, formed in the same way as the gas pockets in his Bahama core.

Localized upwelling of gas in sufficient quantity in muddy sediment produces a blister at the surface. Collapse of this short-lived feature creates a ring, and a concentric pattern may develop if the process is repeated. *Laevicyclus* (e.g., Häntzschel, 1962, Fig. 123,3) is a puzzling structure involving a concentric pattern reminiscent both of the gas expulsion marks figured by Cloud and of feeding traces produced by a recent polychaete. Frey (1970, p. 15) reviewed the pros and cons of both interpretations relative to some Kansas Cretaceous specimens but found the evidence inconclusive.

Beach enthusiasts have long been familiar with small holes produced in the upper part of sand beaches by air expelled from pores as the water table rises with the incoming tide. As described by Emery (1945), these holes range from 1 to 10 mm in diameter and from 1 to 4 cm in depth. As such, they resemble beach burrows produced by the amphipod crustaceans commonly termed "sand hoppers." To compound the problem of distinguishing between the two, Palmer (1928) reported that these practical crustaceans occupy air-escape structures as well as their own burrows. Air-escape passageways descend by a series of steps through sand laminae differing in

grain size. Such a route tends both to follow coarser laminae down dip and to penetrate intervening finer ones. Many of the air holes have ramifying feeder channels at the bottom (Emery, 1945). The surface opening differs in character, depending mainly on whether or not it is covered by water during air expulsion. Where air escapes into water, a crater typically forms because of bubble agitation, and in some instances a raised rim develops. By contrast, the mouth is sharp-edged, having neither crater nor rim, where air escapes subaerially.

Localized areas of contorted and disrupted lamination and pods of nonlaminated sand can result from formation of air pockets in sediment. Resemblance of "bubble-sand" textures (Hoyt and Henry, 1964) to certain bioturbate textures was pointed out by Frey and Mayou (1971, p. 67). When air is driven through beach or tidal-flat sediment by a rising water table, its upward escape can be blocked by downward percolation from faster advancing surface water. If the sediment consists of laminae differing in grain size, air gathering in coarser laminae can be confined beneath overlying laminae of wet fine sand. Domes are produced where overlying laminae are bowed upward above growing air pockets. Heights up to 3 cm and diameters from 2 to 15 cm were recorded by Emery (1945), who noted that distinctive circular patterns can be produced on beach surfaces by truncation of arched laminae. He found that most sand domes lack any trace of a feeder hole.

Contorted laminae and isolated pockets of nonlaminated sand in a modern tidal delta were considered by Stewart (1956) to result from vertically rising air pockets. The sand flat is exposed at low tide, and the air-pump mechanism he postulated is basically that evoked for the beach structures considered above. He supported his interpretation with simulation experiments, and some of the structures illustrated bear notable resemblance to disturbed lamination produced by burrowers. In his interpretation, air pockets, rising through the laminated sequence, rupture laminae and are followed by slumping and inflowing of sand. I observed a similar process through the glass side of a flume, when the dry sand bed was flooded after a period of inactivity. Vertical escape routes are kept open initially by high pressure of the confined air, but most of them are quickly filled with bed-load sediment from above or by wall collapse. In any case, a burrow-like form persists; a sharp contact exists between nonlaminated filling and the laminated sequence it interrupts.

Confusion of inorganic features and biogenic traces is not limited to misidentification of the former as lebensspuren. For example, van der Lingen and Andrews (1969) discussed the possibility of mistaking swash-modified beach tracks of a heavy animal for air-heave structures.

Water

Several of the characteristic sedimentary features produced by burrowers can also result from local upwelling of water or from movement of quicksand. For example, constructional mounds as much as 10 cm high and 30 cm in diameter are produced by some modern infaunal organisms (e.g., Kornicker and Boyd, 1962, Figs. 5 and 11). The structures are similar in size and shape to some of the sedimentary volcanoes described by Gill and Kuenen (1958) from an Irish Carboniferous siltstone and sandstone sequence. These mounds, notably like igneous volcanoes in shape and internal structure, are invariably associated with contorted stratification. Gill and Kuenen thought the volcanoes were built under water by intermittent ejections of sand and silt, and cited experimental evidence to show that either loading or agitation may have caused extrusion of water from the sediment. Modern examples of sedimentary volcanoes were reported by Bondesen (1966) from a glaciofluvial en-

vironment and by Williams and Rust (1969) along channels of a braided river. In both cases, the mechanism involves artesian flow, and water-saturated silt or muddy silt played a significant role in origin of the eruptive structures.

Another lebensspur-like feature thought to be formed by upward movement of water through unlithified deposits is the sandstone-plugged pipe. The term was coined by Allen (1961) for cylindrical structures, as much as 1 m long, piercing fluviatile strata of the Old Red Sandstone. Although the average pipe diameter (8 cm) is larger than that of most burrows, the bedding surface views illustrated by Allen (his Figs. 3 and 4) could have been taken from the ichnological literature so far as form and distribution of the structures are concerned. Moreover, an apparent spreite structure is associated with many of the cylindrical bodies. It consists of nested, curved laminae underlying inclined cylinders. This intriguing feature enhances the resemblance to certain trace fossils, but Allen offered a convincing argument that the sandstone cylinders are plugged ducts, eroded by groundwater rising toward the depositional surface before the deposits were consolidated. The spreiten-like lateral structures were produced by migration of the loci of erosion. His interpretation is strengthened by the observed pinching and swelling of cylinders where they pass through layers of different grain size, and by deviations in inclination coinciding with similar lithologic discontinuities. (Cf. Chapter 16.)

The patterned cones (Fig. 5.2) described by Boyd and Ore (1963) seem to have originated by a process similar to that of the sandstone-plugged pipes. Their notably different form (a low, asymmetric cone) and distinctive surface pattern (branching ridges) are explainable in terms of different grain size and degree of consolidation of host sediment in the two cases. Many of the patterned cones have axes sufficiently inclined to resemble the distal parts of com-

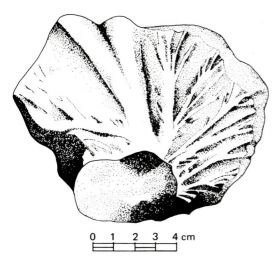

Fig. 5.2 Patterned cone of siltstone, viewed from below. This specimen was a model for cone in Fig. 5.3. Triassic red beds, Wyoming.

pound feeding burrows such as *Phycodes* (e.g., Häntzschel, 1962, Fig. 128,1).

Experiments (Boyd and Ore, 1963) have demonstrated that a water current rising through completely unconsolidated, saturated silt erodes a conical depression at the surface. The diameter of this cone increases until equilibrium is reached. At this point, silt is no longer transported out of the depression but is recycled within it. Downslope movement of grains settling out of suspension, and of groups of grains taking part in small-scale slides, erodes a delicate system of downward-converging gullies. Plaster casts of the experimental depressions are strikingly similar to the natural patterned cones. Some ancient specimens have nested patterned surfaces around one part of the cone (Fig. 5.3). Such specimens may be explained by postulating two or more episodes of excavation, separated by depression-filling sedimentation. This explanation implies that each successive excavation produced a depression of smaller diameter (or with offset center) than the last, and that successive upwelling currents followed essentially the same passageway. The specimen shown in Figure 5.4 illustrates such a sequence of events. The filling of the initial excavation was partly removed, and the resulting space was filled in turn with coarse debris. Finally, a cylindrical duct was excavated through the debris and then plugged with silt.

Irregular but sharply bounded strap-like bodies of siltstone (Fig. 5.5) are typically associated with the patterned cones. Their dimensions, general form, and variable orientation in the enclosing rock are evocative of burrow origin (cf. *Curvolithus*). However, some of the strap-like structures lead upward to the apex of a patterned cone; they probably represent

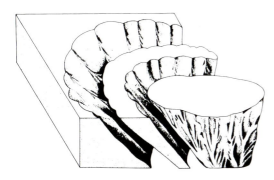

Fig. 5.3 Diagram of mold-cast relationships of nested patterned surfaces in siltstone. Sloping surface on left formed during initial spring action. Depression was filled with silt, but renewed upwelling excavated smaller depression. Patterned cone on right is siltstone cast of the second depression. Triassic, Wyoming.

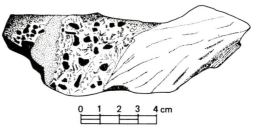

Fig. 5.4 Cross section through asymmetrical patterned cone. Right and left margins represent approximate borders of initial spring pit, subsequently filled by silt. Renewed upwelling produced smaller, steeper walled depression, which then filled with silt chips. Vertical plug of silt piercing coarse debris represents a third stage of upwelling and subsequent filling. Triassic red beds, Wyoming.

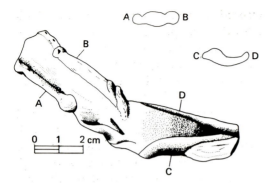

Fig. 5.5 Strap-like siltstone body. Such structures, common in layers bearing patterned cones, are interpreted as casts of partly plugged and collapsed water ducts. Triassic red beds, Wyoming.

partly plugged water passageways, now modified by compaction of the enclosing sediment.

Local disordering of lamination consequent on movement of fluids through sediment (McKee and Goldberg, 1969, Pl. 1, fig. 3) may resemble the disturbance produced by a burrowing mollusk (Schäfer, 1972, Fig. 138). Convolute structures in mud were produced experimentally by McKee and Goldberg (1969). Differential loading caused lateral and upward flowage of mud, accompanied by vertical escape of water or gas, or both. Disruption of laminae by quicksand activity has been produced in experiments, where grain packing was tightened by violent current action (Selley, 1969). This process is apparently one of several ways in which intergranular friction in sand can be reduced. In all cases, the potential result is flowage of the sand, expulsion of water, and localized disruptions of lamination. Possible criteria for recognizing flowage as the cause of disturbance include lateral and vertical relationships of convolute structure to other primary structures (McKee et al., 1967, p. 850), the pattern of distribution in sandstone of fragments of disrupted clay bands, and lateral pinching and swelling of a sandstone bounded by clay bands that are only locally broken (Sellwood, 1972, Fig. 11). In general,

the change from deformed to undeformed lamination is typically more abrupt in a biogenic structure than in one formed by flowage. (See Frey and Howard, 1972.)

Lightning

When lightning strikes a sand dune or an outcrop of sandstone, a burrow-like fulgurite may be formed within the body by fusion of quartz grains. These tubular structures are commonly perpendicular to the surface, and many bear longitudinal ridges and grooves attributable to partial collapse during cooling of a tube initially inflated by expanding gas. Intense magnetization is produced adjacent to the fulgurite by the electrical discharge.

Harland and Hacker (1966) reviewed the literature on characteristics and distribution of modern fulgurites and described ancient fulgurites from the upper Paleozoic Corrie Sandstone of Scotland. The authors even had the unique experience of identifying a derived "fossil" fulgurite in a pebble from the New Red Sandstone. Their case for a lightning origin of the Corrie Sandstone tubes during late Paleozoic deposition of continental dune sands is convincing, even though the structures differ from typical modern fulgurites in having a very thin wall containing little glass and no vesicular texture. Although the sand grains are not completely fused, the Scottish examples were first noticed because of differential weathering. The thin-walled tubes range from 11 to 36 mm in diameter and exhibit an irregular or stellate margin in transverse section. Among the clues to lightning origin are glass fibers connecting grains and intense magnetization immediately surrounding one of the tubes.

Ambient Mineral Grains

Can pyrite grains wander through a solid substrate, producing trails that might be mistaken for structures of organic origin? Tyler and Barghoorn (1963) described such phenomena, even though the minute pyrite grains, possessing "appendages" of quartz or carbonate, are found in Precambrian cherts also possessing well-defined micro-organisms. The authors concluded that the grains advanced through solid chert by growth of the "appendage" mineral on one side of the pyrite grain, while solution affected the chert along the opposite surface of the grain. The carbonate-filled trails are smaller (diameter as much as 0.01 mm; length as much as 0.1 mm) than the quartz-filled trails (diameter as much as 1.4 mm; length as much as 2.95 mm), and the former are more diverse; they include straight, spiral, and irregular forms. A false impression of branching can be produced in thin sections by proximity of two trails. Although a pyrite grain typically occurs at one end of each trail, it is not observable where trails pass obliquely through a thin section.

Size and form of the features illustrated by Tyler and Barghoorn are within the range of borings produced by modern fungi and algae (Bromley, 1970; see Chapter 12), and the Precambrian features are comparable to borings in Cretaceous phosphatic fossils figured by Taylor (1971). Similarity is heightened by the common presence of pyritic spherules at blind ends of the Cretaceous tubes. Taylor attributed the pyrite in his material to in-place formation resulting from bacterial decay of saprotrophic thallophyte filaments, which he believed excavated the tubes. He emphasized the host specificity involved in the Cretaceous occurrences, and his illustrations reveal that the Cretaceous vermiform tubes are longer than any of the tracks behind pyrite grains studied by Tyler and Barghoorn.

Deformation of Unlithified Strata

Faulting

Distinctive relief on the sandstone surface shown in Figure 5.6 appears, on first consideration, to represent sole marks of either organic or inorganic origin. Closer inspection of the specimen, however, indicates that the patterned surface is not the sole; instead, it is at a high angle to internal lamination. I interpret the sculptured face as a fault surface formed when displacement occurred in an unconsolidated sequence, including layers of both sand and cohesive mud. As mud moved past mud, friction resulted in both ripple-like drag features transverse to movement and longitudinal excavations produced by plucking action. As further movement took place, the sculptured mud face was juxtaposed with incompetent sand, which cast the irregularities on the opposing face. The postulated fault presumably resembled one figured by Rettger (1935, Fig. 10), in which clay was faulted against sand during an experiment involving differential loading of unconsolidated sediment. Rettger noted (p. 290) that fault planes in wet clays and sands exhibit irregularities filled with side-wall sediment, but he did not describe three-dimensional sculpture on the fault surfaces.

Compaction and Differential Subsidence

A bed of Carboniferous limestone containing abundant burrow-like features, typically

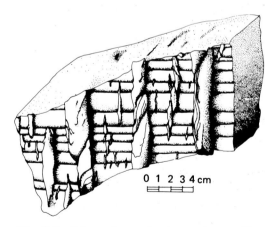

Fig. 5.6 False sole marks in sandstone. Top and bottom surfaces of block are bedding surfaces. Front face interpreted as fault surface; relief on this face apparently formed as sand cast of drag-sculptured mud face, during faulting in unlithified strata. Cretaceous, Wyoming.

several centimeters long and 2 cm or less in diameter, was discussed by Nichols (1966). Although trace fossils are present, he cited reasons for believing that most of the irregular bodies of calcite silt surrounded by calcarenite are boudins produced during compaction.

Some shaly beds exhibit balls of sandstone having concentric lamination. In cross section the aspect of these pseudo-nodules resembles that of exichnial burrow casts. One of many conditions that may be involved in forming such structures was demonstrated experimentally by Kuenen (1958, Pl. 1). He started with a laminated bed of sand deposited over mud. With "earthquake" vibrations as a triggering mechanism, segments of the sand layer sank. As they moved downward, the encircling flow of mud bent their edges upward, resulting in their distinctive form, internal structure, and isolation in the matrix.

The possibility that ellipsoidal bodies having nested internal structure may be misidentified in terms of organic versus inorganic origin is illustrated in another paper by Kuenen (1961). Here he interpreted as burrow fillings or coprolites certain small structures previously considered to have been produced by internal adjustments within unconsolidated strata.

Shrinkage Cracks

The quest for evidence of metazoan life in Precambrian rocks has commonly focused attention on sharply defined, linear features of positive relief on bedding surfaces of sandstone and quartzite. Through the years, various authors have interpreted them as trace fossils, whereas others have called attention to their similarities with undoubted shrinkage-crack patterns. Some of the structures are long and sinusoidal, whereas others are fusiform. The most impressive ones exhibit a central axis and lateral corrugations.

Literature dealing with characteristics and interpretations of such Precambrian bedding-surface features was reviewed by Cloud (1968, p. 29) and Glaessner (1969, p. 370), both of whom concluded that the features are shrinkage-crack casts. Cloud was impressed by their preferred association with ripple troughs, although this alone is not a compelling criterion for inorganic origin. I have observed that trails of some modern intertidal snails are concentrated in the same fashion, because the animals produce furrows in saturated sand while crossing a water-filled trough but leave the sand unmarked as they crawl subaerially over a crest. Glaessner emphasized that some of the unusual characteristics of the structures, especially their curvilinear, sinusoidal, and incomplete patterns, may be the result not of subaerial drying but of cracking under water or during compaction after burial. Young (1969) described evidence for concluding that corrugated vermiform bodies in ripple troughs of Huronian quartzite were produced by both upward and downward injection of silt into fissures developed in mud. Hofmann (1971, p. 36) explained that certain perplexing features of the curved spindles—such as corrugation, interpenetration, and overlap—can be rationalized with an inorganic origin.

Linear features in the Eocene nonmarine Green River Formation are found on both rippled and nonrippled siltstone surfaces; on the former they are present on both crests and troughs. The features were described by Picard (1966), who named them "linear shrinkage cracks." Owing to differential compaction of the Green River sediment, the casts protrude 2 to 4 mm into the overlying bed. Picard attributed the failure to form polygons to a subaqueous origin. Preferred alignment, which characterizes the crack casts in both occurrences, is thought to be the result of slight, downslope gravity movement of the sediment. The resulting pattern, as seen over a small area (e.g., Picard, 1966, Fig. 3), is reminiscent in size and alignment of units to casts of trilobite tracks illustrated by Martinsson (1965, Fig. 14).

The shrinkage cracks discussed above are unique because of their failure to form good polygons. In cases where regular polygons of small dimensions are formed, resulting crack casts resemble the feeding traces assigned to *Paleodictyon* (e.g., Häntzschel, 1962, Fig. 128,5).

CHEMICAL ORIGIN

Differential Precipitation

Formation of aggregate grains by local cementation within carbonate sand is well known in some modern depositional environments. Such composite carbonate particles, rounded during transport, could be mistaken for fecal pellets or burrow-excavation pellets.

Incomplete cementation during formation of beach rock, or during development of a subaqueous hard substrate, is conducive to fragmentation of the product when subjected to undercutting by increased hydraulic activity. (See also Chapter 18.) The resulting lithoclasts may be buried in contemporaneous sediment and subsequently mistaken for coprolites. A similar misinterpretation can involve other products of localized precipitation, such as phosphatic nodules and calcareous concretions (Häntzschel et al., 1968, p. 1).

Color banding that involves alternating zones of enrichment and depletion of a precipitate, such as iron hydroxide, is a common weathering feature of sandstones. The distinctive, commonly concentric pattern is explainable in terms of experimentally produced liesegang rings (Carl and Amstutz, 1958), which form by periodic precipitation from a solution diffusing through an appropriate medium. Ollier (1971), who noted that our knowledge of weathering structures now is about equivalent to the knowledge of sedimentary structures thirty years ago, suggested that patchy and periodic drying up of groundwater may be an important factor in producing color banding in porous rocks.

More rarely, color banding is encountered in limestone or chalk. I know of one Paleozoic limestone exposure where closely spaced concentric zones of iron enrichment are more resistant to chemical weathering than is the intervening gray limestone. The resulting delicate, concentric pattern etched into low relief on bedding and joint surfaces has been mistaken for algal stromatolites by several geologists.

Hofmann (1971, p. 39; Pl. 25, fig. 3) described color patterns simulating the trace fossil *Skolithos* in Precambrian sandstone. Vertical cylinders are visible on a weathered surface because of their brown color, yet bedding passes through them without disruption, and close inspection demonstrates indistinct boundaries. Hofmann interpreted them as chemically formed segregations enriched in iron.

Although resemblance in form between color concentrations and biogenic structures is commonly coincidental, the opposite is true where differential precipitation is localized at the site of prior interaction between an organism and the sediment. An example of the problem of distinguishing between the two situations is Gilmore's (1926, p. 37; Pl. 2, fig. 2) treatment of track-like color markings in sandstone of the Supai Formation. He concluded they are stains resulting from decay of stranded medusae.

Differential Dissolution

In modern littoral-zone limestone outcrops, solution cavities occupied by "homing" gastropods and chitons are difficult to distinguish from other depressions formed by life processes of these organisms. Bromley (1970, p. 62, 64) commented on the problem of differentiating the two. No doubt a spectrum of intermediate situations separates the two end members.

Discontinuity surfaces in carbonate sequences typically show topographic features resulting from differential dissolution during a critical period of either subaerial or subaqueous exposure of the subjacent bed. Dissolution of partly buried shells can produce reentrants in such a surface. In a vertical section the reentrants can easily be mistaken for burrows or borings (Boyd and Newell, 1972, Pl. 4). Narrow anastomosing channelways separating knobs and overhanging platforms characterize a Devonian discontinuity surface described by Koch and Strimple (1968). The channel fillings, in this case argillaceous dolomite, resemble a system of hypichnial ridges.

Selective dissolution of skeletal material from the matrix can be expected in several contexts apart from discontinuity surfaces (e.g., Murray, 1964). Molds, very burrow-like in a three-dimensional sense, remain if this process affects colonies consisting of cylindrical tubes not in contact, such as the Paleozoic coral *Syringopora*.

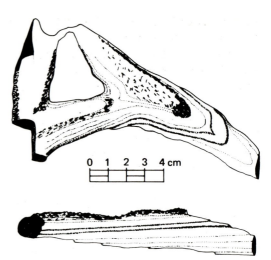

Fig. 5.7 Chert nodule etched from limestone. Lower sketch is side view. Primary lamination of host rock shown on nodule by surface relief and variation in texture. Ordovician, Kentucky.

Silicification

Elongate, branching chert nodules in some limestones form a crude latticework, resembling burrow systems such as *Thalassinoides*; specific instances of the control of crustacean burrows on formation of Cretaceous flint were well documented by Bromley (1967). May we assume that all such silicification patterns follow burrow courses? Specimens such as the nodule shown in Figure 5.7 suggest that each case still requires its own analysis. The figured specimen is typical of many branching, irregularly cylindrical masses of chert liberated together with silicified fossils during laboratory dissolution of Ordovician limestone. Undisturbed lamination of the host sediment, shown by variation in relief and particle size on the surface of the nodule, causes one to question whether a burrow is involved. Other specimens from the same locality show side branches suspiciously smaller in diameter than the segment from which they originate. These branches commonly terminate short distances from the point of origin and appear as protuberances on the main body. However, some

broken ends exhibit a two-part, roughly concentric aspect, having a bluish, more densely silicified outer zone surrounding a less densely silicified core. Proponents of burrow control can argue that silicification followed a burrow, but also affected a narrow peripheral zone of undisturbed sediment. The purpose here is not to quarrel with the theory, but to encourage the reporting of supportive evidence in each case.

Dolomitization

Anastomosing dolomitic bodies form an irregular lattice in some Paleozoic limestones (e.g., Birse, 1928; Griffin, 1942; Swett, 1966). The branching, selectively dolomitized parts stand in relief on weathered surfaces and produce dark, linear designs on polished surfaces. These distinctively mottled limestones pose the same question raised in the preceding section: did chemical alteration follow *Thalassinoides*-like burrow systems or only mimic their pattern?

Arguments against "fucoidal" origin for the mottled pattern in Ordovician carbonates of Manitoba (Birse, 1928) seem to apply also to burrow origin. Birse cited the constancy of development of the pattern through significant vertical and horizontal distances and noted that a typical dolomitic patch contains a tube, in some cases calcite-plugged, 2 to 3 mm in diameter. This is not the definitive evidence of burrow control it first seems, however, because Birse reported the tubular cores within the walls of dolomitized fossils, following their outline, as well as in the branching mottles. A similar anastomosing pattern in a Scottish limestone also reflects incomplete dolomitization, according to Swett (1966). He concluded that the geometry of the dolomitic network precludes a burrow origin.

Crystallization

Radiating, planar crystal growths can mark bedding surfaces with imprints superficially resembling some lebensspuren. Shrock (1948, p. 149–152, Fig. 108) illustrated ice-crystal imprints forming fasciculate patterns in mud, and speculated that some "fucoids" may be casts of such features. The marcasite(?) pattern in Cretaceous rock figured by Cloud (1968, Fig. 4K) is similar in size and form to some specimens of the feeding burrow *Oldhamia* (e.g., Häntzschel, 1962, Fig. 126,4).

ORGANIC ORIGIN

Organic Forms Preserved as Molds

Plants

The root-like aspect of burrows such as *Thalassinoides, Chondrites,* and *Trichichnus* (e.g., Frey, 1970, Figs. 3, 4; Pl. 5) is readily apparent. Is the siliceous latticework shown in Figure 5.8 the cast of burrows or of root molds? I favor the root interpretation because of the crowding, irregularity, and diverse sizes of the components, but I also admit to prejudice from knowl-

edge that the specimen comes from a nonmarine facies. The morphologic aspects noted above can be observed in some burrow complexes, especially those in which large burrows have served one or more "second-occupant" species. Thus, not surprisingly, discrimination between burrows and roots by many authors seems to depend on preconceived ideas concerning depositional environments. A circle is easily completed when, after the structures are identified, they become part of the evidence for the environmental interpretation. The time seems propitious for a major study, from a geological point of view, of modern root systems and their characteristics. The detailed study of root morphology of deep-rooting German marsh plants by Köhnlein and Bergt (1971) is an example of the kind of investigations needed. Until such work provides trustworthy criteria for recognition of fossil roots, authors should include descriptions of the features they interpret as roots (e.g., Walker and Harms, 1971, p. 397) rather

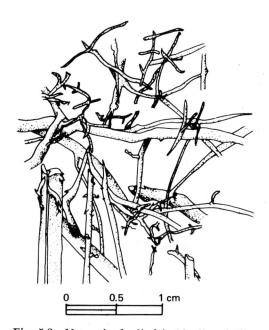

Fig. 5.8 Network of cylindrical bodies of silica, interpreted as root casts, etched from nonmarine, clay-rich limestone. Eocene, Wyoming.

than be content merely to announce their presence. (See Chapter 10.)

An unusually convincing interpretation of root origin for a burrow-like system of calcareous rods was provided by Hoffmeister and Multer (1965). Their conclusion was based not only on analysis of the pattern of roots and pneumatophores of nearby living black mangroves, but also on internal structure of the rods. Some of the latter have a three-part concentric structure analogous to that of roots of the living species. Although no cellular structure was recognized, wood-like texture is replicated by the calcite in such a way as to suggest precipitation of the mineral during decay of the organic matter.

The range of sedimentary environments in which root traces might be preserved is wider than many geologists would expect. For example, plants grow abundantly on coastal dunes in Brazil and their roots commonly penetrate to depths of more than 1 m; McKee and Bigarella (1972, Fig. 5) illustrated deformational structures in the sand caused by root growth. They noted (p. 679) that small roots generally follow lamination, with little or no disturbance, whereas thick roots cause distinctive, irregular warping of adjacent laminae. On the Georgia barrier island coast, low, forested dunes are underlain by a saltwater table. According to R. W. Frey (1972, personal communication), roots penetrate downward until they encounter salt water, then spread laterally. The result is a whorl of roots essentially confined to a thin, horizontal interval, much the way that feeding burrows are concentrated along a carbon-rich layer or dwelling burrows are crowded at some optimum depth. Glennie and Evamy (1968) called attention to the common occurrence of plant-root molds in some Tertiary and younger desert dune sands. Molds are commonly more than 10 cm long, and many are oriented parallel to lamination, apparently in response to maximum permeability and its control on flow of intergranular water.

Molds of subaerial parts of plants can be almost as distracting as roots, to the ichnologist. The expression "plant-like" is recurrent through Häntzschel's (1962) catalog of trace fossil genera, and several of the illustrations (e.g., his Figs. 190,1, 199,3, and 209,1) compare very favorably in form with pictures in the paleobotanical literature. Two familiar patterns shared by both fossil plants and trace fossils are worth special comment. First, some arthropod trackways are simulated by imperfect impressions of slender branches clothed with short, needle-like leaves. Plants possessing such form are as diverse as Paleozoic lycopsids and Mesozoic conifers. Second, joint stems bearing whorls of leaves at each joint, a form typical of sphenopsid plants, can leave imprints strikingly similar to star-like traces produced by various invertebrates. Bandel (1967, p. 2) interpreted *Asterichnus lawrencensis* as feeding traces made by an animal exploiting a circular area. A linear groove, produced as the animal moved from one feeding site to another (his Pl. 1), now connects several rosettes in the same manner that a stem imprint connects leaf whorls in a fossil sphenopsid.

Invertebrates

Molds produced by selective dissolution of skeletal material were discussed previously. Another potential source of confusion between skeletal molds and trace fossils is illustrated by similar planispiral patterns on bedding surfaces, created by imprints of many-whorled, smooth ammonites and by feeding activity of some living polychaetes (e.g., Schäfer, 1972, Fig. 171).

Post-burial decay of a soft-bodied organism can produce a cavity within the sediment, and the surface depression created by collapse of the mold roof constitutes a sediment trap in the manner of an open burrow. Modern examples of molds formed in beach sand by decomposition of jellyfish buried during a storm were described by Kornicker and Conover (1960).

Primary Shell Features

Punctation

To the specialist on endolithic thallophytes, spurious traces should include the minute tubes of primary origin that interrupt the shell structure of many brachiopods and some pelecypods. Brachiopod punctation typically differs in density over the surface of a valve, and punctae have a wide range in diameter throughout the phylum. The punctae are represented in the same size range as borings of both algae and fungi. Although punctae may branch and exhibit some irregularity of course from inner to outer surfaces of a valve, their relatively consistent orientation contrasts sharply with the meandering courses typical of many penetrative thallophytes. Information on borings of algae and fungi, in terms of such characteristics as size, branching, and choice of substrate, was given by Bromley (1970) and Gatrall and Golubic (1970); see also Chapter 12.

Muscle Scars

Muscle scars are familiar features on the inner surfaces of pelecypod valves. Unlike the prominent attachment sites for adductor muscles, the scars of accessory muscles in some species leave small but relatively deep pits in out-of-the-way places. For example, a pedal elevator muscle in several families (e.g., Crassatellidae) is attached in the umbonal cavity of each valve. On the internal mold of such a valve, a sharply defined protuberance rises above the umbo. This feature could be mistaken for the cast of a gastropod boring.

Beveled Umbones

Valve curvature in certain pelecypod species, e.g., the genera *Arca* and *Cardium*, is such that the umbones rub one another when the animal opens its shell. As a result, umbones of old individuals are notably beveled. A truncated umbo of this origin might be mistaken, in dealing with an isolated valve, for the site of an uncompleted attack by a boring predator.

Traces of Modern Organisms in Ancient Host Rock

Some organisms leave traces of their activity on or in hard substrates. Suitability of such surfaces as a substrate commonly bears little or no relation to their geologic age, and a marked chronologic discrepancy may exist between origin of the host material and activity of the organism. Unlike the features discussed in previous sections, these traces are valid lebensspuren. However, they are sources of deception for the ichnologist who assumes an anachronistic trace to be contemporaneous in origin with the host material.

Imprints

When breaking down a ledge of mudstone or siltstone for the purpose of obtaining unweathered samples, one often encounters mat-like living root systems extending several meters along parting surfaces. Opposing rock faces commonly exhibit patterns of the root system they confined. The examples shown in Figure 5.9 represent light tan traces on dark reddish-brown mudstone. Examination of root-infested

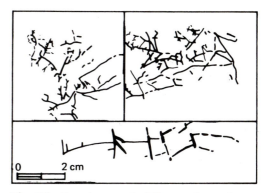

Fig. 5.9 Imprints on bedding surfaces of root systems of three modern plants. Original patterns consist of tan lines on dark reddish-brown mudstone (Triassic, Wyoming).

ledges demonstrates all stages of association between patterns and roots, from that in which the living root system is in place, to that in which the pattern remains after death of the plant and removal of the roots by natural processes. Where the patterns developed along bedding surfaces, they may be mistaken for trails or burrow systems contemporaneous with the host rock. Several features characterize a few patterns that I have studied in detail: (1) linear elements in the pattern have no appreciable depth in the rock face, compared to their width; (2) linear elements both branch and intersect; (3) lines of notably different width are found in any one pattern; (4) both rectilinear and curvilinear segments occur in any one pattern; and (5) interruptions in lines are common. This last feature is explained by the fact that any one rootlet is not necessarily in equally tight contact with the rock face throughout its length.

Some modern bees construct nests on rock surfaces, and a distinctive net-like marking, superficially resembling the trace fossil *Paleodictyon*, can remain after destruction of the nest. Bee-nest patterns described by Sando (1972) are lighter in color than the rock on which they are imprinted; some of the patterns have slight negative relief, others do not. Each pattern, apparently produced by biochemical action of bee saliva on the rock, reflects the outline of anastomosing walls bounding the 10 to 25 capsules comprising a nest. The patterns described by Sando are subovate; maximum dimensions range from 26 to 50 mm. They are on carbonate rock, except for one pattern on quartzite.

Borings and Excavations

Many living organisms are capable of boring into solid substrates. If the medium is an outcrop or an exhumed fossil, then the modern boring may be mistaken for a trace fossil. In cases where either fossil or modern borings could be present on an exhumed surface, cavities should be scrutinized for fossil epizoans and matrix remnants as clues to time of development. (See Chapter 18.)

Organisms capable of producing modern holes in ancient matrix include two terrestrial groups as well as the plethora of more publicized marine forms (Chapter 11). The ubiquitous lichens are important agents of subaerial carbonate erosion (see Chapter 10), but their minute pits are unimportant as sources of deception in the present context. Certain wasps, on the other hand, are reported to excavate holes as much as 5 mm in diameter and 8 cm long in sandstone (Deal, 1963, p. 82). These holes are normal to the outcrop surface, are clustered, and exhibit no control by bedding or jointing.

CONCLUSIONS

False traces are features of sediments or sedimentary rocks that can be mistaken for evidence of behavioral activity.

Many of the physical, chemical, and biological sources of deception are only initially confusing. Most difficult to interpret genetically are certain tool marks, gas-entrapment sites, ducts and surface features formed during expulsion from sediment of gas and water, disruption of lamination due to localized quicksand movement, casts of incomplete shrinkage cracks, concentrations of iron oxide cement, branched chert nodules, and dolomitic mottles.

Recent lebensspuren excavated in, or imprinted on, exhumed surfaces may be mistaken for fossils comparable in age to the substrate.

Organic forms most easily misidentified as burrows are plant roots in the form of casts, molds, and imprints. Modern root systems deserve serious study from a geologic point of view, in order to provide trustworthy criteria for distinguishing their fossil counterparts from animal traces.

ACKNOWLEDGMENTS

The specimens illustrated in Figures 5.2, 5.6, 5.7, and 5.8 were made available, respectively, by Hugh Dresser, Mary Capps, John Pojeta, and Craig Wood. Jeanne Sula helped with library research, and the figures were drawn by Alice Zimmerman. H. J. Hofmann, E. D. McKee, and N. D. Newell took time from busy schedules to provide constructive criticism of the manuscript.

REFERENCES

Allen, J. R. L. 1961. Sandstone-plugged pipes in the lower Old Red Sandstone of Shropshire, England. Jour. Sed. Petrol., 31:325–335.

Bandel, K. 1967. Trace fossils from two Upper Pennsylvanian sandstones in Kansas. Univ. Kansas Paleont. Contr., Paper 18, 13 p.

Barthel, K. W. 1966. Concentric marks: current indicators. Jour. Sed. Petrol., 36:1156–1162.

Birse, D. J. 1928. Dolomitization processes in the Palaeozoic horizons of Manitoba. Royal Soc. Canada, Trans., 22:215–222.

Boekschoten, G. J. 1964. Tadpole structures again. Jour. Sed. Petrol., 34:422–423.

Bondesen, E. 1966. Observations of recent sand volcanoes. Dansk Geol. Foren. Meddr., 16: 195–198.

Boyd, D. W. and N. D. Newell. 1972. Taphonomy and diagenesis of a Permian fossil assemblage from Wyoming. Jour. Paleont., 46:1–14.

———— and H. T. Ore. 1963. Patterned cones in Permo-Triassic redbeds of Wyoming and adjacent areas. Jour. Sed. Petrol., 33:438–451.

Bromley, R. G. 1967. Some observations on burrows of thalassinidean Crustacea in chalk hardgrounds. Geol. Soc. London, Quart. Jour., 123:157–182.

————. 1970. Borings as trace fossils and *Entobia cretacea* Portlock, as an example. In T. P. Crimes and J. C. Harper (eds.), Trace fossils. Geol. Jour., Spec. Issue 3:49–90.

Cameron, B. and R. Estes. 1971. Fossil and recent "tadpole nests": a discussion. Jour. Sed. Petrol., 41:171–178.

Carl, J. D. and G. C. Amstutz. 1958. Three-dimensional liesegang rings by diffusion in a colloidal matrix, and their significance for interpretation of geologic phenomena. Geol. Soc. America, Bull., 69:1467–1468.

Cloud, P. E., Jr. 1960. Gas as a sedimentary and diagenetic agent. Amer. Jour. Sci., 258-A: 35–45.

————. 1968. Pre-metazoan evolution, origins of metazoa. In E. T. Drake (ed.), Evolution and environment. New Haven, Conn., Yale Univ. Press, p. 1–72.

Deal, D. E. 1963. Role of wasps in the erosion of the Pennsylvanian Casper Sandstone along Sand Creek, southeastern Wyoming (abs.). Geol. Soc. America, Spec. Paper 73: 82.

Dżułyński, S. and F. Simpson. 1966. Experiments on interfacial current markings. Geol. Rom., 5:197–214.

———— and E. K. Walton. 1963. Experimental production of sole markings. Edinburgh Geol. Soc., Trans., 19:279–305.

Emery, K. O. 1945. Entrapment of air in beach sand. Jour. Sed. Petrol., 15:39–49.

Ford, T. D. and W. J. Breed. 1970. Tadpole holes formed during desiccation of overbank ponds. Jour. Sed. Petrol., 40:1044–1060.

Frey, R. W. 1970. Trace fossils of Fort Hays Limestone Member of Niobrara Chalk (Upper Cretaceous), west-central Kansas. Univ. Kansas Paleont. Contr., Art. 53, 41 p.

———— and T. M. Chowns. 1972. Trace fossils from the Ringgold road cut (Ordovician and Silurian), Georgia. In T. M. Chowns (Comp.), Sedimentary environments in the Paleozoic rocks of northwest Georgia. Georgia Geol. Surv., Guidebook 11:25–55.

———— and J. D. Howard. 1972. Georgia coastal region, Sapelo Island, U.S.A.: sedimentology and biology. VI. Radiographic study of sedimentary structures made by beach and offshore animals in aquaria. Senckenbergiana Marit., 4:169–182.

———— and T. V. Mayou. 1971. Decapod burrows in Holocene barrier island beaches and washover fans, Georgia. Senckenbergiana Marit., 3:53–77.

Gatrall, M. and S. Golubic. 1970. Comparative study on some Jurassic and recent endolithic fungi using scanning electron microscope. In T. P. Crimes and J. C. Harper (eds.), Trace fossils. Geol. Jour., Spec. Issue 3:167–178.

Gill, W. D. and P. H. Kuenen. 1958. Sand volcanoes and slumps in the Carboniferous of County Clare, Ireland. Geol. Soc. London, Quart. Jour., 113:441–460.

Gilmore, C. W. 1926. Fossil footprints from the

Grand Canyon. Smithsonian Misc. Coll., 77(9):41 p.

Glaessner, M. F. 1969. Trace fossils from the Precambrian and basal Cambrian. Lethaia, 2:369–393.

Glennie, K. W. and B. D. Evamy. 1968. Dikaka: plants and plant root structures associated with eolian sand. Palaeogeogr., Palaeoclimatol., Palaeoecol., 4:77–87.

Griffin, R. H. 1942. Dolomitic mottling in the Platteville limestone. Jour. Sed. Petrol., 12: 67–76.

Häntzschel, W. 1962. Trace fossils and problematica. In R. C. Moore (ed.), Treatise on invertebrate paleontology, Pt. W, Miscellanea. Lawrence, Kan., Geol. Soc. America and Univ. Kansas Press, p. W177–W245.

————— et al. 1968. Coprolites, an annotated bibliography. Geol. Soc. America, Mem. 108, 132 p.

Harland, W. B. and J. F. Hacker. 1966. "Fossil" lightning strikes 250 million years ago. Advmt. Sci., 22:663–671.

Hoffmeister, J. E. and H. G. Multer. 1965. Fossil mangrove reef of Key Biscayne, Florida. Geol. Soc. America, Bull., 76:845–852.

Hofmann, H. J. 1971. Precambrian fossils, pseudofossils, and problematica in Canada. Geol. Surv. Canada, Bull. 189, 146 p.

Hoyt, J. H. and V. J. Henry. 1964. Development and geologic significance of soft beach sand. Sedimentology, 3:44–51.

Koch, D. L. and H. L. Strimple. 1968. A new Upper Devonian cystoid attached to a discontinuity surface. Iowa Geol. Surv., Rept. Invest. 5, 49 p.

Köhnlein, J. and K. Bergt. 1971. Untersuchungen zur Entstehung der biogenen Durchporung im Unterboden eingedeichter Marschen. Zeitschr. Acker- u. Pflanzenbau, 133:261–298.

Kornicker, L. S. and D. W. Boyd. 1962. Shallow-water geology and environments of Alacran reef complex, Campeche Bank, Mexico. Amer. Assoc. Petrol. Geol., Bull., 46:640–673.

————— and J. T. Conover. 1960. Effect of high storm tide levels on beach burial of jellyfish (Scyphozoa) and other organisms. Int. Revue ges. Hydrobiol., 45:203–214.

Kuenen, P. H. 1958. Experiments in geology. Geol. Soc. Glasgow, Trans., 23:1–28.

—————. 1961. Some arched and spiral structures in sediments. Geol. en Mijnbouw, 40: 71–74.

McKee, E. D. and J. J. Bigarella. 1972. Deformational structures in Brazilian coastal dunes. Jour. Sed. Petrol., 42:670–681.

————— and M. Goldberg. 1969. Experiments on formation of contorted structures in mud. Geol. Soc. America, Bull., 80:231–244.

————— et al. 1967. Flood deposits, Bijou Creek, Colorado, June 1965. Jour. Sed. Petrol., 37:829–851.

Martinsson, A. 1965. Aspects of a Middle Cambrian thanatotope on Öland. Geol. Fören. i Stockholm Förhand., 87:181–230.

Murray, R. C. 1964. Preservation of primary structures and fabrics in dolomite. In J. Imbrie and N. D. Newell (eds.), Approaches to paleoecology. New York, John Wiley, p. 388–403.

Nichols, R. A. H. 1966. Petrology of an irregular-nodule bed, Lower Carboniferous, Anglesey, North Wales. Geol. Magazine, 103:477–486.

Ollier, C. D. 1971. Causes of spheroidal weathering. Earth-Sci. Rev., 7:127–141.

Osgood, R. G., Jr. 1970. Trace fossils of the Cincinnati area. Palaeontographica Amer., 41(6):281–444.

Palmer, R. H. 1928. Sandholes of the strand. Jour. Geol., 36:176–180.

Pettijohn, F. J. and P. E. Potter. 1964. Atlas and glossary of primary sedimentary structures. New York, Springer-Verlag, 370 p.

Picard, M. D. 1966. Oriented, linear-shrinkage cracks in Green River Formation (Eocene), Raven Ridge area, Uinta Basin, Utah. Jour. Sed. Petrol., 36:1050–1057.

Rettger, R. E. 1935. Experiments on soft rock deformation. Amer. Assoc. Petrol. Geol., Bull., 19:271–292.

Sando, W. J. 1972. Bee-nest pseudofossils from Montana, Wyoming, and southwest Africa. Jour. Paleont., 46:421–425.

Sarjeant, W. A. S. 1971. Vertebrate tracks from the Permian of Castle Peak, Texas. Texas Jour. Sci., 22:343–366.

Schäfer, W. 1972. Ecology and palaeoecology of marine environments. Edinburgh and Chicago, Oliver & Boyd and Univ. Chicago Press, 568 p.

Schumm, S. A. 1970. Experimental studies on the formation of lunar surface features by fluidization. Geol. Soc. America, Bull., 81: 2539–2552.

Seilacher, A. 1963. Umlagerung und Rolltransport von Cephalopoden-Gehäusen. Neues Jahrb. Geol. Paläont., Mh., 11:593–615.

Selley, R. C. 1969. Torridonian alluvium and quicksands. Scottish Jour. Geol., 5:328–346.

Sellwood, B. W. 1972. Tidal-flat sedimentation in the Lower Jurassic of Bornholm, Denmark. Palaeogeogr., Palaeoclimatol., Palaeoecol., 11:93–106.

Shrock, R. R. 1948. Sequence in layered rocks. New York, McGraw-Hill, 507 p.

Stewart, H. B. 1956. Contorted sediments in modern coastal lagoon explained by laboratory experiments. Amer. Assoc. Petrol., Geol., Bull., 40:153–179.

Swett, K. 1966. Diagenetic mottling in dolomitic limestones, dolostones and cherts, northwest Scotland (abs.). Geol. Soc. America, Spec. Paper 87:171–172.

Taylor, B. J. 1971. Thallophyte borings in phosphatic fossils from the Lower Cretaceous of southeast Alexander Island, Antarctica. Palaeontology, 14:294–302.

Tyler, S. A. and E. S. Barghoorn. 1963. Ambient pyrite grains in Precambrian cherts. Amer. Jour. Sci., 261:424–432.

van der Lingen, G. J. and P. B. Andrews. 1969. Hoof-print structures in beach sand. Jour. Sed. Petrol., 39:350–357.

Walker, R. G. and J. C. Harms. 1971. The "Catskill delta": a prograding muddy shoreline in central Pennsylvania. Jour. Geol., 79:381–399.

Williams, P. F. and B. R. Rust. 1969. The sedimentology of a braided river. Jour. Sed. Petrol., 39:649–679.

Young, G. M. 1969. Inorganic origin of corrugated vermiform structures in the Huronian Gordon Lake Formation near Flack Lake, Ontario. Canadian Jour. Earth Sci., 6:795–799.

PART II

The Geological Significance of Trace Fossils

CHAPTER 6

THE PALEONTOLOGICAL SIGNIFICANCE OF TRACE FOSSILS

RICHARD G. OSGOOD, JR.

Department of Geology, College of Wooster
Wooster, Ohio, U.S.A.

SYNOPSIS

Trace fossils grade imperceptibly into body fossils; indeed, in some cases the two are difficult to distinguish. Although in most instances the organism responsible for a given trace is impossible to identify, trace fossil studies still can add much to our knowledge of the fossil record. Trace fossils provide rudimentary evidence for the morphology of the tracemakers, but the greatest contribution by traces is their demonstration of behavior patterns among extinct organisms. Trilobites and their traces are an excellent example, especially concerning modes of feeding, locomotion, and protection. The contribution of ichnology to the general field of evolution is also important as it pertains to phyletic rates within the Metazoa, particularly during late Precambrian time.

INTRODUCTION

In most respects the study of trace fossils is an integral part of the field of paleontology. Although mentioned only briefly in most paleontology textbooks, trace fossils can make significant contributions to our knowledge of both the morphology and behavior of fossil organisms. My coverage is restricted mainly to biogenic sedimentary structures made by invertebrate organisms. Vertebrate and plant traces, coprolites, and borings are treated in other chapters of this volume.

RELATIONSHIPS BETWEEN BODY FOSSILS AND TRACE FOSSILS

Although for many years trace fossils were largely ignored by paleontologists, a close connection exists between body fossils and trace fossils (Seilacher, 1953a). This relationship is demonstrated by a pie-shaped diagram showing the behavioral classification of trace fossils (see Fig. 3.2). Body fossils occupy the center of the diagram and the various trace fossil categories are represented by "slices" of the pie; but the body fossils and trace fossils merge imperceptibly.

This relationship results from either of two things: the individual author's taxonomic philosophy, or the characteristics of the fossil itself. Howell (1957) described a U-shaped burrow from the Cretaceous of Colorado and named it *Polyupsilon coloradoense*. Even though the burrow exhibited no concrete definitive structures, Howell followed the zoological hierarchy from phylum to species level. His justification was that when dealing with predominately soft-bodied animals, such as the annelids, we are fortunate to find even a U-shaped burrow.

Many trace fossils, especially the Cubich-

nia (resting traces), reflect the general outline of the body of the trace-making
organism. Sea stars are an excellent example
(Seilacher, 1953b, Pls. 7–10). (See Fig. 4.5.)
Also, the trilobite resting trace *Rusophycus*
frequently shows the imprints of the
cephalic and pygidial margins. Nevertheless, the bilobed geometry of this structure
leaves no doubt that it is a trace fossil
rather than the mold of a body fossil; the
lobes result from digging activity by the
legs.

However, even in the case of *Rusophycus*, the trace morphology can sometimes raise crucial questions. A few
examples will illustrate this problem. In a
specimen of *Rusophycus carleyi* from the
Upper Ordovician of Ohio (Fig. 6.1), the

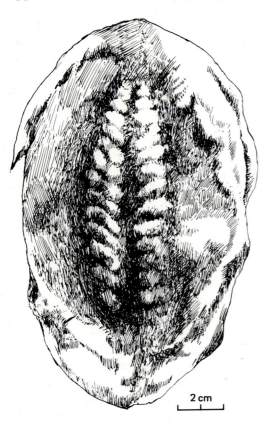

Fig. 6.1 *Rusophycus carleyi*—resting trace of
the trilobite *Isotelus*, preserved as convex hyporelief. Specimen exhibits casts of imprints of
cephalic and pygidial doublures, cephalic spines,
and proximal parts of walking legs. Upper
Ordovician, Ohio. (After Osgood, 1970.)

cephalic doublure, marginal spines, pygidial doublure, and imprints of the
proximal parts of the walking legs are
clearly visible; yet the fossil is bilobed,
demonstrating that animal movement was
involved. Thus, technically, this is a trace
fossil rather than a ventral mold of the
trilobite *Isotelus*.[1] An even better example
is *Walcottia rugosa*, an organism having
an elongate, bilaterally symmetrical body
bearing structures that appear to have been
parapodia (Fig. 6.2). Thus represented is
the segmented body of an annelid, and
apparently a *bona fide* mold. However,
closer examination reveals a large ovate
marking at one end, too indistinct to represent an impression of the head. I postulate
that the organism settled to the bottom,
thereby creating a mold of its ventral
surface, then arched its body and burrowed
almost straight down into the substrate,
without obliterating its earlier imprint. In
a sense, therefore, this specimen is both a
trace fossil and a body fossil.

Another fossil of questionable origin is
the "roostertail" form *Zoophycos* (Fig. 6.3).
Working with specimens from the Cretaceous and Tertiary of Europe and Asia,
Plička (1968; 1970, p. 361) concluded that
Zoophycos represents the ". . . imprints of
abandoned prostomial parts of sedentary
polychaetes." His specimens (1970, Pl. 2)
exhibit hair-like imprints not usually found
in *Zoophycos*. Conversely, most workers—
including myself—consider the fossil to be
the highly complex feeding burrow of a
vermiform organism.[2] Frequently, one
Zoophycos structure may seem to overlap
another. Closer observation reveals that this

[1] As shown subsequently, identifying the originators
of trace fossils is generally impossible. Identification
was possible here because of the large size of the
Rusophycus.

[2] My conclusions on the affinities of *Zoophycos* are
based on observations from the Devonian of east-
central New York State and the Mississippian of
Ohio and Kentucky. Some of the points raised here
are discussed in greater detail by Bischoff (1968).
Also, Bischoff (1968) and Simpson (1970) described
the ethological significance of *Zoophycos*; the interpretation is complex and is not covered here.

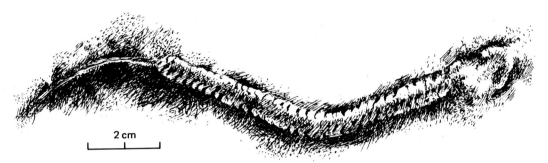

2 cm

Fig. 6.2 *Walcottia rugosa*—convex hyporelief demonstrating the close affinity between trace fossils and body fossils. Parapodia of the vermiform organism are clearly evident. Ovoid marking at right end is interpreted as the place where the organism burrowed into subjacent mud. Upper Ordovician, Ohio. (After Osgood, 1970.)

relation is not overlap—as necessitated if these are prostomial parts of polychaetes—but is instead an interpenetration, where the later formed structure obliterated the one formed earlier. Moreover, many *Zoophycos* specimens exhibit a distinct three-dimensional aspect, in some instances attaining the shape of a cone (e.g., Bischoff, 1968, Pl. 179); this configuration is demonstrated by vertical sections that reveal total disruption of the bedding. Finally, Bischoff (1968, Pl. 179) and Osgood and Szmuc (1972, Pl. 1, fig. 2) illustrated remnants of a cylindrical cast that circumscribes the *Zoophycos* structure (Fig. 6.3). This marginal tube is difficult to reconcile with Plička's interpretation; it is more easily explained as part of a burrow structure.

Zoophycos has also been interpreted recently as a probable umbellulid (Pennatulacea) burrow (Bradley, 1973) and—to complete the circle back to fucoids—as a marine plant (Loring and Wang, 1971).

A somewhat different situation, but one that nevertheless demonstrates the close connection between body fossils and trace fossils, is where the two are actually found together. Probably the most famous examples are the xiphosurid body fossils from the Jurassic Solnhofen Limestone, found at the end of their trackways (Abel, 1935, Figs. 244–247). A similar occurrence was illustrated by Osgood (1970, Pl. 58, figs. 4 and 5); here the Upper Ordovician trilobite *Flexicalymene meeki* was found in situ, the *Rusophycus* attached on the ventral side.

DIFFICULTY IN DETERMINING THE ORIGINATOR OF A GIVEN TRACE FOSSIL

In many cases the maker of a trace cannot be identified even at the phylum level. The reasons for this difficulty are many. First, several phylogenetically distinct groups, e.g., annelids, priapulids, sipunculids, and nemerteans, all have vermiform bodies. These animals can leave sinuous traces that are very similar morphologically. Many such traces exhibit annulations left by peristaltic movements along the body. Even relatively distinct traces, such as U-tubes and the bilobed *Rusophycus* (a genus

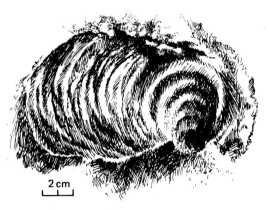

2 cm

Fig. 6.3 *Zoophycos*—surface view of "rooster-tail" feeding burrow. A small part of the marginal tube is preserved near upper margin of the burrow. (After Osgood and Szmuc, 1972.)

generally reserved for trilobite traces), may be produced by several different animal groups. Echiuroids, the enteropneusts *Saccoglossus* and *Balanoglossus*, certain holothurians, many polychaete worms, and even the clam *Solemya velum* and the amphipod *Corophium*, all dwell in U-shaped burrows. Seilacher (1960) showed that the gastropod *Bulla*, the marine polychaete *Aphrodite*, and phyllopods can produce bilobed burrows that, to varying degrees, resemble *Rusophycus*. Similarly, three-dimensional, branching burrow systems such as *Chondrites*—a common genus that was studied in detail by Simpson (1957)—could be excavated by sediment-ingesting polychaetes, sipunculids, or conceivably even by detritus-feeding arthropods.

A second difficulty is that the same organism can produce distinctly different traces under different environmental conditions, different phases of behavior, or during ontogenetic development. As Seilacher (1953a, Fig. 4) demonstrated, *Corophium* constructs a U-shaped burrow in mud, whereas a simple vertical shaft suffices in well-drained sand. Frey and Mayou (1971) reported on burrows of the ghost crab *Ocypode quadrata* at Sapelo Island, Georgia. They found that the crab constructs four different types of burrows, the burrow morphology being linked to the age of the organism and to the environment. These burrows vary from gently inclined cylindrical shafts to J-, Y-, or even U-shaped structures. Both *Corophium* and *Ocypode* leave their burrows on occasion, producing surficial feeding, crawling, and resting traces totally different from their dwelling structures.

A third variable concerns the mechanics of preservation. Seilacher (1953a, Fig. 1) illustrated a bed of mud overlain by coarser grained sand or silt; a fish and a crayfish settled on the bottom, and parts of their ventral surfaces penetrated the sand and extended into the mud. The result was similar traces made by two very different

organisms. The same problem can arise with arthropod tracks, where the walking legs are impressed into the sediment. The configuration of the resulting imprint at a depth of 2 mm can differ markedly from the impression made on the depositional interface (Osgood, 1970, Fig. 19; Goldring and Seilacher, 1971, Fig. 2). The surficial imprints possess sharp morphologic detail, whereas those preserved at depth on cleavage planes are somewhat subdued.

The net result of all these variables is that the identity of the tracemaker generally remains a mystery. One of the main underlying problems is simply our lack of knowledge of the various structures made by invertebrate organisms. Until recently, ichnology was an ill-defined field that lay somewhere on the periphery of zoology, paleontology, and sedimentology. We need many more of the broad works such as Schäfer's (1972) study of recent biogenic structures in the North Sea. Ultimately, detailed studies focusing on the habits of smaller groups will also be required. Examples of these include Caster's (1938) work with *Limulus*; Howard and Elders' (1970) aquarium studies with amphipods at Sapelo Island, Georgia; Stanley's (1970) monograph on the burrowing habits of selected pelecypods; and the work of many authors on *Callianassa* (discussed subsequently). Fortunately, works such as these are being published with increasing frequency. (See Chapter 22.) Naturally, most of them focus on organisms that are easily studied, i.e., hardy animals from the littoral or shallow sublittoral zones. The logistical problems in studying epifaunal—much less infaunal—inhabitants in the remainder of the sublittoral and in the bathyal and abyssal zones are tremendous.

Ewing and Davis (1967) and Heezen and Hollister (1971) have made a magnificent beginning in this field. Heezen and Hollister's book *Face of the Deep* contains remarkable photographs of burrows and trails of bathyal and abyssal organisms. In some instances the pictures were taken while the trails were being made, and the trailmaker can be identified. (See Chapter 21.)

BENEFITS OF TRACE FOSSIL STUDIES TO PALEONTOLOGY

In spite of the difficulties mentioned above, the benefits of trace fossil studies to paleontology are many. For convenience, these are separated into five somewhat artificial categories: contributions to our understanding of (1) taxonomy, (2) the fossil record of soft-bodied invertebrates, (3) the fossil record of invertebrates having hard parts, (4) behavioral patterns of fossil organisms, and (5) evolution.

Taxonomy

Of fundamental importance is the distinction between invertebrate trace fossils, vertebrate traces, and marine algae. As shown in Chapter 1, the descriptive era in paleontology, during the second half of the nineteenth century, resulted in the establishment of a large number of genera of fossil "fucoids" or algae (e.g., Schimper, in Zittel, 1879–1890). Such an array of "fucoids" obviously had its impact on works concerning the phylogeny of plants. Thus, Nathorst (1881, 1886) made a significant contribution to taxonomy, and ultimately phylogeny, when he demonstrated that the "fucoids" were either trace fossils or of inorganic origin.

In a few cases, invertebrate tracks have been assigned to vertebrates. Willard (1935) established *Paramphibius* for trails from the Devonian of western Pennsylvania. He deliberated over the affinity of these tracks for three years and finally concluded, albeit with some hesitancy, that they were made by an amphibian. Because these Devonian specimens predated any amphibian remains known at that time, he erected the Order Icthypoda for his new genus. He postulated that the amphibian possessed

pentadactylous rear legs whereas the forelegs remained fin-like, a carryover from their fish ancestors. He then hypothesized at some length on the early evolution of amphibians. In a classic paper, Caster (1938) demonstrated convincingly that *Paramphibius* actually represents the tracks of xiphosurids. He conducted extensive aquarium studies with *Limulus polyphemus*, which showed the close similarity between the resulting tracks and those from the Devonian. By removing *Paramphibius* from the amphibians, Caster prevented extensive misdirected conjecture on the possible origin and phylogeny of this group of vertebrates.

The Fossil Record of Soft-bodied Invertebrates

The widespread discrepancy in numbers of species and genera between living invertebrates and those in the fossil record is well known to both paleontologists and zoologists. Soft vermiform organisms are important members of the infauna in recent environments, yet their overall fossil record is extremely poor. The main reason for the differences in numbers is obviously that many groups lack hard parts amenable to preservation. Therefore, trace fossils can potentially provide additional documentation of the fossil record of many phyla.

As illustrated by the above discussions, this documentation is a difficult task. To date, trace fossils have told us little about the soft-bodied phyla, other than indicating that some or many animals were present in a given substrate. Nevertheless, this simple fact can be significant: Glaessner (1969) discussed six trace fossils from the Precambrian Ediacara fauna of Australia, the morphology of which could not be reconciled with known Ediacara body fossils. Thus, the trace fossils indicate a more diverse fauna. The same is true of assemblages in the Upper Ordovician of the Cincinnati area (Osgood, 1970). This area is one of the richest fossil localities in the world, yet in most cases I was unable to assign the trace fossils to respective body fossils. However, the trace fossils add to our knowledge in a general way: they indicate that members of the soft-bodied infauna were equally as abundant as their hard-shelled epifaunal counterparts. Also, Frey (1970), studying the Cretaceous Niobrara Chalk of Kansas, determined that the ichnofauna was substantially more diverse than the body fossil fauna. Moreover, he concluded that not one organism now represented by body fossils was a trace-making animal.

In many cases a trace fossil has been attributed to a certain soft-bodied group. An example is Fenton and Fenton's (1934) designation of the straight vertical shaft *Skolithos* as a phoronid dwelling burrow. If this interpretation is true, the range of phoronids is extended back to the Cambrian. Although this assignment is reasonable, no unequivocal evidence supports their case. Living phoronids occupy closely packed, unbranched vertical burrows, as the *Skolithos* animal generally did; but so do some recent polychaetes. Therefore, the evidence is suggestive but could hardly be termed conclusive proof.

Most studies dealing with traces made by soft-bodied invertebrates are analogous to the *Skolithos* situation. Some exceptions are noteworthy, however. Moussa (1970) described sinusoidal trails from the Eocene Green River Formation of Utah, which he attributed convincingly to small nematodes. His argument, buttressed by a considerable amount of literature on the locomotion of recent nematodes, demonstrates that such trails are possible when a mud surface is covered by a thin film of water. This work is a significant contribution because Howell (1962) listed only four genera of these small vermiform organisms known from the fossil record. Three genera are from the Oligocene of Germany, the fourth from the Pleistocene of Siberia.

Another instance where the soft-bodied tracemaker has been identified with some degree of confidence was reported by Kaźmierczak and Pszczołkowski (1969) from the Triassic of Poland. They described, but did not name, several U-shaped tunnels interconnected by horizontal branches. The key to their identification is short, blind galleries, 1 to 3 cm long. Casts taken of these galleries resemble, in both morphology and size, the probosces of recent enteropneusts. The authors conceded that the burrow systems of living acorn worms are not well known, but felt that the galleries (Kaźmierczak and Pszczołkowski, 1969, Fig. 6) provide conclusive evidence for an enteropneust origin.

We must await further studies on recent traces before trace fossils can make many other specific contributions to the fossil record of soft-bodied taxa.

The Fossil Record of Groups Having Hard Parts

In some cases trace fossils can add to our knowledge of the morphology or geologic or geographic range of groups that possess hard parts. Nearly all examples pertain to arthropod anatomy—more specifically, the terminal digits or dactyls of trilobites. In comparatively few specimens of trilobites are the appendages preserved, and even in these rare cases the usually delicate terminal dactyls are missing. [A comparison of Raymond's (1920) monograph with the works of Størmer (1939, 1951) reveals that the former took considerable artistic license in his reconstruction of the dactyls.] Thus, if a certain trilobite can somehow be identified as the maker of a given set of imprints, something of the morphology of the dactyls may be learned. The most comprehensive work in this area is that of Seilacher (1962). He (Pl. 25, fig. 1) identified Illaenid tracks from the Ordovician and Silurian of Europe, North Africa, and the Near East by impressions of the pleurae and also through imprints of the rounded genal angles. The tracks show that these trilobites possessed as many as 11 dactyls on each leg. Seilacher demonstrated that *Phacops* from the Lower Devonian Hunsrück Shale had numerous setae, which surrounded the terminal digits and spread out umbrella-like for added stability when the appendage made contact with the substrate (Fig. 6.4). Osgood (1970, Pl. 71, fig. 6) identified *Isotelus* tracks from the Upper Ordovician by the large width between the two rows of imprints. This study indicated that *Isotelus* had three long, hairlike dactyls. Crimes (1970b, p. 51) showed that the same kind of information may be gained from the study of the bilobed burrowing structures *Cruziana* and *Rusophycus*. (These genera are discussed in detail below.) In the Upper Cambrian of North Wales, he found distinct patterns in the striations on the lobes, made by the legs as the trilobite

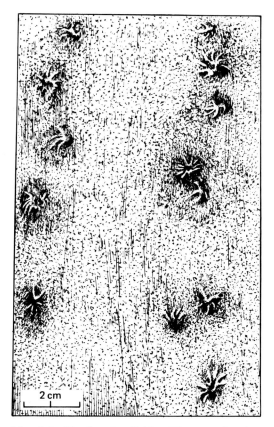

Fig. 6.4 Tracks of trilobite *Phacops*, showing "snowshoe" imprints of the walking legs. Hunsrück Shale (Devonian), West Germany. (After Seilacher, 1962.)

burrowed. He concluded that the organisms, probably olenids, may have had two, three, and even four dactyls.

Xiphosurid tracks from the Devonian of Pennsylvania, the Upper Carboniferous of Great Britain, the Triassic of Arizona and Germany, and the Jurassic of Germany have been studied by various workers: Caster (1938, 1940, 1944), King (1965), and Goldring and Seilacher (1971). The studies generally demonstrate that little change has occurred in the appendages of these "living fossils" since the Devonian.

At least one example exists where the geographic range of an arthropod group has tentatively been extended through the study of trace fossils. Radwański and Roniewicz (1967) established *Aglaspidichnus* for specimens from the Upper Cambrian of Poland. They concluded that the trace was made by an aglaspid rather than a trilobite. If true, this is the first report of these rare merostomes from the rocks of Poland.

The above examples indicate various means by which trace fossils can add information to our knowledge of the morphology of different groups. Fruitful fields for additional studies remain. Goldring and Seilacher (1971) pointed to insect tracks from the Permian of Nierstein, Germany, as one example. An additional possibility is eurypterid trails; they should be highly distinctive but have not received adequate study. Brief discussions may be found in papers by Packard (1900), Sharpe (1932), Størmer (1934), and Allen and Lester (1953).

Behavioral Patterns of Fossil Organisms

Probably the most important contribution that trace fossils make to paleontology is to provide our only direct evidence for the behavior (ethology) of extinct forms. We can infer behavior from hard-part morphology, but we can actually observe its expression in trace fossils. Obviously, this work is most meaningful when it can be tied to a distinct group of animals.

The pioneer work in this field is that

of Rudolf Richter, who, in a series of papers that spanned nearly three decades, reported the results of studies of recent traces from the North Sea "Wattenmeer" and applied this knowledge to the fossil record, especially the Hunsrück Shale. Probably his most dramatic contribution was his interpretation of the tightly packed meanders or spiral patterns of *Helminthoida* (Fig. 6.5) (cf. Raup and Seilacher, 1969). Richter (1928) demonstrated that this pattern originated as the two-dimensional trace of a sediment-eating organism that attempted to "graze" an area with maximum efficiency. Richter noted that such trails, although very dense, rarely crossed over earlier formed parts of the same trace. He attributed this feeding method to a series of behavioral characteristics of the organism:

1. *Phobotaxis*—a chemical or tactile sense that enabled the organism to recognize and thus avoid areas that it had mined-out earlier.
2. *Thigmotaxis*—a guiding sense that allowed the animal to remain in close proximity of its earlier trails, thus preventing large areas of sediment from remaining unexploited.
3. *Strophotaxis*—a sense that periodically caused the organism to reverse its direction by 180°, thus giving rise to the tightly packed meanders.

Phobotaxis can also be demonstrated in other trace fossils. An example is *Chondrites bollensis* from the Lias of Württemburg (see Fig. 1.1); the ends of the burrow branches come to within 1 mm of each other but never interpenetrate—a trait first observed by Richter in 1927. In general, Richter's studies paved the way for Seilacher's (1953a) ethological classification of trace fossils, which in turn allowed Seilacher (1955, 1964) and many subsequent authors to apply trace fossils in facies analyses.

Although significant for paleontology that loosely organized grazing traces analogous to those mentioned above originated at least as early as the Ordovician (Seilacher, 1964, Fig. 7), attributing them to a particular group of organisms is difficult. Seilacher (1953a, Fig. 10b) showed that diverse gastropods can make such traces, and Heezen and Hollister (1971, Fig. 51) illustrated a photograph taken at a depth of 4,871 m that reveals a tight spiral trail having the organism—an enteropneust—still in place.[3] Sediment-ingesting polychaetes could possibly produce such trails also.

For several other trace fossil genera, the behavior of the organism involved is well known even though the identity of the tracemaker remains a mystery. Among these are the bundled feeding burrows called *Phycodes* (Seilacher, 1955, Fig. 3), the hexagonal grazing network *Paleodictyon* (Nowak, 1959; Chamberlain, 1971), and numerous genera of U-tubes such as *Diplocraterion*, *Rhizocorallium*, and *Corophioides* (Osgood, 1970). All these structures were apparently made by highly developed organisms. Recent analogs of *Phycodes* and *Paleodictyon* are unknown, and spreite-bearing U-tubes are uncommon; two examples of the latter are burrows made by the echiuroid *Echiurus echiurus* (Hertweck, 1970) and the amphipod *Corophium volutator* (Schäfer, 1972, Figs. 179 and 180).

[3] This photograph is especially significant because, according to Heezen and Hollister (1971, p. 179), enteropneusts were not previously thought to be significant members of the abyssal community. The only record of the group from deep water consisted of three specimens dredged up by HMS CHALLENGER in 1873. Nearly a century passed before this photograph—taken in 1962—and subsequent ones revealed their presence in large numbers. Heezen and Hollister attributed the discrepancy between numbers of enteropneusts brought up in dredge hauls and the quantity seen in photographs to the animal's fragile body, which broke up and was lost during trawling. (See also Chapter 21.)

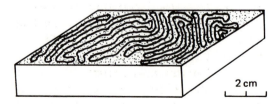

2 cm

Fig. 6.5 *Helminthoida* — two-dimensional, closely meandering trail. Rudolf Richter demonstrated convincingly that it is a grazing trace.

The U-tubes are especially significant because they show that identical burrow morphology can result for at least three different reasons. All these tubes show arcs (a spreite) between the arms of the U (Fig. 6.6), which are the remnants of earlier formed bases of the U. Thus, we know that the organism progressively deepened the U-tube by removing sediment from the floor of the tunnel and plastering it against the ceiling. Yet the reason behind this behavior cannot be determined in many cases. If the animal was vermiform, perhaps an increase in body length with age necessitated lengthening the burrow. However, the spreite indicates that this may be a feeding structure, whereby the organism gleaned nutrients from the sediment as it deepened the burrow. A third possibility is that the animal was forced to elevate or deepen the burrow because of subaqueous deposition or erosion. Goldring (1962, Fig. 3) described such occurrences in the Devonian of Great Britain. In some instances, deposition of sediment buried both openings of the U, and the organism created a spreite in the process of elevating the burrow to the new depositional interface. Conversely, erosion might truncate the burrow and leave both ends of the organism exposed; in this case the animal would lengthen the burrow,

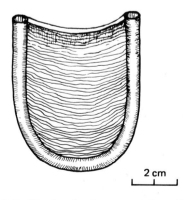

Fig. 6.6 U-tube having a spreite. The arcs between arms of the U indicate that the organism lengthened and deepened the tube by removing sediment from the floor of the burrow and plastering it against the ceiling.

extending it farther down into the substrate. (See Fig. 8.5.)

Several authors (e.g., Häntzschel, 1952; Weimer and Hoyt, 1964; Kennedy and MacDougall, 1969) have studied the burrows of the decapod crustacean *Callianassa* and one of its Mesozoic and Cenozoic analogs, *Ophiomorpha*. Both *Ophiomorpha* and the structures produced by *Callianassa* usually consist of inclined to nearly vertical, cylindrical burrows that have Y- or T-shaped branches. The structures are easily recognized by the nodose exterior of the burrow walls (Fig. 2.2A). *Callianassa* uses the burrow as a domicile but feeds within it. Weimer and Hoyt (1964), through a study of recent *Callianassa* burrows, intensified subsequent interest in *Ophiomorpha* by demonstrating that coastal areas represented by certain ancient sediments may be documented by the presence of abundant *Ophiomorpha*. (Table 2.1.)

Within the last decade, many additional studies have been conducted on *Ophiomorpha*. Those outlined below are a mere sampling. Both Kilpper (1962) and Keij (1965) observed *Ophiomorpha* specimens grading into the vertical corkscrew-shaped burrow *Gyrolithes*. These studies led Kennedy (1967) and Gernant (1972) to regard *Ophiomorpha* and *Gyrolithes* as being synonymous. The corkscrew pattern may be explained by Schmitt's (1965) analysis of the burrowing habits of the soldier crab *Myctris longicarpus*. The organism uses the appendages on only one side of its body for digging; this habit causes the body to rotate and results in a spiral shaft.

Hester and Pryor (1972) reported a different behavioral variant of *Ophiomorpha* from the Eocene of Mississippi. Their work clearly revealed blade-like laminae (i.e., a spreite) within the *Ophiomorpha* burrow system. Hester and Pryor showed that the organism moved upward through the substrate by removing sediment from the roof of the burrow and placing it on the floor. They suggested that the reason for this behavior might have been putrifying conditions in the old burrow.

In summary, *Ophiomorpha* and *Gyrolithes*—burrows mainly of decapod origin —reveal complex morphological variations, which in turn reflect varied behavioral patterns and different environmental conditions. (See also Chapter 2.)

As mentioned previously, xiphosurid traces have also received considerable study. In general, we can be reasonably sure of the results because the extant *Limulus* facilitates application of uniformitarian principles. Of course, this is not true of trilobites. Nevertheless, we know a great deal about the habits of this extinct group. Trace fossils provide us with insight into modes of movement, protection, feeding, and possibly reproduction among trilobites. Because this insight demonstrates just how much we may learn when traces can be assigned to a specific animal group, these categories are discussed in detail. Such coverage provides us with an example of the methodology of palichnology.

Protection

Hard-part morphology indicates that the spinose dorsal surface, where present, as well as the capability of enrollment, afforded trilobites some degree of protection. Trace fossils demonstrate that trilobites could also protect themselves by burrowing, a function that could not be deduced from hard-part morphology. The short, bilobed, buckle-like *Rusophycus*[4] indicates this habit (see Fig. 1.2). Although Fenton (1937) suggested that *Rusophycus* might have served as an egg depository, it

is more logically a shallow, rapidly excavated burrow, where the trilobite temporarily covered itself with a thin layer of sediment. Many recent organisms —e.g., some sea stars, crabs, and amphipods—use this as a method to escape predators. Detailed studies of *Rusophycus* by Seilacher (1953b; 1955; 1970, Figs. 3 and 4), Birkenmajer and Bruton (1971), Radwański and Roniewicz (1967), Crimes (1970b), and Osgood (1970) demonstrated abundant morphological variations that reflect slight behavioral differences. In most cases the walking legs moved the sediment toward the median line of the body and swept it out the rear of the burrow. However, instances are known where the cephalon was employed as a shovel (Seilacher, 1970, Fig. 5b); or, the body was strongly arched, the anterior appendages moving part of the sediment out the front while the posterior ones worked toward the rear (Seilacher, 1970, Fig. 5c). Some *Rusophycus* specimens (Osgood, 1970, Pl. 57, fig. 1; Radwański and Roniewicz, 1967, Pl. 2), in which impressions of the coxae are preserved as nodes between the two lobes, may indicate that the legs were directed more laterally than vertically, thus creating wide, shallow burrows. Crimes (1970b) concluded that these nodes indicate that the legs were attached closer to the midline of the body than is generally shown in reconstructions of ventral surfaces of trilobites.

The depth of burrowing can also affect morphology in other ways. In casts of very shallow burrows, the lobes may not be attached to each other; instead, they resemble two small coffee-bean-shaped bodies. In such specimens the striae are usually weak, and only rarely are imprints of hard parts of the body present (see Birkenmajer and Bruton, 1971, for several examples).

Movement

Modes of trilobite movement are documented both by the elongate, bilobed *Cruziana*—where the organism burrowed

---

[4] Most authors restrict *Rusophycus* to the short, bilobed imprints and employ *Cruziana* for the longer, bilobed trails. Seilacher (1970) preferred to include all forms under *Cruziana*. Here the two are considered as separate genera. Admittedly, the two forms do intergrade. Both Seilacher (1955, Pl. 19) and Osgood (1970, Pl. 66, fig. 3) figured specimens in which the deeper *Rusophycus* (burrows) are connected by the shallower, bilobed bands of *Cruziana*. Nevertheless, *Rusophycus* represents basically downward vertical movement by the organism, whereas *Cruziana* indicates horizontal movement.

half-submerged in the sediment—and by surficial tracks. The precise ethological meaning of *Cruziana* is somewhat in doubt and is discussed below, under Feeding.

At first glance, trilobite trails appear to be so complex as to defy interpretation. However, Seilacher (1955, 1959) devised a valuable method of study.[5] Through comparison with the trails of living arthropods, he reasoned that sets of trilobite tracks would approximate a V, the open end of the V pointing in the direction that the animal moved (Fig. 6.7). Moreover, he postulated that the movement of the legs was similar to that in millipedes, where a wave of motion proceeds along the body from rear to front. In trilobites, a new wave would be initiated at the pygidium before the first one had reached the cephalic legs. This behavioral interpretation was supported by well-reasoned arguments concerning the pattern of the Vs, the acute angle formed by the two arms of the V, and the resulting imprint, which resembles a

[5] Osgood (1970, p. 350–354) gave a more detailed treatment, in English, of Seilacher's work. The terminology for trilobite tracks used in this chapter is that employed by Osgood (1970, p. 351).

sine curve. This curve is the resultant of two vectors, one provided by forward motion of the body and the other by movement of the leg toward the median axis of the body. Utilizing these suppositions, one can determine the net direction of movement, the number of walking legs involved, and the relationship of the body axis to the direction of movement.

Although trails reveal that trilobites could move with the body axis parallel to the direction of movement, in many cases the axis was somewhat oblique; they were "crab-walking." When this occurred, the imprints no longer formed a series of Vs; instead, they became dimorphic (Fig. 6.7B). The imprints on one side are superimposed to form a single row, whereas those on the other side are arranged en echelon. These en echelon imprints (Fig. 6.7B, right) enable us to determine how many legs were actually employed by the trilobite in the movement.

The morphology of individual imprints made by the legs, on both sides of the body, changes as the trail becomes dimorphic. As seen in Figure 6.7C, the body is oriented to the right while movement is toward the

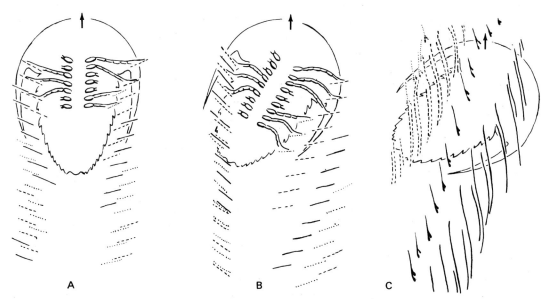

A   B   C

Fig. 6.7 Trilobite tracks, viewed from below. A, movement parallel to body axis. B, movement slightly oblique to body axis. C, movement markedly oblique to axis, with resulting dimorphic imprints. (From Osgood, 1970; Seilacher, 1955.)

top of the page. The imprints of the appendages on the left side of the body (as viewed from below) are short; these appendages supported most of the weight. The imprints of the right side are long scratch marks, resembling to varying degrees the sine curve mentioned above. This trace demonstrates that these appendages were dragged along the surface and could not have borne much weight. Trilobite trails reflecting straight-ahead or moderately oblique movement usually are grouped under *Protichnites* or *Diplichnites*. Seilacher (1955) established *Dimorphichnus* for imprints that are strongly dimorphic. He believed that these were feeding trails, hence they are discussed subsequently. The same is true for *Monomorphichnus* (Crimes, 1970b).

Because many trilobite tracks are preserved in fine-grained sediments, such as shales, they tend to be fragmentary and to present an incomplete picture. Nevertheless, several examples are known that represent an evident change in the type of behavior involved. From the Cambrian of the Salt Range of Pakistan, Seilacher (1955, Pl. lb, fig. 2) illustrated a *Protichnites* (trail) leading from a *Rusophycus*. Similarly, Birkenmajer and Bruton (1971) illustrated several *Rusophycus* associated with *Protichnites*, from Lower Ordovician rocks of Sweden. Crimes (1970b, Pl. 9) mentioned specimens from the Upper Cambrian of Wales that are intermediate between *Cruziana* and *Diplichnites*, and Osgood (1970, Pl. 73, fig. 5) showed a trail exhibiting a change from straight-ahead to oblique movement.

As both Seilacher (1959, p. 391) and Osgood (1970, p. 359–361) pointed out, in some instances the trilobite did not employ all, or even most, of its legs while walking. This behavior was probably employed when movement was rapid. Seilacher hypothesized that, in some cases, the larger and stronger anterior legs could propel the animal while the fleshy caudal cerci acted as skids (Seilacher, 1959, Fig. 2B). Pre-

sumably, paired pygidial spines could achieve the same effect in trilobites not possessing cerci.

Although relatively easy to determine that a given trail is of trilobite origin, identifying the trilobite to the generic level is much more difficult. For example, several genera of trilobites have been reported from the rich Cincinnatian fauna, but in only two cases was I able to assign the many trails studied to a given genus. In one case, *Isotelus*, the large size was diagnostic; I could not identify small isotelid trails. The other example is the "lacy-collared" *Cryptolithus*, one of the rare instances among trilobites where the dactyls are preserved (Raymond, 1920); the dactyls are numerous and bristle-like. Most other trails in this assemblage were of smaller dimensions and were unremarkable in appearance. *Flexicalymene* is by far the most common trilobite in the Cincinnati area, hence most trails were likely produced by members of that genus. Nevertheless, I could not distinguish *Flexicalymene* trails from those of *Triarthrus*, small isotelids, or other trilobites.

Feeding

Hard-part morphology has told us relatively little about trilobite feeding methods. Because trilobites possessed neither special feeding appendages nor hard mouth parts, logic dictates that they were not predators. We may reasonably assume that if they were filter feeders or scavengers, their trace fossils might furnish us some evidence. Not surprisingly, *Cruziana*, *Dimorphichnus*, and *Monomorphichnus* (Crimes, 1970b) have all been interpreted as feeding traces.

Seilacher (1955, Fig. 3) provided a very detailed drawing and analysis of *Dimorphichnus* (Fig. 6.8). The trail has the shape of a large loop, where the organism moved across the substrate in the direction shown, employing strong raking movements. Seilacher stressed that where the animal crossed over the earlier formed part of the

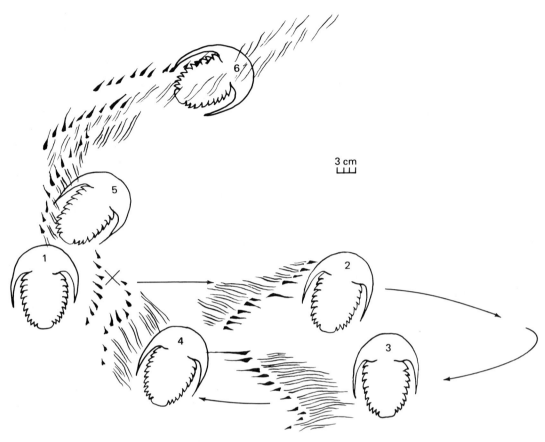

Fig. 6.8 *Dimorphichnus,* interpreted by Seilacher (1955) as a trilobite feeding trace. Legs on "lee" side of body raked the sediment. Direction of movement shown by arrows and numbered trilobites. × indicates place where raking movement ceased briefly, as the trilobite crossed over the previously formed trail. Cambrian; Salt Range, Pakistan. (Simplified from Seilacher, 1955.)

trail (point × of Fig. 6.8), the raking movements briefly ceased, only to begin again once the crossover was made. In the upper left corner of Figure 6.8, the raking ceases entirely, and the trail converts into a *Diplichnites* form—showing movement slightly oblique to the body axis. Seilacher reasoned that this raking stirred up the nutrient-rich mud and that food particles were trapped by filaments on the legs; the food passed to the ventral median axis of the body, and from there forward to the mouth.

Seilacher's thesis has found a champion in Crimes (1970b), who described *Dimorph-ichnus* from the Upper Cambrian of Wales. Also, Whittington (1962) mentioned the idea and did not criticize it. However,

other workers have offered alternative explanations. Banks (1970, p. 28), in describing *Dimorphichnus* from the Lower Cambrian of Norway, inferred that such markings were formed when a dead trilobite was washed across the bottom, a situation in which the limp appendages would drag behind. Martinsson (1965, p. 206–209) reported the genus from the Middle Cambrian of Sweden and, like Banks, attributed it to current activity. Yet in this case a living trilobite, swept by currents, lashed out in an attempt to control its movement. Osgood (1970, p. 353) came to the same conclusion and noted that if *Dimorph-ichnus* represents a feeding method, it must have been inadequate and was discarded by later trilobites. *Dimorphichnus* is restricted to the Cambrian; indeed, Crimes (1970a, p. 117) considered the genus to be an index fossil. (See Chapter 7.)

Monomorphichnus, a trace fossil genus similar to *Dimorphichnus*, was erected by Crimes (1970b) for specimens from the Upper Cambrian of Wales. It is similar to Seilacher's genus in that the long, parallel, raking imprints are present; but Crimes could find no evidence of any body-support imprints. He concluded, therefore, that the trilobite was swimming instead of walking and was raking the bottom in search of food. One could raise the same questions regarding this analysis as those discussed previously for *Dimorphichnus*.

The long, band-like *Cruziana* (see Fig. 1.3), more common than *Dimorphichnus*, has been reported from Paleozoic rocks of nearly every continent. Seilacher (1970) monographed the genus (including *Rusophycus* as a junior synonym of *Cruziana*) and the reader is referred to that work for a detailed description of the variations that can result from behavioral or morphological changes on the part of the trailmaker. Various ethological interpretations have been given to the genus by Seilacher and others. In many instances it probably is a Repichnion (crawling trace); instead of walking, a half-buried trilobite burrowed its way through the sediment.[6] Whether these straight or gently curved trails are feeding burrows is difficult to say. However, some varieties of *Cruziana* suggest more than simple movement. Seilacher (1970, Fig. 6d) illustrated one form from the Ordovician of Iraq in which the pattern of the two-dimensional trail is roughly ovoid, having several secondary arcuate markings within it. Seilacher referred to this as a crude variety of grazing pattern. Although the path is not as well organized as *Helmin-*

thoida (Fig. 6.5), and it ploughs through earlier formed parts, the circular pattern seems to indicate more than random movement.

A second variety illustrated by Seilacher (1970, Fig. 6e) from the Silurian of Chad was designated by Lessertisseur (1956) as *Cruziana ancora*. It is analogous to the bundled feeding burrow *Phycodes* in consisting of a single horizontal shaft having several terminal branches, some of which curve back as much as 170° toward the main axis. The final result is a palmate structure, where the organism first mined out one shaft, then pulled back and mined a second shaft adjacent to the first. This behavior was repeated several times, and almost certainly represents feeding. A similar, although less well developed, form was reported from the Cambrian of Alberta by Fenton (1937).

To me, *Cruziana*, especially the varieties discussed above, seems more likely to be a feeding trace than does *Dimorphichnus*. The latter necessitates the presence of ciliary currents along the animal's leg as well as along the ventral median surface of its body; *Cruziana* does not. However, *Cruziana* is not always present where trilobites are a common constituent of the fauna. The Upper Ordovician of Ohio abounds with both trilobites and *Rusophycus*, but *Cruziana* is very rare. Therefore, although *Dimorphichnus* and *Cruziana* provide possible insights into trilobite feeding mechanisms, some problems remain regarding both.

Reproduction

Very little is known about reproduction in trilobites. Harrington (1959, p. 101–102) referred to various authors who described rare ovoid bodies found near trilobite body fossils. However, the meaning of these is conjectural.

A few authors have suggested that some trace fossils may reflect reproductive activities, e.g., Fenton's (1937) remark that *Rusophycus* might represent trilobite "nests." This view has not gained popular-

[6] Crimes (1970b, p. 62–63, Fig. 6) concluded that the speed of the organisms' pace can be determined from the configuration of the striae on the lobes. This estimate involves a study of the width of the trail, the acute angle formed between the lobes where the striae converge to form a V, and the overall width of the V. Gray (1968) demonstrated the validity of similar principles among recent arthropods.

ity among later authors. A more recent suggestion by Orłowski et al. (1971, p. 346) is that the occurrence of several *Rusophycus* in close proximity (Seilacher, 1955, Pl. 19, fig. 1; Osgood, 1970, Pl. 58, fig. 9) may be connected with the mating process. They refer to King's (1965) study of Jurassic xiphosurid trails from Britain as evidence. King reported on trails in which the impressions of two telsons were superimposed, forming a sinuous, interwoven pattern. He interpreted this pattern as evidence for the mating process, where the male rides on the buckler of the female. However, little evidence supports the theory that trilobite mating was similar to that of *Limulus*. Densely packed or superimposed *Rusophycus* may be explained in other ways, e.g., that the trilobites in question were gregarious; they were threatened by a potential predator and burrowed for protection. Equally probable is that a small number of animals could produce a large number of traces by changes in position, e.g., premature escape attempts.

Crimes (1970a, p. 119–124) utilized *Rusophycus* and *Cruziana* in an entirely different manner in attempting to explain trilobite behavior. He reasoned that both genera, from several different localities, represent faithfully the body width of the organisms that made them. He constructed size-frequency histograms, which revealed no traces less than 5 mm wide. Because many fine, delicate traces are found in the same rocks, he concluded that *Cruziana* and *Rusophycus* of small size

were capable of preservation. His interpretation was that the very young trilobites were planktonic, and only when they reached a size larger than 5 mm did they become benthonic. Moreover, he found that the mean width of *Rusophycus* was significantly smaller than that of *Cruziana*, thus suggesting an early resting phase and a later furrowing habit. This theory could easily be tested at several other localities, with trilobites of different ages.

Evolution

Evolution of Behavior

Because body fossils document evolutionary changes within phyletic groups, one naturally expects trace fossils to show changes in behavioral patterns. Such documentation requires a large bank of data; and because trace fossil studies have lagged until recently, we have relatively few works that illustrate changes within a trace fossil lineage. That trace fossils will ever make a major contribution in this field is unlikely; they reflect responses to an overall set of environmental conditions that presumably have been relatively stable since the Cambrian.

Nevertheless, Seilacher (1967) provided a few examples of evolution in behavior. He interpreted *Dictyodora* (Fig. 6.9) as an essentially horizontal meandering trail formed by an infaunal sediment eater. The organism burrowed vertically until it encountered a nutrient-rich layer, then pro-

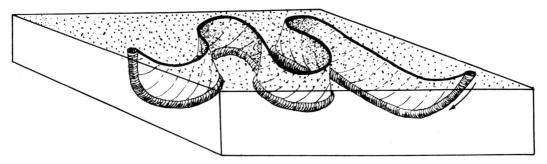

Fig. 6.9 *Dictyodora.* The lower horizontal tube marks the course of a sediment-feeding organism. Upper sinuous line shows where the siphons maintained contact with the depositional interface. (Adapted from Seilacher, 1967.)

ceeded to meander horizontally. Contact with the depositional interface was maintained through thin siphons. The course of the animal's progress can be traced by following the path of the siphons on the surface—analogous to a submarine's periscope breaking the sea surface; the submarine's track parallels the wake of the periscope. Because the earlier examples of *Dictyodora* trails (Cambrian to Devonian) are found at depths of only a few centimeters below the depositional interface, Seilacher concluded that the siphons were very short. He surmised that competition for food at these shallow depths would be intense. Evolutionary changes occurred that lengthened the siphons and enabled the organism to burrow deeper, as found in Mississippian examples. By Mississippian time another change had also taken place; the main axis, instead of being vertical, took on the shape of a corkscrew. In later Mississippian specimens from East Germany, Seilacher reported that the horizontal meandering pattern had been abandoned altogether; the entire trace resembled a massive corkscrew. Although Seilacher did not mention it, this pattern seems to be similar to *Daedalus* from the Silurian of New York State.

Seilacher also discerned a change in habits of the Cretaceous and Tertiary organisms that formed *Zoophycos*. He (1967, p. 80) stated that *Zoophycos* ". . . is the burrow of an unknown wormlike animal that foraged through the sediments by constantly shifting the lobes of its U-shaped tunnel. The two openings of the tunnel remain fixed in the sediment, but thin concentric layers inside the lobes record the tunnel's shift of position." In Cretaceous specimens the distal parts of the burrow have a lobate antler-like shape, recording mined-out areas. The Tertiary counterpart was more efficient; the antler pattern is lacking, and the structure is more compact. Interestingly, Seilacher pointed out that this configuration is true only in the "adult" Tertiary forms. The "juvenile" Tertiary

examples more closely resemble the Cretaceous "adults."[7]

Evolution of Metazoa

To a paleontologist, one of the most fascinating aspects of trace fossil studies is their application to the problem of the apparent sudden increase in diversity of Metazoa at the beginning of Cambrian time. This area is an especially fruitful field for research because trace fossils very commonly occur in siliceous clastic sediments, which do not provide the best medium for preservation of body fossils. As Glaessner (1965b, p. 167) stated, perhaps the study of trace fossils will allow us to ". . . date the first appearance of Metazoa in the geological record." (See also Glaessner, 1965a.)

Our purpose here is not to discuss in detail the Precambrian record of trace fossils but, instead, to see whether they show any evidence for rapid evolution of animals in late Precambrian time. For a discussion of Precambrian traces and pseudotraces, the reader is referred to the following articles: Cloud (1968), Webby (1970), Hofmann (1971), and Glaessner (1962, 1969). (See also Chapter 7.)

Seilacher was the first to apply trace fossils to the Precambrian–Cambrian problem. He showed (1956, Fig. 1) that Precambrian traces are very simple in morphology, whereas Cambrian trace fossils reveal much more sophisticated methods of feeding and movement. He concluded that trace fossils document rapid evolution among the invertebrates in late Precambrian time. Glaessner (1962, p. 481) criticized Seilacher's work on two general points: (1) Seilacher compared Precambrian forms with both Lower and Middle Cambrian trace fossils; he should have deleted the Middle Cambrian from his study, and (2) in Glaessner's opinion, Precambrian traces had not been

---

[7] Seilacher did not mention how he distinguished "juvenile" from "adult" specimens. Perhaps size is the key.

adequately collected—most reports merely referred to "worm trails"—and any conclusions reached in 1956 were bound to be premature.

Since Seilacher's study, several authors have dealt with this problem. The reader is referred to Daily (1972) for a more detailed discussion of several Precambrian–Cambrian sequences and their faunal content. For this chapter I have selected the work of three authors: Glaessner (1969), Webby (1970), and Banks (1970). I believe that these three provide a representative sample of the type of research that is being conducted. Glaessner and Banks discussed sections that embrace a conformable Precambrian–Cambrian sequence. Glaessner,

in a rather comprehensive paper, described but did not formally name six trace fossils from the Ediacara Beds. All are trails left by vermiform sediment or detritus feeders. In addition, he discussed the trace fossils from the Arumbera Formation of central Australia. He determined (Fig. 6.10A) that trace fossils are very rare in the lower two-thirds of the formation but are abundant in the upper one-third. *Phycodes, Diplichnites, Plagiogmus,* and possibly *Rusophycus* are present. Glaessner (1969, p. 382) admitted that detailed systematic studies of the lithology and fauna of the Arumbera Formation had not been carried out. A Precambrian age was assigned to the lower part of the Arumbera on the basis of one

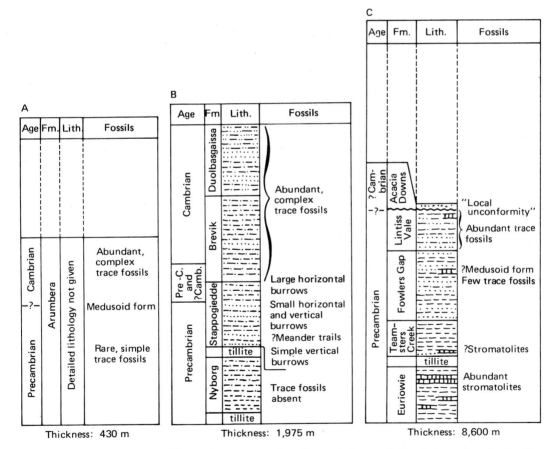

Fig. 6.10 Three stratigraphic sections comparing Precambrian and Cambrian trace fossils. No attempt was made here to correlate the sections; note the differences in scale. A, central Australia (compiled from Glaessner, 1969). B, northern Norway (simplified from Banks, 1970). C, New South Wales, Australia (from Webby, 1970).

specimen of the seapen *Rangea* cf. *R. longa*. Glaessner did not mention Cambrian body fossils and instead drew the Precambrian–Paleozoic boundary at the first occurrence of abundant trace fossils. He concluded (p. 390) that "trace fossils are less common and less varied in the Precambrian than in the Lower Cambrian, but they are neither generally rare nor insignificant."

Banks dealt with a 3,000-m-thick sequence in northern Norway. This section was studied more thoroughly than Glaessner's; both the lithology and environmental analyses are given (Banks, 1970, Figs. 2a, b). Paleontological control, however, is good only for the Middle Cambrian. The Precambrian–Cambrian boundary was based on the occurrence of the serpulid worm tube *Platysolenites antiquissimus* and fragments of the trilobite *Holmia*. Banks documented the trace fossils in appreciable detail. The Precambrian part of the Stappogiedde Formation is characterized only by simple vertical burrows (Fig. 6.10B). The upper part of this unit and the lower part of the overlying Brevik Formation possess a somewhat more diverse trace fossil fauna. The age of this part of the sequence is in doubt. The remainder of the Brevik and the suprajacent Duolbasgaissa contain 18 different kinds of trace fossils. Banks discounted the possibility that this difference is due to environmental changes. As demonstrated in his diagram (Banks, 1970, Fig. 2a, b), the same environments were represented in the lower part of the section, yet they were barren of all traces of life. He concluded that trace fossils provide strong evidence for a rapid increase in numbers and complexity of annelids, arthropods, and mollusks, but not the Metazoa as a whole.

Webby (1970) discussed a very thick sequence in New South Wales, Australia (Fig. 6.10C), which contains an abundant Precambrian trace fossil assemblage. As with the other sections discussed above, picking the base of the Cambrian presents a problem. Webby seriously questioned an Early Cambrian age for the Acacia Downs Beds based on burrows alone, especially because trace fossils are so abundant in the underlying Lintiss Vale. Moreover, an interruption that Webby refers to as a "local unconformity" separates the two units. Equally vexing is that, despite careful study, the Acacia Downs appears to be devoid of body fossils. In addition to uncertainties regarding the boundary, the trace fossils of the Fowlers Gap Beds exhibit a marked change as compared to those of the Lintiss Vale.[8] The Fowlers Gap contains a minimum of 3 and a maximum of 7 "distinctive patterns of activity" whereas the Lintiss possesses a minimum of 10 and a maximum of 21 (Webby, 1970, p. 104); nearly all the traces are horizontal and are interpreted as feeding burrows of soft-bodied organisms. Webby did not discuss whether any Cambrian trace fossils are known from the area. His conclusion is that a definite increase in diversity and complexity of trace-making animals occurred near the end of Precambrian time.

The three authors' documentation of changes in the trace fossil assemblages is impressive and seems to confirm the conclusions of Cloud (1968), Barghoorn (1971), and others that, for one reason or another, a dramatic increase in the rate of evolution occurred late in the Precambrian. The three excellent works cited above should serve as models for others to follow. The contribution of trace fossils to the early history of the Metazoa is a topic that certainly warrants additional study. Yet a disquieting feature of the three sections studied by Glaessner, Banks, and Webby is that none yield good body fossil control with regard to the Precambrian–Cambrian

---

[8] In a more recent work, Wade (1970, p. 102) concluded that the Lintiss Vale trace fossils more closely resembled the "low Lower Cambrian fauna than the Ediacaran." She stated that further collecting of the Lintiss Vale and the subjacent Fowlers Gap is needed to resolve the question.

boundary. Indeed, such control may be difficult if not impossible to obtain; as stated previously, body fossils are rarely preserved in certain rocks that contain abundant trace fossils. Nevertheless, a section that shows a transition from simple to complex trace fossils, and that lies stratigraphically below a well-defined Lower Cambrian unit, would afford compelling evidence for the theory of increased evolutionary rates in late Precambrian time.

ACKNOWLEDGMENTS

I wish to express my gratitude to K. E. Caster, T. P. Crimes, R. W. Frey, A. S. Horowitz, W. J. Kennedy, and C. Teichert for reading the manuscript and offering helpful suggestions. The figures were drawn by Krista Roche of Fredericksburg, Ohio.

REFERENCES

Abel, O. 1935. Vorzeitliche Lebensspuren. Jena, Gustav Fischer, 644 p.

Allen, A. T. and J. G. Lester. 1953. Animal tracks in an Ordovician rock of northwest Georgia. Georgia Geol. Surv., Bull. 60:205–214.

Banks, N. L. 1970. Trace fossils from the late Precambrian and Lower Cambrian of Finnmark, Norway. In T. P. Crimes and J. C. Harper (eds.), Trace fossils. Geol. Jour., Spec. Issue 3:19–35.

Barghoorn, E. 1971. The oldest fossils. Scientific Amer., 224:30–54.

Birkenmajer, K. and D. L. Bruton. 1971. Some trilobite resting and crawling traces. Lethaia, 4:303–319.

Bischoff, B. 1968. *Zoophycos,* a polychaete annelid, Eocene of Greece. Jour. Paleont., 42:1439–1443.

Bradley, J. 1973. *Zoophycos* and *Umbellula* (Pennatulacea): their synthesis and identity. Palaeogeogr., Palaeoclimatol., Palaeoecol., 13:103–128.

Caster, K. E. 1938. A restudy of the tracks of *Paramphibius.* Jour. Paleont., 12:3–60.

————. 1940. Die sogenannten Wirbeltierspuren und die *Limulus*-Fährten der Solnhofener Plattenkalke. Paläont. Zeitschr., 22:12–29.

————. 1944. Limuloid trails from the Upper Triassic (Chinle) of the Petrified Forest National Monument, Arizona. Amer. Jour. Sci., 242:74–84.

Chamberlain, C. K. 1971. Morphology and ethology of trace fossils from the Ouachita Mountains, southeast Oklahoma. Jour. Paleont., 45:212–246.

Cloud, P. E. 1968. Pre-Metazoa evolution and the origin of the Metazoa. In E. T. Drake (ed.), Evolution and environments. New Haven, Conn., Yale Univ. Press, p. 1–72.

Crimes, T. P. 1970a. The significance of trace fossils in sedimentology, stratigraphy and palaeoecology with examples from lower Palaeozoic strata. In T. P. Crimes and J. C. Harper (eds.), Trace fossils. Geol. Jour., Spec. Issue 3:101–127.

————. 1970b. Trilobite tracks and other trace fossils from the Upper Cambrian of north Wales. Geol. Jour., 7:47–68.

Daily, B. 1972. The base of the Cambrian and the first Cambrian faunas. In J. B. Jones and B. McGowan (eds.), Stratigraphic problems of the later Precambrian and Early Cambrian. Univ. Adelaide, Centre Precamb. Res., Spec. Paper 1:13–42.

Ewing, M. and R. A. Davis. 1967. Lebensspuren photographed on the ocean floor. In J. B. Hersey (ed.), Deep-sea photography. Johns Hopkins Oceanogr. Stud., 3:259–294.

Fenton, C. L. 1937. Trilobite nests and feeding burrows. Amer. Midland Natur., 18:446–451.

———— and M. A. Fenton. 1934. *Scolithus* as a fossil phoronid. Pan-American Geol., 61:341–348.

Frey, R. W. 1970. Trace fossils of the Fort Hays Limestone Member, Niobrara Chalk (Upper Cretaceous), west-central Kansas. Univ. Kansas Paleont. Contr., Art. 53, 41 p.

———— and T. V. Mayou. 1971. Decapod burrows in Holocene barrier beaches and washover fans, Georgia. Senckenbergiana Marit., 3:53–77.

Gernant, R. E. 1972. The paleoenvironmental significance of *Gyrolithes* (Lebensspur). Jour. Paleont., 46:735–742.

Glaessner, M. F. 1962. Pre-cambrian fossils. Biol. Review, 37:467–494.

————. 1965a. Biological events and the Pre-cambrian time scale. Canadian Jour. Earth Sci., 5:586–590.

————. 1965b. Pre-cambrian life-problems and perspectives. Geol. Soc. London, Proc., 1626:165–169.

————. 1969. Trace fossils from the Pre-cambrian and basal Cambrian. Lethaia, 2: 369–393.

Goldring, R. 1962. The trace fossils of the Baggy Beds (Upper Devonian) of North Devon, England. Paläont. Zeitschr., 36:232–251.

———— and A. Seilacher. 1971. Limulid under-tracks and their sedimentological implica-tions. Neues Jahrb. Geol. Paläont., Abh., 137:422–442.

Gray, J. 1968. Animal locomotion. Weidenfeld and Nicholson, 479 p.

Häntzschel, W. 1952. Die Lebensspur Ophio-morpha Lundgren im Miozän, ihre welt-weite Verbreitung und Synonymie. Geol. Staatsinst. Hamburg, Mitt. 21:142–153.

Harrington, H. J. 1959. General description of Trilobita. In R. C. Moore (ed.), Treatise on invertebrate paleontology, Pt. O, Arthro-poda 1. Lawrence, Kan., Geol. Soc. Amer. and Univ. Kansas Press, p. O40–O117.

Heezen, B. C. and C. D. Hollister. 1971. Face of the deep. New York, Oxford Univ. Press, 659 p.

Hertweck, G. 1970. The animal community of a muddy environment and the development of biofacies as effected by the life cycle of the characteristic species. In T. P. Crimes and J. C. Harper (eds.), Trace fossils. Geol. Jour., Spec. Issue 3:235–242.

Hester, N. C. and W. A. Pryor. 1972. Blade-shaped crustacean burrows of Eocene age: a composite form of Ophiomorpha. Geol. Soc. America, Bull., 83:677–688.

Hofmann, H. J. 1971. Precambrian fossils, pseudofossils, and problematica in Canada. Geol. Surv. Canada, Bull. 189, 146 p.

Howard, J. D. and C. A. Elders. 1970. Burrow-ing patterns of haustoriid amphipods from Sapelo Island, Georgia. In T. P. Crimes and J. C. Harper (eds.), Trace fossils. Geol. Jour., Spec. Issue 3:243–263.

Howell, B. F. 1957. New Cretaceous scoleciform annelid from Colorado. Palaeont. Soc. In-dia, Jour., 2:149–152.

————. 1962. Worms. In R. C. Moore (ed.), Treatise on invertebrate paleontology, Pt.

W, Miscellanea. Lawrence, Kan., Geol. Soc. America and Univ. Kansas Press, p. W144–W177.

Każmierczak, J. and A. Pszczołkowski. 1969. Bur-rows of Enteropneusta in Muschelkalk (Middle Triassic) of Holy Cross Mountains, Poland. Acta Palaeont. Polonica, 14: 299–318.

Keij, A. J. 1965. Miocene trace fossils from Borneo. Paläont. Zeitschr., 39:220–228.

Kennedy, W. J. 1967. Burrows and surface traces from the Lower Chalk of southern England. British Mus. Nat. Hist. (Geol.), Bull., 15: 127–167.

———— and J. D. S. MacDougall. 1969. Crus-tacean burrows in the Weald Clay (Lower Cretaceous) of southeastern England and their environmental significance. Palaeon-tology, 12:459–471.

Kilpper, K. 1962. Xenohelix Mansfield 1927 aus der miozänen Niederrheinischen Braun-kohlenformation. Paläont. Zeitschr., 36:55–58.

King, A. F. 1965. Xiphosurid trails from the Upper Carboniferous of Bude, North Corn-wall. Geol. Soc. London, Proc., 1626:162–165.

Lessertisseur, J. 1956. Sur un Bilobite Nouveau du Gothlandien de L'Ennedi (Tchad, A.E.F.), Cruziana ancora. Soc. Géol. France, Bull., 6th Ser., 6:43–47.

Loring, A. P. and K. K. Wang. 1971. Re-evalua-tion of some Devonian lebensspuren. Geol. Soc. America, Bull., 82:1103–1106.

Martinsson, A. 1965. Aspects of a Middle Cam-brian thanatotope on Öland. Geol. Fören. i Stockholm Förhand., 87:181–230.

Moussa, M. T. 1970. Nematode fossil trails from the Green River Formation (Eocene) in the Uinta Basin, Utah. Jour. Paleont., 44:304–307.

Nathorst, A. G. 1881. Om spår af nagra everte-brerade djur M.M. och deras paleontolo-giska betydelse. (Mémoire sur quelques traces d'animaux sans vertèbres etc. et de leur protée paléontologique.) Kgl. Svenska Vetensk. Akad. Handl., 18: 104 p.

————. 1886. Nouvelles observations sur les traces d'Animaux et autres phénomènes d'origine purement mécanique décrits comme "Algues fossiles." Kgl. Svenska Vetensk. Akad. Handl, 21: 58 p.

Nowak, W. 1959. Palaeodictyum in the Flysch

Carpathians. Kwartaln. Geol., 3:103–125. (with English summary).

Orłowski, S. et al. 1971. Ichnospecific variability of the Upper Cambrian *Rusophycus* from the Holy Cross Mountains. Acta Geol. Polonica, 21:341–348.

Osgood, R. G., Jr. 1970. Trace fossils of the Cincinnati area. Palaeontographica Amer., 6(41):281–444.

————— and E. Szmuc. 1972. The trace fossil *Zoophycos* as an indicator of water depth. Bulls. Amer. Paleont., 62(271):1–22.

Packard, A. S. 1900. On supposed merostomatous and other Paleozoic arthropod trails with notes on those of *Limulus*. Amer. Acad. Arts Sci., Proc., 36:61–71.

Plička, M. 1968. *Zoophycos,* and a proposed classification of sabellid worms. Jour. Paleont., 42:836–849.

—————. 1970. *Zoophycos* and similar fossils. In T. P. Crimes and J. C. Harper (eds.), Trace fossils. Geol. Jour., Spec. Issue 3: 361–370.

Radwański, A. and P. Roniewicz. 1967. Trace fossil *Aglaspidichnus sanctacrucensis* n. gen., n.sp., a probable resting place of an aglaspid (Xiphosura). Acta Palaeont. Polonica, 12:545–554.

Raup, D. M. and A. Seilacher. 1969. Fossil foraging behavior: computer simulation. Science, 166:994–996.

Raymond, P. E. 1920. The appendages, anatomy and relationships of trilobites. Connecticut Acad. Arts Sci., Mem. 7, 169 p.

Richter, R. 1927. Die fossilen Fährten und Bauten der Würmer, ein Überblick über ihre biologischen Grundformen und deren geologische Bedeutung. Paläont. Zeitschr., 9:193–240.

—————. 1928. Psychische Reaktionen fossiler Tiere. Palaeobiologica, 1:225–244.

Schäfer, W. 1972. Ecology and palaeoecology of marine environments. Edinburgh and Chicago, Oliver & Boyd and Univ. Chicago Press, 568 p.

Schmitt, W. L. 1965. Crustaceans. Ann Arbor, Mich., Univ. Michigan Press, 199 p.

Seilacher, A. 1953a. Über die Methoden der Palichnologie. 1. Studien zur Palichnologie. Neues Jahrb. Geol. Paläont., Abh., 96:421–452.

—————. 1953b. Über die Methoden der Palichnologie. 2. Die fossilien Ruhespuren (Cubichnia). Neues Jahrb. Geol. Paläont. Abh., 98:87–124.

—————. 1954. Die geologische Bedeutung fossiler Lebensspuren. Zeitschr. Deutsche Geol. Gesellsch., 105:214–227.

—————. 1955. Spuren und Lebensweise der Trilobiten; Spuren und Fazies im Unterkambrium. In O. H. Schindewolf and A. Seilacher, Beiträge zur Kenntnis des Kambriums in der Salt Range (Pakistan). Akad. Wiss. u. Lit. Mainz, math.-naturw. Kl., Abh., 10:86–143.

—————. 1956. Der Beginn des Kambriums als biologische Wende. Neues Jahrb. Geol. Paläont., Abh., 103:155–180.

—————. 1959. Vom Leben der Trilobiten. Die Naturwissenschaften, 12:389–393.

—————. 1960. Lebensspuren als Leitfossilien. Geol. Rundschau, 49:41–50.

—————. 1962. Form und Funktion des Trilobiten-Daktylus. Paläont. Zeitschr. (Hermann Schmidt Festband), 218–227.

—————. 1964. Biogenic sedimentary structures. In J. Imbrie and N. D. Newell (eds.), Approaches to paleoecology. New York, John Wiley, p. 296–316.

—————. 1967. Fossil behavior. Scientific Amer., 217:71–80.

—————. 1970. *Cruziana* stratigraphy of "nonfossiliferous" Palaeozoic sandstones. In T. P. Crimes and J. C. Harper (eds.), Trace fossils. Geol. Jour., Spec. Issue 3:447–477.

Sharpe, C. F. S. 1932. Eurypterid trail from the Ordovician. Amer. Jour. Sci., Ser. 5, 24: 355–361.

Simpson, S. 1957. On the trace fossil *Chondrites.* Geol. Soc. London, Quart. Jour., 112:475–499.

—————. 1970. Notes on *Zoophycos* and *Spirophyton.* In T. P. Crimes and J. C. Harper (eds.), Trace fossils. Geol. Jour., Spec. Issue 3:505–514.

Stanley, S. M. 1970. Relation of shell form to life habits of the Bivalvia (Mollusca). Geol. Soc. America, Mem. 125, 296 p.

Størmer, L. 1934. Downtonian Merostomata from Spitsbergen, with remarks on the Suborder Synziphosura. Skrift. Norske Vidensk. Akad. Oslo, 2:1–26.

—————. 1939. Studies on trilobite morphology, Part 1, The thoracic appendages and their phylogentic significance. Norsk Geol. Tidsskrift, 19:143–273.

————. 1951. Studies on trilobite morphology, Part 3, The ventral cephalic structures with remarks on the zoological position of the trilobites. Norsk Geol. Tidsskrift, 29:108–158.

Wade, M. 1970. The stratigraphic distribution of the Ediacara fauna in Australia. Royal Soc. South Australia, Trans., 94:87–104.

Willard, B. 1935. Chemung tracks and trails from Pennsylvania. Jour. Paleont., 9:43–56.

Webby, B. D. 1970. Late Precambrian trace fossils from New South Wales. Lethaia, 3:79–109.

Weimer, R. J. and J. H. Hoyt. 1964. Burrows of *Callianassa major* Say, geologic indicators of littoral and shallow neritic environments. Jour. Paleont., 38:761–767.

Whittington, H. B. 1962. A natural history of trilobites. Smithsonian Inst., Publ. 4489: 405–415.

Zittel, K. A. 1879–1890. Handbuch der Paläontologie. Munich, R. Oldenbourg, 958 p.

CHAPTER 7

THE STRATIGRAPHICAL SIGNIFICANCE OF TRACE FOSSILS

T. P. CRIMES

Department of Geology, University of Liverpool

Liverpool, England

SYNOPSIS

Trace fossils have many uses in stratigraphy. Short-ranging forms can date otherwise unfossiliferous successions, as exemplified here by lower Paleozoic successions in Europe and North America. Trace fossils, including trilobite tracks, also commonly extend beneath the lowest fossiliferous Phanerozoic horizons. This distribution suggests that (1) trilobites may have evolved rapidly at this time from soft-bodied ancestors and (2) trace fossils should be considered in defining a base to the Cambrian system.

Trace fossils are of value in environmental stratigraphy, including forms characteristic of rocky shores, sandy beaches, the neritic zone, and the bathyal–abyssal zone. Trace fossils are also useful in constructing paleogeographical syntheses, as demonstrated by examples from Paleozoic, Mesozoic, and Tertiary strata.

Trace fossils can further aid the stratigrapher in tectonically complex areas, especially in determination of way-up and in quantitative estimation of tectonic compression.

Finally, trace fossils are of value in global tectonics, as illustrated by particular reference to the North Atlantic. An example is the Arenig sequence: at this level, shallow-water Cruziana-*bearing sandstones in Wales give way northwestward, across the Irish Sea fault zone, to deepwater* Nereites-*bearing turbidites; in Newfoundland, a similar change occurs across a fault line between the "Avalon platform" and "central mobile belt." This similarity suggests initial continuity of the faults, and allows equation of the Welsh sequence with that on the "Avalon platform"—not the "central mobile belt," as has generally been claimed.*

INTRODUCTION

Trace fossils may be used in "classical" stratigraphy to define marker beds for short-range correlation, to date otherwise unfossiliferous successions and hence provide regional correlations, and also as geopetals to demonstrate the way-up of strata in structurally complex areas. The first part of this chapter is devoted to these techniques. Modern stratigraphy is, however, moving away from the purely classical concept of dating rocks and elucidating successions and is becoming more concerned with regional paleogeographical syntheses, crustal shortening, and large-scale plate movements on the earth's surface. The potential of trace fossils in these fields has yet to be fully realized, as is discussed later in the chapter.

USE OF TRACE FOSSILS IN BIOSTRATIGRAPHY

As long ago as the late nineteenth century, the Geological Survey of Ireland regarded the radiating worm burrow *Oldhamia* as a Cambrian index fossil (Kinahan, 1878). It was used to date the otherwise almost barren "Bray Series" of eastern Ireland. Recent investigations confirmed the validity of this conclusion for the *Oldhamia*-bearing localities (Crimes and Crossley, 1968; Rast and Crimes, 1969). Similarly, Mägdefrau (1934) recognized that the complex feeding burrow *Phycodes circinatum* is restricted to the Lower Ordovician and, by virtue of its widespread occurrence, could be a useful guide fossil. Little further work was attempted on the use of trace fossils in stratigraphy until Seilacher (1960, 1970) and Crimes (1968, 1969, 1970c) showed that

these fossils could date otherwise unfossiliferous lower Paleozoic successions.

To produce traces of stratigraphical value, an animal should ideally (1) be widely distributed, (2) leave a record of its activities on or within the sediment, (3) belong to a rapidly evolving group, and (4) be facies independent. Few benthic animals are truly facies independent; indeed, the facies control shown by trace fossils is one of their most characteristic features.

In Paleozoic strata, trace fossils occur most abundantly in shallow-water sandstone–shale sequences, many of which are devoid of body fossils. In Mesozoic and

TABLE 7.1   Known Ranges of Selected Trace Fossils.*

	Precambrian	Cambrian	Ordovician	Silurian	Devonian	Carboniferous	Permian	Triassic	Jurassic	Cretaceous	Paleocene	Eocene	Oligocene
Acanthorhaphe													
Asterosoma													
Astropolithon													
Atollites													
Belorhaphe													
Crossopodia													
Cruziana													
Desmograpton													
Dictyodora													
Dimorphichnus													
Diplichnites													
Favreina													
Glockeria													
Helicolithus													
Helicorhaphe													

Tertiary rocks the greatest variety occurs in otherwise relatively fossil-poor, deep-water, distal turbidites. The common occurrence of trace fossils within otherwise unfossiliferous successions means that they have great potential for broad stratigraphical correlations. They might provide the exploration geologist with the first reliable indication of the age of strata. Many trace fossil genera seem to be restricted to one or several systems (Crimes, 1974), so an ichnofauna can often be assigned readily to the correct system on generic identification alone (Table 7.1). Where short-ranging genera are involved, assigning rocks to part of a system may be possible. For example, the occurrence of *Astropolithon*, *Plagiogmus*, and *Syringomorpha* may be taken to indicate Lower Cambrian age. Precise correlations, however, usually require identification to species level. Fortunately, the number of stratigraphically useful species so far described is sufficiently limited that correct identification may often be made in the field by a mapping geologist.

To date, the most detailed work has been attempted on trilobite traces from lower Paleozoic successions. Many trilobites were short ranging and also active, benthic

TABLE 7.1—Continued.

	Precambrian	Cambrian	Ordovician	Silurian	Devonian	Carboniferous	Permian	Triassic	Jurassic	Cretaceous	Paleocene	Eocene	Oligocene
Helminthoida (s.l.)													
Megagrapton													
Oldhamia													
Ophiomorpha													
Plagiogmus													
Rusophycus													
Spirophycus													
Spirorhaphe													
Spongeliomorpha													
Squamodictyon													
Subphyllochorda													
Syringomorpha													
Thalassinoides													
Tomaculum													
Urohelminthoida													

* Compiled from various sources.

CAMBRIAN ORDOVICIAN DEVONIAN

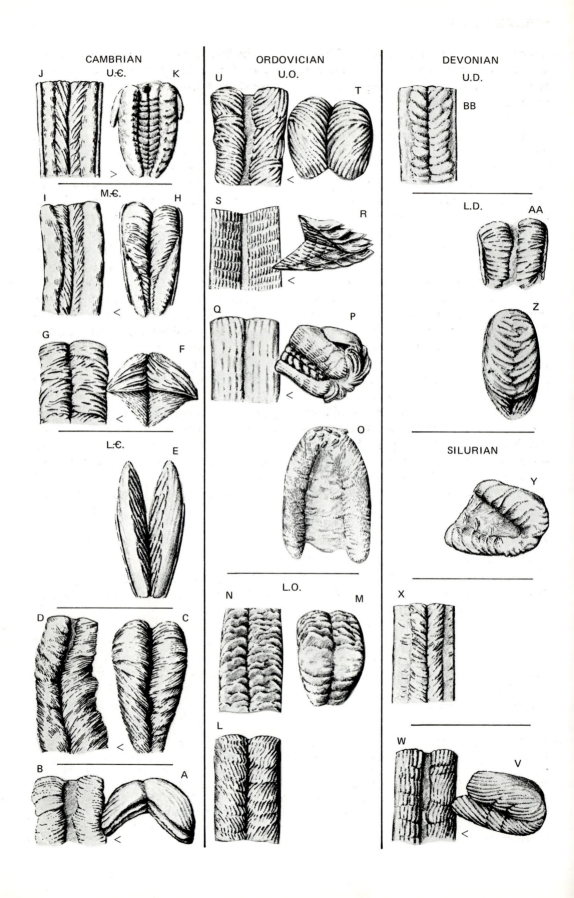

creatures producing copious traces. Normally the most common trilobite trace, the furrow (*Cruziana*), is of cosmopolitan distribution in lower Paleozoic strata. Crimes (1968) demonstrated from material collected within faunally zoned strata in Wales that one species (*C. semiplicata*) is common in the Upper Cambrian but does not range into the Arenig, whereas another (*C. furcifera*) occurs abundantly in the Arenig but not in the Upper Cambrian. The two species are found together only in the intermediate Tremadocian strata (Crimes, 1970c). Rocks of these three series can therefore be distinguished by their contained *Cruziana*. Seilacher (1970) expanded this concept and described 30 species of *Cruziana* having restricted time ranges (Fig. 7.1; Table 7.2). Using this scheme, one should be able to place *Cruziana*-bearing sandstones at least in the correct system and in many cases in the correct series.

In Wales, Spain, and Newfoundland, one can differentiate sequences of "unfossiliferous," lithologically similar sandstones into Upper Cambrian and Arenig on the occurrence of *C. semiplicata* and *C. furcifera*, respectively (Crimes, 1969, 1970c; Seilacher and Crimes, 1969). In Poland, Crimes (1968, p. 363) dated the *C. semiplicata*-bearing quartzites of the Wielska Wisniowka quarry, in the Holy Cross Mountains, as Upper Cambrian. This date was later confirmed by the discovery of olenid trilobites (Orłowski, 1968). Similarly, in North Spain, Schmitz (1971) recognized Upper Cambrian strata by the occurrence of *C. semiplicata*, and separated them from overlying strata containing both *C. semiplicata* and *C. furcifera* and hence assigned to the Tremadoc.

The importance of a trace fossil stratigraphy for rocks spanning the Cambro-Ordovician boundary can be seen from the widespread distribution of thick, usually unfossiliferous sandstones at about this level in South America, Newfoundland, Western Europe, North Africa, Saudi Arabia, and Jordan. In many of these sections the sandstones exceed thicknesses of 500 m and contain abundant trace fossils, providing the easiest or only method of correlation.

The stratigraphical usefulness of trace fossils in upper Paleozoic and Mesozoic rocks is likely to be less than that in the generally less fossiliferous lower Paleozoic strata. Nevertheless, the traces have potential in fossil-poor flysch sequences. Książkiewicz (1970) suggested that the following trace fossil species may have a limited vertical range: *Glockeria glockeri* (lowest Cretaceous), *Gyrophyllites kwassicensis* (upper Senonian), *Taphrhelminthopsis plana* (lower Eocene), *Fucusopsis angulata* (Senonian), and *Scolicia plana* (Albian to Senonian). The abundance of some of these traces may enhance their stratigraphical usefulness, but as yet no Mesozoic turbidite succession has been dated on trace fossil evidence alone.

TRACE FOSSILS AND THE PRECAMBRIAN–CAMBRIAN BOUNDARY

The earliest recorded trace fossil is a burrow system considered to have been made by a worm-like organism, probably an annelid, found in the Grand Canyon Series (U.S.A.) and suggested to be more than 1 billion years old (Glaessner, 1969).

◀ **Fig. 7.1** Paleozoic *Cruziana* species, and approximate stratigraphic distribution. > and < signs indicate whether furrow (= *Cruziana*, left) or resting trace expression (= *Rusophycus*, right) is more common. Forms not separated by horizontal line may occur in same units. See Table 7.2 for further details of stratigraphic distribution. A, B, *Cruziana cantabrica*. C, D, *C. fasciculata*. E, *C. carinata*. F, G, *C. barbata*. H, I, *C. arizonensis*. J, *C. semiplicata*. K, *C. polonica*. L, *C. rugosa*. M, N, *C. imbricata*. O, *C. lineata*. P, Q, *C. almadenensis*. R, S, *C. flammosa*. T, U, *C. petraea*. V, W, *C. acacensis*. X, *C. quadrata*. Y, *C. pedroana*. Z, *C. uniloba*. AA, *C. rhenana*. BB, *C. lobosa*. (From Seilacher, 1970.)

Vertical burrows were claimed in the Buckingham Sandstone and Areyonga Formation in Australia, both reputed to be about 800 million years old (Glaessner, 1969; Ranford et al., 1965). These scattered occurrences indicate that, although more traces of such antiquity will undoubtedly be found, they must be regarded as rare in rocks more than 750 million years old (see also Banks, 1970). More extensive collections were recorded by Webby (1970) from the Torrawangee Group in New South Wales (Australia) and by Glaessner (1969) from the well-known Ediacara Beds in South Australia. Both units are suggested to be about 600 million years old.

Banks (1970) reported that within a continuous succession spanning the Precambrian–Cambrian boundary in Finnmark (Norway), the abundance and diversity of traces increases abruptly in all facies in latest Precambrian and Early Cambrian times (see also Chapter 6). The earliest trilobite traces appear not far above the late Precambrian tillite but before the first preserved trilobites in both Finnmark (Banks, 1970) and Greenland (Cowie and Spencer, 1970). In Sweden, trilobite resting traces (*Rusophycus*) occur in the Hardeberga Sandstone, which is separated by the overlying *Diplocraterion* Sandstone from beds containing the oldest recorded trilo-

TABLE 7.2 Known Ranges of *Cruziana* Species.

	Cambrian			Tremadoc	Ordovician					Silurian		Devonian		
	Lower	Middle	Upper		Arenig	Llanvirn	Llandeilo	Caradoc	Ashgill	Lower	Upper	Lower	Middle	Upper
C. cantabrica														
C. fasciculata														
C. dispar														
C. carinata														
C. barbata														
C. arizonensis														
C. semiplicata														
C. polonica														
C. jenningsi														
C. omanica														
C. imbricata														
C. furcifera														
C. rugosa														
C. goldfussi														
C. grenvillensis														

bites (Bergström, 1970). Near the Barrios de Luna in the Cantabrian Mountains, north Spain, I also found *Rusophycus* and *Cruziana* within the Herreria Formation, immediately above the Precambrian–Cambrian unconformity and about 1,000 m beneath the earliest preserved trilobites. According to Lotze (1961), the trilobites are Early Cambrian and the oldest in the Iberian Peninsula. In Australia, trilobite traces, including *Diplichnites* and *Cruziana*, occur below the first preserved trilobites in the Arumbera Formation (Glaessner, 1969), and Young (1972) also recorded trilobite traces below the earliest recorded trilobites in a late Precambrian–Cambrian succession in Canada. The persistent occurrence, at widespread localities, of trilobite traces stratigraphically below the earliest recorded

trilobites is unlikely to be purely a matter of chance.

The sudden appearance of an animal as complex as a trilobite at the beginning of Cambrian times has been considered something of an enigma. A popular idea is that similar creatures existed for a long time during the Precambrian but did not develop hard skeletons until Early Cambrian times. Soft-bodied trilobites would be expected to leave characteristic traces but no body fossils. The common occurrence of trilobite traces below the first recorded tribolite body fossils might therefore tend to confirm the existence of an early, soft-bodied form.

One objection to the development of a trilobite from a soft-bodied form rests in their complex and powerful musculature.

TABLE 7.2—Continued.

	Cambrian			Tremadoc	Ordovician					Silurian		Devonian		
	Lower	Middle	Upper	Tremadoc	Arenig	Llanvirn	Llandeilo	Caradoc	Ashgill	Lower	Upper	Lower	Middle	Upper
C. petraea														
C. flammosa														
C. perucca														
C. lineata														
C. carleyi														
C. pudica														
C. ancora														
C. acacensis														
C. pedroana														
C. quadrata														
C. dilatata														
C. uniloba														
C. lobosa														
C. rhenana														

Possibly this musculature would have no satisfactory points of attachment without a hard skeleton. The trilobite furrows (*Cruziana*) from earliest Cambrian rocks at the localities mentioned above are, however, markedly different from later types; they are characterized by a simple and relatively shallow overall form having deeply dug, obliquely directed scratch marks made by the front limbs, giving a low V-angle and generally forming a very incomplete *Cruziana* (Fig. 7.1). Such traces are characteristic of slow movement, in which the limbs are directed at a high angle to the body axis, to give the greatest mechanical advantage (Crimes, 1970d, p. 63). The traces depicting more complex types of movement, in which the rear limbs are used to achieve greater speed of locomotion (Crimes, 1970d), are absent in the lowest Cambrian rocks. The traces are therefore consistent with a simple musculature of low efficiency. The observed distribution of trilobite traces and body fossils may therefore be explained by postulating a rapid evolution in late Precambrian–Early Cambrian times, with development of soft-, then hard-bodied forms.

The acquisition of hard parts could have been a normal progression in the rapid evolutionary process, but it may have resulted from oxygen concentration in the primitive sea, rising above a threshold value. Rhoads and Morse (1971) suggested that the change to hard-bodied forms may be reflected in present-day faunal distribution in the Black Sea and other low-oxygen basins, where hard-bodied animals are relatively more common in areas having higher oxygen concentrations (see also Chapter 9).

The above discussion illustrates that trace fossils are more common than body fossils in many late Precambrian–Early Cambrian sequences. This, together with rapid evolution of the animals, means that trace fossils should at least be considered in defining a base to the Cambrian system and may in fact be most valuable in adequately locating that base in many sections.

TRACE FOSSILS IN PALEOGEOGRAPHIC RECONSTRUCTIONS

Trace fossils reflect the behavioral responses of animals. These responses are controlled by such things as energy conditions at the depositional interface, substrate type, and availability of food. In general, the deeper the water, the lower the energy level, the finer the sediment, and the greater the amount of food incorporated in it. One therefore expects these changes to be reflected in the ichnofauna. The sensitivity of many animals to environmental conditions means that trace fossils can be correspondingly more sensitive environmental indicators than inorganic sedimentary structures. Variations in ichnofaunas can therefore be used to infer both vertical and lateral facies changes in ancient sediments, and hence to provide viable paleogeographic reconstructions.

The consistent biological responses of animals to their environment allowed Seilacher (1964, 1967) to suggest that a relatively small number of depth-controlled trace fossil communities recurred throughout Phanerozoic time. These communities, each named after a characteristic trace fossil are, in order of increasing water depth: *Glossifungites* and *Skolithos* (primarily the littoral zone), *Cruziana* (littoral zone to wave base), *Zoophycos* (wave base to zone of turbidite deposition), and *Nereites* (turbidite zone). According to this concept, the occurrence of characteristic members of a community can be used to infer the paleobathymetric level. We shall first consider each depth zone and its trace fossils, and then discuss examples of complete paleogeographical reconstructions using this method. (See also Table 2.1.)

Littoral Zone: the Rocky Coast and Pebbly Shore

Preexisting rocky coastlines can be detected by the presence of abundant rock borings. Pebbles moved to and fro in the littoral zone are also frequently bored. The animals most commonly responsible are sponges, phoronids, sipunculids, polychaetes, gastropods, bivalves, echinoids, and cirripeds. A comprehensive review of borings was recently given by Bromley (1970). (See also Chapter 11.)

Although their abundance can be used to detect ancient shorelines, exposed rock and pebbles may be bored by some organisms at considerable depth. Warme (1970) reported that a variety of rocks exposed down to depths of at least 40 m off southern California have been bored. Clionidae (sponges), for example, are most active and prolific in the uppermost 25 m of the sublittoral zone (Volz, 1939; Hartman, 1954) but occasionally extend to depths as great as 142 m (Topsent, 1904); they are absent from the littoral zone. Nevertheless, according to Bromley (1970), a rich ichnocoenose of sponge borings indicates a depth of less than 100 m, and a varied one should represent shallow water. A few borings are markedly depth restricted. *Cliona viridis* can survive only in shallow water, owing to the presence of symbiotic *Zoochlorellae* that require light. Radwański (1964, 1970) identified Miocene borings from Poland as the work of *Cliona viridis*. Thallophyte borings (algal) are also restricted to relatively shallow water and normally flourish only to depths of 20 to 25 m (Nadson, 1927a), although they have been found at 50 m in the Black Sea (Nadson, 1927b) and 80 m in the Gulf of Naples (Nadson, 1932). In many cases, algal and fungal borings are difficult to differentiate. The distinction is of great importance, however, because of the absence of any photic, and therefore bathymetric, control of fungal borings. Studies by Gatrall and Golubic (1970) suggest that the scanning electron microscope may allow the two to be separated. (See also Chapter 12.)

Littoral Zone: the Sandy Shore

The sandy shore is a very exacting environment, and relatively few benthic animals can fill this niche. The animals must be able to withstand current and wave action, desiccation, and rapid fluctuations in temperature and salinity. Animals that can tolerate such extreme conditions often do so by escaping from the surface into permanent or semipermanent burrows. This response is reflected in the corresponding trace fossils (Fig. 7.2), which show a preponderance of vertical burrows such as the typical *Skolithos* or "pipe rock." U-shaped burrows, such as *Arenicolites* or *Diplocraterion*, are also common. Burrows may be protected and stabilized by a layer of mucous, as in the burrow of the modern lugworm *Arenicola marina*, or by a lining of pellets, as in *Ophiomorpha*, or shells as described for some Eocene burrows by Roniewicz (1970). I also recently found burrows lined with concentrically arranged, imbricated foraminiferal shells in the Cretaceous on the coast north of the Sierra Helada, Benidorm, Spain.

Kennedy et al. (1969) described examples of *Thalassinoides* lined with the microcoprolite *Favreina* in the Jurassic of England. They suggested that the burrows were probably produced by a crustacean living within or just below the intertidal zone. Similarly, *Ophiomorpha* generally seems to be restricted to the littoral or shallow sublittoral zone and normally to occur in neither fresh nor deeper marine water (Hecker, 1956; Hecker et al., 1961; Seilacher, 1964; Kennedy and MacDougall, 1969; Radwański, 1969, 1970; Roniewicz, 1970; Chapter 2). In modern oceans this habitat is occupied by a variety of crustaceans that produce burrows, some of which are lined (Weimer and Hoyt, 1964; Shinn, 1968; Frey, 1970; Farrow, 1971; Braithwaite and Talbot, 1972). The intertidal zone is therefore characterized by an abundance of vertical or U burrows and the use of protective burrow linings. All these traces are intrastratal; surface traces are rarely preserved.

The Neritic Zone

The neritic zone, from low-water mark to the edge of the continental shelf (ca. 200 m),

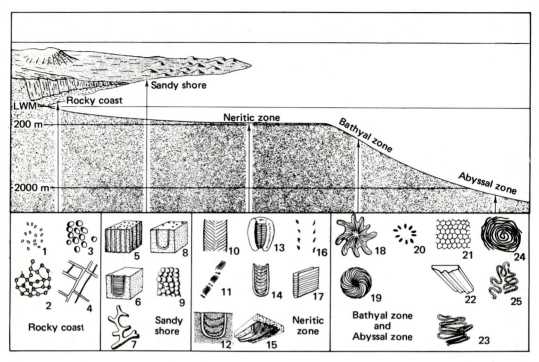

Fig. 7.2 Synoptic diagram illustrating the more common facies and depth–related trace fossil genera. 1, borings of *Polydora* (polychaete); 2, *Entobia*, boring of clionid sponge; 3, echinoid borings; 4, algal borings; 5, *Skolithos*; 6, *Diplocraterion*; 7, *Thalassinoides*; 8, *Arenicolites*; 9, *Ophiomorpha*; 10, *Cruziana*; 11, *Dimorphichnus*; 12, *Corophioides*; 13, *Rusophycus*; 14, *Rhizocorallium*; 15, *Phycodes*; 16, *Diplichnites*; 17, *Teichichnus*; 18, *Zoophycos*; 19, *Spirophyton*; 20, *Lorenzinia*; 21, *Paleodictyon*; 22, *Taphrhelminthopsis*; 23, *Helminthoida*; 24, *Spirorhaphe*; 25, *Cosmorhaphe*. (Cf. Table 2.1.)

is a less exacting environment than the littoral zone. Neither desiccation nor large and sudden fluctuations in temperature and salinity are a problem. Fast-flowing wave and tidal currents can, however, still sweep in and erode significant thicknesses of sediment and disturb sessile benthos and burrowing organisms alike. (See Chapter 8.) With environmental energy often high, the relatively coarse-grained sediment is low in organic carbon. Food for burrowers or browsers is therefore relatively scarce. Protection from environmental fluctuations is less important, consequently vertical burrowing and the construction of protected, U-shaped dwelling burrows is less common. Such burrows, where present, also tend to be shorter (Rhoads, 1966, 1967). Animals are able to move freely on the surface at certain times, and active benthic forms,

which can search for their food, are particularly suited to this environment.

In early Paleozoic seas this part of the ocean floor was dominated by trilobites, and the widespread occurrence of their traces—such as *Cruziana*, *Rusophycus*, *Diplichnites*, and *Dimorphichnus*—led Seilacher (1964) to consider the zone as constituting the "*Cruziana* facies." In later seas other arthropods, especially crustaceans, occupied the niche vacated by the trilobites, and burrows such as *Thalassinoides* and *Rhizocorallium* were common. Crustacean burrows are frequently found in this zone in modern oceans, and subtle differences in burrow types allow distinction between environments, such as back reef and lagoon (Farrow, 1971).

Sediments from this zone also are often characterized by an abundance of ripple marks. The ripples are preserved by the mud deposited during quiet periods; their greater frequency of occurrence in this zone than in the littoral zone therefore reflects higher preservation potential rather than a higher rate of initial production.

In the deeper parts of this zone, the sediments typically contain sufficient organic material for sediment feeders to become established. Here feeding burrows, such as *Corophioides, Teichichnus,* and *Phycodes,* are typical. Increasing water depth tends to be accompanied by a gradation from vertical to horizontal burrowing (Seilacher, 1967). For example, *Rhizocorallium* generally occurs oblique to the bedding in the shallower parts of this zone but is horizontal in the deeper-water, flat-bedded, more muddy sediments (see Ager and Wallace, 1970).

The Bathyal and Abyssal Zones

The bathyal zone occupies the position between wave base and the abyssal zone, which commences at a water depth of 2,000 m. In both zones, conditions are ordinarily quiet for long periods but can be disturbed by the passage of turbidity currents. In the shallower parts, fine-grained sandstones, siltstones, or limestones accumulate. Here complex feeding burrows such as *Zoophycos* and *Spirophyton* are produced. The accumulating sediments are sometimes eroded by turbidity currents, but the amount of erosion, particularly in the more distal areas, may be slight (see Crimes, 1973). In these deeper parts, the character of the sediment often changes. Distinct limestones are fewer and often thinner (see Crimes, 1973). This reduction may be partly the result of dissolution of $CaCO_3$ in deeper water (Berger, 1972). Sediments deposited by turbidity currents increase in importance at the expense of the limestones. However, the farther the turbidity current travels, the finer becomes the grain size of the deposited sediment, and the thinner become the beds. Consequently, the ratio of turbidite to mud gradually decreases: a transition from the "sandy flysch" to the "shaley flysch" of European terminology.

These changes in substrate and environment are reflected in the ichnofauna. Investigations in the lower Paleozoic of Wales (Crimes, 1970c), the Mesozoic and Tertiary of Poland and north Spain (Książkiewicz, 1970; Crimes, 1973) and the Carboniferous of the Ouachita Mountains, U.S.A. (Chamberlain, 1971) suggest that, beneath the zone containing *Zoophycos,* radiating traces such as *Lorenzinia* appear but are soon replaced in more distal beds by spiral, winding, and meandering forms. Finally, in the most distal facies, patterned traces, such as *Paleodictyon,* are added to the ichnospectrum.

Traces showing remarkable resemblance to many of the deep-water flysch types were recently recorded from 6,000 m depth in modern oceans (Ewing and Davis, 1967; Heezen and Hollister, 1971). Excellent photographs of a trace referable to *Spirophycus* are shown (Ewing and Davis, 1967, p. 262–264). Radiating traces are also figured; some can be identified as *Lorenzinia* (Ewing and Davis, 1967, p. 286, Figs. 24–77 and 24–78). Discovery of these modern examples of "deep-water" traces does not in itself prove the bathymetric control of trace fossil distribution, but one is pleased to find in modern deep oceans traces of the type that ichnological thought would predict. (See also Chapter 21.)

Examples of Paleogeographic Reconstruction

The foregoing discussion shows clearly that trace fossils, particularly the facies and bathymetrically influenced types, are valuable tools in paleogeographical reconstruction. Nevertheless, most types are controlled by facies rather than bathymetry. Thus, a given trace fossil can be expected to occur in places outside its normal bathymetric range, as noted by Henbest (1960), Häntzschel (1964), Bandel (1967, p. 9), Frey and Howard (1969), Hattin and Frey (1969), Frey (1970), and Crimes (1970c). Crimes (1970c) and Frey and Howard (1970) contended that mere checklists of trace fossils may not always provide adequate data for delineating facies or bathymetry. The best approach is a complete facies analysis, incorporating evidence from body fossils, trace fossils, lithology, inorganic sedimentary structures, and other features. Such analyses have been attempted on both local and regional scales and have been used (1) to delineate the deposits of a single depth zone, traced laterally; (2) to document paleogeographical changes with time, in a vertical sequence; and (3) to demonstrate the changing paleogeography in both time and space.

For example, in the southern part of the Holy Cross Mountains, Poland, Radwański (1970), from a study of borings made in solid rock surfaces by sponges, polychaetes, and bivalves, accurately delineated the Miocene shoreline over many tens of kilometers. He also followed tracts of continuous rocky seashores for more than 1 km. For the western part of the region, he even drew a detailed paleogeographical map for the time of the transgression. Between the rocky headlands, pebble or sand beaches apparently developed. The pebbles are frequently bored, whereas the littoral sands provided a habitat for burrowers; callianassid decapods produced burrows corresponding to *Ophiomorpha*.

Similarly, in the Kizil-Kum desert (U.S.S.R.) Hecker et al. (1962), Hecker (1970), and Pianovskaya and Hecker (1966) used borings of the bivalve *Lithophaga* and the sponge *Cliona* to map the shores (cliffs) and rocky bottoms (benches) of the Cretaceous (Aptian) sea in perfect exposure. The rock borers covered the surface of the coastal limestones so effectively that no settling place was left for fixed forms such as oysters. The latter are restricted to dense colonies on the granitic dikes not favored by borers. As in the Holy Cross Mountains,

the sandy littoral deposits contain abundant *Ophiomorpha*. These studies are particularly important in that recognition of an ancient coastline provides a datum from which the changing facies and paleogeography may be traced both vertically and laterally.

In other areas, where exposure is less complete, one may be able only to follow facies vertically from an unconformable surface. Nevertheless, this surface usually provides a zero water-depth datum. For example, in Wales, Crimes (1970a, p. 231) demonstrated an abrupt vertical facies change associated with the Arenig transgression. Sandstones contained *Skolithos* at the base, passing upward into *Cruziana*-bearing types, and then to shaly sandstones having abundant feeding burrows such as *Phycodes* and *Teichichnus*.

In many sections, however, unconformities are absent, and paleogeographical deductions must depend solely on the facies analysis. Farrow (1966) attempted to relate trace fossils and facies to bathymetry, in a paleogeographical reconstruction of the shallow-water Jurassic sea of Yorkshire (England). In the Boulonnais (France), Ager and Wallace (1970) demonstrated that the Jurassic sediments show three sequences of shallowing and emergence, each containing a similar trace fossil succession: (1) horizontal *Rhizocorallium*, (2) large *Thalassinoides*, (3) obliquely oriented *Rhizcorallium*, and (4) *Diplocraterion* in intertidal sediments. In the Dakota Formation (Upper Cretaceous, U.S.A.) Siemers (1970) described three trace fossil assemblages, correlatable with the fresh, brackish, and fully marine waters of a deltaic complex.

At the other end of the stratigraphical column, Banks (1970) used trace fossils and physical sedimentary structures to differentiate subtle changes in depositional environments in a conformable succession of dominantly shallow marine sediments of late Precambrian and Early Cambrian age in Finnmark (Norway).

Seilacher (1963) also used trace fossils to demonstrate that the Kharbour Quartzites in Iraq show a vertical sequence of increasing water depth, passing from his littoral *Skolithos* "facies" through *Cruziana* and *Zoophycos* "facies" to the deep-water *Nereites* "facies." He also described lateral transitions between three ichnofacies in the Paleozoic fold belt of the Ouachita Mountains, U.S.A. (Seilacher, 1964, p. 314).

Sediments of the deeper parts of the oceans have in general received less attention. Crimes and Crossley (1968) used trace fossils as part of a facies analysis of the Arenig in southeastern Eire; with a transition from deep to shallow water, first *Nereites* and then *Helminthoida* disappeared from the ichnospectrum, leaving— of the characteristic forms—only radiating traces in the shallower deposits. Tanaka (1971) published a facies analysis of the Cretaceous flysch from Hokkaido, Japan, in which trace fossils were used to infer alternations of shallower and deeper water; *Zoophycos* and radiating traces were present in the shallower deposits swept by transverse paleocurrents but absent in the deepest deposits swept by lateral paleocurrents; here *Helminthoida* and *Paleodictyon* were the most common forms.

Crimes (1973) reported that in one section from Cretaceous to Eocene flysch of north Spain, the trace fossils are closely facies controlled. *Zoophycos*-rich limestones are overlain by turbidites bearing radiating traces, and then by more distal turbidites containing meandering, spiral, and patterned traces.

To date, few attempts have been made at facies analysis involving complete eras, or even systems. This situation makes the pioneering work of Seilacher and Meischner (1964) even more remarkable. They used all available bathymetric criteria to reconstruct the sedimentary history and crustal movements in the Oslo district during almost the complete span of early Paleozoic time. The study revealed the existence of two phases of basin development and

crustal subsidence, separated by an episode of stability. Subsequently, Crimes (1970b) presented a facies analysis of the Cambrian of Wales, incorporating paleocurrent and lithological maps as well as vertical facies columns from the three main outcrops. Biogenic and inorganic sedimentary structures were consistently in agreement and indicated that the sediments did not all accumulate in either shallow or deep water, as had been previously claimed, but in fluctuating water depths at all localities. The analysis allowed a detailed account of vertical and lateral facies changes to be given, thus providing a picture of continuously changing paleogeography during the 70-million-year period of deposition.

In summary, one can have little doubt that many trace fossils provide valuable, and in some cases indispensable, evidence for paleogeographical reconstructions in most Phanerozoic successions. For this purpose, identification to generic level is usually sufficient and can readily be made by the mapping geologist. To assist in this work, Figure 7.2 shows some common facies and probable depth-related genera.

TRACE FOSSILS AS GEOPETALS

Trace fossils have been virtually ignored as geopetals (way-up indicators), even though the possibility was discussed by Shrock (1948) and Seilacher (1954), and they were applied to this purpose in the flysch of north Spain (Gomez de Llarena, 1954). In many successions where suitable inorganic sedimentary structures are absent, trace fossils can quickly provide an unequivocal reading of way-up of the rocks.

Traces useful as geopetals may be divided into those initially formed: (1) within beds, (2) on the top surface of beds, and (3) on the lower surface of beds. Of traces formed within beds, the way-up can be determined easiest from U burrows such as *Arenicolites*, which are dwelling burrows and consequently hang down from the depositional interface (Seilacher, 1954); the surface against which the limbs of the U terminate must therefore be the initial top (Fig. 7.3A). Some U burrows, such as *Corophioides* (Fig. 7.3B), are similar but also have upwardly concave septa (spreite) between the limbs of the U. Others, such as *Diplocraterion*, have upwardly concave backfill laminae within the limbs themselves. This concave-upward backfilling is also evident in the wall-like trace fossil *Teichichnus* (Fig. 7.3C). Some burrowers produce downward deflections of the surrounding sediment, and *Monocraterion* shows a multiple funnel structure, like a series of stacked filter funnels (Fig. 7.3D). Here the funnel opening and concavity of the laminae point to the top surface of the bed. Specimens of the simple vertical burrow *Skolithos* may penetrate a complete bed, but where several end against only one surface, that must be the top (Fig. 7.3E). Borings in hardgrounds similarly terminate against the previously exposed upper surface,[1] and, at burrowed disconformable surfaces, overlying sediments are piped downward into subjacent beds (Frey, 1971; Chapter 18).

Traces formed and preserved on the upper surface of beds are rare. Post-depositional burrows formed between beds sometimes adhere to the upper surface of the bed and can be confused with the upper surface traces. (See Chapter 2.) Tracks of arthropods are sometimes preserved as a series of minute depressions, and even double-lobed depressions corresponding to trilobite resting excavations have been reported (Crimes, 1970d). Where traces of this type occur as depressions on a bedding plane, it is clearly the upper surface. More important are traces that are formed on the sea floor as depressions but are now preserved as sand casts and therefore occur as downward depressions on the bases of the coarser grained rocks. In shallow-water sandstone–shale sequences, trilobite furrows

---

[1] Where the discontinuity surface is highly irregular, however, the borings may be inclined or even inverted in original configuration (see Chapter 18).

(*Cruziana*), resting excavations (*Rusophycus*), and walking traces (*Diplichnites*) are normally preserved this way and can be used to determine the way-up (Fig. 7.3F). In deep-water flysch successions, a host of traces are preserved in this manner; examples include *Helminthopsis*, *Spirophycus*, *Paleodictyon*, and *Cosmorhaphe* (Fig. 7.3G).

Some animals burrowed between or within beds and produced traces having similar upper and lower surfaces, so that their exposed presence on a surface cannot indicate the way-up; this applies to the supposed gastropod trace *Scolicia* (Fig. 7.3H). Despite this difficulty, the geopetal significance of most traces is readily apparent, and they often provide the easiest method of determining way-up.

TRACE FOSSILS AS INDICATORS OF COMPACTION AND DEFORMATION

Quantitative estimates of compaction or deformation depend on the existence in the rock of objects of simple, known, initial shape (cf. Chmelik, 1970). Most sediment-filled worm burrows are initially approximately cylindrical in form. After compaction, those oriented perpendicular to bedding remain cylindrical. Burrows parallel to bedding suffer a change in shape such that, in cross section, they become elliptical rather than circular. If the burrow is composed of essentially the same material as the surrounding sediment, then measurements on the two axes of the ellipse can provide a measure of the amount of compaction suffered by the sediment as a whole after the burrow was produced (Fig. 7.4A–C). In some cases, however, the burrow is defined by coarser material within a fine-grained matrix. Here such measurements reveal only the compaction suffered by the burrow; the surrounding sediment will have suffered more. This difference can sometimes clearly be seen by the deflection of laminae around the burrow (Fig. 7.4D).

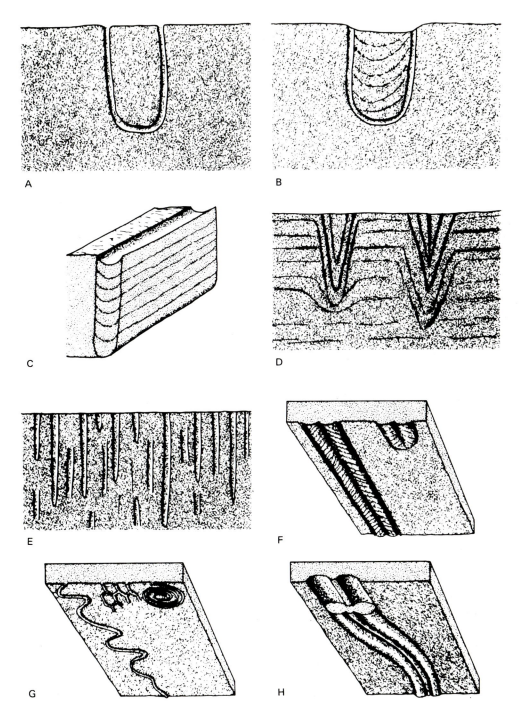

Fig. 7.3 Trace fossils useful as geopetals. In each case, top of bed is upward. A, *Arenicolites* hanging down from top surface. B, *Corophioides* hangs down and also has concave-up laminae. C, *Teichichnus* also has concave-up laminae. D, *Monocraterion* having upwardly concave stacked funnels. E, *Skolithos* ending against top surface. F, *Cruziana* and *Rusophycus* preserved as ridges on lower surface of sandstones. G, *Helminthopsis*, *Paleodictyon*, and *Spirophycus* also preserved as ridges, on lower surface of turbidites. H, *Scolicia* formed between beds, showing confusingly similar upper and lower surfaces.

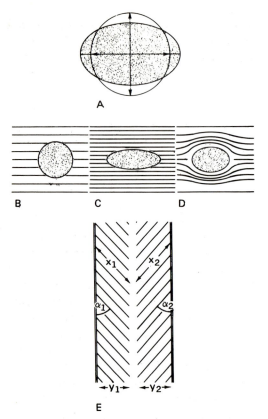

Fig. 7.4 Use of trace fossils in deformational studies. A, elliptical outline of burrow (stippled) may be used to determine percentage compaction by construction of circle of equal area to represent outline of undeformed burrow, and then comparing axial lengths. Quantitative significance depends on comparative behavior of burrow and matrix. In simplest case, B, no compaction after burrowing, the burrow remains circular in outline and adjacent laminae undeformed. If burrow infill and matrix compact equally after burrowing, C, measurement of burrow compaction gives correct value for matrix. In D, matrix compacts more, the laminae deflected around burrow; compaction values for burrow therefore are less than for matrix. Because many burrows have coarse-grained infill, this case is common. E, effects of tectonic deformation on *Cruziana*. In undeformed state, normally $X_1 = X_2$, $\alpha_1 = \alpha_2$, $y_1 = y_2$. Any increment of compression or shear disturbs the equality and can be used to quantify deformation, two-dimensionally in plane of trace.

In some fine-grained sediment, burrows can be completely flattened. Where the ramifying burrow system *Chondrites* occurs in muds, it is usually preserved as a flattened film on the surface of laminae. Occasionally, it is found within limestone and is truly three-dimensional, consisting of cylindrical burrows. Similarly, within concretionary mudstones *Chondrites* can sometimes be clearly seen within the concretions but cannot be discerned easily in the mudstones. Here early diagenetic cementation protected the burrow system within the concretion from the effects of compaction that obliterated the structure within the surrounding sediment. Some concretions show laminae that come closer together at their edges, thus demonstrating growth during compaction. In these cases, progressive changes in burrow shape should be detectable. Trace fossils can therefore be used to determine the amount of compaction suffered by a sediment, but that same compaction can lead to their modification and eventual destruction.

Trace fossils can also be used in deformational studies. *Skolithos* forms a cylinder at right angles to the bedding plane. Deviations from this shape and orientation may be used to calculate increments of compressive distortion. In undeformed rocks, straight *Cruziana* usually show equal lengths of the two limbs of the V and constant angular relationships between the Vs and the edges of the trace (Fig. 7.4E). Any variations can therefore be used to calculate the amount of compressive distortion within the plane containing the *Cruziana*. Similarly, *Rusophycus* of a given species has a constant ratio of length to breadth, regardless of orientation (see Crimes, 1970d); compressional distortion results in variations in this ratio, the values dependent on the relationship between *Rusophycus* orientation and principal stress.

Perhaps the most useful burrows are U types, such as *Diplocraterion* and *Arenicolites*. Variations in shape between the vertical and horizontal parts of the tube can be used to calculate compaction and further increments of strain across the horizontal plane. Taken together with measurements in two perpendicular planes in the vertical tubes, the data allow reconstruction of strain ellipsoids. This method was attempted by Plessmann (1966) on material from the Upper Cretaceous of Germany and the San Remo flysch (Italy). He also studied U burrows from the Lower Carboniferous of Edersee (Germany), and differentiated quantitatively the increments of precleavage compaction, compression, and rotation, as well as cleavage distortion (Plessmann, 1965). Amounts of finite strain were calculated by Brace (1961) and P. Geiser (1972, personal communication) from studies of *Skolithos*.

Body fossils and many other strain indicators are more rigid than the surrounding sediments and hence have a different response to stress. Most trace fossils consist of the same or nearly the same sediment as the surrounding matrix. Strain data from trace fossils are then likely to be more typical of the rock as a whole and hence more useful.

TRACE FOSSILS AND PLATE TECTONICS

The contributions made by paleontologists to plate tectonics stem from the provinciality shown by some fossils. Differences in faunas are used to define faunal provinces, and the arrival of species at different locations can define migration paths.

Trace fossils reflect the morphology of the animals that produced them. For example, the width of trilobite furrows (*Cruziana*) compare closely with the width of the trilobites, whereas the bunching of scratch marks making the Vs is a function of the claw pattern of the animals' limbs. Trilobite resting excavations (*Rusophycus*) accurately reflect the shape of the animals' body and also show impressions of segments, genal spines, and other diagnostic morphological features. Thus, if trilobites show provinciality and can be used to define migration paths, so can trilobite traces.

Seilacher and Crimes (1969) thereby showed from *Cruziana* and *Rusophycus* that during Late Cambrian and Arenig times, north Spain, Wales, and Newfoundland were linked together in the "Atlantic" province, clearly separated from mainland North America, which possessed trilobite traces of very different affinities, presumably constituting the "Pacific" province.

In addition to their provinciality, trace fossils have another useful attribute in global studies: their marked facies control. This relationship allows broad facies provinces to be defined. The traces are particularly useful because in many less-well-known regions, such as north Africa, trace fossils, by virtue of their abundance and mystique, have been recorded from many sections where the available sedimentological data could not alone provide an unequivocal paleogeographical interpretation. As an example, consider the Lower Ordovician strata from Iraq to Newfoundland. The basal Ordovician beds consist of sandstones yielding *Cruziana* (usually *C. furcifera*, *C. rugosa*, and *C. goldfussi*) at the following localities:

1. the Ora Anticline, Iraq (Seilacher, 1963, 1964)
2. the northern part of the Arabian Shield (Helal, 1965)
3. south Jordan (Bender, 1963; Selley, 1970)
4. Turkey (Frech, 1916)
5. Libya (Desio, 1940)
6. Algeria (Seilacher, 1970)
7. Portugal (Delgado, 1886)
8. Spain (Färber and Jaritz, 1964; Schmitz, 1971; Tomain *et al.*, 1970, and my own observations)
9. France (Peneau, 1946)
10. England and Wales (Crimes, 1968, 1970d)
11. Newfoundland (Seilacher and Crimes, 1969)

Much of the sedimentological data available from the above localities is sketchy, but when considered with the abundant occurrence of *Cruziana*, leaves little doubt that a shallow sea stretched over much of the area from Iraq to Newfoundland (Fig. 7.5). The extent of this sea implies that the area was a stable shelf. To the northwest, no *Cruziana* are recorded.

For example, in the British Isles none occurs in the Lower Ordovician northwest of Wales. Extensive searching by me in a sandstone–shale sequence in County Wexford, Eire, only several hundred kilometers from south Wales, failed to reveal any trilobite traces. Instead, the rocks yielded an abundant ichnofauna consisting of deep-water forms, such as *Helminthopsis, Lorenzinia,* and *Nereites* (Crimes and Crossley, 1968). The change in ichnofaunas represents the abrupt passage from shelf sea to rapidly subsiding eugeosynclinal trough; it demonstrates the importance of the intervening, fault controlled, northeast–

southwest trending, Irish Sea landmass as a fundamental fracture zone separating areas having radically different crustal response (Crimes, 1970a).

A zone of such importance should be traceable on the west side of the Atlantic Ocean. The Lower Ordovician rocks of southeastern Newfoundland yield abundant trilobite tracks on Bell Island (Seilacher and Crimes, 1969), thus defining the area as part of the stable shelf; it is known locally as the "Avalon platform." To the northwest, a northeast–southwest trending fault zone separates it from the turbidites and volcanics of the eugeosynclinal "cen-

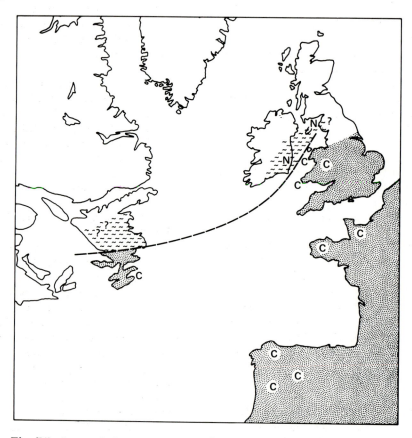

Fig. 7.5 Lower Ordovician facies relationships in North Atlantic. Shallow-water *Cruziana*-bearing Arenig sandstones (stippled) and *Cruziana* localities (C), and deep-water *Nereites* facies sediments and main trace fossil localities (N). Facies are separated in United Kingdom by Irish Sea fault zone and in Newfoundland by Avalon boundary fault system, which can be equated. Fault trace is diagrammatic only; it may be effected by dextral transcurrent faulting in the Atlantic (see Kay, 1969).

tral mobile belt" (Williams, 1969), which have nowhere yielded *Cruziana* but which can be expected to contain deep-water traces. This sharp ichnofaunal and facies break allows equation of the two fault zones across the Atlantic Ocean. The Welsh Ordovician succession is thus the equivalent of the sequence on the "Avalon platform," not that of the "central mobile belt" as claimed by many authors (see Kay, 1969).

Clearly, the full potential of trace fossils in global studies has yet to be realized. This is another field where common occurrence, easy recognition, and marked facies control may be particularly useful attributes of trace fossils.

ACKNOWLEDGMENTS

I thank J. Lynch for redrawing the diagrams.

REFERENCES

Ager, D. V. and P. Wallace. 1970. The distribution and significance of trace fossils in the uppermost Jurassic rocks of the Boulonnais, northern France. In T. P. Crimes and J. C. Harper (eds.), Trace fossils. Geol. Jour., Spec. Issue 3:1–18.

Bandel, K. 1967. Trace fossils from two Upper Pennsylvanian sandstones in Kansas. Univ. Kansas Paleont. Contr., Paper 18, 13 p.

Banks, N. 1970. Trace fossils from the late Precambian and Lower Cambrian of Finnmark, Norway. In T. P. Crimes and J. C. Harper (eds.), Trace fossils. Geol. Jour., Spec. Issue 3:19–34.

Bender, F. 1963. Stratigraphie der "Nubischen Sandsteine" in Süd-Jordanien. Geol. Jahrb., 81:237–276.

Berger, W. H. 1972. Dissolution facies and age-depth constancy. Nature, 236:392–395.

Bergström, J. 1970. *Rusophycus* as an indication of Early Cambrian age. In T. P. Crimes and J. C. Harper (eds.), Trace fossils. Geol. Jour., Spec. Issue 3:35–42.

Brace, W. F. 1961. Mohr construction in the analysis of large geologic strain. Geol. Soc. America, Bull., 72:1059–1080.

Braithwaite, C. J. R. and M. R. Talbot. 1972. Crustacean burrows in the Seychelles, Indian Ocean. Palaeogeogr., Palaeoclimatol., Palaeoecol., 11:265–285.

Bromley, R. G. 1970. Borings as trace fossils and *Entobia cretacea* Portlock, as an example. In T. P. Crimes and J. C. Harper (eds.), Trace fossils. Geol. Jour., Spec. Issue 3: 49–90.

Chamberlain, C. K. 1971. Bathymetry and paleoecology of Ouachita geosyncline of southeastern Oklahoma as determined from trace fossils. Amer. Assoc. Petrol. Geol., Bull., 55:34–50.

Chmelik, F. B. 1970. An investigation of changes induced in macrostructures in pelitic sediments during primary consolidation. Texas A & M Univ., Dept. Oceanogr., Tech. Rept. 70–8–T, 131 p.

Cowie, J. W. and A. M. Spencer. 1970. Trace fossils from the late Precambrian/Lower Cambrian of East Greenland. In T. P. Crimes and J. C. Harper (eds.), Trace fossils. Geol. Jour., Spec. Issue 3:91–100.

Crimes, T. P. 1968. *Cruziana*: a stratigraphically useful trace fossil. Geol. Magazine, 105: 360–364.

————. 1969. Trace fossils from the Cambro-Ordovician rocks of north Wales and their stratigraphic significance. Geol. Jour., 6: 333–337.

————. 1970a. A facies analysis of the Arenig of western Lleyn, north Wales. Geologists' Assoc., Proc., 81:221–240.

————. 1970b. A facies analysis of the Cambrian of Wales. Palaeogeogr., Palaeoclimatol., Palaeoecol., 7:113–170.

————. 1970c. The significance of trace fossils in sedimentology, stratigraphy and palaeoecology, with examples from lower Palaeozoic strata. In T. P. Crimes and J. C. Harper (eds.), Trace fossils. Geol. Jour., Spec. Issue 3:101–126.

————. 1970d. Trilobite tracks and other trace fossils from the Upper Cambrian of north Wales. Geol. Jour., 7:47–68.

————. 1973. From limestones to distal turbidites: a facies and trace fossil analysis in the Zumaya flysch (Paleocene-Eocene), north Spain. Sedimentology, 20:105–131.

————. 1974. Colonisation of the early ocean floor. Nature, 248(5446):328–330.

———— and J. D. Crossley. 1968. The stratigraphy, sedimentology, ichnology and structure of the lower Palaeozoic rocks of northeastern County Wexford. Royal Irish Acad., Sect. B, Proc., 67:185–215.

Delgado, J. F. N. 1886. Étude sur les *Bilobites* et autres fossiles des quartzites de la base du système silurique du Portugal. Lisbon, Soc. Trab. Geol. Portugal, Mem., 113 p.

Desio, A. 1940. Vestigia problematiche paleozoiche della Libia. Mus. Libico sotira Natur., Ann., 2:47–92.

Ewing, M. and R. A. Davis. 1967. Lebensspuren photographed on the ocean floor. In J. B. Hersey (ed.), Deep-sea photography. Johns Hopkins Oceanogr. Stud. 3:259–294.

Färber, A. and W. Jaritz. 1964. Die Geologie des westasturischen Kustengebietes zwischen San Esteban de Pravia und Ribadeo (NW–Spanien). Geol. Jahrb., 81:679–738.

Farrow, G. E. 1966. Bathymetric zonation of Jurassic trace fossils from the coast of Yorkshire, England. Palaeogeogr., Palaeoclimatol., Palaeoecol., 2:103–151.

————. 1971. Back-reef and lagoonal environments of Aldabra Atoll, distinguished by their crustacean burrows. In D. R. Stoddart and C. M. Yonge (eds.), Regional variations in Indian Ocean coral reefs. Zool. Soc. London, Symp., 28:455–500.

Frech, F. 1916. Geologie Kleinasiens im Bereich der Bagdadbahn. Zeitschr. Deutsche Geol. Gesell., 68:1–30.

Frey, R. W. 1970. Environmental significance of recent marine lebensspuren near Beaufort, North Carolina. Jour. Paleont., 44:507–519.

————. 1971. Ichnology—the study of fossil and recent lebensspuren. In B. F. Perkins (ed.), Trace fossils, a field guide. Louisiana State Univ., Misc. Publ. 71-1:91–125.

———— and J. D. Howard. 1969. A profile of biogenic sedimentary structures in a Holocene barrier island—salt marsh complex, Georgia. Gulf Coast Assoc. Geol. Socs., Trans., 19:427–444.

———— and J. D. Howard. 1970. Comparison of Upper Cretaceous ichnofaunas from siliceous sandstones and chalk, western interior region, U.S.A. In T. P. Crimes and J. C. Harper (eds.), Trace fossils. Geol. Jour., Spec. Issue 3:141–166.

Gatrall, M. and S. Golubic. 1970. Comparative study on some Jurassic and recent endolithic fungi using scanning electron microscope. In T. P. Crimes and J. C. Harper (eds.), Trace fossils. Geol. Jour., Spec. Issue 3: 167–178.

Glaessner, M. F. 1969. Trace fossils from the Precambrian and basal Cambrian. Lethaia, 2:369–393.

Gomez de Llarena, J. 1954. Observaciones geologicas en el Flysch cretacico-numulitico de Guipuzcoa. Inst. "Lucas Mallada" Invest. Geol., Monogr. 13, 98 p.

Häntzschel, W. 1964. Spurenfossilien und Problematica im Campan von Beckum (Westf.). Fortschr. Geol. Rheinld. u. Westf., 7:295–308.

Hartman, O. 1954. Marine annelids from the northern Marshall Islands. U.S. Geol. Surv., Prof. Paper 260Q:Q619–Q644.

Hattin, D. E. and R. W. Frey. 1969. Facies relations of *Crossopodia* sp., a trace fossil from the Upper Cretaceous of Kansas, Iowa, and Oklahoma. Jour. Paleont., 43:1435–1440.

Hecker, R. T. 1956. Ecological analysis of crustacean decapods in Fergana Gulf of the Palaeogene sea of central Asia. Byull. mosk, obshch. Ispyt. Prir., 61:77–98. (In Russian).

————. 1970. Palaeoichnological research in the Palaeontological Institute of the Academy of Sciences of the USSR. In T. P. Crimes and J. C. Harper (eds.), Trace fossils. Geol. Jour., Spec. Issue 3:215–226.

———— et al. 1962. Fergana Gulf of Palaeogene sea of Central Asia, its history, sediments, fauna and flora, their environment and evolution, 1 and 2. Moscow, Akad. Nauk SSSR, 332 p. (in Russian) See also Amer. Assoc. Petrol. Geol., Bull., 47:617–631.

Heezen, B. C. and C. D. Hollister. 1971. The face of the deep. New York, Oxford Univ. Press, 659 p.

Helal, A. H. 1965. Stratigraphy of outcropping Paleozoic rocks around the northern edge of the Arabian shield (within Saudi Arabia). Zeitschr. Deutsche. Geol. Gesell., 117:506–543.

Henbest, L. G. 1960. Fossil spoor and their environmental significance in Morrow and Atoka Series, Pennsylvanian, Washington County, Arkansas. U.S. Geol. Surv., Prof. Paper 400–B:B383–B385.

Kay, M. (ed.). 1969. North Atlantic—geology and continental drift. Amer. Assoc. Petrol. Geol., Mem. 12, 1082 p.

Kennedy, W. J. and J. D. S. MacDougall. 1969. Crustacean burrows in the Weald Clay (Lower Cretaceous) of south-eastern England and their environmental significance. Palaeontology, 12:459–471.

———— et al. 1969. A *Favreina-Thalassinoides* association from the Great Oolite of Oxfordshire. Palaeontology, 12:549–554.

Kinahan, G. H. 1878. Manual of the geology of Ireland. London, Kegan Paul, 444 p.

Książkiewicz, M. 1970. Observations on the ichnofauna of the Polish Carpathians. In T. P. Crimes and J. C. Harper (eds.), Trace fossils. Geol. Jour., Spec. Issue 3:283–322.

Lotze, F. 1961. Das Kambrium Spaniens. I Stratigraphie. Akad. Wiss. u. Lit. Mainz, math.–naturw. Kl., Abh., 6:283–498.

Mägdefrau, K. 1934. Über *Phycodes circinatum* Reinh. Richter aus dem thuringischen Ordovicium. Neues Jahrb. Geol. Paläont., Abh., 72:259–281.

Nadson, G. A. 1927a. Les algues perforantes de la Mer Noire. Compt. rend. Hebd. Seane. Acad. Sci., Paris, 184:896–898.

————. 1927b. Les algues perforantes, leur distribution et leur role dans la nature. Compt. rend. Hebd. Seane. Acad. Sci., Paris, 184:1015–1017.

————. 1932. Contribution a l'etude des algues perforantes. Izv. Akad. Nauk SSSR (7), Math. Nat. classe, 1932(6):833–855.

Orłowski, S. 1968. Upper Cambrian fauna of the Holy Cross Mts. Acta Geol. Polonica, 18:257–291.

Peneau, J. 1946. Étude sur l'Ordovicien Inferieur (Arenigien-Grès Armoricain) et sa faune (specialement en Anjou). Soc. Étude scient. Angers, N. Ser., Bull., 74–76:37–106.

Pianovskaya, I. A. and R. T. Hecker. 1966. Rocky shores and hard ground of the Cretaceous and Palaeogene seas in central Kizil-Kum and their inhabitants. In Organisms and environment in the geological past, a symposium. "Nauka." (in Russian)

Plessmann, W. 1965. Laterale Gesteinsverformung vor Faltungsbeginn im Unterkarbon des Edersees (Rheinisches Schiefergebirge). Geol. Mitt., 5:271–284.

————. 1966. Diagenetische und Kompressive Verfermung in der Oberkreide des Harz-

Nordrandes sowie im Flysch von San Remo. Neues Jahrb. Geol. Paläont., Abh., 8:480–493.

Radwański, A. 1964. Boring animals in Miocene littoral environments of southern Poland. Acad. Pol. Sci. (Sér. Sci. Geol. Geogr.), Bull., 12:57–62.

————. 1969. Lower Tortonian transgression on to the southern slopes of the Holy Cross Mountains. Acta Geol. Polonica, 19:137–143.

————. 1970. Dependence of rock-borers and burrowers on the environmental conditions within the Tortonian littoral zone, southern Poland. In T. P. Crimes and J. C. Harper (eds.), Trace fossils. Geol. Jour., Spec. Issue 3:371–390.

Ranford, L. C. et al. 1965. The geology of the central part of the Amadeus basin, Northern Territory. Australia Bur. Min. Resourc., Rept. 86, 48 p.

Rast, N. and T. P. Crimes. 1969. Caledonian orogenic episodes in the British Isles and north-western France and their tectonic and chronological interpretation. Tectonophysics, 7:277–307.

Rhoads, D. C. 1966. Missing fossils and paleoecology. Discovery, Yale Peabody Museum, 2:19–22.

————. 1967. Biogenic reworking of intertidal and subtidal sediments in Barnstable Harbor and Buzzards Bay, Massachusetts. Jour. Geol., 75:461–476.

———— and J. W. Morse. 1971. Evolutionary and ecologic significance of oxygen-deficient marine basins. Lethaia, 4:413–428.

Roniewicz, P. 1970. Borings and burrows in the Eocene littoral deposits of the Tatra Mountains, Poland. In T. P. Crimes and J. C. Harper (eds.), Trace fossils. Geol. Jour., Spec. Issue 3:439–446.

Schmitz, U. 1971. Stratigraphie und sedimentologie im Kambrium und Tremadoc der Westlichen Iberischen Ketten nördlich Ateca (Zaragoza), NE-Spanien. Münster Forsch. Geol. Paläont. 22, 123 p.

Seilacher, A. 1954. Die geologische Bedeutung fossiler Lebensspuren. Zeitschr. Deutsche Geol. Gesell., 105:114–227.

————. 1960. Lebensspuren als Leitfossilien. Geol. Rundschau, 49:41–50.

————. 1963. Kaledonischer Unterbau der

Irakiden. Neues Jahrb. Geol. Paläont., Abh., 10:527–542.

———. 1964. Biogenic sedimentary structures. In J. Imbrie and N. D. Newell (eds.), Approaches to paleoecology. New York, John Wiley, p. 296–316.

———. 1967. Bathymetry of trace fossils. Marine Geol., 5:189–200.

———. 1970. *Cruziana* stratigraphy of "nonfossiliferous" Palaeozoic sandstones. In T. P. Crimes and J. C. Harper (eds.), Trace fossils. Geol. Jour., Spec. Issue 3:447–476.

——— and T. P. Crimes. 1969. "European" species of trilobite burrows in eastern Newfoundland. In M. Kay (ed.), North Atlantic—geology and continental drift. Amer. Assoc. Petrol. Geol., Mem. 12:145–148.

——— and D. Meischner. 1964. Faziesanalyse im Paläozoikum des Oslo-Gebietes. Geol. Rundschau, 54:596–619.

Selley, R. C. 1970. Ichnology of Palaeozoic sandstones in the southern desert of Jordan: a study of trace fossils in their sedimentologic context. In T. P. Crimes and J. C. Harper (eds.), Trace fossils. Geol. Jour., Spec. Issue 3:477–488.

Shinn, E. A. 1968. Burrowing in recent lime sediments of Florida and the Bahamas. Jour. Paleont., 42:879–894.

Shrock, R. R. 1948. Sequence in layered rocks. McGraw-Hill, 507 p.

Siemers, C. T. 1970. Facies distribution of trace fossils in a deltaic environmental complex: upper part of the Dakota Formation (Upper Cretaceous), central Kansas (abs.).

Geol. Soc. America, Abs. Prog., 2(7):683–684.

Tanaka, K. 1971. Trace fossils from the Cretaceous flysch of the Imushumbetsu area, Hokkaido, Japan. Geol. Surv. Japan, Rept. 242, 31 p.

Tomain, G. et al. 1970. L'Ordovicien de la Sierra Morena Orientale (Espagne). Compt. Rend. 94th Congr. Nat. Socs. savantes, Pau, 1969. Sci. II, p. 275–292.

Topsent, E. 1904. Spongiaires des Açores. Result. Camp. Scient., Prince Albert I., Fasc, 25 p.

Volz, P. 1939. Die Bohrschwämme (Clioniden) der Adria. Thalassia, 3:1–39.

Warme, J. E. 1970. Traces and significance of marine rock borers. In T. P. Crimes and J. C. Harper (eds.), Trace fossils. Geol. Jour., Spec. Issue 3:515–526.

Webby, B. D. 1970. Late Precambrian trace fossils from New South Wales. Lethaia, 3:79–109.

Weimer, R. J. and J. H. Hoyt. 1964. Burrows of *Callianassa major* Say, geologic indicators of littoral and shallow neritic environments. Jour. Paleont., 38:761–767.

Williams, H. 1969. Pre-Carboniferous development of Newfoundland Appalachians. In M. Kay (ed.), North Atlantic—geology and continental drift. Amer. Assoc. Petrol. Geol., Mem. 12:32–58.

Young, F. G. 1972. Early Cambrian and older trace fossils from the southern Cordillera of Canada. Canadian Jour. Earth Sci., 9:1–17.

CHAPTER 8

THE SEDIMENTOLOGICAL SIGNIFICANCE OF TRACE FOSSILS

JAMES D. HOWARD

Skidaway Institute of Oceanography

Savannah, Georgia, U.S.A.

SYNOPSIS

In sedimentologic and stratigraphic studies we should consider most trace fossils for what they are: sedimentary structures. Used in this context, trace fossils can furnish valuable information concerning (1) general depositional processes, (2) episodes of local deposition and erosion, and (3) characteristics of currents, substrate consistency, and in some instances, causes of sediment sorting. This information is important in and for itself, and it should also be utilized more fully by paleoecologists and others concerned with the reconstruction of ancient depositional environments.

INTRODUCTION

The significance assigned to trace fossils by a specific investigator is usually related to the nature of the study being undertaken and to the investigator's background in geology. As discussed in other chapters of this book, traces commonly receive a paleontologic or zoologic connotation. Because of this aspect, traces are often given short shrift by sedimentologists. This situation is unfortunate, and indeed unfair, to the study of sediments because the contained lebensspuren *are* sedimentary structures (albeit biologically formed) and should receive attention equal to that devoted to structures developed by physical processes. In fact, these traces often supply evidence of sedimentological conditions that is superior to information gained only by the study of physical structures.

If the foregoing is not sufficient reason for sedimentologists to be concerned with the study of ichnology, perhaps they can be prodded into it by virtue of the fact that the nefarious beasts creating the biogenic structures have a nasty habit of destroying their beloved physical structures, and they should at least attempt to identify the enemy! Indeed, unless he knows something about his foe, it may be rather difficult for the sedimentologist to distinguish between physical and biogenic sedimentary structures (e.g., Kuenen, 1961).

In this chapter, lebensspuren are considered as sedimentary structures, and the animal's activities are considered as (1) processes that form new, or destroy preexisting, fabrics and structures; (2) mechanisms for sediment concentration, reworking, or modification; and (3) devices for determining rates of sediment deposition or erosion. Other facets of these topics are considered in Chapter 9.

DESTRUCTIONAL AND CONSTRUCTIONAL FEATURES

To discuss independently the destructional and constructional aspects of burrowing organisms and the resulting sedimentary structures is essentially impossible. With few exceptions, the two processes occur simultaneously. Any organisms moving in or on the substrate invariably leave traces of their activity; but in so doing, the organisms inevitably disturb, modify, or even completely obliterate the preexisting texture or fabric. This fact, coupled with the knowledge that some type of plant or animal life is able to survive under nearly any environmental condition, should make an investigator rather suspicious of any sedimentary rock sequence completely devoid of a biogenic record.

From the moment sediment is deposited, under most conditions, it becomes subject to the possibility of biogenic reworking. In many situations a "contest" begins, to determine whether the geologic record will consist mostly of physical or biogenic sedimentary structures, or a combination of the two, according to the relative dominances of the two processes. The ultimate outcome of this contest depends on the conditions of the environment, including such things as physical energy, rate of sedimentation, and the density, adaptations, and variety of organisms.

In a very general way, the above relationship is shown in Figure 8.1; decreasing wave or current energy is correlative with an increase in the degree of biogenic reworking for a particular depositional environment. This general condition has been widely reported, and was succinctly stated by Purdy (1964). Detailed studies of beach-to-offshore facies sequences, and other compound facies relationships in modern and ancient sediments, have repeatedly shown the interrelationships between biogenic activity and physical processes (e.g., Moore and Scruton, 1957; Evans, 1965; Ager and Wallace, 1970; Reineck, 1970; Reineck and Singh, 1971; Howard, 1971a, 1972; Howard and Reineck, 1972; Clifton and Hunter, 1973; Chapters 7 and 9).

This relationship can be of significant value to the stratigrapher in attempting to trace shorelines from widely scattered outcrops or from drill cores. It can be of further use in the interpretation of transgressive and regressive deposits, as well as erosional contacts or reversals in deposi-

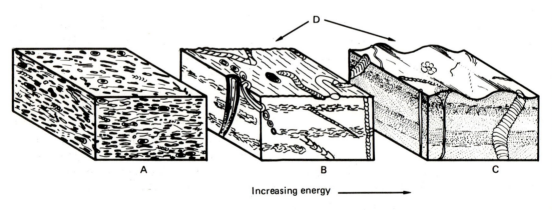

Fig. 8.1 General grouping of Upper Cretaceous trace fossils in the Book Cliffs and Wasatch Plateau of Utah. A, siltstone having highly mottled textures and very few taxonomically identifiable traces. B, very-fine-grained sandstone having larger but less abundant mottles, and containing individually recognizable, well-developed burrows. C, cleaner, coarser sands showing near absence of mottled texture but numerous recognizable burrows. D, tracks and trails developed on bedding plane surfaces. An increase in grain size and distinctiveness of individual burrows is evident from left to right, accompanied by a decrease in organic matter, fine-grained matrix, and mottled textures; these changes represent an increase in wave and current transport energy, to the right. (From Howard, 1966.)

tional conditions, especially in nearshore situations. In nearly any sedimentary sequence, one should expect to find an energy gradation that is reflected in the biogenic as well as physical sedimentary structures. Specifically, as in the case of a nearshore sequence, this gradation is a change of facies in which the abundance of physical sedimentary structures increases landward, commensurate with the increase in wave energy. Furthermore, the *rate* of facies change increases in a landward direction; the sedimentary facies are more closely spaced near shorelines. Thus, in tracing a depositional unit laterally, from offshore landward, one would expect a high degree of bioturbation offshore, where sedimentation rates are lower and where physical processes are less (or less frequently) important. This regime is succeeded by an intermediate zone exhibiting an interplay between physical and biogenic processes. The transition is culminated finally by the nearshore or inshore area, where physical processes dominate. In outcrops or cores, these relations are reflected similarly in the vertical sequence.

Obviously, many subtle interruptions and exceptions run counter to this ideal situation; but the model is valid, and should be kept in mind as answers are sought to explain the exceptions. For instance, in the Holocene of the Georgia coast, an interruption seen in the fine-grained nearshore sequence is due to the presence of palimpsest sediments, composed of medium to coarse sand (Howard and Reineck, 1972), and by outcrops of lithified Miocene sandy marl (Howard et al., 1973).

Similarly, local interruptions of the lateral sequence can be due to topographic irregularities. In the Georgia nearshore area, we see both higher energy and lower energy interruptions in the lateral beach-to-offshore sequence. For example, the intertidal estuary-entrance shoals, which commonly extend several miles offshore, are lenses of clean, well-sorted, trough-cross-bedded, ripple-laminated sand, surrounded

by otherwise fine-grained, commonly muddy, highly bioturbated sediments (Howard and Reineck, 1972). Another example is the incorporation of highly bioturbated, and in places muddy, sand of locally restricted tidal-flat deposits (see Fig. 8.6), some of which are contained in the main body of otherwise clean, well-laminated, fine sand of the beach (Howard and Dörjes, 1972). Preserved in the rock record, deposits of these small tidal flats would appear as lense-shaped units representing relatively low energy, surrounded by sediments deposited under considerably higher energy. (See also Chapter 20.)

An increase of biogenic activity within a laterally continuous depositional sequence can also be seen in situations where the conditions of life change even though the physical energy remains constant. An example of this change may occur within an estuary, where the physical processes and resulting physical sedimentary structures remain similar but the gradient in marine conditions permits the support of more organisms near the mouth of the estuary (e.g., Howard and Frey, 1973).

Various authors have attempted to characterize bioturbation and bioturbate textures from a sedimentological point of view (e.g., Reineck, 1967; Hanor and Marshall, 1971, and references cited). Such studies are of course important but are largely beyond the scope of this chapter.

The important point in the foregoing discussion is that within a depositional sequence, the biogenic structures—equally as much as the physical structures—can be used as subtle indicators of change in facies, which in turn indicate a change in the overall depositional environment.

LEBENSSPUREN AS INDICATORS OF GENERAL DEPOSITIONAL PROCESSES

Trace fossils can be of special help in the interpretation of conditions of erosion and sedimentation that occurred when the rock interval being examined was part of a dynamic sedimentary environment. Specific clues to these processes are the abrupt truncations of burrows and bioturbate textures, even in units that may otherwise appear to be thick sequences of completely bioturbated sediments. Failure to recognize these breaks may result in the impression of slow, continuous sediment accumulation, when in fact the sequence resulted from intermittent deposition and erosion. Perhaps the best means of illustrating the ways in which lebensspuren aid in the interpretation of depositional environments is to consider various conditions under which sediments can accumulate, and to discuss them in terms of the record left by the trace-making organisms. Scale of course plays an important role in this discussion, because essentially any depositional process is sooner or later interrupted by cessation of deposition, commonly accompanied by erosion. Thus, the features considered here are seen within a bedding unit or a local sequence of bedding units, and the break in deposition may represent little or much time.

Continuous Deposition

Slow Rate of Deposition

In the event of continuous, slow, uninterrupted sedimentation, the record is one of complete biogenic reworking (Fig. 8.2), generally by a variety of organisms. This record is the "fossitextura deformativa" of Schäfer (1972, p. 404). Recognizing the traces of specific organisms is commonly impossible because the record is that of continued working and reworking of the substrate by numerous individuals. In examining such a unit (Fig. 8.1A), one commonly gets the impression that this texture must record an environment in which myriad organisms were crawling in, on, and through the substrate, because of

Fig. 8.2 Continuous slow deposition, represented by siltstone unit in which biogenic activity was so intense that virtually all physical sedimentary structures were destroyed. Vertical face of sawed rock; Panther Sandstone Member, Star Point Formation (Upper Cretaceous), east-central Utah.

the complete biogenic destruction of primary fabrics and structures. The examination of recent sediments, however, indicates that this implied density is seldom the actual case. Typically, the high degree of bioturbation has more to do with the amount of time available for biogenic activity per unit accumulation of sediment than with animal density or "frenzy" of activity.

An example of this relationship is in protected areas of estuaries and offshore areas below storm-wave base, where sedimentation is so slow that infaunal reworking of deposited sediment far exceeds the rate of introduction of new sediments. Even though the sediments become bioturbated throughout, most of the macroinvertebrate organisms do not move extensively through the substrate but use it only as an anchoring medium; these animals are dependent —directly or indirectly—on the overlying waters for respiration, nutrients, and food. The substrate in which they live is characterized by highly reduced sediments, in places beginning only a few millimeters below the sediment–water interface. When investigated in detail, even those organisms found alive at depths of 0.5 m or more in a core are almost invariably related to the surface by some sort of permanent burrow opening or siphon. [Although exceptions are known—such as *Sipunculus nudas* (Schäfer, 1972, p. 274), which is able to live in the sediment free of surface contact— such animals are relatively uncommon.] The resulting sedimentary structures record multiple overprints in which a confused assemblage of truncations and interruptions of animal traces are superimposed one upon another (see Fig. 17.1).

Fast Rate of Deposition

Next, consider the case of continuous rapid deposition on a large scale, such as transport and outfall of sediments by long-lived turbidity currents or splay processes, as well as more local conditions, such as laterally accreting point bars in tidal estuaries, or abrupt buildups of beaches and shoals associated with storms. In this context, an essential provision is that, following deposition, little or no erosion occurs, i.e., a "catastrophic" event, after which conditions return more or less to what they were before the event occurred. Following these criteria, the next important consideration in this part of the model is the thickness of the bed. If this thickness is greater than about 30 cm, one commonly sees little bioturbation or biogenic reworking, except in the upper few centimeters of the bed (Fig. 8.3). If bed thickness is less than 30 cm or so, the sequence of rapid sedimentation may be less obvious because of organisms that reestablish themselves in the bed following deposition, or move upward from below; given sufficient time, these completely destroy the original physical structures, as noted above.

Escape structures made by organisms caught below the rapidly deposited interval should be sought (e.g., Fig. 8.5), but the chance for organisms to succeed in penetrating the new unit of sedimentation is probably inversely related to the thickness of the material laid down above them. That very many macroinvertebrates are able to penetrate appreciably more than 30 cm of rapidly deposited sand is unlikely, although Schäfer (1972, p. 372) reported the ability of the polychaete *Aphrodite* to escape through 25 to 30 cm of sediment, and certain pelecypods have even greater prowess (Kranz, 1970). Frequently, however, the paucity of animals within lower parts of such a unit is made more obvious by dense burrowing activity recorded in the upper few centimeters of the bed. Examples of this situation in Holocene sediments of the La Jolla deep-sea fan were described by Piper and Marshall (1969).

In such situations of rapid deposition, the abruptly deposited units are commonly interbedded with units of considerably finer grained texture. Because of this contrast, two additional factors affect the ichnolog-

Fig. 8.3 Continuous rapid deposition, represented by clean, well-sorted, fine- to medium-grained sandstone interbedded with thin beds of highly bioturbated gray siltstone. Sandstone beds are large foresets deposited as a river mouth bar, and are as much as 1.5 m thick. Individual beds consist of parallel to subparallel laminations, bioturbate textures being very rare. Only at top of bedding units does one see any significant biogenic record, but here bioturbation was intense. Panther Sandstone Member, Star Point Formation (Upper Cretaceous), east-central Utah.

ical record. First, the contrasting textures permit preservation of many delicate biogenic structures that were present on the preexisting subjacent unit and on the upper surface of the depositional unit (see Chapter 4). This preservation potential is in strong contrast with situations in which new additions of sediment have approximately the same textural parameters as the old. Because of this differential preservation, and the necessity of most burrowing animals to maintain a connection with the sediment surface, trace fossils may be considered as essentially "interface phenomena" (Martinsson, 1970). The study of trace fossils, especially in flysch basins characterized by turbidite sedimentation, has received considerable impetus because of the many textural contrasts and good preservation in this type of depositional sequence. The abundance and environmental implications of trace fossils at these interfaces also has had significant impact on the interpretation of flysch facies, as exemplified especially by the classic work of Seilacher (1958, 1962, 1964) and more recently by Książkiewicz (1970) and Chamberlain (1971). (See also Chapter 7.)

A second aspect of this type of sedimentation is the influence that it may have in providing new substrates—unexploited —in a particular depositional environment. An example of this situation was cited by Howard (1972) from the Cretaceous of Utah, where thin beds of clean sand are found within a gray siltstone sequence.

Here the clean sands contain a strikingly different trace fauna than do the inter-bedded siltstones, and it is believed that the character of the fauna was affected almost exclusively by sediment texture; all other physical and chemical aspects of the environment evidently remained essentially unchanged.

Discontinuous Deposition

Variable Rates of Deposition: No Erosion

Under circumstances of discontinuous slow deposition, but no erosion, one probably could discern relatively little in the biogenic record differing from that resulting from slow, continuous sedimentation, because the organisms in either case would be more than able to keep pace with the addition of new sediment.

If thick beds were deposited rapidly, separated by intervals representing slow deposition but not erosion, one would find a stacking of essentially the same bedding types as those resulting from rapid continuous sedimentation. Under these conditions, the breaks in time are recorded by bedding planes that form because of the settling out of finer materials between times of major sediment accumulation. In the absence of such silt or shale breaks, one should look for features such as those pointed out by Seilacher (1964): horizons of concentrated burrows, or fecal pellet accumulations in units otherwise lacking any evidence of changes in sedimentation. Also to be documented very carefully is the *absence* of erosional indicators, such as truncated terminations of vertical burrows and bioturbation structures modified by scour.

Variable Rates of Deposition: With Erosion

Probably the most common depositional type is that in which discontinuous periods of sediment accumulation are preceded and followed by erosion (Fig. 8.4). Based on examinations of both modern (Howard and Reineck, 1972) and ancient examples (Howard, 1972; Goldring and Bridges, 1973), this situation results in one of the most common kinds of bedding unit found in the nearshore area of beach-to-offshore sequences; the cycle of erosion and deposition produces a very specific type of stratification. This sequence is discussed in detail in the papers referenced above and thus is only briefly recounted here. The bedding type referred to as "parallel-laminated to burrowed" (Howard, 1971b)

Fig. 8.4 Discontinuous sedimentation, represented by muddy, fine-grained sandstone consisting of beds recording variable amounts of biogenic reworking. Beds range in thickness from 30 to 45 cm and have erosional upper and lower contacts; in each, bioturbate textures increase in abundance from base to top. Upper surface characterized by erosional truncation of burrow structures. This type of sedimentation seems to be typical of nearshore areas at or just below wave base. Spring Canyon Member, Blackhawk Formation (Upper Cretaceous), east-central Utah.

is characterized by erosional upper and lower contacts. The lower part of the bed lacks, or is impoverished in, recognizable biogenic features and is characterized by physical sedimentary structures—generally in the form of parallel-laminated sand. Bioturbate textures generally increase upward within the bed, and gradually become the dominant sedimentary structures. The top of the bed is recognized by erosional truncations of distinct burrows. The origin of this bed type is believed to be storm activity, which erodes the sea floor and resuspends sediments during the storm build-up, followed by redeposition as the storm subsides. Following this episode, the disturbed infauna is reestablished, as is a slower rate of sedimentation.

The example just cited represents the median of two end points: greater and lesser rates of sedimentation, erosion, and biogenic reworking. To find less rapidly deposited discontinuous sequences interrupted by erosion, one has only to move farther seaward; here the bedding units are also separated by erosional planes, but the beds are completely reworked biogenically. Farther landward, one sees evidence of more abrupt, discontinuous sequences, and here the parallel laminations comprise a greater percentage of the bed—at the expense of biogenic structures.

Burrows and borings in hardground (carbonate) discontinuity surfaces are also distinctive and important, as recounted in Chapter 18.

RELATIVE RATES OF SEDIMENTATION AND EROSION

In addition to serving as indicators of overall or general environmental conditions and subtle changes in these depositional conditions, trace fossils can give important clues concerning local episodes of deposition and erosion.

Examples of animal response to both sedimentation and erosion are frequently seen in the sedimentary record; some structures record a combination of both events through the apparent lifetime of an individual. The best, most sensitive indicators of these processes are individual structures and distinct burrow systems, or "Fossitextura Figurativa" of Schäfer (1972, p. 404), as opposed to general patterns of bioturbation; as discussed above, the latter illustrate broad, general conditions of deposition and erosion.

In recent years, numerous geologists have recognized the responses of invertebrate organisms to local conditions of deposition and erosion. An excellent brief summary is included in Seilacher's 1964 paper, in which he cited several examples of vertical movement by organisms in response to sedimentation, a classic example being that of ophiurans from the Lower Triassic Seisic Beds of Tyrol, Austria (Fig. 8.5D).

Some specific structures have received particular study with regard to their record of sedimentation because they are very commonly found in an upward building configuration. One thus assumes that animals constructing these forms were particularly adapted to environments of rapid deposition. These examples include *Teichichnus*, as discussed by Chisholm (1970) from the Lower Carboniferous of Scotland, and *Skolithos* from Cambrian beds of Scotland (Hallam and Swett, 1966) and Wyoming (Boyd, 1966).

Of particular interest has been the attention devoted to the study of vertical response movements of invertebrate animals in modern sediments. Studies of this type are summarized in detail in the writings of Schäfer (1962, 1972). In his two books, the first a German edition and the second an English translation that includes an update of more recent work, Schäfer cited dozens of examples of how organisms have responded to conditions of deposition and erosion. Although these studies are primarily from examples in the North Sea, they must stand as basic references for examples of escape structures made by burrowing organisms. Other recent work, which en-

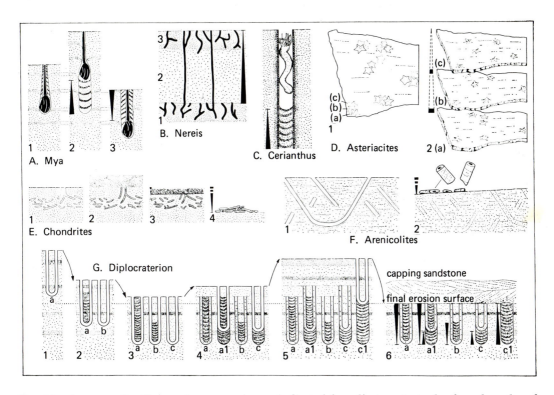

Fig. 8.5 Amount of sedimentation or erosion as indicated by adjustments to depth and modes of preservation of various lebensspuren. Heights of solid arrows show amount of deposition or erosion. A, movement pattern of pelecypod *Mya*, which has a single siphon. With stationary sedimentary surface (1), growing organism gradually burrows deeper; bottom of structure wider than top. With rapid sedimentation (2), organism migrates toward surface, leaving infilled burrow the width of the shell. With degradation of surface (3), organism migrates downward, leaving burrow of same width but having different internal structure. B, movement pattern of polychaete worm *Nereis*. Older colonized surface (1) is rapidly covered by sediment (2), and during deposition, paths of escape are directed upward. With stabilization, new colonization surface (3) has irregular "normal" burrows. Structures in (1) and (3) are generally mucus lined; in (2) they are unlined. C, movement pattern of anemone *Cerianthus*, an organism dwelling in a single tube. With sedimentation, animal moves upward, leaving an unfilled or passively filled burrow. A similar pattern might be expected in traces such as *Skolithos* and *Monocraterion*. D, movement represented by trace fossil *Asteriacites lumbricalis*, resting place of a stelleroid. With sedimentation, animal migrated upward, in stages a–c; combined (1) and separate (2) plan of all impressions. E, preservation patterns of trace fossil *Chondrites*. Tunnel system (1) is infilled (2), following a change in type of sediment being deposited (bed-junction preservation). Slight degradation of surface (3) removes the proximal shafts before further sediment, of a different type, accumulates (concealed bed-junction preservation). Renewed degradation of surface winnows away sediment, leaving mucus-lined infilled tunnels as burial preservations (4). F, preservation pattern of trace fossil *Arenicolites curvatus*. Sedimentary surface containing open U-tubes (1) has been degraded (2), the mucus-cemented tube fragments accumulating in an intraformational conglomerate; sediment has filled the tubes. G, movement pattern represented by trace fossil *Diplocraterion yoyo*. In Upper Devonian Baggy Beds of North Devon, this trace occurs in various configurations shown in (6); all have been truncated to common erosion surface. Repeated phases of erosion and sedimentation (1–5) evidently led to development of the various types. (1), development of burrow (a). With degradation of surface, this tube migrates downward, and at intervals, new tubes (b, c) are constructed (2, 3). Sedimentation follows (4, 5), but some tubes are abandoned. (6), all tubes are abandoned, and erosion reduces them to a common base. (From Goldring, 1964.)

compasses other environments, includes the studies of reef lagoons by Shinn (1968) and Farrow (1971), both of whom showed examples of vertical movement in carbonate sediments by organisms in response to conditions of rapid deposition. Shinn also cited parallel examples from ancient carbonates.

Examples of combined erosion and deposition reflected by specific trace fossils are not abundant in the literature but have been noted by several writers. Perhaps this phenomenon has been encountered but overlooked because of (1) the difficulty of recognizing the evidence for this combined activity, (2) the relatively short life-span of many invertebrate organisms relative to the duration of erosion and deposition, and (3) the paucity of specific environments that possess the physical and biogenic conditions necessary to permit this sequence of events to be recorded clearly.

The best description and summary of this combined process is that published by R. Goldring in two papers (see Fig. 8.5); he described responses to erosion and deposition represented by the aptly named *Diplocraterion yoyo* from the Upper Devonian Baggy Beds of England (Goldring, 1962), and then (1964) compared these with other examples in the literature on erosional and depositional responses. Another example is that cited by Howard (1971b) of the episodes recorded by an *Ophiomorpha* burrow in a rapidly prograding Pleistocene beach sequence; the burrow reflects five periods of erosion and deposition within a vertical sequence of 20 cm.

One point to bear in mind is that the scour horizon does not necessarily coincide with the top end of truncated burrows. Some burrows have reinforced walls that make them more resistant to erosion than the adjacent substrate. These burrows do become truncated by scour, but their tops may still project slightly above the main substrate surface (Frey, 1968, Text-fig. 2). After gentle scour, truncated burrows of *Callianassa major* have been seen to stand

5 to 8 cm above the surface of protected sand flats in Georgia. In the rock record, particularly where seen in thin, poor exposures of rock, these could be mistaken for post-erosion burrows because they do, in fact, "cross-cut" the main erosional plane. This phenomenon is, of course, more common in low-energy environments than in high-energy ones.

CURRENT INDICATORS

The role of trace fossils as indicators of relative current strength and direction has been noted by several writers. Examples of this relation are shown as (1) direct evidence—as where tracks, trails, and burrows acted as baffels to form crescent or shadow deposits in the lee of the structure, and (2) indirect evidence—as where animal tracks and trails show interruption or deflection of the animal by currents.

One of Seilacher's earlier papers (1955) beautifully illustrated the diversion of animals (and thus their trails) by currents. In this example, Seilacher reconstructed the sinuous path followed by a trilobite moving across a current-swept substrate. (See also Fig. 6.8.) Subsequently, the lebensspuren of current-affected trilobites have been mentioned in several papers, the most noteworthy of which is that by Birkenmajer and Bruton (1971); they discussed in detail several current-induced trilobite responses and the resulting markings.

The development of sediment accumulations in the lee of positive and negative features at the sediment–water interface have long been recognized as indicators of current activity. In many instances, such projections were formed biogenically. Several examples of specific organisms which, due to the effect of currents, left markings in the substrate were given by Seilacher (1953, 1961, 1964), Schäfer (1972), and Bandel (1967).

Indirect evidence of currents can also be derived from trace fossils in other ways. These include the fillings of abandoned burrows by sediments of coarser grain size than that in which the burrows are developed, as cited by Frey (1970b, Pl. 8, fig. 2) from the Fort Hays of Kansas. Also, as pointed out by Frey (1970b, p. 33), the very presence of burrows of filter-feeding organisms indicates the occurrence of some type of water movement in order to supply these organisms with nutrients necessary for their survival.

TRACE FOSSILS AS INDICATORS OF SUBSTRATE CONSISTENCY

The nature of the original fluid content of ancient sediments is frequently indicated by careful study of the associated trace fossils and bioturbation structures. Although long recognized that animal distribution is in large part controlled by the character of the substrate, only in recent years has attention been focused on the significant influence that organisms have on determining the mass properties of the substrates that they inhabit. One of the principal contributors to this subject is Rhoads (1970; Rhoads and Young, 1970; see also Chapter 9).

In the examination of ancient sedimentary sequences, one has numerous opportunities to estimate the relative consistency of sediments at the time the organisms were present, by observing the diversity and preservation of burrows present and the nature of biogenic activity. The type of burrowing itself can indicate something about the nature of the substrate. As discussed by Purdy (1964), in his analyses of the geologic significance of studies by Sanders (1956, 1958), a "textural continuum" exists in depositional environments, from substrates composed completely of sand to those composed completely of clay-size material. Two examples of this range can be drawn from the "ends" of the textural continuum. At the relatively high-energy end, we see clean, well-sorted sands and conditions of frequent erosion and deposition. Under these circumstances,

the environment is characterized by few species and few individuals, most of which are suspension feeders that commonly build deep, well-constructed burrows. At the opposite end of the spectrum, the communities are composed almost exclusively of deposit feeders; but here also are relatively few species and few individuals, because of the problems of locomotion and other activities in this type of muddy substrate. Examples of this relation applied to modern and ancient sedimentary sequences were cited in the first part of this chapter.

More direct evidence of substrate consistency is seen in the specific tracks, trails, and biogenically formed impact structures preserved on bedding planes. Examples include instances in which flow structures formed before burrowing (Howard, 1971b) and after burrowing (Kennedy and MacDougall, 1969), as well as the process of subaerial burrowing in dry sand (Hanley et al., 1971). Frey (1968, 1970a) cited examples from recent sediments suggesting that where filter feeders are found in thixotropic sediments, they overcome sediment instability by building thick burrow walls. Even in the study of deep-sea sediments, the consistency of the substrate can often be discerned readily by examination of photographs, many of which show the surface traces and burrow openings of deep-water organisms (Ewing and Davis, 1967; Heezen and Hollister, 1971). (See Chapter 21.)

In one way or another, substrate coherence is probably one of the main factors controlling the distribution of trace-making organisms—even more so than sediment composition or grain size (Frey and Howard, 1970).

SEDIMENT SORTING BY BIOGENIC ACTIVITY

Although reworked sediment does not itself constitute a trace fossil, the burrowing activity of organisms has been shown in several instances to be responsible for grain sorting—an important sedimentological process that deserves mention here. Animals that are sediment ingestors pass vast amounts of sediment through their digestive systems and in the process extract whatever nutrients they can from the substrate. As a result of this activity, a large amount of material is transported and redeposited, especially locally. One has only to look at a modern tidal flat to appreciate the magnitude of the process (Fig. 8.6). Commonly such organisms are oriented in such a way that they move material from depths of several centimeters to the sediment–water interface. Although many of these organisms are relatively nonselective feeders, they are restricted by the size of materials that they are able to ingest. Thus, in sediment containing a wide variety of grain sizes, they are forced to reject particles too large to be manipulated by their jaws or guts. These large sedimentary fragments are left behind, and if sufficient burrowing activity is accomplished, the particles eventually become concentrated at a specific depth—the phenomenon of biogenic graded bedding. An excellent example of this phenomenon is the shell beds of the North Sea tidal flats, described by van Straaten (1952, 1954, 1956), in which valves of the mollusk *Hydrobia* have been concentrated in beds 2 to 3 cm thick, at depths of 20 to 30 cm below the sediment surface, owing to burrowing activity of the sand worm *Arenicola*. Similar situations, but involving other organisms, were cited by Rhoads and Stanley (1965), Rhoads (1967), and Warme (1967).

Shinn (1968) pointed out a different kind of sediment-sorting action, initiated

Fig. 8.6 Muddy sand flat at low tide, showing profusion of fecal castings by the sediment-ingesting animals *Balanoglossus* (enteropneust) and *Leptosynapta* (holothurian). Castings of the two cannot be distinguished, from the surface. This is the ecological niche of *Arenicola* (polychaete) on North Sea tidal flats. Cabretta Island, Georgia. (Photo by R. W. Frey; see Howard and Dörjes, 1972.)

by the burrowing shrimp *Callianassa*. In the process of building a complex system of burrow galleries, *Callianassa* pumps considerable sediment out of the burrow to the sediment–water interface. In the presence of even slight currents, this excavated material is sorted during transport; the coarser fraction is deposited near the burrow, and the finer, clay fraction remains longer in suspension and is transported away from the immediate area of the burrow.

Another way in which organisms may be responsible for concentrations of a particular sediment size or composition has to do with organisms that build their burrows of specific, selected particles and thus incorporate exotic detritus into the depositional sequence. The resulting record of these habits may be local concentrations of unusual sediment where abandoned burrows formerly existed, or concentrations where numerous burrows have been eroded, transported, and concentrated by wave or current action.

The ability of animals to selectively pick detrital materials and cement them to their burrow walls has been noted in recent sediments by several workers (e.g., Fager, 1964; Frey and Howard, 1969; Myers, 1970; and Schäfer, 1972). Recognition of such features in the fossil record has been less common, although Frey (1971) suggested this as a possible origin of a trace fossil reported by Howell (1953), and Roniewicz

(1970) described a similar tube built of foraminiferal tests.

CONCLUSIONS

The sedimentologic significance of trace fossils rests in large part on the ability of the investigator to look beyond the paleontologic character of lebensspuren and to consider traces as the result of biogenic processes that are contemporaneous or penecontemporaneous with physical processes. As such, they add another useful dimension to the repertory of facts that the investigator can gather in his attempt to reconstruct geologic history.

Trace fossils thus give direct evidence of biogenic activities that are fundamental processes operating in depositional environments. In addition, biogenic activity can act upon, or be modified by, associated physical processes and thereby substantially increase our ability to recognize indications of long- and short-term conditions of local and regional depositional regimes.

ACKNOWLEDGMENTS

The ideas and interpretations expressed in this chapter have been influenced and strengthened by discussions with numerous colleagues, especially by R. W. Frey and H.-E. Reineck, whom I wish to thank. Research support by the Oceanography Section, National Science Foundation, grant GA-39999X, is gratefully acknowledged.

REFERENCES

Ager, D. V. and P. Wallace. 1970. The distribution and significance of trace fossils in the uppermost Jurassic rocks of the Boulonnais, France. In T. P. Crimes and J. C. Harper (eds.), Trace fossils. Geol. Jour., Spec. Issue 3:1–18.

Bandel, K. 1967. Isopod and limulid marks and trails in Tonganoxie Sandstone (Upper Pennsylvanian) of Kansas. Univ. Kansas Paleont. Contr., Paper 19:1–10.

Birkenmajer, K. and D. L. Bruton. 1971. Some

trilobite resting and crawling traces. Lethaia, 4:303–319.

Boyd, D. W. 1966. Lamination deformed by burrows in Flathead Sandstone (Middle Cambrian) of central Wyoming. Contr. to Geol., 5:45–54.

Chamberlain, C. K. 1971. Bathymetry and paleoecology of Ouachita geosyncline of southeastern Oklahoma as determined from trace fossils. Amer. Assoc. Petrol. Geol., Bull., 55:34–50.

Chisholm, J. I. 1970. *Teichichnus* and related trace fossils in the Lower Carboniferous, St. Monance, Scotland. Geol. Survey Great Britain, Bull. 32:21–51.

Clifton, H. E. and R. E. Hunter. 1973. Bioturbation rates and effects in carbonate sand, St. John, Virgin Islands, U.S.A. Jour. Geol., 81:253–268.

Evans, G. 1965. Intertidal flat sediments and their environments of deposition in the Wash. Geol. Soc. London, Quart. Jour., 121:209–245.

Ewing, M. and R. A. Davis. 1967. Lebensspuren photographed on the ocean floor. In J. B. Hersey (ed.), Deep sea photography. Johns Hopkins Oceanogr. Stud., 3:259–294.

Fager, E. W. 1964. Marine sediments: effects of a tube building polychaete. Science, 143:356–359.

Farrow, G. E. 1971. Back-reef and lagoonal environments of Aldabra Atoll distinguished by their crustacean burrows. Zool. Soc. London, Symp., 28:455–500.

Frey, R. W. 1968. The lebensspuren of some common marine invertebrates near Beaufort, North Carolina. I. Pelecypod burrows. Jour. Paleont., 42:570–574.

————. 1970a. The lebensspuren of some common marine invertebrates near Beaufort, North Carolina. II. Anemone burrows. Jour. Paleont., 44:308–311.

————. 1970b. Trace fossils of Fort Hays Limestone Member of Niobrara Chalk (Upper Cretaceous), west-central Kansas. Univ. Kansas Paleont. Contr., Art. 53, 41 p.

————. 1971. Ichnology—the study of fossil and recent lebensspuren. In B. F. Perkins (ed.), Trace fossils, a field guide. Louisiana State Univ., School Geosci., Misc. Publ. 71-1:91–125.

———— and J. D. Howard. 1969. A profile of biogenic sedimentary structures in a Holocene barrier island—salt marsh complex. Gulf Coast Assoc. Geol. Socs., Trans., 19:427–444.

———— and J. D. Howard. 1970. Comparison of Upper Cretaceous ichnofaunas from siliceous sandstones and chalk, western interior region, U.S.A. In T. P. Crimes and J. C. Harper (eds.), Trace fossils. Geol. Jour., Spec. Issue 3:141–166.

Goldring, R. 1962. The trace fossils of the Baggy Beds (Upper Devonian) of North Devon, England. Paläont. Zeitschr., 36:232–251.

————. 1964. Trace fossils and the sedimentary surface in shallow water marine sediments. In L. M. J. U. van Straaten (ed.), Deltaic and shallow marine deposits. Developments in Sedimentology, 1:136–143.

———— and P. Bridges. 1973. Sublittoral sheet sandstones. Jour. Sed. Petrol., 43:736–747.

Hallam, A. and K. Swett. 1966. Trace fossils from the Lower Cambrian Pipe Rock of the north-west Highlands. Scottish Jour. Geol., 2:101–106.

Hanley, J. H. et al. 1971. Trace fossils from the Casper Sandstone (Permian), southeastern Laramie Basin, Wyoming and Colorado. Jour. Sed. Petrol., 41:1065–1069.

Hanor, J. S. and N. F. Marshall. 1971. Mixing of sediment by organisms. In B. F. Perkins (ed.), Trace fossils, a field guide. Louisiana State Univ., School Geosci., Misc. Publ. 71-1:127–135.

Heezen, B. C. and C. D. Hollister. 1971. The face of the deep. New York, Oxford Univ. Press, 659 p.

Howard, J. D. 1966. Characteristic trace fossils in Upper Cretaceous sandstones of the Book Cliffs and Wasatch Plateau: Utah Geol. Mineral. Survey, Bull. 80:35–53.

————. 1971a. Comparison of the beach-to-offshore sequence in modern and ancient sediments. In J. D. Howard et al., Recent advances in paleoecology and ichnology. Short Course Lect. Notes, Amer. Geol. Inst., p. 148–183.

————. 1971b. Trace fossils as paleoecologic tools. In J. D. Howard et al., Recent advances in paleoecology and ichnology. Amer. Geol. Inst., Short Course Lect. Notes, p. 184–211.

————. 1972. Trace fossils as criteria for recognizing shorelines in stratigraphic record. In J. K. Rigby and W. K. Hamblin (eds.), Recognition of ancient sedimentary environments. Soc. Econ. Paleont. Mineral., Spec. Publ. 16:215–225.

———— and J. Dörjes. 1972. Animal–sediment relationships in two beach-related tidal flats; Sapelo Island, Georgia. Jour. Sed. Petrol., 42:608–623.

———— and R. W. Frey. 1973. Characteristic physical and biogenic sedimentary structures in Georgia estuaries. Amer. Assoc. Petrol. Geol., Bull., 57:1169–1184.

———— and H.–E. Reineck. 1972. Georgia coastal region, Sapelo Island, U.S.A.: sedimentology and biology. IV. Physical and biogenic sedimentary structures of the nearshore shelf. Senckenbergiana Marit., 4:81–123.

———— et al. 1973. Physical and biogenic characteristics of sediments from the outer Georgia continental shelf (abs.). Amer. Assoc. Petrol. Geol., Bull., 57:784.

Howell, B. F. 1953. A new terebellid worm from the Carboniferous of Texas. Wagner Free Inst. Sci., Bull., 28:1–4.

Kennedy, W. J. and J. D. S. MacDougall. 1969. Crustacean burrows in the Weald Clay (Lower Cretaceous) of south-eastern England and their environmental significance. Palaeontology, 12:459–471.

Kranz, P. M. 1970. Bivalve escape behavior as an indication of sedimentary rates and environments (abs.). Geol. Soc. America, Abs. Prog., 2(7):599.

Książkiewicz, M. 1970. Observations on the ichnofauna of the Polish Carpathians. In T. P. Crimes and J. C. Harper (eds.), Trace fossils. Geol. Jour., Spec. Issue 3:283–322.

Kuenen, P. H. 1961. Some arched and spiral structures in sediments. Geol. en Mijnbouw, 40:71–74.

Martinsson, A. 1970. Toponomy of trace fossils. In T. P. Crimes and J. C. Harper (eds.), Trace fossils. Geol. Jour., Spec. Issue 3: 323–330.

Moore, D. G. and P. C. Scruton. 1957. Minor internal structures of some recent unconsolidated sediments. Amer. Assoc. Petrol. Geol., Bull., 41:2733–2751.

Myers, A. C. 1970. Some palaeoichnological observations on the tube of *Diopatra cuprea* (Bosc): Polychaeta, Onuphidae. In T. P. Crimes and J. C. Harper (eds.), Trace fossils. Geol. Jour., Spec. Issue 3:331–334.

Piper, D. J. W. and N. F. Marshall. 1969. Bioturbation of Holocene sediments on La Jolla deep sea fan, California. Jour. Sed. Petrol., 39:601–606.

Purdy, E. G. 1964. Sediments as substrates. In J. Imbrie and N. D. Newell (eds.), Approaches to paleoecology. New York, John Wiley, p. 238–271.

Reineck, H.–E. 1967. Parameter von Schichtung und Bioturbation. Geol. Rundschau, 56: 420–438.

———— (ed.). 1970. Das Watt, Ablagerungs– und Lebensraum. Frankfurt, W. Kramer, 142 p.

———— and I. B. Singh. 1971. Der Golf von Gaeta (Tyrrhenisches Meer). III. Die Gefüge von Vorstrand und Schelfsedimenten. Senckenbergiana Marit., 3:185–201.

Rhoads, D. C. 1967. Biogenic reworking of intertidal and subtidal sediments in Barnstable Harbor and Buzzards Bay, Massachusetts. Jour. Geol., 75:461–476.

————. 1970. Mass properties, stability and ecology of marine muds related to burrowing activity. In T. P. Crimes and J. C. Harper (eds.), Trace fossils. Geol. Jour., Spec. Issue 3:391–406.

———— and D. J. Stanley. 1965. Biogenic graded bedding. Jour. Sed. Petrol., 35:956–963.

———— and D. K. Young. 1970. The influence of deposit-feeding organisms on sediment stability and community trophic structure. Jour. Marine Res., 28:150–178.

Roniewicz, P. 1970. Borings and burrows in the Eocene littoral deposits of the Tatra Mountains, Poland. In T. P. Crimes and J. C. Harper (eds.), Trace fossils. Geol. Jour., Spec. Issue 3:439–446.

Sanders, H. L. 1956. Oceanography of Long Island Sound, 1952–1954. X. The biology of marine bottom communities. Bingham Oceanogr. Coll., Bull., 15:345–414.

————. 1958. Benthic studies in Buzzards Bay. I. Animal–sediment relationships. Limnol. Oceanogr., 3:245–258.

Schäfer, W. 1962. Aktuo-paläontologie nach Studien in der Nordsee. Frankfurt, W. Kramer, 666 p.

————. 1972. Ecology and palaeoecology of marine environments. Edinburgh and Chicago, Oliver & Boyd and Univ. Chicago Press, 568 p.

Seilacher, A. 1953. Der Brandungssand als Lebensraum in Vergangenheit und Vorzeit. Natur u. Volk, 83:263–272.

————. 1955. Spuren und Lebensweise der Trilobiten; Spuren und Fazies im Unterkambrium. In O. H. Schindewolf and A. Seilacher, Beiträge zur Kenntnis des Kambriums in der Salt Range (Pakistan). Akad. Wiss. u. Lit. Mainz, math.–naturw. Kl., 10:342–399.

―――――. 1958. Zur ökologischen Charakteristik von Flysch und Molasse. Eclogae Geol. Helvetiae, 51:1062–1078.

―――――. 1961. Krebse im Brandungssand. Natur. u. Volk, 91:257–264.

―――――. 1962. Paleontological studies on turbidite sedimentation and erosion. Jour. Geol., 70:227–234.

―――――. 1964. Biogenic sedimentary structures. In J. Imbrie and N. D. Newell (eds.), Approaches to paleoecology. New York, John Wiley, p. 296–316.

Shinn, E. A. 1968. Burrowing in recent lime sediments of Florida and the Bahamas. Jour. Paleont., 42:878–894.

van Straaten, L. M. J. U. 1952. Biogenic textures and the formation of shell beds in the Dutch Wadden Sea. Koninkl. Nederl. Akad. Wetensch., Proc., 55:500–516.

―――――. 1954. Composition and structure of recent marine sediments in the Netherlands. Leidse Geol. Meded., 19:1–110.

―――――. 1956. Composition of shell beds formed in the tidal flat environment in the Netherlands and in the Bay of Arcachon (France). Geol. en Mijnbouw, 18:209–226.

Warme, J. E. 1967. Graded bedding in the recent sediments of Mugu Lagoon, California. Jour. Sed. Petrol., 37:540–547.

CHAPTER 9

THE PALEOECOLOGICAL AND ENVIRONMENTAL SIGNIFICANCE OF TRACE FOSSILS

DONALD C. RHOADS

Department of Geology and Geophysics, Yale University
New Haven, Connecticut, U.S.A.

SYNOPSIS

Trace fossils have great paleoecologic utility because they are (1) widespread in space and time, (2) found in place, and (3) largely the record of animal behavior and response, making them ideal indicators of environmental conditions.

Traces may be used together with body fossils to increase knowledge of taxonomic richness in an ancient biotic assemblage. The morphology of traces may be used to reconstruct modes of feeding by many trace-producing organisms. The orientation of traces within sediments is sensitive to such depth-related variables as salinity, temperature, and food supply. Adaptation of many shallow-water benthos for vertical burrowing reflects the effectiveness of sedimentary cover as a temperature and salinity "buffer." Gradients in dissolved oxygen may be detected by size gradients in burrow diameter. Low-oxygen marine environments are charac-terized by small-diameter horizontal feeding traces.

Primary deformational features of traces may be used to reconstruct the original water content of the sea floor (bottom hardness). In addition, the presence or absence of shallow burrows, the preserved evidence of vertical movement reflected by specific types of traces, and gradients in intensity of bioturbation may be used to reconstruct relative and, in some cases, absolute rates of sedimentation or erosion.

The greatest value of trace fossils in paleoecologic reconstruction is realized when they are used as independent evidence brought together with other sources of paleoecologic data. An interpretive, environmental model is proposed here in which information gained from the relative abundance of body and trace fossils is integrated; in this model, eight biolithofacies are recognized.

INTRODUCTION

Trace fossils are more widely distributed through the geologic column than are invertebrate body fossils. Traces commonly accompany body fossils and are also encountered in sedimentary units devoid of benthic body fossils. Historically, paleontologists commonly made only passing reference to such "barren" intervals, while concentrating their efforts on body-fossil-bearing strata. In other instances, trace fossils may be overlooked in the field because they are inconspicuous. Although many traces are easily observed on weathered outcrop surfaces, special techniques—such as staining, etching, or radiography—may be required to reveal others (see Chapter 23); furthermore, small-diameter traces common to many lutites may be observed only in rock thin-sections, acetate peels, or prepared rock surfaces.

The widespread distribution of traces in both space and time, and their in situ nature,[1] makes this category of fossils attractive in paleoecology. In addition, because most organic traces are records of the behavior of their producers, these preserved responses are useful in environmental reconstruction, especially in sediments containing few, or no, body fossils. The purpose of this chapter is thus to demonstrate the utility of trace fossils in "community" paleoecology and environmental reconstruction. Autoecologic (ethologic) information derived from trace morphology is discussed in Chapter 6 and elsewhere.

USE OF TRACE FOSSILS IN RECONSTRUCTING BIOTIC ASSEMBLAGES

Taxonomic Diversity

A fundamental ecologic parameter in recent benthic research is the number of individuals per taxon per unit sample of sea floor. This parameter is usually expressed as a diversity index, and is used to characterize the taxonomic structure and temporal stability of a benthic community (Chapter 20; see Sanders, 1968, for a discussion of diversity measures).

Soft muddy seafloors are populated by abundant soft-bodied invertebrates. These organisms, if preserved at all, are recorded as tracks, trails, or burrows. The accuracy of estimating species abundance in a fossil community may be increased significantly by including soft-bodied taxa (represented by trace fossils) with the list of body fossils. Unpublished work by me on a recent sandy, intertidal area in Barnstable Harbor, Massachusetts, shows that the ratio of total living invertebrate species to the number of dead shell-bearing taxa ($\geqq$ 1 mm in size) is 10:7, i.e., by observing only dead shell material, the accuracy of reconstruc-

tion is 70 percent. If traces are also included, the accuracy may be increased to about 90 percent.

Johnson (1964) summarized data on the range of percentage of organisms possessing hard parts on different bottom types. These data are a compilation of 534 samples from boreal shallow-water environments. The percentages range from 10 to 15 percent for mud, 7 to 35 percent for muddy sand, 21 to 50 percent for clean sand, and 7 to 67 percent for gravel. Although the range in percentage is great, muds tend to have fewer numbers of preservable taxa than do coarser grained bottoms. Traces are preserved well in lutites, providing the paleontologist an opportunity to increase his resolution even in those environments having the poorest potential body fossil record. (See also Chapter 2.)

Some care in interpretation must be taken in this approach. Size, behavior, and morphology may change throughout the ontogeny of an organism, and these changes may produce a marked difference in the trace made by a particular organism during its life-span (Hertweck, 1970). Organisms represented by body fossils can also leave traces that may not be readily ascribed to them. Different species can produce traces so similar that differentiation may not be possible. In addition, a single species may produce different traces; the morphology of ghost crab burrows has been shown to depend not only upon the age of the crab but also on the location of the burrow in foreshore, backshore, or dune environments (Frey and Mayou, 1971). All these factors can complicate reconstruction of an ichnocoenose.

Faunal and Floral Density

The meaning of absolute abundance of trace or body fossils, in terms of the abundance of living organisms at any particular time (standing crop), is elusive. For this reason, measures of biotic density—numbers of individuals per taxon—are

---

[1] Dwelling tubes may be exhumed and transported intact or as fragments but are generally recognized as such (see Chapter 2).

usually not calculated for fossil assemblages. Relative abundances may be used cautiously, however, to estimate common and rare species during life of the assemblage. Opportunistic species tend to dominate the fossil record because birth and death rates are high (e.g., Levinton and Bambach, 1970). The high turnover rate of these populations of characteristically short-lived species may produce shell layers and (or) abundant tubes and burrows. Such populations are usually associated with new or recently changed habitats. Highly variable or unpredictable environments · are commonly populated by these adaptive types.

An intensively burrowed stratigraphic unit may tempt speculation about the original densities of the burrow producers. However, the density of traces, like body fossils, is a function of both turnover rate of the population and sedimentation rate. Recent studies by Rhoads (1967) also indicate that the rate of bioturbation is generally poorly correlated with faunal density but is closely related to the mobility of the burrowing or grazing organisms.

Trophic Diversity

Another important ecologic parameter in benthic work is the proportioning of feeding types among the constituent species (Walker, 1972; Rhoads et al., 1972; Walker and Bambach, 1974). The distribution of herbivores, carnivores, or scavengers, especially suspension-feeding and deposit-feeding benthos, can provide useful information about food resources, relative sedimentation rates, water turbidity, and sea-floor stability (Rhoads et al., 1972). Trace morphology may be used to reconstruct the general mode of feeding of the organism producing the trace. Schäfer (1972) summarized examples of feeding traces produced by recent trophic types from the North Sea.

Walker and Laporte (1970) provided an excellent example of how trace fossil data, integrated with body fossil data, can increase accuracy in paleoenvironmental re-

construction. In a comparative study between Ordovician and Devonian carbonate sequences in upper New York State, the authors recognized 15 common invertebrate fossils that represented the bulk of the fauna. Five of these taxa are trace fossils. Analysis of the morphology of the traces indicates that two were produced by infaunal, sessile suspension feeders and three by vagile, infaunal deposit feeders or scavengers. Frey (1972), in his environmental reconstruction of the Fort Hays Member of the Niobrara Chalk, augmented sparse body fossil data with information from abundant and diverse traces. Most of the animals present were apparently soft-bodied, living in a soft coccolith ooze.

USE OF TRACE FOSSILS IN RECONSTRUCTING ANCIENT ENVIRONMENTS

In both terrigenous and carbonate sedimentary settings, the distribution of trace fossils in marine environments seems, in most cases, to be strongly related to depth, i.e., the most striking changes in trace fossil suites occur perpendicular to depositional strike, except along estuarine salinity gradients (Fig. 9.1A). Examples were given by Seilacher (1964, 1967), Farrow (1966), Mc-Alester and Rhoads (1967), and Rodriguez and Gutschick (1970).

Causes for this bathymetric zonation may be related to several depth-dependent factors (Fig. 9.1B). Seilacher (1967) attributed the depth zonation to partitioning of feeding types as related to food resources. Figure 9.1A shows Seilacher's bathymetric zonation of trace fossil "facies." These ichnocoenoses are named after characteristic trace genera. The *Scoyenia* "facies" is associated with red beds and other nonmarine deposits. Vertical tubes also dominate the *Skolithos* "facies" in very shallow, if not intertidal, water. *Glossifungites* traces are borings or excavations into "hardgrounds," erosion surfaces, or semiconsolidated mud. *Glossifungites* assemblages are

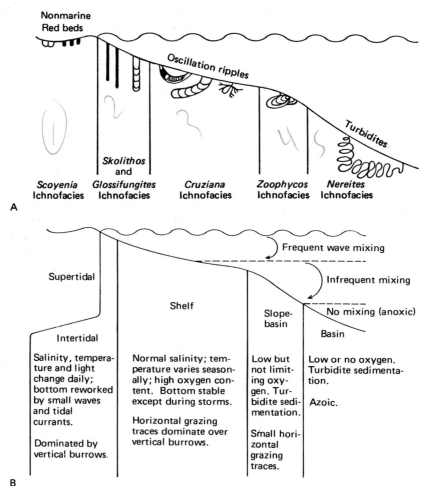

Fig. 9.1 Cross section of geosynclinal shelf and basin, showing (A) trace fossil assemblages (after Seilacher, 1967), and (B) gradients in important ecologic parameters. (Compare with Table 2.1 and Fig. 7.2.)

not, therefore, restricted to this depth zone. Recognition of boring tunnels of endolithic algae can assist in documenting shallow-water carbonate environments because these algae are distributed in a relatively narrow zone near the strand line (Golubic and Schneider, 1972; Chapter 12). Bromley (1970) and Warme (Chapter 11) summarized recent and fossil boring taxa. The *Cruziana* "facies" is associated with the wave-influenced shelf and consists of shallow resting, feeding, and grazing traces. The *Zoophycos* and *Nereites* "facies" are characteristic of deeper water and consist primarily of systematic grazing traces.

Mixed facies may also be found, in which traces of two or more of the above assemblages are present (Frey and Chowns, 1972). (See Chapter 7.) Hecker (1970) documented a shallow-water habitat for the *Zoophycos* animal in the Carboniferous of the Russian Platform. Similarly, Osgood and Szmuc (1972) found that *Zoophycos* in the Mississippian of Ohio occurs in shallow-water deposits, and suggested that the form requires closer examination regarding its range of both morphologic variations and depth distribution.

This bathymetric zonation of burrow types was suggested by Seilacher (1967) to reflect primarily a transition in feeding types, from nearshore suspension feeders to offshore deposit feeders. His conclusion is based on the idea that food is in suspension in the nearshore, wave-agitated zone, whereas sedimentary particles are deposited in quiet, deeper water in offshore areas. The spatial separation of suspension and deposit feeding benthos in recent shallow seas is well documented, but the cause is not related solely to food partitioning. Recent studies in Cape Cod Bay and Buzzards Bay, Massachusetts, and Long Island Sound show that suspension feeders (tube-dwelling amphipods and filtering bivalves) are excluded from deeper water muds, in some cases, by physical instability of the bottom. Reworking of muds by deposit feeders (primarily errant polychaetes and protobranch bivalves) can produce a fluid sediment surface that is easily resuspended, even by weak tidal currents. High turbidity at the sediment surface effectively excludes suspension feeding organisms that are sensitive to clogging (Rhoads and Young, 1970).

Suspension feeders may also be excluded from deeper waters by decreasing concentrations of dissolved oxygen. Deposit feeders are better adapted than suspension feeders for exploiting reducing aqueous or sedimentary environments (Theede et al., 1969). In my opinion, abundant deposited food is important to maintain large populations of deposit feeders in shallow or deep water, but suspended food is not a major limiting factor for suspension feeders in epeiric seas. Suspended particulate detritus may be limiting to abyssal organisms, however. A full discussion of ecologic factors known to limit the distribution of bivalve feeding types was given by Rhoads et al. (1972).

The nearshore zone is associated with high variability of light, heat, and salinity, compared with deeper water. Dissolved oxygen and substratum mobility are also depth related (surface wave mixing); see Figure 9.1B.

With so many depth-related variables, how may trace fossils be used to identify specific parameters? These variables are considered separately, below.

Salinity and Temperature

The behavioral responses of burrowing organisms to steep gradients in salinity and temperature are similar; therefore, these two variables are considered together. The intertidal zone, shallow lagoons, estuaries, and delta platforms experience extremes in temperature and (or) salinity. Deep-living infaunal organisms are extremely common in these types of high-stress[2] environments. The deep infaunal habitat is a refugium from the highly variable sediment surface. Sediment is effective in slowing down the exchange of pore water. For example, Figure 9.2 gives the range of salinity with

[2] "Stress" is used here in the sense of physiological stress. Stressful environments are unpredictably variable and require physiologic regulation or accommodation by the inhabitants. Burrowing fulfills, in part, the mechanism of regulation.

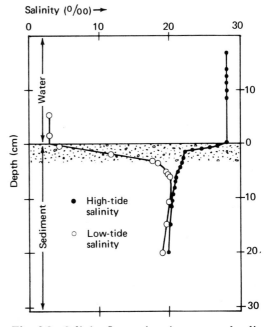

Fig. 9.2 Salinity fluctuations in water and sediment at high and low tides, Pocasset River estuary, Massachusetts. Deep vertical burrowers experience uniform salinity conditions, whereas near-surface dwellers experience large salinity fluctuations. (After Sanders et al., 1965.)

sediment depth during a tidal cycle in the Pocasset River, Massachusetts (Sanders et al., 1965). Organisms living at depths $\geqq$ 5 cm experience essentially an isohaline environment. Significantly, marine animals penetrating the farthest landward into an estuary are infaunal species; this was documented by Alexander et al. (1955) in the Tees and Tay estuaries of Great Britain and is depicted here in Figure 9.3.

Sediment cover is also effective in dampening temperature variations. Figure 9.4 shows the range of sediment temperatures at three different depths in an intertidal sand flat in Tomales Bay, California. Johnson (1965) found that 70 percent of the infaunal organisms within this flat live at depths greater than 5 cm. These intertidal organisms experience temperature variations equal to subtidal populations, whereas intertidal epifaunal organisms experience temperature variations that are three times as great. Deep vertical burrows are characteristic of, but not unique to, nearshore environments. The uniformity of the infaunal environment may have evolutionary significance because this appears to be a likely path for the initial colonization of the intertidal and supertidal zones by marine species.

Frey (1971) indicated that deep vertical burrows (having reinforced walls or linings) are also a means of successfully populating shifting or otherwise unstable bottoms. Unpublished work by me on recently depopulated subtidal mud bottoms in Buzzards Bay, Massachusetts, and Long Island Sound, Connecticut, indicates that early benthic colonizers are tube dwellers. Dense populations of tube-dwelling "pioneer" species may initiate physical stability in a new or newly changed bottom, providing opportunities for settlement of species characterized as later colonizers.

To distinguish whether temperature, salinity, or bottom stability is of major importance in a particular ancient environment requires supporting evidence from sedimentary or geochemical data. Ideally,

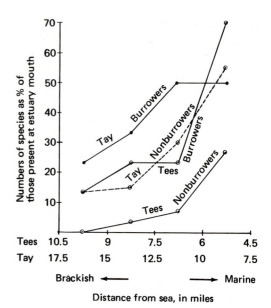

Fig. 9.3 Distribution of burrowing and non-burrowing organisms along salinity gradients in the Tees and Tay estuaries. (After Alexander et al., 1955.)

one wishes to be able to recognize unique suites of trace fossils, which can be related to specific salinity regimes. Seilacher (1963) suggested that this may be done by studying in detail those trace fossil suites from lithotopes of limited vertical and hori-

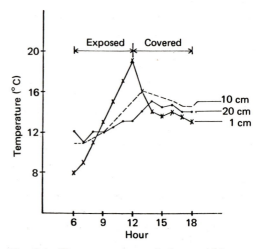

Fig. 9.4 Temperature variation within an intertidal sand flat during a tidal cycle, Tomales Bay, California. Deep-burrowing species experience less thermal variation than near-surface organisms. (After Johnson, 1965.)

zontal extent (such as a Carboniferous cyclothem). By relating trace morphologies to independent evidence from lithologic, mineralogic, geochemical, and body fossil data, one may be able to reconstruct relative salinity or temperature gradients. Once a correlation is established between trace morphology and a relative salinity or temperature scale, indicator species may be identified. Care must be taken to avoid circular reasoning in this type of analysis, however.

Dissolved Oxygen

Dissolved oxygen decreases with increasing depth in many recent marine basins, such as the Black Sea, California borderland basins, and the deeper depressions of the Gulf of California. Many Paleozoic and Mesozoic geosynclines also seem to have been low in oxygen. Fossil evidence of benthic life in the centers of these basins is sparse. Shell-bearing invertebrates, especially large ones, are today confined to well-mixed shelf waters, where oxygen levels exceed 1 ml/l (Rhoads and Morse, 1970). Descent from the well-mixed shelf waters to the upper or midslope regions of poorly aerated basins is accompanied by a marked change in benthos. As oxygen falls below 1 ml/l, shell-bearing invertebrates decrease in abundance. This part of the sea floor is dominated by small infaunal deposit feeders. Gas diffusion becomes a problem at low oxygen tension. The surface area-to-volume ratio is high in these organisms (Raff and Raff, 1970). As oxygen falls below 0.1 ml/l, most metazoans disappear.

One would predict that, if this recent model holds for ancient settings, somewhere between the shelly shelf facies of Paleozoic and Mesozoic geosynclines and the deep, azoic parts of these basins one should find an intermediate facies; it represents a transitional area of low, but not limiting, oxygen concentration (<1.0 and >0.1 ml/l). Such transitional areas would possess only small-size traces; few or no body fossils

should be present. This facies may well be represented by Seilacher's *Zoophycos* and *Nereites* ichnofacies (Figure 9.1A). Figure 9.5 shows a reconstruction of the Upper Devonian marine facies off the Catskill delta, New York State. Reconstruction of dissolved oxygen levels in this Paleozoic basin is based upon a combined trace and body fossil analysis (Byers, 1972; Bowen et al., 1974).

Mass Properties of the Bottom

One of the most important variables controlling the distribution of bottom-dwelling organisms is the nature of the substratum, such as grain-size distribution, organic content of sediment, bottom compaction, and sedimentation rate. Sediment grain size and organic content are directly observable in lithified deposits. The more elusive features, syndepositional bottom hardness and sedimentation rate, may be approached by using trace fossils.

Bottom hardness of marine sediments (measured as cohesion or water content) ranges from hard lithified rock to "soupy" muds rich in organic matter. Mass properties of the bottom control, in part, the distribution of burrowing and sessile epifaunal species, as well as deposit feeding and suspension feeding types. Uncompacted muds are easily resuspended by tidal currents. Such unstable bottoms are dominated by infaunal deposit feeders (Rhoads and Young, 1970). These syndepositional sedimentary properties can be reconstructed after lithification of the bottom.

The initial state of sediments may be depicted by deformational structures associated with traces. Shallow-burrowing organisms (grazing or foraging traces) moving through an uncompacted sediment, high in water content ($\geqq 50$ percent by weight), ordinarily produce a narrow zone of deformation around their burrows. Water-lubricated grains slide past one another in the loosely packed matrix, to accommodate volume displacement by the burrower. In

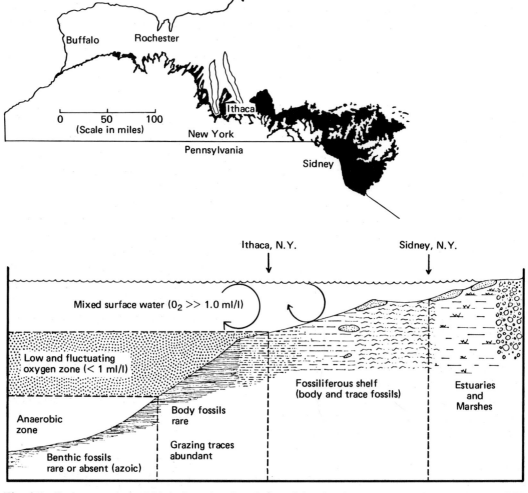

Fig. 9.5 Reconstruction of biofacies related to inferred levels of dissolved oxygen on the Catskill marine shelf and basin (Upper Devonian; Sonyea Group), New York State. Dark pattern on index map represents area of exposure. (After Bowen et al., 1974.)

contrast, an organism burrowing through a firm bottom, low in water content (≤ 50 percent), deforms the bottom plastically; i.e., each burrow is surrounded by a relatively larger zone of deformation, extending several grain diameters away from the burrow wall. In highly indurated or lithified sediments, boring takes place. In the latter case, no soft-sediment deformation occurs (*Glossifungites* "facies"). Excavation may be mechanical or chemical. Rhoads (1970) summarized criteria for recognizing soft-sediment deformational structures pro-

duced by burrowers, and gave both recent and ancient examples.

Bottom Stability, Sedimentation Rate, and Currents

The frequency of accretion or erosion of the bottom is important for relatively sedentary benthos, which must maintain a connection with the sediment surface for respiration and feeding. Trace fossils may be used to recognize erosional or accretionary sequences, and in some cases, quantitative estimates may be made. Goldring (1964) summarized some ways in which trace fossils may be used to identify changes in position of the sedimentary surface. (See Fig. 8.5.) The occurrence of shallow and deep burrows suggests accretion, whereas sediments containing only deep, truncated burrows indicate erosion. The quantity of material removed or accreted may, in turn, be estimated by the requisite behavior and preservational pattern of specific trace fossils. U-shaped burrows (cf. the genus *Diplocraterion*) lend themselves to such analysis because vertical movement of the organism can be recognized by the preserved, abandoned, horizontal base of the U (spreite). Vertical displacement of the tube leaves evidence of earlier construction below the new tube (response to sedimentation). Similarly, remnants of old, abandoned tube segments above sections of new tube construction indicate downward movement (response to erosion).

The record of storm events in marine shelf sediments may be recognized with the help of trace fossil data. Rapid erosion of the sea floor results in resuspension of bottom sediments. The erosion surface may be recognized by the truncation of endostratal traces at a common horizon. After storm conditions subside, resuspended (and newly introduced) sediments are rapidly deposited, producing a laminated post-storm unit. Organisms buried below, or within, this newly deposited sediment may burrow upward to reestablish connection with the sediment surface (cf. Fig. 9.6). Escape or "trauma" traces are characteristic of such short-period sedimentary events and have been illustrated for the inner part of the German Bay (Reineck et al., 1968); additional fossil examples include Siluro-Devonian deposits from Arsaig, Nova Scotia (Rhoads, 1966) and Upper Devonian rock of New York (Bowen et al., 1974). (See Chapter 22.)

Traces may also be used to infer the presence of currents, and in some cases, actual current direction may be determined.

Rheotropic responses of benthic organisms may produce a common alignment of burrow or tube openings, siphon openings, and crawling, grazing, and resting traces (Seilacher, 1964; Frey, 1971). This type of information is routinely used via oriented bottom photographs to determine relative intensity and direction of currents, in recent sea-bottom surveys (Fig. 9.7). Examples of current orientation of Lower Cambrian and Lower Devonian trilobite tracks were summarized by Seilacher (1964). (See also Chapter 8.)

Moore and Scruton (1957) used the intensity of bioturbation to characterize deltaic and delta-influenced shelf environments of the Mississippi Delta region. Sediment layering is produced by high sedimentation rates and wave and current reworking of the bottom. Destruction of these layers is accomplished by burrowing activity. Moore and Scruton found that the production of layering exceeded destruction by burrowing when sedimentation rates exceed about 4 cm/yr. Gradients in bioturbation are sensitive indicators of relative rates of sedimentation, current strength, and

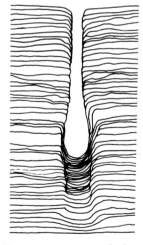

Fig. 9.6 Escape structure made by the bivalve *Mya arenaria* in tidal flat sediments of the North Sea. (Adapted from Reineck, 1970.) Escape structures must be interpreted with caution, because some are easily confused with certain physical collapse features (e.g., Howard, 1971, p. 200, Fig. 9).

Fig. 9.7 Photographs of two different mud bottoms in Long Island Sound (20 m), showing current alignment of (A) bivalve siphon openings (*Pitar morhuana, Mulinia lateralis*) and (B) polychaete tubes. Arrows indicate current direction. (Photos by W. Sacco.)

wave activity; Reineck (1967) demonstrated this phenomenon in intertidal and non-deltaic marine shelf areas of the North Sea.

ENVIRONMENTAL RECONSTRUCTION BASED UPON RELATIVE ABUNDANCES OF TRACE FOSSILS AND BODY FOSSILS

The strength of any environmental reconstruction is determined by the number of pieces of complementary independent evidence used in the reconstruction. A major advantage of using trace fossils in paleoecology is their value as complementary evidence, together with information derived from body fossils, sedimentary structures, and stratigraphic relations. This approach, illustrated in Figure 9.8, is a modification of Schäfer's (1972) biofacies concept. My students have found this conceptual model helpful in thinking about the relationships between assemblages of marine body fossils and trace fossils. I caution the reader not to consider this model as all inclusive or a panacea for environ-

mental reconstruction. Perhaps the major value of the model is in pointing out our lack of understanding of the formation of specific kinds of assemblages or associations of trace and body fossils. The scheme is used here to demonstrate a concept, rather than as a precise working model.

Two important facts must be considered in this analysis. First, not all traces are apparent in the field. Documentation of the abundance of traces within a rock normally requires thin-section or acetate-peel study. Second, the level of resolution of the examples in Figure 9.8 is on the order of tens of feet of section. Inferences made from observations on units of smaller scale can result in spurious conclusions. Application of this model by other workers will hopefully result in refinements and additions to the range of environments listed under each example discussed below.

Trace Fossils Rare

Example 1. The presence of abundant in situ body fossils, together with a paucity of

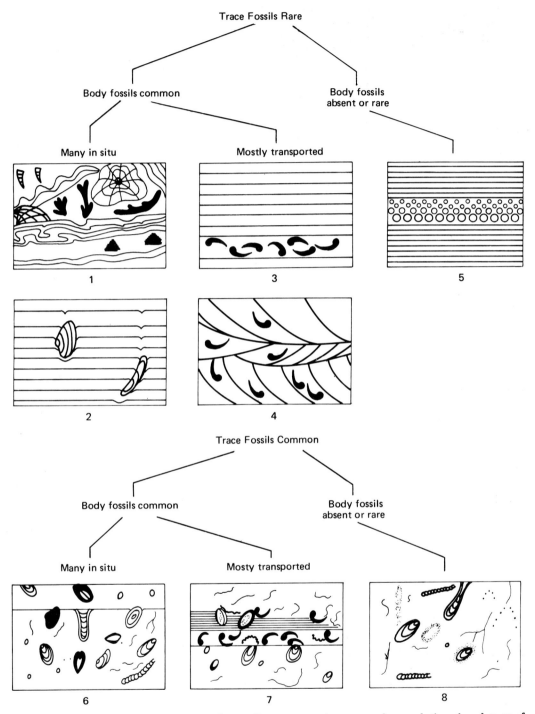

Fig. 9.8 Interpretation of ancient marine sedimentary environments from relative abundances of trace fossils and body fossils.

traces, may suggest hard or otherwise impenetrable bottoms. Traces that may be present are borings into the substratum (*Glossifungites* "facies"). Body fossils are primarily represented by epifaunal suspension feeders (reefs, bio-

stromes, erosional or nondepositional hard bottoms).

Example 2. Very high sedimentation rates may produce sediments in which traces are rare. Occasional layers containing in situ body fossils may represent periodic settlement of opportunistic species in a low-diversity setting (high turbidity and sedimentation rates, low and variable salinity). Body fossils are also rare, usually due to dilution by high rates of sediment influx (delta platform).

Example 3. Transportation of body fossils into anoxic basins by turbidity currents or by epiplanktonic rafting may produce shell concentrations in otherwise barren sediments. Bedding units are thin and essentially horizontal. Storm transport of body fossils into an area of high sedimentation rate may produce a similar association (turbidite basin or delta platform).

Example 4. An association like that of Example 3, but occurring with large-scale cross bedding, represents a shell-bank type of bottom frequently reworked by tidal currents or waves (tidal channels and high-energy beaches).

Example 5. The absence of both body and trace fossils (excluding diagenetic factors) may define an anoxic sea floor (poorly ventilated basins) or sandy foreshore deposits (high energy). (Cf. Fig. 21.19.)

Trace Fossils Common

Example 6. In situ body fossils indicate low energy in the water. Complete bioturbation of the bottom indicates a low sedimentation rate (oxygenated deep-water or protected shallow-water shelf). Most fossiliferous marine shelf deposits are of types 6 and 7, often in close proximity, one alternating with the other.

Example 7. Alternating with conditions in Example 6, one commonly finds layers of exhumed and disarticulated body fossils bounded above and below by a type 6 lithology. This distribution represents occasional reworking of the bottom by shoaling waves (storm generated), preceded and followed by quiet-water conditions (open marine shelf).

Example 8. Highly bioturbated sediments essentially devoid of benthonic body fossils are not uncommon. (The absence of body fossils must, of course, be determined not to be the result of diagenesis.) This association, characteristic of low-oxygen environments or hypersaline lagoons, is common in the rock record but is not well understood. Not all the environments in which this association is formed can be identified by present criteria.

SUMMARY

Trace fossils are used in paleoecology to add information about overall fossil community structure and diversity and to provide data about environmental conditions. Consideration of distinctive traces of soft–bodied organisms may yield a more accurate picture of species richness and feeding-type diversity within fossil communities. The density of traces in rocks may, however, contribute little, or misleading, information about original standing crop densities; burrow density is a function of population turnover rate, mobility of the burrowing species, and sedimentation rate.

Trace suites are known to undergo distinctive changes related to increasing water depth. This phenomenon may be caused by a variety of depth-related parameters, such as gradients in water temperature, salinity, dissolved oxygen, and distribution of food resources. Successful population of the highly variable intertidal environment has been accomplished by many invertebrates that construct deep vertical burrows. Sedimentary cover is an effective "buffer" of short-period changes in surface temperature and salinity. Many vertical burrowers are suspension feeders, although deposit feeders, carnivores, and scavengers are also represented.

Gradients in dissolved oxygen may be detected by the transition from a dominance of large calcified body fossils in shallow, well-aerated shelf water, to a dominance of soft-bodied metazoans in deeper, oxygen-poor water. Low-oxygen marine environments are characterized by small-diameter horizontal feeding traces. Subtidal mud bottoms having normal levels of dissolved oxygen in the water column contain both body fossils and trace fossils.

Mass properties of the sea floor (water content) may be reconstructed from primary deformational structures adjacent to infaunal feeding traces. Sedimentation or erosion surfaces may be determined from the presence or absence of shallow traces, truncated deep burrows, behavioral responses reflected by certain traces, and gradients in bioturbation. Traces may be used to recognize storm events. In some cases, quantitative estimates of erosion or sedimentation rates may be made. Lineation of dwelling tubes and surface tracks can yield information about current direction.

A model is presented here in which trace and body fossil data are integrated with sedimentologic information; eight biolithofacies are recognized. The most ecologically interesting biofacies are those consisting solely of trace fossils. The spectrum of possible environments of formation of this type is not well known, nor are all the physiological or functional reasons for such assemblages; yet these interesting assemblages are common in the fossil record.

The adaptive significance of associations of benthic species lacking hard parts may have fascinating ecological, physiological, and evolutionary significance. Answers to this set of problems will involve the joint efforts of biologists and geologists in field and laboratory studies.

ACKNOWLEDGMENTS

I wish to thank R. W. Frey, University of Georgia, for his suggestions and criticisms of this manuscript in its formative stages and for securing Figure 9.6. I also benefitted from the comments of T. P. Crimes, University of Liverpool, J. E. Warme, Rice University, and K. M. Waage, Yale University.

REFERENCES

Alexander, W. B. et al. 1955. Survey of the Tees. Part II—the estuary—chemical and biological. Dept. Sci. Indust. Res., Water Pollution Res. Tech. Paper 5, 171 p.

Bowen, Z. et al. 1974. Marine benthic communities in the Upper Devonian of New York. Lethaia, 7:93–120.

Bromley, R. G. 1970. Borings as trace fossils and *Entobia cretacea* Portlock, as an example. In T. P. Crimes and J. C. Harper (eds.), Trace fossils. Geol. Jour., Spec. Issue 3:49–90.

Byers, C. 1972. Analysis of paleoenvironments in Devonian black shale by means of biogenic structures (abs.). Geol. Soc. America, Abs. Prog., 4(7):464.

Farrow, G. E. 1966. Bathymetric zonation of Jurassic trace fossils from the coast of Yorkshire, England. Palaeogeogr., Palaeoclimatol., Palaeoecol., 2:103–151.

Frey, R. W. 1971. Ichnology—the study of fossil and recent lebensspuren. In B. F. Perkins (ed.), Trace fossils, a field guide. Louisiana State Univ., School Geosci., Misc. Publ. 71-1:91–125.

———. 1972. Paleoecology and depositional environment of Fort Hays Limestone Member, Niobrara Chalk (Upper Cretaceous), west-central Kansas. Univ. Kansas Paleont. Contr., Art. 58, 72 p.

——— and T. M. Chowns. 1972. Trace fossils from the Ringgold road cut (Ordovician and Silurian), Georgia. In T. M. Chowns (ed.), Sedimentary environments in the Paleozoic rocks of northwest Georgia. Georgia Geol. Surv., Guidebook 11:25–55.

——— and T. V. Mayou. 1971. Decapod burrows in Holocene barrier island beaches and washover fans, Georgia. Senckenbergiana Marit., 3:53–77.

Goldring, R. 1964. Trace fossils and the sedimentary surface in shallow–water marine sediments. In L. M. J. U. van Straaten (ed.), Deltaic and shallow marine deposits. Developments in Sedimentology, 1:136–143.

Golubic, S. and J. Schneider. 1972. Relationship between carbonate substrate and boring patterns of marine endolithic microorganisms (abs.). Geol. Soc. America, Abs. Prog., 4(7):518.

Hecker, R. T. 1970. Palaeoichnological research in the Palaeontological Institute of the Academy of Sciences of the U.S.S.R. In T. P. Crimes and J. C. Harper (eds.), Trace fossils. Geol. Jour., Spec. Issue 3:215–226.

Hertweck, G. 1970. The animal communities of a muddy environment and the development of biofacies as affected by the life cycle of the characteristic species. In T. P. Crimes and J. C. Harper (eds.), Trace fossils. Geol. Jour., Spec. Issue 3:263–282.

Howard, J. D. 1971. Trace fossils as paleoecological tools. In J. D. Howard et al., Recent advances in paleoecology and ichnology. Amer. Geol. Inst., Short Course Lect. Notes, p. 184–212.

Johnson, R. G. 1964. The community approach to paleoecology. In J. Imbrie and N. D. Newell (eds.), Approaches to paleoecology. New York, John Wiley, p. 107–134.

————. 1965. Temperature variation in the infaunal environment of a sand flat. Limnol. Oceanogr., 10:114–120.

Levinton, J. S. and R. K. Bambach. 1970. Some ecological aspects of bivalve mortality patterns. Amer. Jour. Sci., 268:97–112.

McAlester, A. L. and D. C. Rhoads. 1967. Bivalves as bathymetric indicators. Marine Geol., 5:383–388.

Moore, D. G. and P. C. Scruton. 1957. Minor internal structures of some recent unconsolidated sediments. Amer. Assoc. Petrol. Geol., Bull., 41:2723–2751.

Osgood, R. G., Jr. and E. J. Szmuc. 1972. The trace fossil Zoophycos as an indicator of water depth. Bulls. Amer. Paleont., 62:1–22.

Raff, R. A. and E. C. Raff. 1970. Respiratory mechanisms and the fossil record. Nature, 228:1003–1005.

Reineck, H.–E. 1967. Layered sediments of tidal flats, beaches, and shelf bottoms of the North Sea. In B. Lauff (ed.), Estuaries. Amer. Assoc. Advmt. Sci., Publ. 83:191–206.

———— (ed.). 1970. Das Watt. Frankfurt, W. Kramer, 142 p.

———— et al. 1968. Sedimentologie, Faunenzonierung und Faziesabfolge vor der Ostküste der inneren Deutschen Bucht. Senckenbergiana Leth., 49:261–309.

Rhoads, D. C. 1966. Missing fossils and paleoecology. Discovery, Yale Peabody Mus., 2:19–22.

————. 1967. Biogenic reworking of intertidal and subtidal sediments in Barnstable Harbor and Buzzards Bay, Massachusetts. Jour. Geol., 75:461–476.

————. 1970. Mass properties, stability, and ecology of marine muds related to burrowing activity. In T. P. Crimes and J. C. Harper (eds.), Trace fossils. Geol. Jour., Spec. Issue 3:391–406.

———— and J. Morse. 1970. Evolutionary and ecologic significance of oxygen deficient marine basins. Lethaia, 4:413–428.

———— and D. K. Young. 1970. The influence of deposit-feeding benthos on bottom sediment stability and community trophic structure. Jour. Marine Res., 28:150–178.

———— et al. 1972. Trophic group analysis of Upper Cretaceous (Maestrichtian) bivalve assemblages from South Dakota. Amer. Assoc. Petrol. Geol., Bull., 56:1100–1113.

Rodriguez, J. and R. C. Gutschick. 1970. Late Devonian—Early Mississippian ichnofossils from western Montana and northern Utah. In T. P. Crimes and J. C. Harper (eds.), Trace fossils. Geol. Jour., Spec. Issue 3: 407–438.

Sanders, H. L. 1968. Marine benthic diversity: a comparative study. Amer. Naturalist, 162: 243–282.

———— et al. 1965. Salinity and faunal distribution in the Pocasset River, Massachusetts. Limnol. Oceanogr., 10:216–229.

Schäfer, W. 1972. Ecology and palaeoecology of marine environments. Edinburgh and Chicago, Oliver & Boyd and Univ. Chicago Press, 568 p.

Seilacher, A. 1963. Lebensspuren und salinitätsfazies. Fortschr. Geol. Rheinhld. u. Westf., 10:81–94.

————. 1964. Biogenic sedimentary structures. In J. Imbrie and N. D. Newell (eds.), Approaches to paleoecology. New York, John Wiley, p. 296–316.

————. 1967. Bathymetry of trace fossils. Marine Geol., 5:413–429.

Theede, H. et al. 1969. Studies on the resistance of marine bottom invertebrates to oxygen-deficiency and hydrogen sulphide. Marine Biol., 2:325–337.

Walker, K. R. 1972. Trophic analysis: a method for studying the function of ancient communities. Jour. Paleont., 46:82–93.

———— and R. K. Bambach. 1974. Feeding by benthic invertebrates: classification and terminology for paleoecological analysis. Lethaia, 7:67–78.

———— and L. F. Laporte. 1970. Congruent fossil communities from Ordovician and Devonian carbonates of New York. Jour. Paleont., 44:928–944.

PART III

Selected Groups
of Trace Fossils

CHAPTER 10

PLANT TRACE FOSSILS

WILLIAM ANTONY S. SARJEANT

Department of Geological Sciences, University of Saskatchewan
Saskatoon, Saskatchewan, Canada

SYNOPSIS

Discriminating between plant body-fossils and biogenic structures resulting from plant activity is often difficult. This difficulty is a consequence not only of their modes of life but also of semantic problems and differences in viewpoint among various workers. However, structures resulting from plant activity afford important information both to ichnologists and to paleobiologists in general, and merit fuller study than they have hitherto received.

Four major groups of plant traces—borings and attachment traces, phytoliths, root molds and casts, and stromatolites—are discussed here, including an evaluation of their ecological and biological significance and their status as trace fossils. Only stromatolites have hitherto attracted a significant degree of attention from geologists.

INTRODUCTION

The terms "trace fossil" and "biogenic structure" have many published definitions and are comprehended by geologists in an even greater variety of ways. (See Chapter 3.) Since the ending of the "fucoid" period in the history of trace fossil study (see Chapter 1), research in this field has been largely undertaken by paleozoologists; and as a consequence, both of the above terms have come to be applied primarily to structures resulting from the activity of animals.

Structures produced by animals, in general, display a marked component of motion: the animal producing the traces most often moved completely away from the position where the traces were made, and only in certain instances is any dispute likely to arise concerning their character. In contrast, most multicellular plants live and die in one place, so that deciding whether the structures surviving in sediments are to be regarded as body fossils or as trace fossils may be far from easy.

Of the structures considered in this chapter, perhaps only one group—plant borings and attachment traces—is likely to be accepted without question by most paleoichnologists as constituting true plant trace fossils; the others selected for discussion may well provide a focus for dispute. However, I consider that the stance adopted in discriminating plant trace fossils must be slightly different from that assumed in discriminating traces made by animals. If a structure directly reflects the morphology of the plant at the time of its death, or consists of a component part of the complete plant (e.g., bark, leaf, pollen, or spore), it must be interpreted as a body fossil. If, in contrast, the structure results from the activity of the plant (or one of its component parts), reflecting the morphology of the plant (or that part) only indirectly or partially, then we are dealing with a biogenic structure and not with a body fossil.

In the accounts that follow, an attempt is made in each case to explain why I consider the structures to be classifiable as trace fossils. The reader must, in each instance, make up his own mind as to whether he accepts this opinion.

BORINGS AND ATTACHMENT TRACES

Treatment of plant borings as trace fossils needs little justification; the parallels between them and borings produced by animals (e.g., bryozoans, clionid sponges) is self-evident. Attachment traces generally correspond more to the degree of penetration of organically secreted acids than to the form of the attachment structure of the plant in question; clearly, these excavations also must be regarded as trace fossils, although they have as yet attracted little study by paleoichnologists.

Fungi

Members of numerous groups of plants bore into hard substrates, as a means for obtaining food or shelter, or to secure attachments. (See Chapter 12.) Fungi bore into shell, dentine, and bone, and into the scales, armor, and carapaces of fish and reptiles, as a means of obtaining food. Fungal borings, which may penetrate any accessible substrate surface, are consistently small (usually not larger than 25 μm in diameter, rarely larger than 50 μm). The borings tend to be straight or gently curved; they may fork dichotomously, or may develop side branches at a consistent angle (60–90°) with the main tunnel. Swellings in the tunnels accommodate reproductive cells.

Fungi are important agents in the decay of wood; however, the evidence of their activity is not always separable from other types of borings, because fungi often enter through animal borings (which provide access for infestation) and are themselves often entered and enlarged by wood-boring bivalves and crustaceans.

Evidence of fungal boring within fossil pollen and spores also is often encountered (Fig. 10.1). Elsik (1966, p. 515) noted three distinct patterns in spores and pollen from Mississippian, Lower Cretaceous, and Tertiary sediments: (1) simple, circular to slightly irregular perforations, some ±0.25 to 1 μm in diameter, others 1 to 2 μm in diameter; (2) wedge-shaped perforations arranged in basic rosettes of four or more wedges, their narrow ends joined, often modified by further branching from the broad end of each wedge (dimensions not stated by Elsik); and (3) a meandering, branched groove, ±0.5 μm wide, having no particular symmetry.

Some groups of living fungi, the Chytridiales and Blastocladiales (Phycomycetes), absorb food from their host by means of bud-like or tufted rhizoids. These rhizoids may penetrate the spore or pollen wall by absorption or enzyme solution, rather than by physical abrasion; this produces distinct patterns of very smooth-sided perforations, so closely analogous to those described by Elsik as to leave no room for reasonable doubt concerning the origin of the latter (Gaumann, 1952, p. 32, 42; Elsik, 1966, p. 516). Indeed, Moore (1963) described coccoid, ovoid, or bacilli-like cells associated with aseptate filaments (*Palynomorphites diversiformis*), considered to be a fungus, investing and penetrating Carboniferous spores; in some instances, the spore wall had been entirely destroyed and was represented only by a "pseudomorph" formed by the presumed "fungal hyphae" (Moore, 1963, Pl. 54, fig. 11).

Algae

Members of the groups of algae that bore into hard substrates probably do so only as a means of obtaining shelter; their borings are found only on illuminated surfaces. Although some borings are as small as 4 μm in diameter, the tunnels are generally larger than those produced by fungi; algal borings typically have diameters greater than 25 μm and frequently greater than 50 μm. The main tunnel usually is by no means straight,

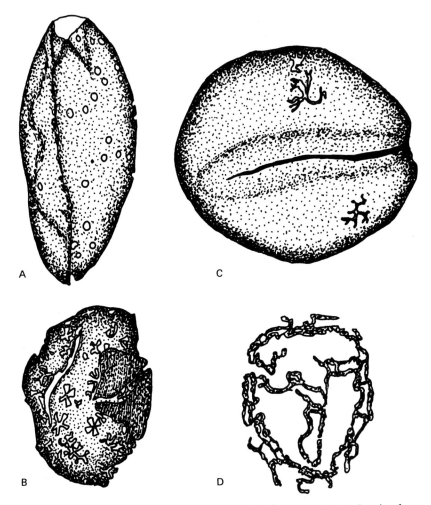

Fig. 10.1 Traces of biological degradation of fossil spores. A, simple, circular perforations in walls of laevigate spore; Eocene, Texas. B, wedge-shaped perforations, arranged in rosettes, in wall of laevigate spore (*Laevigatosporites* sp.); Eocene, Texas. C, irregularly meandering and branching grooves in wall of unidentified spore; Mississippian, Oklahoma. D, a filamentous organism, *Palynomorphites diversiformis* (probably a fungus), "pseudomorphing" a trilete spore that it has entirely destroyed; Upper Mississippian, Scotland. [Drawings based on photographs in Elsik (1966) (A–C) and Moore (1963) (D).]

and may even exhibit contortions; it may be occupied by several threads, the individual ones diverging from a "false ramification," or converging with the tunnel at various points, so that the tunnel thins and thickens at intervals. Branching of algal tunnels is highly irregular; thus, although the tunnels resemble those of fungi in the occurrence of regular swellings to accommodate reproductive cells, they differ in almost all other particulars. Fuller discussions of marine algal and fungal borings are given in Chapters 11 and 12.

Some algae seem to have a chemotactic affinity for iron (Ellis, 1914, p. 119); indeed, several species, mainly members of the Cyanophyceae, deposit iron in or upon their filaments and occur abundantly within iron oolites. Other algae, only partially endolithic, produce densely spaced, minute, pin-prick holes 50 to 80 μm in diameter, in shell surfaces (see

Bromley, 1970). The larger seaweeds, in particular the brown algae (Phaeophyceae), often affix themselves to hard substrates (rock, shell, bone, or wood) by means of attachment structures (holdfasts), with or without root-like branches (hapterae), at the base of the thallus; these also may produce indentations or holes in substrate surfaces, although I have not discovered any accounts of fossil examples. Algal borings into wood are much less frequent than those of fungi, and scarcely merit detailed comment.

Lichens

On rock surfaces exposed in the terrestrial realm, most plants are unable to become established because of the pronounced deficiency of water and nutrients, extremes of temperature experienced, and exposure to solar radiation. Crustose lichens, produced by the symbiosis of a crust-like fungus that enmeshes and protects the terrestrial algae, are almost the only plants able to survive in such situations. The lichens are spongelike and able to absorb water rapidly; they obtain mineral nutrients by secreting carbon dioxide, which mixes with water to form dilute carbonic acid and thus attacks the underlying rock surface, into which the rhizoids of the lichen may penetrate to several millimeters. The rock surface is thus corroded and decomposed, not only at the position of contact between thallus and rock but also as far beyond the margins of the thallus as the carbonic acid may penetrate. In general, the scars produced by crustose lichens correspond to the lichens in shape but are markedly larger. The rate at which the rock surface is attacked depends on its mineral character; crustose lichens may persist for centuries on the surface of quartzite or basalt in a dry climate, whereas those developing on limestones or sandstones in a moist climate may destroy the surface— to the point at which a little soil is produced—within only a few years.

As soon as a little soil is available, foliose lichens start to develop, at first in depressions or other sheltered positions. Their expanding, leaf-like thalli soon overshadow the crustose lichens, which die and decay as a result. The structure of the foliose lichens makes them more efficient collectors of water; they are attached to the substrate at a single point or along a single margin, and at these positions, the acid secreted by them eats more deeply into the rock, producing scars very dissimilar to those resulting from crustose lichens. When yet more soil has accumulated, other plants—in particular, xerophytic mosses and frutiose lichens—begin to develop; their secretions attack the rock surface over a wider area, and do not produce such characteristic scars (Weaver and Clements, 1938, p. 66–69).

Lichens are thus important agents of rock erosion in the terrestrial realm. Where extreme desiccation resulting from exceptionally dry weather, or the effects of fires or other processes, has removed the lichen flora, the stage reached in the plant succession before its interruption can readily be determined by study of the lichen scars.

In the marine realm, lichens occur to a much more restricted degree; the most important boring species, *Arthropyrenia sublittoralis*, forms pits in limestone or shell surfaces similar to, but larger than, those produced by algae (Santesson, 1939).

PHYTOLITHS

Many different structures produced by plants are composed of opaline silica—the external skeletons or shells of silicoflagellates, frustules of diatoms, etc. In most instances, these structures clearly have a protective or supporting function and must be regarded as body fossils. However, phytoliths (also called "opal phytoliths"), although produced within the cell walls of higher plants, cannot be so regarded, and in my view, merit treatment here.

Characteristics of Phytoliths

Phytoliths[1] are small particles of opaline silica, ranging from a few tens to several hundreds of microns in maximum cross measurement, and are of highly variable shape (Fig. 10.2). They are not present in all plants, and may not be present even in all individuals of species in which phytoliths are typically developed. Wilding and Drees (1968, p. 97) noted that:

> ... *the silica content of a plant is a function of two major factors, soil and plant species, with a third factor, transpiration rate, which reflects climate, also important in some species. The concentration of soluble silica, monosilicic acid, available for plant uptake, is dependent on soil pH and sesquioxide content. With increases*

in *pH, soluble silica in the soil decreases, reaching a minimum between pH 8 and 9, which is attributed to an absorption reaction of monosilicic acid to sesquioxides (particularly aluminum oxide) by H-bonding. Soil temperature, moisture regime, and nutrient status also affect the concentration of monosilicic acid and/or efficiency of water use by plants, and thus silica uptake.*

> *Plant species differ in their uptake of silica, and monocotyledons generally contain considerably more plant opal than dicotyledons . . . The non-uniform distribution of silica concentrations toward the uppermost leaves, blades, and sheaths of mature plants and a systematic increase from the basal to apical region of the same leaf support the mechanism that silica uptake is a passive process, with most silica concentrated where greatest water is lost. It has also been demonstrated that silica contents of the plant increase markedly with age and maturity.*

[1] Maslov (1960, p. 56), apparently unaware of an existing definition, applied the term "phytolith" to stromes, oncolites, and katagraphs. Usage of the term in this very different sense can only cause confusion, and should be discontinued.

Thus, although silica sometimes secondarily fulfills a strengthening function (see

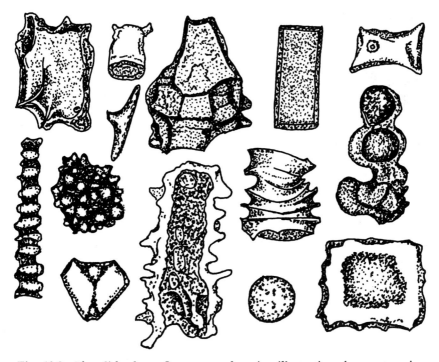

Fig. 10.2 Phytoliths from Quaternary deposits, illustrating the great variation in morphology. [Redrawn after Ehrenberg (1841, 1846) and Deflandre (1963).] Not to constant scale, but magnifications are in the range ×700–×1,300.

later discussion), phytoliths seemingly are generated as a means of disposing of the soluble silica present in absorbed water—in other words, they are accumulations of waste, comparable with the calcium oxalate crystals so widely present in plants. Phytoliths certainly cannot be regarded as body fossils, because they are not uniformly present in plants, or even in different individuals of the same species. Because phytoliths are waste materials accumulated in particular situations as a consequence of physiological processes, they are in many respects comparable with the coprolites of animals; and considering that plants are unable to excrete waste directly, to treat them in a manner similar to coprolites—i.e., as trace fossils—is reasonable.

Phytoliths are formed in the walls of epidermal cells of numerous groups of plants, notably the Equisetales (see Scagel et al., 1965, p. 404) and various deciduous trees (Wilding and Drees, 1968, p. 102); however, plants of the Order Graminales (the grasses and their allies) produce phytoliths in greatest abundance. In a recent study of cellular differentiation in the oat *Avena sativa*, Kaufman et al. (1970, p. 304–306) showed that silica accumulates in the epidermis, in special silica cells, each of which is paired with a cork cell. At an early stage of differentiation, the silica cell begins to protrude out from the plane of the adjacent epidermal cells; its nucleus breaks down, and all cellular membranes and organelles disappear. The cell then becomes filled with fibrillar material (considered to be breakdown products of the membranes and organelles), containing a few osmiophilic droplets; the lumen becomes filled by bead-like chains of amorphous silica bodies (about 83 percent silica). The silica is thus accumulated in a dead cell; the authors were unable to determine whether this habit had any adaptive significance, and reasonably regarded the silica as stored waste.

After decay of the plant tissue, the chains of bead-like silica bodies break down,

and the phytoliths accumulate in soil as particles of very variable shape (see later discussion), depending on the shape of the cell within which they formed. They withstand weathering and can be eroded and transported, to accumulate in riverine, aeolian, telmatic (marsh), lacustrine, or marine sediments; rounding during transport may render them indistinguishable from other opaline silica particles or siliceous matter originating from rock breakdown.

Silica is also deposited in the entire cell wall of the trichomes of the oat (Kaufman et al., 1970, p. 306) and is most dense in the apical region; it clearly provides structural support, adding to the "bristliness" of the trichome. In this instance, the lumen does not fill up with silica bodies, and the resultant bodies would have a morphology dissimilar to that of the typical phytolith.

Fossil Phytoliths

Fossil phytoliths were first described by Christian Gottfried Ehrenberg in 1841, from "edible clays" of the Amazon basin, Brazil. In a later paper, describing similar siliceous particles from the Atlantic island of Ascension, Ehrenberg (1846) designated them "Phytolitharien." A full discussion of the history of phytolith studies, and of the classifications applied to them, was given by Deflandre (1963), who thus summarized their morphology (p. 242, transl.):

> First of all, one finds types absolutely classic and characteristic of the epidermal cells, having diverse forms—small cushions, crescent-shaped cells, two-bladed axes, small batons with straight or undulose sides or with saw-teeth, etc. Other types, sometimes very large (150–250μ), probably belong to the stems of Graminae . . . we consider also a fairly large number of bodies of rounded shape, measuring from 3 to 8–10μ, to be from the same source.

Phytoliths have attracted very little attention from geologists to date; a review by Wetzel (1967), although not entirely comprehensive in coverage [the paper by Baker (1960) is perhaps the most glaring omission], gives a good impression of the meagerness of known fossil records. Published records are almost entirely from the Quaternary (e.g., Fig. 10.2), and no pre-Tertiary records have yet been documented. Most literature available on phytoliths is thus found in botanical journals, or journals concerned with soil science.

In the belief that particular phytoliths are rarely, if ever, amenable to assignment to a particular plant group, Deflandre (1963) advocated a parataxonomic classification for them. His premise was in part contested by numerous later workers. Wilding and Drees (1968, p. 102) found no problems in distinguishing between phytoliths from tree leaves and phytoliths from grasses; they considered that black, opal phytoliths were of forest origin, and found that phytoliths from the sugar maple, *Acer saccharum*, the white ash, *Fraxinus americana*, and the slippery elm, *Ulmus rubra*, were very distinct from those of grasses, although those of the white oak, *Quercus alba*, were much less readily distinguished. Lutwick and Johnston (1969) recognized phytoliths characteristic of two genera of Alberta foothill grasses (*Festuca, Calamogrostis*) and used this criterion in elucidating the origin of cumulic soils. Twiss et al. (1969), studying phytoliths from a variety of sources (atmospheric dust, soils, paleosols, Pleistocene loess, and deep-sea sediments), found that a particular grass species might contain phytoliths of several different types; but the authors nonetheless were able to distinguish morphological groups characteristic of three subfamilies of grasses—Festucoid, Chloridoid, and Panicoid. A fourth phytolith group ("elongate") was uniformly distributed in all species, however, and could not be used in taxonomy. Blackman (1971), examining phytoliths from range grasses in southern Alberta, showed that phytoliths of different shapes characterized different positions in the leaf sheaths of festucoid grasses, although no

such association could be demonstrated in the leaf blades. Phytoliths of 26 grass species, attributable to 12 genera, were studied; whereas the majority of forms occurred widely in a variety of different species, some were sufficiently restricted in occurrence to be taxonomically useful. Apparently, however, particular phytoliths do not generally characterize individual plant genera, and phytolith classifications must remain largely morphographic.

Nevertheless, the fact that broad groups of plants may be recognized from phytoliths gives the prospect that, in future work, they may be used not only to determine whether grassland conditions prevailed at a particular time in an area but also to discriminate between different types of forests or grasslands. Certainly these microfossils merit more attention from geologists than they have hitherto received, and they should prove to be a means of elucidating the spread of the grasses through geological time.

ROOT MOLDS AND CASTS

In many Paleozoic plants, the character of the root (or more correctly, the rhizophore) was extremely regular. The organ genus *Stigmaria*, perhaps the most common fossil of the Carboniferous coal measures in both Europe and North America, comprises the rhizophore systems of a group of tree-size plants (*Sigillaria, Lepidodendron*, and probably *Lepidophloios* and *Bothrodendron*; see Crookall, 1929, p. 33). From the central trunk, four main branches extended downward; each branch normally forked three times, giving 16 ultimate branches. Such root systems may fairly be treated as "body fossils," because the surviving molds and casts exactly correspond to the structure of the living plant.

In more advanced plants, however, the correspondence between the root molds and casts, preserved in the sediment in which the plant grew, and the structure of the actual root at the time of the death of the plant, is much less close. The root develops primarily in quest for water, secondarily in order to absorb minerals from the soil. Minerals are absorbed in the zone of the root that is still growing. Water is absorbed through root hairs which, being physiologically very active, are relatively short lived; they normally shrivel up and die very quickly, usually becoming detached; if they are retained, suberin (a water-proof, cork-like material) is deposited in them so that they become inactive. In consequence, only the tip of any given root is likely to be physiologically active; root branches regularly die, shrivel, and decay, new root branches being developed as replacements. The root molds and casts produced by biannual and perennial plants, and even by many annuals, thus represent plant activity over a period of time and are correspondingly more complex and often more extensive than the root system itself was at any given time during the life of the plant. (For example, in the leafy spurge, *Euphoria exculta*, the roots continuously grow and die.) For this reason, I disagree with the view that these traces are to be regarded as body fossils, and consider instead that root molds and casts merit treatment as trace fossils.

Characteristics of Plant Roots

Although the character of the root developed by a particular plant is determined initially by hereditary factors, the range of variation in root structures within a particular genus, or even within a particular species, is surprising. Thus, one species of blazing star, *Liatris punctata*, has a strong, much-branched taproot reaching depths as great as 13 ft (4 m), whereas another species, *L. scariosa*, has a large number of fibrous roots originating from the base of a corm. Also, field corn usually has three primary roots, whereas sweet corn usually has only one (Weaver and Clements, 1938, p. 293).

The form of the root of any individual plant, however, is closely controlled by the environment of growth and by any changes in that environment during the life of the plant. Of prime importance are the quantity of water in the soil and the levels at which the water is concentrated. Where water content is relatively low—provided enough is available to enable good growth—root development is stimulated to a greater degree than in soils where water is more abundant. Weaver and Clements (1938, p. 296) recorded that corn grown for five weeks in a moist, rich, loess soil—having an available water content of 19 percent—had a total root area 1.2 times greater than that of the transpiring surfaces of leaves and stems; in contrast, corn grown in comparable soil—having an available water content of only 9 percent—had a total root area 2.1 times greater than that of the transpiring surfaces. In both instances, the main roots developed to the same degree; but in the drier soil, the need to seek out more water evoked a much more substantial development of secondary and tertiary root branches.

Beyond a certain point, however, the above rule ceases to apply. Perhaps this difference may be more readily comprehended if expressed in terms of ratio of biomass (dry weight). Results of the grassland studies under the International Biological Program in the United States (R. T. Coupland, 1973, personal communication) show that, in humid grassland,[2] 1,111 g of dry roots per m² were observed, constituting 84 percent of the total biomass; in semiarid grasslands,[3] having a range of 1,025 to 2,540 g/m², 88 to 95 percent of the total biomass was underground; however, in truly arid conditions,[4] the biomass was at a minimum—204.6 g/m²—having only 68 percent underground. Both moist and dry conditions thus restrict root development in grasslands.

---

[2] Osage, Oklahoma.
[3] Summation of values from Hays, Kansas; Cottonwood and Bison, South Dakota; Dickinson, North Dakota; Pawnee, Colorado.
[4] Hornada, New Mexico.

Where soils are very dry, the development of roots depend very much on the way in which water may be most readily obtained: by tapping water at depth, or by rapid collection of percolating water after rains. (See Fig. 10.3.) According to Cannon (1911, p. 194), three broad groups of root systems may be distinguished in desert environments:

1. Plants having taproots as the most prominent features, e.g., junco or allthorn (*Koeberlina*

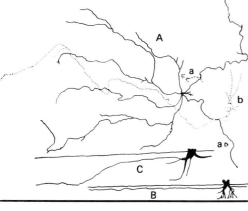

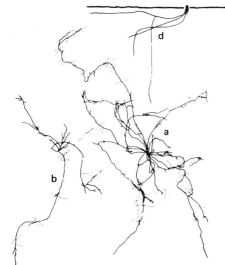

Fig. 10.3 Root systems of desert plants near Tucson, Arizona. Upper: roots of *Opuntia fulgida* in vertical view (A) and in section (B, C), intergrown with those of *Acacia constricta* (b, c) and *Riddellia cooperi* (a). Lower: roots of *Cavillen tridentata* shown in section (d) and in vertical view, the upper system (a) being separated—for clarity—from the lower (b). (From Cannon, 1911.)

spinosa), Mormon tea (*Ephedra*), etc. These plants are found only very locally, in places where the soil is fairly deep and the taproot can reach down to an available water supply (mostly in floodplains).

2. Plants having essentially lateral root systems. These plants include nearly all cacti and most annual plants of other groups; the roots of the latter rarely extended deeper than 20 cm, and their largest development occurs within 4 to 5 cm of the surface. Such root systems often extend laterally for great distances—sometimes as much as 50 ft (15 m). The root systems of summer annuals tend to be better developed and more complex than those of winter annuals; the lower winter temperatures (a) reduce the rate of water loss by transpiration and evaporation, so that water remains present in the soil somewhat longer, and (b) directly inhibit root development. Where sand is especially loose, a dense mat of shallow roots may be produced to stabilize it, e.g., in the sand grass (*Ammonphila*).

3. Plants having generalized root systems, the taproots and laterals both well developed. Most perennials, other than cacti, have root systems of this kind (e.g., acacia—*Acacia*; hackberry—*Celtis*; ratama or Jerusalem thorn —*Parkinsonia*; and mesquite—*Prosopis*). Relative development of laterals and taproots is dependent on local circumstances. [R. T. Coupland (1973, personal communication) found plants having both an extensive shallow lateral root system and very deep taproots, extending down 12 to 15 ft (*ca.* 3 to 4 m) in sandhills near Beaver Creek, Saskatchewan.]

Many perennials of arid conditions (in groups 2 and 3) developed deciduous roots: short filamentous roots, arising in groups of six or more from the larger laterals. These roots are formed during the rainy season, especially in the summer, and serve to greatly and quickly increase the absorption surface without adding much to the distance of water transport—a matter of great importance when transpiration rates are high. At the end of the rainy season, they quickly die and shrivel (Cannon, 1911, p. 96).

Where water content of the soil is particularly high, the surface soil wet, and the water table shallow, plants generally root shallowly; this habit is primarily a response to lack of aeration. Trees growing in bogs are more shallowly rooted than trees of the same species growing in drier circumstances; some marsh-dwelling trees, e.g., the tamarack *Larix laricina*, develop no taproot. In certain palms and mangroves, the roots may grow not merely horizontally, but actually upward! (Weaver and Clements, 1938, p. 299). The roots of many marshland plants contain large internal air spaces, through which gas exchange occurs; in some species, e.g., the willow *Salix negra*, roots continue growing even in the absence of atmospheric oxygen. Such roots would be expected to collapse in a special way when they decay, and their root casts should thus to be readily distinguishable; but no studies on this point have yet been published.

Soil particle size also serves as a control on root growth, in part because moisture adheres to exterior surfaces of particles (more surfaces are available in fine soil), and in part because coarser soils tend to be more permeable. Coupland and Johnson (1965), in a study of grassland plants of Saskatchewan, found that, in general, the coarser the soil, the deeper the roots; the pasque flower (*Anenome patens* var. *wolfgangiana*) was exceptional, rooting most deeply in moist soils. (See Fig. 10.4.) Orientation of root molds in sediments from semiarid or arid conditions may reflect soil permeability; Glennie and Evamy (1968) found root-molds directed parallel to sediment lamination in dune sands, apparently along levels where the supply of intergranular water was maximal.

In saline environments, roots are developed preferentially at levels above or below the layers of maximum salt concentration. (See Chapter 5.) Some plants can withstand salt concentrations encountered at depth (when the plant already has an extensive root system at shallower depths) that would be fatal to seedlings and younger plants. Excessive soil acidity likewise may be adverse to root development, resulting in short roots, often very much thickened and having abrupt ends; these produce very

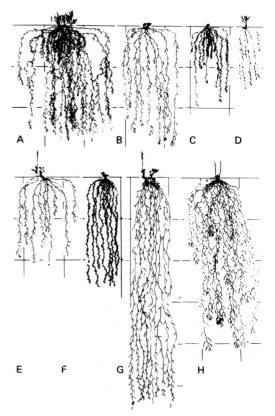

Fig. 10.4 Variations in root system of grassland plant (*Artemisia frigida*) developed in different environments of Saskatchewan. A, stabilized sand. B, south-facing loam slope in the dark brown soil zone. C, north-facing loam slope in the dark brown soil zone. D, top of loamy knoll in the dark brown soil zone. E, level black loam soil. F, level dark brown loam soil. G, lower loamy slope in the dark brown soil zone. H, level brown loam. (From Coupland and Johnson, 1965.)

distinctive casts (Weaver and Clements, 1938, p. 303). The presence of "hardpan" layers in soil may entirely prevent root penetration, and result in shallow root systems.

Roots may also directly reflect the amount of light available to the shoot. Plants adapted to shade or diffuse light develop more elaborate root systems when in full sunshine, where transpiration demands are greater. White pine seedlings, where grown in darkness, are tall and have poorly developed roots; in diffuse light, they are shorter, having long, branching roots; in full light, they are short and stocky, and again have long, branching roots (Weaver and Clements, 1938, p. 309).

Fossil Root Molds and Casts

The stratigraphically earliest report of fossil root molds is perhaps Barrell's (1913) account of rootlet impressions on fracture surfaces in Upper Devonian argillaceous sandstones of the Appalachians. Fossil roots, root casts, and molds have since been reported in abundance from younger continental sediments, especially in underclays (seat earths).

In simplest form, the roots may consist of vertical cylindrical tubes, which may be infilled with sand or other sediments to form reed casts (see Osborne, 1948). The terms "root mark" or, where appropriate, "root cast," are applied to more complex structures, ramifying downward from an original central tube or group of tubes. If the root shrinks away from the sides of the hole that it has made—after the death of the plant—the surrounding cavity may be infilled by mineral matter, and a structure having a central cavity may be produced after the roots have completely decayed away. Jenkins (1925, p. 241) found calcareous tubes formed around carbonized rootlets in a lacustrine sand of the Pend Oreille River, Washington, U.S.A.; Claxton (1970) described cylindrical tapering structures, formed by ferruginous material, in the alluvial sands of the Trent Valley, England, believed to have formed around taproots of the thistle *Cirsium lanceolatus*, but which were actually markedly larger than the original root mold would have been.

Despite the abundance of preserved records, fossil root marks and casts have received little attention from geologists, and their paleoecological potential has scarcely even begun to be examined.

STROMATOLITES

Unlike the other groups of plant trace fossils treated in this chapter, stromatolites (Fig. 10.5) have long been a focus for attention by paleontologists and are the subject of an abundant literature. Late Precambrian and early Paleozoic forms have attracted special attention; the history of their study was reviewed by Hofmann (1969, 1973) and, more briefly, by Walter (1972). Most specialists would agree that stromatolites are, at very least, biogenic sedimentary structures; whether they are strictly "trace fossils," or whether they should be regarded merely as biostratification structures, however, is a matter for argument (see Table 3.1). Because of the doubt concerning their admissibility as trace fossils, only a very brief treatment of stromatolites is attempted here.

Hofmann (1969, p. 6) gave the following definition:

> . . . a millimetre- to decametre-sized organosedimentary structure whose growth is recorded by a succession of laminae. These laminae represent intervals of accumulation of fine sediment on surfaces presumed to have been populated by a community of microorganisms. The sedimentary material is accumulated by trapping or agglutination of particles from suspension on the organic film, or by direct or indirect precipitation resulting from the metabolic activity of elements of the microbiota.

Characteristics of Stromatolites

Although present-day stromatolites are considered to be formed in different settings by several different groups of organisms, they seem to be formed predominantly by blue-green algae. Logan et al. (1964, p. 69) considered that the organic film is probably a complex of filamentous and unicellular algae, composed of both green and blue-green algae (Chlorophyta and Cyanophyta). Stromatolites are most typical of the shallow subtidal, intertidal, and supratidal zones, although they may also occur at greater depths. Their gross morphology (Fig. 10.6) is determined by the characteristics of the environment in which they develop: depth, degree of turbulence, salinity, and carbon-

Fig. 10.5 Modern stromatolites exposed at low tide, Carbla Point, Shark Bay, Western Australia. The smaller, light-colored mounds at center and lower front are active; the numerous dark mounds are inactive, except for growth at the sides. Ruler is 15 cm long. (Photo courtesy H. J. Hofmann.)

ate concentration of the seawater, nature and slope of the seafloor, the direction of current flow or tidal onset, and numerous other factors. The duration of the process of accretion of laminae profound affects not only the size but also the shape of the stromatolite. As a result of these factors, the character of stromatolites is highly variable, from essentially horizontal structures to convex mounds, columns, or ramifying groups in which component columns may be similar or markedly dissimilar in height.

Some stromatolites are moved passively by tides or currents, growing by peripheral encrustation; these structures are termed oncolites and typically exhibit a rounded form. Other subtidal stromatolites may be

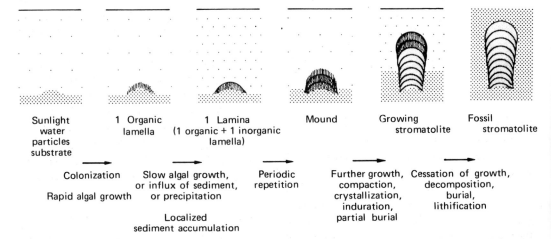

| Sunlight water particles substrate | 1 Organic lamella | 1 Lamina (1 organic + 1 inorganic lamella) | Mound | Growing stromatolite | Fossil stromatolite |

Colonization — Rapid algal growth
Slow algal growth, or influx of sediment, or precipitation — Localized sediment accumulation
Periodic repetition
Further growth, compaction, crystallization, induration, partial burial
Cessation of growth, decomposition, burial, lithification

Fig. 10.6 Sequence of events in the development of a marine algal stromatolite. (From Hofmann, 1969.)

torn up by unusually strong currents, transported downcurrent, and buried in sediment; these may be distinguished from oncolites, however, by the character and orientation of the laminae.

In intertidal environments, stromatolites do not grow in runoff channels, where the scouring effects—not only of backwash but also of translatory wave motion—are concentrated and thus destroy the organic film. The algae also avoid the bottoms of splash pools, apparently because permanent or semipermanent wetting prohibits growth (see Logan, 1961). Under these conditions, and in the supratidal environment· where the algal mats suffer especially prolonged periods of exposure to the atmosphere, recurrent desiccation may convert the algal mats into a mosaic of polygonal plates. In any environment, unusually heavy sedimentation may halt the development of stromatolites temporarily or permanently.

Stromatolites are not formed only in marine environments, however. Monty (1972) described in detail the development of present-day stromatolites in seasonal lakes on Andros Island, Bahamas, formed entirely by algae and having a fine-grained, homogeneously laminated structure; the laminations are produced not by sedimentary influx but by seasonal halts in growth. Walter et al. (1972) described stromatolites, formed by algae and photosynthetic flexibacteria, from hot spring and geyser effluents in Yellowstone National Park, Wyoming, U.S.A. These stromatolites differ from most others hitherto described in that they are composed of nearly amorphous silica, not of calcareous material; in other respects, they are so strikingly similar in morphology to certain Precambrian marine stromatolites that the authors suggested that the latter might also have had (in part at least) a bacterial origin.

The form of the stromatolite (Fig. 10.7) is thus an expression of its environment and not of precise biological affinity. Particular algal species (or groups of species) can produce markedly dissimilar

stromatolites under different conditions; conversely, unrelated species or markedly different groups of species may produce closely similar stromatolites where determined by the character of the environment. Thus, although evolution must certainly have occurred within the algal groups that gave rise to the structures, trying to elucidate evolutionary lineages from changes in the overall shape of stromatolites is extremely hazardous. Unless the algae themselves are preserved in the sediment fabric (a rare occurrence), the only direct indication of taxonomic affinity is the microfabric—and often this is destroyed by diagenetic and postdiagenetic processes. In consequence, although workers are generally agreed that a binomial nomenclature should be applied to stromatolites, they have expressed profound disagreements about the choice of characteristics for delimiting form genera and species (see discussion in Hofmann, 1969).

Fossil Stromatolites

Stromatolites have been reported from all levels of the geological column, from Precambrian (Proterozoic or earlier)[5] to the present. Because they are the largest and most conspicuous fossils of the Proterozoic, stromatolites at this level have attracted the most attention from stratigraphers. Although the dangers inherent in using such variable structures in stratigraphical correlation are obvious, stromatolites have been successfully employed for this purpose in Asia, Australia, and northwest Africa. [For assessments of their use in Precambrian and Cambrian stratigraphy, see Cloud and Semikhatov (1969), Awramik (1971), Preiss (1972), and Walter (1972).] At other stratigraphic levels, their greatest importance is likely to be as indicators of environment

---

[5] Schopf et al. (1971) reported stromatolites from the Bulawayan Group of Rhodesia, dated radiometrically at more than 2.6 billion years old. (See also Button, 1973.)

Laminae		
Components	**Laminae: light and dark laminae**	
Microstructure* — Microfabric	▨ Ribboned ▨ Distinct ▨ Striated ▨ Diffuse ▨ Lumpy Massive	
Microstructure* — Cellular biofabric	▨ Porostromatid	

Configuration — Curvature type and order	⌒ Even
	Crinkled
	   ⌒ Wavy 1st ⌒ 1 Order
	   ∿ Corrugate 1st 2nd ⌒ 2 Orders
	   ∿ Crenate 2nd 3rd
	   ⌄⌄⌄ Dentate 1st ⌒ 3 Orders	

Configuration — Profile	— Flat
	⌒ Convex
	⌣ Concave
	Inflexed { Very acute
	    ∧ Angulate Acute } Symmetrical
	    ∧ Geniculate Obtuse Asymmetrical
	    ∧ Cuspate Very obtuse
	Globoidal
	   ○ Penecinct
	   ○ Plenicinct
	   ▨ Obscure	

Configuration — Plan outline	○ Round, circular, elliptical, ovate
	⬭ Oblong
	▽ Scutate
	▽ Crescentic
	Lobate
	   ⧈ Laxilobate
	   ⧈ Densilobate
	   ♡ Brevilobate
	   ○ Polygonal
	   ⬯ Lanceolate	

Linkage	∿∿ Linked ∿∿ Partly linked ∿∿ Unlinked	
Spacing	∿∿ Contiguous (P = 0) ∿∿ Very close (P ⩽ r) ∿∿ Close (P ⩽ 2r) ∿∿ Open (P > 2r) ⌒ Isolated (P > 20r)	
Relief (relative)	⌒⌒ Low (2r ≫ h) ∿ ∩ Moderate (2r ≈ h) ∩ ⋂ High (2r ≪ h)	
Degree of inheritance	▨ Low ▨ Moderate ▨ High	

*Only very few varieties of microstructure are illustrated here

Fig. 10.7 Graphical summary of the morphological attributes of stromatolites. (From Hofmann, 1969.)

Stromatolites		
Habit of accretion vector	Growth factor	Columnar: Terete, Cylindrical, Turbinate; Bulbous; Nodular; Stratiform; Spheroidal
Habit of accretion vector	Variability	Uniform; Constringed; Ragged; Crustose; Stubby; Slender
Habit of accretion vector	Attitude	Straight; Curved; Centrifugal; Erect; Recumbent; Inclined; Decumbent; Horizontal; Sinuous
Habit of accretion vector	Branching style	Furcate; Umbellate; Digitate; Dendroid; Coalescent; Anastomosed
Surface ornamentation		Smooth; Tuberculate; Fimbriate; Rugate
Internal features		Axial zone; Ribs (plan view); Wall; Mantle; Nucleus

Other attributes		
Dimensional	Thickness of laminae (T)	μ, mm, constant, tapering
Dimensional	Span of laminae (2r) (width of stromatolite)	μ-sized (< 1mm); mm-sized (1-10 mm); cm-sized (1-10 cm); (etc.)
Dimensional	Relief of laminae (h)	μ, mm, cm, dm, etc.
Dimensional	Height of stromatolite (H)	μ, mm, cm, dm, m, etc.
Dimensional	Orientation	Azimuth of elongation; Azimuth and amount of inclination of accretion vector
Material	Petrology, mineralogy	Carbonate: calcite, dolomite, etc.; Silica; Silicates; Oxides, hydroxides (Fe, Mn); Organic matter; Open pore space; Fluids
Positional	Geographic position	Latitude, longitude, elevation
Positional	Geologic setting	Stratigraphic unit; environment
Positional	Geologic age	Years
Taxonomic		Name of originating organism(s): genus, species
Nomenclatorial		Name of stromatolite: group, form

Fig. 10.7 *See facing page for caption.*

[perhaps even nonmarine ones (Button, 1973)].

Although stromatolites are widely represented in Phanerozoic sediments, they are certainly much less prominent than their Precambrian counterparts; indeed, Fischer (1965, p. 1205) considered that they were in steady decline throughout the Paleozoic, playing only a minor role in Mesozoic marine sediments, and virtually disappearing from the geological record in the Cenozoic. Their restriction was attributed by Garrett (1970) and Stanley (1973) to the progressive diversification during the Phanerozoic of grazing animals, which feed on surface algal mats, and of burrowing animals, which destroy sedimentary structures. Their persistence at the present day apparently results primarily from the unusual ecologic tolerance of the blue-green algae, which still inhabit a very wide range of marine environments; had their tolerance been less wide, stromatolites would in all probability be features of the geologic past and might have defied interpretation by geologists today.

ACKNOWLEDGMENTS

In writing this chapter (as a result of discussions with R. W. Frey), I have strayed far outside the fields where I have any research expertise; for these reasons, I am indebted for the guidance on particular points by T. A. Steeves (Biology), R. T. Coupland (Plant Ecology), V. L. Harms (Biology), R. E. Redmann (Plant Ecology), and G. M. Simpson (Crop Science), all of the University of Saskatchewan, and also to Steeves and Coupland for their critical reading of parts of the manuscript. I am also indebted to the various reviewers of the first draft of the manuscript, for their constructive comments.

In addition, I would like to thank the following for permission to reproduce figures: Carnegie Institution of Washington (Fig. 10.3), R. T. Coupland and Messrs. Blackwell Scientific Publications (Fig. 10.4), and H. J. Hofmann and the Director of the Geological Survey of Canada (Figs. 10.6 and 10.7).

REFERENCES

Awramik, S. M. 1971. Precambrian columnar stromatolite diversity: reflection of metazoan appearance. Science, 174:825–827.

Baker, G. 1960. Fossil opal-phytoliths. Micropaleontology, 6:79–85.

Barrell, J. 1913. The Upper Devonian delta of the Appalachian geosyncline. Pt. I: The delta and its relations to the interior sea. Amer. Jour. Sci., 36:429–472.

Blackman, E. 1971. Opaline silica bodies in the range grasses of southern Alberta. Canadian Jour. Bot., 49:769–781.

Bromley, R. G. 1970. Borings as trace fossils and *Entobia cretacea* Portlock, as an example. In T. P. Crimes and J. C. Harper (eds.), Trace fossils. Geol. Jour., Spec. Issue 3: 49–90.

Button, A. 1973. Algal stromatolites of the early Proterozoic Wolkberg Group, Transvaal Sequence. Jour. Sed. Petrol., 43:160–167.

Cannon, W. A. 1911. The root habits of desert plants. Carnegie Inst. Wash., Publ. 131, 96 p.

Claxton, C. W. 1970. Cylindrical tapering structures in the alluvial sands of the Trent Valley. Mercian Geol., 3:265–267.

Cloud, P. C. and M. A. Semikhatov. 1969. Proterozoic stromatolite zonation. Amer. Jour. Sci., 267:1017–1061.

Coupland, R. T. and R. E. Johnson. 1965. Rooting characteristics of native grassland species in Saskatchewan. Jour. Ecol., 53: 475–507.

Crookall, R. 1929. Coal Measure plants. London, Arnold, 80 p.

Deflandre, G. 1963. Les Phytolithaires (Ehrenberg). Protoplasma, 57:234–259.

Ehrenberg, C. G. 1841. Über Verbreitung und Einflüss des mikroskopischen Lebens in Süd- und Nordamerika. Monatsber. Preuss. Akad. Wiss., Berlin:139–144.

————. 1846. Über die vulkanischen Phytolitharien der Insel Ascension. Monatsber. Preuss. Akad. Wiss., Berlin:191–202.

Ellis, D. 1914. Fossil micro-organisms from the Jurassic and Cretaceous rocks of Great Britain. Royal Soc. Edinburgh, Proc., 35: 110 et seq.

Elsik, W. C. 1966. Biologic degradation of fossil pollen grains and spores. Micropaleontology, 12:515–518.

Fischer, A. G. 1965. Fossils, early life, and atmospheric history. Nat. Acad. Sci. U.S., Proc., 53:1205–1213.

Garrett, P. 1970. Phanerozoic stromatolites: noncompetitive ecologic restriction by grazing and burrowing animals. Science, 169:171– 173.

Gaumann, E. A. 1952. The fungi. (Transl. by F. L. Wynd.) New York, Hafner, 420 p.

Glennie, K. W. and B. D. Evamy. 1968. Dikarka; plants and plant root structures associated with aeolian sand. Palaeogeogr., Palaeoclimatol., Palaeoecol., 4:77–87.

Hofmann, H. J. 1969. Attributes of stromatolites. Geol. Surv. Canada, Paper 69-39:1–58.

————. 1973. Stromatolites: characteristics and utility. Earth-Sci. Rev., 9:339–373.

Jenkins, O. P. 1925. Clastic dikes of eastern Washington and their geological significance. Amer. Jour. Sci., 10:234–246.

Kaufman, P. B. et al. 1970. Ultrastructural studies on cellular differentiation in internodal epidermis of *Avena sativa*. Phytomorphology, 20:281–309.

Logan, B. W. 1961. *Cryptozoon* and associated stromatolites from the recent, Shark Bay, Western Australia. Jour. Geol., 69:517–533.

———— et al. 1964. Classification and environmental significance of algal stromatolites. Jour. Geol., 72:68–83.

Lutwick, L. E. and A. Johnston. 1969. Cumulic soils of the rough fescue–prairie poplar transition region. Canadian Jour. Soil Sci., 49:199–203.

Maslov, V. P. 1960. Stromatolity . . . Akad. Nauk S.S.S.R., Trudy Geol. Inst., 41, 188 p. (in Russian)

Monty, C. L. V. 1972. Recent algal stromatolitic deposits, Andros Island, Bahamas. Preliminary report. Geol. Rundschau, 61:742–783.

Moore, L. R. 1963. Microbiological colonization and attack on some Carboniferous miospores. Palaeontology, 6:349–372.

Osborne, G. D. 1948. A review of some aspects of the stratigraphy, structure and physiology of the Sydney Basin. Linnean Soc. New South Wales, Proc., 73:vii.

Preiss, W. V. 1972. Proterozoic stromatolites— succession, correlations and problems. In J. B. Jones and B. McGowan (eds.), Stratigraphic problems of the later Precambrian and Early Cambrian. Univ. Adelaide Centre Precamb. Res., Spec. Paper 1:53–62.

Santesson, R. 1939. Amphibious pyrenolichens. I. Arkiv. Bot., 29A(10): (not seen).

Scagel, R. F. et al. 1965. An evolutionary survey of the plant kingdom. Belmont, Calif., Wadsworth Publ. Co., 658 p.

Schopf, J. W. et al. 1971. Biogenicity and strati-graphic significance of the oldest known stromatolites. Jour. Paleont., 45:477–485.

Stanley, S. M. 1973. An ecological theory for the sudden origin of multicellular life in the late Precambrian. Nat. Acad. Sci. U.S.A., Proc., 70:1486–1489.

Twiss, P. C. et al. 1969. Morphological classification of grass phytoliths. Soil Sci. Soc. America, Proc., 33:109–115.

Walter, M. R. 1972. Stromatolites and the biostratigraphy of the Australian Precambian and Cambrian. Palaeont. Assoc., Spec. Papers Palaeontology 11, 190 p.

———— et al. 1972. Siliceous algal and bacterial stromatolites in hot spring and geyser effluents of Yellowstone National Park. Science, 178:402–405.

Weaver, J. E. and F. E. Clements. 1938. Plant ecology (2nd ed.). New York, McGraw-Hill, 601 p.

Wetzel, W. 1967. Die geologische Bedeutung der Phytolitharien, ein besonderes Kapitel der Mikropaläontologie. Neues Jahrb. Geol. Paläont., Mh., 12:731–735.

Wilding, L. P. and L. R. Drees. 1968. Biogenic opal in soils as an index of vegetative history in the Prairie Peninsula. In E. Bergstrom (ed.), The Quaternary of Illinois. Univ. Illinois, Coll. Agriculture, Spec. Publ. 14:96–105.

CHAPTER 11

BORINGS AS TRACE FOSSILS, AND THE PROCESSES OF MARINE BIOEROSION

JOHN E. WARME

Department of Geology, Rice University
Houston, Texas, U.S.A.

SYNOPSIS

Marine borers are nearly ubiquitous in the modern seabed. Their distinctive excavations provide abundant potential trace fossils, and their general erosional activities (bioerosion) are important factors in marine sedimentation and benthic ecology.

Species of excavators include protozoans, plants, and animals. Those best studied are boring fungi, algae, sponges, sipunculids, polychaetes, gastropods, bivalves, and echinoids; however, several other groups also contain species of borers.

Many excavations in the fossil record are distinctive enough to be identified and named as trace fossils. Rock borings mark ancient shorelines, hardgrounds, and unconformities.

Shell borings are useful for paleoecologic and taphonomic reconstructions.

Most borers penetrate for their protection, but in the process they sculpt and significantly bioerode exposed substrates. This process is especially evident on coral reefs and limestone outcrops, because these substrates are very susceptible to biochemical attack.

The mechanisms of penetration are understood for very few borers. Certain borers use solely mechanical processes, whereas others supplement or replace these with chemical means. Although bioerosion is an important natural process, little is known of the rates at which it proceeds.

INTRODUCTION

The wide pervasion and importance of marine borers and excavators generally is underestimated by biologists, paleontologists, and sedimentologists. The excavators attack rocks, shells, calcareous tests, wood, peat, carbonate sand grains, and even man-made materials. They belong to diverse taxonomic groups, and through their activities they produce abundant and varied borings that are potential trace fossils. Many such excavations are indeed recognized in the fossil record.

The term "bioerosion" was proposed by Neumann (1966) for "erosion of substrate by means of biological procedures." However, in addition to borers that produce distinct holes, bioerosion is accomplished by forms that nestle, scrape, rasp, gnaw, or otherwise cause attrition of the seabed or associated materials. Thus, a spectrum of bioerosion processes exists that is significant for marine sedimentation and ecology, as well as for the study of trace fossils.

At least 12 phyla of animals, as well as several groups of plants and protozoans, contain marine borers. On a worldwide basis, the following are the most obvious and probably the most important excavators: microscopic fungi and algae, clionid sponges, sipunculid worms, polychaete annelids, pholadid and mytilid bivalves, amphineurans, gastropods, crustaceans, echinoids, and fishes. Different groups may be the chief excavators in local situations, such as clionid sponges on limestone shores, spionid polychaetes in oyster shells, parrot fishes in reef areas, or pholadid bivalves on mudstone or sandstone shoreline benches. As yet, not enough is known about most of these groups to allow prediction of which species or higher taxonomic units will be the dominant excavators, or even be present, in a given area.

Because some groups of simple organisms—such as algae, fungi, and foraminifers—contain species that clearly are borers, no group can be excluded as potential borers simply because this habit seems to be an unlikely circumstance. Certain borers are well known because they are obvious and abundant (e.g., pholadid bivalves), some are well-known borers but are poorly understood (e.g., boring sipunculids), and others generally are not recognized as species that bore (e.g., certain foraminifers). Some large groups, such as the coelenterates, have not been studied as eroders, but they probably contain many species that bore or at least corrode the substrate (e.g., holdfasts of gorgonians).

Marine bioeroders play important ecological and sedimentological roles. Vast intertidal exposures, subtidal outcrops, and carbonate reefs harbor a myriad of borers belonging to many groups of animals and plants. Beneath the often-crowded surface of attached and encrusting organisms is an endolithic assemblage that may be fully as diverse as the more obvious epilithic species; the excavations are apparent only where the substrate is naturally broken or eroded, or specially prepared (Fig. 11.1).

Fig. 11.1 Encrusted and bored rock. Left side encrusted with corals, bryozoans, vermetid gastropods, attached bivalves, etc., somewhat protecting the underlying rock. Right side shows several generations of borings, mostly by bivalves, in various stages of exposure as the rock was worn away. A, irregular, scratched openings into rock, many of which are interconnected within. B, exposed posterior end of calcareous sheath secreted by the boring bivalve *Lithophaga plumula* and bored by clionid sponges. 20 m depth; Scripps Submarine Canyon, California.

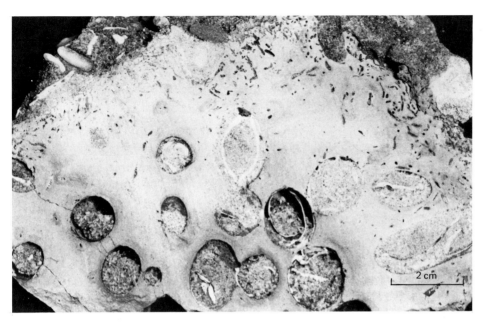

Fig. 11.2 Miocene siltstone containing borings by Pleistocene bivalves (*Lithophaga* sp.) and polychaetes (probably Spionidae). Rock was exposed on flank of a marine terrace, bored, then covered with sand that filled the larger borings. Newport Bay, California.

Outcrops on sandy shores, in submarine canyons, or elsewhere may not become fully colonized, or they may be colonized sporadically because of shifting sediment, ice formation, or other natural events; nevertheless, the exposures are bored occasionally and thus contain potential trace fossils (Fig. 11.2).

At present, the thrust of biological research on marine borers is to document the species that bore and to study their ecology, physiology, and functional morphology as related to methods of penetration. Much effort has been directed toward understanding the destructive wood-boring bivalves (*Teredo* and *Xylophaga*) and arthropods (*Limnoria* and *Sphaeroma*) for application to their control (e.g., Ray, 1959). Some work has been accomplished on calcareous reef and limestone borers (e.g., Otter, 1937; Yonge, 1963a), but little attention was paid to the physical borings themselves, which constitute potential trace fossils. Fossil borings have been used notably to interpret the life habits and taphonomy of fossil shells (Boekschoten, 1967) and bones (Boreske et al., 1972), the environmental history of ancient indurated seabeds (Rose, 1970, 1972;

Boyd and Newell, 1972), and to document uplifted shorelines (e.g., Bradley, 1956; Higgens, 1960).

This review is limited to marine excavators, although various substrates in freshwater and terrestrial environments also are attacked. (See Chapter 10.) Marine borers are best known from shallow water because of the technical difficulties of studying them in deeper habitats. Herein, more emphasis is placed on recent borers than on fossil ones, because recent ones have been studied more thoroughly, and with few exceptions, such studies have not been applied to the geological record. Other chapters in this volume contain more detailed discussions of algal and fungal microborers (Chapters 10 and 12), records of shell drillers and other predators (Chapter 13), and ancient limestone borers (Chapters 17 and 18).

Comments on Terminology

The term "bioerosion," as used herein, includes distinctive borings as well as other processes of biologic erosion, such as rasping and scraping.

Confusion and lack of agreement exist over the terms "borer, boring, burrower, burrow," and such synonyms as "penetrant, driller, excavator," etc. "Boring" is used herein simply for penetration of any hard substrate and "burrow" for penetration of soft ones; in neither case is a particular method of penetration implied. Choice of this convention follows the suggestions of Bromley (1970a), Alexandersson (1972), Frey (1973), and Golubic et al. (Chapter 12), and stems from traditional use by early workers. Osler (1826) used both terms without defining them, but confined boring to excavations in hard rock, such as those made by the bivalves *Pholas* and *Saxicava*. Carriker and Smith (1969), however, recommended that the term "burrow" be used for excavation of a domicile, in which much or all of the excavator resides; they reserved "boring" for peneration in quest of food, such as gastropod drillings in bivalve shells. These definitions do not distinguish between penetration of a hard versus a soft substrate (Warme et al., 1971), which is of paleoecological importance. R. G. Bromley (1973, personal communication) pointed out that the definitions by Carriker and Smith require that fungal excavations be classed as borings whereas very similar algal excavations be classed as burrows.

McLean (1972) proposed that any rock eroder be termed "lithophagic," following the terminology of some previous workers (e.g., Raynaud, 1969), and that surficial, shallow, and deep excavations be designated as formed by epilithophagic, mesolithophagic, and endolithophagic species, respectively. Such a scheme is useful, but the terms give the undesirable connotation that the rock is eaten. Golubic et al. (Chapter 12) adopted similar prefixes, using the terms "epi-, meso-, endo-," and "cryptolithic," the latter referring to cavity dwellers. In this chapter, I have not attempted to assign all marine borers to such categories.

Designation of an excavation as a soft-substrate burrow or a hard-substrate boring is a difficult question in some ancient settings because (1) contemporaneous substrate layers may vary in hardness, especially in carbonate terrains (Shinn, 1969), and (2) some burrows are very similar to borings, especially where separate but related taxa of excavators have different substrate-hardness thresholds (Warme, 1970). Certain species of pholadid bivalves penetrate rocks that contain constituents harder than their shells; others penetrate soft rock, shells, wood, or semiconsolidated or loose sediment, yet all species possess similar basic shell morphology. Furthermore, species of the same genus, or even individuals of the same species, may tolerate a wide range of hardnesses, e.g., the common polychaete *Polydora*, different species of which bore into cherty limestone, shell, soft mudstone, and even burrow into very soft, watery mud (Boekschoten, 1966, p. 353). In each case, basic geometry and size of domiciles are similar.

Bromley (1970a) used "drill" for penetration for food, such as the drilling gastropods. He also employed "embed" for borers that reside on the surface of living corals, etc., and are overgrown and embedded by their host; to differentiate between these concepts is important.

Carriker and Smith (1969) also suggested using the term "substratum" for the surface that is penetrated. Discounting the widespread use of the term "substrate" in the literature, they suggested that it should be restricted to bacterial culture media, and that "substratum" be used in benthic ecology. I have retained the word substrate, because of long-standing use by geologists and biologists (see Frey, 1973).

Past Work

The literature on marine excavators is scattered, and most papers treat only one group or occurrence. Some of the more important published reviews that have appeared during the last 15 years are summarized in Table 11.1. Other important papers are stressed at appropriate places in the text.

ECOLOGY AND PALEOECOLOGY OF MARINE EXCAVATORS

Why Excavate?

Borers, like burrowers, penetrate the substrate; but boring usually requires more time. To my knowledge, most rock borings are protective domiciles; none are constructed for the primary purpose of obtaining food, except for (1) boring fungi—seeking organic skeletal matrix, and (2) scraping gastropods, amphineurans, and echinoids—grazing intertidal surfaces for epilithic and endolithic floras, and perhaps subsequently resting in the scraped excavations. Most borings in skeletons also are protective, but some gastropods, octopus, and perhaps other groups bore into live shells in order to devour the contents. In addition, certain borers construct domiciles within shells of living animals, as do spionid worms in oyster shells, orienting their excavations so as to take advantage of feeding or other currents created by the host, and receiving both protection and food. Grazing forms, such as echinoids, amphineurans, and gastropods, commonly can move away from their excavations; but most borers remain in their respective excavations and eventually become trapped. Few are attached physically to the walls of their borings; known exceptions are *Lithotrya*, a rock-boring barnacle, and certain byssate bivalve borers.

Extensively excavated rock and shell surfaces commonly are the result of several generations of inhabitants that may have included different species; the product is a cavernous substrate teeming with many species of borers and nonborers alike (Warme et al., 1971).

Fossil Borings

Considering the widespread boring in rocks and shells on modern coasts, disappointingly few fossil examples are documented.

TABLE 11.1 Some of the Chief Bibliographic Sources on Marine Borers.

Authors	Date	Orientation	No. of Pages	Remarks
Menzies	1957	Marine borers	6	50 references; annotated bibliography only
Ray	1959	Symposium on marine borers and foulers	536	34 papers; largely restricted to wood borers
Clapp and Kenk	1963	Marine borers	1,136	4,045 references; comprehensive annotated bibliography, cross-referenced
Sognnaes	1963	Symposium on destruction of hard tissue	764	26 papers total; 3 papers on boring; medical and dental aspects mainly
Boekschoten	1966	Shell borers	46	110 references
Carriker et al.	1969	Symposium on borers in calcareous substrates	391	37 papers; useful and current
Bromley	1970a	Marine borers—recent and fossil	41	303 references; concise and comprehensive
Crimes and Harper	1970	Symposium on trace fossils	547	35 papers; 11 papers on borers; good guide to European literature
Alexandersson	1972	Carbonate microborers	35	115 references; excellent review of microborers

However, certain bored unconformities, hardgrounds, and boulder beds are well understood (Hecker, 1970; Radwański, 1970; Roniewicz, 1970; Rose, 1970, 1972; Perkins, 1971; McCrevey, 1974; see Chapters 17 and 18), and the paleontological record of some shell borers is known (e.g., Cameron, 1969; Seilacher, 1969; Sohl, 1969; Gatrall and Golubic, 1970).

Shell borers probably are better studied than rock borers because shells provide a durable substrate more often preservable in the rock record, and generally are investigated more carefully by paleontologists than are rock surfaces or clasts (e.g., Clarke, 1921; Elias, 1956; Rodda and Fisher, 1962; Boekschoten, 1967; Richards and Shabica, 1969; Rodriguez and Gutschick, 1970). Borings in shells are usually small, distinctive, and seldom apt to be confused with soft-sediment burrows. Problems in interpretation arise because some borings in rock are nearly identical in size and geometry with excavations of soft-sediment burrowers (e.g., Seilacher, 1969, Pl. 1; Warme, 1970, Pl. 4).

Radwański (1968), Perkins (1971), and Kennedy and Klinger (1972) described bored boulders that were excavated on all sides (see Fig. 11.3). This distribution of penetrations does not necessarily indicate that the boulders were turned over on the seabed; pieces of limestone debris on modern seabeds often have underlying hollows that permit free circulation of water and thus harbor a variety of life. The hollows are maintained by the activity of fishes and invertebrates, as well as by physical processes such as wave surge.

Methods of Penetration

The precise method of excavation is known for only a few borers. Based on what is known, however, many of them very likely use strictly mechanical methods, and all borers probably use some amount of mechanical action, which may or may not be aided by biochemical processes. In early studies, litmus paper or other chemical indicators commonly were used to demonstrate acid as a chemical aid, with equivocal results (see Clapp and Kenk, 1963).

Probably the best understood borer is the gastropod *Urosalpinx cinerea*, which drills other shelled mollusks and devours the contents (Carriker, 1969; Smarsh et al., 1969). This species utilizes carbonic anhydrase as a catalyst to form carbonic acid and soften the prey's shell, which then is more susceptible to radular rasping. Biologists

Fig. 11.3 Limestone boulder bored on top and bottom by bivalves and other borers. Lower Cretaceous, Texas.

seek carbonic anhydrase in biochemical systems as a possible aid in boring; it probably fulfills this function in several species (Carriker and Smith, 1969), and is suspected in many others (e.g., Tomlinson, 1969b, p. 840). Other chemical methods also are employed and perhaps should be expected among various animals not yet studied (Jaccarini et al., 1968; Carriker and Smith, 1969).

In most cases where chemically aided boring is documented or suspected, the biochemical activity clearly is accompanied by mechanical action. For instance, boring sponges (Warburton, 1968) and mytilid bivalves (as discussed below) excavate sand-size chips or finer powder, and do not completely dissolve even pure calcium carbonate substrates. *Urosalpinx* alternates biochemical softening with radular attack. This combination of methods seems to be metabolically economical.

Habitat Distribution

Rock borers exist in most marine habitats, and are especially obvious in soft rocks exposed in the intertidal and shallow subtidal zones. In general, more opportunities for chemical borers exist in carbonate terrains of the tropical belt, but strictly mechanical borers can extend into temperate and boreal seas. Boring bivalves exhibit this segregation: borers that penetrate with the aid of chemical means, such as *Lithophaga* and probably *Gastrochaena*, are most abundant and diverse in low latitudes, whereas primarily mechanical borers, such as the pholadids and *Hiatella*, predominate in higher latitudes. Clionid sponges are restricted to pure, calcareous substrates, but exist at all latitudes. The dearth of formation of modern shallow-water limestone in higher latitudes restricts clionids to shells, other calcareous skeletons, or outcropping ancient limestone. So little is known about the methods of penetration of members of geographically widespread groups, such as the polychaetes and sipunculids, that

further statements about the habitat selection of mechanical and chemical borers would be premature.

Tropical Reefs and Coasts

The importance of borers on modern coral-algal reefs and limestone shores was emphasized in the comprehensive surveys by Gardiner (1902, 1903) in the Indian Ocean, Otter (1937) on the Great Barrier Reef, Ginsburg (1953) in Florida, Cloud (1959) on Saipan, Neumann (1966) on Bermuda, and the reports of Newell (1954) and numerous others in the Atoll Research Bulletin (U.S. National Museum), and in widely scattered papers of tropical biological stations and elsewhere. The subject was reviewed by Yonge (1963a). On some limestone shores, excavators play important if not commanding roles in attacking and molding such large-scale geographic features as organic reefs and seacliffs, as well as reducing substrates to sand- and mud-size particles.

Using SCUBA, one can observe and collect in situ borings on reefs. Our work in Yucatan and Jamaica suggests that all exposed surfaces in water depths to 30 m—not occupied by living corals, sponges, etc.—are infested with borers.

The depth limits of reef borers are not known. Neumann (1966) clearly documented boring clionid sponges as being responsible for the shallow subtidal nip in Bermuda. Goreau and Hartman (1963) emphasized erosion by sponges to 60 m depth on the north coast of Jamaica, and recent collecting to 300 m there by submersible (L. S. Land, 1973, personal communication) yielded limestone riddled with live sponge.

Comparison of bored substrates from the open reef and forereef with those at similar depths (0 to 30 m) in sheltered bays of Jamaica suggests that the latter habitats harbor a more diverse fauna of borers. Cloud (1959, p. 422) found more sipunculids in the lagoon at Saipan than offshore. This may be a general relationship in many areas, and is worthy of investigation. In quiet water, dead coral fragments of all sizes teem with borers and epifauna. On the reef crest and in the surge-washed habitats seaward of it, only coarser coral fragments remain and are attacked by the borers.

Ecologically, borers are important on reefs because they greatly increase species diversity, as well as biomass, and they sculpt reef morphology. Geologically, the borers destroy any available surface—on all scales: e.g., algae bore sand-size grains; sipunculids, polychaetes, echinoids, and other borers help level reef flats; and bivalves, sponges, and other animals help to topple coral colonies by excavating the dead bases.

Limestone and coral borers also create and enlarge cavities within the substrate, resulting in a framework that is porous or even cavernous (Storr, 1964; Reidl, 1969). Reef cavities become centers for cementation and related diagenetic processes after burial. This important role of borers in the "fossilization" of some reefs now is becoming clear (Zankl and Schroeder, 1972; Schroeder, 1972; Scoffin, 1972).

Terrigenous Coasts

Borers can be exceedingly numerous in terrigenous sedimentary rocks along coasts.

Fig. 11.4 Eocene siltstone broken along bedding plane to expose: calcareous sheaths of the bivalve *Lithophaga plumula* (A), part of a pholadid bivalve boring having shell scratches (B), and irregular tunnels probably sculpted by arthropods and annelids (C). Small polychaete boring (D), probably a spionid, branches from a main tunnel. Scripps Submarine Canyon, California.

Jehu (1918) reviewed the early literature on borers around Great Britain, and suggested that their activity aids serious coastal erosion in some localities (rates are discussed below). Sandstones and mudstones along the California–Oregon coast are extensively eroded by pholadids (Evans, 1968a, 1968b), as well as other borers that crowd in favorably exposed rocks (Fig. 11.4) (Warme and Marshall, 1969; Warme, 1970; Warme et al., 1971; McHuron, 1972). Even in these places, however, calcareous shells and clasts are sought by chemical borers. Borings in wave-cut terraces (Bradley, 1956), as well as in boulders and sea stacks (Higgens, 1960), were used to document relative sea-level changes on the California coast. In the former case, pholadid shells within the borings gave carbon-14 dates to document the time of movement.

Deep-Water Habitats

On the forereef slopes of reefs (30 to 300 m), sponges actively bore exposed limestones (see above), and additional samples from these depths would probably reveal other taxa of borers. Bivalves, polychaetes, sipunculids, and other groups bore terrigenous rocks on the upper rims of submarine canyons of western North America (15 to 45 m—Warme and Marshall, 1969; Warme et al., 1971) and galatheid crabs excavate mudstone near the canyon floors (300 m—Warme, 1971, p. 70). Similar borings occur in outcrops exposed on Atlantic coast canyons at depths of about 800 m (Dillon and Zimmerman, 1970).

Sedimentary clasts dredged off California commonly show polychaete and bivalve borings (see section on polychaetes, below). Samples that we have collected there by submersible, at depths to 2,000 m, contained empty borings of these groups, as well as a small, live sipunculid. Only sedimentary rocks were bored; igneous and metamorphic fragments appeared untouched.

A fascinating problem is the widespread occurrence of bored mudstone and chalk

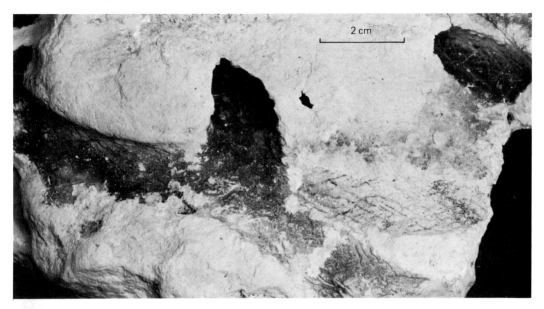

Fig. 11.5 Deep-sea chalk boulder, riddled with large tunnels ornamented by scratch marks. Dredged from Caribbean.

recovered from bathyal and abyssal depths; the circular or irregular borings exhibit scratch marks (Fig. 11.5). They are known from the deeper parts of submarine canyons in the Bering Sea (Fig. 11.6) (Scholl et al., 1970); from Caribbean basins; from the flanks of the Atlantis Fracture Zone (2,900+ m) near the Midatlantic Ridge (Scott et al., 1972); and from other widespread localities (J. L. Worzel, 1972, personal communication). The cross-hatched "scratch marks" on the tunnel walls are similar to those ascribed to crustaceans in chalk–limestone sequences now exposed on land (Bromley, 1967). Because some or all of the excavations are coated or completely blocked by manganic crusts, we are not sure when the holes were made or what might be their paleobathymetric significance. They are very likely being created by some unknown excavator(s) today in the deep sea.

SEDIMENTOLOGICAL SIGNIFICANCE OF MARINE BORERS

Physical Processes Versus Bioerosion

Knowledge of the extent of erosive activities and of the sedimentological importance of marine borers is meager. We know that in many marine locations virtually every exposed surface shows some sign of boring, and that bioerosion is a significant erosional process. Countless clasts in the sedimentary record were produced by the excavating activities of marine borers.

Erosion of rock attacked by borers is aided by waves and tidal action. Commonly,

Fig. 11.6 Scratch marks ornamenting bored tunnels that are plugged with sediment and fecal pellets. Diatomaceous mudstone; submarine canyon, Bering Sea. (Sample courtesy D. W. Scholl.)

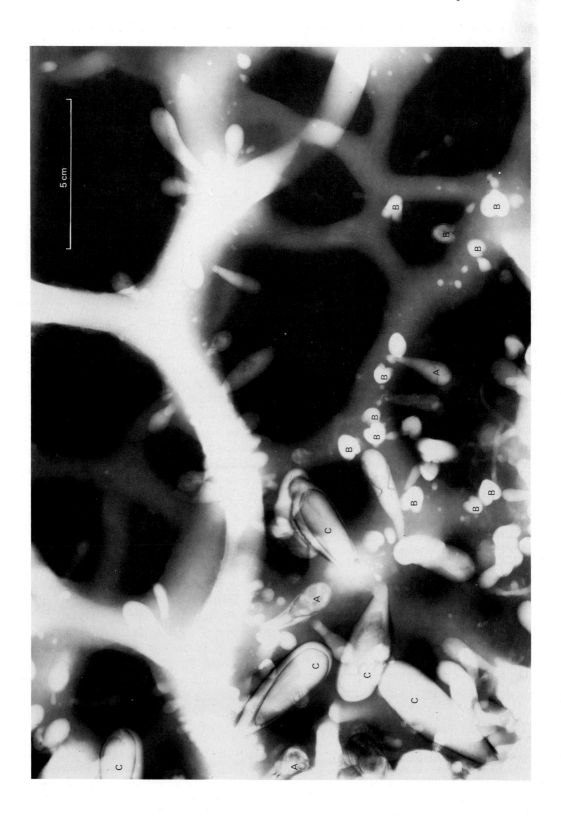

rocks exposed on parts of reefs, submerged pinnacles, submarine canyon walls, etc., are not visibly affected by physical or chemical processes, but are deeply bored (Warme and Marshall, 1969). On a small scale, the work of borers can be a quantitatively greater factor in erosion than that of other processes. For instance, Bertram (1936) inspected 45 coral colonies broken during a storm on the Red Sea; 60 percent were weakened by sponges and 20 percent by mollusks, whereas only 20 percent had growth defects or had grown on inadequate bases.

Rock samples collected from Scripps Submarine Canyon, when broken out of water, commonly are unweathered—even dry— only a few millimeters from the surface. Biological penetration, however, extends inward one or more decimeters (Fig. 11.7). Thus, the *relative* importance of bioeroders, at least in some cases, is greater than other erosive agencies. The *absolute* rate of erosion is more difficult to ascertain (see below).

In general, the sizes of materials excavated by borers and added to the local sedimentary regime are not known. Undoubtedly much, or perhaps most, of the loose sediment on reefs is produced in this way (Matthews, 1966; Hoskin, 1966).

The activity of shallow marine borers can help account for the prominent limestone reef flats, "pedestal rocks" (Umbgrove, 1947, Pl. 5, figs. 1 and 2), and wave-cut terraces so frequently seen on shorelines of the world. Intertidal or shallow subtidal benches, flats, or terraces commonly culminate landward in intertidal or shallow subtidal nips. Perhaps the flats are partially a product of bioerosion accompanying the Holocene sealevel rise, and the nip is a result of excavation at the present stillstand. Many investigators have discussed the relative importance of chemical solution, wave action, and bioerosion in the genesis of this nip, as well as associated flats and "solution" basins (Macfadyen, 1930; Bertram, 1936; Emery, 1946; Umbgrove, 1947; Newell and Imbrie, 1955; Newell, 1956; Revelle and Fairbridge, 1957; Kaye, 1959; Yonge, 1963a).

Rates of Bioerosion

Many organisms cease to bore under laboratory conditions; in situ monitoring and measurements are difficult, and have yielded equivocal results. In an effort to make such measurements, Evans (1970a) devised a detachable "drawing table" that can be set up periodically in the same position on intertidal rocks and thereby used to measure erosion rates.

Neumann (1966, p. 93) reviewed published estimates for rates of bioerosion; they seem to be low compared with the abundance of borers in many localities, and generally are based on one taxon rather than on the entire biota of excavators. Neumann's (p. 106) calculations for boring sponges in Bermuda, however, range as high as 1 m/70 yr, and probably are realistic. Hodgkin (1964) marshalled evidence for a general rate of intertidal erosion of about 1 mm/yr for several carbonate localities in Australia.

For any given taxon, an estimate of potential bioerosion can be made by measuring the volume of the average excavation, estimating densities of the borers, and choosing an average life-span or generation time. Typical densities for shallow-water species are 10 spionid polychaetes/cm^2, 50 to 500 pholadid bivalves/m^2, and 100 echinoids/m^2. Evans (1968c) constructed such a model, in two dimensions, for pholadids.

The most difficult data to collect are records of longevity; other problems involve

---

◄ **Fig. 11.7** Radiograph of siltstone showing extensive tunnels bored parallel to bedding by unknown excavators, and abundant borings by the bivalves *Nettastomella rostrata* (A), *Adula californiensis* (B), and *Lithophaga plumula* (C). 15 m depth; Scripps Submarine Canyon.

patchy distribution of the live populations and spatial shifts of population abundances with time. Nevertheless, just the standing crop of borers counted in many localities has produced significant volumes of excavated material.

IDENTIFICATION AND CLASSIFICATION OF EXCAVATIONS

Fossil and modern excavations encompass a spectrum ranging from simple pits, scrapings, or holes to complex tunnel-and-gallery systems. Most fossil borings are identified and named simply by comparing them with modern counterparts or with other known fossil examples (Häntzschel, 1962, p. W228–W232). Certain borings are unique and rather easily ascribed to given taxonomic groups. Others are of general form and can be ascribed to more than one group. Possible bases for classification of excavations are phylogenetic relationships, substrates penetrated, or size-geometry categories. No classification is attempted here, however, because of the following problems:

1. Many families, genera, and even species of borers and other excavators exhibit variations in the size and geometry of penetrations that they make. The sipunculids, for example, are a small but widespread phylum of marine worms containing many species that bore a variety of substrates. Some of their borings are solitary, having only one individual per penetration, which may have single or multiple openings. Other substrates yield many sipunculids in a single boring or system of borings. Another example is the bivalve family Pholadidae. In normal population densities of such pholadids as *Penitella penita*, the animals construct circular, downward expanding, club-shaped borings that exhibit little or no curvature. However, in crowded conditions, they may be stunted (Evans, 1968c); when restricted to small or heterogeneous substrates, their borings may curve to avoid other individuals (Evans, 1970b), to circumvent impenetrable obstacles, or to prevent piercing another exterior surface (Fig. 11.8). In extreme cases, the borings

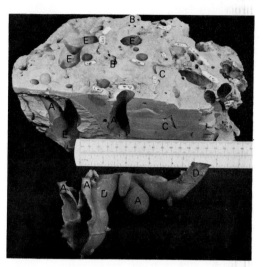

Fig. 11.8 Bedded Eocene mudstone boulder penetrated by large pholadid bivalves (A), small, circular, flabelligerid polychaetes (B), and irregular borings of uncertain origin (C). Large tube (D), cast in Latex, represents a single boring of the pholadid *Chaceia ovoidea*, which was confined to the boulder and changed direction to avoid exterior surfaces and other borers. Parts of its boring are labeled E. (Rock was sawed on front face, and a 2-cm slab was removed from the top). 15 m depth; Scripps Submarine Canyon. (Sample courtesy E. J. McHuron.)

of some pholadids become elongate and sinuous, and even the siphons may be twisted to allow the animal to continue growing and still feed at the surface of its excavation (Fig. 11.8). Pholadid borings and shells also change length–diameter proportions in different substrate hardnesses, becoming relatively longer in soft substrates and more tumid in hard ones (Evans, 1970b).

2. Similarly, penetrants may vary in their choice of substrates. Knowledge of substrates that are attacked by different groups is helpful in identifying some excavators, although relationships are too haphazard among different groups to be useful for a phylogenetic classification of borings. Separate species of pholadids, for instance, prefer calcareous or noncalcareous rocks, shells, wood, or unconsolidated sand or mud.

3. Separate groups can produce very similar borings. Some species of sipunculids, bivalves, barnacles, and polychaetes produce similar finger-like, solitary borings (Fig. 11.9).

4. Within groups, the size and shape of borings may have evolved through time, as docu-

mented by Seilacher's (1969) study of acrothoracican barnacles.

5. Adequate description of borings is lacking, as well as ecological documentation for many groups of modern borers.

These problems cause difficulties in developing size/shape categories that would arrange borings according to phylogeny or any other meaningful biological interrelationships (see Chapter 3). A purely descriptive (nonphylogenetic) scheme, based on size and geometry of excavations, is now used in part to name and classify borings (Häntzschel, 1962; Bromley, 1970a).

NOMENCLATURE

Bromley (1970a, 1972) argued for formal names of borings, and urged that these be kept separate from those of the borers, even if the latter are known; this system conforms to traditional ichnological practice (see Chapter 3). As examples, sponge borings are distinctive, yet identity of the animal species is based on spicules that are rarely if ever fossilized in the borings. Bromley (1970a, p. 70–80) recommended the trace fossil genus *Entobia* for ancient sponge borings resembling those by modern clinoid sponges. He (1972) also recommended the trace genus *Trypanites* to cover a spectrum of more generalized pouch-like, single-entrance excavations that are made by sipunculids, polychaetes, bivalves, thoracican barnacles, and probably other groups. Figure 11.9 shows some of the variation possible within his redefinition of *Trypanites* (1972, p. 95). This designation is useful because a gradation probably exists between most forms shown. However, it covers such a wide range of sizes and shapes that it eventually may harbor an immense number of trace fossil species. Also, Bromley's synonymy includes taxa that many workers consider to be body fossils (e.g., the *Treatise on Invertebrate Paleontology*).

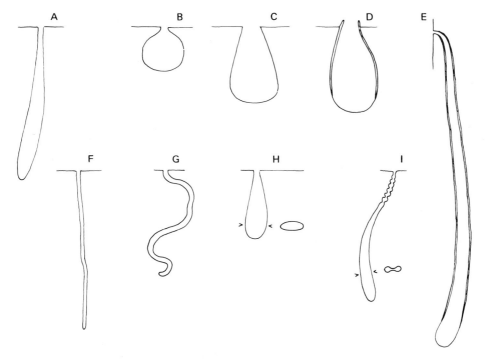

Fig. 11.9 Some variations included under the designation *Trypanites*, by Bromley (1972). A, general form. B, C, forms ascribed to bivalve borings lacking linings; D, E, those having calcareous linings. F–I, forms ascribed to polychaetes, sipunculids, or phoronids.

TAXONOMIC SURVEY OF
MARINE EXCAVATORS

Bacteria

Bacteria may be very important in carbonate breakdown, especially by attacking the organic skeletal matrix of such organisms as corals and mollusks. DiSalvo (1969) identified bacteria from discolored and weakened areas in corallites, whereas adjacent healthy regions yielded none. Several bacteria on reefs were chitin-digesting forms (10 to 20 percent). These microfloral borers may precede sponges, algae, or other borers by penetrating and weakening skeletons on a microscopic scale, or perhaps succeed them in rotten pockets within the substrate. Little additional work has been done on this group.

Plants

Chapter 12 is devoted to endolithic algae and fungi, and should be consulted for details of the morphology and ecologic distributions of these penetrants (see also Chapter 10). Endolithic algae and fungi are probably ubiquitous at the shoreline and in shallow-shelf depths, especially in calcareous sediments and limestones. Boring fungi also are present in deep water. Although they are individually small, they have an important—perhaps dominating—aggregate effect on size and shape of grains by producing clasts, and on shoreline and shallow submarine topography by weakening and sculpting the substrate (Perkins and Halsey, 1971). Modern and fossil algal and fungal borings were compared by Gatrall and Golubic (1970).

Fungal Microborers

Both living and dead substrates are penetrated by marine boring fungi. Cavaliere and Alberte (1970) documented tube-like fungal borings in shells, where the slender hyphae are 1 to 3 μm in diameter, and

bulbous sporangea 20 to 50 μm in diameter. The borings may also be root- and filament-like outgrowths in various shapes and sizes. Bonar (1936) described fungal borings in shells of intertidal limpets (*Acmaea*) in California. Most individuals were infected, as were other mollusks and barnacles. *Acmaea* responds by thickening its infested shell, further confusing the taxonomy of an already difficult group of limpets; this fact, of course, has important paleontological implications.

Kohlmeyer (1969) stated that marine fungi are widespread in shells of mollusks, plates of barnacles, calcareous tubes of wood-boring bivalves (teredinids), and calcareous algae. Carbonate sands from widespread localities are reportedly infested by fungi (Zebrowski, 1936; Porter and Zebrowski, 1937), the fragments being derived from mollusks, foraminifers, ostracods, and calcareous sponges.

The work of boring fungi begins as a roughening of the surface, and continues until a spongy, pitted substrate is developed (Kohlmeyer, 1969, p. 745). As penetrating hyphae colonize the shell in a branching network, the matrix becomes fragile and more readily susceptible to breakage and disintegration.

Kohlmeyer (1969) suspected that a symbiotic relationship exists between some species of marine algae and fungi. Santeson (1939) described intertidal marine lichens having worldwide distribution, living in salinities above 15 or 20 ‰. He suggested that some of the organisms discussed by Bornet and Flauhault (1889) and Bonar (1936) were in fact lichens. Boekschoten (1966, p. 346–347) found the lichen *Arthropyrenia* to be very common on the Dutch coast, in shells of the bivalve *Mya* and in all but the youngest specimens of the gastropod *Littorina littorea*. It was present in *Mya* and other burrowing bivalves washed onto the beach, but not in fresh, articulated valves, suggesting that infestation and boring occurred after death and exposure of the host shells.

Algal Microborers

Golubic (1969) stated that endolithic algae are the main constituents of the dark-colored band so prominent on cliffed carbonate coasts. Destruction of carbonate within the spray zone was documented by Purdy and Kornicker (1958) in the Bahamas, and this phenomenon is very evident on many coastlines and shallow-water localities (see Neumann, 1966). The algae weaken the rock and make it susceptible to gastropod and amphineuran rasping (Newell and Imbrie, 1955; Newell, 1956; Yonge, 1963a).

Examples of algal penetration in fossils were provided by Hessland (1949) from the Ordovician, and Taylor (1971) from the Cretaceous. Duncan (1876) mentioned that the dead, lower parts of coral skeletons were intensely perforated by microfloral borers, but Duerden (1902) pointed out the presence of algal borers also just beneath living tissue of reef corals in Jamaica; they had colonized very recently deposited skeletal material. He stated that decalcification of a coral skeleton leaves a filamentous model, showing such morphological details as septa. According to Duerden (1902, p. 327–328):

> *Among the thirty different species of West Indian corals examined, the only places where filaments have not been found are at the growing apices of branching corals like* Madrepora *and* Oculina. *Here the calcareous deposition apparently takes place a little in advance of the upward growth of the algal filaments, for the latter are always to be found in the older parts of the corallum.*

If this activity is widespread, as Duerden indicated, it is an important factor in coral ecology and in diagenesis and lithification of coral rock.

Odum and Odum (1955) emphasized the importance of boring algae in reef trophic structure, and Stephenson (1961) documented the susceptibility of different calcareous substrates to algal attack.

Petrographic Implications of Microborers

Floral microborers are important in the disintegration of molluscan shells and other carbonate fragments, especially in quiet bays where the clasts are not moved by waves and currents (Bornet and Flauhault, 1889). These relations were substantiated by Swinchatt (1969), and thoroughly documented by Alexandersson (1972) in his work in the Mediterranean and elsewhere. Alexandersson pointed out the marked difference in surface texture of sand-size and smaller grains in agitated environments, where they are smooth, and in non-agitated environments, where they are riddled with holes and other biogenically produced depressions (Fig. 11.10).

Another process of importance to carbonate petrography is the preservation of molluscan bioclasts by formation of peripheral algal borings and subsequent formation of a resistant aragonitic micrite coating. After death of the algae, the void is filled with aragonite mud that becomes cemented. With successive generations, the aragonite coating may become complete, and act as a resistant shield against physical erosion of the fragment (Bathurst, 1966).

The importance of endolithic penetrations in diagenesis and lithification of voids in shallow-water carbonates, such as reef cavities, was shown by Schroeder (1972). He studied the endolithic alga *Ostreobium*, which penetrates the substrate, continues into cavities within the substrate, and forms a meshwork within these cavities; the meshwork then becomes calcified. This phenomenon takes place in the dark, and raises the possibility of heterotrophic green algal borers, as discussed by Nadson (1927a, 1927b; 1932) and Frémy (1945). Schroeder (1972) pointed out that algae very likely bore, and at the same time (or alternately), precipitate calcium carbonate, just as do many invertebrates.

Macroscopic Algae

Barnes and Topinka (1969) documented the strength with which the common alga

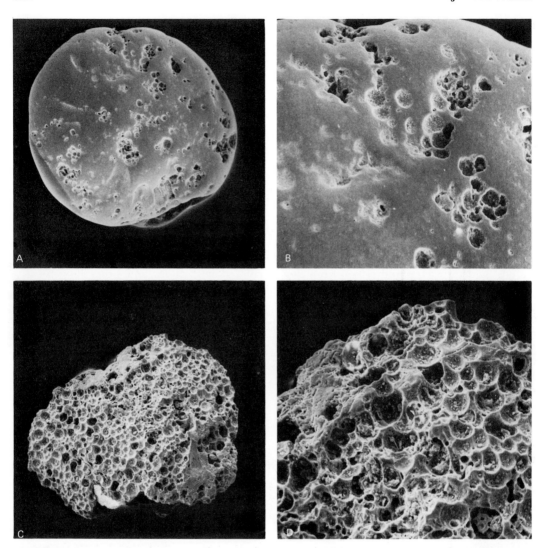

Fig. 11.10 Bored foraminifera (*Amphistegina*). A, B, specimen from agitated environment, 1 m depth, smoothed and polished and having few borings. C, D, specimen from quiet environment, 30 m depth, intensely bored. Most pits are probably sponge microborings, but microfloral borers produce similar results. Tests ca. 0.5 mm in diameter. Barbados. (SEM photos courtesy T. Alexandersson.)

Fucus can attach itself to intertidal rocks, and showed dissolution of the calcareous substrate beneath the holdfasts of the alga as it becomes established. Much of the calcareous material under *Fucus* holdfasts is thoroughly penetrated by rhizoids of the plant; when the plant is pulled loose, a thin layer of substrate also is removed. Similarly, Bertram (1936) described removal of corallite material where larger "soft algae" were detached from coral bases in the Red Sea.

Other macroscopic algae clinging by holdfasts, and perhaps even boring into intertidal rocks and subtidal outcrops, were discussed by Emery (1963). He documented the ability of the giant kelp *Macrocystis* to float rock fragments as large as boulders free from the seabed, owing to the buoyancy of the rapidly growing plant. The effect commonly is enhanced because of the progressive weakening of rock beneath the holdfasts by invertebrate borers.

Foraminiferida

Foraminifers are documented as borers in bivalve shells, calcareous algae, and other materials. Shells of the bivalve *Lima* dredged off Africa show pits and scars bored by *Rosalina carnivora* (Todd, 1965). Apparently, the host was alive at the time of penetration, because the holes were sealed from within. Another species in California, *R. globularis*, was observed resting in pits on calcareous algae; the pits are believed to be formed by the protist (Sliter, 1965; 1972, personal communication). Delaca and Lipps (1972) confirmed that *R. globularis* can pit calcareous substrates at points of attachment. *Cymballopora tabellaformis* rests in similar pits on dead coral rock; the pits are made by the protist, but not all specimens excavate them (Bertram, 1936). Banner (1971) documented endoparasitization of one foraminifer by another, producing small pits in the test of the infested host.

Poag (1969, 1971) provided a fossil example in which the middle Tertiary species *Vasiglobulina alabamensis* partially or completely penetrated bivalve shells, barnacle fragments, and bryozoans in order to anchor its spines. Poag envisaged pseudopodial penetration followed by spine secretion; borings produced in this way are 10 to 20 μm in diameter, and occur in clusters where the foram was attached (Fig. 11.11). Some of the borings completely penetrate the shell (Poag, 1971, Pl. 14, fig. 1), leaving a circular or hexagonal hole.

Porifera

About 100 species of sponges are borers or potential borers; most belong to about 14 genera in the Family Clionidae. Both recent and fossil clionids have been studied extensively (Goreau and Hartman, 1963; Bromley, 1970a). The sponges are present in abundance in tropical, temperate, and boreal seas, and several species may exist sympatrically, apparently by finely dividing

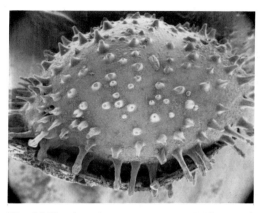

Fig. 11.11 Attached Eocene foraminifer (*Vasiglobulina alabamensis*), its spines penetrating mollusk shell. Specimen ca. 0.8 mm wide. (SEM photo courtesy C. W. Poag.)

the niches available to them (Hartman, 1957). Their abundance and widespread distribution make them among the most important marine borers.

Clionids make microscopic to small macroscopic excavations in shells and pure limestones. A less known coral borer, *Siphonodictyon*, falls in a different family (Adociidae) and creates large cavities within coral heads, in tropical seas (Bergquist, 1965; Rützler, 1971). Species in both families have siliceous spicules, but I am aware of no evidence that spicules are used in boring. Clionid colonies may be confined to part of a shell, or extend in a thin layer as much as 1 m in diameter across limestone outcrops, penetrating one to several centimeters inward. The borings initially are a series of scallop-shaped excavations, gradually interconnected to become sculptured galleries, penetrating and anastamosing through the substrate (Fig. 11.12). Tubes rise to the substrate–water interface, providing water exchange through the canal system of the sponges. The borings are distinctive for certain species, but may also vary, depending on the age and size of the sponge and the substrate under attack. Boring sponges are largely confined within their substrate, but they may excavate until it is entirely destroyed and then stand free on the bottom (Rützler, 1971, p. 1).

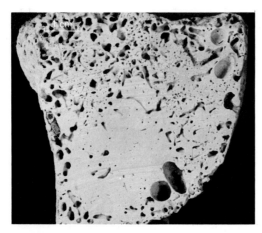 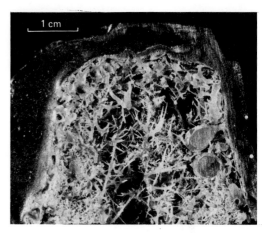

Fig. 11.12 Cretaceous limestone showing peripheral attack by bivalves and perhaps polychaetes, and extensive penetration by clionid sponges. A, cobble sectioned to reveal internal voids. B, similar rock fragment impregnated with plastic and the limestone matrix dissolved, to show clionid networks. Jetties; Galveston, Texas. (Sample courtesy E. J. McHuron and J. A. McCrevey.)

Evans (1969, Fig. 1) used time-lapse radiography to show sponge growth in shells. In a period of six months, the sponges extended laterally a few centimeters and widened and ramified their galleries, apparently excavating on all surfaces.

Goreau and Hartman (1963) showed that settling clionid larvae produce etched lines almost immediately on a calcareous substrate, and Cobb (1969) described "etching-type" cells that penetrated around the part of their periphery in contact with the substrate surface, then continued downward and inward to isolate small chips. The chips are irregular, having numerous curved facets and long axes up to 80 μm across. They are passed through the mesenchyme and excurrent canal system and expelled through an osculum onto the surface, giving rise to the calcareous debris described by Nassanow (1883), Topsent (1887), Warburton (1958), Goreau and Hartman (1963), and Cobb (1969). This mechanism shows a precise adaptation for the boring mode of life. Cobb (p. 787) also showed that *Cliona celata* can penetrate the shell periostracum, cutting a mosaic similar to that produced in calcareous layers. To my knowledge, the mechanism of isolation and removal of the

chips is unknown, but is likely a combination of chemical and mechanical processes.

Sponges are classified on the basis of spicule size and shape. Spicules are rarely found in fossil clionid borings (Bromley, 1970a), thus the unlikelihood of assigning generic and specific names to fossil clionid borings. Bromley recommended assigning all fossil sponge borings resembling those of clionids to the genus *Entobia*, even though the given borings might be very distinctive and assignable to a clionid animal species.

The ecology of clionids is known to some extent. They exist from the lower intertidal zone to depths of at least several hundred meters, and they apparently bore for protection only. At least one species of *Cliona* has symbiotic algae (Sàra and Liaci, 1964), and another bores into calcareous algae (Cotte, 1914); distribution of these species presumably is dependent on light (Bromley, 1970a).

Hartman (1957) showed vertical zonation of nine species of clionids down a steep underwater face on an Adriatic shore. Each species had different depth preferences, and also different oscular papillae openings (which are useful paleoecologically). The distribution was especially well developed for the more abundant species, and suggested Gaussian competitive exclusion. Hartman postulated that the sizes of incurrent openings bored in the substrate reflect food particle sizes available at different depths.

In estuarine clionids, the sizes of papillae openings at the surface of oyster shells vary with salinity, as demonstrated by Old (1941). This relationship was applied by Hopkins (1956, 1962) in determining the average salinity, or variation in salinity, on oyster banks at different locations within American Atlantic and Gulf Coast estuaries. Larger openings were found in species living under normal or near-normal marine conditions, whereas smaller openings were dominant under freshened or unstable salinities. These relationships should prove valuable in deciphering paleosalinities in Pleistocene or older oysters, and were applied to an Oligocene oyster assemblage by Lawrence (1969); he also reviewed the biology and paleoecology of clionids. (See also Wiedemann, 1972.)

Three species of *Siphonodictyon* are known: *S. muscosum* was described by Bergquist (1965) from the southwest Pacific, and *S. cachacrouense* and *S. coralliphagum* by Rützler (1971) from the Caribbean. In contrast with millimeter-size excavations by clionids, *Syphonodictyon* makes large cavities—up to several hundred cubic centimeters in volume, some being fist-size or larger (Fig. 11.13). These are connected to the surface of corals by one or many finger-like extensions, and may be topped by a cone- or chimney-like sponge growth having an apical oscular opening. Rützler (1971) described four varieties of *S. coralliphagum*, representing different growth forms of the sponge, all of which have similar spicules. *Siphonodictyon* from the Bahamas and Caribbean was collected at depths of 1.5 to 20 m, and seemed to prefer quiet-water habitats (Rützler, 1971, p. 16). Other sponges reported to bore large cavities in coral are *Adocia pumila* and *Oceanopia* sp., in the Red Sea (Bertram, 1936).

The ecological significance and erosional work of sponges was documented by Goreau and Hartman (1963) and Neumann (1966). On the northern coast of Jamaica, Goreau and Hartman showed that clionids apparently dominate the boring fauna on the deeper forereef slope, owing to a paucity of other borers. The clionids are able to bore at a pace that is significant compared with the growth of individual coral colonies. Flat, horizontal branches of colonies in 30 to 60 m depths on the forereef are bored at their bases, some to such an extent that they are completely cut off and collapse in an imbricated pile on the steep reef face. Large colonies commonly are broken off from the slender eroding bases by divers simply swimming by and barely touching them. Goreau and Hartman (1963) suggested that boring sponges are a leading factor, if not a dominating force, in sculpting parts of these reefs; sponges play an exceedingly important role in the growth, geometry, and concurrent destruction of coral colonies.

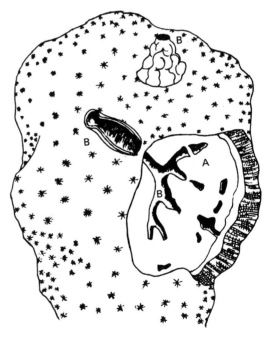

Fig. 11.13 Drawing of sponge *Siphonodicyton coralliphagum*, showing animal (A) in excavated interior of *Montastrea* coral, and sponge canal system and chimney (B), opening to the exterior. (From Rützler, 1971.)

Clionids in Bermuda are of primary importance in generating a notch at the subtidal fringe. Neumann (1966) showed that *Cliona lampa* penetrates to depths of 10 cm but is concentrated in the upper few centimeters of substrate. This species can excavate 6 to 7 kg of carbonate substrate/ m^2/100 days, which equals approximately 1 m of penetration each 70 years. Most (90 percent) of this material is excavated as chips. Neumann showed weight losses of 20 percent by some limestone fragments within 100 days.

In his thorough review of boring sponges, Bromley (1970a) indicated their usefulness in determining the history of the substrate having been attacked. Bored shells having pore openings on both sides were empty and exposed to the water on both sides. Openings only on one side indicate that the shell was exposed to the sea only on one side, or that the borings were made in the shell of a living animal. Such relationships were used by Boek-schoten (1966, 1967) to interpret the taphonomic history of bored shells from both Holocene and Pliocene localities in northern Europe.

Sponge borers are also important in sculpting ancient hardgrounds and discon-formable surfaces (see Chapters 17 and 18). Bromley (1970b) and Bromley and Nordmann (1971) illustrated encrusting foraminifera that encircled clionid pores, giving an interesting example of com-mensalism, in Cretaceous chalk. Bromley (1970b) also showed scratch marks around clionid papillae, suggesting attack by some unknown predator.

Coelenterata

Attached coelenterates of all kinds, espe-cially the larger true corals and alcyonar-ians, likely alter their substrate and erode it in the process. Considering the immense diversity and abundance within the phylum, records of penetration are meager. Only one example is known to me: Storr (1964,

Pl. 5, figs. 1 and 2) showed that after death, a species of the zoanthid anemone *Palythoa*, from the Bahamas, left pits about 5 mm in diameter and equally as deep in the cal-careous substrate. The possibility remains, however, that living calcareous algae or other encrusting forms grew around the bases of the already established *Palythoa* polyps, creating relief where each sat (J. Lang, 1972, personal communication).

Turbellaria and Nematoda

For information on borings by these two groups of animals, see references cited by Yonge (1963b) and Bishop (Chapter 13).

Phoronida

Phoronids comprise a separate phylum of lophophorate worms. Two small species bore shells, and when they reproduce asexually, they thoroughly infest the substrate. Marcus (1949) reported densities up to 150 borings/ cm^2 in shells of *Thais* and *Mytilus* and plates of balanid barnacles. Tube openings are circular (0.2 to 0.3 mm in diameter). *Phoronis ovalis* is documented from Brazil (Marcus, 1949) and Scandinavia (Lönöy, 1954), apparently working chemically. Phoronid borings as trace fossils were re-viewed by Bromley (1970a, p. 58–59), who also illustrated the borings of *P. ovalis* colonies (1970a, Fig. 2b). Voigt (1972) re-viewed the present knowledge of phoronid borings, and ascribed the common trace fos-sil boring *Talpina ramosa* to phoronids (see Bromley, 1970a, Fig. 3).

Bryozoa

Several groups of ectoprocts penetrate hard substrates, leaving more or less distinctive patterns of borings that represent zoid openings and the system of connective stolons between them. Thus, the borings are small, not more than 1 mm across zoid openings; stolonal canals are only 10 to 20 μm wide. Some colonies excavate a dendritic or otherwise ramifying pattern, whereas others are less regular and leave a haphazard meshwork of tubes and cavities (Fig. 11.14). Zoid openings always reach the surface, whereas stolons may be completely or only partially covered, or merely slightly impressed into the substrate surface. Each zoid of certain cheilostome species is clearly etched into the substrate beneath the colony (Fig. 11.14a).

Soule and Soule (1969) discussed the systematic position of bryozoan borers within the difficult classification of the phylum. Borers fall within at least four families—Immergentiidae, Terebriporidae, Penetrantiidae, and Hypophorellidae—and belong in two of the three existing orders—Ctenostomata and Cheilostomata. Some living species of bryozoans were described from borings only, creating taxonomic problems because the borings are not specifically unique (Silén, 1948), and the practice is contrary to recommended convention (Chapter 3). Voigt and Soule (1973), however, argued for naming fossil bryozoan borings as zoological taxa rather than ichnotaxa.

Methods of boring by bryozoans are not fully understood; very likely they are aided by chemical means. Silén (1946) suggested that phosphoric acid is employed by *Penetrantia*; perhaps certain other forms use it too (Bromley, 1970a, p. 57). Some species within the peculiar Hypophorellidae have a rasping apparatus associated with the aperture, and bore into other bryozoans or worm tubes. The other three families have no obvious boring organs (Soule and Soule, 1969, p. 794). Molluscan shells are the most common material attacked, but boring bryozoans also excavate worm tubes, coniferous twigs, and both calcareous and noncalcareous rock. The borings are described from South America, the North Sea, and other widespread localities, in depths ranging from intertidal to at least 400 m. As with most borers, they seem to bore for protection (Soule and Soule, 1969).

The sizes of borings are relatively uniform because of the rather equal size of individuals in colonies of different species. Geometry of the pattern of borings is highly variable, however, as illustrated in Figure 11.14.

Positions where the zoids resided are commonly better exposed in worn shells than in fresh ones. For the three most common families, Soule and Soule (1969)

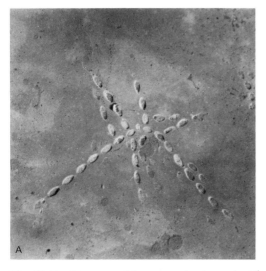

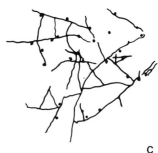

Fig. 11.14 Patterns of bryozoan borings. A, *Electra* sp. on bivalve *Spisula*. (Photo courtesy R. G. Bromley.) B, *Terebripora comma* (after Bromley, 1970a). C, *Iramena* sp. (after Boekschoten, 1970). Each colony ca. 2–3 mm across.

gave data on the shape of the openings and their cross-sectional outline where exposed on a worn substrate. The immergentids make elongate, fusiform borings, which are ovoid in cross section because the penetration is curved. The terebriporids have zoecial cavities that are teardrop-shaped and elongate suboval in cross section. The penetrantids have circular or subcircular zoecial borings, which may or may not have a calcareous projection or "tooth" at the rim; they are round in cross section when worn down, except for gonozoids, which are double holes separated by a thin septum.

Brachiopoda

Bromley (1970a, p. 61) reviewed the scant data on brachiopods that use the finely divided distal ends of their pedicles to bore into skeletal fragments within loose sediment. At least three orders of brachiopods contain members that have (or had) this habit: two species of living terebratulids, a Permian strophomenid, and a Devonian spiriferid (Bromley, 1970a, p. 61, Rudwick, 1965, p. H200–H201).

Bromley and Surlyk (1973) described pedicle borings by modern rhynochonellids, terebratellids, and terebratulids, and compared them with very similar Cretaceous borings. The inarticulate brachiopod *Lingula* occurs in borings in Silurian corals, but was judged to have occupied holes abandoned by borers such as annelids or bivalves (Newall, 1970).

Sipunculida

Many species of sipunculids bore pencil-size or smaller excavations in calcareous and noncalcareous rocks and in coral. Their borings are variable; most are simple, blind, straight to gently curved, or even highly sinuous tubes containing a single specimen (Rice, 1969); however, several closely jammed specimens have been observed in a branching tunnel system in subtidal mudstone (Warme, 1970, Pl. 1b). The latter seemed to be in excavations of their own making, but which could have been enlarged or modified from polychaete or other borings.

Sipunculids are major borers in many localities, being uniquely adapted for excavating hard substrates; they possess a variety of hooks, spines, papillae, or other structures embedded in their leathery skin. These structures probably anchor the worms in their domiciles, and function as tools for species that bore. They are deployed at the tip, base, or along the length of the introvert, along the side of the body, or at the posterior extremity (Fig. 11.15). Their size, shape, and position are criteria in classification of the group, but their function seldom is discussed in taxonomic work.

Gardiner (1902, p. 336) regarded sipunculids to be important in breaking up coral blocks in various habitats on Indian Ocean reefs; he stated that the sizes of borings seldom exceed one-third or one-half of the diameter that the animal assumes when freed from the substrate. In attempting to remove sipunculids from a partially exposed boring, our experience shows that the result usually is a burst sipunculid; they have high body-fluid pressures that keep them anchored firmly against excavation walls. Hyman (1959) doubted that sipunculids bore, but mentioned Sluiter's work (1891); he noted that species boring in calcareous rock have shorter and more muscular introverts, bearing stronger hooks, than those from soft substrates. The worms have been noted in very hard rocks that other groups do not bore, suggesting that the sipunculids made the excavations (Rice, 1969).

Sipunculids are prominent endolithic animals in many reef areas, boring into live and dead coral, but also excavating hard, well-cemented limestones exposed· between and below tide levels. Utinomi (1953) described multiple openings of sipunculid excavations in live corals, and suggested that the worms were at least in part embedded as the host corals grew. Gardiner (1902, p. 336) listed five genera of borers from Indian Ocean reefs, which were dis-

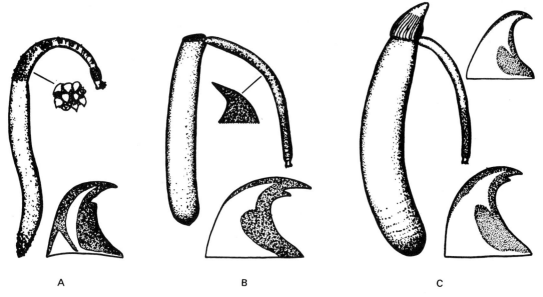

A   B   C

Fig. 11.15 Sipunculid borers. A, *Phascolosoma dentigerum*. Enlargement shows hook from anterior part of introvert (right end) and tubercular papillae from near base of introvert. B, *Aspidosiphon elegans*, having smaller spinelet and larger hook, from the introvert. C, (?)*Lithacrosiphon alticonum*, having one- and two-pointed hooks, from the introvert. Worms ca. 20 mm long; spines, hooks, etc., range from ca. 0.03–0.06 mm, maximum dimension. (From Robertson, 1963.)

cussed by Shipley (1902). Otter (1937) recounted three species of coral borers on the Great Barrier Reef; Cloud (1959, p. 392) listed several species taken in coral rock in Saipan and stated that most fit their borings precisely, this being taken as evidence that they are primary borers and are not simply nestling in readymade excavations. Seven species were clearly illustrated by Robertson (1963) from southern Florida, where they were a major component of the boring biota. Most species are present both in intertidal and shallow subtidal habitats; *Phascolosoma dentigerum* (Fig. 11.15A) is the dominant intertidal sipunculid.

Sipunculids also excavate terrigenous rocks. Both calcareous and noncalcareous sandstones and mudstones collected from intertidal to 50 m depths off southern California contained abundant specimens. One sample from 30 m depth in Scripps Submarine Canyon off La Jolla, California, yielded numerous specimens closely packed in a branching network; the tunnels were oriented primarily parallel to bedding planes, but some also crossed them (Warme, 1970). Other rocks collected from the same locality and kept in aquaria showed sipunculid activity only at night, when they extended their introverts over the rock surface, apparently feeding; nocturnal habits likely enable endolithic animals to avoid predation. A mudstone fragment collected by manned submersible from about 1,800 m depth off California contained one small, boring specimen, together with polychaete borings.

Sipunculids that we have placed in drilled holes in rock, bounded on one side by glass observation plates, did not bore actively; but some specimens doubled over, placed their introverts downward into the hole, and became tightly appressed in an upside-down position. This position perhaps is assumed by some species while boring with the embedded hooks and spines of the introvert and upper part of the trunk. We lack details on which species bore, their habits, and the magnitude, variation, and importance of their activity.

Echiurida

Some echiurida worms have been reported as modern and fossil borers (Farran, 1851; Jehu, 1918, p. 6; Joysey, 1959, p. 398; Bromley, 1970a, p. 59). Lamy and André (1937) were not certain whether two species reported by several authors were in fact borers (in Clapp and Kenk, 1963, p. 584). Little documentation of boring eciurids exists.

Annelida: Polychaeta

Most marine annelid worms are polychaetes, a very diverse group consisting of about 70 families, 1,600 genera, and 10,000 species (Hartman, 1968, p. 3). Families such as the Spionidae and Cirratulidae contain well-known borers (Evans, 1969); others, such as the Flabelligeridae and Capitellidae, contain borers but are poorly documented. Commonly, identification of genera and even species of polychaetes found boring into various substrates is not difficult; mention of their habits, however, seldom is given in the literature.

Polychaete bioerosion is probably vastly underestimated. Many submerged and intertidal rocks, or coral heads and limestone platforms, are so riddled with a variety of polychaete borings that to sort out which worms are responsible for which excavations in the substrate is a bewildering task (see Bertram, 1936). The task is further complicated because many polychaetes not only bore but also secrete chitinous or calcareous tubes at the entrance of their boring, extending the domicile into the water column as well as into the substrate.

Some polychaete borings are single, blind tubes. Others are basically U-shaped, representing a double trackway. The latter may sometimes lengthen and meander, the two limbs more or less parallel, being U-shaped at the base and having closely spaced dual openings at the substrate–water interface (Fig. 11.16; see Fig. 11.4D for cross-sectional view).

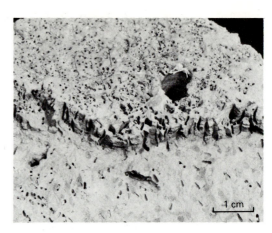

Fig. 11.16 Cretaceous mudstone bored by living spionid polychaetes (probably *Boccardia uncata*). Each boring is U shaped, and space between limbs is filled with mud. Abundance of borings caused a layer of rock to loosen and fall away. Larger hole is a pholadid boring. Intertidal zone near San Diego, California.

By far the best known family of boring polychaetes is the Spionidae, which includes the oyster-shell borer *Polydora ciliata* and numerous related species. However, five families of shell borers were described by Blake (1969) from New England: Spionidae, Cirratulidae, Capitellidae, Terebellidae, and Sabellidae. Primary rock borers in sandstones, mudstones, and limestones of the Pacific coast of North America alone belong to at least three families, and species that modify other borings to a greater or lesser extent belong to at least five additional families, perhaps to nine or more. Another six families have species that nestle in abandoned borings (E. J. McHuron, 1972, personal communication).

Some polychaete borings are very regular and characteristic whereas others are not. From my experience, the diameter of relaxed worms usually matches that of their borings, regardless of the length and geometry of the excavation. Because the sizes of polychaetes vary so, the sizes of their borings also vary greatly. Diameters range from less than 1 mm—resembling pin holes (Fig. 11.16), for the smaller forms such as *Polydora*—up to large tunnelways

5 mm or more in diameter and commonly many decimeters in length, for the eunicid *Marphysa sanguinea* (Fig. 11.17). Numerous polychaete borings are lined with smooth, chitinous or mucoid substances that apparently ease the passage of the animals. Several species have been observed doubled up in their borings, even inside one limb of a U-shaped excavation; they reverse direction in the middle of the tube simply by doubling back on themselves and squeezing by their own bodies.

Polychaete borings are protective, but they also serve as centers for food-gathering activities. Several species of foraging polychaetes bore long, slender holes through sandstone and shale boulders on the California coast, using exit positions to snare food. Aquarium observations show that detritus feeders (e.g., *Neanthes*) behave similarly, foraging in a radius around their domicile's opening, which may be a boring or simply a nestling place. Marsden (1962) described similar feeding behavior for *Hermodice carunculata*, which devours coral polyps near the opening of its dwelling.

Species of the spinoid *Polydora* commonly make a U-shaped domicile having a spreite-like filling between the limbs, parallel to the base (e.g., Seilacher, 1969, Pl. 1); this structure may wash away when the worm dies. The worms can occur in great abundance (cf. Fig. 11.16). They excavate a variety of materials, including carbonates (both shell and limestone), mudstones, and calcareous-cemented sandstones; they even invade wood, and make tubes in soft, gelatinous mud (Korringa, 1951, 1952; Dorsett, 1961; Dean and Blake, 1966; Boekschoten, 1966, p. 353; Cheng, 1967; Evans, 1969; Blake, 1971). One species, *P. commensalis*, lives in the columella of live gastropods; this boring is a common trace fossil (Kern et al., 1974). In this case, borings and burrows of many species must be studied carefully before attempting to characterize typical traces for the Spionidae.

Any classification of polychaete borings remains artificial until more is learned about the variation of excavations made by given species, genera, and families. Furthermore, not even a widely accepted phylogenetic classification of polychaetes above the family level exists. For instance, in Hartman's (1968, 1969) description of polychaetes from the west coast of North

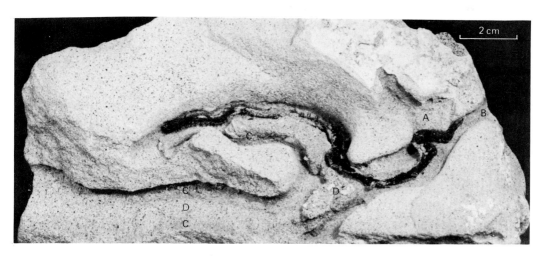

Fig. 11.17 Large, meandering, U-shaped boring of the polychaete *Marphysa sanguinea*, in sandstone. Two entrances (A, B) lead to a pathway (C) at least 50 cm long, the limbs being separated by a sand plug (D) that has largely been removed. Position of boring was revealed by x-radiography before overlying sandstone was removed; worm was still in its boring. Intertidal exposure near San Diego, California.

America, including 701 species, 312 genera, and 61 families, she did not attempt to arrange them in a superfamilial phylogenetic scheme. For this reason, we may have little chance to classify polychaete borings relative to any phylogenetic system for the group.

Boring polychaetes are present at most latitudes, and evidence indicates that at least some species live at relatively great depths. For instance, phosphatic pebbles dredged from banks in 500 to 1,000 m of water off California are commonly riddled with spionid-like borings. A large boulder collected by the submersible DEEP QUEST at 1,800 m in the San Clemente basin off southern California exhibited large, spionid-like borings in a mudstone stratum. The boulder was probably not transported from very much shallower depth. No live borers were found when the rock was recovered. In contrast, a sabellid (*Caobangia*) bores freshwater snail shells in the streams of southeast Asia (Jones, 1969).

Most polychaetes are equipped with a variety of potential mechanisms for eroding rocks. Mechanically, they can use setae, parapodia, and other structures associated with the lateral margins of each segment, as well as jaws and other mouth structures. Elliott and Lindsay (1912) thought that the giant setae of *Polydora* were used in boring, but Haigler (1969) removed the giant setae of *P. websteri* and the worms still were able to etch the substrate. A chemical means utilizing "acid" was employed, but the details were not determined. Annulations on sculpted tunnels in intensely eroded rock suggest that polychaetes play an important role in forming or maintaining the passages (Fig. 11.4), but the method for accomplishing the erosion is unknown.

The importance of boring polychaetes in limestone and coral regions was recognized long ago by Gardiner (1902, p. 336):

> *Polychaeta are perhaps really the most important boring animal in coral rock, although the actual forms are inconspicuous and of small diameter. In coral*

> *reefs at least some specimens can be obtained from every rock below tide marks. All large coral masses are bored into and penetrated by their tubes, which bend and twist in every direction. The surface at the edge of the reef is made rotten for some inches by their borings, and the section of broken base of a coral often appears as a regular sieve from their holes . . . From their prevalence in the rock, be it coral, sand, or millipore, the total effect of the Polychaeta must be enormous, and they must certainly be regarded as the prime and most effective agent in the breaking down of coral rocks.*

Gardiner felt that the Eunicidae were the most important. They extend up the skeleton and even into areas on coral branches that are covered with live polyps. They are the chief cause of "rotting" on the reef flat, and they also are found in depths up to 75 m in Indian Ocean atoll lagoons.

Gardiner's statement is cited by subsequent writers (e.g., Hartman, 1954; Ebbs, 1966), but few detailed studies are available on polychaete borers or boring in carbonates. Ebbs (1966, p. 546–549) stated that eunicids are important and abundant borers at the base of coral heads in Florida, and noted the possible sedimentological effect of the Atlantic palolo worm *Eunice schemacephala*, an important borer:

> *On dissecting the guts of a number of these eunicids, they were found to contain a rather coarse sand anteriorly, as it might be immediately after rasping, but the texture of the gut contents graded to a smooth, thick mud posteriorly. This mud differed only in its color (brownish white) from that gray mud found in the galleries [of the worm (p. 546)].*

Fossil borings ascribed to polychaetes are numerous. Cameron (1969) reviewed evidence for Paleozoic shell borers; he also described (1967) a remarkable occurrence of soft-part preservation in a Devonian spionid.

Slender, straight to highly sinuous penetrations and U-shaped spionid-like penetrations are described as fossils in cobbles and boulders (Radwański, 1968; Voigt, 1970) and in associated hardgrounds (Mägdefrau, 1932; Ellenberger, 1947; Radwański, 1964, 1965; Hölder and Hollmann, 1969; Perkins, 1971; and McCrevey, 1974). Belemnites also were commonly bored by sinuous penetrations that are confined to the skeleton (Mägdefrau, 1932; Gripp, 1967); these probably are polychaete borings. Interesting examples of borings and associated tubes on fossil brachiopods and other shells were discussed by Clarke (1921), Teichert (1945), Cameron (1969), and many others. They can be used to infer feeding currents and life positions of the hosts, together with other borings and encrusting animals on the host shells. They also can be used, with other borings, to infer steps in the taphonomy of shell beds (Boyd and Newell, 1972).

Arthropoda

Cirripedia

Three groups of barnacles contain species of borers. The elongate borings of the thoracican *Lithotrya* and the scar-like impressions by certain thoracican acorn barnacles are large compared with those of

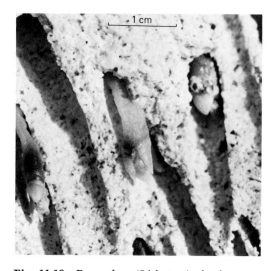

Fig. 11.18 Barnacles (*Lithotrya*) boring upward beneath intertidal ledges. A common habit. Beachrock; Bahamas. (Photo courtesy W. M. Ahr.)

the millimeter-size acrothoracicans and ascothoracicans.

Lithotrya (Fig. 11.18) is a prominent coral and carbonate rock borer that makes pencil-size or larger (as much as 1 cm maximum diameter), oval-shaped borings 10 cm or more long. At least eight species occur in the tropics, worldwide (Sewell, 1926). They commonly hang upside down under intertidal limestone ledges (Ahr and Stanton, 1973), but are capable of boring at all angles to the substrate surface, as I have observed in dead coral-algal blocks in the intertidal zone on Curaçao. They occur in great numbers, from the low- to mid-tide zones on carbonate coasts, and are important rock eroders (Gardiner, 1902; Sewell, 1926; Cannon, 1935; Otter, 1937; Ginsburg, 1953; Lewis, 1960; Robertson, 1963). Some species also bore shell (Cannon, 1935). Within the boring, a calcareous attachment is secreted along the side and near the base, clearly illustrated by Bromley (1970a, Fig. 4g) and Ahr and Stanton (1973). The animals may use spicules that are embedded along their stalk in order to mechanically widen their borings as they grow; but these structures are an unlikely aid in deepening the hole. *Lithotrya* is gregarious, densely populating available outcrops in its habitat (Fig. 11.19), and occupying as much as 50 percent of the substrate to depths of 5 to 7 cm (Newell and Imbrie, 1955; Newell, 1956). Newell (1954, p. 23) mentioned another "gooseneck" barnacle, *Drupa*, occupying 20 percent of the substrate to a depth of 5 to 8 cm at mean sea level on Raroia Island.

Acrothoracican barnacles comprise a group of small, specialized forms that make sac-like excavations only a few millimeters deep and having aperatures about 1 mm in diameter. The aperatures are drop- to slit-shaped; they are similar to bryozoan borings in dimension, but are solitary (although commonly gregarious) and are not connected by a network of stolons. The occurrences of several zoologic and trace fossil genera of acrothoracicans were dis-

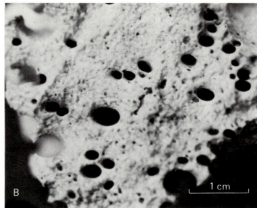

Fig. 11.19 Lateral (A) and surface (B) views of dense *Lithotrya* borings. Bahamas. (Specimen courtesy W. M. Ahr.)

cussed and reviewed by Newmann et al. (1969), Boekschoten (1966), and Bromley (1970a, p. 67–70); their biology discussed by Tomlinson (1969a); and their evolutionary trends and paleoecology by Newmann et al. (1969) and Seilacher (1969). As a group, they infest a variety of skeletal materials, as well as carbonate pebbles; but some forms have definite substrate preferences.

The position of these borings with respect to mantle openings or other possible sites of water-current generation (and hence food-bearing currents) has been used to infer the soft anatomy of hosts as well as the requirements of the barnacles. Seilacher (1968) showed that the preferred orientation of borings on belemnites indicated that the cephalopods were bored while still alive, and suggested that their normal direction of motion was head forward. Some species require that their substrate be occupied; an example is the modern *Trypetesa*, a commensal of hermit crabs in gastropod shells (Tomlinson, 1953).

Seilacher (1969) illustrated an evolutionary trend in acrothoracican borings, from narrow, parallel-sided penetrations in Paleozoic shells to the development of side pockets near the base of the borings (for gonad accommodation) in Mesozoic and Cenozoic shells.

A third group, the ascothoracicans,

today are parasitic in some echinoids and corals, making characteristic excavations through the skeleton to the exterior (Bromley, 1970a, p. 67). The borings have been identified in Cretaceous corals and echinoids (Newmann et al., 1969).

Decapoda

Several groups of decapods, including snapping shrimp, ghost shrimp, porcelain crabs, and brachyuran crabs, contain species that are suspected to bore—or at least to modify—rock surfaces. They possess adaptations such as pincers, pointed walking legs, or strong mouth parts for use in excavating hard materials. Evidence for these animals being borers generally is circumstantial; such species appear repeatedly within excavations of the same character, rather than occupying haphazard nooks and crannies, and commonly the walls of their abodes are ornamented with scratch marks.

Examples from California include snapping shrimp (*Crangon*) that live deep inside tunnels and galleries of excavated rock, and true crabs (*Cancer* spp., *Pilumnus spinohirsutus*, *Paraxanthias taylori*) living within bored boulders on tidal flats (Ricketts and Calvin, 1968, p. 144). Some species of the Caribbean ghost shrimp *Callianassa* occur regularly in cavities of dead corals (Biffar, 1971).

Evidence for excavation by porcelain

crabs is more direct; in California, *Pachycheles rudis* and *Petrolisthes* spp. live in slot-shaped tunnels that exhibit scratch marks and seem to be modified from smaller, round excavations made by bivalves or polychaetes. Porcelain crabs feed with mouth parts modified for fanning the water to obtain suspended food, yet they possess relatively large (*Petrolisthes*) to massive (*Pachycheles*) pincers; these are very likely used for excavating as well as for defense. Distinctive sculpture on tunnels and galleries may be attributable to these arthropods (Fig. 11.20).

Isopoda

Isopods are well-known excavators in wood and rock. Damage to wooden pilings has prompted research on the small (approximately 1 mm wide) isopod *Limnoria*: its excavations resemble those of termites (see Ray, 1959; Clapp and Kenk, 1963). Activities of the larger but related rock excavator *Sphaeroma* were described by Barrows (1919) and Higgins (1956). On the shore of San Pablo Bay, California, they presumably bite sandstone and tuff with their mandibles, making holes 2 to 9 mm in diameter and 6 to 35 mm long. Thousands of individuals are present, and

Fig. 11.20 Delicate, feathery sculpting parallel to bedding. Such scratches (or etches) are common, and may be attributable to porcelain crabs or polychaete worms. Eocene rock; Scripps Submarine Canyon, California.

empty holes are scarce. They excavated chalk blocks in aquaria, leaving scratch marks on the boring walls. Barrows (1919, p. 302–303) stated that some species of *Sphaeroma* also attack wood and that they are present in other localities, notably in Australia.

Mollusca: Bivalvia

The habit of boring into rock, coral, shell, or wood is supremely developed within three families of bivalves: the Pholadidae are exclusively borers or deep burrowers, using primarily mechanical means of penetration; the Gastrochaenidae and some species of the Mytilidae penetrate predominantly calcareous substrates, using chemical means—at least in part. Several other genera of boring bivalves belong to families that are not mainly borers: *Petricola* (Petricolidae), *Hiatella* (Hiatellidae), *Platydon* (Myidae), *Tridacna* (Tridacnidae) (Yonge, 1963b), as well as some other lesser-known taxa. Many bivalves not belonging to the Pholadidae, Mytilidae, or Gastrochaenidae are nestlers, residing in ready-made holes or modifying and enlarging them as they grow. In places, several generations of nestlers occupy the same hole, leaving several pairs of shells, one inside the other; these subsequently may be packed with mud or otherwise modified by polychaetes, sipunculids, or other non-molluscan residents, and in the fossil state thus represent burrows inside borings. The three major families of borers mentioned above are clearly primary borers, however, and occur in vast numbers under favorable conditions.

Each species of bivalve borer normally produces a typical or "model" excavation that is characteristic and easily attributable to that species, or at least to a few closely related species. During growth, however, this ideal form may be modified, owing to unfavorable circumstances such as encountering other specimens, impenetrable obstacles, or exterior surfaces. As an example, Figure 11.8 shows the flexibility of certain pholadids' behavior under crowded or restricted conditions. This block of Eocene mudstone, collected in Scripps Submarine Canyon, southern California, was attacked from all sides by pholadid bivalves as well as nonmolluskan borers. Many rock ledges in the same locality show countless pinhole-size penetrations, where larval settlement of pholadids resulted in preliminary boring and then death of most of the individuals. Thus, pholadid holes range from less than 1 mm to several centimeters in diameter, and vary from flask- or club-shaped borings having straight axes, to cylindrical holes making one or more bends away from the initial axis of penetration.

Pholadidae

The pholadids are clearly adapted for endolithic life, possessing a shell and foot modified for boring activities. About 42 species represent the family on North American shores; the range of certain east coast species also extends to Europe. Pholadids are the subject of an extensive literature, summarized in the excellent monographs by Turner (1954, 1955)

MacGinitie (1935) described the rotary movement of *Zirfaea pilsbryi* boring into firm mud in a California estuary. Rotary motions on a vertical axis, each encompassing a small fraction of a revolution and followed by a pause, are accomplished by manipulation of the foot and valves. The process is accompanied by intermittent expulsion of excavated debris, which is gathered into the mantle cavity and forcefully propelled out the siphons with a jet of water. Yonge (1964) stated that pholadids rotate in their burrow, first in one direction and then the other; similar observations have been made on several species by other workers. Pholadids thus are primarily mechanical borers, excavating in sandstone, mudstone, and wood, as well as schist and other apparently very hard substrates (Turner, 1954, p. 6). *Penitella conradi*,

however, is confined to abalone (*Haliotis*) shells (Smith, 1969). *Diplothyra smithi* bores oysters (Turner, 1954, p. 2) as well as limestone. These species most likely bore by using chemical means, at least in part, suggesting that many species of pholadids are perhaps equipped to use both mechanical and chemical means, one or the other being accentuated in various taxa or under diverse conditions.

The longest bivalve borings are probably those of some larger pholadids that do not secrete a callum over the foot, as adults, and thus continue growth and boring throughout life. *Zirfaea pilsbryi* attains shell heights of at least 72 mm, and thereby creates a boring in soft mudstones at least this diameter, or larger. Bioerosion by such large pholadids is exhibited on mudstone outcrops in shallow water off many California beaches, where closely spaced siphonal holes and interspaced empty borings, many containing shells still in place, are in all degrees of exposure, owing to wave erosion and bioerosion. In such places, the sea floor is literally riddled by pholadid borings.

The highly specialized *Xylophaga* (Fig. 11.21) and related *Teredo* are especially adapted for wood boring (Turner, 1967), and some of the more typical piddocks, such as *Martesia*, also bore into wood (Turner, 1954, 1955).

Sculpture on pholadid borings varies directly with hardness of substrate. In soft

Fig. 11.21 Wooden test panel infested with *Xylophaga washingtona* and *X. duplicata*, after submersion for three years at 1,600 m off California. Small holes are openings to the large borings revealed by sawed face. (Sample courtesy J. S. Muraoka.)

Fig. 11.22 Fragment of diatomaceous mudstone (A) bounded and penetrated by pholadid borings; Malibu Beach, California. Rock has been grooved by prominent growth lines and spines of pholadid shells, such as *Zirfaea pilsbryi* (B).

mudstones, spines of the shell ornament produce deep, parallel grooves as the animal rotates, demonstrating the effectiveness of the spines as tools (Fig. 11.22). In hard substrates, the animal rotates repeatedly without making much progress downward, and the borings tend to be smooth, even polished, as well as less deep—compared with borings made by the same species in softer substrates (Evans, 1970b).

Mytilidae

The most abundant and best known mytilid borers are species within the genus *Lithophaga*, largely confined to calcareous substrates. Kühnelt (1930) listed and described 46 species, mostly from tropical localities. However, at least one species, *L. plumula*, bores into mudstones that were judged by Haas (1942) to be noncalcareous; other rocks from the same locality in California contain a trace (<1 percent) to more than 50 percent carbonate (Warme and Marshall, 1969). *Lithophaga plumula* also can penetrate quartzose sandstones containing about 35 percent carbonate cement. They have

no prominences or other obvious shell structures to aid in mechanical boring; they are believed to use chemical means to loosen fragments (Hodgkin, 1962) and ciliary currents to remove them (Yonge, 1955). Jaccarini et al. (1968) showed that the chemical utilized for boring by *L. lithophaga* is not a free acid but rather a neutral mucroprotein having calcium-binding ability. The animals clearly do not dissolve all of the substrate that is removed. *L. plumula* in aquaria deposits a fine mud outside its borings where the substrate is mudstone. Undoubtedly, the sediment removed is at least part sand when this species is boring into sandstone composed of noncalcareous clasts.

Several species of *Lithophaga* secrete a heavy calcareous encrustation on the surface of their shells, especially near the siphonal or posterior end, where these deposits merge with the posterior prolongations also present in valves of certain species (Fig. 11.23) (Turner and Boss, 1962). The latter are pointed tips at the posterior end of the shell that are shoved into the exit of the boring, and effectively seal it when the siphons are withdrawn and the

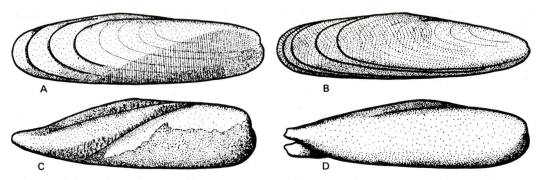

Fig. 11.23 Shells of four species of *Lithophaga*. *L. nigra* (A) and *L. antillarum* (B) have rounded posterior margins. *L. bisulcata* (C) and *L. aristata* (D) have well-developed posterior prolongations. From Florida; length ca. 8 cm. (From Robertson, 1963.)

shell is closed. *Lithophaga* shells are covered with a dark-brown periostracum which, together with their shape, give rise to the common name "date shell" (Turner and Boss, 1962, p. 81). Several authors suggest that this covering protects the shells during chemical boring. However, freshly collected specimens of several species exhibit periostraca partly worn away by movement of the clam in its boring (Soliman, 1969; my personal observation). This movement is accomplished by pulling on byssal threads attached in a midventral position to the inside of the boring (Yonge, 1955).

The mytilid *Adula* is at least partially a mechanical borer, moving by the use of byssal threads attached to the boring wall. In the specimens that I have collected in situ, *A. californiensis* is oriented in a vertical or near vertical position (Fig. 11.7). *Adula* frequently exhibits valves that are worn through the periostracum, owing to friction with the boring wall, and in some specimens surficial layers of shell are also worn away, particularly on the umbones. Whether *Adula* is a chemical borer, in part, is not clear (Haas, 1942; Yonge, 1955). Its excavation is heart-shaped in cross section (Fig. 11.7) because the animal does not rotate when boring. In contrast, some individuals of certain species of *Lithophaga* apparently do rotate, forming spindle-shaped borings of circular cross section (Figs. 11.4 and

11.7). Because *Lithophaga* also is suspended by byssal threads, these circular borings suggest either that the mantle or some other structure can be extended around the circumference of the boring to erode it, or that the threads can be detached or dissolved and resecreted to permit rotation to a new position. Kühnelt (1933) stated that *Lithophaga* demonstrates two metabolic stages in its growth. One phase is a period of normal respiration, feeding, growth, and other metabolic activity, and the other a phase of reduced metabolic activity, corrosion of the internal shell layer next to the mantle, and concurrent enlargement of the boring. A chip of $CaCO_3$ placed between the mantle and the shell retarded shell corrosion during the second phase. Because at least some species of *Lithophaga* secrete not only a shell but also an internal lining or sheath on their boring (see below), shell growth and sheath secretion probably coincide, as do shell corrosion, sheath dissolution, and borehole enlargement.

Several species of *Lithophaga* secrete sheaths that line part or all of their boring. These linings are typically thickest around the siphonal opening. In juveniles of *L. plumula*, the sheath is incomplete and very thin, but larger specimens may be completely encased by sheaths 1 to 2 mm thick. The sheaths may be exposed by erosion of the substrate around them and then bored by clionid sponges and polychaetes (Fig. 11.1). Premature precipitation of the lining occurs when the animal is annoyed by another borer. In such cases, the bivalve seals off the intruder by thickening its lining at the point of irritation, and excavates on the opposite side of the boring (Fig. 11.7C). Some individuals of *L. plumula* can extend the calcareous lining anteriorly and completely encapsulate their shells, still being freely suspended inside by byssal threads. *Lithophaga* sheaths in Red Sea species seem especially resistant to sponge borings (Bertram, 1936).

Similar deposits were described by Soliman (1969) for species of *Lithophaga* in rapidly growing corals; they must bore and enlarge their holes upward in order to maintain communication with the surface, flooring their excavation with meniscus-shaped partitions as they go. Three species have both firm calcareous deposits and calcareous paste in their borings, the latter occurring in various stages of hardening. Soliman (1969) argued for dominantly mechanical boring by these species.

An ultimate accommodation of boring mytilids associated with corals is exhibited by *Fungiacava eilatensis,* which lives commensally with *Fungia scutaria;* the bivalve bores into the coral skeleton and feeds within the coelenteron of the live coral (Goreau et al., 1969, 1971). Caribbean mytilid borers are *Botula fusca* and *Gregariella coralliophaga* (Warmke and Abbott, 1962, p. 164; Robertson, 1963). Species of *Modiolus* (e.g., *M. cinnamoneus*) and *Mytilus* [e.g., *M. (Brachidontes) exustus*] are mentioned as coral borers (e.g., Bertram, 1936; Ginsburg, 1953); their taxonomic status and documentation as borers are not known to me.

Gastrochaenidae

Species of *Gastrochaena* are confined to calcareous substrates, primarily coral skeletons and mollusk shells. The borings of two sympatric Caribbean species, *G. hians* and *G. ovata,* differ in that the excavation of the former commonly is three or more times the length of the shell [similar to *G. cuniformis* from the Great Barrier Reef (Otter, 1937, Pl. 2)], and of the latter only about two times the shell length (Fig. 11.24); they also differ slightly in other respects (Robertson, 1963).

In addition to boring, certain species also build calcareous, club-shaped domiciles in loose sediment, made from gravel- and sand-size material cemented with carbonate. These hollow vessels resemble the shape of gastrochaenid borings in solid substrates, and commonly occur beneath the shells through which they bored after settling as larvae (Fig. 11.25). With growth, the animals must partially dissolve, enlarge, and resecrete these casements from the inside.

Another coral borer, *Spenglaria rostrata,* constructs an unique boring; the adult has widely separated siphons that extend to the substrate–water interface via separate tubes that are lined with a smooth, calcareous secretion (Fig. 11.26A). The siphons probably erode the outer margin of the siphonal holes and secrete the calcareous lining on the inner margins, creating a V-shaped wedge of material deposited between them (Fig. 11.26B). Both *Gastrochaena* and *Spenglaria* secrete calcareous chimneys, raising their siphonal exits above the substrate, and together with *Lithophaga,* exhibit the ability to dissolve calcareous substrates and also to secrete calcareous shells, linings, and siphonal tubes.

Other Bivalvia

A venerid heterodont genus well known as a borer is *Petricola*. *P. pholadiformis* from the west coast of North America has a shell elongated posteriorly, and is probably a primary borer, whereas *P. carditoides* nestles and enlarges other borings. *P. lapicida* bores into calcareous rocks in the Indopacific region (Otter, 1937; Yonge, 1963b, p. 12), and the Caribbean (Robert-

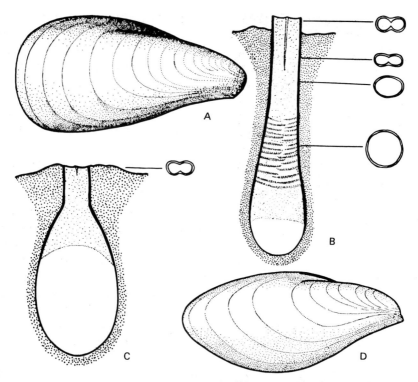

Fig. 11.24 Shells and borings of two common Caribbean species of *Gastrochaena*. A, B, *G. hians*. C, D, *G. ovata*. Thickened areas on borings represent calcareous sheaths. Irregularities on borings of *G. hians* are ridges that seemingly correspond to wrinkles on the joined siphons. The delicate shells lack ornament for boring mechanically. Shells ca. 2 cm long; borings ca. 8 cm (*G. hians*) and 4 cm (*G. ovata*) long. (From Robertson, 1963.)

son, 1963); its shell exhibits no obvious adaptations for boring. Another petricolid borer, *Hiatella*—occurring almost worldwide in temperate and boreal regions (Rowland and Hopkins, 1971)—either bores into soft rocks or nestles. The biology and habits of these two widespread species were described by Hunter (1949). Shells of *Hiatella*, as well as some species of *Petricola*, are distorted in their borings, varying among specimens of the same species.

Another northeastern Pacific borer is the myid *Platyodon cancellatus*, related to the softshell clam *Mya arenaria* (Yonge, 1963b, p. 9–10). It is an active borer, excavating 15 cm or more into mudstones. It does not rotate, and probably bores mechanically because it occurs in noncalcareous rocks.

Yonge (1963b, p. 16–17) summarized data on *Tridacna crocea*, which bores into coral by rocking (with the aid of byssal attachments) until the ventral margin is flush with the substrate surface, creating an excavation as much as 10 cm long and 7 cm deep.

Numerous other bivalve species are suspected to be borers. *Arca* cf. *A. umbonota* is byssally attached to, and partially imbedded in, rocks collected at 15 m depth off Texas. Although bryozoans, algae, and other encrusters build up around them, they scrape at the substrate by continually pulling on their byssus, and their excavations become enlarged laterally as their shells grow. Otter (1937, p. 328–329) described the indentations by *A. imbricata* from the Great Barrier Reef, and Ginsburg (1953) noted similar activity by *A. barbata* from Florida. Undoubtedly, many other species of bivalves bore or bioerode to some extent.

Fig. 11.25 *Gastrochaena* sp. encapsulated in cemented tube (arrow) after boring through gastropod shell. Siphonal opening on opposite side resembles raised chimney illustrated in Figure 11.24B. Attica, Greece.

Ecology of Bivalve Borers

The boring habit of bivalves is clearly protective and is supplemented in many species by special features mentioned above, such as posterior prolongations and very deep borings several times the shell length.

Rock-excavating bivalves occur at most latitudes, but a general segregation exists: primarily chemical borers inhabit the tropical and warm-temperate belts, where calcareous substrates are abundant, and primarily mechanical borers inhabit the terrigenous rocks of temperate and boreal zones.

Boring bivalves are most abundant in shallow water, but their depth limitations are not known. Both mytilid and pholadid borers are so abundant as to completely occupy all available outcrop space at depths of 30 to 50 m or more in Scripps and La Jolla Submarine Canyons off southern California (Warme et al., 1971). The pholadid *Jouanettia quillingi* occurs in vast numbers, the shells almost touching one another, on

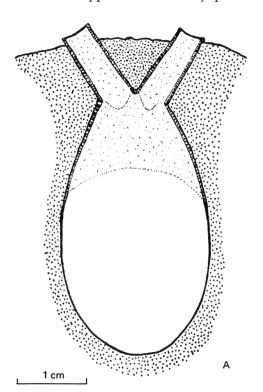

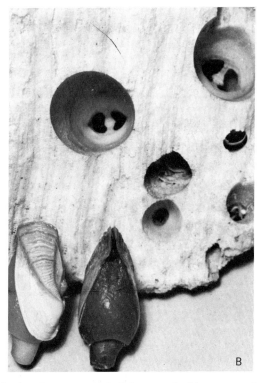

Fig. 11.26 *Spenglaria rostrata* borings. A, sketch showing separated siphons and chimneys, and calcareous lining (thickened). (From Robertson, 1963.) B, *S. rostrata* specimens (below), showing well-developed foot and pedal gape; broken coral (*Acropora palmata*) reveals posterior end of borings, having separate siphonal exits. *Spenglaria* can shove its posterior shell margins into posterior end of boring, so that siphonal exits are blocked.

mudstone banks 100 km off the Texas coast, at depths from the bank tops (20 m) to at least 30 m. Cobbles and boulders dredged from shelf, slope, and basin depths off California commonly show bivalve excavations, most likely made by large pholadids, but presently covered with phosphorite or ferromanganic coatings. Neither the age of the borings nor the water depth where they were formed are known. That the depth was much different from that at which the rocks were recovered seems unlikely. Larvae of pholadid bivalves, particularly the wood borers *Xylophaga* spp., are abundant at depths as great as 1,600 m off California (Muraoka, 1965); they can settle and grow, to occupy about 75 percent of the volume of wooden test panels placed at these depths, within a period of three years (Fig. 11.21). *Xylophaga* borings in fossil wood indicate that it was once driftwood exposed to the sea, as do similar borings of the shallow-water "shipworm," *Teredo*.

Other Mollusca

Gastropoda

In addition to shell-drilling gastropods (see Chapter 13), various species of snails accomplish four kinds of destruction of hard substrates: they graze across algae-covered rocks, nestle and excavate at resting positions, penetrate as larval vermetids, and bore into corals.

Grazing of rock for feeding on epilithic and endolithic algae was documented for two species of littorines from the west coast of North America by Emery (1946, 1960, p. 16) and North (1954); these abundant gastropods can erode significant volumes of intertidal sandstones and mudstones. Similarly, the limpet-like gastropod *Siphonaria pectinata* erodes calcareous beachrock initially weakened by intertidal endolithic algae, in Florida (Craig et al., 1969). Neritids play a similar role on tropical limestone shores, cited by numerous writers (e.g., Ginsburg, 1953; Newell, 1956).

In many cases, the effect of rasping is modified by chemical solution and development of microkarst in the upper part of the intertidal zone (Emery, 1946, 1960, p. 16–17). Lowenstam (1962a) showed that the radulae of some, but not all, intertidal gastropods is mineralized and hardened with goethite, probably aiding their grazing activities. Thus, even in noncalcareous or slightly calcareous rocks, intertidal grazers scrape grains from the surface that are first loosened physically, chemically, or biologically (bored by algae or fungi).

The erosive and sedimentological effect of rasping is significant, but fossil rasp traces—indicating intertidal or shallow subtidal conditions—are not likely to be preserved.

Nestling, particularly by intertidal limpets, creates oval depressions on rocks that closely fit their shells, and suggests either that the animals do not move or that they return to the same position and orientation after grazing.

Another form of gastropod excavation is the penetration of initial parts of vermetid shells. Bromley (1970a, p. 63, Pl. 1c) illustrated the seating of juvenile vermetid tubes on a *Trydacna* shell, and mentioned other examples. Vermetids off California grow gregariously in masses that break loose and wash ashore, together with fragments of the underlying substrate to which they were attached. Slightly calcareous or noncalcareous sandstone and mudstone is penetrated initially by juveniles, which then grow outward and entwine.

Snails that bore into live corals are unique gastropods because their entire shells are contained within their excavations. Three species of coral-boring gastropods are present in the Red Sea (Soliman, 1969), boring into separate kinds of coral. The shells are thin, roughened on the exterior, and commonly encrusted with coral fragments and other debris (Soliman, 1969, Fig. 4; Cloud, 1959, Pl. 129). Soliman (1969) believed that penetration is primarily by mechanical means. The animals can partially rotate in their borings, probably enlarging their excavations intermittently; periods of activity are represented by ridges in the bored hole. The borings must be lengthened outward as the surface of the coral colony grows, as well as deepened and widened as the helical shell expands. Characteristic excavation by some species is deeper than that of others. Biology of coralliophilid gastropods was summed by Gohar and Soliman (1963), Bertram (1936), and Cloud (1959), and other workers mention these coral borers from widespread Indopacific localities.

Fig. 11.27 Chitons grazing on algae-covered rocks in intertidal zone. Each individual fits closely into a slot-shaped depression formed during its continued habitation there. Southern California.

Cephalopoda

Shell-drilling octopus are discussed in Chapter 13.

Amphineura

Amphineurans are rock eroders, especially in the intertidal zone, where they rasp algae with the aid of their radula. As with gastropods, the radulae of some chitons contain mineral compounds such as magnetite (Lowenstam, 1962b) and opal (Lowenstam, 1971) that hardens the apparatus and perhaps aids in foraging activities.

Most chitons nestle in shallow rock depressions, but some make very deep holes, which may be slot shaped (Fig. 11.27) or circular, and five or more times deeper than the height of the animal (Doran, 1955). One example recorded in southern California was a hole 5 cm in diameter and 7 cm deep containing a large chiton (cf. *Mopalia mucosa*) securely attached at the bottom, but covered with loose rock and algal fragments.

Because many species live on rocks in the direct path of breaking waves, they must have a clean surface on which to maintain pedal grasp. Constant gripping perhaps removes weathered grains and deepens the holes.

Otter (1937, p. 331–332) stated that amphineurans within their borings on the Great Barrier Reef expelled feces that were calcareous, reflecting the composition of the substrate that was rasped and ingested. That some or all chitons may also employ chemical means of penetrating rock for endolithic flora or for nestling purposes should not be discounted. The abundance of intertidal rock-grazing chitons and gastropods makes them important eroders on many shorelines. To a lesser extent, the activities of chitons may promote bioerosion in deeper water; they are known to exist at depths of 700 m or more, although certainly at reduced densities. The activities of primarily intertidal amphineurans were described by Otter (1937) on the Great Barrier Reef; Cloud (1959) in the Indopacific area; Kaye (1959) in Puerto Rico; and Emery (1960, p. 15–16) in California. (See also Bromley, 1970a, p. 62–70.)

Echinodermata: Echinoidea

The Subclass Regularia of the Echinoidea includes species of borers that are common on many tropical and temperate coasts, especially intertidally, but also to depths of 10 m or more. Boring by echinoids was summarized by Fewkes (1890) and Otter (1932).

Echinoids excavate substrates ranging in hardness from mudstone, sandstone, and limestone to schist and granite. Many species excavate to some degree; some species probably complete their borings in one generation whereas others require many generations, especially where living on harder rock.

Regular echinoids may occur in great numbers, honeycombing lower intertidal to shallow subtidal substrates in the direct path of breaking waves. *Strongylocentrotus purpuratus* and *S. franciscanus* populate the California coast in this manner, and numerous species occupy similar habitats on tropical shores. In California, *S. franciscanus* reoccupied the top of a large, square, sandstone boulder in 10 m of water, sculpting it with about eight hemispheres, each 10 to 12 cm in diameter, forming an underwater landmark familiar to divers. Commonly, not all the hemispheres contain live urchins.

In softer shale and limestone, echinoid borings may extend deeper than the height of their tests, or may be slot-shaped runways with or without branches. Tropical long-spined echinoids commonly excavate 10 or 15 cm—depths greater than the combined height of their test and spines.

Hunt (1969) observed that *Echinometra lacunter* in Bermuda tended to have circular tests and occupy borings that were circular in cross section when young, and to have slightly elongated tests and occupy slot-shaped, wandering trackways as adults. McLean (1967) demonstrated that this species scrapes algae from the boring walls with the teeth, as well as with spines, and thus erode it. They apparently do not leave their borings, and in fact little or no movement occurred in several specimens during an observational period of three weeks (Hunt, 1969).

Certain species of urchins living in the quieter waters of tropical bays and lagoons occupy deep borings or crevices and forage at night (Thornton, 1956; Ormond and Campbell, 1971, p. 451). Although Otter (1932) suggested that specimens in deep, flask-shaped holes could not leave them, Bromley (1970a, p. 67) cited Märkel and Maier (1967) as having documented echi-

noids squeezing out of small orifices by manipulating their spines. Several writers remark that echinoid-bored limestone exhibits few or no empty borings, suggesting that the individuals either made the borings in which they occur or compete for abandoned borings.

Echinoids intensely attack intertidal and subtidal carbonates in some localities (e.g., Kaye, 1959, Fig. 34; Storr, 1964, p. 68; Neumann, 1966; Hunt, 1969). By analyzing gut contents and using in situ measurements, McLean (1967) showed a rapid rate of erosion—averaging about 1 cm/yr—in the borings of *Echinometra lacunter* in Barbados. Certain species can excavate glaucophane schist, granite, and similar durable rocks. Echinoid borings on rocks of uplifted wave-cut terraces in California (Fig. 11.28) give positive proof of rather recent tectonic movement.

Echinoids can excavate by use of both spines and Aristotle's lantern (McLean, 1967), the latter being a very flexible feeding device consisting of five radially arranged, calcified teeth. The lantern apparatus can be extended downward through the mouth opening, some distance below the test, in order to forage on vegetation, primarily algae. Neumann (1966, Fig. 7) illustrated echinoid "browse marks" on limestone. These traces indicate rock erosion and are potential trace fossils as well. In hard substrates, such as granites or silicified shales, the echinoid excavations are polished by spine movement. The tube feet can be extended beyond the length of the spines in most species, and probably are used to keep the borings clean of excavated material. Very hard surfaces are perhaps excavated by removal of softer cements, as in calcareous-cemented sandstones, or of softer minerals, such as mica, in igneous and metamorphic rocks.

Pisces

A very important group of bioeroders and destroyers of living reef animals in tropical

Fig. 11.28 Spherical holes excavated by echinoids, probably *Strongylocentrotus purpuratus*, in up-lifted metamorphic rock about 15 m above sea level. (Coin is 2.5 cm in diameter.) Similar borings by *S. purpuratus* presently occur in granitic and metamorphic rocks exposed in present intertidal zone nearby. Near Bodega Bay, California.

carbonate terrains is the scarid parrot fishes. They consume vast quantities of carbonate material, much of which passes through their intestinal tracts and is scattered as sand- and perhaps silt-size materials over the reefs. This group may be the most significant bioeroders in terms of volume of material removed, as discussed by Stephenson (1961) and especially by Cloud (1952, 1959). They defecate a fine stream of rock material, especially when frightened. Some species feed alone or in pairs or small groups, and others work in schools of hundreds of individuals, stirring the bottom and cleaning the substrate. They consume live coral and also algae and perhaps other organisms, on living and non-living surfaces. Bertram (1936) emphasized that coral colonies damaged by parrot fish are susceptible to entrance by other borers, particularly sponges, and that the corals

may never recover once infested. "Beak traces" of parrot fishes are common on Indian Ocean reefs; here they cause considerable destruction, to water depths of 10 m (Barnes et al., 1971, p. 98–99). Bakus (1966) documented intense predation in the tropics by fishes of all kinds—a compelling reason for prey organisms to bore.

ACKNOWLEDGMENTS

Research on marine bioerosion, at Rice University, was supported by National Science Foundation grant GB-14321 and the Henry L. and Grace Doherty Charitable Foundation. I am indebted to numerous people, especially E. J. McHuron and J. A. McCrevey of Rice University, for gathering and contributing data and drawing my attention to references, and to R. G. Bromley, J. W. Evans, and R. W. Frey for reviewing drafts of the manuscript.

REFERENCES

Ahr, W. M. and R. J. Stanton. 1973. The sedimentologic and paleoecologic significance of *Lithotrya*, a rock boring barnacle. Jour. Sed. Petrol., 43:20–23.

Alexandersson, T. 1972. Micritization of carbonate particles: processes of precipitation and dissolution in modern shallow-marine sediments. Geol. Inst. Univ. Uppsala, Bull. (N.S. 3), 7:201–236.

Bakus, G. J. 1966. Some relationships of fishes to benthic organisms on coral reefs. Nature, 210:280.

Banner, F. T. 1971. A new genus of the *Planorbulinidae*, an endoparasite of another foraminifer. Rev. Española Micropaleont., 3:113–128.

Barnes, H. and J. A. Topinka. 1969. Effect of the nature of the substratum on the force required to detach a common littoral alga. Amer. Zool., 9:753–758.

Barnes, J. et al. 1971. Morphology and ecology of the reef front of Aldabra. Zool. Soc. London, Symp., 28:87–114.

Barrows, A. L. 1919. The occurrence of a rock-boring isopod along the shore of San Francisco Bay, California. Univ. California Publ. Zool., 19:299–316.

Bathurst, R. G. C. 1966. Boring algae, micrite envelopes, and lithification of molluscan biosparites. Geol. Jour., 5:15–32.

Bergquist, P. R. 1965. The sponges of Micronesia, Part I: the Palau Archipelago. Pacific Sci., 4:123–204.

Bertram, G. C. L. 1936. Some aspects of the breakdown of coral at Ghardaga, Red Sea. Zool. Soc. London, Proc. 1936, p. 1011–1026.

Biffar, T. A. 1971. The genus *Callianassa* (Crustacea, Decapoda, Thalassinidea) in south Florida, with keys to the western Atlantic species. Bull. Marine Sci., 21:637–715.

Blake, J. A. 1969. Systematics and ecology of shell-boring polychaetes from New England. Amer. Zool., 9:813–820.

———. 1971. Revision of the genus *Polydora* from the east coast of North America (Polychaeta: Spionidae). Smithsonian Contr. Zool., 75, 32 p.

Boekschoten, G. J. 1966. Shell borings of sessile epibiontic organisms as paleoecological guides (with examples from the Dutch

Coast). Palaeogeogr., Palaeoclimatol., Palaeoecol., 2:333–379.

———. 1967. Palaeoecology of some Mollusca from the Tielrode Sands (Pliocene, Belgium). Palaeogeogr., Palaeoclimatol., Palaeoecol., 3:311–362.

———. 1970. On bryozoan borings from the Danian at Fakse, Denmark. In T. P. Crimes and J. C. Harper (eds.), Trace fossils. Geol. Jour., Spec. Issue 3:43–48.

Bonar, L. 1936. An unusual *Ascomycete* in the shells of marine animals. Univ. California Publ. Bot., 19:187–194.

Boreske, J. R. et al. 1972. A reworked cetacean with clam borings: Miocene of North Carolina. Jour. Paleont., 46:130–139.

Bornet, E. and C. Flauhault. 1889. Sur quelques plantes vivant dans le test calcaire des mollusques. Soc. Bot. France, Bull., 36:147–176.

Boyd, D. W. and N. D. Newell. 1972. Taphonomy and diagenesis of a Permian fossil assemblage from Wyoming. Jour. Paleont., 46:1–14.

Bradley, W. C. 1956. Carbon-14 date for a marine terrace at Santa Cruz, California. Geol. Soc. America, Bull., 67:675–677.

Bromley, R. G. 1967. Some observations on burrows of thalassinidean Crustacea in chalk hardgrounds. Geol. Soc. London, Quart. Jour., 123:157–182.

———. 1970a. Borings as trace fossils and *Entobia cretacea* Portlock, as an example. In T. P. Crimes and J. C. Harper (eds.), Trace fossils. Geol. Jour., Spec. Issue 3:49–90.

———. 1970b. Predation and symbiosis in some Upper Cretaceous clionid sponges. Dansk Geol. Foren. Meddr., 19:398–405.

———. 1972. On some ichnotaxa in hard substrates, with a redefinition of *Trypanites* Mägdefrau. Paläont. Zeitschr., 46: 93–98.

——— and E. Nordmann. 1971. Maastrichtian adherent foraminifera encircling clionid pores. Geol. Soc. Denmark, Bull., 20:362–368.

——— and F. Surlyk. 1973. Borings produced by brachiopod pedicles, fossil and recent. Lethaia, 6:349–365.

Cameron, B. 1967. Fossilization of an ancient

(Devonian) soft-bodied worm. Science, 155: 1246–1248.

———. 1969. Paleozoic shell-boring annelids and their trace fossils. Amer. Zool., 9:689–703.

Cannon, H. G. 1935. On the rock-boring barnacle, *Lithotrya valentiana*. British Mus. (Nat. Hist.), Sci. Repts., 5:1–17.

Carriker, M. R. 1969. Excavation of boreholes by the gastropod, *Urosalpinx*: an analysis by light and scanning electron microscopy. Amer. Zool., 9:917–934.

——— and E. H. Smith. 1969. Comparative calcibiocavitology: summary and conclusions. Amer. Zool., 9:1011–1020.

——— et al. (eds.) 1969. Penetration of calcium carbonate substrates by lower plants and invertebrates. Amer. Zool., v. 9, no. 3, ed. 2, 391 p.

Cavaliere, A. R. and R. S. Alberte. 1970. Fungi in animal shell fragments. Jour. Elisha Mitchell Sci. Soc., 86:203–206.

Cheng, T. C. 1967. Marine molluscs as hosts for symbiosis. In F. S. Russell (ed.), Advances in marine biology, vol. 5. New York, Academic Press, 424 p.

Clapp, W. F. and R. Kenk. 1963. Marine borers: an annotated bibliography. Office Naval Res., Dept. Navy, ACR–74, 1136 p.

Clarke, J. M. 1921. Organic dependence and disease: their origin and significance. New York State Mus., Bull. 221–222, 113 p.

Cloud, P. E., Jr. 1952. Preliminary report on the geology and marine environments of Onotoa Atoll, Gilbert Islands. Atoll Res. Bull. 12, 73 p.

———. 1959. Geology of Saipan, Mariana Islands. Part 4. Submarine topography and shoalwater ecology. U. S. Geol. Surv., Prof. Paper 280–K: K361–K445.

Cobb, W. R. 1969. Penetration of calcium carbonate substrates by the boring sponge, *Cliona*. Amer. Zool., 9:783–790.

Cotte, J. 1914. L'association de *Clionia viridis* (Schmidt) et de *Lithophyllum expansum* (Phillippi). Compt. Rend. Séanc. Soc. Biol., 76:739–740.

Craig, A. K. et al. 1969. The gastropod, *Siphonaria pectinata*: a factor in destruction of beach rock. Amer. Zool., 9:895–901.

Crimes, T. P. and J. C. Harper (eds.). 1970. Trace fossils. Geol. Jour., Spec. Issue 3, 547 p.

Dean, D. and J. A. Blake. 1966. Life history of *Boccardia hamata* (Webster) on the east and west coasts of North America. Biol. Bull., 130:316–330.

Delaca, T. E. and J. H. Lipps. 1972. The mechanism and adaptive significance of attachment and substrate pitting in the foraminiferan *Rosalina globularis* d'-Orbigny. Jour. Foraminiferal Res., 2:68–72.

Dillon, W. P. and H. B. Zimmerman. 1970. Erosion by biological activity in two New England submarine canyons. Jour. Sed. Petrol., 40:542–547.

DiSalvo, L. H. 1969. Isolation of bacteria from the corallum of *Porites lobata* (Vaughn) and its possible significance. Amer. Zool., 9:735–740.

Doran, R., Jr. 1955. Land forms of the southeast Bahamas. Univ. Texas Publ. 5509: 1–38.

Dorsett, D. A. 1961. The behavior of *Polydora ciliata* (Johnst.). Tube-building and burrowing. Jour. Marine Biol. Assoc. United Kingdom, 41:577–590.

Duerden, J. E. 1902. Boring algae as agents in the disintegration of corals. Amer. Mus. Nat. Hist., Bull., 16:323–332.

Duncan, P. M. 1876. On some thallophytes parasitic within recent Madreporaria. Royal Soc. London, Proc., 25:238–257.

Ebbs, N. K. 1966. The coral-inhabiting polychaetes of the northern Florida reef tract. Part 1. Aphroditidae, Polynoidae, Amphinomidae, Eunicidae, and Lysaretidae. Bull. Marine Sci., 16:485–555.

Elias, M. K. 1956. Recent and ancient penetrants. Jour. Paleont., 30:1001.

Ellenberger, F. 1947. Le problème lithologique de la craie durcie de Meudon. Bancs-limites et "contacts par racines": lacune sousmarine ou émersion? Soc. Géol. France, Bull., 17:255–274.

Elliott, W. T. and B. Lindsay. 1912. Remarks on some of the boring Mollusca. British Assoc. Advmt. Sci., Rept., 81:433.

Emery, K. O. 1946. Marine solution basins. Jour. Geol., 54:209–228.

———. 1960. The sea off southern California. New York, John Wiley, 366 p.

———. 1963. Organic transportation of marine sediments. In M. N. Hill (ed.), The sea. New York, Wiley-Interscience, p. 776–793.

Evans, J. W. 1968a. The effect of rock hardness and other factors on the shape of the burrow of the rock-boring clam, *Penitella penita*. Palaeogeogr., Palaeoclimatol., Palaeoecol., 4:271–278.

――――. 1968b. The role of *Penitella penita* (Conrad 1837) (Family Pholadidae) as eroders along the Pacific coast of North America. Ecology, 49:156–159.

――――. 1968c. A theoretical consideration of crowding and its effects on the biology of the rock-boring clam, *Penitella penita*. 2nd Internat. Congr. Marine Corrosion, Fouling, 1968, Athens, Greece.

――――. 1969. Borers in the shell of the sea scallop, *Placopecten magellanicus*. Amer. Zool., 9:775–782.

――――. 1970a. A method for measurement of the rate of intertidal erosion. Bull. Marine Sci., 20:305–314.

――――. 1970b. Palaeontological implications of a biological study of rock-boring clams (Family Pholadidae). In T. P. Crimes and J. C. Harper (eds.), Trace fossils. Geol. Jour., Spec. Issue 3:127–140.

Farran, C. 1851. *Thalassema neptuni*. Ann. Mag. Nat. Hist., 7:156.

Fewkes, J. W. 1890. On excavations made in rocks by sea urchins. Amer. Naturalist, 21:1–21.

Frémy, P. 1945. Contribution à la physiologie des thallophytes marins perforant et cariant les roches calcaires et les coquilles. Ann. Inst. Océanogr., 22:107–144.

Frey, R. W. 1973. Concepts in the study of biogenic sedimentary structures. Jour. Sed. Petrol., 43:6–19.

Gardiner, J. S. 1902. The action of boring and sand-feeding organisms. In J. S. Gardiner (ed.), The fauna and geography of the Maldive and Laccadive Archipelagoes. New York, Cambridge Univ. Press, 1:333–341.

――――. 1903. The origin of coral reefs as shown by the Maldives. Amer. Jour. Sci., 16:203–213.

Gatrall, M. and S. Golubic. 1970. Comparative study on some Jurassic and recent endolithic fungi using scanning electron microscope. In T. P. Crimes and J. C. Harper (eds.), Trace fossils. Geol. Jour., Spec. Issue 3:167–178.

Ginsburg, R. N. 1953. Intertidal erosion on the Florida Keys. Bull. Marine Sci., 3:55–69.

Gohar, H. A. F. and G. N. Soliman. 1963. On biology of three coralliophilids boring in living corals. Mar. Biol. Sta. Al-Ghardaga, Red Sea, Publ., 12:99–126.

Golubic, S. 1969. Distribution, taxonomy, and boring patterns of marine endolithic algae. Amer. Zool., 9:747–751.

Goreau, T. F. and W. D. Hartman. 1963. Boring sponges as controlling factors in the formation and maintenance of coral reefs. In R. F. Sognnaes (ed.), Mechanisms of hard tissue destruction. Amer. Assoc. Advmt. Sci., Publ. 75, p. 25–54.

―――― et al. 1969. On a new commensal mytilid (Mollusca, Bivalvia) opening into the coelenteron of a *Fungia scutaria* (Coelenterata). Jour. Zool. London, 158:171–195.

―――― et al. 1971. On feeding and nutrition in *Fungiacava eilatensis* (Bivalvia, Mytilidae), a commensal living in funguid corals. Jour. Zool. London, 160:159–172.

Gripp, K. 1967. *Polydora biforans* n. sp., ein in Belemniten—Rostrum bohrender Wurm der Kreidezert. Meynicana, 17:9–10.

Haas, E. 1942. The habits and life of some west coast bivalves. Nautilus, 55:109–113.

Haigler, S. A. 1969. Boring mechanism of *Polydora websteri* inhabiting *Crassostrea virginica*. Amer. Zool., 9:821–828.

Häntzschel, W. 1962. Trace fossils and problematica. In R. C. Moore (ed.), Treatise on invertebrate paleontology, Pt. W, Miscellanea. Lawrence, Kan., Geol. Soc. America and Univ. Kansas Press, p. W177–W245.

Hartman, O. 1954. Marine annelids from the northern Marshall Islands. U.S. Geol. Surv., Prof. Paper 260-Q:Q619–Q644.

――――. 1968. Atlas of the errantiate polychaetous annelids from California. Allan Hancock Found., 828 p.

――――. 1969. Atlas of the sedentariate polychaetous annelids from California. Allan Hancock Found., 812 p.

Hartman, W. D. 1957. Ecological niche differentiation in the boring sponges (Clionidae). Evolution, 11:294–297.

Hecker, R. T. 1970. Palaeoichnological research in the Palaeontological Institute of the Academy of Sciences of the USSR. In T. P. Crimes and J. C. Harper (eds.), Trace fossils. Geol. Jour., Spec. Issue 3:215–226.

Hessland, I. 1949. Investigations of the lower

Ordovician of the Siljan District, Sweden. II. Lower Ordovician penetrative and enveloping algae of the Siljan District. Geol. Inst. Univ. Uppsala, Bull., 33:409–428.

Higgens, C. G. 1956. Rock-boring isopod. Geol. Soc. America, Bull., 67:1770.

————. 1960. Ohlson Ranch Formation, Pliocene, northwestern Sonoma County, California. Univ. California Publ. Geol. Sci., 36:197–232.

Hodgkin, E. P. 1964. Rate of erosion of intertidal limestone. Zeitschr. Geomorph., 8:385–392.

Hodgkin, N. M. 1962. Limestone boring by the mytilid *Lithophaga*. Veliger, 4:123–129.

Hölder, H. and R. Hollmann. 1969. Bohrgänge mariner Organismen in jurassischen Hart- und Felsböden. Neues Jahrb. Geol. Paläont., Abh., 133:79–88.

Hopkins, S. H. 1956. Notes on the boring sponges in Gulf Coast estuaries and their relation to salinity. Bull. Marine Sci., 6:44–58.

————. 1962. Distribution of species of *Cliona* (boring sponge) on the eastern shore of Virginia in relation to salinity. Chesapeake Sci., 3:121–124.

Hoskin, C. M. 1966. Coral pinnacle sedimentation, Alacran Reef Lagoon, Mexico. Jour. Sed. Petrol., 36:1058–1074.

Hunt, M. 1969. A preliminary investigation of the habits and habitat of the rock-boring urchin *Echinometra lacunter* near Devonshire Bay, Bermuda. In R. N. Ginsburg and P. G. Garrett (eds.), Seminar on organism–sediment interrelationships. Bermuda Biol. Sta. Res., Spec. Publ. 2, 153 p.

Hunter, W. R. 1949. The structure and behavior of *Hiatella gallicana* (Lamarck) and *H. arctica* (L.) with special reference to the boring habit. Royal Soc. Edinburgh, Proc., (B), 63:271–289.

Hyman, L. H. 1959. The invertebrates (vol. 5). Smaller coelomate groups. New York, McGraw-Hill, 783 p.

Jaccarini, V. et al. 1968. The pallial glands and rock boring in *Lithophaga lithophaga* (Lammellibranchia, Mytilidae). Jour. Zool. London, 154:397–401.

Jehu, T. J. 1918. Rock-boring organisms as agents in coast erosion. Scottish Geogr. Magazine, 34:1–11.

Jones, M. L. 1969. Boring of shell by *Coabangia* in freshwater snails of Southeast Asia. Amer. Zool., 9:829–835.

Joysey, K. A. 1959. Probable cirripede, phoronid, and echiuroid burrows within a Cretaceous echinoid test. Palaeontology, 1:397–400.

Kaye, C. A. 1959. Shoreline features and Quaternary shoreline changes, Puerto Rico. U.S. Geol. Surv., Prof. Paper 317-B, 140 p.

Kennedy, W. J. and H. C. Klinger. 1972. Hiatus concretions and hardground horizons in the Cretaceous of Zululand. Palaeontology, 15:539–549.

Kern, J. P. et al. 1974. A new fossil spionid tube, Pliocene and Pleistocene of California and Baja California. Jour. Paleont., 48:978–982.

Kohlmeyer, J. 1969. The role of marine fungi in the penetration of calcareous substances. Amer. Zool., 9:741–746.

Korringa, P. 1951. The shell of *Ostrea edulis* as a habitat: observations on the epifauna of oysters, living in Oosterschelde, Holland, with some notes on polychaete worms occurring there and in other habitats. Arch. Nëerl. Zool., 10:32–152.

————. 1952. Recent advances in oyster biology. Quart. Rev. Biol., 27:266–308, 339–365.

Kühnelt, W. 1930. Bohrmuschelstudien. I. Palaeobiologica, 3:53–91.

————. 1933. Bohrmuschelstudien. II. Palaeobiologica, 5:371–408.

Lamy, E. and M. André. 1937. Annélides perforants les coquilles de mollusques. Compt. Rend. 12th Internat. Congr. Zool., Lisbon 1935, p. 946–948.

Lawrence, D. R. 1969. The use of clionid sponges in paleoenvironmental analyses. Jour. Paleont., 43:539–543.

Lewis, J. B. 1960. The fauna of the rocky shores of Barbados, West Indies. Canadian Jour. Zool., 38:391–435.

Lönöy, N. 1954. A comparative study of *Phoronis ovalis* Wright from Norwegian, Swedish, and Brazilian waters. Univ. Bergen, Naturv. Rekke, 1953, 2:1–29.

Lowenstam, H. A. 1962a. Goethite in radular teeth of recent marine gastropods. Science, 137:279–280.

————. 1962b. Magnetite in denticle capping in recent chitons (Polyplacophora). Geol. Soc. America, Bull., 73:435–438.

————. 1971. Opal precipitation by marine gastropods (Mollusca). Science, 171:487–490.

Macfadyen, W. A. 1930. The undercutting of coral reef limestone on the coast of some islands in the Red Sea. Geogr. Jour., 75:27–34.

MacGinitie, G. E. 1935. Ecological aspects of a California marine estuary. Amer. Midland Natur., 16:629–765.

Mägdefrau, K. 1932. Über einege Bohrgänge aus dem Unteren Muschelkalk von Jena. Paläont. Zeitschr., 14:150.

Marcus, E. du B. R. 1949. *Phoronis ovalis* from Brazil. Zoologia, 14:157–171.

Märkel, K. and R. Maier. 1967. Über die Beweglichkeil von Seeizeln. Naturwissenschaften, 53:535.

Marsden, J. R. 1962. A coral-eating polychaete. Nature, 193:598.

Matthews, R. K. 1966. Genesis of recent lime mud in southern British Honduras. Jour. Sed. Petrol., 36:428–454.

McCrevey, J. A. 1974. Fossil traces of the Whitestone Limestone and associated strata of the Walnut Formation, Lower Cretaceous, south-central Texas. Unpubl. M.A. Thesis, Rice Univ., 105 p.

McHuron, E. J. 1972. Characteristic morphology of modern invertebrate borers and borings: paleontologic implications. Geol. Soc. America, Abs. Prog., 4(4):285.

McLean, R. F. 1967. Erosion of burrows in bedrock in the tropical sea urchin, *Echinometra lucunter*. Canadian Jour. Zool., 45:586–588.

————. 1972. Nomenclature for rock-destroying organisms. Nature, 240:490.

Menzies, R. J. 1957. Marine borers (annotated bibliography). In J. W. Hedgpeth (ed.), Treatise on marine ecology and paleoecology, vol. I, Ecology. Geol. Soc. America, Mem. 67(1):1029–1034.

Muraoka, J. S. 1965. Deep-ocean boring mollusk. BioScience, 15:191.

Nadson, G. A. 1927a. Les algues perforantes de la mer Noire. Compt. Rend. Acad. Sci., Paris, 184:896–898.

————. 1927b. Les algues perforantes, leur distribution et leur rôle dans la nature. Compt. Rend. Acad. Sci., Paris, 184:1015–1017.

————. 1932. Contribution à l'étude des algues perforantes. Acad. Sci. U.S.S.R., Bull., 7:833–845.

Nassonow, N. 1883. Zur Biologie und Anatomie der *Cliona*. Zeitschr. Wiss. Zool., 39:295–308.

Neumann, A. C. 1966. Observations on coastal erosion in Bermuda and measurements of the boring rate of the sponge, *Cliona lampa*. Limnol. Oceanogr., 11:92–108.

Newall, G. 1970. A symbiotic relationship between *Lingula* and the coral *Heliolites* in the Silurian. In T. P. Crimes and J. C. Harper (eds.,), Trace fossils. Geol. Jour., Spec. Issue 3:335–345.

Newell, N. D. 1954. Reefs and sedimentary processes of Raroia. Atoll Res. Bull., 36:1–35.

————. 1956. Geological reconnaissance of Raroia (Kon Tiki) Atoll, Tuamoto Archipelago. Amer. Mus. Nat. Hist., Bull., 109:311–372.

———— and J. Imbrie. 1955. Biological reconnaissance in the Bimini area, Great Bahama Bank. New York Acad. Sci., Trans., 18:3–14.

Newmann, W. A. et al. 1969. Cirripedia. In R. C. Moore (ed.), Treatise on invertebrate paleontology, Pt. R, Arthropoda. Lawrence, Kan., Geol. Soc. America and Univ. Kansas Press, p. R206–R295.

North, W. J. 1954. Size distribution, erosive activities, and gross metabolic efficiency of the marine intertidal snails, *Littorina planaxis* and *L. scutulata*. Biol. Bull., 106:185–197.

Odum, H. T. and E. P. Odum. 1955. Trophic structure and productivity of a windward coral reef community on Eniwetok Atoll. Ecol. Monogr., 25:291–320.

Old, M. C. 1941. The taxonomy and distribution of boring sponges (Clionidae) along the Atlantic coast of North America. Chesapeake Biol. Sta., Publ. 44, 30 p.

Ormond, R. F. G. and A. C. Campbell. 1971. Observations on *Acanthaster planci* and other coral reef echinoderms in the Sudanese Red Sea. In D. R. Stoddart and C. M. Yonge (eds.), Regional variation in Indian Ocean coral reefs. Zool. Soc. London, Symp., 28:433–454.

Osler, E. 1826. On burrowing and boring marine animals. Royal Soc. London, Philos. Trans., 116:342–371.

Otter, G. W. 1932. Rock-burrowing echinoids. Biol. Rev. (Biol. Proc. Cambridge Philos. Soc.), 7:89–107.

————. 1937. Rock-destroying organisms in

relation to coral reefs. British Mus. (Nat. Hist.), Great Barrier Reef Expedit. 1928–1929, Sci. Rept. 1(12):323–352.

Perkins, B. F. 1971. Traces of rock-boring organisms in the Comanche Cretaceous of Texas. In B. F. Perkins (ed.), Trace fossils, a field guide. Louisiana State Univ., School Geosci., Misc. Publ. 71–1:137–147.

Perkins, R. D. and S. D. Halsey. 1971. Geologic significance of microboring fungi and algae in Carolina shelf sediments. Jour. Sed. Petrol., 41:843–853.

Poag, C. W. 1969. Dissolution of molluscan calcite by the attached foraminifer *Vasiglobulina*, new genus (Vasiglobulininae, new subfamily). Tulane Stud. Geol. Paleont., 7:45–72.

————. 1971. Notes on the morphology and habit of *Vasiglobulina alabamensis* (Foraminiferida). Jour. Paleont., 45:961–962.

Porter, C. L. and G. Zebrowski. 1937. Lime-loving molds from Australian sands. Mycologia, 29:252–257.

Purdy, E. G. and L. S. Kornicker. 1958. Algal disintegration of Bahamian limestone coasts. Jour. Geol., 66:97–99.

Radwański, A. 1964. Boring animals in Miocene littoral environments of southern Poland. Acad. Pol. Sci. (Sér. Sci. Géol. Géogr.), Bull., 12:57–62.

————. 1965. Additional notes on Miocene littoral structures of southern Poland. Acad. Pol. Sci (Sér. Sci. Géol. Géogr.), Bull., 13:167–173.

————. 1968. Tortonian cliff deposits at Zahorska Bystrica near Bratislava (southern Slovakia). Acad. Pol. Sci. (Sér. Sci. Géol. Géogr.) Bull., 16:97–102.

————. 1970. Dependence of rock-borers and burrowers on the environmental conditions within the Tortonian littoral zone of southern Poland. In T. P. Crimes and J. C. Harper (eds.), Trace fossils. Geol. Jour., Spec. Issue 3:371–390.

Ray, D. L. (ed.) 1959. Marine boring and fouling organisms. Seattle, Univ. Washington Press, 536 p.

Raynaud, J.–F. 1969. Lamellibranches lithophages. Application à l'étude d'un conglomérat à cailloux perforés du Miocène du midi de la France. Travaux Lab. Paléont. Orsay, Univ. Paris, 75 p.

Reidl, J. M. R. 1969. Sea caves. Oceans, 1(4):32–43.

Revelle, R. and R. W. Fairbridge. 1957. Carbonate and carbon dioxide. In J. W. Hedgpeth (ed.), Treatise on marine ecology and paleoecology, vol. 1, Ecology. Geol. Soc. America, Mem. 67(1):239–296.

Rice, M. E. 1969. Possible boring structures of sipunculids. Amer. Zool., 9:803–812.

Richards, R. P. and C. W. Shabica. 1969. Cylindrical living burrows in Ordovician dalmanellid brachiopod beds. Jour. Paleont., 43:838–841.

Ricketts, E. F. and J. Calvin. 1968. Between Pacific Tides (4th ed., revised by J. W. Hedgpeth). Stanford, Calif., Stanford Univ. Press, 614 p.

Robertson, P. B. 1963. A survey of the marine rock-boring fauna of southeast Florida. Unpubl. M. S. Thesis, Inst. Mar. Sci., Univ. Miami, 167 p.

Rodda, P. U. and W. L. Fisher. 1962. Upper Paleozoic acrothoracic barnacles from Texas. Texas Jour. Sci., 14:460–479.

Rodriguez, J. and R. C. Gutschick. 1970. Late Devonian–Early Mississippian ichnofossils from western Montana and northern Utah. In T. P. Crimes and J. C. Harper (eds.), Trace fossils. Geol. Jour., Spec. Issue 3:407–438.

Roniewicz, P. 1970. Borings and burrows in the Eocene littoral deposits of the Tatra Mountains, Poland. In T. P. Crimes and J. C. Harper (eds.), Trace fossils. Geol. Jour., Spec. Issue 3:439–446.

Rose, P. R. 1970. Stratigraphic interpretation of submarine versus subaerial discontinuity surfaces: an example from the Cretaceous of Texas. Geol. Soc. America, Bull., 81:2787–2798.

————. 1972. Edwards Group, surface and subsurface, central Texas. Bur. Econ. Geol., Univ. Texas, Rept. Invest. 74, 198 p.

Rowland, R. W. and D. M. Hopkins. 1971. Comments on the use of *Hiatella arctica* for determining Cenozoic sea temperatures. Palaeogeog., Palaeoclimatol., Palaeoecol., 9:59–64.

Rudwick, M. J. S. 1965. Ecology and paleoecology. In R. C. Moore (ed.), Treatise on invertebrate paleontology, Pt. H, Brachiopoda. Lawrence, Kan., Geol. Soc. America and Univ. Kansas Press, p. H199–H214.

Rützler, K. 1971. Bredin-Archbold-Smithsonian biological survey of Dominica: burrowing sponges, genus *Siphonodictyon* Bergquist, from the Caribbean. Smithsonian Contr. Zool., 77, 37 p.

Santeson, R. 1939. Die Geologische Bedeutung von Bohr-Organismen in tierischen Hartteilen, aufgezeigt an Balaniden-Schill der Innenjade. Senckenbergiana Leth., 20:304–313.

Sàra, M. and L. Liaci. 1964. Symbiotic association between zooxanthellae and two marine sponges of the genus Cliona. Nature, 203:321.

Scholl, D. W. et al. 1970. The structure and origin of the large submarine canyons of the Bering Sea. Marine Geol., 8:187–210.

Schroeder, J. H. 1972. Calcified filaments of an endolithic alga in recent Bermuda reefs. Neues Jahrb. Geol. Paläont., Mb., 1972:16–33.

Scoffin, T. P. 1972. Fossilization of Bermuda patch reefs. Science, 178:1280–1282.

Scott, R. B. et al. 1972. Manganese crusts of the Atlantic fracture zone. EOS, Amer. Geophys. Union, Trans., 53:529.

Seilacher, A. 1968. Swimming habits of belemnites—recorded by boring barnacles. Palaeogeogr., Palaeoclimatol., Palaeoecol., 4:279–285.

————. 1969. Paleoecology of boring barnacles. Amer. Zool., 9:705–719.

Sewell, R. B. S. 1926. A study of Lithotrya nicobarica Reinhardt. Records, Indian Mus. (Calcutta), 28:269–330.

Shinn, E. A. 1969. Submarine lithification of Holocene carbonate sediments in the Persian Gulf. Sedimentology, 12:109–144.

Shipley, A. E. 1902. Sipunculoidea, with an account of a new genus Lithocrosiphon. In J. S. Gardiner (ed.), The fauna and geography of the Maldive and Laccadive Archipelagoes. New York, Cambridge Univ. Press, 1:131–140.

Silén, L. 1946. On two new groups of Bryozoa living in shells of molluscs. Ark. Zool., 38B:1–7.

————. 1948. On the anatomy and biology of Penetrantiidae and Immergentiidae (Bryozoa). Ark. Zool., 40A:1–48.

Sliter, W. V. 1965. Laboratory experiments on the life cycle and ecologic controls of Rosalina globularis d'Orbigny. Jour. Protozool., 12:210–215.

Sluiter, C. P. 1891. Die Evertebraten aus der Sammlung des Königlichen Naturwissenschaftlichen Vereins in Niederländisch Indien in Batavia. III. Die Gephyreen. Natuurk. Tijdschr. Nederlandsch-Indië, 50:102–125.

Smarsh, A. et al. 1969. Carbonic anhydrase in the accessory boring organ of the gastropod, Urosalpinx. Amer. Zool., 9:967–982.

Smith, E. H. 1969. Functional morphology of Penitella conradi relative to shell penetration. Amer. Zool., 9:869–880.

Sognnaes, R. F. (ed.) 1963. Mechanisms of hard tissue destruction. Amer. Assoc. Advmt. Sci., Publ. 75, 764 p.

Sohl, N. F. 1969. The fossil record of shell boring by snails. Amer. Zool., 9:725–734.

Soliman, G. N. 1969. Ecological aspects of some coral-boring gastropods and bivalves of the northwestern Red Sea. Amer. Zool., 9:887–894.

Soule, J. D. and D. F. Soule. 1969. Systematics and biogeography of burrowing bryozoans. Amer. Zool., 9:791–802.

Stephenson, W. 1961. Experimental studies on the ecology of intertidal environments at Heron Island. II. The effect of substratum. Australian Jour. Mar. Freshw. Res., 12:164–176.

Storr, J. F. 1964. Ecology and oceanography of the coral-reef tract, Abaco Island, Bahamas. Geol. Soc. America, Spec. Paper 79, 98 p.

Swinchatt, J. P. 1969. Algal boring: a possible depth indicator in carbonate rocks and sediments. Geol. Soc. America, Bull., 80:1391–1396.

Taylor, B. J. 1971. Thallophyte borings in phosphatic fossils from the Lower Cretaceous of southeast Alexander Island, Antarctica. Palaeontology, 14:294–302.

Teichert, C. 1945. Parasitic worms in Permian brachiopod shells in Western Australia. Amer. Jour. Sci., 243:197–209.

Thornton, I. W. B. 1956. Diurnal migrations of the echinoid Diadema setosum (Leske). British Jour. Animal Behavior, 4:143–146.

Todd, R. 1965. A new Rosalina (Foraminifera) parasitic on a bivalve. Deep-Sea Res., 12:831–837.

Tomlinson, J. T. 1953. A burrowing barnacle of the genus Trypetesa (Order Acrothoracica). Jour. Washington Acad. Sci., 43:373–381.

————. 1969a. The burrowing barnacles

(Cirripedia, Order Acrothoracica). U. S. Nat. Mus., Bull., 296:1–162.

———. 1969b. Shell-burrowing barnacles. Amer. Zool., 9:837–840.

Topsent, É. 1887. Contribution à l'étude des Clionides. Arch. Zool. Expér. Gén., (2)5, Sup., Mem. 4, 165 p.

Turner, R. D. 1954. The family Pholadidae in the western Atlantic and the eastern Pacific, Part I—Pholadinae. Johnsonia, 3:1–63.

———. 1955. The family Pholadidae in the western Atlantic and eastern Pacific, Part II—Martesiinae, Jouannetiinae and Xylophaginae. Johnsonia, 3:65–160.

———. 1967. The Xylophaginae and Teredinidae—a study in contrasts. Amer. Malacol. Union, Ann. Rept. 1967:46–48.

——— and K. J. Boss. 1962. The genus Lithophaga in the western Atlantic. Johnsonia, 4:81–116.

Umbgrove, J. H. F. 1947. Coral reefs of the East Indies. Geol. Soc. America, Bull., 58:729–778.

Utinomi, H. 1953. Coral-dwelling organisms as destructive agents of corals. Pan-Pacific Sci. Congr., Auckland, Proc., Pt. 4, Zool., 533 p.

Voigt, E. 1970. Endolithische Wurm-Tunnelbauten (Lapispecus caniculus n. g. n. sp. und Dodecaceria [?] sp.) in Brandungsgeröllen der oberen Kreide im nördlichen Harzvorlande. Geol. Rundschau, 60:355–380.

———. 1972. Über Talpina ramosa v. Hagenow 1840, ein wahr scheinlich zu den Phoronidea gehöriger Bohr Organismus aus der oberen Kreide. Nachr. Akad. Wiss. Göttingen II. Math.-Phys. Kl., 1972: 93–126.

——— and J. D. Soule. 1973. Cretaceous burrowing bryozoans. Jour. Paleont., 47:21–33.

Warburton, F. E. 1958. The manner in which the sponge Cliona bores into calcareous objects. Canadian Jour. Zool., 36:555–562.

Warme, J. E. 1970. Traces and significance of marine rock borers. In T. P. Crimes and J. C. Harper (eds.), Trace fossils. Geol. Jour., Spec. Issue 3:515–526.

———. 1971. Biological energy in erosion and sedimentation, and animal-sediment interrelationships. In J. D. Howard et al., Recent advances in paleoecology and ichnology. Amer. Geol. Inst., Short Course Lect. Notes, p. 55–72.

——— and N. F. Marshall. 1969. Marine borers in calcareous terrigenous rocks of the Pacific Coast. Amer. Zool., 9:765–774.

——— et al. 1971. Submarine canyon erosion: contribution of marine rock burrowers. Science, 173:1127–1129.

Warmke, G. L. and R. T. Abbott. 1962. Caribbean seashells. Wynnewood, Pa., Livingston Publ. Co., 348 p.

Wiedemann, H. U. 1972. Shell deposits and shell preservation in Quaternary and Tertiary estuarine sediments in Georgia, U.S.A. Sediment. Geol., 7:103–125.

Yonge, C. M. 1955. Adaptation to rock boring in Botula and Lithophaga (Lamellibranchia, Mytilidae) with a discussion on the evolution of this habit. Micros. Sci., Quart. Jour., 96:383–410.

———. 1963a. The biology of coral reefs. In F. S. Russell (ed.), Advances in marine biology, v. 1. New York, Academic Press, p. 209–260.

———. 1963b. Rock-boring organisms. In R. F. Sognnaes (ed.), Mechanisms of hard tissue destruction. Amer. Assoc. Advmt. Sci., Publ. 75:1–24.

———. 1964. Rock borers. Sea Frontiers, 10:106–116.

Zankl, H. and J. H. Schroeder. 1972. Interaction of genetic processes in Holocene reefs off North Eleuthera Island, Bahamas. Geol. Rundschau, 61:520–541.

Zebrowski, G. 1936. New genera of Cladochytriaceae. Ann. Missouri Bot. Garden., 23:553–564.

CHAPTER 12

BORING MICROORGANISMS AND MICROBORINGS IN CARBONATE SUBSTRATES

STJEPKO GOLUBIC
Department of Biology, Boston University
Boston, Massachusetts, U.S.A.

RONALD D. PERKINS
Department of Geology, Duke University
Durham, North Carolina, U.S.A.

KAREN J. LUKAS
Harbor Branch Foundation Laboratory
Fort Pierce, Florida, U.S.A.

SYNOPSIS

Boring or endolithic microorganisms discussed herein are photosynthetic cyanophytes, eucaryotic green and red algae, and heterotrophic fungi that actively penetrate carbonate substrates. Although their existence has been known since the mid-nineteenth century, new techniques for preparation and study developed within the last decade have brought about significant progress in our understanding of them. Boring microorganisms have been studied in a variety of carbonate substrates, including the shells and skeletons of living organisms or their fragmented remains, and within coastal limestones.

Endolithic microorganisms exhibit a wide but discontinuous, horizontal and vertical distribution, and accordingly indicate geographical, climatic, and ecological conditions. Their activity within carbonate substrates produces characteristic boring patterns, which permit taxonomic identification. Precision of this identification is restricted by the variability and convergent morphology of endolithic taxa. Biologically specific boring behavior is often modified by environmental factors, such as light

intensity and water supply in the habitat; such modifications may be used as indicators of local ecological conditions. Orientation of the borings and their surface outlines are partly a function of biologically specific modes of penetration; they are also partly controlled by properties of the bored substrate. Microboring activity has a cumulative destructive effect on coastal limestones and on carbonate sediments.

Post-mortem changes within endolithic microborings include (1) evacuation of the boring; (2) secondary carbonate precipitation or secondary leaching within the boring; (3) diagenetic alteration of host substrates, such as replacement, dolomitization, or silicification; and (4) infilling of microborings by phosphatized chalk or limonite to form natural casts.

Study of recent boring microorganisms strongly suggests that microborings may be suitable as paleoecological indicators. Paleoecological application of endoliths must be tested initially within stratigraphically and paleoecologically well known ancient environments, proceeding from Quaternary or Tertiary deposits backward in geologic time.

INTRODUCTION

Previous Work

Penetration of hard substrates by micro-organisms was noted in the mid-nineteenth century by Carpenter (1845), who first thought that borings were part of the initial substrate morphology. The microbial origin of borings was established by Wedl (1859), who considered them to be algal; Kölliker (1860) assumed a fungal identity. In 1876 Duncan compared recent and fossil endoliths in corals and gave the name *Palaeochlya perforans* to the fossil forms; Roux (1887) provided the name *Mycelites ossifragus* for microbial borings within vertebrate bones. The taxonomic affinities of both names, whether algal or fungal, have never been clarified, and subsequently they have been used as ichnotaxa, without specifying the origin of the borings.

In 1885 and 1886 Lagerheim taxonomically described the first boring algae. Bornet and Flahault followed with additional algal descriptions (1888) and also described the first boring fungi (1889). Nadson (1902, 1927) recognized the depth distribution and biogeochemical significance of boring algae, although many of his taxonomic identifications were unreliable. Ercegović (1932) described an array of new intertidal boring cyanophyte species and studied their community structure and zonal distribution. Many of his new species descriptions have not been accepted by all subsequent authors (Geitler, 1932; Frémy, 1945). However, these descriptions characterize boring morphologies rather precisely, and have therefore proved useful for our present purposes. Frémy (1945), working principally in the Mediterranean Sea and the English Channel, reported the characteristics of algal and fungal borings from brackish-water, intertidal, and deep-marine environments.

In 1937 Pia reviewed the extensive literature that had accumulated on the subject of recent and fossil boring microorganisms and microborings. He included among recent forms those that indirectly contribute to substrate disintegration but which do not leave recognizable boring patterns. Pia objectively reported conflicting opinions, which makes his contribution particularly valuable. His review also presented an extensive inventory of fossil borings, including known forms that had not been named. This important paper also summarized the then-contemporary state of knowledge concerning ecology, mechanisms of penetration, distribution, and geological significance of boring microorganisms.

In 1949 Hessland reported on fossil microborings from the Ordovician of Sweden. Unlike Pia, Hessland opposed the use of ichnofossil names for microborings and recommended instead a classification according to size, as suggested earlier by Wetzel (1938).

Bathurst (1966) proposed that the widely distributed micrite envelopes so commonly found on fossil skeletal fragments were initiated by the activity of boring microorganisms. Boekschoten (1966) and Swinchatt (1969) suggested the application of microborings as paleoecological guides and paleobathymetric indicators. Le Campion-Alsumard (1969, 1970) studied the taxonomic relationships and microzonation of modern intertidal boring algae of the Mediterranean coast at Marseille, and Gatrall and Golubic (1970) presented a direct comparison between fossil and recent boring patterns, utilizing the scanning electron microscope (SEM). Studies of microboring organisms within modern sediments were also conducted by Perkins and Halsey (1971), Rooney and Perkins (1972), Golubic (1972), and Golubic and Schneider (1972).

At present, research on microborings encompasses a wide, interdisciplinary field of study that is significantly aided by modern instrumentation, as exemplified by the work of Alexandersson (1972); he examined the diagenetic alteration of recent and sub-recent borings in relation to environmental chemistry, utilizing the SEM and microprobe.

Terminology

Boring microorganisms are defined here by their activity, character of the substrate, size, and biological identity. The products of their activity are termed "microborings." The term "boring" refers to penetration into hard substrates by chemical or mechanical means, in contrast to "burrowing," which refers to activities in loose or soft substrates (Bromley, 1970; Frey, 1973). This terminology does not conform with the usage recommended by Carriker and Smith (1969). (See Chapter 11.)

The term "endolith" has been used to describe organisms that bore into hard substrates, specifically in reference to algal borings (Hessland, 1949; Golubic, 1969), and more broadly (Gary et al., 1972) to refer to "an organism that lives within rock or other stony matter." However, in studying marine boring microorganisms, distinction should be made between (1) epiliths, which live on the surface of the substrate; (2) chasmoliths, which adhere to the surfaces of fissures and cavities within the substrate; and (3) endoliths, which actively penetrate into carbonate substrates (Fig. 12.1). Friedmann et al. (1967) and Friedmann (1971, 1972), studying the microflora of desert rocks, used these terms in somewhat different contexts. According to the scheme proposed here, organisms colonizing existing internal spaces, with or without an apparent connection between them and the surface of the substrate, should be referred to as "chasmoliths" rather than as "endoliths."

Distinctions between these boring habits are not always clear; some organisms may belong to more than one category, or may alter their habit during their life cycle. Several boring algae initially colonize shell surfaces and mature, during which time part of their thallus remains epilithic while endolithic filaments penetrate the substrate (Ercegović, 1932). Shallow surface grooves continuing as endolithic tunnels were illustrated by Alexandersson (1972). The endo-

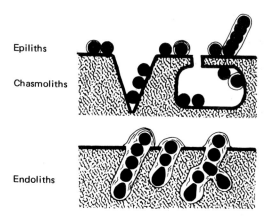

Fig. 12.1 Types of microorganisms in relation to a hard, carbonate substrate. Bold line indicates surface.

lithic green alga *Ostreobium* evidently leaves its boreholes and assumes a chasmolithic habit in larger cavities within the substrate (Schroeder, 1972). The red algae *Porphyra* and *Bangia* are endolithic during the early stages of their life cycle ("*Conchocelis*" phase), but epilithic during their adult stages (Drew, 1949, 1954, 1958).

Boring or endolithic microorganisms mostly include cyanophytes (blue-green algae) and microscopic eucaryotic algae and fungi that actively penetrate hard carbonate substrates and produce boreholes conforming to the shapes of their thalli. The size of these microorganisms and their borings range from less than 1 to about 100 μm in diameter. In their upper size range, these borings overlap with similar structures formed by boring sponges and bryozoans (Chapter 11).

Methods for Study of Endolithic Microorganisms

Direct examination of recent boring microorganisms poses methodological problems and limitations, owing to their relative inaccessibility within translucent hard substrates. However, the following basic techniques are employed, which permit analyses of endolithic microorganisms and their effects on calcareous substrates: (1)

isolation of endoliths through dissolution of the substrate, (2) observation of endoliths and their borings in situ, and (3) a resin-embedding and acid-etching technique that produces three-dimensional casts of microboring networks.

Isolation of Endoliths

Most agents applied to remove the rock or skeletal substrate cause structural damage to enclosed endolithic organisms. This damage is reduced by fixing the entire sample in 3 percent formaldehyde solution prior to substrate removal. Fixation in 5 percent glutaraldehyde, followed by 0.5 to 1.0 percent osmium tetroxide, is required if transmission electron microscopy is planned. Carbonate substrate is usually removed by dissolution in dilute acids. Hydrochloric acid (2 percent), acetic acid (10 percent), and concentrated lactic acid are used frequently. The least damage to cellular structures is incurred with dilute EDTA (ethylenediaminetetraacetic acid; Prud'homme van Reine and van den Hoek, 1966), but its dissolution capacity is very low. Fast dissolution with relatively little damage is achieved by Perényi solution (0.5 percent chromic acid:10 percent nitric acid:70 percent ethanol, as 30:40:30), a combined fixing and dissolving agent. However, all these methods remove the hard matrix that supports endolithic microorganisms, and the entire organic component of the boring collapses. Consequently, spatial relationships and growth arrangements are difficult to reconstruct utilizing this method of study.

In Situ Observation

Microboring structures also may be studied in standard petrographic thin sections. Limitations of this technique are the restricted transparency of the section and an incomplete presentation of the overall boring network.

In order to observe microorganisms and their boring patterns in original three-dimensional arrangements, optically clear substrates have been chosen for study. Small fragments and thin chips of bivalve shells are usually sufficiently transparent to be studied directly by transmission light microscopy. Some bivalve shells, even though thicker (e.g., *Pinna nobilis*), are equally transparent due to the favorable optical orientation of large calcite crystallites (T. Le Campion–Alsumard, unpublished). The best results have been achieved by embedding such fragments in optical media (such as immersion oil, 1515) having refractive indices similar to those of the substrate (Rooney and Perkins, 1972) (see Fig. 12.8). This method minimizes dispersion of light within the skeletal substrate and renders the shell essentially transparent. Kendall and Skipwith (1969) applied 40 percent hydrofluoric acid for the same purpose.

Large Iceland spars (calcite cleavage rhombohedrons) are optically well suited for endolith analysis, although they are rarely found in natural environments undergoing endolithic penetration. However, clear Iceland spars have been successfully "planted" and colonized by endolithic algae (Golubic, 1969; Golubic et al., 1970).

External surface aspects of boreholes may be studied using incident light microscopy for lower magnifications and scanning electron microscopy for higher magnifications (Golubic, 1973, Fig. 21.19A). Similarly, internal borehole surfaces may be examined after samples have been fractured. Endolithic organisms in recent substrates may be removed through Clorox (sodium hypochlorite) treatment, to provide clean borehole surfaces for study (see Fig. 12.5A). Early diagenetic dissolution and precipitation within recent and subrecent borings have been successfully studied in this manner, using the SEM with an attached microprobe (Alexandersson, 1972).

Fossil borings are sometimes filled with such materials as phosphatized chalk (Bromley, 1970, Pl. 1a) or limonite (Hessland, 1949; Wendt, 1969), which are less soluble in acids than is the matrix. These fillings, where exposed by natural dissolution of the surrounding aragonitic or calcitic matrix, are observed as casts of the boring tunnels (Kennedy, 1970, Pl. 4c, d). (See Fig. 12.9.) Such casts also may be exposed by careful etching of thin-sections and polished surfaces of the rock, and observed with the light microscope (Schindewolf, 1962) and the SEM (Gatrall and Golubic, 1970, Fig. 20).

Casts of Microboring Networks

Spatial arrangements of boring tunnels having the included microorganisms preserved in situ have been rendered accessible to study by combined embedding and casting in polymerized resins, and subsequent dissolution of the calcareous matrix (Golubic et al., 1970). Achieving complete resin penetration is difficult, owing to the extremely small diameters of the microbial borings and the depth of their extension in the substrate. The most complete penetration is achieved if subaqueous marine samples are collected and transferred through liquid phases of dehydration and medium replacements without being dried. If the specimens dry at any stage in the procedure, air becomes trapped within the fine boreholes, and penetration of the resin is hindered. The procedure is further complicated if applying either a vacuum or increased pressure, or boiling the samples in liquids, becomes necessary. Similar problems are encountered in examining empty or partially filled fossil microborings collected on land. The embedding–casting technique is easily combined with methods for fixation and preservation of cellular structures, as described above. For routine work, the following simplified procedure is recommended:

1. Fix small samples of substrate containing microorganisms as described above. Solutions should be prepared in filtered seawater.
2. Rinse in distilled water.
3. Dehydrate by transferring specimens through a series of ethanol (acetone) solutions in distilled water, along a gradient of increasing concentration, ending with several baths in pure ethanol followed by propylene oxide (pure acetone). Total dehydration time ranges from 2 to 3 hours.
4. Begin the infiltration phase in a covered container, with a 50:50 solution of *complete* Araldite mixture (Durcupan ACM mixture No. 2, which includes 964 accelerator) and propylene oxide (acetone), at room temperature, for 6 hours. After 3 hours, uncover samples to allow propylene oxide (acetone) to evaporate.
5. For final infiltration, place samples in rubber molds (used for SEM blocks) filled with pure, *complete* Araldite mixture, and leave them at room temperature overnight. For *both* infiltration phases, use Araldite Durcupan ACM mixture No. 2.
6. Cure (polymerize) at 60°C for 48 hours. To assure complete resin infiltration in particularly fine, deep borings, polymerization may be delayed by conducting one or both infiltration steps under refrigeration.

After complete polymerization, the substrate, boreholes, and microorganisms are contained within a solid resin block. These blocks are cut open (*caution*: avoid friction heating, which destroys polymer bonds) and the substrate is etched using dilute acid, EDTA, or Perényi solution. Resin casts of the borings are thus revealed in their original position. Such casts, partially or completely exposed by etching, may be coated with metal under a vacuum and then observed with a SEM (e.g., Figs. 12.3, 12.4, and 12.5B to 12.7). Partial etching is often advantageous in that it facilitates a comparative study of the boring patterns and bored substrate. Substrates containing organic components, such as lamellae between crystals in bivalve shells, are best studied in partially etched, oriented sections having the lamellae running perpendicular to the cut.

Parts of specimens from which the substrate was removed by etching may be reembedded in the same resin. In this way the substrate is replaced by a homogeneous, optically clear medium, rendering the organisms visible in their original positions (see Figs. 12.4A, 12.5A, and 12.6A). Specimens that have been used for scanning electron microscopy may also be re-embedded. The metal coating marks the outline of the original substrate surface, but it is thin enough to permit undisturbed observation with a light microscope. Ultra-fine sections of both substrate and organisms may also be made, using an ultra-microtome having a diamond knife, and preparations made for study of the cellular ultrastructure with the transmission electron microscope. In the manner described, practically every structure observed and photographed as a cast, and also its biological content, can be subsequently reanalyzed and interpreted.

DISTRIBUTION OF RECENT BORING MICROORGANISMS AND MICROBORINGS

The activity of boring microorganisms has been reported from fresh water as well as from marine environments (Pia, 1937). (See also Chapter 10.) Lacustrine algal boring in limestone was used to explain the origin of "furrowed" stones ("Furchensteine") in European lakes (reviewed by Kann, 1941). These structures, however, were later found to be a result of predominantly bacterial influence on the chemistry of the micro-environment, and are not associated with boring patterns (Golubic, 1961, 1962). Algal boring of subaerial limestones reported by Bachmann (1915) was questioned by Jaag (1945, p. 433), although Ercegović (1925) in part corroborated Bachmann's observations. Subaerial algal boring activity from the freshwater realm was recently reported by Folk et al. (1971). No question remains of the existence of freshwater boring algae penetrating shells of mollusks (Huber and Jadin, 1892; Geitler, 1927). However, to distinguish between algae that actively bore, and those that become encrusted and buried within precipitated calcium carbonate, is often difficult (Geitler, 1932, p. 77).

The highest diversity and greatest ecological significance of boring micro-organisms are found in the marine environment. The organisms have been reported from a wide variety of habitats, depths, salinities, and substrates, and their cosmopolitan distribution was stressed by Nadson (1927) and Frémy (1945). Although these generalities are valid for endoliths as a whole, the group is nevertheless highly diversified, and many species are characterized by narrow ecological niche specialization. Therefore, at the species level one may validly apply endoliths as ecological and paleoecological indicators.

Although information available at present is limited, evidence suggests that at least some endolithic microorganisms are restricted regionally. Among endolithic siphonaceous chlorophytes, Lukas (1974) found Ostreobium quekettii dominant in all corals studied. O. constrictum, however, was found only in corals from the Atlantic (Bahamas, Bermuda, Jamaica), and O. brabantium only in those from the Pacific (Palau). The regional distribution of different species of Porphyra is presumably accompanied by a similar distribution in their endolithic Conchocelis stages, but the morphology of their boring filaments does not permit species distinction (Kurogi, 1953; Drew, 1954). The difference seems more significant when endoliths from different latitudes are compared. A distinctive, large siphonaceous green alga (Ostreobium sp.), found abundantly in skeletal fragments from the temperate coasts of the Carolinas, was extremely rare in the tropical waters of Puerto Rico and Australia, and was not found in the boreal waters of New England (Perkins and Halsey, 1971; Rooney and Perkins, 1972; M. Carreiro, unpublished). These findings suggest a need for further regional comparisons.

Knowledge of the vertical distribution of marine boring microorganisms is necessary for application of endoliths as paleobathymetric indicators (Fig. 12.2). Light penetration in the sea and the compensation depth of algae (depth at which photosynthesis equals respiration)

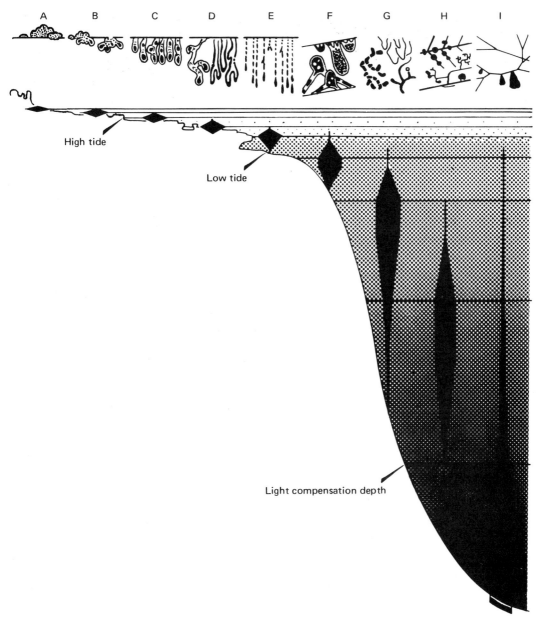

Fig. 12.2 Relative vertical distribution of recent marine microboring assemblages, as controlled by light penetration and water supply in the habitat, superimposed on an idealized coastal profile. From the coast seaward: A, coccoid cyanophytes. B, *Hormathonema luteobrunneum, H. viola-ceonigrum.* C, *Hormathonema paulocellulare.* D, *Solentia foveolarum, Kyrtuthrix dalmatica.* E, *Hyella tenuior.* F, chlorophytes *Eugomontia sacculata, Codiolum polyrhizum.* G, *Hyella caespitosa, Plectonema terebrans, Mastigocoleus testarum.* H, *Conchocelis*-stages of *Porphyra,* and *Ostreobium* spp. I, fungi.

set the limits for the vertical distribution of photosynthetic microorganisms, whereas boring heterotrophs continue into abyssal depths, lacking an apparent lower limit. Frémy (1945) observed that borings in dark, deep waters (to 1,113 m) are the exclusive product of fungal activity, but he did not describe the organisms. Höhnk (1969) studied nine forms of boring fungi in the North Sea that penetrated shell fragments collected from shallow waters down to a depth of 70 m. He found that the percentage of fragments attacked by fungi was highest in nearshore areas and decreased gradually with depth. In a recent comparison of materials from more than 300 collecting sites, ranging from the intertidal zone to a depth of 780 m, fungal borings were found in all samples (Perkins and Halsey, 1971; Rooney and Perkins, 1972; Perkins, 1972; Fig. 12.3A).

The compensation depth, and accordingly the absolute thickness of the photic zone in the sea, varies considerably from region to region, depending upon clarity of the water. Within the photic zone, light changes gradually with depth, both in intensity and in spectral composition. The response of photosynthetic organisms that evolved corresponding pigment systems is known as "chromatic adaptation" (see Levring, 1960) and applies also to many of the boring algae. Light limitation determines floral composition and depth of penetration of endolithic algae into the substrate (M. Carreiro and Golubic, unpublished).

The lower part of the photic zone is characterized by predominantly long, straight, rectangularly branched borings 2 to 5 μm in diameter, interrupted occasionally by globose swellings (Fig. 12.3B). The borings are distributed mainly between depths of 18 and 30 m in Australia (Rooney and Perkins, 1972). Conchocelis-stages of Porphyra, and Ostreobium quekettii, are found in the same depth range in coastal waters at Woods Hole, Massachusetts, often accompanied by two chromatically adapted cyanophytes, Hyella caespitosa and Plectonema terebrans (M. Carreiro and Golubic, unpublished). In deeper samples, algal borings spread horizontally, immediately below the substrate surface. Also characteristic of depths greater than 18 m are microboring sponges, found in sediments of the Arlington Reef Complex, Australia (Rooney and Perkins, 1972) and the southwestern Puerto Rico shelf (Perkins, 1972). In clear waters off Jamaica, Ostreobium quekettii (see Fig.

12.7A) and Plectonema terebrans were found in corals and calcareous sponges to a depth of at least 75 m (Lukas, 1974).

In the upper part of the photic zone, where illumination is greater, borings are dominated by the septate chlorophytes Eugomontia, Phaeophila, and Entocladia (Wilkinson and Burrows, 1972), but the cyanophytes Hyella caespitosa, Plectonema terebrans, and Mastigocoleus testarum are abundant locally from the intertidal zone to a depth of 30 m (T. Le Campion–Alsumard, unpublished).

In shallow marine waters, other environmental factors become critical for the distribution of boring microorganisms. Currents and wave action mechanically affect sedimentary particles and interfere with colonization by the microorganisms (Swinchatt, 1965; Alexandersson, 1972), so that the more intensive boring is found in protected environments. This observation refers particularly to borings within carbonate sediments. In high-energy environments, boring algae are exposed to frequent burial by moving sediment. Borings have been observed within ooid grains in actively shifting oolite shoals of the Bahamas (Newell et al., 1960) and the Persian Gulf (Shearman and Skipwith, 1965); however, the boring community is reduced (usually to one species: Hyella caespitosa), and the percentage of grains containing algae is very low (Margolis and Rex, 1971). In discussing bathymetric implications of boring microorganisms, Frémy (1945) expressed concern for possible sediment transport and hence an allochthonous origin for borings in the sediment samples. Rooney and Perkins (1972) discussed the same possibility, but stressed the potential application of well-defined borings as markers for sediment tracing or provenance.

The study of boring microorganisms in sedimentary particles is generally hindered by a low proportion of living organisms in relation to the total number of borings. Alexandersson (1972) estimated that, at any one time, fewer than 5 percent of the

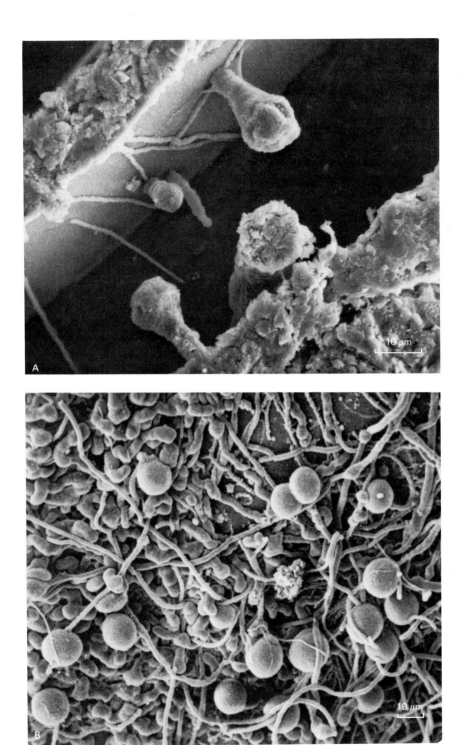

Fig. 12.3 Fungal and algal microborings in deeper waters. A, resin casts of fungal borings in etched pteropod shell; the "bags" are probably reproductive bodies growing from a network of thinner hyphae (fungal filaments). Recent sediment sample; 312 m, southwestern Puerto Rico shelf. B, casts of mixed endolith assemblage within a molluskan substrate. At least four types of borings are visible in photo: (1) very small filaments (<1 μm), (2) long, straight filaments 2–3 μm wide, connecting (3) globulose structures 12–15 μm in diameter, and (4) short, knobby filaments 3–6 μm wide. Recent sediment sample; ca. 20 m, Arlington Reef Complex, Australia. A, B = Epon-915 casts; SEM-Jeolco, Duke Univ.

borings are inhabited; these microborings thus represent a cumulative record of several generations of endolithic colonizers. In addition, poor preservation of endoliths within these borings, noted earlier by Frémy (1945), has been the main obstacle in efforts at identification. The organic production of endoliths in sediment grains has not been measured; however, endoliths should be considered a potential food source for sediment-ingesting animals. The effects of sediment feeders on endoliths, and the extent to which they may be responsible for the small number of living microorganisms present within sediment borings, are unknown (Alexandersson, 1972).

In the nearshore environment one sees a higher incidence of hard, rocky bottoms, which provide a firm substrate for endoliths. Limestones and dolomites, as well as encrusting calcareous organisms, are heavily bored. Wave action and associated abrasion restrict colonization by boring microorganisms only on extremely soft, friable rocks (e.g., chalk cliffs of Dover, England, and Rügen, Germany). Unlike sediment particles, carbonate rocks contain a very high proportion of living microorganisms within the borings.

In the lower intertidal and subtidal zone (immediately above and below the low-water mark) septate green algae are found commonly; they dominate in protected embayments, within both limestones and shell fragments. In these environments,

Eugomontia sacculata forms dense, complicated boring patterns of tunnels combined with bag-shaped swellings of various sizes and shapes (Fig. 12.4B). *Codiolum polyrhizum* produces simple, elongate, bag-shaped perforations, generally circular in cross section and variable in size. Initially, *C. polyrhizum* is connected to the substrate surface with only small rhizoidal extensions (Fig. 12.4C). More mature "bag" structures have been observed to penetrate completely through smaller shell fragments. The lower intertidal zone of agitated rocky coasts is also bored by the cyanophytes *Mastigocoleus testarum*, *Hyella tenuior*, and *Kyrtuthrix dalmatica*. *Mastigocoleus* is dominant locally but shows patchy distribution, whereas *Kyrtuthrix* and *Hyella* frequently cover large surfaces of the rock in almost pure populations (Le Campion–Alsumard, 1969, 1970). (Figs. 12.4A and 12.5).

In areas of increased exposure to air and light in the upper intertidal zone, the endolithic flora is dominated by blue-green algae having shorter boring filaments and pigmented sheaths. *Scopulonema hansgirgianum*, *Hormathonema luteobrunneum*, and surface parts of *Hyella*, *Mastigocoleus*, and *Kyrtuthrix* produce a yellow-brown pigment, "scytonemin." A parallel series of forms evolved blue pigment, "gloeocapsin," within their sheaths: *Solentia foveolarum*, *Hormathonema paulocellulare* (Fig. 12.6), and *H. violaceonigrum* (Ercegović, 1932). Although these forms may represent varia-

Fig. 12.4 Subtidal and lower intertidal microborings. A, *Kyrtuthrix dalmatica*, a dominant boring cyanophyte within limestone [resin (Araldite) cast of borings]. Pitted cast surface replicates microcrystalline texture of the rock; fine filaments probably represent borings of parasitic fungi. *Inset*: the organism in situ (double resin embedding; same scale). Characteristic loop-type false branching is visible on light microscope photo as well as on SEM photo of the cast. Lower intertidal zone; Marseille, France, SEM-Cambridge, (H. Scholz) Univ. Göttingen; Zeiss, Nomarski contrast. B, resin casts of borings of *Eugomontia sacculata* (Chlorophyta) in limestone. *Inset*: bag-like swelling of the algal filament, showing chloroplast having two pyrenoids and a thickened cross wall (septum) characteristic of the organism. 1 m depth; Marseille, France. SEM-Jeolco, Univ. Napoli; Zeiss, Nomarski contrast. C, *Codiolum polyrhizum* (Chlorophyta) "bags" penetrating a shell (Epon-812 cast). *Inset*: organism in position of penetration; rhizoidal appendages are in contact with shell surface. Lower intertidal zone; Barnstable Harbor, Massachusetts. SEM-Cambridge, Univ. Bonn; Zeiss, Nomarski contrast. Scale same as in B.

tions of fewer biological species (Frémy, 1945), they are morphologically and ecologically well defined, and the names have proved useful ·in the characterization and identification of boring patterns. The uppermost wave-spray zone is dominated by shallow boring and epilithic coccoid forms.

Fig. 12.5 Deep cyanophyte microborings in lower intertidal zone. A, characteristic straight borings of *Hyella tenuior* in limestone (fractured, Clorox-treated rock). *Inset*: organisms in situ (double resin embedding). Marseille, France. SEM-Cambridge, (H. Scholz) Univ. Göttingen; Zeiss, Nomarski contrast. Scale same as in B. B, *H. tenuior* (partially etched Araldite cast preparation), same location as A. Terminal cells produce club-shaped boring tips, visible in both A and B.

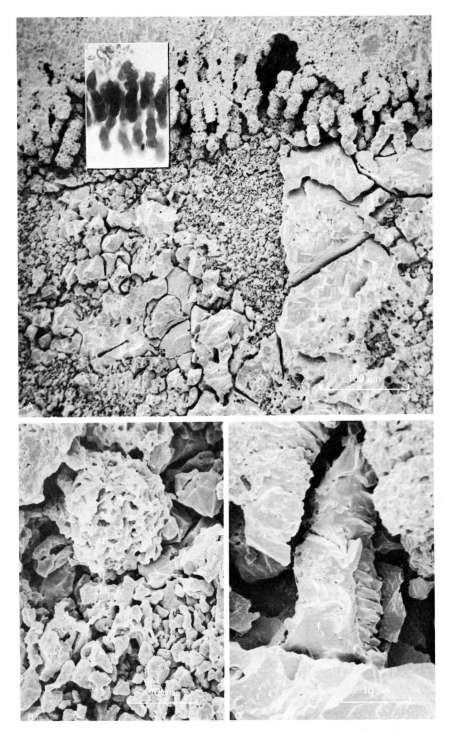

Fig. 12.6 Shallow cyanophyte microborings in upper intertidal zone. A, *Hormathonema paulocellulare* (partially etched Araldite cast preparation). Short, direct borings penetrate limestone of heterogeneous crystal composition. *Inset*: organism in situ (double resin embedding). Marseille, France. SEM-Cambridge, (H. Scholz) Univ. Göttingen; Zeiss, Nomarski contrast. B, C, detail of filaments in center of A. Surface sculpture of cast conforms with microcrystalline texture of the substrate (a sparry crystal in C). (*f* = filament; *s* = substrate.) Cf. boring cast in A, upper right.

Along a gradient of diminishing water supply, the general tendencies in algal growth are an overall reduction in average cell size and a reduction in depth of carbonate penetration (Ercegović, 1932; Le Campion–Alsumard, 1970). Depth of substrate penetration also decreases as the lower limit of light is approached. The degree of ecological variability to which endolithic organisms are subjected increases upward, from stable conditions at depth to highly fluctuating conditions in nearshore environments. Accordingly, biological zones for algal endoliths are wider toward the lower end of their distributional spectrum and more compressed toward their upper limit (Fig. 12.2).

BIOLOGICAL SPECIFICITY OF MICROBORINGS

In discussing borings as environmental indicators, Bromley (1970) postulated: ". . . the more precisely the boring organism can be identified the more useful is the fossil boring as an environmental indicator." Important information in this context is the degree of specificity of boring behavior, because borings are generally the only trace of the organism that can be readily preserved and fossilized. The problem may be stated by posing these fundamental questions: (1) to what degree does the morphology of borings reflect constant, genetically determined properties of the boring organism, (2) how well does boring morphology reflect the response of the boring organism to critical environmental influences, and (3) to what extent is this morphology determined by the properties of the bored substrate? Number (3) is a special case of number (2), but because of its particular importance to this discussion, it is here singled out.

Different species penetrating into the same substrate under uniform ecological conditions produce distinctive boring patterns, as a result of their genotypic differences. A single species, when subjected to varying ecological conditions or to different substrates, may exhibit wide variations in morphology of boring patterns. To be useful paleoecological indicators, these patterns should be morphologically characteristic and recognizable, regardless of the conditions that determined their form. At present, only a few microorganisms and their boring behavior are sufficiently studied to permit an identification based on boring patterns or a practical application for paleoecological interpretations.

Boring behavior of the intertidal endolithic cyanophytes *Hormathonema luteobrunneum*, *H. violaceonigrum*, *H. paulocellulare*, *Solentia foveolarum*, and *Hyella tenuior* is modified markedly by water supply, and this modification is expressed in the morphology of their borings. Depth of boring decreases for each of these species in areas having reduced water supply, in the upper range of their distribution. Short, unbranched borings of *H. paulocellulare* are shown in Figure 12.6A. Boring tunnels are straight on exposed rocky coasts (e.g., Fig. 12.5A) and tortuous in small rock-pools containing stagnant water. If recognized, such boring modifications identify the organism on an infraspecific level, and can provide paleoecological indicators of extreme subtlety.

Recognition of the organism at the species level, based on boring pattern morphology, proved possible for two heterocystous, filamentous cyanophytes: *Mastigocoleus testarum* and *Kyrtuthrix dalmatica* (Fig. 12.4A) (Golubic and Le Campion–Alsumard, 1973). Among green algae, the bag-shaped boreholes typical of *Codiolum polyrhizum* may permit species determination, although the variability of growth forms within this species may result in confusion with similar structures in other algae and fungi. For example, borings identified initially as the fungus *Dodgella inconstans* (Gatrall and Golubic, 1970, Plate 3b, d, f) are now known to belong to the chlorophyte *Codiolum polyrhizum* (Fig. 12.4C).

Other microorganisms produce boring

patterns that are either too variable or too simple, and therefore do not allow precise identification of the organism to be made. Many different, even unrelated, species produce similar borings. Small boring tunnels, 1 to 4 μm in diameter, were assumed to be fungal (Perkins and Halsey, 1971), but unless accompanied by characteristic reproductive bodies or other distinctive structures, fungal borings are difficult to distinguish from equally delicate borings of the blue-green algae *Plectonema terebrans* and *P. endolithicum*. Descriptions of marine endolithic fungi (see Fig. 12.3A) are few (Bonar, 1936; Zebrowski, 1936; Porter and Zebrowski, 1937; Johnson and Anderson, 1962; Höhnk, 1969; Cavaliere and Alberte, 1970) and relatively little is yet known about these organisms. A recent critical evaluation of such findings was given by Kohlmeyer (1969). Bromley (1970) showed that various criteria employed for distinguishing between fungal and algal borings are unreliable.

Branched borings generally ranging from 2 to 5 μm (but up to 8 μm) are produced by a siphonaceous green alga, *Ostreobium quekettii* (Fig. 12.7A), and by *Conchocelis*-stages of the red alga *Porphyra*. Similarity of their branching patterns led to an early confusion between these two algae (Nadson, 1902). *Ostreobium* is nonseptate, whereas *Conchocelis* is septate; but the septa (cross walls) are not always conspicuous. Contrary to claims in the literature (Pia, 1937, p. 291, 301), we found that *Ostreobium* does not turn red in deeper water.

Branched boring tunnels 6 to 12 μm in diameter can be produced by a variety of blue green and green algae: *Hyella caespitosa*, *Solentia foveolarum*, *Eugomontia sacculata* (Fig. 12.4B), and some endolithic species of *Phaeophila* and *Entocladia*. However, the paths of boring filaments and their mode of branching appear different in these forms, and a refined characterization may improve the possibility for differentiation of their borings.

When complete thalli of *Ostreobium quekettii* are observed in corals, fine filaments (2 to 5 μm) can be traced back to a few large, branching trunks having diameters as much as 25 μm. *O. constrictum* has either constricted, cylindrical filaments as much as 20 μm, or dilated filaments as much as 60 μm, in diameter (Lukas, 1974). *O. brabantium* was reported, from Pacific corals, to form very large, branched filaments 40 to 60 μm in diameter; near bifurcations, some of these reach a size of 160 μm (Weber-van Bosse, 1932).

Bag-shaped boreholes and swellings represent a common boring pattern (Figs. 12.3, 12.4B, 12.4C, 12.8B, and 12.8C), which evolved separately in various unrelated taxa of chlorophytes (*Ostreobium*, *Codiolum*, *Eugomontia*), rhodophytes (*Conchocelis*-stages of *Porphyra*), and fungi (*Ostracoblabe*, *Lithopythium*, *Dodgella*, *Pharcidia*).

Taxonomic identity of most species is at present tentative because our knowledge of life cycles and their morphological variability is incomplete. A significant contribution toward clarification of taxonomic problems came from laboratory studies on organisms in cultures, although applicability of these results to natural conditions still needs evaluation. *Conchocelis* was described initially as a separate species, *C. rosea*, by Batters (1892); but culture studies by Drew (1949, 1958) revealed that *C. rosea* represents a microscopic, endolithic stage in the life cycle of the familiar macroalgae *Porphyra* and *Bangia*. In subsequent studies of different species of *Porphyra*, differences were found between their corresponding *Conchocelis* forms in culture, although they had been virtually indistinguishable when growing within shells (Kurogi, 1953; Kornmann, 1962a; Krishnamurthy, 1969). Culture studies have also clarified the taxonomic status of a complex of septate, boring green algae often cited under the name "*Gomontia polyrhiza*," which is now known to include several separate species having complicated life cycles involving alternation

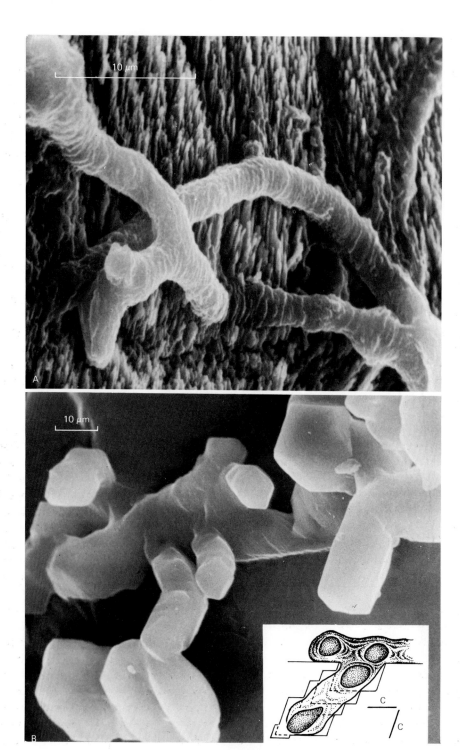

Fig. 12.7 Substrate control of algal boring patterns: at scale of scanning electron microscope. A, borings of *Ostreobium quekettii* (Chlorophyta) in *Montastrea annularis* (coral) (partially etched Epon-812 cast). Arrangement of aragonite crystals of coral skeleton is replicated on the cast surface. 17 m, Discovery Bay, Jamaica. SEM-Cambridge, Univ. Rhode Island. B, borings of *Hormathonema paulocellulare* and *Hyella tenuior* within Iceland spar (calcite). Direction and surface outlines of boring tunnels conform largely with cleavage planes: *Inset*: diagrammatic presentation of filament of *Hormathonema paulocellulare* penetrating spar (*c* = cleavage planes); gelatinous sheath of alga is more pigmented in upper part. Upper intertidal zone; La Parguera, Puerto Rico. SEM-Jeolco, Univ. Napoli.

of morphologically different generations (Kylin, 1935; Kornmann, 1959, 1960, 1962b; Wilkinson and Burrows, 1970, 1972; Nielsen, 1972). Because of difficulties with techniques, fewer culture studies on boring fungi and cyanophytes have been reported (Bonar, 1936; Kohlmeyer, 1969; Le Campion–Alsumard, 1972).

SUBSTRATE CONTROL OF BORING PATTERNS

Boring activity represents an ecological interrelationship between boring microorganisms and the bored substrate. The extent to which each of these components influences the morphology of the resulting boring pattern is important in its interpretation.

The principal differences between algae and fungi in their relationship to substrate were discussed briefly by Gatrall and Golubic (1970) and Bromley (1970, p. 55). Algae are basically light-dependent organisms limited to environments where light intensity permits an excess of photosynthesis over respiration. They are independent of organic matter in the substrate, although the existence of substrate-specific algal floras may imply special nutritional requirements in some endolithic algae. Such specificity or preference was found for boring chlorophytes in corals (Weber-van Bosse, 1932; Lukas, 1973), and in the calcareous red alga *Tenarea tortuosa* (T. Le Campion-Alsumard, unpublished). Both substrates are attacked by only a few species of boring chlorophytes, which follow closely the growing front of the organism where carbonate substrate is being deposited.

Fungi are basically independent of light, but being heterotrophs, they depend upon a supply of usable organic matter for food. Whereas algae can colonize and bore into inorganic substrates such as limestones and calcite, fungi can invade such substrates only secondarily, after colonization by algae. Fungi regularly accompany endolithic algae within limestones and are frequently observed to parasitize them. For example,

fungal hyphae often run parallel to filaments of *Hyella caespitosa* and *Mastigocoleus testarum*, sending haustoria (short lateral hyphae that absorb food from the host) into each cell of the algal filament. They have been seen also to penetrate the rock in a straight line connecting two algal borings, which perhaps indicates a chemotactic ability to detect algae within the rock. Fungal filaments often penetrate deeper than those of algae, exploring the interior of the rock and possibly intercepting the boring algal filaments. Fungal infestation of endolithic algae is particularly intensive in the stagnant waters of intertidal rock pools (J. Schneider, T. Le Campion-Alsumard, and Golubic, unpublished). A special relationship between algae and fungi is their symbiosis as lichens, many of which are known to be endolithic (see Pia, 1937, p. 316–328), and some of which occur in marine environments (J. Schneider, unpublished). Because fungi depend on algae for food, and algae are excluded from environments that are in permanent darkness, *inorganic* substrates below the photic zone should be free of microbial borings. However, substrates containing organic matter, such as skeletal fragments and shells, can be invaded directly by fungi; in environments without sufficient light, such substrates are bored exclusively by fungi (Fig. 12.3A).

Invasion of organic substrates and mineralized tissues by microorganisms falls into two categories: (1) invasion during the life of the substrate-producing animal, and (2) *post mortem* invasion. Odum and Odum (1955) reported a change in the flora within corals of Eniwetok Atoll, following the death of the animal, from one containing exclusively chlorophytes to a mixed flora including other endolithic algae. Based on this observation, they suggested a symbiotic relationship to exist between the green algae and coral host. Shells of living bivalves are often bored, but the most intensive boring takes place after the animal's death. In some cases, boring within living hosts resembles parasitism, particularly where damage is severe and the host is forced to protect itself. Shells of living mollusks are sometimes bored as far as the mantle surface. In such cases the mantle repairs the shell by forming pearl-like "blisters" and other structures (Chodat, 1898; Kessel, 1937). Barnacles repair parts of their plates that have disintegrated from algal boring (Parke and Moore, 1935). *Littorina* snails were observed to scrape endolithic algae from the shells of others of their own species, with their sharp radulas, thus contributing to shell destruction—an unusual ecological circularity (J. Schneider, unpublished). Presumably less damaging to the host are endoliths present in skeletons of living corals and sclerosponges (Lukas, 1969, 1973) and in spines of sea urchins (Schmidt, 1962a).

Post mortem invasion of shells and skeletal fragments of different sizes and composition—including vertebrate teeth and bones, fish scales, and spicules of diverse animals—by both boring algae and fungi is universal and well documented (Pia, 1937; Frémy, 1945; Schmidt, 1962b).

Boring patterns are also influenced by characteristics of the bored substrate. The influence of a mineralogically well-defined substrate, such as calcite spar, is expressed in the degree to which the boring pattern conforms to the regularity of molecular arrangement of the crystal lattice. Analyses of a bored Iceland spar (calcite) (Golubic, 1969) and of resin casts of the borings (Golubic et al., 1970) revealed an impressive degree of such conformity (Fig. 12.7B). Initial borings resemble the etching pattern (="Etzmuster") that inorganic acids leave on surfaces of crystals. Removal of carbonate by boring algae proceeds layer by layer along the main cleavage planes, most dissolution occurring at the tips of boring filaments. In this way, the excavated pit assumes a rhombohedral shape controlled by the three perfect cleavages of calcite. Size of the excavation is determined by the size of the boring algal or fungal filament (Fig. 12.7B). Advancement of the filament

follows a direction predominantly diagonal to the main cleavage planes, starting with the removal of successive carbonate layers at the intersection of these planes. The bored tunnel thus represents a series of rhombohedral cavities arranged along the direction of least resistance to the etching process. A similar observation of the relationship between *Conchocelis* growth and calcite cleavage planes was recorded earlier by Ogata (1955).

Less substrate control over the shape and direction of microbial boring is found in microcrystalline limestones. Where grains are smaller than the boring filaments, the corrosive action affects many grains simultaneously, and may extend along grain surfaces and between grains. A resin replica of such relationships has a honeycombed, pitted surface, on which each depression marks the position of a small grain along the boring interface (Fig. 12.4A). Preparations in which the substrate is only partially removed (Fig. 12.6A) show a correlation between grain size and the surface pattern of the boring cast (cf. Fig. 12.6B, C). Impressions of aragonite needles composing coral skeletons are frequently found on casts of *Ostreobium* (Lukas, unpublished) (Fig. 12.7A). This surface pattern, however, is not universal. Certain boring microorganisms, such as fungi and larger boring chlorophytes, exhibit smooth cast surfaces side by side with the highly pitted casts of boring cyanophytes. The differences in mechanisms of penetration that this observation suggests are as yet unresolved, but the characteristic morphology of these two types of borings are diagnostic (Golubic and Schneider, 1972).

Alexandersson (1972) noted the tendency for microborers to avoid emerging from the substrate during boring activity, either at the exposed substrate surface or by intersecting other borings. A similar observation was recorded earlier for fossil borings in ammonite shells (Schindewolf, 1962). However, Drew (1954) demonstrated fusion of *Conchocelis* borings to form anastomoses, and Schroeder (1972) found calcified *Ostreobium* filaments within larger cavities of corals. These seemingly contradictory reports may refer to species-specific differences in boring behavior, which may have bearing on water circulation and microchemistry within boring networks.

Skeletal substrates, such as shells of mollusks, plates of barnacles, or sea urchin shells and spines, contain different amounts of organic matter, which is often organized in a well-defined manner (Travis and Gonsalves, 1969; Kobayashi, 1969). For endolithic algae, these organic lamellae present an obstacle to penetration. *Ostreobium quekettii* was observed densely filling a single crystallite in a shell of the bivalve *Pinna nobilis*, but was unable to penetrate across an organic lamella into adjacent crystallites (T. Le Campion–Alsumard, unpublished). A similar case was illustrated by Rooney and Perkins (1972, Fig. 8). Figure 12.8A, B seems to indicate different degrees of control by the same prismatic layer of a molluscan shell upon two types of borings. Boring behavior of fungi in relation to organic lamellae of shells has not yet been demonstrated. Many borings appear irregular regardless of substrate architecture (Fig. 12.8C).

Boring filaments of *Hyella* within barnacle plates show evidence of mechanical perforation of the organic lamellae. Alternating compressed and constricted parts of the boring filaments are reflected in resin casts (J. Schneider, unpublished). Mechanical perforation by a boring green alga of organic lamellae in shells of the freshwater bivalve *Anodonta* was observed as early as 1898 by Chodat. *Codiolum* borings in marine bivalve shells reveal that carbonate is removed faster than are the organic lamellae. This mode of boring leaves concave depressions on the bored substrate surfaces, outlined by ridges of organic lamellae (Golubic and Schneider, 1972; cf. Gatrall and Golubic, 1970, Pl. 3d, f). No comparable data for endolithic fungi are presently available.

Among various factors that determine the morphology of boring patterns, we thus conclude that specific biological properties, responses of boring microorganisms to environmental variables, and characteristics of the bored substrate must be considered and evaluated. Size, shape, and mode of branching of boring tunnels are generally determined by the biological properties of the microorganism. Orientation of the boring tunnels and their surface relief are controlled in part by the bored substrate and in part by the biologically determined mechanism of penetration. The relative extent to which each of these controls may participate in boring morphology varies from organism to organism and from substrate to substrate. Boring depth is modified strongly by environmental factors, such as light intensity and water supply for endolithic algae, and food supply for the fungi.

ECOLOGICAL AND SEDIMENTOLOGICAL SIGNIFICANCE OF MICROBORING ACTIVITY

The unusual habitat and mode of life of boring microorganisms has attracted the attention of both geologists and biologists, and different aspects of their ecology have been discussed repeatedly in the literature (see Pia, 1937). These discussions, which include the possible mutual relationships between borers and their hosts, trophic relationships, symbiosis and antibiosis, and chemotactic attraction and repulsion, are largely speculative in substance and show the need for more directed studies. The two basic questions—why do microorganisms bore, and how do they accomplish it—also remain unanswered. Our main interest here is addressed to (1) the rate of microbial boring activity and (2) its cumulative effect on the substrate.

The colonization and attack of boring microorganisms proceeds rapidly. In laboratory studies, Kornmann (1959) recorded the release and settlement of spores of *Codiolum polyrhizum* on exposed shell frag-

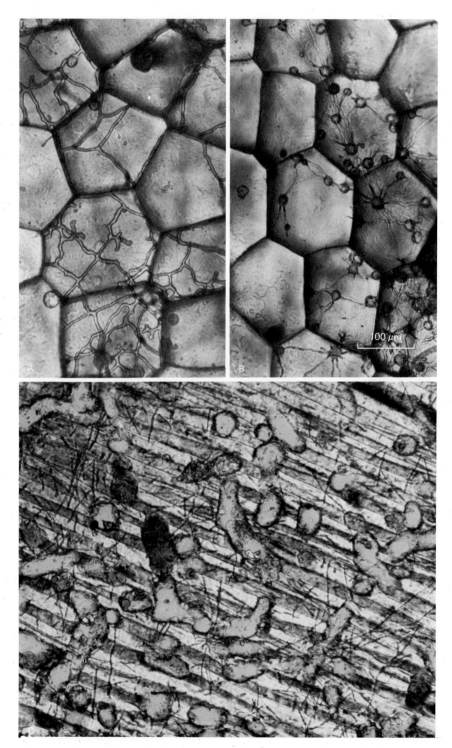

Fig. 12.8 Substrate control of algal borings: at scale of light microscope. A, endoliths within prismatic layer of bivalve shell, showing varying degrees of control by shell structure. Borings of algae (mostly *Ostreobium quekettii*) show deflections along organic lamellae separating crystals. Recent sediment sample; 30 m, Arlington Reef Complex. Australia. Mounted in oil: plane-polarized light. Scale as in B. B, endolithic fungi(?) within same specimen as A, showing no apparent deviation in boring patterns at organic lamellae-crystal contacts. C, endolithic borings in molluskan substrate, showing no apparent control of boring pattern by shell-wall architecture. Recent sediment sample; 20 m, North Carolina continental shelf. Partially polarized light. Scale as in B.

ments within 1.25 hours, which was followed by germination within 2 to 3 days. A marked penetration into the shell was noticed after 12 days. Germination and penetration by the *Conchocelis*-phase of *Porphyra* were carefully monitored by Kurogi (1953) and Drew (1954). Kurogi noticed the initial penetration 3 days after germination. Boring activity continued at an increasing rate by branching endolithic filaments. Within 30 days, main filaments originating from a single spore had bored 0.5 mm from the point of penetration, forming a dense network of boring tunnels having a total length of more than 2 mm. The penetration recorded by Drew proceeded more than twice as fast, producing a tunnel system 5.5 mm long, at a maximal distance of 0.5 mm from the point of penetration, within 2 weeks [calculated from data of Kurogi (1953) and Drew (1954)].

Under natural conditions, colonization by endoliths of exposed carbonate rock surfaces competes with a dense flora of epilithic organisms. Within 1 to 2 weeks, newly exposed surfaces are completely covered by algae; but less than 5 percent of these are endoliths. The few spores of the endolith *Hyella* bore immediately, however, so that within 3 to 4 weeks their filaments penetrate 30 to 50 μm into the substrate. Early colonization stages are the most unstable, and frequent removal of microorganisms by grazers or mechanical erosion repeatedly exposes new settling areas. Selection clearly favors the endolithic species, which rapidly increase in density. (T. Le Campion–Alsumard, unpublished). Le Campion–Alsumard (1969, 1970) also demonstrated that recolonization of denuded rock surfaces does not necessarily repeat taxonomically the community that originally occupied the surface. Succession occurred in two or more steps, although the sequence was not completed during five years of observation. This aspect suggests that pioneer colonizers (e.g., *Hyella*), which are capable of efficient substrate occupation under competitive conditions, are only slowly replaced by other, more specialized boring microorganisms.

Parke and Moore (1935) recorded the colonization of barnacle plates by endolithic algae. They observed the first endoliths on young animals 4 to 6 months after settlement of the larvae. A colonization started by *Plectonema terebrans* was followed by *Hyella* and "*Gomontia*" 3 months later. The penetration of barnacle plates proceeded at a rate of about 100 μm/yr, but did not exceed a depth of about 300 μm.

After successful colonization and penetration below the surface, the further course of boring activity within exposed carbonate rock is determined by gradients in critical ecological factors within the endolithic microenvironment. Penetration, population density, and ultimately, carbonate removal by endolithic algae are restricted to the photic zone, which seems to be a thin layer within the relatively opaque rock substrate. Approaching the compensation light intensity, penetration slows down and the algal population stabilizes. Thus, on their own, endolithic algae represent a "surface phenomenon." The relationship between habitation and destruction of rock substrates is comparable, on a smaller scale, with that of larger animals that penetrate carbonate substrates in search of shelter (clionid sponges, bivalves, and worms). (See Chapter 11.) With no external interference, the cumulative effect of carbonate rock destruction by a stabilized subtidal endolithic community is minimal.

Shallow, predominantly horizontal algal borings in the supratidal zone, however, have been observed to result in the chipping of conchoidal fragments from the rock surface, in contrast to less destructive perpendicular-to-surface "anchoring" of boring algae in the lower intertidal zone (Frémy, 1945). Similar destruction of oolite limestone in the Bahamas, where removal of cement by endolithic algae loosened the oolite grains, was observed by Purdy and Kornicker (1958).

Boring microorganisms provide food for numerous animals. Positively correlated with water supply in the intertidal zone, gastropods, chitons, and sea urchins are present in sizable populations and can exercise strong grazing pressure on both epiliths and endoliths (J. Schneider, unpublished; Chapter 11). Carbonate substrates, once perforated and loosened by boring microorganisms, are easily removed by rasping of mollusk radulas and scraping of sea urchin teeth. Removal of the surface layer of the substrate results in a dynamic situation in which the photic zone is displaced toward the interior of the rock, and algal penetration can resume. Under persistent grazing pressure, continuous algal penetration is maintained. This is a homeostatically regulated system, within which rock destruction depends upon mutually adjusted rates of algal penetration and grazing of the animals. Grazing rate is controlled by the presence of algae as a food source, which in turn depends on rate of algal growth and boring. If grazing exceeds the microbial boring velocity, it is halted by exhaustion of the resource, and the denuded surface must be recolonized. Because algal boring is light-controlled, and the boring front advances concomitantly with substrate removal, grazing rates lower than the maximal boring velocity determine the rate of substrate destruction. Relative differences in carbonate removal rates are expressed in the morphology of coastal (intertidal) profiles. Carbonate removal is often modified by heterogeneity of the substrate, physical forces of destruction, and counteracting forces fortifying the substrate, such as cementation, and incrustation by calcareous algae. The cumulative effect of microbial boring on carbonate coastal erosion, as viewed within its ecosystem context, is considered to be geologically significant (J. Schneider and Golubic, unpublished).

The cumulative effect of endolith activity on skeletal debris is a relative increase in the finer size carbonate fraction (Klement and Toomey, 1967), and ultimately, a decrease in total carbonate content of the sediment. Such selective removal of carbonate components results in a shift toward higher clastic-to-carbonate ratios, where admixtures are present (Perkins and Halsey, 1971). Within the carbonate fraction, certain skeletal components (particularly molluskan) are more susceptible to endolithic attack than others (Boekschoten, 1966; Perkins and Halsey, 1971; Alexandersson, 1972; Rooney and Perkins, 1972; Schroeder, 1972). Selective removal of the molluskan fraction yields a carbonate–grain assemblage depleted of its most susceptible components, and results in a "residual lag" of sediments more resistant to biological erosion—a "maturity index."

PALEOECOLOGICAL POTENTIAL OF MICROBORINGS

Boring microorganisms leave grooves, perforations, and tunnels within hard carbonate substrates, which easily can be preserved and fossilized. Whereas larger fossil remains are often preserved as unidentifiable fragments, having little or no value as paleoecological guides, microborings can be preserved in their entirety due to their small size. Rocks containing only fragmented skeletal debris, having enclosed endolithic structures, may therefore be useful for paleoecological analysis.

Several workers have previously suggested that algal borings may be useful in defining ancient photic zones (Mägdefrau, 1937; Hessland, 1949; Boekschoten, 1966; Golubic, 1969; Swinchatt, 1969; Gatrall and Golubic, 1970; Perkins and Halsey, 1971; Rooney and Perkins, 1972). Swinchatt (1969), in his work on sediments of Arlington Reef Complex, Australia, postulated the possible value of microborings as paleobathymetric indicators of water depths less than 15 to 20 m (50 to 60 ft), and proposed that micritic envelopes may indicate such depths in ancient sediments. Taken out of its environmental context, the limitations of this proposal lie in (1) the neglect of fungal borings, which are light independent (Friedman et al., 1971); (2) the neglect of low-light-adapted boring algae, which occur in deeper waters (reported herein); and (3) the lack of evidence that all ancient micrite envelopes are of algal or even microbial origin [see Alexandersson (1972) for discussion of shell-residue micrite].

Perkins and Halsey (1971) used a qualitative and quantitative comparison of borings from recent nearshore areas and deeper sediments to detect ancient shoreline positions. Two offshore zones paralleling the present coastline were found to have a high incidence of borings similar to those in active, nearshore sediments, and were interpreted as relict nearshore deposits associated with lower stands of Pleistocene sea level.

In order to refine an application of microborings to paleoecology, the following points must be considered: the morphology and ecology of recent boring microorganisms and their borings, the effects of post mortem changes, fossilization and diagenetic changes, and the systematic treatment of recent versus fossil microborings. In preceding sections we discussed qualitative and quantitative aspects of microbial boring activity in recent environments that provide information for interpretation of equivalent fossil situations. These aspects can be summarized as follows:

1. Recent boring microorganisms exhibit a wide horizontal (regional and latitudinal) and vertical (depth) distribution. Boring patterns can be assigned a geographical, climatic, and bathymetric affinity with respect to their origin, and they can indicate corresponding ecological conditions.
2. Boring microorganisms produce specific boring patterns that permit taxonomic identification. Precision of this identification is restricted by variability within some endolithic taxa and by convergent morphologies of unrelated microorganisms.
3. Specific boring behavior can be modified by environmental factors, such as light and water supply. Conversely, these morphological modifications of boring patterns have the potential of indicating critical ecological parameters of the habitat.

4. Orientation of microborings and their resultant borehole surfaces are in part controlled by biologically specific modes of penetration and by the properties of the bored substrate.
5. Microbial boring activity has a cumulative effect on both lithified and skeletal carbonate substrates, and in time produces recognizable changes in gross coastal morphology and in sediment composition.

In studying fossil borings and in comparing them with recent forms, post mortem changes prior to fossilization, as well as later diagenetic changes, must be considered. Ancient endolithic microorganisms are known from their preserved borings in rocks ranging in age from Cambrian to Holocene—reviewed by Pia (1937), Hessland (1949), and Bromley (1970). These studies indicate that microborings are preserved despite diagenetic alteration. Isolation of well-preserved organic remains of algae, through acid dissolution of carbonate rocks as old as Cambrian, was reported by De Meijer (1969). Whether these remnants belong to sediment-building or to boring algae was not established; this distinction is important, however, because boring microbial activity is not necessarily contemporaneous with substrate formation. Boring algae can invade a substrate at any time after lithification, including the present day (cf. Chapter 5).

Many of the recent borings found in skeletal grains are empty (Alexandersson, 1972), and several generations of borers are thought to participate in the boring of a single fragment, often interrupted by periods of carbonate precipitation and infilling of older borings (Bathurst, 1966). Newell et al. (1960) discussed the possibility of secondary changes induced by bacterial decomposition of endolithic algae in oolites, but Alexandersson (1972) did not detect any such effects in well-aerated, oligotrophic environments.

Alexandersson (1972) described two different ways that bored carbonate fragments change in response to the carbonate saturation state of the ambient sea water.

In shallow, warm seas, supersaturated with respect to CaCO₃, vacated borings are filled with acicular aragonite and rhombohedral Mg-calcite. This process corresponds to the mechanism of micrite envelope formation as described by Bathurst (1966; 1971, p. 388). In cold and deeper waters undersaturated with respect to carbonates, the borings remain empty. In addition to destruction of skeletal grains by microboring activity, Alexandersson (1972) described inorganic "selective leaching" that results in "shell residue micrite." Exaggerated surface sculpture of fossil borings shown by Gatrall and Golubic (1970) can be explained by secondary leaching of the walls of borings. (Fig. 12.10A).

A well-known fact is that skeletal aragonite and Mg-calcite undergo diagenetic alteration to calcite when exposed to meteoric water during subaerial exposure (Friedman, 1964; Land, 1967). The fate of microborings contained within these substrates is then controlled largely by the manner and degree of "host" alteration during progressive stages of diagenesis. Where the original substrate composition is aragonitic, and mineralogical changes proceed through processes of replacement or inversion (see Bathurst, 1971, p. 239, 347, 486), the prospects of preservation for enclosed borings are fairly good. If, however, the aragonitic substrate undergoes dissolution, leaving an empty shell mold that is later filled with calcite, preservation is possible only where microborings have been filled with (or altered to) calcite prior to leaching. In Miocene sediments from North Carolina, endoliths or casts of their borings are found in growth position along the margins of shell molds of mollusks, apparently preserved as insoluble residues of the carbonate leaching process (Nease and Wolf, 1971). The walls of such shell molds, which have not been filled completely with secondary calcite, may therefore provide another site for preservation of endolithic organisms (Fig. 12.9); however, complete filling undoubtedly would result in obliteration of such delicately preserved structures.

Borings are often filled by less-soluble minerals, such as phosphatized chalk (Bromley, 1970, Pl. 1a) and limonite. The ability to concentrate iron was attributed to both living algae and fungi, and possible mechanisms were discussed by Cayeux (1914), Frémy (1945), and Hessland (1949). Iron-rich fillings of microborings are usually well preserved as natural casts of borings (Hessland, 1949; Wendt, 1969), which can be exposed and studied by acid etching (Schindewolf, 1962; Gatrall and Golubic, 1970). (Fig. 12.10.)

Alteration of high-magnesium calcite to low-magnesium calcite through the process of exsolution (Land, 1967) apparently permits good preservation of borings, such as those illustrated by Wolf (1965) in fossil crinoids. Shells that are composed initially of calcite, such as those of oysters and brachiopods, generally preserve microborings very well, as shown by the work of Hessland (1949). In many cases, such remarkable preservation is maintained throughout later diagenesis, including subsequent dolomitization or silicification (Fig. 12.10B).

Recent endoliths are a morphologically and physiologically diverse group of microorganisms acting in a peculiar microenvironment. From all evidence, their activity extended throughout the Phanerozoic, associated with complex phenomena that we have only begun to understand. Therefore, the recognizable, morphologically characterized fossil borings seemingly have high paleoecological potential.

Classification of fossil and recent borings according to size proved inadequate, and the protagonists of this system had to use additional information—such as direction and ramification of tunnels, and presence or absence of swellings—to complete their descriptions (Wetzel, 1938; Hessland, 1949). Providing that the study of recent microorganisms yields a basis for reliable ecological and taxonomic identifications, the

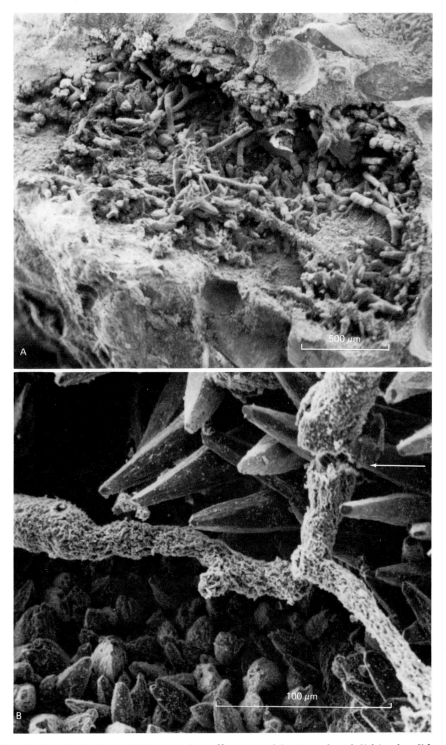

Fig. 12.9 Fossil microborings: Miocene. A, well-preserved boring of endolithic alga [identified as *Ostreobium quekettii* by Nease and Wolf (1971)] exposed within leached shell void. Fractured surface of a sample collected near Aurora, North Carolina. B, well-preserved boring of endolithic alga within leached mollusk mold, showing rupturing of algal filament (arrow) by actively growing calcite crystal. Same sample locality as A. A, B, SEM-Jeolco, Duke Univ.

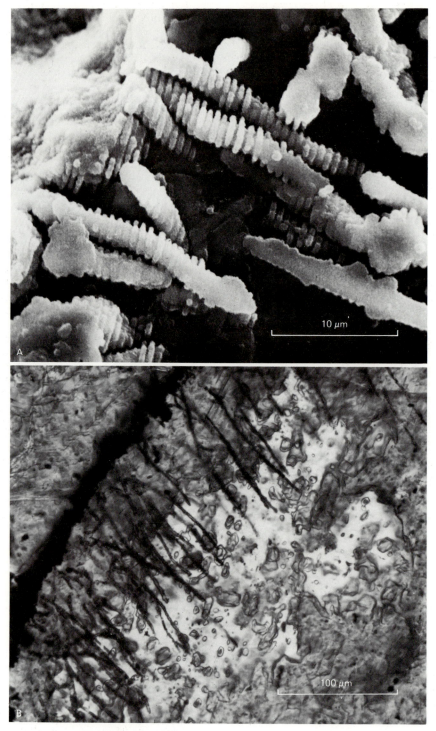

Fig. 12.10 Fossil microborings: Jurassic (A) and Silurian (B). A, limonite cast of microborings from polished and etched shell fragment. Casts replicate the substrate crystallite arrangement, possibly exaggerated by subsequent leaching. Twinhoe Beds; Bath, England. SEM Cambridge, Univ. Leicester (courtesy M. Gatrall). B, endoliths within partially silicified brachiopod. Thin section, Reynales Limestone; near Rochester, New York. Cross-polarized light.

question arises: how can these correlations be extrapolated for ancient situations, and accordingly, to what extent is the use of the nomenclature of extant forms applicable to fossils? Virtually nothing is known about the evolution and phylogenetic relationships among endoliths. These considerations prompted Pia (1937, p. 341) to endorse the use of trace genera ("artificial genera"), but he did not support the description of corresponding trace species. Limitations in the recognition of microorganisms exclusively on the basis of their borings, as discussed in this chapter, do extend to fossil forms (adding to the problems of preservation and diagenesis), and we expect that some types of borings inevitably represent heterogeneous groups; accordingly, these have little diagnostic value. This situation justifies the use of ichnotaxa, together with the familiar system of binominal nomenclature. (See Chapter 3.) At present, however, to evaluate the ichnological descriptions of fossil microborings that have already accumulated in the literature (e.g., Mägdefrau, 1937; Pia, 1937; Schindewolf, 1962; Klement and Toomey, 1967; Bromley, 1970; Kennedy, 1970) is premature.

Progress in the study of endoliths is needed along the following lines.

1. A better understanding of the distribution and environmental requirements of endoliths.

2. Refinement of endolith identification based on boring patterns.
3. The study of fossil forms from stratigraphic units that are well defined and paleoecologically well understood.
4. The study of progressively older stratigraphic units, from those of Quaternary and Tertiary times, which presumably contain many extant forms, to older units containing fewer forms having living relatives.
5. Refinement of the taxonomic and environmental interpretation of fossil microborings and ichnotaxa.

ACKNOWLEDGMENTS

The work reported here resulted from close collaboration between us and T. Le Campion-Alsumard, Marseille, France; J. Schneider, Göttingen, West Germany; B. d'Argenio, Napoli, Italy; and M. Carreiro, Boston, Massachusetts, U.S.A. In addition to previously unpublished data that they made available for this chapter, many ideas originated from fruitful discussions with them. Scanning electron microscopes used in our studies were those of the Universities of Bonn, Göttingen, Napoli, and Rhode Island, and also Duke University. We thank these institutions for providing instrument time and for the technical assistance of their personnel, particularly H. Scholz, whose excellence in SEM operation and photomicrography deserves special acknowledgment. The manuscript was read critically by B. Cameron, M. Carreiro, I. Friedmann, L. Margulis, and J. E. Warme. This research was supported by NSF grants GB-6543, BO-25271, and GA-31168 to Golubic, and GA-22690 to Perkins.

REFERENCES

Alexandersson, T. 1972. Micritization of carbonate particles: processes of precipitation and dissolution in modern shallow-water sediments. Geol. Inst. Univ. Uppsala, Bull., N. Ser., 3:201–236.

Bachmann, E. 1915. Kalklösende Algen. Ber. Deutsch. Bot. Gesell. Jahrb., 33:45–47.

Bathurst, R. G. C. 1966. Boring algae, micrite envelopes and lithification of molluscan biosparites. Geol. Jour., 5:15–32.

————. 1971. Carbonate sediments and their diagenesis. Developments in Sedimentology, 12, 620 p.

Batters, E. A. 1892. On *Conchocelis*, a new genus of perforating algae. Phycological Mem., Pt. I., p. 25–29.

Boekschoten, G. J. 1966. Shell borings of sessile epibiontic organisms as paleoecological guides (with examples from the Dutch coast). Palaeogeogr., Palaeoclimatol., Palaeoecol., 2:333–379.

Bonar, L. 1936. An unusual ascomycete in the shells of marine animals. Univ. California Publ. Bot., 19:187–193.

Bornet, E. and C. Flahault. 1888. Note sur deux nouveaux genres d'Algues perforantes. Jour. Botanique, 2:161–165.

———— and C. Flahault. 1889. Sur quelques plantes vivant dans le test calcaire des mollusques. Soc. Bot. France, Bull., 36:147–176.

Bromley, R. G. 1970. Borings as trace fossils and *Entobia cretacea* Portlock, as an example. In T. P. Crimes and J. C. Harper (eds.), Trace fossils. Geol. Jour., Spec. Issue 3:49–90.

Carpenter, W. 1845. On the microscopic structure of shells. British Assoc. Advmt. Sci., Rept., 14:1–24.

Carriker, M. R. and E. H. Smith. 1969. Comparative calcibiocavitology: summary and conclusions. Amer. Zool., 9:1011–1020.

Cavaliere, A. R. and R. S. Alberte. 1970. Fungi in animal shell fragments. Jour. Elisha Mitchell Sci. Soc., 86:203–206.

Cayeux, L. 1914. Existence de nombreuses traces d'algues perforantes dans les minerais de fer öolithiques de France. Compt. Rend. Acad. Sci. Paris, 158:1539–1541.

Chodat, R. 1898. Sur les algues perforantes d'eau douce. Études de Biologie lacustre. Herb. Boiss., Bull., 6:431–476.

De Meijer, J. J. 1969. Fossil non-calcareous algae from insoluble residues of algal limestones. Leidse Geol. Med., 44:235–263.

Drew, K. M. 1949. *Conchocelis*-phase in the life-history of *Porphyra umbilicalis* (L) Kütz. Nature, 164(4174):748–749.

————. 1954. Studies in the Bangioideae III. The life-history of *Porphyra umbilicalis* (L) Kütz. var. *laciniata* (Lightf.) J. Ag. Ann. Botany, N.S., 18:183–211.

————. 1958. Studies in the Bangiophycidae. IV. The *Conchocelis*-phase of *Bangia fuscopurpurea* (Dillw.) Lyngbye in culture. Stazion Zool. Napoli, Publ., 30:358–372.

Duncan, P. M. 1876. On some thallophytes parasitic within recent Madreporaria. Royal Soc. London, Proc., 24:238–257.

Ercegović, A. 1925. La vegetation des lithophytes sur les calcaires et les dolomites en Croatie. Acta Botanica (Zagreb), 1:64–114. (Croatian with French summary.)

————. 1932. Études ecologiques et sociologiques des Cyanophycées lithophytes de la côte yougoslave de l'Adriatique. Acad. Youg. Sci. Arts Classe Sc. Math. Nat., Bull. Internat., 26:33–56.

Folk, R. L. et al. 1971. Black phytokarst from Hell. Geol. Soc. America, Abs. Prog., 3(7):569–570.

Frémy, P. 1945. Contribution à la physiologie des Thallophytes marines perforant et cariant les roches calcaires et les coquilles. Ann. Inst. Oceanogr., 22:107–144.

Frey, R. W. 1973. Concepts in the study of biogenic sedimentary structures. Jour. Sed. Petrol., 43:6–19.

Friedman, G. M. 1964. Early diagenesis and lithification in carbonate sediments. Jour. Sed. Petrol., 34:777–813.

———— et al. 1971. Micrite envelopes of carbonate grains are not exclusively of photosynthetic algal origin. Sedimentology, 16:89–96.

Friedmann, E. I. 1971. Light and scanning electron microscopy of the endolithic desert algal habitat. Phycologia, 10:411–428.

————. 1972. Ecology of lithophytic algal habitats in Middle Eastern and North American deserts. In Eco-physiological foundation of ecosystems productivity in arid zone. U.S.S.R. Acad. Sci., "Nauka," p. 182–185.

———— et al. 1967. Desert algae of the Negev (Israel). Phycologia, 6:185–200.

Gary, M. et al. (eds.) 1972. Glossary of geology. Washington, D.C., Amer. Geol. Inst., 805 p.

Gatrall, M. and S. Golubic. 1970. Comparative study on some Jurassic and recent endolithic fungi using scanning electron microscope. In T. P. Crimes and J. C. Harper (eds.), Trace fossils. Geol. Jour., Spec. Issue 3:167–178.

Geitler, L. 1927. Neue Blaualgen aus Lunz (Neue oder wenig bekannte Mikro–organismen aus der Umgebung von Lunz). Arch. Protist., 60:440–448.

————. 1932. Cyanophyceae. In L. Rabenhorst (ed.), Kryptogamen-Flora von Deutschland, Österreich, und der Schweiz. Leipzig, Akad. Verlagsgesell, 14, 1196 p.

Golubic, S. 1961. Der Vrana–See an der Insel Cres—ein Chara-See. Int. Verein. Theor. Angew. Limnol., Verh., 14:846–849.

————. 1962. Zur Kenntnis der Kalkinkrustation und Kalkkorrosion im Seelitoral. Schweiz. Zeitschr. Hydrol., 24:229–243.

————. 1969. Distribution, taxonomy, and boring patterns of marine endolithic algae. Amer. Zool., 9:747–751.

————. 1972. Scanning electron microscopy of recent boring Cyanophyta and its possible paleontological application. In T. V.

Desikachary (ed.), Taxonomy and biology of blue-green algae. Univ. Madras, India, p. 167–170.

————. 1973. The relationship between blue-green algae and carbonate deposits. In N. Carr and B. A. Whitton (eds.), The biology of blue-green algae. London, Blackwell Sci. Publ., p. 434–472.

———— and T. Le Campion-Alsumard. 1973. Boring behavior of marine blue-green algae Mastigocoleus testarum Lagerheim and Kyrtuthrix dalmatica Ercegović, as a taxonomic character. Schweiz. Zeitschr. Hydrol., 35:157–161.

———— and J. Schneider. 1972. Relationship between carbonate substrate and boring patterns of marine endolithic microorganisms (abs.). Geol. Soc. America, Abs. Prog., 4(7):518.

———— et al. 1970. Scanning electron microscopy of endolithic algae and fungi using a multipurpose casting-embedding technique. Lethaia, 3:203–209.

Hessland, I. 1949. Investigations of the Lower Ordovician of the Siljan District, Sweden, II. Lower Ordovician penetrative and enveloping algae from the Siljan District. Geol. Inst. Univ. Uppsala, Bull., 33:409–428.

Höhnk, W. 1969. Über den pilzlichen Befall kalkiger Hartteile von Meerestieren. Deutsch. Wiss. Komm. Meeresforsch., Ber., 20:129–140.

Huber, J. and M. F. Jadin. 1892. Sur une nouvelle algue perforante d'eau douce. Jour. Botanique, 6:278–286.

Jaag, O. 1945. Utersuchungen über die Vegetation und Biologie der Algen des nackten Gesteins in den Alpen, im Jura und im Schweizerischen Mittelland. Beitr. Kryptogamenflora Schweiz, 9:1–560.

Johnson, T. W., Jr. and W. R. Anderson. 1962. A fungus in Anomia simplex shell. Jour. Elisha Mitchell Sci. Soc., 78:43–47.

Kann, E. 1941. Krustensteine in Seen. Arch. Hydrobiol., 37:504–532.

Kendall, C. G. S. C. and P. A. d'E. Skipwith. 1969. Holocene shallow-water carbonate and evaporite sediments of Khor al Bazam, Abu Dhabi, southwest Persian Gulf. Amer. Assoc. Petrol. Geol., Bull., 53:841–869.

Kennedy, W. J. 1970. Trace fossils in the chalk environment. In T. P. Crimes and J. C. Harper (eds.), Trace fossils. Geol. Jour., Spec. Issue 3:263–282.

Kessel, E. 1937. Schalenkorrosion bei lebenden Strandschnecken (Littorina littorea) und ihre Ursachen. Deutsch. Zool. Ges. Anz., Verh., 39:69–77.

Klement, K. W. and D. F. Toomey. 1967. Role of the blue-green alga Girvanella in skeletal grain destruction and lime mud formation in the Lower Ordovician of west Texas. Jour. Sed. Petrol., 37:1045–1051.

Kobayashi, I. 1969. Internal microstructure of the shell of bivalve molluscs. Amer. Zool., 9:663–672.

Kohlmeyer, J. 1969. The role of marine fungi in the penetration of calcareous substances. Amer. Zool., 9:741–746.

Kölliker, A. 1860. On the frequent occurrence of vegetable parasites in the hard tissues of the lower animals. Microscop. Soc., Quart. Jour., 8:171–188.

Kornmann, P. 1959. Die heterogene Gattung Gomontia. I. Der sporangiale Anteil. Codiolum polyrhizum. Helgoländer Wiss. Meeresuntersuch., 6:229–238.

————. 1960. Die heterogene Gattung Gomontia. II. Der fädige Anteil. Eugomontia sacculata nov. gen. nov. sp. Helgoländer Wiss. Meeresuntersuch., 7:59–71.

————. 1962a. Die Entwicklung von Monostroma grevillei. Helgoländer Wiss. Meeresuntersuch., 8:195–202.

————. 1962b. Zur Kenntnis der Porphyra-Arten von Helgoland. Helgoländer Wiss. Meeresuntersuch., 8:176–192.

Krishnamurthy, V. 1969. The Conchocelis-phase of three species of Porphyra in culture. Jour. Phycol., 5:42–47.

Kurogi, M. 1953. Study of the life-history of Porphyra. I. The germination and development of carpospores. Tohoku Reg. Fish. Res. Lab., Bull., 2:67–103. (Japanese with English summary.)

Kylin, H. 1935. Über einige kalkbohrende Chlorophyceen. Kungl. Fysiogr. Sallsk, i Lundforh., 5:186–204.

Lagerheim, G. 1885. Codiolum polyrhizum n. sp., Ett Bidrag. Till Kännedomen om Slägtet Codiolum A. Br. Övers. Kgl. Vetensk. Akad. Forhandl., 42:21–31.

————. 1886. Notes sur le Mastigocoleus, nouveau genre des algues marines de l'ordre des Phycochromacées. Notarisia, 1:65–69.

Land, L. S. 1967. Diagenesis of skeletal carbonates. Jour. Sed. Petrol., 37:914–930.

Le Campion-Alsumard, T. 1969. Contribution à l'étude des cyanophycées lithophytes des étages supralittoral et mediolittoral (region de Marseille). Tethys, 1:119–172.

———. 1970. Cyanophycées marines endolithes colonisant les surfaces rocheuse denudees (Étages supralittoral et Mediolittoral de la region de Marseille). Schweiz. Zeitschr. Hydrol., 32:552–558.

———. 1972. Quelques remarques sur la culture et le cycle de developpement de la Cyanophycée lithophyte *Hyella caespitosa* Bornet et Flahault. Thétis, 4:391–396.

Levring, T. 1960. Submarines Licht und die Algenvegetation. Botanica Marina, 1:67–73.

Lukas, K. J. 1969. An investigation of the filamentous, endolithic algae in shallow–water corals from Bermuda. In R. N. Ginsburg and P. Garrett (eds.), Reports of research— the 1968 seminar on organism-sediment interrelationships. Berumuda Biol. Station Res., Spec. Publ., 2:145–152.

———. 1973. Taxonomy and ecology of the endolithic microflora of reef corals, with a review of the literature on endolithic microphytes. Unpubl. Ph. D. Dissert., Univ. Rhode Island, 159 p.

———. 1974. Two species of the chlorophyte genus *Ostreobium* from skeletons of Atlantic and Carribbean corals. Jour. Phycol., 10:331–335.

Mägdefrau, K. 1937. Lebensspuren fossiler "Bohr"-Organismen. Beitr. Naturk. Forsch. Südwdtl., 2:54–67.

Margolis, S. and R. W. Rex. 1971. Endolithic algae and micrite envelope formation in Bahamian oolites as revealed by scanning electron microscopy. Geol. Soc. America, Bull., 82:843–852.

Nadson, G. A. 1902. Die perforienden Algen und ihre Bedeutung in der Natur. Scripta Hort. Bot. Univ. Imperialis Petropol., 1900–1902:35–40.

———. 1927. Les algues perforantes de la mer Noire. Compt. Rend. Acad. Sci. Paris., 184:896–898.

Nease, F. R. and F. A. Wolf. 1971. A nonpetrified fossil alga from a phosphate mine site in eastern North Carolina. Jour. Elisha Mitchell Sci. Soc., 87:51–52.

Newell, N. D. et al. 1960. Bahamian oolitic sand. Jour. Geol., 68:481–497.

Nielsen, R. 1972. A study of the shell-boring marine algae around the Danish island Læsø. Bot. Tidsskrift, 67:245–269.

Odum, H. T. and E. P. Odum. 1955. Trophic structure and productivity of a windward coral reef community on Eniwetok Atoll. Ecol. Monogr., 25:291–320.

Ogata, E. 1955. Perforating growth of *Conchocelis* in calcareous matrices. Bot. Magazine, 68:371–372.

Parke, M. and H. B. Moore. 1935. The biology of *B. balanoides*. II. Algal infection of the shell. Jour. Marine Biol. Assoc. United Kingdom, 20:49–56.

Perkins, R. D. 1972. Microboring organisms as environmental indicators and sediment tracers: SW Puerto Rico Shelf (abs.). Geol. Soc. America, Abs. Prog., 4(7)·624.

——— and S. D. Halsey. 1971. Geologic significance of microboring fungi and algae in Carolina shelf sediments. Jour. Sed. Petrol., 41:843–853.

Pia, J. 1937. Die kalklösenden Thallophyten. Arch. Hydrobiol., 31:264–328, 341–398.

Porter, C. L. and G. Zebrowski. 1937. Lime-loving molds from Australian sands. Mycologia, 29:252–257.

Prud'homme van Reine, W. F. and C. van den Hoek. 1966. Isolation of living algae growing in the shells of molluscs and barnacles with EDTA (Ethylene-diamine-tetra-aceticacid). Blumea, Pays-Bas, 14:331–332.

Purdy, E. G. and L. S. Kornicker. 1958. Algal disintegration of Bahamian limestone coasts. Jour. Geol., 66:96–99.

Rooney, W. S. and R. D. Perkins. 1972. Distribution and geologic significance of microboring organisms within sediments of the Arlington Reef Complex, Australia. Geol. Soc. America, Bull., 83:1139–1150.

Roux, W. 1887. Über eine im Knochen lebenden gruppe von Fadenpilzen (*Mycelites ossifragus*). Zeitschr. Wiss. Zool., 45:227–254.

Schindewolf, O. H. 1962. Parasitäre Thallophyten in Ammoniten-Schalen. Paläont. Zeitschr., H. Schmidt-Festband: 206–215.

Schmidt, W. J. 1962a. *Mycelites*-Befall an Stacheln von lebenden Seeigeln. Zool. Anzeig., 169:245–252.

———. 1962b. Über *Mycelites*-Befall an Zähnen fossiler Haie. Internat. Rev. Gesamt. Hydrobiol., 47:587–601.

Schroeder, J. H. 1972. Calcified filaments of an endolithic alga in recent Bermuda reefs. Neues Jahrb. Geol. Paläont., Mh., 16–33.

Shearman, D. J. and P. A. d'E. Skipwith. 1965. Organic matter in recent and ancient limestones and its role in their diagenesis. Nature, 208:1310–1311.

Swinchatt, J. P. 1965. Significance of constituent composition, texture, and skeletal breakdown in some recent carbonate sediments. Jour. Sed. Petrol., 35:71–90.

———. 1969. Algal boring: a possible depth indicator in carbonate rocks and sediments. Geol. Soc. America, Bull., 80:1391–1396.

Travis, D. F. and M. Gonsalves. 1969. Comparative ultrastructure and organization of the prismatic region of two bivalves and its possible relation to the chemical mechanism or boring. Amer. Zool., 9:635–661.

Weber-van Bosse, A. 1932. Algues. In Résultats Scientifiques du Voyage aux Indes Orientales Néerlandaises de LL. AA. RR. le Prince et la Princesse Léopold de Belgique. Mus. Roy. Hist. Natur. Belgique, Mém., Hors Sér.:1–28.

Wedl, G. 1859. On the significance of the canals found in many mollusc and gastropod shells. Sitzungsber. Kl. Akad. Wiss., 33:451–472.

Wendt, J. 1969. Stratigraphie und Paläogeographie des Roten Jurakalks im Sonnwendgebirge (Tirol, Österreich). Neues Jahrb. Geol. Paläont., Abh., 132:219–238.

Wetzel, W. 1938. Die Schalenzerstörung durch Mikroorganismen, Erscheinungsform, Verbreitung und geologische Bedeutung in Gegenwart und Vergangenheit. Kieler Meeresforsch. 2:255–266.

Wilkinson, M. and E. M. Burrows. 1970. Eugomontia sacculata Kornm. in Britain and North America. British Phycol. Jour., 5:235–238.

——— and E. M. Burrows. 1972. An experimental taxonomic study of the algae confused under the name Gomontia polyrhiza. Jour. Marine Biol. Assoc. United Kingdom, 52:49–57.

Wolf, K. H. 1965. Petrogenesis and palaeoenvironment of Devonian algal limestones of New South Wales. Sedimentology, 4:113–178.

Zebrowski, G. 1936. New genera of Cladochytriaceae. Ann. Missouri Bot. Garden, 23:553–564.

CHAPTER 13

TRACES OF PREDATION

GALE A. BISHOP

Department of Geology, Georgia Southern College
Statesboro, Georgia, U.S.A.

SYNOPSIS

Predation, a normal activity in most recent communities, is a way of life for many animals and may exert control over the population density of other members of the community. Our knowledge of ancient examples of this important function is meager. This dearth stems mainly from (1) destruction of the evidence itself during predation, (2) collecting biases toward complete specimens and away from incomplete and broken specimens that might show predation damage, and (3) our inability to recognize the evidence of predation.

Each instance of predation may be considered as a chronological series of activities: search, capture, penetration, ingestion, digestion, and defecation. Each activity may result in preservable evidence of predation. The actual chance of preservation, however, is extremely small.

To resolve this deficiency in our knowledge, we must increase our observations on predation in the recent and on evidence for probable instances preserved in the fossil record.

INTRODUCTION

Predation is the relationship between individuals whereby one (the predator) eats or partly eats another (the prey). This relationship, an important part of nearly all recent communities, has long been fundamental to animal life (Stanley, 1973). Predation is a normal but antagonistic activity that results in the death of the prey to maintain the life of the predator (Ager, 1963, p. 247).

The goal of this chapter is twofold: (1) to summarize the evidence for predation known in the fossil record, as well as instances reported from the recent that probably could be found in the fossil record, and (2) by this review of processes and traces of predation, to stimulate further searches for evidence of this important way of life among both recent and fossil animals.

Not all the phenomena discussed here qualify as trace fossils in the strict sense, but all are included in order to place the actual traces of predation in their proper perspective. (Cf. Müller, 1963.)

PREDATION

Predation in the Community

The structure of a community may be depicted by a food pyramid, illustrating relative abundances of organisms that acquire their food in different ways and that constitute different trophic levels. At successively higher levels in the food pyramid, the quantity of detritus and living organisms (biomass) decreases because of energy losses involved in consumption, digestion, assimilation, and growth. The food pyramid does not summarize community functions as accurately as does a functional diagram, however. Figure 13.1 more closely depicts the total scheme of relationships of predators to the community. Of course, even this diagram is a gross simplification of community structure. (See Odum, 1971; Valentine, 1973.)

The amount of energy flow actually involving predators in a community is relatively unimportant (Odum, 1963, p. 100). To the predator, however, this energy flow is vital. The amount of energy expended in predation must result in an equal or greater amount of energy return (profit), or the predator dies. Within the community, predation may be one of the more important factors controlling the populations of primary, secondary, and tertiary consumers (Odum, 1963, p. 103; Seed, 1969, p. 336).

Opportunistic Predation

Errington (1967), in his studies of predation in the north-central part of the United States, cited much evidence indicating that predation is opportunistic and nonspecific. Populations of animals that increase beyond the carrying capacity of the ecosystem become vulnerable prey to scores of predators. When the surplus is reduced, predation decreases. Animals that become most vulnerable are those that are diseased, careless, or are under other stresses due to overpopulation.

The diet of many predators is highly varied and reflects seasonal population fluctuations of prey animals (Errington, 1967). Other predators seem to be tied more closely to a single or to a few prey species. The common mud crab *Panopeus herbstii*, for example, is exceedingly common under oyster clumps in estuaries on the Georgia coast. The crab is adapted to burrowing among masses of oyster shells and has stout claws (see Fig. 13.3C) that are useful for cracking the shells of live oysters. Where this crab lives in close association with oysters, it probably preys almost exclusively on the oysters (Menzel and Nichy, 1958).

"Predation" is defined as the eating of living animals. The eating of dead animals is called "scavenging." This distinction becomes less well defined, however, when we consider predation on dying or diseased animals. Errington (1967, p. 5) pointed out that much of the evidence cited for predation is "sign," consisting of tracks, trails, blood smears, patches of fur or feathers, accumulations of prey remains outside dens, and undigested material in feces. Yet one often cannot determine whether these "signs" are evidence of predation or of scavenging.

Predation as a Function

The activity of predation may be subdivided into a chronological series of events: (1) search, (2) seizure, (3) penetration, (4) ingestion, (5) digestion, and (6) defecation. In many cases a part of this series of events is essentially a single act, as when a fish gulps down a smaller fish in one bite. Many times, however, each event is a distinct activity, and in some instances one or more parts of it may be exceedingly complex.

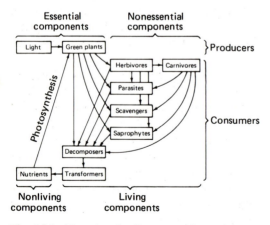

Fig. 13.1 Functional diagram of ecosystem, illustrating relationship of carnivores (some are predators) to other functional units in system. [From Clarke (1954) by permission of John Wiley & Sons, Inc.]

Search

Search for prey involves whatever sensory devices are available to the predator. Sight, smell, and hearing are probably most useful to terrestrial animals, and touch, chemoreception, and sight are most useful to marine animals.

Seizure

Once the prey is located by search, it is captured in some manner. Entrapment may involve gulping the prey down with jaws, killing it by biting, immobilizing it by weight of the predator, poisoning, or by moving along on or with the prey animal in a contest for supremacy.

Penetration

In order to reach the flesh contained in animals protected by exoskeletal armor, often the predator must force an entry. Gaining access to the edible parts may be done by cracking or crushing, abrading or boring, or by weakening the prey by poisoning or by extended periods of muscular stress or exertion.

Ingestion

The captured animal is then eaten. It may be ingested intact, or torn apart and eaten piecemeal. The predator may selectively eat parts of the prey (such as flesh only, or only a "hindquarter"), which biases the record of predation.

Digestion

The process of digestion varies with the type of predator. Ingested animals may be digested as eaten, or may be further mascerated by various intestinal organs such as gizzards, gastric teeth, or auxiliary stomachs. Undigested remains may be passed on through the intestine, to be excreted as feces, or the undigested material may accumulate in the gut and be regurgitated periodically as pellets or masses of these remains.

Defecation

The remains of prey that are passed through the intestine usually are partly digested and fragmented (or otherwise rendered small). These particles may be preserved in the fossilized excrement (coprolites) of the predator.

Tools of Predation

Structures used to capture, dismember, and ingest prey may be single organs (jaws of fish) or several organs (walking legs, claws, and mouth parts of crabs).

Appendages

Appendages not directly associated with the mouth that are used to capture prey are legs (arthropods, vertebrates), arms (echinoderms), or tentacles (cephalopods). These mobile appendages are used to grasp, entangle, or otherwise immobilize the prey.

If the appendage has a mineralized exoskeleton or bears mineralized tissues (claws or hooks), it may also be used to dismember the prey, either by piercement of the prey's exoskeletal armor or by tearing large prey animals to pieces of a size that can be eaten.

The appendages of many carnivorous arthropods are used actively to search for and capture prey. Many arthropods (horseshoe crabs, eurypterids, scorpions, stomatopods, crabs, and lobsters) have one or more pairs of appendages modified into a grasping organ (claw) capable of pinching, tearing, or crushing.

The claws may be unspecialized, or highly specialized, in their function. The shape of crab claws is closely correlated with their function (Schäfer, 1954). In many taxa, each claw is different morphologically and is used for different functions by the animal (cutting and crushing claws of the lobster *Homarus americanus*, the blue crab *Callinectes sapidus*, and the mud crab *Panopeus herbstii*).

Most vertebrates and scores of invertebrates, such as echinoderms and cephalopods, have soft appendages useful for capturing prey. The appendages may be modi-

fied with hooks (vertebrate claws, cephalopod hooklets) or suction cups (tube feet of
echinoderms, sucker cups of cephalopods)
to hold or dismember the prey.

Mouth organs

Jaws and beaks are mouth-associated structures consisting of opposed mineralized tissues that are used to capture, dismember,
and devour prey (Fig. 13.2F, G).

Most vertebrates have toothed jaws. The
shape of the tooth is closely related to its
function (or to the type of food that it
must prepare for digestion): piercing teeth
(Fig. 13.2A) are cone- or blade-shaped (reptile teeth, canine teeth of dogs); shearing
or cutting teeth (Fig. 13.2B) are long and
blade-shaped (carnassial teeth of the cat);
crushing and grinding teeth (Fig. 13.2C)
are low, flat, and button-shaped (skates and
rays); digging and stabbing teeth (Fig. 13.2
D) are rod-shaped and large (sabertooth
cats, walruses).

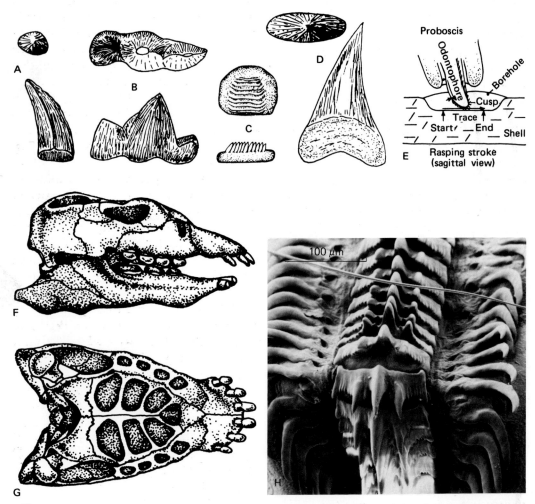

Fig. 13.2 Mouth organs used for predation. A–D, functional shapes of vertebrate teeth. A, cone-
shaped piercing tooth. B, blade-shaped carnassial tooth. C, button-shaped crushing tooth. D, knife-
shaped stabbing tooth. E, structure and use of gastropod radula (after Carriker and Van Zandt,
1972). F, G, skull of placodont reptile (*Placodus*) having crushing teeth (after Romer, 1945). H,
radula of gastropod *Urosalpinx cinerea follyensis* at bending plane, showing worn and unworn
teeth (scanning electron micrograph; from Carriker, 1969).

Beaks are jaw-like mouth organs that do not carry teeth but which can be used to rip and tear food items apart before ingestion. Beaked vertebrates include most birds, turtles, and a few other taxa. Among the beaked invertebrates are cephalopods and many echinoids, the latter having the beak-like Aristotle's lantern.

Small rasping organs called "radulae" are found in gastropods, cephalopods, and turbellarians (Woelke, 1957). These small tooth-like masses of mineralized tissue (Fig. 13.2E, 13.2H) are used to gather food by scraping or rasping. Carnivorous gastropods (Muricidae, Naticidae) use the radula and an accessory boring organ to bore holes into shells, in order to get at the flesh of the prey (Carriker, 1969, p. 920).

Boring is accomplished by alternately dissolving or softening the prey shell by chemical action and then partially removing the weakened material with the rasp-like radula. Shell material is dissolved by the accessory boring organ, which secretes an unidentified acidic substance. This acid dissolves both the mineral and organic constituents, and accounts for most of the removal of shell material (Carriker, 1969, p. 928; Carriker and Van Zandt, 1972). (See Chapter 11.)

The radula of large specimens of the muricid gastropod *Urosalpinx* (Fig. 13.2H) is about 400 μm wide, and is composed of a medial multicusped rachidian tooth and a pair of lateral, single-cusped marginal teeth. Abrasion is accomplished by moving the membrane bearing the hard radular teeth, as well as by moving the organ carrying the membrane. As the teeth wear, they are continuously replaced by new teeth. Once the shell of the prey is penetrated, the gastropod feeds upon the exposed flesh by tearing off small pieces with the radula.

The octopus also possesses a radula (in addition to its beak), which is often used to bore into other mollusks (see Fig. 13.7E, F). The process of feeding on gastropods involves (1) selection of the prey, (2) hole boring, (3) secretion of a substance to relax the prey, (4) removal of the prey from its shell, and (5) feeding by means of the tearing and nipping actions of the beak (Pilson and Taylor, 1961; Wodinsky, 1969).

A few predators use part of their exoskeleton as an abrading tool. The gastropod *Busycon* (Fig. 13.3A) attacks and opens bivalves by rubbing and chipping the edge of the commissure of the clam against the outer edge of its own aperture (Carriker, 1951; Paine, 1962). This action, usually attempted only by thick-shelled members of *Busycon*, chips the predator's shell as well as the prey's. A Miocene *Busycon* figured by Fagerstrom (1961, Pl. 60, fig. 1) shows evidence of the animal having periodically repaired its chipped apertural margin (presumably chipped during predation). *Murex fortispina* (Fig. 13.3B) has a spine that functions as a device to force shells open (Wells, 1958; Schäfer, 1972, p. 150). Certain crabs (*Callinectes, Calappa, Panopeus, Menippe*) have heavily mineralized claws useful for crushing or prying open prey (Fig. 13.3C–E).

The use of objects as tools to capture or open prey is practiced by the sea otter of California, which carries a rock on its stomach as it floats on its back and hammers the prey's shell against the rock. Birds are known to carry shells aloft and drop them onto rocks exposed near the beach, thus cracking the shells open.

Drinnan (1967) described predation by oyster catchers (*Haematopus ostralegus*) on edible cockles (*Cardium edule*). The birds penerate feeding clams by stabbing the beak between the gaping valves, and wedging the valves farther apart by a lateral twist of the head. If the bird's first attempt is unsuccessful, or if the clams are not feeding, a clam is dug out and pecked at along its commissure until the shell fractures, exposing the flesh inside (Drinnan, 1967; Carter, 1968, p. 32).

Poisoning and Stunning

A few predators have specialized structures used for poisoning or stunning their

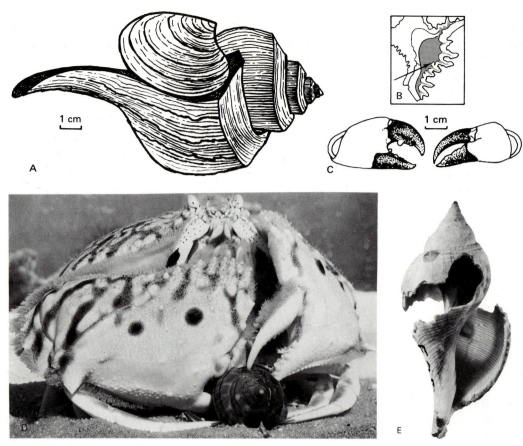

Fig. 13.3 The exoskeleton as a tool for predation. A, gastropod *Busycon* using edge of aperture to abrade commissure of clam *Mercenaria* (after Carriker, 1951). B, tooth (arrow) on aperture of gastropod *Murex fortispina*, used to pry bivalves apart [from Schäfer (1972), courtesy of Oliver and Boyd]. C, crushing and cutting claws of common mud crab *Panopeus herbstii*. D, the crab *Calappa* opening a gastropod shell containing a hermit crab (length of *Calappa* carapace, 91 mm). E, a gastropod shell that contained a hermit crab and was opened by *Calappa* [from Shoup (1968); copyright 1968, American Association for the Advancement of Science].

prey, e.g., the gastropod *Conus* (Purchon, 1968, p. 68) and the electric eel, respectively.

Manufactured tools

A few animals manufacture tools to aid them in predation. Man has thoroughly utilized his ability to make tools, to his obvious advantage, and has thus become one of the world's foremost predators. The relatively young age of possible evidence of man's predatory acts removes them from consideration by most paleontologists and places them in the fields of anthropology and archeology. The literature on man's predatory nature can be approached by reading *Man the Hunter* by Lee and De-Vore (1968).

EVIDENCE OF PREDATION

Preservation of the Evidence

Because the object of predation is to eat, most evidence for instances of predation is literally eaten up. The activities of predation, particularly mastication and digestion, tend to destroy the evidence or at least make it difficult to recognize. Cadee (1968, p. 84) listed major predators of bivalves as being fish, crustaceans, gastropods, and birds; he pointed out that body fragments resulting from fish predation are finer than those resulting from crustacean predation. This conclusion might be challenged (to emphasize the difficulty in recognition of evidence): decapod stomach contents consisting of fragments of bivales 1 mm in size were described by Ropes (1969, p. 186).

The evidence most probably preserved involves hard parts that are marked or broken during seizure or ingestion of prey; these remains are not digested, but are regurgitated or excreted and preserved within pellets or coprolites, often accumulating in unusual aggregations or other arrangements.

Evidence of unsuccessful attempts at predation may have a better chance of preservation (and later recognition) than successful attempts. Traces left on hard parts are not so apt to be destroyed if the prey continues living, or if the prey escapes and then dies. If the prey lives, the scars of predation are usually preserved as healed areas.

Examples of Evidence for Predation

Evidence of predation in the fossil record may be referred to as "traces of predation." The examples cited here will help illustrate the kinds of evidence that various workers have found and interpreted as traces of predation. Some instances are well documented, whereas others are based on circumstantial evidence. This range of substantiation is to be expected because of the biases inherent in successful predation, in addition to more traditional kinds of taphonomic "information losses" (Ager, 1963, p. 185; Lawrence, 1968).

Search

The evidence of predators searching for food must be inferred from such things as trackways, faunal associations, and dis-

rupted bedding laminations. Much of the evidence is circumstantial and can be explained in one or several different ways.

Colbert (1961, p. 14) cited one instance of possible searching by a predator:

> A series of dinosaur trackways . . . discovered at Glen Rose, Texas, in sediments of Lower Cretaceous age, gives clear evidence that a large sauropod comparable to Brontosaurus in size was followed or trailed by a carnosaur of Allosaur-like proportions. Here is what seems to be the evidence of an ancient hunt [see Chapter 2; Fig. 14.17].

Evidence of the search for prey by the asteroid *Astropecten* was described by Seilacher (1953) and Schäfer (1972, p. 394). This sea star crawls over the sediment surface until it locates buried prey, then digs into the substrate to capture the prey. The activity of digging into the substrate may result in a trace having fivefold symmetry (Fig. 13.4B). The animal also burrows to rest. However, the association of a five-rayed trace and the remains of a possible prey animal would be good evidence for predation.

Clarke (1912) described an accumulation of 400 Devonian sea stars (*Devonaster eucharis*) in close association with shells of the bivalves *Grammysia* and *Pterinea* (Fig. 13.4A); some of the sea stars are preserved in feeding positions (Fig. 13.4C) over the clams (Clarke, 1912, p. 117; Ladd, 1957, p. 35). The association of predators and possible prey may thus be good evidence for predation.

In search for food, many crabs use two or more of their eight walking legs as chemical and tactile probes (Schäfer, 1972, p. 252). This activity may disrupt laminations in the near-surface layers, causing a zigzag pattern of disrupted laminae called "under tracks" (Fig. 13.5A). However, the same sort of bedding traces are made when the crabs are merely walking across the sea bottom (see Table 3.5, cleavage relief).

Moulton and Gustafson (1956, p. 992) described the role of the green crab

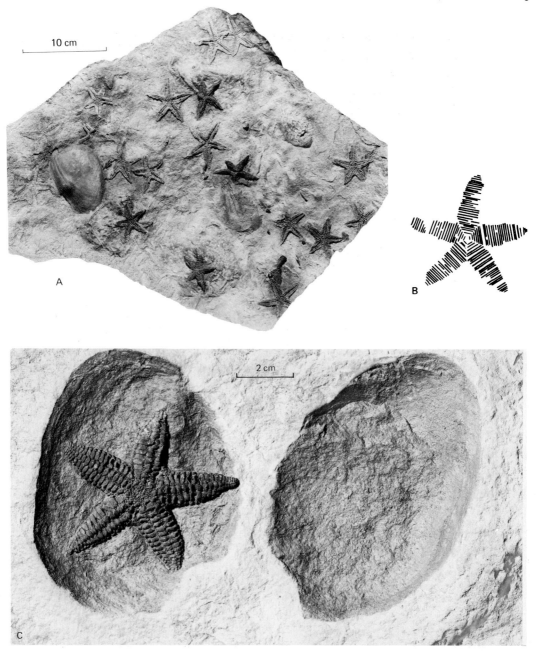

Fig. 13.4 Search for prey by asteroids. A, C, sea star *Devonaster eucharis* associated with clams upon which they were probably preying (Mt. Marion Beds; Hamilton; Middle Devonian): A, slab displaying associated sea stars and clams; C, sea star preserved on clam (courtesy D. W. Fisher; N.Y. State Mus. and Sci. Service, Albany, N.Y.). B, diagrammatic trace (*Asteriacites*) left on soft bottom by sea star burrowing down to prey upon infaunal animal [after Schäfer (1972) and my personal observations].

Carcinides maenas as a vector in the redistribution of quahogs (*Mercenaria*) (Fig. 13.5B). The green crab is a predator on these clams in northern New England. After

the authors discovered a quahog pinched onto the tip of a crab's walking leg, they picked up the crab; the quahog snapped shut, pinching the crab's leg off and leaving

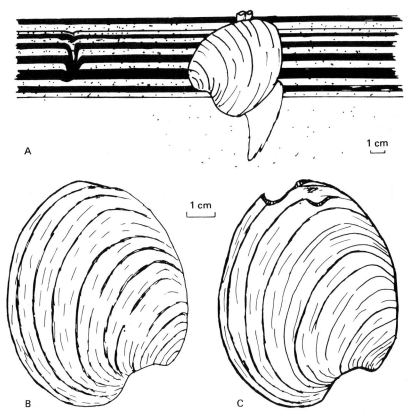

Fig. 13.5 "Search traces" left by crabs. A, probe "track" (left) made in thinly laminated sediment by tip of leg of crab searching for prey, and a quahog (*Mercenaria*) in living position in the sediment (modified from Schäfer, 1972). B, shell of quahog showing no evidence of predation. C, quahog shell bearing two sets of "nicks," possibly made by crushing of crab legs inserted between valves while clam was feeding.

a pair of small circular nicks or spall marks on the commissure of the clam shell. Further examination of the clam population disclosed many other clams having similar nicks. Moulton and Gustafson assumed that these clams also had clamped onto the crabs' legs when the crabs inadvertently placed their leg tips between the gaping valves of quahogs. Perhaps the act is inadvertent, but the activity might well be involved in the search for prey. A quahog showing this kind of nick was collected by me from a salt marsh behind Jekyll Island, Georgia (Fig. 13.5C). Carriker (1951, p. 79) described the same sort of damage in quahogs due to blue crabs (*Callinectes sapidus*) trying either to break pieces off the shell or to force entry by pushing a cheliped between the valves. Specimens of fossil *Mercenaria* showing this sort of predation trace probably could be found in many paleontological collections.

Bite Traces

Bite traces provide some of the best evidence for ancient predation. Kauffman and Kesling (1960) described an ammonite bitten by a mosasaur (Fig. 13.6A). The use of teeth in the seizure of a crab by a fish (Fig. 13.6B) was described by Bishop (1972). Gripp (1929, p. 243) described echinoids bitten by fish. Boyd and Newell (1972) analyzed and illustrated Permian clams preyed upon by shell-crushing fish (Fig. 13.6D–13.6F). Kauffman (1972) described Cretaceous *Inoceramus* clams damaged by a shell-crushing shark, *Ptychodus*. Speden (1971) and Frey (1972) also described possible cases of *Inoceramus* being bitten by fish.

Kier and Grant (1965, p. 54) described the predation of a small burrowing fish on the thin edges of sand dollars. The observation of predation was made after the sand dollars reached a near vertical position; the fish appeared and began nipping at the exposed edges of the test. Kier and Grant reported seeing many specimens of sand dollars having edges that seem to have been nipped off in this manner. This type of injury to sand dollars is common in populations of *Mellita* examined by me on the Georgia coast (Fig. 13.6C). The circumstantial nature of this evidence was recognized by Kier and Grant (1965, p. 54), who pointed out that predation is only one of many possible ways in which the edges of sand dollars might be broken.

Zangerl and Richardson (1963) described a fossil fauna composed largely of fishes that were killed by predators. The occurrence of biting is clearly shown by amputated parts of bodies and by zones of disrupted scales and skeletal elements (Fig. 13.6G).

Green (1961) described evidence of biting on the radius of a rhinocerotid, and a possible tooth puncture near the left orbit of a large pig-like mammal, *Archaeotherium*. Scott (1937, p. 618, Figs. 381–383) described and figured the skull of a "false sabre-tooth cat," *Nimravus bumpensis*, which bears a partly healed wound (presumably from a coexisting sabre tooth cat, *Eusmilus*) on its left frontal region. Although this record is a debatable piece of evidence for predation (it might represent a fight over food, etc.), it undoubtedly is a preserved bite trace.

Miller (1969, p. 24) studied the traces on bones made by zoo animals, and compared them with traces preserved on fossil bones. He demonstrated that many "rodent gnawed" bones could have been marred by carnivores the size of coyotes.

The evidence of predation by shell-crushing animals can sometimes be preserved as chips and crush impressions along shell margins. More often, the predatory attempt is successful and the shell is fragmented. The problem then becomes one of distinguishing predatory breakage from breakage due to tumbling about on the bottom, diagenetic compaction, and so on. Boyd and Newell (1972) were confronted by this problem in their taphonomic and diagenetic study of a bed of Permian rock (Park City Formation) in Wyoming, but were able to distinguish pre-lithification damage from post-lithification damage. Most valves were disarticulated and scattered. Pre-lithification breakage was determined and plotted on triangles representing the shapes of clams (Fig. 13.6F). In addition to the preferred positions of these breaks, the authors noted that the hinge teeth were commonly broken, too. Several specimens exhibiting a pattern of crescentic reentrants along the shell margin, comparable to modern examples of predation, were seen (Fig. 13.6D, E). This evidence, together with the presence of crushing teeth of holocephalan fishes from rocks of the same age in the same area, strongly indicates predation on the clams by shell-crushing fish.

Fig. 13.6 Evidence of predatory biting on prey. A, Cretaceous ammonite (*Placenticeras*) bearing ▶ tooth impressions of a mosasaur (from Kauffman and Kesling, 1960). B, Cretaceous crab (*Raninella*) containing tooth impressions of a fish (from Bishop, 1972). C, possible bite traces along edge of test of recent sand dollar (*Mellita*). D, valve of recent clam (*Placuna placuna*) having crescentic chips removed from margin by fish. E, valve of Permian clam having crescentic chips removed from margin. F, analysis of position of breakage on shell of Permian bivalve, interpreted as predation. (D–F after Boyd and Newell, 1972.) G, Pennsylvanian fish having bite marks across body (1) and behind now dismembered head (2) (from Zangerl and Richardson, 1963).

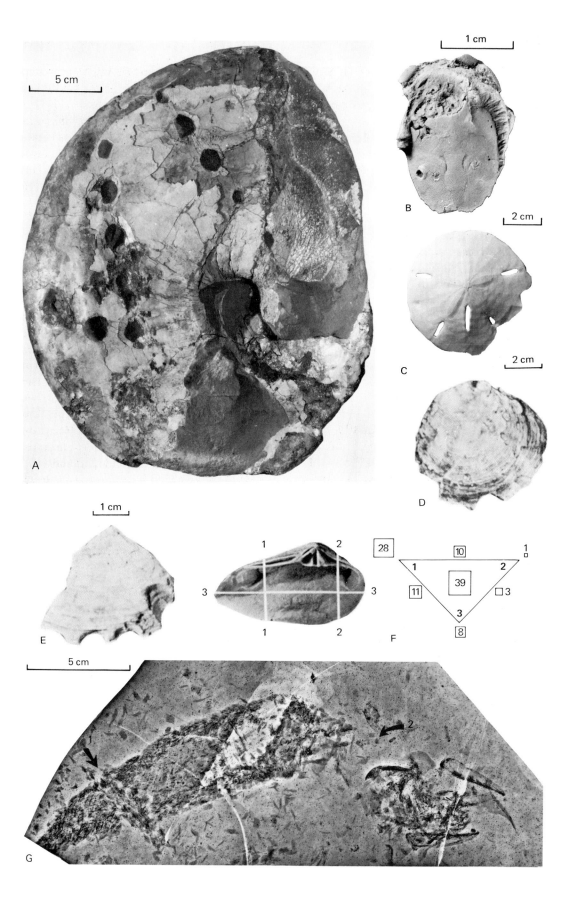

Biting, Nipping, Scraping, and Boring

Evidence of abrasion by the predator's exoskeleton to gain entry into prey is commonly described for recent occurrences (Carriker, 1951; Galtsoff, 1964) but is seldom reported from the fossil record (see Fig. 13.7B, C). Abrasion on the commissures of bivalves (or on the predator) is probably disregarded by most paleontologists, or considered as effects of postmortem transportation.

The record of biting by animals lacking teeth (or having teeth so small that they can be used only as scrapers) is even more obscure. Bromley (1970, p. 403) described evidence of scrape marks made by a predator trying to eat the papillae of boring clionid sponges. The sponges were living in an *Inoceramus* shell, and the scrape marks (Fig. 13.7A) were usually (but not exclusively) found around the holes where the clionid sponge was exposed at the surface of the shell. The predatory damage might have been caused by the mouth of a small fish, a regular echinoid, or by the claw of a decapod crustacean.

Kier and Grant (1965, p. 49) observed a sea star (*Oreaster*) preying upon a sea urchin (*Meoma*). Upon removal of the sea star from the urchin, the authors could see a smooth area where the spines and test had been dissolved by the everted stomach of the sea star. The dissolution of the test completely erased the evidence of spine bases and made that part of the test appear smooth.

Borings by predators can be preserved either as the boring itself or as molds of the interior of the holes. Preservation of the hole itself depends upon preservation of the prey's shell, and seems to be the most common mode.

The recent record of boring predators is considerable, and because of the impact that boring predators have on commercial fisheries, this mode of predation has been studied extensively (Moore, 1956; Boek-schoten, 1966; Carriker and Yochelson, 1968; Arnold and Arnold, 1969; Wodinsky, 1969; Carriker and Van Zandt, 1972).

The fossil record of predatory boring is well represented back through the Tertiary and into the Cretaceous (Carriker and Yochelson, 1968; Sohl, 1969; Taylor, 1970). Pre-Mesozoic borings are uncommon and problematical. [The inconsistent use of nomenclature to describe borings was discussed by Carriker and Yochelson (1968, p. B2). They suggested that usage should be consistent; summarized their nomenclature in a drawing; and provided a glossary for use by subsequent workers.]

Borings by gastropods (1) tend to penetrate perpendicular to the prey's shell; (2) are parabolic, cylindrical, or cone-shaped in axial cross section; and (3) are usually circular in cross sections normal to the axis of penetration (Fig. 13.7G–I). The borings are usually symmetrical and smooth. Cephalopod boreholes are known only from the recent but should be present in the fossil record. These borings tend to penetrate the prey's shell at oblique angles, and decrease rapidly in diameter with depth. They are highly variable in size and shape but do tend to be somewhat cone-shaped (Fig. 13.7E, F). The point of penetration may or may not be at the apex of the cone.

Borings made by gastropods of the families Muricidae and Naticidae are similar if the prey's shell is thin, but can be differentiated if the shell is thick. Parameters important in distinguishing borings by these gastropods are (1) borehole location on the prey, (2) axial cross section, (3) type of countersinking, and (4) presence or absence of a shelf. Cross-sectional shape perpendicular to the axis is usually circular, but may be the shape of a crescent or heart, or may be irregular. The diameter of the hole decreases with depth. Differences in hardness in layers and ornamentation of the shell may cause variations in borehole morphology.

Muricid gastropod borings are randomly

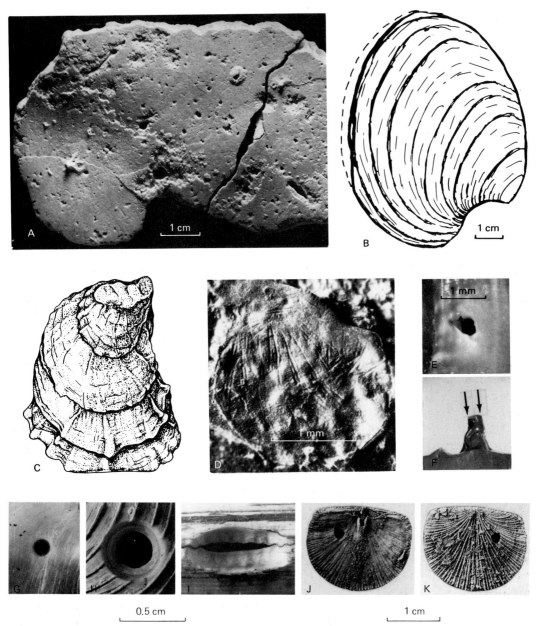

Fig. 13.7 Predation damage due to rasping and abrading by predators. A, predatory rasp traces around papillae of boring sponge *Cliona* in shell of Cretaceous bivalve *Inoceramus*. (From Bromley, 1970.) B, C, abrasion along bivalve commissures by gastropod *Busycon*. B, abraded quahog (*Mercenaria*) (after Carriker, 1951). C, abraded oyster valve (bottom edge) (after Galtsoff, 1964). D, G–I, gastropod borings. D, rasping trace in shallow, incomplete borehole made by *Urosalpinx* boring into *Mya* (from Carriker, 1969). G, cylindrical borehole of muricid gastropod *Eupleura*. H, parabolic borehole of naticid gastropod *Lunatia* in *Mercenaria*. I, borehole of muricid gastropod *Murex* on commissure of *Mercenaria*. E, F, *Octopus* boring in gastropod shell. E, borehole at outer surface. F, cast or impression of borehole shown in E (from Arnold and Arnold, 1969). J, interior and K, exterior of brachial valve of brachiopod *Dalmanella* perforated by polychaete borehole (Lexington Limestone, Middle Ordovician; Kentucky). [G–K from Carriker and Yochelson (1968), courtesy U.S. Geol. Survey.]

distributed on the shell of prey animals. Rarely a boring is made on the inside of a disarticulated valve, but only where the shell lies in close proximity of live bivalves whose effluent confuses the predator. The axial cross section of the borehole tends to be cylindrical or cone-shaped (Fig. 13.7G–I). Because of the shape of the accessory boring organ, the bottoms of incomplete boreholes tend to be shallowly concave (Fig. 13.7D). When the inner side of the shell is penetrated, drilling stops, commonly leaving a shelf. A few muricid gastropods (e.g., *Thais haemastoma*) select the commissure opposite the hinge as their boring site. The resultant borehole is then an elliptical, parabola-shaped hole affecting both valves (Fig. 13.7I).

Naticid gastropod boreholes are usually excavated within certain limited areas on a valve, and sometimes one specific valve is selected for boring. The edge of the hole on the outside of the valve is usually countersunk. Incomplete boreholes may have a raised central boss. The interior of naticid borings usually converge toward the center of the inner opening, describing a graceful, pronounced, parabolic curve (Fig. 13.7H), a feature that is diagnostic of all naticid boreholes (Carriker and Yochelson, 1968, p. B7).

Boreholes of the octopus (Fig. 13.7E, F) have been described in several recent papers (Fugita, 1916; Pilson and Taylor, 1961; Arnold and Arnold, 1969, Fig. 2, p. 994). Holes vary with the octopus species, hardness of prey shell, and size of the octopus (Wodinsky, 1969, p. 1002). The outer diameter of a typical hole is about 1 mm; diameter of the inner part is about 0.5 mm; and the depth (depending on prey shell thickness) is a little less than 1 mm. The hole is roughly cone-shaped, bored obliquely to the prey shell surface. The point or points of penetration of the inner surface of the shell may be at the apex of the cone, or tangential to it. Once initial penetration is attained, boring stops, be-

cause no feeding is done through the borehole; the hole is used as an avenue for transmitting a secretion to relax the prey, so that it may be pulled from its shell. A strong preference is exhibited for boring within the first 90° or 180° of the lip of the aperture on the spire of large gastropods. No instances of fossil cephalopod boreholes cited in the literature have yet come to my attention.

Nematode worms have been described boring into, and apparently preying upon, two species of foraminifers in both laboratory and natural populations (Sliter, 1971). One to three holes completely penetrated the walls of chambers in which live nematodes were coiled. The penetrations are round to oval in plan view, about 10 μm in diameter, and are "distinguished from gastropod boreholes by their minute size, irregular shape, roughened inner periphery, oblique orientation, and lack of smoothly beveled outer edges" (Sliter, 1971, p. 21). Nematode borings could be confused with brachiopod pedicle borings, which also occur in foraminifers (Bromley and Surlyk, 1973).

Carriker and Yochelson (1968) summarized the Paleozoic record of "boreholes." The older "boreholes" are found most commonly on brachiopods (Fig. 13.7J, K); show considerable variation in cross sections perpendicular to the axis of penetration; commonly penetrate obliquely; and are commonly chamfered in one or two quadrants. The origin of these borings is not predatory; they represent the dwelling tubes of polychaetes (Richards and Shabica, 1969).

Ingestion

Preservation of evidence for ingestion was discussed by Jepsen (1967), who illustrated (my Fig. 13.8A) a 14-cm perch (*Mioplosus*) that died while trying to eat a small herring (*Knightia*); he proclaimed it "an example of terminal voracity of a fossilized moment

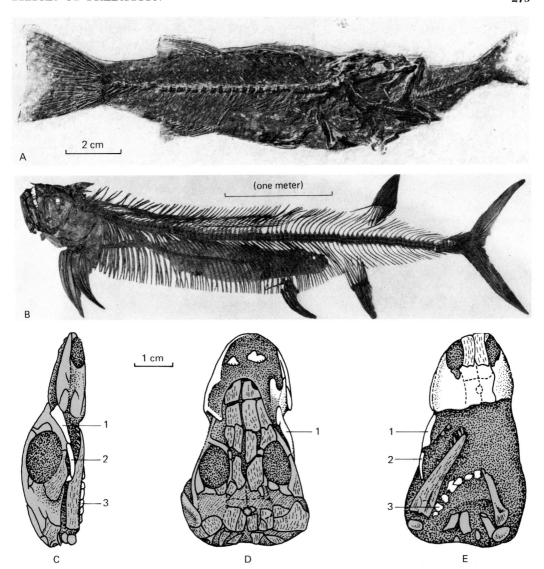

Fig. 13.8 Ingestion of prey. A, B, fish preying upon fish. A, perch (*Mioplosus*) that died while trying to eat a herring (*Knightia*). Early Eocene; Green River Formation, Wyoming (from Jepsen, 1967). B, giant Kansas Cretaceous fish (*Xiphactinus*) having smaller, ingested fish (*Gillicus*) within the abdominal cavity (Sternberg Memorial Mus., Ft. Hays Kansas State College; photo by E. C. Almquist.) C–E, captorhinomorph reptile preserved with partly ingested younger reptile in its mouth (prey in bolder outline). C, right lateral view of predator and prey. D, dorsal view of predator skull and ventral view of prey skull. E, ventral view of predator skull, showing dorsum of prey's skull, vertebrae (3), and leg elements (1, 2) (from Eaton, 1964).

of fishy gluttony." An ingested fish (*Gillicus*) is preserved intact inside the predator (*Xiphactinus*), displayed at the Sternberg Memorial Museum, Hays, Kansas (Fig. 13.8B). Bradack (1965) listed 7 of 18 specimens of *Xiphactinus* preserved with in-gested prey. Eastman (1911) reported reptiles preserved in fish from the Jurassic of France. Eaton (1964) described a reptile preserved as it had been ingesting another reptile, from the Permian of Texas (Fig. 13.8C–E).

Gastric Residues

As digestion progresses, the chance of preservation of identifiable remains decreases. Pollard (1968) nevertheless described an ichthyosaur from the Lias of Lyme Regis, Dorset, containing within its abdominal area a gastric mass (pre-coprolite) consisting of thousands of tentacle hooklets of dibranchiate cephalopods. Lehmann (1971) described preserved crops of ammonites that contained foraminiferans, ostracodes, and fragments of other ammonites.

Regurgitates

Many animals separate the digestible part of their food from the indigestible part. The latter may be regurgitated through the mouth (Schäfer, 1972, p. 408). Recognition of these regurgitated masses may be possible, and may be important because of the potential biases that they can introduce into the fossil record.

Birds of prey commonly separate the digestible and indigestible parts, and regurgitate the latter as "pellets" (Fig. 13.9B). Grim and Whitehouse (1963), using x-

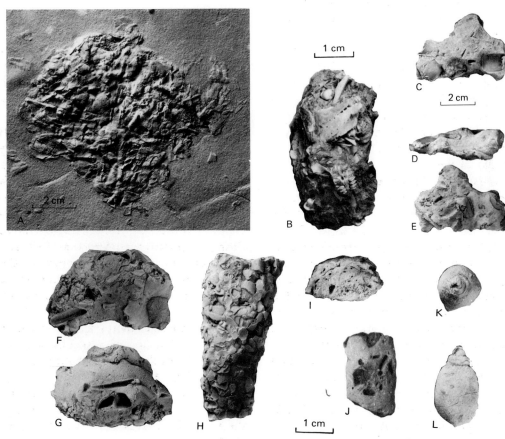

Fig. 13.9 Gastric residues, regurgitates, and coprolites. A, gastric residue mass of crushed bivalve fragments (*Myalina*); Pennsylvanian, Indiana (from Zangerl and Richardson, 1963). B, recent owl pellet containing numerous rodent remains (collected in Montana by R. M. Petkewich). C, top, D side, and E bottom views of slab of concretionary apatite around a possible regurgitated mass of *Callianassa* claws. F, G, two views of concretion containing disarticulated skeletal elements of decapod *Dakoticancer*. H–L, coprolites of predators. H, side and I, end of coprolite containing numerous fish scales. J, coprolite containing prismatic shell fragments of bivalve *Inoceramus*. K, top and L, side view of spiral coprolite of shark or dogfish. (C–L from Upper Cretaceous Pierre Shale, South Dakota.)

radiography to study a captured great horned owl, described the formation of pellets. These pellets may form enormous accumulations at the roosting or nesting site of the birds. If fossilized, the large accumulation of pellets, together with the concentration of skeletal debris contained in them, could give us an erroneous impression of prey population density, particularly if the predator selected particular prey animals. Reed (1957) reported 1,018 skulls from 275 barn owl pellets collected from three recent owl burrows in Colorado. These pellets were situated in an environment that should provide them a relatively high possibility of preservation. (See Chapter 15.)

Meanley (1962) discussed pellets of king and clapper rails and the fragments of crayfish, fiddler crabs, and clams that they contained. The pellets were found on the ground and in muskrat houses within the marsh. Sparks and Soper (1970) cited the occurrence of fossilized barn owl pellets in the Caribbean.

Masses of unconsolidated material are often regurgitated by aquatic animals, such as sharks. The regurgitated material described by Zangerl and Richardson (1963, p. 140), from the Pennsylvanian of Indiana, consisted of whole to partially disarticulated and disoriented regions of the bodies of prey animals; the regions of the bodies nevertheless remained in their correct relative positions. The area surrounding these stomach ejecta are usually strewn with "tufts of brownish material." Zangerl and Richardson (p. 135) also interpreted "concentrations of finely minced shell material" of the pelecypod *Myalina* (my Fig. 13.9A) to be regurgitated, indigestible animal remains. Wright and Wright (1940, p. 233) considered a mass containing the ossicles of 10 species of asteroids as "forming, almost certainly, a pellet of indigestible parts ejected by a starfish-eating fish or other animal."

. Speden (1971, p. 57) described patches of *Inoceramus* shell fragments rendered similar to those of *Myalina*, by virtue of their being reduced to a hash. Pieces of the prismatic shell of *Inoceramus* may remain intact, if regurgitated soon after ingestion, or they may be broken down into individual prisms or felted masses of prisms, if regurgitated at a later stage of digestion.

Disarticulated crab fragments that are preserved together (Fig. 13.9F, G) and accumulations of numerous claws of mud shrimp (*Callianassa*) (Fig. 13.9C-E) from the Pierre Shale are considered by me to be regurgitated material.

Coprolites

Häntzschel et al. (1968) reviewed the literature on coprolites (fossilized dung or excrement) in their annotated bibliography. Sarjeant also discusses coprolites in Chapter 14.

Coprolites containing the preserved parts of ingested and undigested remains of prey may be the only "sign" or evidence for an instance of predation (Fig. 13.9H–L). This kind of evidence is probably fairly common in the fossil record, but is not commonly reported because of problems in clearly identifying either the defacator (predator) of a particular coprolite, or the enclosed remains (prey). In many cases, however, prey may be identified from remains left in the feces; thus, even if the predator remains unknown, clear evidence exists for the action of predation and for the affinities of the prey animal (cf. Barthel and Janicke, 1970).

Some predaceous animals do have distinctive and recognizable fecal masses, e.g., those of the sharks and dogfishes (Fig. 13.9K, L) are sometimes preserved before excretion, in the form of a spiral (Buckland, 1841; Zangerl and Richardson, 1963, p. 142; Williams, 1972), and are often mistaken for coprolites.

Coprolites containing fragments of fish and an ichthyosaur described by W. Buckland were refigured by Häntzschel et al.

(1968, Pl. 4, figs. 17 and 18). Predation is clearly demonstrated by coprolites collected by me from the upper Pierre Shale of north-central South Dakota. Of 38 coprolites examined, 34 contained fish scales (Fig. 13.9H, I); one contained numerous small, subrounded plates of prisms and a few isolated prisms of the bivalve *Inoceramus* (Fig. 13.9J); and three had no preserved hard parts. Wetmore (1943, p. 441) described bird feathers preserved in a coprolite from the Miocene of Maryland.

CONCLUSIONS

Predation, a functional part of most recent communities, is so important as a way of life and as a possible controlling factor of populations that we should search for additional evidence of it in the fossil record.

The references cited herein clearly show that numerous workers have recognized and described instances of recent and fossil predation. With a few exceptions, however, the coverage has consisted of short reports on a few specimens. This kind of inventory is necessary to build a reservoir of knowledge, but many more observations and detailed analyses are needed.

Collection of fossils in the past has been biased toward complete, well-preserved specimens. The recent trend toward analysis of assemblages of fossils, and their significance, will undoubtedly continue to turn up evidence for acts of predation. The recognition and publication of these details should produce a further awareness of the importance of predation in paleoecological studies. If we know *what* to look for and that *we ought to be looking for it*, our knowledge of this important function in fossil communities will be greatly enhanced.

ACKNOWLEDGMENTS

I am grateful to R. W. Frey for inviting me to review evidence of predation that might be expected in the fossil record. The manuscript was improved by critical reviews rendered by M. R. Carriker, R. W. Frey, M. R. Voorhies, and E. L. Yochelson. The many colleagues contacted while I compiled data, ideas, and figures, and who graciously provided their comments and materials, are gratefully acknowledged. The manuscript was typed by Sue Colson, Georgia Southern College. The initial literature search was done by students in Invertebrate Paleontology (1972–73; Georgia Southern College) as a practical library exercise.

REFERENCES

Ager, D. V. 1963. Principles of paleoecology. New York, McGraw-Hill, 371 p.

Arnold, J. M. and K. O. Arnold. 1969. Some aspects of hole–boring predation by *Octopus vulgaris*. In M. R. Carriker et al. (eds.), Penetration of calcium carbonate substrates by lower plants and invertebrates. Amer. Zool., 9:991–996.

Barthel, K. W. and V. Janicke. 1970. Aptychen als Verdauungsrückstand. Neues Jahrb. Geol. Paläont., Mh., 1970:65–68.

Bishop, G. A. 1972. Crab bitten by a fish from the Upper Cretaceous Pierre Shale of South Dakota. Geol. Soc. America, Bull., 83:3823–3826.

Boekschoten, G. J. 1966. Shell borings of sessile epibiontic organisms as palaeoecological guides (with examples from the Dutch Coast). Palaeogeogr., Palaeoclimatol., Palaeoecol., 2:333–379.

Boyd, D. W. and N. D. Newell. 1972. Taphonomy and diagensis of a Permian fossil assemblage from Wyoming. Jour Paleont., 46:1–14.

Bradack, D. 1965. Anatomy and evolution of chirocentrid fishes. Univ. Kansas Paleont. Contr., Art. 10:1–88.

Bromley, R. G. 1970. Predation and symbiosis in some Upper Cretaceous clionid sponges. Dansk Geol. Foren., Meddr., 19:398–405.

———— and F. Surlyk. 1973. Borings produced by brachiopod pedicles, fossil and recent. Lethaia, 6:349–365.

Buckland, W. 1841. Geology and mineralogy considered with reference to natural theology. Treatise VI of the "Bridgewater treatises on the power, wisdom and goodness of God as manifested in the creation." Lea and Blanchard, 468 p.

Cadee, G. C. 1968. Molluscan biocoenoses and thanatocoenoses in the Ria de Arosa, Galicia, Spain. Zool. Verhand., 95:1–121.

Carriker, M. R. 1951. Observations on the penetration of tightly closing bivalves by Busycon and other predators. Ecology, 32:73–83.

————. 1969. Excavation of boreholes by the gastropod Urosalpinx: an analysis by light and scanning electron microscopy. In M. R. Carriker et al. (eds.), Penetration of calcium carbonate substrates by lower plants and invertebrates. Amer. Zool., 9:991–996.

———— and D. Van Zandt. 1972. Predatory behavior of a shell-boring muricid gastropod. In H. E. Winn and B. L. Olla (eds.), Behavior of marine animals. New York, Plenum Publ. Co., p. 157–244.

———— and E. L. Yochelson. 1968. Recent gastropod boreholes and Ordovician cylindrical borings. U.S. Geol. Survey, Prof. Paper 593-B:B1–B26.

———— et al. (eds.) 1969. Penetration of calcium carbonate substrates by lower plants and invertebrates. Amer. Zool., 9:629–1020.

Carter, R. M. 1968. On the biology and palaeontology of some predators of bivalved molluscs. Palaeogeogr., Palaeoclimatol., Palaeoecol., 4:29–65.

Clarke, G. L. 1954. Elements of ecology. New York, John Wiley, 560 p.

Clarke, J. M. 1912. Early adaptation in the feeding habits of star-fishes. Acad. Nat. Sci. Philadelphia, Jour., Ser. 2, 15:113–118.

Colbert, E. H. 1961. Dinosaurs. New York, Dutton, 300 p.

Drinnan, R. E. 1967. The winter feeding of the oystereater (Haematopus ostralegus) on the edible cockle (Cardium edule). Jour. Animal Ecol., 26:441–469.

Eastman, C. R. 1911. Jurassic saurian remains ingested within fish. Ann. Carnegie Mus., 8:182–187.

Eaton, T. H. 1964. A captorhinomorph predator and its prey. Amer. Mus. Nat. Hist., Novitates, 2169:1–3.

Errington, P. L. 1967. Of predation and life. Ames, Iowa, Iowa State Univ. Press, 277 p.

Fagerstrom, J. A. 1961. Busycon (Busycon) tritone Conrad redescribed and reillustrated. Jour. Paleont., 35:448–450.

Frey, R. W. 1972. Paleoecology and depositional environment of Fort Hays Limestone Member, Niobrara Chalk (Upper Cretaceous), west-central Kansas. Univ. Kansas Paleont. Contr., Art. 58:1–72.

Fujita, S. 1916. On the boring of pearl oysters by Octopus (Polypus) vulgaris Lamarck. Dobytsugaku Zasshi, 28:250–257. (in Japanese)

Galtsoff, P. S. 1964. The American oyster, Crassostrea virginica Gmelin. U.S. Fish Wildlife Serv., Fish. Bull., 64:1–480.

Green, M. 1961. Pathologic vertebrate fossils and recent specimens. South Dakota Acad. Sci., Proc., 40:142–148.

Grim, R. J. and W. M. Whitehouse. 1963. Pellet formation in a great horned owl: a roentgenographic study. The Auk, 80:301–306.

Gripp, K. 1929. Über Verletzungen an Seeigeln aus der Dreide Norddeutschlands. Paläont. Zeitschr., 11:238–245.

Häntzschel, W. et al. 1968. Coprolites: an annotated bibliography. Geol. Soc. America, Mem. 108, 132 p.

Jepsen, G. L. 1967. Notable geobiologic moments. Geotimes, 12(6):16–18.

Kauffman, E. G. 1972. Ptychodus predation in a Cretaceous Inoceramus. Palaeontology, 15:439–444.

———— and R. V. Kesling. 1960. An Upper Cretaceous ammonite bitten by a mosasaur. Univ. Michigan Mus. Paleont. Contr., 15:193–248.

Kier, P. M. and R. E. Grant. 1965. Echinoid distribution and habits, Key Largo Coral Reef Preserve, Florida. Smithsonian Misc. Coll., 149(6):1–68.

Ladd, H. S. 1957. Paleoecological evidence. In H. S. Ladd (ed.), Treatise on marine ecology and paleoecology, V. 2, Paleoecology. Geol. Soc. America, Mem. 67(2):31–66.

Lawrence, D. R. 1968. Taphonomy and information losses in fossil communities. Jour. Paleont., 79:1315–1330.

Lee, R. B. and I. DeVore (eds.) 1968. Man the hunter. Chicago, Aldine-Atherton, 415 p.

Lehmann, U. 1971. Jaws, radula, and crop of Arnioceras (Ammonoidea). Palaeontology, 14:338–341.

Meanley, B. 1962. Pellet casting by king and clapper rails. The Auk, 79:253–254.

Menzel, R. W. and F. W. Nichy. 1958. Studies

of the distribution and feeding habits of some oyster predators in Alligator Harbor, Florida. Bull. Marine Sci., 8:125–145.

Miller, G. J. 1969. A study of cuts, grooves, and other marks on recent and fossil bones. I. Animal tooth marks. Tebiwa, Jour. Idaho State Univ. Mus., 12:20–26.

Moore, D. R. 1956. Observation on predation of echinoderms by three species of Cassididae. Nautilus, 69:73–76.

Moulton, J. M. and A. H. Gustafson. 1956. Green crabs and the redistribution of quahogs. Science, 123:992.

Müller, A. H. 1963. Lehrbuch der Paläzoologie. I. Allgemine grundlagen. Jena, Gustav Fisher, 387 p.

Odum, E. P. 1963. Ecology. New York, Holt, Rinehart and Winston, Modern Biol. Ser., 152 p.

————. 1971. Fundamentals of ecology. Philadelphia, Saunders, 574 p.

Paine, R. T. 1962. Ecological diversification in sympatric gastropods of the genus *Busycon*. Evolution, 16:512–522.

Pollard, J. E. 1968. The gastric contents of an ichthyosaur from the Lower Lias of Lyme Regis, Dorset. Palaeontology, 11:376–388.

Pilson, M. E. Q. and P. B. Taylor. 1961. Hole drilling by *Octopus*. Science, 134:1366–1368.

Purchon, R. D. 1968. The biology of the Mollusca. Pergamon Press, 560 p.

Reed, E. B. 1957. Mammal remains in pellets of Colorado barn owls. Jour. Mammal., 38:135–136.

Richards, R. P. and C. W. Shabica. 1969. Cylindrical living burrows in Ordovician dalmanellid brachiopod beds. Jour. Paleont., 43:838–841.

Romer, A. S. 1945. Vertebrate paleontology. Chicago, Univ. Chicago Press, 687 p.

Ropes, J. W. 1969. The feeding of the green crab, *Carcinus maenas* (L.). U.S. Fish Wildlife Serv., Fish. Bull., 67:183–203.

Schäfer, W. 1954. Form und Funktion der Brachyuren-Schere. Senckenberg. Naturf. Gesell., Abh., 489:1–65.

————. 1972. Ecology and palaeoecology of marine environments. Edinburgh and Chicago, Oliver & Boyd and Univ. Chicago Press, 568 p.

Scott, W. B. 1937 (1962 reprint). A history of land mammals in the Western Hemisphere. New York, Hafner, 786 p.

Seed, R. 1969. The ecology of *Mytilus edulis* L. (Lamellibranchiata) on exposed rocky shores. 2. Growth and mortality. Oecologia, 3:317–350.

Seilacher, A. 1953. Studien zur Palichnologie. II. Die fossilen Ruhespuren (Cubichnia). Neues Jahrb. Geol. Paläont., Abh., 98:87–124.

Shoup, J. B. 1968. Shell opening by crabs of the genus *Calappa*. Science, 160(3830):887–888.

Sliter, W. V. 1971. Predation on benthic foraminifers. Jour. Foram. Res., 1:20–29.

Sohl, N. F. 1969. The fossil record of shell boring by snails. In M. R. Carriker et al. (eds.), Penetration of calcium carbonate substrates by lower plants and invertebrates. Amer. Zool., 9:725–734.

Sparks, J. and T. Soper. 1970. Owls: their natural and unnatural history. New York, Taplinger Publ. Co., 206 p.

Speden, I. A. 1971. Notes on New Zealand fossil Mollusca. 2. Predation on New Zealand Cretaceous species of *Inoceramus* (Bivalvia). New Zealand Jour. Geol. Geophys., 14:56–60.

Stanley, S. M. 1973. An ecological theory for the sudden origin of multicellular life in the late Precambrian. Nat. Acad. Sci. USA, Proc., 70:1486–1489.

Taylor, J. D. 1970. Feeding habits of predatory gastropods in a Tertiary (Eocene) molluscan assemblage from the Paris basin. Palaeontology, 13:254–260.

Valentine, J. W. 1973. Evolutionary paleoecology of the marine biosphere. Englewood Cliffs, N.J., Prentice-Hall, 511 p.

Wells, H. W. 1958. Feeding habits of *Murex fulvescens*. Ecology, 39:556–558.

Wetmore, A. 1943. The occurrence of feather impressions in the Miocene deposits of Maryland. The Auk, 60:440–441.

Williams, M. E. 1972. The origin of "spiral coprolites." Univ. Kansas Paleont. Contr., Paper 59:1–19.

Wodinsky, J. 1969. Penetration of the shell and feeding on gastropods. In M. R. Carriker et al. (eds.), Penetration of calcium carbonate substrates by lower plants and invertebrates. Amer. Zool., 9:997–1010.

Woelke, C. E. 1957. The flatworm *Pseudostylochus ostreophagus* Hyman, a predator of oysters. Nat. Shellfish. Assoc., Proc., 47:62–67.

Wright, C. W. and E. V. Wright. 1940. Notes on Cretaceous Asteroidea. Geol. Soc. London, Quart. Jour., 94:231–248.

Zangerl, R. and E. S. Richardson, Jr. 1963. The paleoecological history of two Pennsylvanian black shales. Chicago Mus. Nat. Hist., Fieldiana, Geol. Mem., 4:1–352.

CHAPTER 14

FOSSIL TRACKS AND IMPRESSIONS OF VERTEBRATES

WILLIAM ANTONY S. SARJEANT

Department of Geological Sciences, University of Saskatchewan
Saskatoon, Saskatchewan, Canada

SYNOPSIS

The study of vertebrate trace fossils constitutes the oldest branch of ichnology, although several problems have hampered its development and ultimate usefulness: obscure publications, inadequate illustrations and text descriptions, poorly preserved specimens that are easily overlooked or misinterpreted, and simply the lack of interest in the field shown by most vertebrate paleontologists. Nevertheless, vertebrate ichnology has much to offer paleontology, stratigraphy, and facies analysis, and these fields would benefit greatly by its increased use and refinement.

Although the methods of interpretation and classification of fossil footprints have not been settled unequivocally, the present approaches have already yielded valuable data on footprints as evidence of (1) animals unknown as body fossils; (2) geographic and geologic ranges of vertebrate species, including biostratigraphic zonations by trace fossils; (3) behavior and habitat adaptations of vertebrates; and (4) prevailing environmental conditions. Other vertebrate traces and impressions—such as swimming and resting traces of fishes and amphibians, coprolites, various traces reflecting the death throes of animals, and vertebrate burrows (discussed in the following chapter)—are also important and deserve increased attention.

INTRODUCTION

Of all branches of paleoichnology, the study of fossil vertebrate footprints is most venerable. The date of the first human observation of tracks in stone cannot be established with certainty. Kirchner (1941) suggested that observation of Triassic reptile footprints at Siefriedsburg (Siegfriedsburg), Germany, may have been the origin of the legend of Siegfried and the dragon; but as Lessertisseur (1955, p. 89) drily commented, "this incursion of a paleontologist into medieval folklore appears more curious than convincing!" (transl.).

The earliest authenticated observation of fossil traces is the discovery by Pliny Moody, a New England farm boy, of dinosaur tracks impressed in red sandstone near South Hadley in the Connecticut Valley, U.S.A., in 1802. Because the tracks were strikingly bird-like and dinosaur remains were then unknown, the oddities were understandably taken to be tracks of gigantic birds—by some, as the tracks of Noah's raven! However, these tracks were not made the subject of scientific study for some 35 years thereafter, nor was their discovery known to geologists.

The history of paleoichnology must therefore be said to begin in 1828, with the description of tracks from red sandstones

283

in a quarry at Corncockle Muir, Dumfries-shire, Scotland. Although these tracks were first discovered in 1824, the first published record of them was an anonymous note in the *London & Paris Observer* for 1828, followed in the same year by an account presented by Henry Duncan to the Royal Society of Edinburgh (published in 1831). The discovery caused great excitement among geologists and was reported in several other British and foreign journals. From the outset, the tracks were recognized as those of a quadrupedal reptile. Buckland (1828), noting the broad trackway and short stride, very reasonably concluded that they were tracks of tortoises; but they are now considered to be tracks of amphibians (Labyrinthodonts).

During the remainder of the nineteenth century, and indeed, until about 1930, reports of fossil vertebrate footprints were published in profusion: from Scotland, England, Germany, the United States, and eventually from most countries in the world.

Other types of vertebrate traces also came to be recognized. As early as 1829, William Buckland demonstrated that some nodular structures in the English Lias were the fossil feces of vertebrates; he proposed the name coprolites for them, and showed them to be a potential source of phosphates for agricultural purposes. Teeth and claw marks were first recognized on bones from cave deposits, later from more ancient geological strata. Vertebrate burrows, some of spectacular size, were discovered; so were resting impressions of fishes, traces of birds in process of taking flight, taphoglyphs (imprints of the bodies of dead or dying animals) in sands and muds and, most spectacularly, in volcanic ash. Even supposed "tadpole nests" have been reported.

Although some presumed traces have subsequently been shown to be either inorganic in origin or produced by invertebrates, a diverse array of evidence for the activity of vertebrates has now been discovered in the geological column and is available for interpretation by paleoecologists.

PROBLEMS AND ATTRACTIONS OF VERTEBRATE ICHNOLOGY

Literature Problems

Since about 1930, the study of fossil vertebrate footprints has fallen into disfavor. Curry in 1957 accurately commented that, by that date, the study of fossil tracks and trackways had become "a minor, neglected, and somewhat disreputable branch of paleontology." This comment may be applied with equal strength to all branches of vertebrate paleoichnology; as a consequence, scientific papers on this topic have most often been written by amateur geologists and have tended to be published in journals difficult of access to most paleontologists. The *Proceedings of the Liverpool Geological Society* which, arguably, has contained more major early papers on this topic than any other journal, is relatively easy to obtain; but much more difficult to obtain are some equally relevant papers in the *Proceedings of the Liverpool Literary & Philosophical Society* and the *Transactions of the Liverpool Geological Association*. Gaining access to papers published in German high school publications (Sickler, 1834) and "works newspapers" (Wolansky, 1952; Weingardt, 1961) or by Austrian touring clubs (Kittl, 1901) is virtually impossible; yet such places of publication are scarcely even exceptional. In Kuhn's (1963) invaluable bibliographic compilation, *Ichnia tetrapodorum*, the words "Nicht gesehen" (*not seen*) and "Nicht auffindbar!" (*not discoverable!*) occur frequently; one can entirely sympathize with his problems. Literature on coprolites was ably reviewed by Häntzschel et al. (1968), but no guides to papers on other aspects of vertebrate paleoichnology are available.

The problems of literature on fossil footprints have been greatly accentuated by inadequate illustrations. Many papers, even those proposing new taxa, are illustrated only by outline sketches; when the footprint cast or mold has a sharp angle at its deepest or highest part and grades smoothly at its margins into the surrounding sediment surface, where does one place the line of demarcation? In museums I have seen type specimens in which India-ink lines, supposedly bounding a print, actually traverse the impression of a digit; drawings based on such outlines would be very misleading. Photographs, when published at all, have tended to be of low quality and are almost always much too small; they are often taken in overhead lighting, the worst possible arrangement for showing a cast or mold. If oblique lighting was employed, the direction of incidence of the light was rarely indicated and the result sometimes highly misleading.

Text descriptions also are often inadequate; frequently they are overly brief, and adequate measurements are rarely given. In English texts, the words "step" and "stride" are frequently interchanged. Interdigital angles are sometimes quoted, but because digits almost always show some degree of curvature, are valueless unless the positions of reading of the angles are indicated. Adequate comparisons with previously described forms are rarely made, a consequence in part of the particular literature difficulties in this field. The formal lodgement of specimens is only rarely indicated, and with depressing frequency, locating holotypes and figured specimens proves impossible. Even when originally deposited in a named collection or museum, numerous footprint slabs are so large as to be a nuisance to curators; many have been shifted outside to yards or gardens, to be damaged or destroyed by rain, wind, and human feet or hands, or even simply thrown away.

Problems of Preservation

Not all the problems to be faced, however, are the fault of the describer; many diffi-culties result from the characteristics of the fossils themselves. Footprints are formed as molds—indentations in the surface over which the animal passed. If the surface is fairly fine grained and cohesive, neither too wet nor too dry, and if the animal progressed sufficiently slowly, a perfect impression of the whole undersurface of the foot (pes) and, in quadrupeds, of the hand (manus), may be formed; details of claws or nails, shape of pads, or patterns of scales may be shown clearly. If the sediment is too coarse, the resultant mold will not show such details. If too wet, the shape of the mold may be distorted, slightly or markedly, especially in the deeper hollows, which might well fill with water; if under water, the print may be entirely washed out (Fig. 14.1). In any case, prints impressed into beaches or in shallow estuaries are likely to be destroyed by the next rising tide; prints impressed into sand dunes are normally destroyed quickly by wind and shifting sands. Prints made by animals traveling at greater speed are unlikely to be complete, often comprising digit impressions only.

The most favorable circumstances for preservation are at the end of a phase of water advance, when fine-grained sediments are gradually drying out. If a subsequent inrush of suspension-laden waters follows storms, the molds may then be filled (cast) before they are destroyed. Such casts, being formed by a potentially more solid and less friable sedimentary material, have a greater survival potential than do the original argillaceous molds; even so, they occur on the undersides of blocks and are thus unlikely to be seen except where screes are forming, or during quarrying operations. The quarrying of sandstone for building purposes has virtually ceased in Europe and North America since about 1910; for example, the Triassic sandstones that were once so popular for building purposes in Britain (cf. Fig. 14.2) and Germany are in almost entire disuse now. Quarries that do continue in production tend to be highly mechanized, and blocks containing foot-

Fig. 14.1 Section from borehole core (A), showing trackway (?*Aetosauripus*) impressed into wet sand and partly washed out by swash of receding tide. B, C, details of manus and pes molds. Bunter Pebble Beds (Triassic); Worcestershire, England; at ca. 304 m depth. (From Wills and Sarjeant, 1970.)

prints must frequently escape notice; only in the smaller quarries, still hand worked, are they apt to be noticed.

Fig. 14.2 Large, clear, well-formed footprint casts on slab propped up to weather. A trail—attributable to *Chirotherium storetonense*—and some isolated prints are seen. Keuper Sandstone (Triassic); quarry at Storeton, Cheshire, England. (Photo: W. H. Rock, ca. 1906; now in Beasley Coll., Liverpool Geol. Soc.)

Light conditions are also a critical factor in footprint collecting. Molds on bedding planes, and even shallow casts on the undersides of overturned blocks, are likely to escape attention when the sun is overhead, especially in the intense sunshine of desert conditions. At the other extreme, completely overcast skies also cause difficulties; shadows are not prominent, and color contrasts are reduced. Ideal conditions for footprint collection are the early morning and late afternoon of summer days or sunny days in winter, when slanting sunlight creates long shadows and causes casts or molds to stand out prominently.

Even under the most favorable conditions, the prints of small animals, lightly impressed and producing only shallow casts (cf. Fig. 14.3) readily escape notice; individual prints of such animals rarely show much detail. Moreover, small reptiles have minimal water requirements and might rarely need to traverse wet sediment surfaces. A direct relationship thus exists between the size of the animal and the prob-

Fig. 14.3 Poorly impressed small prints (molds) —attributed to *Varanopus* aff. *V. curvidactylus* and *Microsauripus* aff. *M. acutipes*—forming a confusing pattern. Footprints are clearly visible only in oblique illumination. Keuper Waterstones (Triassic); Nottingham, England. (From Sarjeant, 1969.)

ability of its footprints ever being studied: the larger the animal, the better the chance of its footprints being preserved, the higher the probability that critical details will be evident, and the greater the chance of the track being noticed and collected or described.

Other sorts of vertebrate traces are equally unlikely to receive the attention of geologists in the field. Vertebrate burrows (see Chapter 15) generally occur as disturbances of stratification and, if noted at all, may be misinterpreted as structures produced by cryoturbation (freeze–thaw action

in cold conditions, where the ground freezes in winter), slumping, or other inorganic effects. Coprolites are not always easily distinguished from inorganic concretions. Resting and flight traces are not readily appreciated; vertebrate taphoglyphs, where associated with skeletal remains, are likely to be destroyed in the extraction of the bones, and where not so associated, are not readily perceived. (See also Fig. 14.18.)

Attractions of Vertebrate Paleoichnology

Despite the various handicaps discussed above, the study of fossil vertebrate tracks and other traces remains an extremely worthwhile discipline.

The sediments in which fossil footprints are preserved very often suggest formation under continental climatic conditions, including a period of summer aridity; indeed, they seem very often to have formed under semiarid, or even arid, conditions. In such environments, members of animal populations can never have been particularly numerous; the number of skeletons available for fossilization must have been low, and even when preserved, the odds against discovering the remains of any individual animals must be extremely high. During its lifetime, however, each animal might well produce many thousands, even millions, of footprints; the larger animals, in particular, would tend to range widely in search of food, and the need for water would draw them to the drying-out margins of waterways and pools. Thus, although the odds against the survival of any individual print are vast, a very much higher chance remains that a few, out of the multitude of tracks produced during a lifetime of activity, might survive and be discovered and that some of these few might be of adequate quality for scientific study.

In the Permian and Triassic "new red sandstones" of Europe, finds of vertebrate skeletons have been very few; indeed, occurrences of fossils of any kind, with the single exception of pollen and spores, are meager.

Vertebrate footprints, in contrast, have been discovered in abundance and have been used successfully in both local and long-distance correlation. Moreover, the prints indicate an infinitely more abundant and varied vertebrate fauna than can be deduced from skeletal remains, in both local and absolute terms.

Fossil tracks and other traces are also of special interest in that they provide a dynamic record of the animal in action, allowing a partial determination of its way of life and, in some instances, even affording a glimpse of the dramas of the past. In the ensuing pages, special attention is devoted to fossil footprints, and examples of their stratigraphical and paleontological uses are given. A brief review is presented of other types of vertebrate impressions. Accounts of vertebrate burrows and fossil evidence of predation may be found in Chapters 13 and 15.

FOOTPRINTS

Methods of Interpretation

For study purposes, the paleoichnologist hopes to have a series of successive footprints—a trackway, usually taken to consist of a minimum of three sequential sets of impressions (six footprint casts or molds)—so that the progressive motion (the gait) of the trackmaker may be determined. Under ideal circumstances, indeed, he hopes to have a series of trackways, illustrating slow and fast movement by related trackmakers. Sometimes, however, he must be concerned with (1) an isolated track—the adjacent hand and foot impressions of the same (right or left) side, or at worst, (2) only a single print—possibly only a single mold (impress, imprint, or hyporelief) or a single cast (epirelief). The study of a single print has serious limitations; one cannot always be sure whether it is the track of a quadruped or biped, or even whether it represents manus or pes.

In quadrupeds, the sequence of walking

or running movements always begins with one pes, followed by the manus on the same side, then the pes and manus of the opposite side. In general, two or three feet are simultaneously in contact with the ground during fast movement, three or four in slower movement. Walking and running tracks always have right manus near left

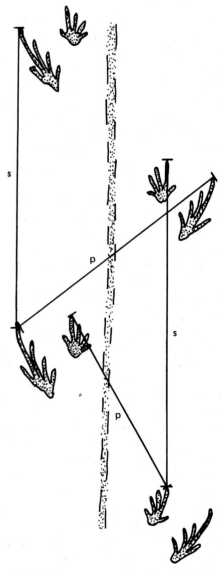

Fig. 14.4 Measurement of stride (s) and pace (p), shown on trackway *Rhynchocephalichnus pisanus*. Although length of stride is inevitably the same for manus and pes, the pace may be markedly different. Central stippled groove is tail-drag trace. From Keuper (Triassic); Italy. (After von Huene, 1941.)

pes, and vice versa; the impressions may be well separated, or the pes impression may approach or even overlap that of the manus. Jumping tracks have all four impressions set close together (but not overlapping), the groups of impressions being widely spaced. In bipeds, left and right feet alternate; the impressions are of constant form and are never side by side.

The stride is the approximate unitary forward movement of the feet on the right or on the left side; it is measured from some fixed point on the impression of one foot to the same point on the next impression (Fig. 14.4). The pace is the distance between the impression of the right manus or right pes and the left manus or left pes; measurement of pace length is thus oblique to the midline of the trackway (Fig. 14.4). The step is the overall progress during which each foot and each hand are moved forward one stride length. Although step-length measurements are not normally quoted, knowing the step angle (pace angulation) is useful: the angle formed by lines joining the midpoint of three successive manus or pes (right-left-right or left-right-left) imprints (Fig. 14.5).

Measurements of the track breadth should always be given. A broad trackway and short stride indicate an inefficient walker; a narrow trackway and long stride indicate an efficient walker moving at appreciable speed; but moderate trackway breadth and stride can indicate either a walker of moderate efficiency or an efficient walker moving relatively slowly.

In quadrupeds, a matter of great interest is the glenoacetabular distance, measured from the midpoint between two manus impressions and the midpoint between two pes impressions. This parameter more or less corresponds to the glenoid and acetabular axis (the body length), although in some animals, notably the amphibians, the body motion in locomotion is rather sinuous and the equivalence is therefore not quite exact. (Cf. Dalquist, 1964.)

Also of interest in interpretation of

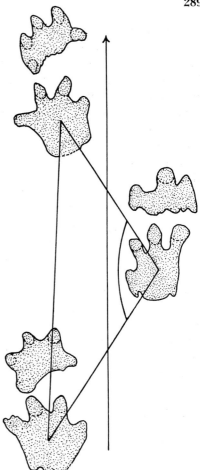

Fig. 14.5 Measurement of step angle (pace angulation), taken from three successive imprints in trackway *Tetrapodosaurus borealis*. Cretaceous; western Canada. (After Sternberg, 1930.)

tracks is the degree of overlap of the impression of the pes upon that of the manus. Peabody (1959, p. 6) distinguished between (1) primary overlap, in which the pes is emplaced on part or all of the manus imprint immediately after the manus has been retracted; (2) secondary overlap, in which the pes is one full stride behind the manus at the time of its retraction; and (3) tertiary overlap, in which—as in some long-bodied salamanders—the pes is two full strides behind the manus at the time of its retraction (Fig. 14.6). In terms of the footprints produced, the imprints of manus and pes in primary overlap show a consistent positional relationship; this is characteristic of

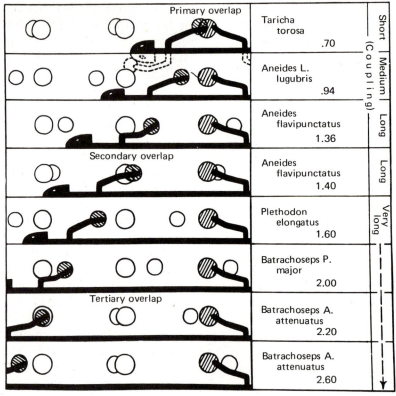

Fig. 14.6 Degree of overlap of pes upon manus (coupling) in tracks of living salamanders. (From Peabody, 1959.)

animals in which the distance between the pectoral and pelvic girdles is not great (short-coupled forms). In secondary and tertiary overlap, the footprints do not show a constant positional relationship; the relative spacing of manus and pes impressions changes with each step, the manus impression only periodically having that of the pes superimposed on it. Tertiary overlap occurs only in extremely long-bodied (long-coupled) forms (e.g., some salamanders) and, in the fossil state, is indicated by the wide interval between overlapping prints and a high, even obtuse, step angle (Fig. 14.6).

The length of the dorsal region is sometimes termed the "degree of coupling" and may be expressed in terms of the coupling value, a number obtained by dividing the glenoacetabular distance by the sum of the lengths of fore and hind limb. This value is readily determined in studies of living forms, such as the salamanders investigated

by Peabody (1959); but in the case of fossil tracks, serious inaccuracies in calculation can arise and the resultant numbers are of questionable usefulness.

The position of prints in relation to the midline of the trackway is also of interest in determining gait: (1) the angle of the long axis of the whole print, taken from the foremost tip of one of the three central digits to the rearmost point on palm or heel, and (2) the direction in which the digits are pointed, whether toward the midline (inward or positive rotation) or away from it (outward or negative rotation). (Fig. 14.7.)

No standard method has yet been established for quoting detailed measurements of each impression. The length and breadth of impressions have been measured in whatever direction an author saw fit. Haubold's suggestion (1971a, p. 6) that length be measured parallel to digit III and breadth

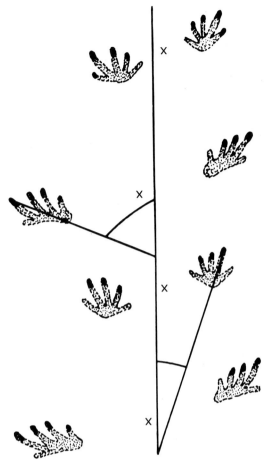

Fig. 14.7 Position of prints in relation to midline, giving rotation of manus and pes, shown on trackway *Phalangichnus simulans*. The midline here does not correspond exactly to midpoints between paired imprints (indicated by ×) because the gait is somewhat sprawling. Permian; Germany. (After Schmidt, 1959.)

at right angles to it, has great merit; but even on this basis, digit curvature can cause serious problems in deciding the direction for measurement. The length of each digit should also be cited, but here again digit curvature causes difficulties, and, especially where digits are relatively broadly based, deciding where to terminate the measurement is difficult (Fig. 14.8).

Digits may be parallel but more often are divergent to varying degrees, sometimes markedly so; opposition of a particular digit (its placement at right angles to the main axis) is a pronounced feature in a few

vertebrate groups, notably the primates—where the thumb (digit I) is opposed—and the diapsid reptiles—where digit V may be opposed. Digit divarication is frequently quoted in terms of the interdigital angle (the angle of divergence of adjacent digits), but here again, digit curvature causes problems, and the choice of positions for axial lines is so subjective that a diagram showing positions of readings should always be appended (Fig. 14.9). Moreover, because digits spread out to gain a better footing on soft sediments and close up on harder surfaces, small differences in interdigital angle are unlikely to be of taxonomic importance.

Footprints are, of course, three-dimensional objects. In thin-bedded sediments or in laminated sediments, such as shales, which acquire fissility as a consequence of diagenetic effects, the upper sediment layer containing the true footprint mold may be separable from an underlying layer in which original deformation caused subsurface impressions of the most deeply impressed parts of the footprint (Heyler and Lessertisseur, 1963). The resultant subtrace ("under track") may be preserved separately and may be taken at first sight as a footprint of a type very different from the true footprint mold (Fig. 14.10).

Wherever possible, both mold and cast should be examined, not only because the character of the cast enables ready distinction between genuine imprints and subtraces, but also simply because the morphology of a cast is more readily appreciated (see Sarjeant, 1971). For this reason, where only molds are available for examination, artificial casts are frequently made (in plaster of paris, rubber, or various plastics) to facilitate study and description.

In attempting skeletal restorations, one can normally assume that joints between the bones of the digit (phalanges) correspond in position to nodes and pads of the imprint; the innermost pad, in instances where the front part of the palm or sole is impressed, corresponds to the junction between the basal phalange and a meta-

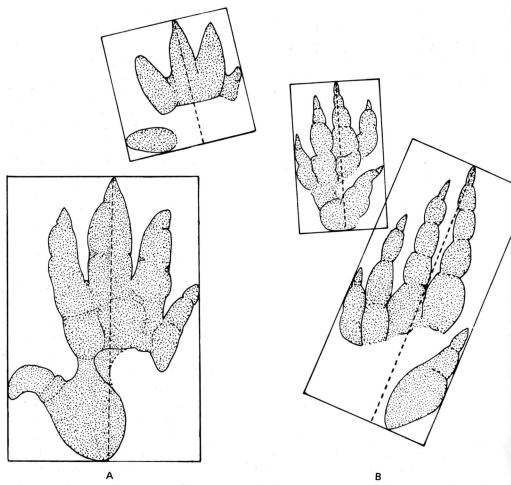

Fig. 14.8 Contrasting methods of length–breadth measurement. A, *Chirotherium barthi*. Here Haubold's method is applied; length is measured parallel to digit III, breadth at right angles to it. The method works well for the manus, less well for the pes where this digit is plump and slightly curving. B, *Chirotherium pseudosuchoides*. Here measurements are made parallel, and at right angles, to a midline; choice of midline is arbitrary, made more so by digit curvature. Both methods pose problems, but A is preferable. (After Haubold, 1971a.)

carpal or metatarsal. Reconstruction of the structure of the skeleton of the manus or pes is thus possible in many instances, although this should be attempted only where warranted by a reasonable presumption that the impression is complete; partial impressions formed by a fast-moving animal can be very misleading.

Having obtained the glenoacetabular length of a quadrupedal trackmaker and the length of its stride, attempting a restoration of the animal becomes feasible. This interpretation is most readily done for rep-

tiles and amphibians, where limb movement is essentially in phase, the forelimb and hind limb movement of opposite sides occurring more or less simultaneously. The relative sizes of manus and pes should be noted; where the pes is significantly larger, bipedal (diapsid) reptilian affinities may be assumed (Fig. 14.11). By drawing parallels in the modes of locomation between the fossil tracks and the body proportions of living reptile and amphibian species that move in a similar fashion, a reasonable estimate can be made of limb lengths, and a recon-

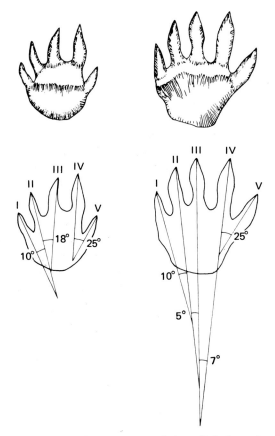

Fig. 14.9 Measurements of interdigital angle. Always involving an arbitrary choice of midlines for digits, methods should be diagrammatically explained. On the trackway drawn here, *Chelichnus hicklingi*, other lines could easily have been selected. Permian; Nottinghamshire, England. (From Sarjeant, 1967.)

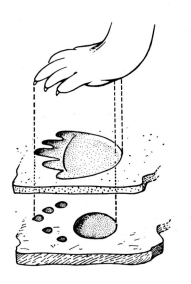

struction of the trackmaker can be attempted. Tail length cannot, of course, be determined with exactitude. Tail-drag impressions, in the form of sinuous grooves, are sometimes found and indicate a tail of appreciable (but indeterminable) length; however, because many long-tailed animals carry their tail entirely clear of the ground, the absence of such impressions does not conclusively indicate a short tail. For most reptiles and amphibians, one reasonably expects a tail length markedly greater than the glenoacetabular distance.

Many mammals have developed an out-of-phase pattern of limb movements in gaits designed for speed, but such species tend to revert to an alternating gait when progressing at a more leisurely pace (Fig. 14.12). The structure of the feet is more variable, including development of hooves, pads, and retractible claws in different species; to determine affinities is more or less possible, but to estimate the length of limbs and of tails is less easy, so that reconstructions from tracks are more difficult to make. The tracks of bipeds in general and birds in particular, although they may display special features permitting identification of the taxonomic group to which the trackmakers belong, afford insufficient direct evidence for making reconstructions. Bird tracks are only rarely sufficiently individualistic to permit generic, or even familial, identification of the trackmaker; even to link bird tracks to a particular order is often difficult.

Human tracks are commonly encountered on cave floors (see Vallois, 1931; Casteret, 1939) and on former "living surfaces" exposed in archeological excavations. I have seen a "living surface" dating from about 2500 B.C., beaten flat and shiny by naked feet, exposed from beneath a barrow

◀ **Fig. 14.10** Formation of footprint subtrace ("under track") on bedding plane beneath the footprint itself. (From Heyler and Lessertisseur, 1963.)

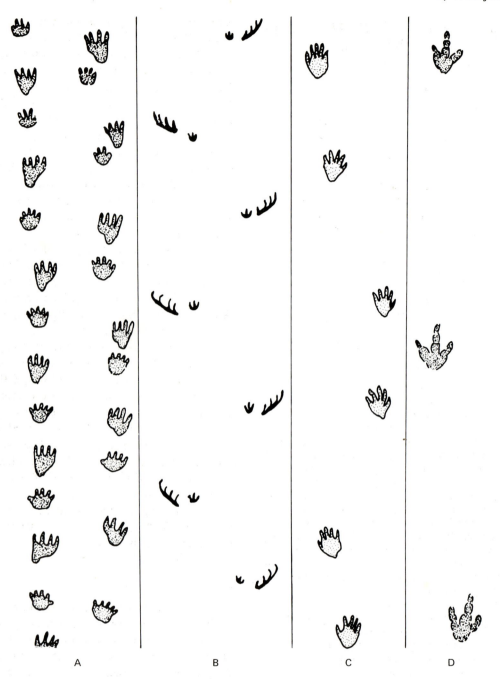

Fig. 14.11 Different reptile gaits, as illustrated by tracks. A, *Megapezia praesidentis*. Plantigrade tracks of a quadruped: trackway broad and rather irregular; stride short—an inefficient walker. Upper Carboniferous; Germany. B, *Hamatopus wildfeueri*. Digitigrade tracks of a quadruped: track-way broad but stride quite long and much more regular—a running track evidencing greater effi-ciency in movement. Triassic; Germany. C, *Dicynodontipus geinetzi*. Plantigrade tracks of a quadruped: trackway narrow; stride long and regular—a highly efficient walker. Triassic; Germany. D, *Anomoepus* sp. Semidigitigrade tracks of a biped: trackway very narrow; stride long—also an efficient walker. Triassic; France. [After Schmidt (1956) (A) and (1959) (B); Haubold (1971a) (C); Ellenberger (1965) (D).] Not to scale.

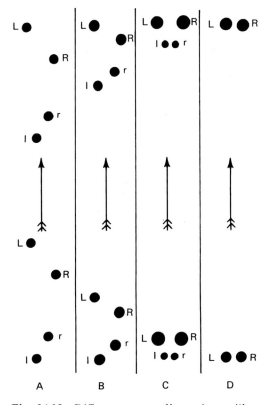

Fig. 14.12 Different mammalian gaits, as illustrated by tracks. l,r = left and right manus. L,R = left and right pes. A, rotary gallop of mammal having strong forelimbs: intervals of suspension of the body separate the moment when the second forelimb leaves the ground and the hind limb on the same side makes contact, and the corresponding moment when the second hind limb leaves the ground; never are all four limbs simultaneously in contact. B, rotary gallop of mammals having weak forelimbs: only the hind limbs propel suspension, and the four imprints show a closer grouping; all four limbs are repeatedly in contact with the ground. C, a springing gait: movement of all four limbs is synchronized, and the hind limbs provide propulsion. D, a bounding gait: the hind limbs provide total propulsion, and the forelimbs—used only in stabilization—may not be represented in short sections of the trackway. (After Casamiquela, 1964.)

(tumulus) in an excavation at Pitnacree, Perthshire, Scotland, in 1964.

The use of Cartesian diagrams in the study of vertebrate footprints was initiated by Lull (1953), in his study of the ichno-fauna from the Triassic of the Connecticut Valley. The method was first applied by Sir D'Arcy Thompson in evolutionary studies. Lull used it instead to show the relations of different trace species within the same genus. For example, in the case of the carnosaur footprint genus *Eubrontes*, a basic rectangular grid was prepared, having the parallel coordinates, both vertical and horizontal, placed at intervals of 0.5 inch (1.25 cm). Upon this grid was traced a full-size outline of the type species, *E. giganteus*, which was used as a basis for comparison with other species.

The points of intersection of coordinates with track outline, having been thus determined, are then transferred to the track outlines of other species and become the points through which the corresponding coordinates must be drawn. The latter are no longer straight or parallel and no longer have rectangular intersections, but show instead varying degrees of distortion of the basic grid. Approximation to the original grid illustrates closeness of relationship, and vice versa (Fig. 14.13). The method thus aids in determining relationships; it is not influenced by size alone. An excellent means for analyzing the validity of related species in a single genus or a group of related genera, it readily helps expose synonyms. The method has not attained widespread use, however.

Classification

The question of classification of vertebrate footprints forms part of the general problem of trace fossil classification, on which I have recently expressed my views at length (Sarjeant and Kennedy, 1973); because of this, and because the matter is treated in Chapter 3 as well, a full discussion of the problem would be inappropriate here. Instead, the history of vertebrate footprint classification is briefly reviewed and the most recent approaches discussed.

The earliest application of a formal

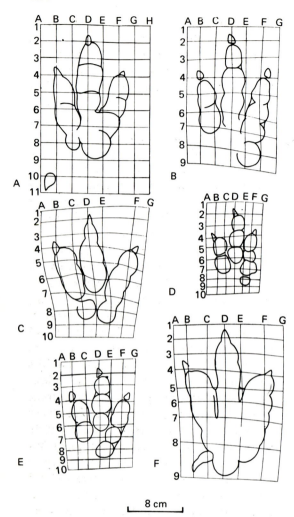

Fig. 14.13 Cartesian diagrams applied to members of the trace genera *Eubrontes* and *Gigandipus*. A, the basic form, *E. giganteus*. B, *E. approximatus*. C, *E. divaricatus*. D, *E. tuberatus*. E, *E. platypus*. F, *Gigandipus caudatus*. Triassic; Connecticut Valley, U.S.A. (From Lull, 1953.)

taxonomic name to vertebrate tracks was by Kaup (1835), who applied the appelation *Chirotherium* or *Chirosaurus barthii* to some hand-like imprints in German Triassic sandstones; the provision of alternative names indicated his doubt concerning their affinity.

The earliest classification was advanced by Hitchcock (1836) who, still then convinced that the Connecticut Valley tracks were those of birds, formulated a discipline that he named "ornithichnology"; he classed

the tracks as Ornithichnites and divided them into two groups, Pachydactyli and Leptodactyli, on the form of their digits. By the following year, however, the progress of geological knowledge had forced him to formulate categories for other groups; he termed the impressions "Ichnites" and recognized reptile tracks (Sauroidichnites), bird tracks (Ornithichnites), and tetrapod, essentially mammalian tracks (Tetrapodichnites). Between 1841 and 1844, he formalized his classification, erecting a Class Ichnolithes; this incorporated invertebrate as well as vertebrate tracks and was divided into four orders (Polypodichnites, Tetrapodichnites, Dipodichnites, and Apodichnites), based on the number, or absence, of foot impressions. The existence of bipedal reptiles was by then recognized; the Dipodichnites was thus divided into two suborders (Sauroidichnites and Ornithoidichnites). Not until 1845, however, did Hitchcock commence to apply familial and generic names.

The question of footprint classification continued to engage Hitchcock's attention, and he was not averse to altering or abandoning names as his views changed. In 1865, when his final classification was published, very few relics remained of his earlier ideas; footprints were by then placed in the Lithichnozoa, and the seven groups he recognized from the Connecticut Valley were designated by Roman numbers (I, marsupialoid animals; II, pachydactylous, or thick-toed birds; III, leptodactylous, or narrow-toed birds; IV, ornithoid lizards or batrachians; V, lizards; VI, batrachians; and VII, chelonians). Perhaps because of his instability in approach, the classifications proposed by Hitchcock never secured widespread acceptance, despite the reverence with which his work was viewed.

The Belgian paleontologist Louis Dollo, who studied in detail the large group of *Iguanodon* skeletons from Bernissart, showed in 1883 that some tracks from the Wealden (Lower Cretaceous) accorded with the foot skeleton of that dinosaur and therefore applied the name *Iguanodon* to the tracks themselves. Although this assignment was reasonable, even laudable, its unfortunate effect was to cause other geologists to describe as "*Iguanodon* tracks" *any* Lower Cretaceous tridactyl or tetradactyl tracks that they happened upon, regardless of the tracks' detailed form; thus occasioned were taxonomic problems that still have to be resolved. The precedent has only rarely been followed at other geological levels, in part because few generic names are so well known as *Iguanodon*, in part because affinity between tracks and skeleton can rarely be shown so convincingly.

The renowned German paleontologist W. Pabst, in contrast, accepted the principles behind Hitchcock's approach but introduced his own modifications. In a study of the German Rotliegenden (Lower Permian) published in 1900, he attributed all tracks encountered to the single genus *Ichnium* (see Table 14.1).

His system was both (1) much too cumbrous to be a convenient method for classification of tracks at one stratigraphic level and (2) too specific to be a good basis for any wider classification of vertebrate tracks. Although employed for a while in Germany and occasionally elsewhere [e.g., by Hardaker (1912) in England], Pabst's classification passed swiftly into oblivion.

Most subsequent approaches attempted to incorporate vertebrate tracks into the general system of vertebrate classification. This approach was employed by Richard Swann Lull, in his restudy of the abundant material from the Connecticut Valley (e.g., Fig. 14.14), commenced in 1904 and finished in 1953; tracks were firmly assigned to reptile classes or subclasses except where their affinity was not determinable.

His procedure was, to some extent, a compromise; within the vertebrate classes and subclasses were placed families based upon trace genera. For example, the genus *Grallator*, originally defined by Hitchcock, was made type for the family Grallatoridae, which was in turn placed into the osteologically based Infraorder Coelurosauria of the saurischian Suborder Theropoda. This approach gained many adherents (e.g., Baird, 1957) and finds its fullest expression in Haubold's (1971a) admirable review of amphibian and reptile tracks. Haubold, however, preferred to distinguish families based on trace genera as "morphofamilies."

The Hungarian paleontologist Ferenc Nopcsa, attempting a comprehensive review of all amphibian and reptile tracks (1923), used no formal hierarchical groupings but produced six categories relating closely to existing taxonomic units: (1) salamandroid and stegocephaloid, (2) lacertoid, (3) theromorphoid (dicynodontoid, theriodontoid), (4) rhynchosauroid, (5) crocodiloid, and (6) dinosauroid. Unfortunately, each grouping proved on scrutiny to be heterogeneous, bringing together footprints of

TABLE 14.1   Genus *Ichnium*.

I. Brachydactylichnia (short-toed tracks)	
1. Pachydactylichnia (heavy-toed tracks)	*Ichnium pachydactylum*
2. Brachydactylichnia (short-toed tracks)	*I. brachydactylum*
3. Anakolodactylichnia (shortest-toed tracks)	*I. anakolodactylum*
4. Sphaerodactylichnia (lump-toed tracks)	*I. sphaerodactylum*
5. Rhopalodactylichnia (club-toed tracks)	*I. rhopalodactylum*
II. Dolichodactylichnia (long-toed tracks)	
6. Akrodactylichnia (pointed-toed tracks)	*I. acrodactylum*
7. Tanydactylichnia (elongate-toed tracks)	*I. tanydactylum*
8. Dolichodactylichnia (long-toed tracks)	*I. dolichodactylum*
9. Gampsodactylichnia (curved-toed tracks)	*I. gampsodactylum*

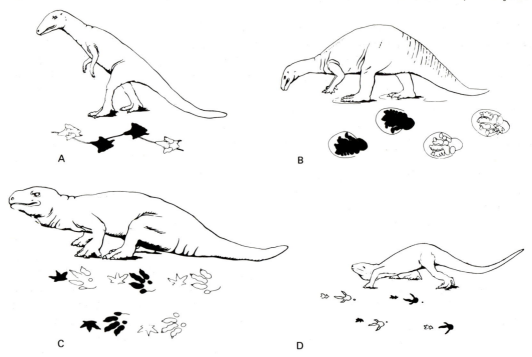

Fig. 14.14 Visualizing the trackmakers: four reconstructions by Richard S. Lull of reptiles forming tracks in Triassic of Connecticut Valley, U.S.A. A, *Hyphepus fieldi,* a small coelurosaur. B, *Otozoum moodii,* probably a prosauropod. C, *Cheirotheroides pilulatus,* considered by Lull to be a quadrupedal pseudosuchian. D, *Ammopus marshi,* a quadruped of uncertain affinity. (From Lull, 1953.)

morphologically convergent, but unrelated, types (see Abel, 1935). Despite these disadvantages, Nopcsa's approach remains current, being adopted, after critical discussion, by Lessertisseur (1955) and subsequently used by several other French paleoichnologists.

Most subsequent authors, even though they may not have consciously advocated or even formulated any classificatory philosophy, have tried to find osteological analogs for the footprints they have studied; accordingly, the result has been an increasing tendency to place vertebrate trace genera into systematic categories based upon skeletal remains.[1] In some instances, fossil tracks have even been assigned directly to families based on living genera; for example, Peabody (1959) assigned some Pliocene tracks from California to the

salamander family Ambystomidae. In other instances, especially for fossil mammalian tracks, informal names have been applied; for example, Alf (1959) listed camel tracks, antelope tracks, and carnivore tracks from the California Miocene.

Kuhn (1958), in a review of fossil reptile and amphibian tracks, placed post-Permian reptile tracks into osteologically based suborders but listed all Carboniferous and Permian tracks alphabetically by trace genus; he made no attempt to assign amphibian tracks to systematic categories. In his bibliography (1963), the pre-Triassic tracks were again listed alphabetically, but an attempt was made at a systematic assignment for all later tracks. This cautious approach epitomizes that of most recent workers, who wish to express systematic relationships as accurately as possible but who are reluctant to assign trace genera to osteologically based families; the absurdity is manifest in giving equal status within

---

[1] The proposal by Faul (1951) being a conspicuous exception.

a family to a vertebrate genus and a trace genus—based on the sedimentary impression of the foot of one of its members.

Not all paleoichnologists have adopted this approach, however. The most vigorous proponent of alternative schemes is a Soviet writer, O. S. Vialov. In 1966 he proposed a hierarchy comparable in many respects to that advocated by Hitchcock (1841–1844), embracing all tracks and trails—the Zooichnia (Vivichnia), divided into Invertebratichnia and Vertebratichnia. The vertebrate traces were subdivided as follows:

Amphibipedia
    Order Labyrinthopedida
    Order Caudipedida
        Suborder Salamandripedoidei
Reptilipedia
    Superorder Theromorphipedii
    Order Therapsidipedida
    Superorder Cotylosauripedii
    Order Procolophonipedida
    Superorder Chelonomorphipedii
    Order Testudipedida
    Superorder Lepidosauripedii
    Order Rhynchocephalipedida
    Order Lacertipedida
    Order Sauropterygipedida
    Order Pterosauripedida
    Order Saurischipedida
        Suborder Coelurosauripedoidei
    Order Ornithischipedida
        Suborder Ornithopedoidei
    Order Thecodontipedida
        Suborder Pseudosuchipedoidei
        Suborder Parasuchipedoidei
Mammalipedia
    Order Carnivoripedida
    Order Perissodactipedida
    Order Artiodactipedida
        Suborder Pecoripedoidei
Avipedia

Vialov's approach is entirely logical, and the classification, designed to stand alongside (and yet be independent from) that based on living animals and body fossils, is conceptually excellent. The new generic names he proposed are based on morphological comparability (e.g., *Avipeda, Hippipeda, Gazellipeda*) and yet are not so precise as to occasion taxonomic embarrassment. This classification has much

to recommend it as a means for the coherent organization and comprehension of data.

More recently, Vialov (1972) offered a system of nomenclature, for both invertebrate and vertebrate traces, based on function; here he distinguished between Vivichnia (direct traces of the organism's body, such as tracks and trails) and Vivisignia (indications of physiological functions), noting that, in the vertebrates, distinction between these taxa is easy. Traces of the latter type are distinguished as Vertebratisignia. They include Veterovata (eggs), reptile eggs (Reptiliava) being distinguished from those of birds (Aviava); egg capsules of fishes (Pisciava) and fossil embryos (Vetembrionae) are also distinguished. Other proposed categories include Vertebratocopria (vertebrate coprolites), Gastrolithia (gastric stones), various types of injuries—Corruptisignia, subdivided into Dentisignii (bites), Fractisignii (fractures), and Vulneratisignii (wounds), and categories for the various modes of death—including even Voratimortia (death from gluttony!) (See Fig. 13.8A).

This last classification hitherto has not been regarded with any enthusiasm by paleoichnologists and appears unlikely to prove helpful in the organization of data; but the overelaboration and other weaknesses evident in Vialov's latest scheme should not obscure the validity and good sense of his earlier approach (1966, 1968).

At present, however, one must recognize that no single system for the classification of vertebrate footprints has secured general acceptance. Although several arguments have been advanced for the assimilation of trace genera into the general system of vertebrate classification, the fact that footprints and impressions are, in the last analysis, nothing more than sedimentary structures is surely a major reason for *not* classing them alongside body fossils in any unified classification. Whether the level of separation between trace genera and animal genera should be at familial level only, the "morphofamilies" and ordinary families placed together into vertebrate infraorders and suborders, or whether an entirely separate classificatory hierarchy for vertebrate traces should be maintained, can only be a matter of opinion.

Evidence for Unknown Animal Species

Because of the much greater potential abundance of vertebrate footprints compared to body fossils, the probability is high that skeletons of the trackmakers may be unknown when their footprints are discovered. Footprints are thus an extremely valuable supplement to the osteological fossil record and may be used in reconstructing evolutionary lineages (e.g., Haubold, 1969a). Space permits citation of only two examples:

The Elusive Chirotherium

The most famous instance of representation of an unknown animal by fossil footprints is that of *Chirotherium*. These specimens were first found as large prints about the size of a man's hand and very like it, having four forward-directed digits comparable to fingers and a markedly opposed "thumb"; but the traces reflected distinct claws instead of nails. Often the specimens were found alone, in single blocks or in trackways; less frequently, they were found associated with a much smaller but essentially similar print. Workers recognized from the outset that the large print was that of the pes, the smaller print that of the manus.

The prints were first recorded from the Triassic red sandstones of Hildburghausen in Thüringia, Germany, in a published letter from F. K. L. Sickler (1834) to the eminent anatomist Blumenbach. Sickler did not attempt to assess their affinity, but several other paleontologists, excited by the discovery, soon offered their opinions; indeed, nine papers discussing the tracks were published during 1835. Voigt (1835) initially considered them to be tracks of a giant ape, but an ape was certainly unlikely in the desert Triassic environments;

thus he (1836) quickly revised his views and decided, on the basis of a track not exhibiting the thumb, that these were footprints of a cave bear! More reasonable hypotheses were advanced by the great scientist Alexander von Humboldt (1835) and by a French geologist, Link (1835), who respectively considered the tracks to be those of a kangaroo-like marsupial and of a toad. (The bipedal character of some diapsid reptiles remained unknown at this time.) The tracks were christened by another German, J. F. Kaup (1835), who guardedly proposed the alternative names *Chirotherium* or *Chirosaurus*, "hand beast" or "hand reptile"; unfortunately, the former and less appropriate generic name has taxonomic priority, although the latter name also continued in intermittent use for the remainder of the century. The spelling was shown to be incorrect, the correct Greek transliteration being *Cheirotherium*; both spellings continue in use even today, although the earlier spelling is taxonomically correct.

Similar tracks were reported from equivalent horizons in the English Triassic, from quarries at Storeton, near Liverpool, and elsewhere (Fig. 14.15). Within the ensuing century, reports of these tracks came from many other parts of Europe, many more localities in Germany, and eventually from the United States and South America. Some 40 species (not all of them valid) have been proposed on the bases of variations in size and proportions of the footprints.

Interpretation of the tracks continued to cause headaches. The anatomist Owen (1842) offered the hypothesis that these were labyrinthodont tracks; but the opposed digit, actually digit V, continued to be misinterpreted as a thumb, and reconstructions were published showing labyrinthodonts, depicted as toad-like animals, laboriously walking cross-legged! (Fig. 14.16A). In 1874, however, the zoologist L. C. Miall, in an address to the British Association, showed this attribution to be

Fig. 14.15 *Chirotherium storetonense* manus (upper) and pes, showing skin patterns. (Photo: H. C. Beasley, ca. 1906; now in Beasley Coll., Liverpool Geol. Soc.) *Note*: Haubold (1971b) considered *C. storetonense* to be a junior synonym of the type species, *C. barthii*.

fallacious, both on morphological and stratigraphic grounds, and suggested instead that these were dinosaur footprints. (An alternative suggestion, by the Manchester botanist W. C. Williamson in 1867, that the footprints· were those of crocodiles, curiously attracted little attention.) Although Owen's authority was such that many geologists continued to speak of "hand-footed labyrinthodonts," workers gradually came to accept that these were indeed reptile footprints. Acceptance came easier after Watson (1914) stressed the relative narrowness of the trackway; an amphibian trackway would unquestionably have been broad. Moreover, no fossil evidence suggested that amphibians were ever bipedal.

The most profound study of *Chirotherium* footprints was made by another German, Wolfgang Soergel, in 1925. He demonstrated beyond doubt that these were reptile tracks, on the basis of the scale patterns discernible on some well-preserved prints (cf. Fig. 14.15). By comparison with the foot skeleton of the South African Triassic reptile *Euparkeria*, he also showed that the supposed "thumb" was digit V; however, a reconstruction of the probable footprint of *Euparkeria*, made from the skeleton, ruled out that reptile as a possible trackmaker. The depth and pattern of tracks assigned to *C. barthii*, the type species, suggested that the trackmaker was basically a quadruped, carrying the principal weight on its hind limbs; to achieve the correct balance, a large heavy tail, carried clear of the ground, had to be postulated. The presence of strong claws suggested a carnivore, or at least a habit in part carnivorous. The body length was computed as roughly 3 ft (1 m); the tail was considered probably of similar length; with due allowance for head and neck, a total length of 8 ft (2.5 m) seemed reasonable. Other species were of markedly dissimilar dimensions; the smallest is perhaps *C. sickleri*, estimated at only 14 inches (35 cm). All in all, Soergel concluded that the trackmakers were Pseudosuchian reptiles—ancestral to both dinosaurs and crocodiles—having gaits varying from quadrupedal to

Fig. 14.16 Possible *Chirotherium* trackmakers. A, "hand-footed labyrinthodont" visualized by Sir Richard Owen. (From Lyell, 1855.) B, reconstruction by Soergel (1925). C, *Ticinosuchus ferox*, as reconstructed from skeleton in Triassic of Switzerland. (From Krebs, 1965.)

bipedal (Fig. 14.16B); but he was unable to link them to any animal yet known from skeletal remains, and indeed, no members of that reptile group then known were of a length even approaching 8 ft (2.5 m).

This size difficulty did not long persist; 10 years later (in 1935), the German verte-brate paleontologist Friedrich von Huene discovered in Brazil the remains of a pseudo-suchian, *Prestosuchus*, which not only had a foot structure comparable to that estab-lished for *Chirotherium* but was actually much bigger, fully 15 ft 6 in (4.7 m) long. The affinity presumed for *Chirotherium* was thus perfectly possible. Both Peabody (1948), working on tracks from Arizona and

Utah, and Swinton (1960), studying British tracks, arrived at the same conclusions as Soergel regarding the probable morphology and affinity of *Chirotherium*; but still no skeletal remains were found that accorded exactly with the hypothetical trackmaker.

However, in 1965 the discovery of an appropriate "candidate" was at last re-ported by a Swiss paleontologist, Bernard Krebs. The skeletal remains, found in the Swiss Middle Triassic and named *Ticino-suchus*, were those of a slim-bodied Pseudo-suchian, about 8 ft (2.5 m) long. The forelimbs were only about half as long as the hindlimbs and the pes much larger than the manus. In all four feet, the fifth digit was opposed; all digits bore strong claws. The teeth confirmed a carnivorous habit. The tail was long, and the body balanced on the pelvic girdle. In all particulars (Krebs, 1966), *Ticinosuchus* qualifies as trackmaker for footprints of the type species, *Chirotherium barthii* (Fig. 14.16C). To account for all the other forms, however, workers still necessarily presume the existence of yet undescribed pseudosuchians, some dominantly quadrupedal in gait, some dominantly or exclusively bipedal (cf. Fig. 14.14C). Haubold (1971a), although he considered many existing species to be synonyms, still distinguished at least 23 species of *Chirotherium*; the fossil evidence from footprints for animal species of this reptile group thus remains overwhelmingly more ample than that from osteological fossils.

The Ancestry of Salamanders

The amphibian order Urodela (newts and salamanders) is first represented in the fossil record in Upper Cretaceous strata. Probable direct ancestors of the order are a group of lepospondyls, inappropriately named Microsauria; the latest occurrence of these in the fossil record is in Upper Permian rocks. Thus, a major stratigraphic and evolutionary hiatus separates the salamanders from their presumed ancestors.

This gap is in part bridged by trace fossils. Footprints of a four-toed quadrupedal animal of appropriate size, having digits II and III longest, the outer digits shorter and lacking claws, occur in the Triassic of Tuscany, Italy; placed in the monotypic genus *Cryptobranchichnus*, they were considered by von Huene (1941), their discoverer, to be amphibian footprints comparable to those of living Urodeles. Similarly, in the upper Bunter Sandstone of Germany, footprints placed by Kuhn (1958) in the monotypic genus *Ruecklinichnium* also appear to be those of small amphibians, comparable with salamanders; they exhibit three digits linked by webs

and were imprinted by an animal about 15 cm long. No footprints of salamander-like animals have yet been reported, however, either from the Jurassic or from the Lower Cretaceous.

Evidence for Vertebrate Distribution

In virtually every instance in which a footprint ichnofauna has been described, it immensely supplements our knowledge of the distribution of vertebrates. For example, my own studies (1967, 1969) of fossil footprints from the Triassic of Nottinghamshire and Derbyshire, England, demonstrate the existence in this region of a fauna including two species of small carnivorous dinosaurs (coelurosaurs), possibly an early bipedal plant-eating dinosaur (ornithopod), a pseudosuchian of the *Chirotherium* group, two species of lepidosaurs, and a quadruped of undetermined nature (perhaps also a pseudosuchian?); no osteological remains have yet been found in the Triassic of either county. The footprint fauna reported by Gilmore (1926–1928) and others from the Grand Canyon, Arizona, is immensely more varied than that known from body fossils. Fossil evidence from the Permian of Texas shows cursorial bipedal reptiles to have been present, which can only have been Younginiformes—a group having a definite osteological record only from South Africa (Sarjeant, 1971). The recent discovery of tracks of the large herbivorous dinosaur *Iguanodon* on the cliffs of Spitzbergen (Colbert, 1964) is the first evidence for dinosaurs in the present Arctic and poses interesting paleogeographic problems.

Such instances of the value of vertebrate footprints as evidence for animal distribution could be multiplied almost indefinitely. The preparation of faunal distribution maps, based on footprint evidence, would be an invaluable addition to paleontological, environmental, and paleogeographical knowledge.

Evidence for Vertebrate Behavior

As in the documentation of unknown animal species, the value of footprints in determining animal behavior is best illustrated by selection of several examples.

Sauropod Mobility on Land

When the giant sauropod dinosaurs were first discovered and described, and their immense bulk appreciated, workers felt that——despite the immense strength of these skeletal structures—the animals must have been habitually aquatic, relying on water buoyancy to compensate much of their weight. Many paleontologists considered that, if the dinosaurs did indeed move on land, they must have had a crawling gait like that of lizards and crocodiles, their elbows and knees well splayed out, the belly close to the ground. Other workers seriously questioned the ability of these animals to move on land at all.

A series of trackways discovered by Bird (1944, 1954) at Paluxy Creek and West Verde Creek, Texas, gave a clear solution to these problems. The footprints were undoubtedly those of sauropods. Bird considered them to be those of brontosaurs (*Apatosaurus*), although the age of the rocks (Aptian–Albian) makes this identification questionable. The trackway was very narrow and clearly that of an efficient walker, progressing upright and taking steps fully 6 ft (ca. 2 m) long. Such footprints could have been formed only on inshore mudflats, either covered by shallow water—retreat of which would allow hardening of the mud—or lying entirely exposed. At Paluxy Creek, no sign of tail drag was found; the water cover may have been sufficient to float the tail, or the tail may have been simply carried clear of the ground. At West Verde Creek, however, a continuous deep groove associated with the footprints seems to represent tail drag and indicates that the mudflats were indeed fully exposed. Clearly the sauropods *were* capable of walking on land, unsupported by water buoyancy. (See also Chapter 2.)

A touch of drama was added to one of the trackways from Paluxy Creek by the superposition of footprints of a large carnivorous dinosaur, which may well have been in pursuit of a possible victim. (Chapter 13.) This trackway is now displayed in the American Museum of Natural History; the skeleton of an *Apatosaurus* is placed into the sauropod tracks, with the skeleton of a carnivorous dinosaur (*Allosaurus*)—comparable to the presumed pursuer—alongside (Fig. 14.17). [These Texas trackways have also stimulated considerable interest among local residents, sometimes to the detriment of science (Fig. 14.18).]

Recently, Bakker (1971), on the basis of detailed studies of the skeleton (in particular the structure of the feet and limbs), suggested that the sauropods were not aquatic at all. Characteristics of the sediments in which their remains are found, and also of the associated fauna, support his contention. The wheel has thus gone full circle!

Soft–Part Morphology of the Vertebrate Foot

Indications of skin and scale pattern are frequent in imprints preserved in fine-grained sediments, as mentioned previously for *Chirotherium* (Fig. 14.15). The presence of tissue pads is also reflected frequently in footprints of the larger vertebrates. For example, Langston (1960) described a footprint of a plant-eating dinosaur (hadrosaur) that indicates three digital pads and a massive tissue pad beneath the metatarsal bundle (Fig. 14.19).

Reports of the existence of webbing between digits in fossil footprints are often based on misinterpretations of the ridges produced by mud displacement. Nonetheless, several instances of the preservation of webbing are known. Three digits in the footprint of a coelurosaur (*Swinnertonichnus*) from the Triassic of Nottinghamshire,

Fig. 14.17 Trackways from Paluxy Creek, Texas, on display in American Museum of Natural History. Skeleton of brontosaur (*Apatosaurus*) is placed in its tracks. Tridactyl prints of the pursuing carnivorous dinosaur are superposed (a skeleton of *Allosaurus*, for comparison, is seen at right). (Photo: R. F. Sisson, Nat. Geogr. Soc.)

England (Fig. 14.20), are linked by a delicate line that can only have been produced by webbing, suggesting that the trackmaker was semiaquatic (Sarjeant, 1967). More broadly spreading tridactyl reptile footprints (*Talmontopus*), also thought to be webbed, were reported from the Lower Jurassic of France (Lapparent and Montenat, 1967).

Footprints of birds, showing partly or

Fig. 14.18 Dinosaur footprints in commerce! Slabs on sale near Paluxy Creek, Texas. The piecemeal extraction of footprints for sale to tourists has resulted in the destruction of many potentially important trackways before scientific studies could be made. (From Bird, 1954.)

completely developed webs, considered to be those of early members of the Order Charadriiformes (wading birds and gulls),

were reported from the Eocene of Utah (Curry, 1957; Moussa, 1968). Erickson (1967) described from these beds webbed tracks having "dabble" patterns, which he attributed to the Anseriformes (ducks and geese). (Fig. 14.21.)

Running Behavior of Reptiles

A well-known fact is that many present-day longtailed lizards become temporarily bipedal in gait while running at great speed. I have several times seen whiptails in Arizona rise up onto their hind feet in this fashion. Also well known is that reptiles, like mammals, rest only on their digits—in particular, on the central digits—while running; a walking track having clear palm prints (plantigrade) is thus transformed to one in which only the front of the palm is impressed (semidigitigrade) and ultimately into a track showing only toeprints (digitigrade). However, all three conditions are rarely manifested in a single fossil track;

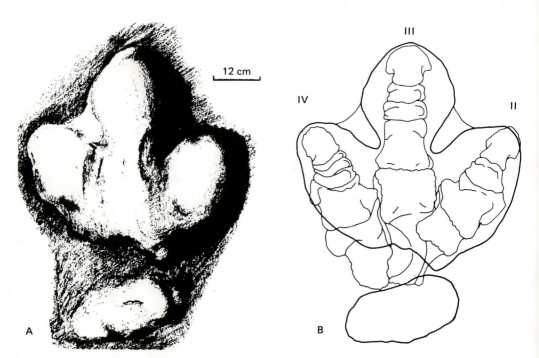

Fig. 14.19 Hadrosaur footprint (A), with right pedal bones of the hadrosaur *Hypacrosaurus* superimposed (B), indicating the presence of sole and phalangeal pads. Upper Cretaceous; Alberta, Canada. (From Langston, 1960.)

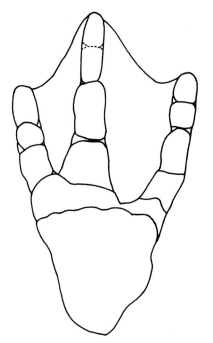

Fig. 14.20 A webbed tridactyl print, *Swinner-tonichnus mapperleyensis.* Keuper Waterstones (Triassic); Nottingham, England.

to identify the makers of digitigrade imprints with any degree of confidence is not usually possible.

Tracks of such variable character add to the problems of footprint classification: should they be placed together under a single generic name or, following the precedent set for trilobites (different generic names are given to burrows, furrows, and walking and swimming traces, even where clearly made by the same individuals—see Chapters 6 and 7), does the evidence of different behavior merit different generic designations?

An unexpected feature of certain early tracks of bipedal reptiles is that they anticipate a trend familiar in the horse lineage—extreme reduction in the number of digits functional in running. These tracks, described from Permian sandstones of Castle Peak, Texas (Moodie, 1930), show the imprint of only two digits; one (probably digit IV) is much larger and more deeply impressed than the other (probably digit III). The tracks (Fig. 14.22) are now

distinguished as a separate genus, *Moodie-ichnus* (Sarjeant, 1971). I stress that these are tracks of a pes which may well have had four or five toes, all of which might have been impressed in a walking gait (when the reptile may have been quadrupedal). In running, however, the reptile certainly became bipedal, and the pes was functionally didactyl; principal stress focused on one central digit, the adjacent digit perhaps aiding in balance and thus being impressed also. The earliest reptiles apparently capable of adopting a bipedal gait—the Younginiformes—occur in the Permian. Not only remarkable is the fact that such a running gait was developed so early in vertebrate history, but also that it was so quickly abandoned; didactyl reptile tracks are unknown from Triassic or later sediments.

Pterodactyls on the Ground

Most pterodactyl remains were found in marine or lagoonal sediments, in some of which (such as the Solnhofen lithographic limestones of south Germany) impressions of wing membranes are clearly evident. The position and structure of the wing membranes caused many paleontologists to question whether pterodactyls were capable of walking at all on relatively flat surfaces; some paleontologists instead preferred to visualize them at rest, hanging bat-like from rocks or trees, clinging to the face of cliffs, or at best, being capable only of laborious clambering on more or less vertical surfaces (see illustrations in Augusta and Buryan, 1961).

In 1952, pterodactyl footprints were discovered by William L. Stokes in the Morrison Formation (Jurassic) near Four Corners, Arizona. The tracks were impressed into sandstones showing current lineations and faint current bedding; the environment was interpreted as a sandbar lying parallel to a flowing stream, in a semiarid alluvial plain. The degree of indentation of the tracks and unsteadiness

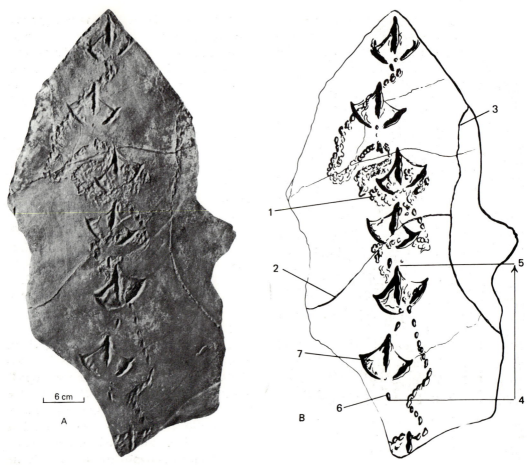

Fig. 14.21 Bird tracks (A) and interpretative drawing (B), showing interrupted dabble pattern (1), mud cracks (2, 3), stride (4, 5), impression of hallux only (6), and webbing on left pes (7). Eocene; Utah, U.S.A. (From Erickson, 1967.)

of the gait indicated that the sands were moist. The manus was small and tridactyl; the outer digits were short and bore blunt claws, the central one (digit III) very much larger and supporting the wing membrane, evidenced by a short trailing impression. The pes was large, having a broad sole, narrow heel, and short digits (Fig. 14.23). Poorly formed grooves in the trackway may represent the external edges of wings. The combination of short digits on all four feet and an impression trailing from the manus identified the trackmaker; the tracks were placed in the monotypic genus *Pteraichnus* (Stokes, 1957).

The tracks indicate that at least one species of pterodactyl could walk fairly freely on the ground. Although the pes is directed slightly outward, the tracks show no awkwardness of gait, the pes overstepping the manus. Walking seems to have been made possible by the bending backward of the third, wing-supporting digit; a deep central depression seems to correspond to the joint between the heavy metacarpal and the first wing-bearing phalange. The digits of the pes are turned downward (inward); they were clearly flexible and suited to climbing or clinging. This track is also of interest from another viewpoint: it provides only the second record of pterodactyls in the Jurassic of the Western Hemisphere (one incomplete skeleton is known, also from the Morrison Formation).

Fig. 14.22 Didactyl footprint, *Moodieichnus didactylus*, showing one long, deeply impressed digit and a shorter, divergent digit. Permian; Texas, U.S.A.

Dinosaurs in Herds or Packs

With the discovery of accumulated skeletons of *Iguanodon* in a fossil limestone ravine at Bernissart, Belgium, in 1878 (see Casier, 1960), workers have reason to suppose that herbivorous dinosaurs may have lived in herds. Similarly, one interpretation of the puzzling accumulation of skeletons of the coelurosaur *Coelophysis* at Ghost Ranch, New Mexico, discovered in 1947 (see Whitaker and Meyers, 1965), was that these carnivorous dinosaurs may have hunted in packs. Direct evidence for gregarious behavior cannot, however, be derived from osteological remains.

The first direct suggestion from footprint studies that dinosaurs were gregarious was made by Bird (1944, 1954), who re-garded the abundance of overlapping track-ways of sauropods (23 of them) in Paluxy Creek and West Verde Creek, Texas, and the variable size of the individual track-makers, as an indication that they lived in herds. The first full analysis of the evidence provided by fossil trackways was made, however, by Ostrom (1972), who brought together several records. Most notable are his records of (1) the Triassic of Mount Tom, Massachusetts, where 19 parallel tracks referable to the genus *Eubrontes* (considered to be footprints of a carnivorous dinosaur) were found, all of them going in the same direction (Fig. 14.24), and (2) specimens from Lake Ennes, Texas, comprising 25 northwest-trending trails of a large ornithopod (possibly *Camptosaurus* or *Tenontosaurus*).

Thus, both herbivorous and carnivorous dinosaurs seemingly were gregarious. Bakker (1968) even suggested, on the basis of Bird's observations, that the sauropods traveled in a "structured" herd, the juveniles at the center and the adults around the periphery as protection; but further evidence is needed on this point.

Did some dinosaurs live in herds all the time, or did they band together only during climatically induced migrations? As long ago as 1928, Friedrich von Huene, using evidence both from footprints and from the distribution of skeletons in the Triassic of Germany, suggested that the prosauropod *Plateosaurus* undertook annual migrations induced by summer droughts. One can easily imagine plant-eating dinosaurs living in herds all the time, or coming together only during migrations. Should a statistical analysis of trackway patterns in an area reveal that the dominant paths were along the same compass line, but in two opposite directions, a strong case could certainly be made for migration. But flesh-eating dinosaurs, at least some species, perhaps more likely hunted in packs all the time; carnivorous animals that were normally solitary would be unlikely to associate in migratory groups.

Hibernation of Cave Bears

Tracks comparable to those of living bears, but much larger, occur on the floors of many European caverns; the footprints are frequently associated with bones of the cave

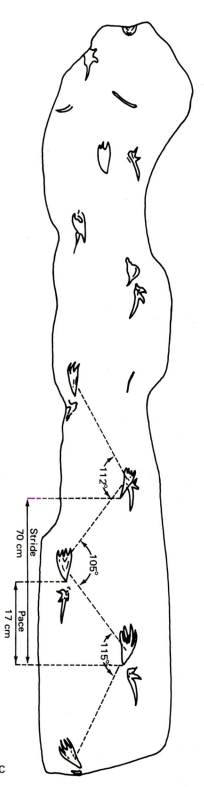

Fig. 14.23 Pterodactyl tracks, *Pteraichnus saltwashensis*. A, three slabs showing four consecutive sets of footprints. B, closeup of best pair of tracks. C, interpretative drawing, with measurements. Jurassic; Arizona, U.S.A. (From Stokes, 1957.)

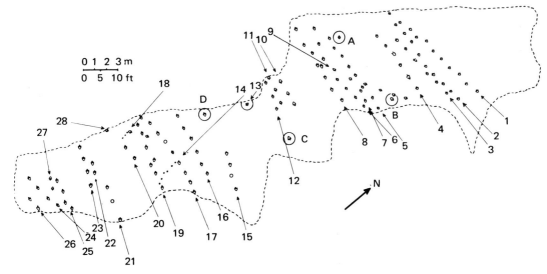

Fig. 14.24 Tracks of ?gregarious dinosaurs. Tracks 1–12, 15–26 are those of carnivorous dinosaurs (*Eubrontes* sp.); the solitary prints A–D (circled) may represent continuations of these tracks, but the isolated footprint (13) cannot be so interpreted. A smaller track of dissimilar character (14) is also noted. Of 26 tracks represented, 19 are more or less parallel; arrows indicate trackway directions. Triassic; Mount Tom, Massachusetts, U.S.A. (From Ostrom, 1972.)

bear (*Ursus spelaeus*) and can confidently be attributed to that animal. In some caverns, the tracks are associated with vertical and horizontal scratch marks of claws, found to a considerable height on the cavern walls. Even polished walls or crusts of hardened clay may be present, indicating that the bears rubbed against the walls while scratching or cleaning themselves (Bachofen-Echt, 1931). These details suggest prolonged residence and have been interpreted as indicating hibernation (Lessertisseur, 1955, p. 88); the general absence of gnawed bones or other food remnants tends to suggest a hibernaculum rather than a lair.

In this instance, although the evidence for hibernation is derived from other features, the presence of footprints shows that the cave bears actively inhabited the caverns and thus forms an essential prelude to the interpretation of that evidence.

Environmental Interpretation

Conditions for Preservation

The presence of vertebrate footprints generally indicates that sediments accumu-

lated in a terrestrial environment. By indicating the character of the fauna and providing evidence for the way of life of particular animals, the tracks allow reasoned deductions to be made concerning the general environment (see previous sections). Footprint preservation shows that the sediment must have dried out shortly after they were imprinted, although this desiccation may not have happened immediately. Beasley (1904) observed imperfect *Chirotherium* prints in the Triassic of Cheshire, England, having ripple marks superposed on them, indicating that "a thin layer of water was still present over the mud, just sufficient to form the ripples represented" (p. 222). A footprint slab that I illustrated from the Triassic of Worcestershire shows washing out of tracks at one side (Fig. 14.1), indicating onlap of waters —presumably wind-driven waves—before drying-out finally occurred (Wills and Sarjeant, 1970).

Tracks that Go Uphill

An excellent example of how vertebrate footprints can aid in environmental interpretation is furnished by footprints in the

Coconino Sandstone (upper Paleozoic) of the Grand Canyon region, Arizona. In his original description of these tracks, Gilmore (1926) noted that they almost always led uphill, having been imprinted into sand surfaces deposited at angles of about 30°. This configuration was observed in all but three of hundreds of trackways studied; none of the tracks led downhill.

Because the environment of deposition of the Coconino Sandstone was not clearly understood, the footprints proved of critical significance. Matthes (in McKee, 1934) noted the sharp definition, even margins, and generally excellent preservation of the tracks, and reasoned that they could not have been formed in loose sand or under water:

> If it is suggested that the tracks were made at times of low tide, when the front of the delta was partly above water, then the question arises why the tracks were not wiped out at the next high tide. It is much easier to account for the preservation of animal tracks in dune sand. True, they are readily destroyed when the sand is dry, but the chances of their preservation with sharp definition are particularly good if they are made after a shower, when the sand is damp enough to possess fair cohesion, and then are buried by dry sand blown over them.

The lack of downhill tracks, assuming Matthes' deduction of conditions to be correct, was accounted for by Reiche (1938) from observations of tracks on present-day dunes:

> On the lee slopes in the dune field near Laguna, New Mexico, recognizable tracks of coyotes, rabbits and lizards formed after the rain. Without exception they led up the slopes, those directed downwards having been obscured by sliding of surface layers of sand.

He considered the sliding of sand to result from the inertia of the animal, transmitted to the sand by efforts to maintain its balance and control its speed on the poor footing of the steep sands of the windward slope.

A series of experiments conducted with living lizards by McKee (1944) in part contradicted Matthes' ideas and in part supported them. He found that tracks similar to the fossils were developed neither in wet sands nor in dried sands having a thin crust. Instead, he concluded that the fossil tracks must have been formed in *dry* sand, to account for the depth of implantation of tracks of relatively lightweight animals, and must have been preserved by subsequent dampening by mist or dew, which would form a crust on the surface without destroying any tracks. His experiments endorsed Reiche's opinion that slumping would destroy downhill tracks.

These results indicate that the Coconino environment was one of sand dunes, sufficiently mobile to rapidly cover and preserve any wettened surface. Climatic conditions, at least at times, must have been moist enough to permit the formation of dew or mist; and the direction of wind onset must have been relatively constant, considering the development of dunes having such consistent form.

Chronostratigraphic Indices

Recognition that vertebrate footprints might provide a useful means for stratigraphical correlation of terrestrial sediments has been slow; the great majority of published studies emphasize the trace's paleontological interest, and even when comparisons are made with prior studies of similar forms, little comment is made on their possible stratigraphic value. However, several exceptions to this general rule are known; for example, Hickling (1909) suggested a Permian date for the problematic red sandstones of Dumfries-shire, Scotland, based on comparisons of the vertebrate ichnofauna with those of localities elsewhere in Scotland and England. A great handicap to stratigraphic work has been the obscurity of much of the literature and the inadequacy of descriptions and illustrations (see previous discussion).

Within the last five years, however, a resurgence of interest in the chronostratigraphic potential of vertebrate footprints has occurred, as a result of initially independent work by Georges Demathieu and others in France and of Hartmut Haubold in Germany (see in particular Courel et al., 1968; Haubold, 1969b, 1971b). Recently, Demathieu and Haubold (1972) outlined their conclusions concerning the trace fossil biostratigraphy of the European Triassic (Fig. 14.25) and showed that this scheme can be used to obtain correlations not only within Europe but also among localities as far away as the southern United States and Lesotho. Their work may be expected to form a basis for further and more detailed chronostratigraphic studies of vertebrate ichnofossils.

OTHER VERTEBRATE TRACES AND IMPRESSIONS

Fish Impressions

The mode of life of most fishes is such that these animals are unlikely to leave traces of any kind (other than coprolites) in sediments. Only benthic species, or species living close to the bottom, are at all likely to leave fossilizable traces. (See Chapters 15 and 19.)

Häntzschel (1935) noted that a small goby (*Gobias minutus*), which inhabited the marine strandline, formed furrows in the sand surface. In particular, after laying its eggs onto the inner side of its "nest" (formed by a shell of the bivalve *Mya*), it would cover the nest with sand by several

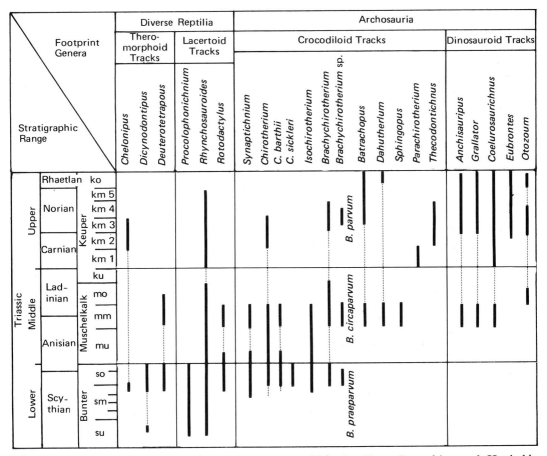

Fig. 14.25 Footprint biostratigraphy of the European Triassic. (From Demathieu and Haubold, 1972.)

times swimming radially in the direction of the "nest," sand being shoveled inward by the movement of the fish's pectoral fins and tail. The result is distinctive star-like traces in the sand; these have not yet been reported in fossil form.

Resting traces of various fish occur in sandy sediments of the nearshore zone and in sandflats exposed at low tide. Flatfish (Order Pleuronectiformes) may form oval impressions; skates and rays (Order Batoidea) may form impressions of roughly triangular or trapezoidal outline. More often, however, the fishes form rounded depressions, 15 to 100 cm in diameter, having concave bottoms 2 to 5 cm deep, which give no exact impression of the animals that made them; such traces are frequently associated with oscillation ripple marks. The traces were first reported by Lessertisseur (1955); later, Cook (1971) independently recorded such impressions, noting that they were made by the California halibut (*Paralichthys californicus*), the bat ray (*Myliobates californicus*), and the skate (*Raja erinacea*). At somewhat greater depths, Lessertisseur also noted the occurrence of trough-like impressions formed by weaverfish (*Trachinus*) and of unstable burrows formed by sand eels (*Ammodytes*). Few of these types of fish impressions have yet been found as fossils (see Chapter 15).

Schäfer (1972, p. 398) listed 38 species of present-day North Sea fish that frequently rest on or in loose sediments; these include, in addition to the orders named above, representatives of the Anguilliformes (*Anguilla anguilla, Conger conger*), Gadiformes (*Zoarces viviparus*), Perciformes (*Trachinus draco, Callionymus lyra, Anarchiches lupas*, etc.), Synbranchiformes (*Trigla corax, T. gurnardus, Agonus cataphractus*) and Lophiiformes (*Lophius piscatorius*). In supplementing Lessertisseur's observation, Schäfer noted that the weaverfish use their anal fins to burrow into sand and that their trough-like burrows may exhibit transverse grooves; the fish lie more or less horizontally in these

burrows, their breathing apertures uncovered. Sand eels and eels (*Anguilla*, etc.), which excavate burrows by undulations of their body, lie obliquely in the sediment—their body hidden but their mouth and gill slits exposed; after collapse, their burrows are represented by narrow sag zones in the sediment. In general, indeed, resting fishes tend to destroy the original sediment texture and do not leave structures of well-defined shape; perhaps for this reason, resting traces have not hitherto been reported in the fossil state.

Locomotory traces of fish are known to be formed in present-day sediments (Fig. 14.26). Swimming traces of three different types were noted by Schäfer (1972, p. 259–263), in his study of North Sea phenomena:

1. Swimming by horizontal undulation of an elongate body (e.g., the eels) produces a continuous, undulating trace when the whole body is in contact with the sediment and a series of curves and semicircles where only a part of the body touches bottom.
2. Swimming by vertical undulation of a flattened body (e.g., the flatfishes) produces —when the fish swims parallel to the substrate—a rhythmic series of impressions of the anterior part of the fin-fringe; arcuate impressions cross one another, and some have a central, elongate–arcuate impression resulting from the downward flip of the tailfin.
3. Swimming by the beating of enlarged pectoral fins sometimes produces motion in a series of rather clumsy jumps (e.g., the bullhead, *Agonus cataphractus*, and the gobies, *Gobius* spp.). The beats of the pectoral fins then produce a series of isolated, crescentic impressions—the horns pointing in the direction of movement; these may be associated with similarly isolated lens-shaped tail traces aligned with the direction of movement. Where movement is a continuous glide (e.g. *Callionymus lyra*), the sediment is scarcely disturbed.

Some fish, in particular the gurnards (Cottidae), "walk" along the sea floor, the three ventral rays of the pectoral fins functioning as limbs and the tail fin dragging to produce a continuous trail. The fossilization of traces of this type was first

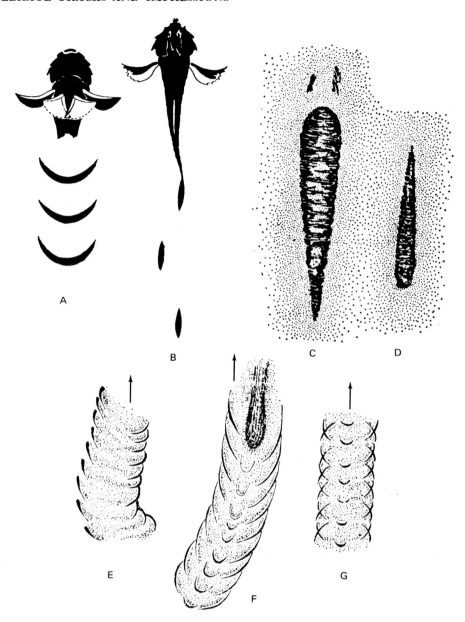

Fig. 14.26 Fish impressions. A, B, swimming traces of the bullhead *Agonus cataphractus*. A, beating of the enlarged pectoral fins produces crescentic impressions in sediment. B, lens-shaped trace produced by the tailfin. C, D, "burrows" excavated into sediment by the weaverfish *Trachinus draco* (C), and by *Callionymus* (D). E–G, surface traces of flatfishes. E, body oriented slightly tail-downward, so that during downward beats, the fins on dorsal and ventral body edges at rear brush the sea floor. F, body oriented more strongly tail-downward, the dorsal and ventral fins and the tailfin brushing the sea floor during downward beats. G, body horizontal, the anterior and posterior parts of dorsal and ventral fins and the tailfin successively brushing the sea floor. (After Schäfer, 1972.)

discussed by Fraipont (1915). Schäfer (1972, p. 262–263) described "walking trails" produced in the present-day North Sea. Fin traces from the Jurassic lagoon at Solnhofen, Germany (Walther, 1904), seem to be walking traces and are thought to have been produced by a coelacanth, *Holophagus penicillatus*.

Tadpole Holes

In 1858 Edward Hitchcock described some polygonal network structures from the Lockport Dolomite (Silurian) of New York and from red Triassic shales of Massachusetts. He compared these with the "nests" formed on the bottoms of ponds by tadpoles and placed them in a new genus, *Batrachioides*; the species were named *B. antiquior* and *B. nidificans*. In 1867, however, the latter group of structures were reexamined by Charles U. Shepard, who considered them to be wind-induced wave patterns and not the result of animal action. This conclusion was later given emphatic support by the German paleontologist Abel (1926, p. 53–54, Figs. 27 and 28), after reexamination of the type specimen. The character of the older specimens remained a matter for discussion. Kindle (1914) considered them to be interference ripple marks modified by the motion of tadpoles, thus in general supporting Hitchcock's concepts. Eventually the Russian ichnologist Vialov (1964) concluded that these and similar polygonal structures in the geological record were definitely *not* tadpole "nests" and that all the structures described by Hitchcock were of inorganic origin.

Tadpole holes are by no means uncommon today. The entire bottom of ponds may be dimpled by such depressions. In ponds in Quebec, Dionne (1969) noted some that were 10 to 20 mm in diameter, having bowl-shaped or concave bottoms 10 mm deep. He noted that they were occupied by resting tadpoles and formed by the initial circular motion of the tail of the tadpole when commencing to swim. Bragg (1965), in a study of spadefoot toads, concluded that such depressions were formed only when water level was low; the tadpoles gathered in the cooler waters at the bottom of evaporating pools, stopped feeding, and

moved their tails so as to form a depression. The holes so formed would effectively deepen the pond and give the tadpoles a better survival prospect.

Accepting *Batrachioides* as a name based on sedimentary structures and thus invalid, Boekschoten (1964) proposed instead the new generic name *Benjaminichnus* for modern tadpole holes. Arguably, this name was inadvertently made a junior synonym of Hitchcock's invalid "genus": unfortunately, Boekschoten phrased his statement to the effect that the new name was designed "to replace *Batrachioides*" (p. 423). On this basis, the new name was rejected by Cameron and Estes (1971). Better reasons for the rejection of *Benjaminichnus* are (1) the philosophic objection that generic names should not be proposed for sedimentary structures made by living animals and (2) the technical objection that no type specimen was designated.

Although tadpole holes or nests were at that time unknown in the fossil state, Ford and Breed (1970) pointed out cogently that those formed in drying out ponds would have a fairly high fossilization potential. Subsequently, Bhargava (1972) described some bowl-shaped structures, approximately 0.5 cm deep and from 0.7 to 1.1 cm in diameter, formed in a thin shale layer and cast by an overlying orthoquartzite, in the Upper Jurassic of Simla Himalaya, India. These structures are separated by a roughly polygonal pattern of raised ridges; they correspond well with modern tadpole nests, yet the environment (shallow marine) is not one in which amphibian larvae would normally be found, and the record must be viewed with doubt. (See also Chapters 5 and 15.)

Hoof Prints

In examining the sediments of a New Zealand beach, Van der Lingen and Andrews (1969, Figs. 1–7) discovered some structures that, at the surface, were seen as roughly circular hollows surrounding a central symmetrical or asymmetrical dome-like structure of low relief, protruding barely above general beach level. In section, these structures were found to be part of a disturbance of stratification down to about 25 cm, where the structure faded out. The lower part consisted of parallel laminae bowed down to form a smooth, concave-up stratification. The upper part showed a roughly circular, steep-sided cavity truncating the general beach stratification, infilled by homogeneous sediment or by concave-up sand laminae; at the center, the sand layers were convex-upward, forming the dome-like structure.

By direct experimentation, Van der Lingen and Andrews demonstrated that such structures were hoof prints of a horse ridden across the beach, backfilled by sand from the swash of waves. The weight of the horse and the sharpness of its hooves thus caused a disturbance of beach stratification that survived even after wave action had rendered the horse's footprints wholly unrecognizable.

Similar structures must have been produced in the past by sharp-hooved animals moving across poorly consolidated sands; but they have not yet been reported in the fossil state.

Coprolites

The fossilized excrement of animals is commonly encountered in marine and terrestrial sediments, both modern and ancient. Some groups of reptiles—the lizards and snakes, and probably some related extinct reptilian groups—produce nonliquid urinary secretions which, as Duvernoy (1844) pointed out, are perfectly capable of fossilization and should correctly be distinguished as "urolites." The majority of fossil excrement, however, consists of alimentary feces (coprolites); the latter term is now generally applied to all types of fossil excrement.

Fossil excrement may form entire layers of sediment. The accumulated droppings of seabirds occur locally in such quantity and relative purity as to form thick deposits (guano) of great commercial importance as fertilizers; the droppings of bats, which may form thick deposits (chiropterite) in some caves, are also exploited sometimes for agricultural purposes.

Most often, however, coprolites occur individually, or as isolated masses, in sediments. Their recognition is not always a straightforward matter, because their form may be closely similar to that of pellets and concretions of entirely inorganic origin. Occasionally "pre-coprolites" are found in place within the fossil remains of an animal; for example, Buckland (1836, p. 199)[2] reported that "Lord Greenock had discovered, between the laminae of a block of coal, from the neighbourhood of Edinburgh, a mass of petrified intestine distended with coprolite, and surrounded with the scales of a fish, which Professor Agassiz refers to the *Megalichthys*." Such unambiguous discoveries are rare, however, and other means of identification must normally be used. In his pioneer description of coprolites, Buckland (1829) employed two criteria —their form, which in some instances reflects the intestinal structure of the organism that excreted them, and the presence of included matter (fish scales, bones, otoliths, and other indigestible or undigested food remains) of a characteristic kind. A third criterion is their chemical composition; coprolites are typically rich in organic compounds and in calcium phosphate, derived from the phosphatization of part of the organic foodstuffs. However, both the form and the composition of coprolites may be modified so much by diagenetic and post-diagenetic processes that confident recognition becomes impossible.

Even when coprolites *have* been recognized, to decide on the particular animal group to which they should be ascribed

[2] Häntzschel et al. (1968) noted only the American edition of this work, published in 1841.

usually is by no means easy. Although the character of the droppings of land-living mammals is fairly well known in general (largely as a consequence of their importance to hunters), much less information is available concerning the excretory products of other groups of land-living animals; the excretory processes of aquatic animals generally have attracted little study. That animals of very different systematic position may produce feces of closely similar, even identical, morphology is known; usually, distinction can readily be made between the coprolites of invertebrates and vertebrates, yet extreme difficulty may be encountered in distinguishing between the excretory products of the major vertebrate groups. For example, the English Lias coprolites have at different times been ascribed to fish, amphibians, and reptiles.

In instances where coprolites may be attributed with confidence to particular groups—or better still, to particular genera or species—they become of great interest to the vertebrate paleontologist from the following three principal viewpoints.

Evidence of Feeding Habits

Dry feces of the moa, *Dinornis robustus*, of New Zealand, found together with feathers in a cave, show the bird to have been a vegetable feeder (Buick, 1931, p. 340). Coprolites of ground sloths, found in New Mexico, indicate that these animals, too, were vegetable feeders; the feces contain fragments of a range of different plants— Compositae (*Gutierrezia*), Chenopodiaceae (*Atriplex*), Malvaceae (*Sphaeralcea*, ?*Sida*), and Polypodiaceae (see Eames, 1930). Eocene coprolites, considered to be those of crocodiles, contain stomach stones and bones of frogs—and also bones of young crocodiles! (Nürnberger, 1934). Coprolites from the Alberta Upper Cretaceous, examined by Waldman and Hopkins (1972), yielded an extensive spore-pollen microflora suggestive of an aquatic environment, but no macrofloral material; the authors concluded that the feces were derived from

small to medium-size aquatic carnivorous reptiles (crocodiles, champsosaurs, or chelonians). Coprolites of a fossil dog from the California Pliocene contain teeth and bones of rabbits, pocket mice, and other small mammals (Stirton, 1959). Coprolites have even been used as a means for determining ancient human diet; for example, Callen (1963) showed that the diet of the inhabitants of Tamaulipas, Mexico, from 7000 B.C. to A.D. 1700, was largely a vegetable one but also included mice, snakes, lizards, grasshoppers, and perhaps other insects, and (1965, in Häntzschel et al., 1968, p. 30) that the diet of ancient inhabitants of northern Peru (2500 to 1200 B.C.) was entirely vegetarian. (See also Chapter 13.)

Indications of Intestinal Structure

Dean (1894) examined a "coprolite" in position in the visceral region of the cladodont fish *Cladoselache newberryi*; it gave clear indication of the presence of low intestinal septa and of a spiral valve. Neumayer (1904) described two types of amphibian coprolites from the Texas Permian, the larger forms attributed to *Eryops* and smaller forms to *Diplocaulus*; both indicated a digestive tract having a spiral flap. Walter (1928) noted that "twisted coprolites . . . found with the bones of ichthyosaurs, indicate that these extinct reptiles might have been equipped with a spiral-valve device that molded the feces into a twisted shape." Spiral coprolite-like structures are indeed found widely; their geological occurrence was reviewed recently by Williams (1972). In a microscopical study of thin sections of specimens from the Lower Permian of Kansas, Williams (p. 17) demonstrated the presence of well-preserved mucosal folds, including *tunica propria*, indicating that these specimens are in fact not true coprolites but rather fossilized intestines—probably of sharks, paleoniscoids, or lungfishes. In consequence, he preferred to call them "enterospicae"— a term originally proposed by Fritsch (1907).

Indications of Parasitism

Samuels (1965, in Häntzschel et al., 1968, p. 103–104) noted the presence of ova of the parasite *Enterobius vermicularis* in a human coprolite from Wetherill Mesa, Colorado, U.S.A. The coprolites are 700 to 900 years old.

Taphoglyphs

Traces of death throes, etc., furnish information on the manner of death of the animals represented and thus, indirectly, also on their mode of life.

Taphoglyphs—imprints in sediments of the bodies of dying or dead animals—may be produced (1) in a position attained by the animal during the final movements of its life, (2) by the dead body after downward settlement from water or air immediately after death, or (3) by the dead body after it has been transported to a new position by a predator or scavenger or by an inorganic transporting medium, such as water or ice. Although taphoglyphs are in all instances likely to prove of great paleontological or paleoecological importance, only in the first instance can they justly be classed as trace fossils. If the animal died relatively passively (as was the case with the stranded ichthyosaur illustrated by Thenius, 1973, Fig. 16), the taphoglyph may reproduce, with fair or complete exactitude, the animal's morphology and may thus be as much a *body* fossil as a *trace* fossil.

Taphoglyphs of vertebrates almost always enclose skeletal remains; the osteology of the animal represented is thus also determinable. As a direct indication of the soft morphology of the vertebrate body, these traces are of great importance in vertebrate paleontology. They may be three-dimensional, where accumulation of sediments or volcanic ash was rapid, followed by correspondingly rapid lithification. The most familiar examples are the gruesome casts of men and dogs writhing in death under hot ashfall from the eruption of Vesuvius, prepared from molds in the consolidated ashes at Pompeii, Italy. Much more often, taphoglyphs occur as imprints, little more than two-dimensional, in bedding-plane surfaces; the Solnhofen Stone of Germany has furnished many fine examples, including specimens of (1) *Archaeopteryx* showing the cover of feathers that identifies it as a fossil bird (the skeleton, taken alone, would probably not), and (2) a pterodactyl, *Rhamphorhynchus*, in which not only the wing membranes are present but also a tail flap, not predictable from the skeleton (see Augusta and Buryan, 1961).

Sometimes bacterial decay was incomplete, perhaps as a result of euxinic or anaerobic conditions, and the taphoglyph takes the form of a carbonized film coating the skeleton and the surrounding bedding plane, reproducing the outline of the animal. The fine series of ichthyosaur specimens prepared by Bernhard Hauff from materials in quarries in the Lias (Lower Jurassic) of Holzmaden, south Germany, are of this kind; the carbonized layer represents the skin.

Because they are, in many instances, capable of being treated as body fossils, taphoglyphs have hitherto attracted little attention from paleoichnologists. This dearth of interest is regrettable, because taphoglyphs give information not only on the morphology but also on the manner of death of the animal and, directly or indirectly, their way of life also. The Holzmaden ichthyosaurs, for example, not only exhibited the shark-like sweep of an arcuate tail, predictable from a downward flexure in the caudal vertebrae, but also a dorsal fin, lacking skeletal support and not predictable at all. The attitude of the taphoglyphs and skeletons indicates that they died as a result of stranding on a shoreline, as do many porpoises, dolphins, and whales in the present day. One female of the genus *Stenopterygius* was apparently stranded when approaching the time of giving birth to her young, the birth pangs perhaps being expedited by the stranding; three young remain in the body cavity, a fourth was just in process of being born (see Augusta and Buryan, 1964; Thenius, 1973, Fig. 74a). The decisive evidence in this instance was osteological, however, the taphoglyph being extremely poor.

Taphoglyphs, although certainly rare, may be somewhat more common than is currently recognized; they must often pass unnoticed in the field, and may also be removed, without being detected, during specimen preparation. The Holzmaden taphoglyphs, for example, were not recognized until about 1890, although skeletons of ichthyosaurs were extracted from the quarries forty years earlier.

ACKNOWLEDGMENTS

I am indebted to the Council of the East Midlands Geological Society, for permission to reproduce Figs. 14.1, 14.3, and 14.20; to the regents of the University of California, for permission to reprint Fig. 14.6; to the Muséum National d'Histoire Naturelle, Paris, and D. Heyler and J. Lessertisseur, for permission to reproduce Fig. 14.10; to the Director of the Connecticut Geological and Natural History Survey, for permission to reproduce Figs. 14.13 and 14.14; to the Directors of the American Museum of Natural History, for permission to reproduce Fig. 14.18; to the National Geographic Society—particularly its Art Editor, A. Poggenpohl—for furnishing the print used for Fig. 14.17 and permission to reproduce it and Fig. 14.18; to W. Langston, Jr., and the National Museum of Natural Sciences of Canada for permission to reproduce Fig. 14.19; to B. R. Erickson and the Directors of the Science Museum of Minnesota, for permission to reproduce Fig. 14.21; to W. L. Stokes and editors of the *Journal of Paleontology*, for permission to reproduce Fig. 14.23; to J. H. Ostrom, for his loan of the original artwork for Fig. 14.24 and his permission, and that of the editors of *Palaeogeography, Palaeoclimatology, Palaeoecology*, for its reproduction. Figures 14.2 and 14.15 are reproduced from photographs in the Beasley Collection, for the loan of which I am indebted to the Council of the Liverpool Geological Society. I am further indebted to Wann Langston, Jr., Texas Memorial Museum, Austin, for pointing out (*in litt.*) the stratigraphic inconsistencies in R. T. Bird's account of the Paluxy Creek trackways. The help in photography by S. Mizinski, library research by A. Morgan, and the patient deciphering of the manuscript by N. Allan are also gratefully acknowledged.

REFERENCES

Abel, O. 1926. Amerikafährt. Eindrücke, Beobachtungen und Studien eines Naturforschers auf einer Reise nach Nordamerika und Westindien. Jena, Gustav Fischer, 462 p.

————. 1935. Vorzeitliche Lebensspuren. Jena, Gustav Fischer, 644 p.

Alf, R. M. 1959. Mammal footprints from the Avawatz Formation, California. South. California Acad. Sci., Bull., 58:1–7.

Anonymous. 1838. Antediluvian remains at Stourton Quarry. Liverpool Mechanics' Inst. Liverpool Mercury, 14 Aug. 1838:273.

Augusta, J. and Z. Buryan. 1961. Prehistoric reptiles and birds. London, Paul Hamlyn, 104 p.

———— and Z. Buryan. 1964. Prehistoric sea monsters. London, Paul Hamlyn, 64+50 unnumb. p.

Bachofen-Echt, A. von. 1931. Fährten und andere Lebensspuren. In O. Abel and G. Kyrle (eds.), Die Drachenhöhle bei Mixnitz. Speläolog. Monogr., 7–9. (*not seen.*)

Baird, D. 1957. Triassic reptile faunules from Milford, New Jersey. Mus. Comp. Zool. Harvard, Bull., 117:449–520.

Bakker, R. T. 1968. The superiority of dinosaurs. Discovery, Yale Peabody Mus., 3:11–22.

————. 1971. Ecology of the brontosaurs. Nature, 229:172–174.

Beasley, H. C. 1904. Report on footprints from the Trias, Pt. I. British Assoc. Advmt. Sci., Rept. (Southport, 1903):219–230.

Bhargava, O. N. 1972. A note on structures resembling molds of tadpole nests in the Upper Jurassic Tal Formation, Simla Himalaya, India. Jour. Sed. Petrol., 42:236–245.

Bird, R. T. 1944. Did *Brontosaurus* ever walk on land? Natural History, 53:63–67.

————. 1954. We captured a "live" brontosaur. Nat. Geogr. Magazine, 105:707–722.

Boekschoten, G. J. 1964. Tadpole structures again. Jour. Sed. Petrol., 34:422–423.

Bragg, A. N. 1965. Gnomes of the night: the spadefoot toads. Philadelphia, Univ. Pennsylvania Press, 127 p.

Buckland, W. 1828. Note sur les traces de tortues observées dans le grès rouge. Ann. Sci. Natur., 13:85–86.

————. 1829. On the discovery of coprolites, or fossil faeces, in the Lias at Lyme Regis, and in other formations. Geol. Soc. London, Trans., Ser. 2, 3:223–236.

————. 1836. Geology and mineralogy considered with reference to natural theology. The Bridgewater Treatises, VI. (2 vols.) London, Pickering, 599+128 p.

Buick, T. L. 1931. The mystery of the moa, New Zealand's avian giant. Avery, New Plymouth, N. Z., 357 p.

Callen, E. O. 1963. Diet as revealed by coprolites. In E. S. Higgs (ed.), Science in archaeology. London, Thames & Hudson, p. 186–196.

Cameron, B. and R. Estes. 1971. Fossil and recent "tadpole nests": a discussion. Jour. Sed. Petrol., 41:171–178.

Casamiquela, R. M. 1964. Estudios icnológicos. Problemas y métodos de la icnología con aplicación al estudio de pisadas Mesozoicas (Reptilia, Mammalia) de la Patagonia. Gobierno Prov. Rio Negro, Buenos Aires, 229 p.

Casier, E. 1960. Les Iguanodons de Bernissart. Inst. Roy. Sci. Natur. Belgique, 134 p.

Casteret, N. 1939. Ten years under the earth. London, Dent, 200 p.

Colbert, E. H. 1964. Dinosaurs of the Arctic. Natural History, 73(4):20–23.

Cook, D. O. 1971. Depressions in shallow marine sediments made by benthic fish. Jour. Sed. Petrol., 41:577–578.

Courel, L. et. al. 1968. Empreintes de pas de vertébrés et stratigraphie du Trias. Soc. Géol. France, Bull., Ser. 7, 10:275–281.

Curry, H. D. 1957. Fossil tracks of Eocene vertebrates, southwestern Uinta Basin, Utah. Intermountain Assoc. Petrol. Geol., Proc. 8 Ann. Field Conf., 42–47.

Dalquist, W. W. 1964. Large amphibian ichnites from the Permian of Texas. Texas Jour. Sci., 15:220–224.

Dean, B. 1894. A new cladodont from the Ohio Waverly, Cladoselache newberryi n. sp. New York Acad. Sci., Trans., 13:115–119.

Demathieu, G. and H. Haubold. 1972. Stratigraphische Aussagen der Tetrapodenfährten aus der terrestrischen Trias Europas. Geologie, 21:806–840.

Dionne, J.-C. 1969. Tadpole holes; a true biogenic sedimentary structure. Jour. Sed. Petrol., 39:358–360.

Dollo, L. 1883. Troisième note sur les dinosauriens de Bernissart. Mus. Roy. Hist. Natur. Belgique, Bull., 2:85–120.

Duncan, H. 1831. An account of the tracks and footprints of animals found impressed on sandstone in the quarry of Corncockle Muir in Dumfries-shire. Royal Soc. Edinburgh, Trans., 11(1830):194–209.

Duvernoy, G. 1844. Sur l'existence des urolithes fossiles, et sur l'utilité que la science des fossiles organiques pourra tirer de leur distinction d'avec les coprolithes, pour la détermination des restes fossiles de Sauriens et d'Ophidiens. Compt. Rend. Hébd., Acad. Sci., Paris, 19:255–260.

Eames, A. J. 1930. Report on ground sloth coprolite from Dona Ana County, New Mexico. Amer. Jour. Sci., Ser. 5, 20:353–356.

Ellenberger, P. 1965. Découverte de pistes de vertébrés dans le Permien, le Trias et le Lias inférieur, aux abords de Toulon (Var) et d'Anduze (Gard). Compt. Rend. Hébd., Acad. Sci., Paris, 260:5856–5859.

Erickson, B. R. 1967. Fossil bird tracks from Utah. Museum Observer, 5:1–6.

Faul, H. 1951. The naming of fossil footprint "species." Jour. Paleont., 25:409.

Ford, T. D. and W. J. Breed. 1970. Tadpole holes formed during desiccation of overbank pools. Jour. Sed. Petrol., 40:1044–1046.

Fraipont, C. 1915. Essais de paléontologie expérimentale. Geol. Fören. i Stockholm Förhänd., 37: 435.

Fritsch, A. 1907. Miscellanea palaeontologica. I. Palaeozoica. Prague, 23 p.

Gilmore, C. W. 1926. Fossil footprints from the Grand Canyon, Arizona. Smithsonian Misc. Coll., 78:45–48.

————. 1927. Fossil footprints from the Grand Canyon: second contribution. Smithsonian Misc. Coll., 80(3):1–78.

———— 1928. Fossil footprints from the

Grand Canyon: third contribution. Smithsonian Misc. Coll., 80(8):1–16.

Häntzschel, W. 1935. Ein Fisch (*Gobius microps*) als Erzeuger von Sternspuren. Natur u. Volk, 65:562.

———— et al. 1968. Coprolites: an annotated bibliography. Geol. Soc. America, Mem. 108:1–132.

Hardaker, W. H. 1912. On the discovery of a fossil-bearing horizon in the 'Permian' rocks at Hamstead quarries, near Birmingham. Geol. Soc. London, Quart. Jour., 68:631–683.

Haubold, H. 1969a. Die Evolution der Archosaurier in der Trias aus der Sicht ihrer Fährten. Hercynia, 6:90–106.

————. 1969b. Parallelisierung terrestrischer Ablagerungen der tieferen Trias mit Pseudosuchier-Fährten. Geologie, 18:836–843.

————. 1971a. Ichnia amphibiorum et reptiliorum fossilium. In O. Kuhn (ed.), Handbuch der Paläoherpetologie, Pt. 18, 121 p.

————. 1971b. Die Tetrapodenfährten des Buntsandsteins in der Deutschen Demokratischen Republik und in Westdeutschland und ihre Äquivalente in der gesamten Trias. Paläont. Abhandl., Ser. A, 4:395–660.

Heyler, D. and J. Lessertisseur. 1963. Pistes de tétrapodes permiens de la région de Lodève (Hérault). Mus. Nat. Hist. Natur., Paris, Mém., New Ser., 11(2):125–221.

Hickling, G. 1909. British Permian footprints. Manchester Lit. Phil. Soc., Proc., Mem., 53:1–28.

Hitchcock, E. 1836. Ornithichnology. Description of the footmarks of birds (Ornithoidichnites) on New Red Sandstone in Massachusetts. Amer. Jour. Sci., 29:307–340.

————. 1841. Final report on the geology of Massachusetts, v. 3. Amherst and Northampton, p. 301–714.

————. 1843. Description of five new species of fossil footmarks, from the red sandstone of the valley of the Connecticut River. Assoc. Amer. Geol. Natural., Trans., p. 254–264.

————. 1844. Report on ichnology, or fossil footmarks, with a description of several new species. Amer. Jour. Sci., 47:292–322.

————. 1845. An attempt to name, classify and describe the animals that made the fossil footmarks of New England. Assoc. Amer. Geol. Natural., Proc. 6th Mtg., 23–25.

————. 1858. Ichnology of New England. A report on the sandstone of the Connecticut Valley, especially its fossil footmarks. Boston, W. White, 220 p.

————. 1865. Supplement to the ichnology of New England. Boston, Wright & Potter, 96 p.

Kaup, J. F. 1835. [Mitteilung über Tierfährten von Hildburghausen.] Neues Jahrb. Geogr. Geol. Petrol. (1835):327–328.

Kindle, E. M. 1914. An enquiry into the origin of "Batrachioides Antiquior" of the Lockport Dolomite of New York. Geol. Magazine, Ser. 6, 1:58–161.

Kirchner, H. 1941. Versteinerte Reptilfährten als Grundlage für den Drachenkampf in einem Heldenlied. Zeitschr. Deutsche Geol. Gesell., 93:309.

Kittl, E. 1901. Saurierfährten von Bozen. Mitt. Sekt. Naturkund. Osterreich. Touristenklub, 3. (*not seen*)

Krebs, B. 1965. *Tichinosuchus ferox* n.g. n.sp., ein neuer Pseudosuchier aus der Trias des Monte San Giorgio, Schweizer. Paläont. Abhandl., 81:1–411.

————. 1966. Zur Deutung der *Chirotherium*–Fährten. Natur u. Museum, 96:389–396.

Kuhn, O. 1958. Die Fährten der vorzeitlichen Amphibien und Reptilien. Bamberg, Meisenbach, 64 p.

————. 1963. Ichnia tetrapodorum. Fossilium Catalogus I: Animalia. Pars. 101, 176 p.

Langston, W., Jr. 1960. A hadrosaurian ichnite. Natur. Hist. Papers, Nat. Mus. Canada, 4:1–9.

Lapparent, A. de and C. Montenat. 1967. Les empreintes de pas de reptiles de l'Infralias du Veillon (Vendée). Soc. Géol. France, Mém., n. s., 107:1–44.

Lessertisseur, J. 1955. Traces d'activité animale et leur signification paléobiologique. Soc. Géol. France, Mem., N. Ser., 74, 150 p.

Ley, W. 1951. Dragons in amber. London, Sidgwick & Jackson, 328 p.

Link, —. 1835. Notes sur les traces des pieds d'animaux inconnus à Hildburghausen. Ann. Sci. Natur., Ser. 2, 4:139–141.

Lull, R. S. 1915. Triassic life of the Connecticut Valley. Connecticut Geol. Natur. Hist. Surv., Bull. 24, 285 p.

————. 1953. Triassic life of the Connecticut Valley. Connecticut Geol. Natur. Hist. Surv., Bull. 81, 336 p.

Lyell, C. 1855. Manual of elementary geology. London, 512 p.

McKee, E. D. 1934. The Coconino Sandstone— its history and origin. Carnegie Inst., Publ., 440:1–97.

———. 1944. Tracks that go uphill. Plateau, 16:61–72.

Miall, L. C. 1874. Report of the Committee . . . on the Labyrinthodonts of the Coal Measures. British Assoc. Advmt. Sci., Rept. (Bradford, 1873):225–249.

Moodie, R. L. 1930. Vertebrate footprints from the red beds of Texas. II. Amer. Jour. Sci., 38:548–565.

Moussa, M. T. 1968. Fossil tracks from the Green River Formation (Eocene) near Soldier Summit, Utah. Jour. Paleont., 42:1433–1438.

Neumayer, L. 1904. Die Koprolithen des Perms von Texas. Palaeontographica, 51:121–128.

Nopcsa, F. 1923. Die Fossilen Reptilien. Förtschr. Geol. Paläont., 2:1–210.

Nürnberger, L. 1934. Koproporphyrin im Tertiären Krokodilkot. Nova Acta Leopoldina, Halle, N. Ser., 1:324–325.

Ostrom, J. H. 1972. Were some dinosaurs gregarious? Palaeogeogr., Palaeoclimatol., Palaeoecol., 11:287–301.

Owen, R. 1842. Description of parts of the skeleton and teeth of five species of the genus Labyrinthodon . . . with remarks on the probable identity of the Cheirotherium with this genus of extinct Batrachians. Geol. Soc. London, Trans., Ser. 2, 6:515–543.

Pabst, W. 1900. Beiträge zur Kenntis der Tierfährten in dem Rothliegenden "Deutschlands." Zeitschr. Deutsche. Geol. Gesell., 52:48–63.

Peabody, F. E. 1948. Reptile and amphibian trackways from the Lower Triassic Moenkopi Formation of Arizona and Utah. Univ. California Publ. Zool., 27:295–468.

———. 1959. Trackways of living and fossil salamanders. Univ. California Publ. Zool., 63:1–72.

Reiche, P. 1938. An analysis of cross-lamination: the Coconino Sandstone. Jour. Geol., 46:918.

Sarjeant, W. A. S. 1967. Fossil footprints from the Middle Triassic of Nottinghamshire and Derbyshire. Mercian Geol., 2:327–341.

———. 1969. Fossil footprints from the Middle Triassic of Nottinghamshire and the Middle Jurassic of Yorkshire. Mercian Geol., 3:269–282.

———. 1971. Vertebrate tracks from the Permian of Castle Peak, Texas. Texas Jour. Sci., 22:343–366.

——— and W. J. Kennedy. 1973. Proposal of a code for the nomenclature of trace fossils. Canadian Jour. Earth. Sci., 10:460–475.

Schäfer, W. 1972. Ecology and palaeoecology of marine environments. Edinburgh and Chicago, Oliver & Boyd and Univ. Chicago Press, 568 p.

Schmidt, H. 1956. Die grosse Bochumer Oberkarbon-Fährte. Paläont. Zeitschr., 30:199–206.

———. 1959. Die Cornberger Fährten in Rahmen der Vierfüssler–Entwicklung. Hess. Landes-Amt. Bodenforsch., Abh., 28:1–137.

Shepard, C. U. 1867. On the supposed tadpole nests, or imprints made by Batrachoides nidificans Hitchcock, in the Red Shale of the New Red Sandstone of South Hadley, Mass. Amer. Jour. Sci., Ser. 2, 43:99–104.

Sickler, F. K. L. 1834. Sendschreiben an Dr. J. F. Blumenbach über die hochst merckwürdigen, vor einigen Monaten erst entdecken Reliefs der Fährten urweiltlicher grosser und unbekannter Thiere in den Hessberger Sandsteinbrüchen bei der Stadt Hildburghausen. Schulprogramm Gymnasiums Hildburghausen, 16 p.

Soergel, W. 1925. Die Fährten der Chirotheria. Eine paläobiologische Studie. Jena, Gustav Fischer, 92 p.

Sternberg, C. M. 1930. Dinosaur tracks from Peace River, British Columbia. Nat. Mus. Canada, Ann. Rept. (1930):59–85.

Stirton, R. A. 1959. Time, life and man. New York, John Wiley, 558 p.

Stokes, W. L. 1957. Pterodactyl tracks from the Morrison Formation. Jour. Paleont., 31:952–954.

Swinton, W. E. 1960. The history of Chirotherium. Liverpool Manchester Geol. Jour., 2:443–473.

Thenius, E. 1973. Fossils and the life of the past. New York, Springer-Verlag, 194 p.

Vallois, H. V. 1931. Les empreintes de pieds humains des grottes préhistoriques du Midi de la France. Palaeobiologia, 4:79.

Van der Lingen, G. J. and P. B. Andrews. 1969. Hoofprint structures in beach sand. Jour. Sed. Petrol., 39:350–357.

Vialov, O. S. 1964. Network structures similar to those made by tadpoles. Jour. Sed. Petrol., 34:665–666.

————. 1966. Sledy zhiznedeyatelnosti organizmow i ikh paleontologicheskoe znachenie. (The traces of the vital activity of organisms and their paleontological significance.) Akad. Nauk. Ukraine, S.S.R., 219 p.

————. 1968. Materialii k klassifikatsii iskolaemix sledov i sledov zhiznedeyatel'nosti organizmov. (Materials for the classification of fossil tracks and of traces of vital activity.) Paleont. Sbornik, 5:125–129.

————. 1972. The classification of the fossil traces of life. 24 Internat. Geol. Congr., Montreal, Proc., 7:639–644.

Voigt, F. S. 1835. [Thierfährten im Sandstein von Hildburghausen.] Neues Jahrb. Mineral. Geogr. (1835):322–326.

————. 1836. (Weitere Nachrichten über die Hessberger Thierfährten.) Neues Jahrb. Mineral. Geogr. (1836):165–174.

von Huene, F. 1928. Lebensbild des Saurischier-Vorkommens im obersten Keuper von Trossingen. Palaeobiologica, 1:103.

————. 1935. Die fossilen Reptilien des Südamerikanischen Gondwanalands I. München, Becksche Verlagsbuchhandl. (not seen)

————. 1941. Die Tetrapoden-Fährten im toskanischen Verrucano und ihre Bedeutung. Neues Jahrb. Mineral. Geol. Paläont., Ser. B, 86:1–34.

von Humboldt, A. 1835. Note sur les empreintes de pieds d'un quadrupède dans les grès bigarrés de Hildburghausen. Ann. Sci. Natur., Ser. 2, 4:134–138.

Waldman, M. and W. S. Hopkins, Jr. 1972. Coprolites from the Upper Cretaceous of Alberta, Canada, with a description of their microflora. Canadian Jour. Earth Sci., 7:1295–1303.

Walter, H. E. 1928. Biology of the vertebrates. New York, Macmillan, 788 p.

Walther, J. 1904. Die Fauna der Solnhofener Plattenkalke bionomisch betrachtet. Denkschr. Med. Naturw. Gesell. Jena, 11:135–214.

Watson, D. M. S. 1914. The *Cheirotherium*. Geol. Magazine, N. Ser., 61:395–398.

Weingardt, H. W. 1961. Tierfährten-Funde aus der Steinkohlzeit. Schacht u. Heim, Werkzeitung Saarbergwerke AG, 7 (Apr. 1961).

Whitaker, G. O. and J. Meyers. 1965. Dinosaur hunt. New York, Harcourt, Brace, 94 p.

Williams, M. E. 1972. The origin of "spiral coprolites." Univ. Kansas Paleont. Contr., Paper 59, 19 p.

Williamson, W. C. 1867. On a *Cheirotherium* footprint from the base of the Keuper Sandstone of Daresbury, Cheshire. Geol. Soc. London, Quart. Jour., 23:56–57.

Wills, L. J. and W. A. S. Sarjeant. 1970. Fossil vertebrate and invertebrate tracks from boreholes through the Bunter Series (Triassic) of Worcestershire. Mercian Geol., 3:399–414.

Wolansky, D. 1952. Spuren von Landwirbeltieren aus der Steinkohlzeit. Werkzeitung Gruppe Hamborn Gelsenkirchen A. G. (Jan. 1952).

CHAPTER 15

VERTEBRATE BURROWS

M. R. VOORHIES

Department of Geology, University of Georgia
Athens, Georgia, U.S.A.

SYNOPSIS

A large proportion of living vertebrates, particularly mammals, excavate burrows. At least half of the extant mammalian species are partially fossorial, and the habit has also been developed by many fishes, amphibians, reptiles, and even birds. Yet vertebrate burrows have rarely been reported as fossils. Possible explanations for this dearth of information or occurrences are: (1) lack of detailed observations, (2) tendency of burrowers to avoid areas of active sedimentation and (or) (3) an evolutionary increase in the burrowing habit as a result of Cenozoic climatic changes.

The census of modern and fossil burrows and burrowers given herein emphasizes the potential paleoecological, paleoclimatic, and phylogenetic importance of fossil burrows.

INTRODUCTION

Fossil marine invertebrate burrows have received considerable attention from paleontologists and stratigraphers, as is evident in numerous other chapters of this volume. However, except for such spectacular types as *Daimonelix* (see Fig. 15.4), fossil burrows of vertebrates—or indeed of any organisms—are mentioned in very few publications dealing with nonmarine strata. (See Chapter 19.) In the following pages I attempt to: (1) speculate on reasons for this paucity of information, (2) summarize the widely scattered data on fossil and modern burrows and burrowers, in each of the vertebrate classes, and in passing, (3) describe some new discoveries of vertebrate burrows from the Cenozoic of the Great Plains of North America.

Useful summaries of data on modern vertebrate burrowers, especially as related to soil formation, were given by Kühnelt (1950, p. 186–191) and Kevan (1962, p.

13–18). Ager (1963, p. 112, 114) briefly discussed fossil examples. Vertebrate tracks and trails are discussed in Chapter 14.

THE ?PAUCITY OF FOSSIL VERTEBRATE BURROWS

Only a few kinds of fishes excavate burrows[1], but among the tetrapod classes—amphibians, reptiles, mammals, and even birds—many species are habitual hole diggers. In the mammals, most species lead at least partly subterranean lives, and all the terrestrial classes except Aves contain many species that rarely come to the surface at all. Thus, one may legitimately ask why the published record of ancient vertebrate burrows is so scanty.

[1] For purposes of this discussion, any hole in the ground made by a vertebrate—whether dug in hard or soft sediment, and whether serving for concealment, shelter, feeding, etc.—is considered a burrow. Bromley's (1970) useful distinction between borings and burrows, and Seilacher's (1964) ethological categories are difficult if not impossible to apply to vertebrate burrows. (See Chapter 3.)

On first consideration, the burrowing habit might seem to increase an organism's chances for fossilization; but this expectation is not borne out by discoveries. Such fully committed burrowers as the worm-like caecilians (Amphibia:Apoda) and the "marsupial moles" (Mammalia:Notoryctidae) have virtually no fossil record, and the equally fossorial amphisbaenians (Reptilia:Annulata) and African mole rats (Mammalia:Bathyergidae) are nearly unknown as fossils (Romer, 1966). Moles (Mammalia:Talpidae), in contrast, are known from literally thousands of fossil specimens [Hutchison (1968) listed a minimum of nearly 100 individuals from the Miocene and Pliocene of Oregon alone], yet none seems to have been discovered in its burrow. The same is true for the commonly fossilized pocket gophers.

Some possible explanations for these apparent anomalies include the following:

1. Fossil burrows may not really be uncommon, but:
   (A), *they have gone unreported* (publication bias). My own limited observations of nonmarine Cenozoic strata suggest that fossil vertebrate burrows, although not obvious at every outcrop, are not nearly so rare as the literature would suggest. Apparently, numerous workers simply choose not to report the burrows that they observe.
   (B), *they have not been recognized as such* (preservation-interpretation bias). The distinction between fossil burrows and enclosing strata may be very subtle, except where the burrows:
      (1), have a distinctive shape (e.g., the spiral *Daimonelix*),
      (2), are filled with material that strongly contrasts with the surrounding sediment (e.g., burrows containing volcanic ash, described below),
      (3), are lined (e.g., with vegetation— as in *Daimonelix*, or mucus—as in lungfish burrows), or
      (4), actually contain the remains of the animals that made them (e.g., *Daimonelix*, and others described below).
   Probably, most burrows lack any obvious development of the above features and thus go unrecognized. An additional problem is the tendency for populations of burrowers (moles, prairie dogs) to inhabit a favored locality for many decades or centuries. Complete mixing (bioturbation) of the substrate undoubtedly occurs in such circumstances, and distinguishing between a "massive" stratum never burrowed at all, and one completely homogenized by organic activity, might be difficult. Ideally, for field recognition, only a few generations of burrows should be present. As stated elsewhere in this volume (e.g., Chapters 8 and 23), one should be automatically suspect of marine strata that seem to be completely devoid of biogenic structures; does the same premise hold true for continental deposits?

2. Fossil vertebrate burrows may be truly rare in the record, because:
   (A), *environments favored by burrowers are not well represented in the stratigraphic record* (sedimentation bias). Most nonmarine sedimentary deposits accumulated in lowland habitats— stream channels, floodplains, etc.— whereas the most common burrowers (rodents, moles, lizards) prefer better drained soils in upland areas. McNab (1966), for example, found that in Florida a correlation exists between the distribution of the gopher *Geomys pinetis* and soils of low water-retention capacity (< 20 percent). He also noted that in areas subject to seasonal flooding, e.g., the lowlands southwest of Lake Okeechobee, such characteristic burrowers as moles, gophers, and burrowing tortoises were absent. Only those animals that make their burrows in swamps and riverbanks (lungfish, alligators, beavers) would seem to have better-than-average chances of having their excavations preserved in the fossil record.
   (B), *fossorial habits were less widespread among vertebrates in the geologic past than at present* (evolutionary bias). Many, or perhaps most, fossorial vertebrates burrow to avoid environmental extremes, especialy during hibernation or estivation. Progressive climatic deterioration during the Cenozoic, climaxing in the Pleistocene, may have led to increased burrowing by many vertebrate groups. Some anatomical evidence exists among such groups as

moles for increased fossorial specialization during the Cenozoic (e.g., Hutchison, 1968, p. 80). That fossil vertebrate burrows are known only from the late Paleozoic and late Cenozoic—times of extreme fluctuations in both temperature and humidity—may not be coincidental. I have found no references to vertebrate burrows in either the Mesozoic or early Cenozoic, which were characterized by comparatively equable nonglacial climates.

Perhaps a combination of these biases is responsible for our lack of knowledge about the burrowing habits of ancient vertebrates, as discussed below.

CLASS AGNATHA

Larval lampreys (*Lampetra, Petromyzon*) burrow in soft sediment in freshwater streams, until they metamorphose into mobile, parasitic adults. (See Chapter 19.) Hagfishes (*Myxine*) also spend much time buried in mud, having only the head exposed; but unlike lampreys, they are strictly marine and they burrow as adults (Bigelow and Schroeder, 1948). The inability of these modern cyclostomes to burrow in any but the least consolidated sediment nearly precludes any possibility that their fossorial behavior might be "fossilized."

In contrast, the "snowplow"-shaped cephalic shields of the osteostracan and heterostracan ostracoderms (armored Paleozoic jawless fishes) suggest that these animals may have effectively burrowed, or plowed, through sediment in search of the organic detritus that probably formed their diet. Halstead (1968, p. 26), in his discussion of ostracoderm ecology, concluded that they were "in general semi-sessile bottom feeders" that spent "much of their time buried in sediment." This habit seems particularly likely in such nearly eyeless types as *Eglonaspis* (Obruchev, 1964, p. 54).

Unfortunately, no unequivocal record of ostracoderm burrowing activity has yet been found in the rock record. Large sand-filled tubes occur abundantly in several ostracoderm collecting localities within the Old Red Sandstone in Britain; Allen (1961), who described the structures, discussed the possibility that they might be burrows, but marshalled more convincing evidence for an inorganic (fluid-injection) origin for them.

CLASS PLACODERMI

The placoderms, a heterogeneous assemblage of Paleozoic jawed fishes, are extinct; thus, habits of the widely disparate members of this group are conjectural. The dorsoventrally flattened, skate-like shape of such forms as the rhenanids suggests that they may have preyed upon benthic invertebrates, and perhaps excavated depressions in the sea floor while doing so; but the chances seem unlikely that these or any other known placoderms were true burrowers.

CLASS CHONDRICHTHYES

Most modern sharks (Order Selachii) are streamlined predators that not only do not burrow but are physiologically incapable even of resting on the bottom. The closely related skates and rays (Order Batoidea), however, are primarily benthic feeders on invertebrates (Bigelow and Schroeder, 1953); their shape and unique respiratory adaptations enable them to pursue their prey by burrowing through bottom sediments. Cook (1971) described and illustrated the sedimentary structures formed by modern batoids, noting that in some the outline of the fish is preserved, but usually not.

No sediment-penetration structures definitely ascribed to chondrichthyans seem to have been reported in ancient strata. R. W. Frey and J. D. Howard (1973, personal communication) have, however, observed structures strikingly similar to stingray holes in the Pleistocene of Georgia–Florida and the Cretaceous of Utah. These occurrences have not yet been described, but the same workers have published an excellent illustration of a modern ray (probably *Dasyatis* sp.) excavation on the Georgia coast (Frey and Howard, 1969, Pl. 3, fig. 5).

CLASS OSTEICHTHYES

Much more varied in shape and habits than any of the preceding groups are the bony fishes, which include numerous true burrowers; some of these (lungfish) demonstrably developed the habit as far back as the Paleozoic.

Modern Examples

Burrowing is shallow in such forms as the flounders (Pleuronectidae and Soleidae) but is considerably deeper in the mudminnow (*Umbra*) and the Japanese weatherfish (*Misgurnus*), which "on alarm, literally swim and dive into soft bottom material" (Lagler et al., 1962, p. 198); certain marine eels (Ophichthidae) can burrow in moments, tail first, to depths of a meter. American freshwater catfish (Ictaluridae) often occupy burrows in rather firm, riverbank sediment, but I have been unable to determine whether they excavate the holes themselves or merely occupy old beaver tunnels. None of the forms so far mentioned seems likely to have left an important ichnological record.

Certain members of the Order Dipnoi (lungfishes), however, construct permanent burrows quite capable of being preserved and recognized in ancient strata. The African (*Protopterus*) and South American (*Lepidosiren*) lungfish both live in swamps on floodplains of tropical rivers subject to seasonal drought. During the dry season they estivate in flask-shaped burrows, which they excavate themselves and sometimes line with mucus to retard drying out. Specimens of *Protopterus aethiopicus* have been known to survive for 18 months or more before being released from their sun-baked burrows by the return of the rainy season (Smith, 1931, p. 180).

The burrowing technique of *Protopterus annectens* in Gambia was studied comprehensively by Johnels and Svensson (1955). As water levels fall at the end of the rainy season, the fish literally begin chewing into the compact red clay bottom. The mud is forced into the mouth, where it is diluted with water and expelled through the gills. At the proper depth (as much as 1 m, depending on the size of the fish) the end of the burrow is widened, and the fish reverses its position and assumes a tail-down orientation. The animal periodically rises to the surface to breathe, until the water level drops below the estivating chamber, at which time the mucus cocoon is secreted and the remarkable physiological modifications accompanying "summer sleep" begin. Burrow densities of as many as 5 burrows per m² over many hectares were reported for adult lungfish by Johnels and Svensson, and burrows of immature specimens are even more numerous. A segregation into size groups—the young ones estivating in shallower waters at the edge of the floodplain, and larger individuals congregating in topographically lower areas—was reported by the same authors.

Fossil Examples

Romer and Olson (1954) were the first to describe dipnoan burrows in the rock record. The authors recognized the peculiar cylinders weathering out of Lower Permian shales at several localities in Texas as being natural casts of burrows remarkably similar to those made by modern lungfish (Fig. 15.1). The presence of skeletal remains of the dipnoan genus *Gnathorhiza* within some of the burrows confirmed this identification. In more recent years, similar burrows have been reported from Lower Permian beds on Prince Edward Island (Langston, 1963), the Sangre de Cristo Formation (also Lower Permian) in northern New Mexico (Vaughn, 1964), Middle Pennsylvanian strata in Michigan (Carroll, 1965), and the Wellington Formation

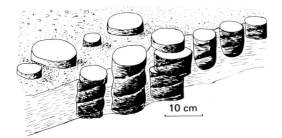

Fig. 15.1 Natural casts of lungfish burrows. Arroyo Formation, Clear Fork Group, Lower Permian: Reed Ranch Locality, Willbarger County, Texas. (From photographs in Romer and Olson, 1954.)

(Lower Permian) in north-central Oklahoma (Carlson, 1968). The last occurrence is particularly noteworthy in that well-preserved, articulated skulls and skeletons of *Gnathorhiza* were found in typical tail-down estivating position.

These discoveries have paleozoological and geological importance beyond the obvious but often overlooked fact that, in these and other fossil burrows, we have conclusive proof of authochthonous burial —rare instances of fossil vertebrates known to have lived and died where they are found. First, the burrows establish the great antiquity of the estivating habit—a behavior pattern that apparently has remained essentially unchanged for more than 300 million years. In this regard a particularly revealing discovery would be to find still older burrows and thus to determine the time of origin of estivation among lungfish—perhaps during the Devonian, when dipnoan morphological evolution was proceeding most rapidly (Westoll, 1949, p. 172). Second, important behavioral evidence is added to the skeletal evidence that *Gnathorhiza* is phylogenetically related to the modern lepidosirenids (*Lepidosiren* and *Protopterus*), which estivate, rather than to the living Australian lungfish (*Neoceratodus*), which does not (see Carlson, 1968, p. 654). Third, strong support is added to the interpretation of Permian climates (largely based on sedimentological evidence) as being seasonally arid over large areas of

North America. The Middle Pennsylvanian burrows from Michigan are intriguing in this respect. Further such discoveries may require that traditional ideas about uniformly wet tropical climates in the Carboniferous of eastern North America be modified.

CLASS AMPHIBIA

Amphibians as a group are more sensitive than other tetrapods to environmental fluctuations in temperature and, especially, humidity. Not surprisingly, therefore, many of them seek the relative constancy of the underground habitat.

Modern Examples

Most salamanders (Order Urodela) and frogs (Order Anura) have relatively weak limbs not suited for digging, and although they may burrow for short distances into mud to hibernate, they probably contribute little to sediment dynamics. Several permanently aquatic salamanders of the Family Sirenidae are reported to possess the lungfish-like ability to withstand several months of complete drying out of their habitat, by burrowing into bottom mud (Freeman, 1958). The more terrestrial salamanders often utilize burrows made by other animals; the tiger salamander *Ambystoma tigrinum*, for instance, is often seen emerging from prairie dog holes (Smith, 1956, p. 41).

Several groups of terrestrial anurans ("toads") have become expert burrowers, notably members of the Breviceptidae, Bufonidae, and Pelobatidae. The last group includes the highly specialized "spadefoot" toads (*Pelobates* in Europe, *Scaphiopus* in America), in which broad, sharp-edged epidermal tubercles ("spades") are present on the inner edges of the hind feet. Bragg (1944, 1945), in his investigation of the unusual life cycle of *Scaphiopus* in Oklahoma, noted that they dig to any depth necessary to remain properly moist—often as much as several meters.[2]

The amphibians most highly specialized for burrowing are certainly the caecilians (Order Apoda)—curious limbless creatures superficially resembling the earthworms upon which they feed. Taylor, in his monograph on the group (1968), noted that of all the larger ordinal divisions of vertebrates, these animals are the least well known—except possibly for some of the deep-sea fishes. Caecilians occur in the tropics, on all major land masses except Australia and Madagascar, and are highly efficient burrowers, even in rather tough lateritic soils. They probably are not rare animals, but because of their habits and tropical distribution, great effort is required to collect them. Some species (e.g., *Caecilia thompsoni*) exceed 1.3 m in length.

Fossil Examples

Except for a single fossil caecilian specimen (Estes and Wake, 1972), only the Urodela and Anura have a fossil record. Unfortunately, the geologic setting even of the few relatively complete skeletons that have been found is rarely described in sufficient detail for determining whether the animals may have died in burrows. This lack of attention to lithology is a trait widespread among even modern workers in vertebrate

paleontology. Not only are highly pertinent sedimentologic data seldom included in published descriptions, but they may not even be noticed during the scrupulous removal of hateful "matrix" from the bones. The hygenic zeal of a good preparator may result in invaluable clues about the habit and habitat of unique specimens being lost forever.

Notable exceptions to the general disregard of lithologic clues in paleozoological work can be found in E. C. Olson's studies of the nonmarine Permian rocks and vertebrate faunas in the southern Great Plains. Pertinent here is Olson's (1956, 1967, 1970, 1971) interpretation of the estivating habits of the long, slender amphibian *Lysorophus*. Individuals found in burrows are invariably tightly coiled and, at any one locality, are all approximately the same size; those not in burrows are uncoiled and are of many sizes. These are interpreted, respectively, as estivating swarms and populations of free-swimming individuals.

Among the larger Paleozoic amphibians (Subclass Labyrinthodontia), nothing has been observed that would indicate habitual burrowing. Halstead (1968, p. 87) suggested that labyrinthodonts were less susceptible to water loss than are the modern amphibians, because the dermal armor of the former acted as waterproofing. Some labyrinthodonts (e.g., many embolomeres and plagiosaurs; see Romer, 1966, p. 92–94) have degenerate limbs reminiscent of modern burrowers; but this condition has been interpreted, probably correctly, as indicating a return to strictly aquatic life. Somewhat more terrestrial forms, such as *Eryops* —which apparently filled the "alligator" niche in late Paleozoic swamps (Colbert, 1969, p. 97)—may well have excavated the equivalent of "gator holes," but published evidence on this point is lacking. No large burrows are known in Permian rocks.

Although hardly qualifying as burrows, "tadpole nests" should be briefly mentioned here, because they are the only amphibian-produced structures to have received exten-

---

[2] This habit of *Scaphiopus* may lead to confusion of paleontologists. On several occasions I have found spadefoot toads, both live ones and skeletons, embedded in unconsolidated fossiliferous Tertiary strata in the Great Plains (Gering Formation, Valentine Formation, etc.). In some cases the tunnel to the surface was obvious, but in others—especially in loose, "channel" sands—it was not. Subtle differences in preservation were sometimes the only clue to the Holocene age of the toad remains. Such "stratigraphic leakage" could lead to serious errors in interpretation, especially considering that lower vertebrates generally have evolved more slowly than mammals during the Cenozoic. Intrusive fossorial mammals are likely to be recognized as heterochthonous; but would, for example, a Pliocene toad that burrowed into a Miocene sand be so easy to recognize as an entity not coeval with the older sediments?

sive attention in the geological literature (see Cameron and Estes, 1971, for bibliography). To quote from Cameron and Estes:

> "Tadpole nests" are true biogenic sedimentary structures produced by the activities of tadpoles and are known only from Recent occurrences. They are composed of circular to nearly hexagonal interconnecting ridges with central depressions and are found only in shallow, low-energy, freshwater pools at the sediment–water interface.

The likelihood of such ephemeral structures being preserved seems very small, because very slight disturbance sends the sediment into suspension. Most "tadpole structures" in the paleontologic record are probably intereference ripples. Especially suspect are the Silurian "tadpole nests," which not only predate the oldest amphibians by tens of millions of years but are also in marine strata—a habitat most unlikely for larval amphibians. Boekschoten (1964) suggested criteria for distinguishing between true tadpole nests and interference ripples, including the observation that the latter frequently possess convex slopes on the ridges and have a pattern directed along two axes, whereas the former consist of irregular polygons having concave slopes. (See Chapters 5 and 14.)

CLASS REPTILIA

Modern Examples

All four orders of existing reptiles (Chelonia—turtles, Rhynchocephalia—tuataras, Crocodilia—crocodiles and alligators, and Squamata—lizards and snakes) include members that burrow to some extent. Carr (1963, p. 83) described three ethological categories of burrowing reptiles:

1. Those that dig and inhabit a permanent burrow, e.g., monitor lizards, gopher tortoises; in these cases, feeding is not subterranean.
2. Those that burrow through loose sand in search of food—a behavior aptly described as "sand-swimming"; this category includes

many snakes and numerous lizards in the skink family.
3. Permanent underground foragers—primarily the amphisbaenians.

To this list might be added those, like most crocodiles and turtles, that dig deep nests for laying their eggs but that are otherwise not fossorially inclined. A review of the anatomical specializations in burrowing reptiles was given by Bellairs (1970, p. 110–115).

Chelonia

Although most turtles excavate nests, few dig permanent burrows; members of the American genus Gopherus, however, are true burrowers. G. polyphemus of the southeastern United States commonly excavates tunnels exceeding 5 m in length. One burrow was followed laterally for 8 m and to a depth of 3 m, and no end was in sight when the investigators gave up (Pope, 1939, p. 249).

An account of the burrow morphology and substrate preferences of the gopher tortoise was given by Hallinan (1923). The elevation of the habitat must be high enough that the burrows can be kept above the water table, but not so high that the turtles cannot end their tunnels in slightly damp sediment. G. polyphemus is colonial, regularly occurring in densities of several hundred individuals per km^2.

Gopherus agassizi of arid southwestern North America nicely illustrates the effect that climate has on burrowing. Through most of its range, it digs shallow (about 1 m) burrows for both hibernation and estivation; but Auffenberg and Weaver (1969) reported that in Utah, at the northern edge of its range, G. agassizi excavates communal winter quarters ("hibernacula"), which may exceed 10 m in length. In contrast, Bogert and Oliver (1945, p. 399) found that in the vicinity of Alamos, Sonora, which has a comparatively equable climate, the tortoises seem not to burrow at all.

Rhynchocephalia

Because *Sphenodon*, the tuatara, invariably inhabits islands frequented by petrels and is sometimes seen taking refuge in burrows made by the birds, observers once believed that both species shared the same burrow. More detailed observations, however, showed that the tuatara usually excavates its own burrow (Bellairs, 1957, p. 125).

Crocodilia

The American alligator (*Alligator mississipiensis*) digs three kinds of dens (Neill, 1971, p. 272–276). The first two types— open depressions in soft peat accumulations in ponds ("gator holes"), and enlarged undercut holes in soft riverbank mud—are the usual domiciles in the southern parts of the alligator's range. Actual burrows—long horizontal tunnels having an expansion at the end to permit turning around—are found in seasonally inundated floodplains having a hard clay pan, primarily in the northern parts of the species' range. The burrows are apparently used exclusively for hibernation, a behavioral adaptation that may allow the alligator to range much farther north than do crocodiles and caimans. Some tropical crocodilians, notably *Crocodylus niloticus*—the Nile crocodile— digs deep (as much as 12 m) riverbank burrows in the drier parts of its range. As many as 15 estivating crocodiles have been found in a single burrow (Guggisberg, 1972, p. 106). Surprisingly, the jaws rather than the feet are used in digging the burrows (Pooley, 1969).

Squamata

Burrowing by lizards (Suborder Lacertilia) was reviewed by Bellairs (1970, p. 91–95), who stated that approximately half of the existing families contain one or more members in which the limbs have become reduced or eliminated in connection with fossorial habits. Examples are the slow worms (Anguidae), scale-foot lizards (Pygopodidae), and the footless lizards (Annielidae). The most impressive lizard burrows are undoubtedly those of the very large monitor lizards (*Varanus komodoensis* and *V. salvator*) of the Indonesian archipelago (Hoogerwerf, 1970).

The highly fossorial, worm-like amphisbaenids are a taxonomic puzzle. First classed with the snakes because of their lack of limbs, upon discovery of the two-legged genus *Bipes* they were transferred to the lizard suborder. They now are often regarded as an independent category distinct from both snakes and lizards. Gans, the principal recent student of the group, regards the amphisbaenids as ". . . the only true burrowers among the reptiles . . . they can make their own tunnels in compact soils," and they very rarely come to the surface (1969, p. 147).

Burrowing has been regarded as playing a crucial role in the origin of the snakes (Suborder Serpentes; Walls, 1942; see discussion in Bellairs, 1970). Two families of snakes regarded as primitive ones are burrowers (Typhlopidae and Leptotyphlopidae), as are the Aniliidae and many boids. A fossorial habit has also been secondarily regained by some members of the "advanced" Family Colubridae (e.g., in the hog-nosed snake *Heterodon*).

Fossil Examples

As mentioned in the introduction, the "Age of Reptiles" is notable for the absence of any known vertebrate burrows. None of the known dinosaurs, large or small, seems to have been adapted for digging. Various workers (e.g., Hotton, 1963, p. 169) have suggested that this lack of adaptation may have been a factor in the extinction of the dinosaurs, whereas lizards, turtles, etc., survived into the Cenozoic. Even in Cenozoic rocks, no reptile burrows seem to have been reported; the interesting Pleistocene box turtle hibernaculum described by Auffenberg (1959) is not actually a burrow but apparently a crevice into which the animals crawled. Cox (1972) described an apparently fossorial Late Permian dicynodont (an aberrant group of "mammal-like" reptiles) from the Karroo of Tanzania. The skeleton of *Kawingasaurus fossilis*, a small toothless animal, strongly suggests that it was a burrower. Yet no burrow of this, or any other Karroo reptile, seems to have been discovered.

CLASS AVES

Modern Examples

Birds would seem at first glance to be preeminently unsuited to a fossorial existence. Essentially all familiar birds, and even the unfamiliar flightless types (with the exception of the kiwi) seek their food entirely above ground. A surprisingly large number, however, make underground nests, some being of startlingly large dimensions. Van Tyne and Berger's (1959, p. 376–552) summary of the biology of the 168 families of birds indicates that 33 families include at least one species that constructs its own burrow.[3]

Most of the obligatory avian burrowers are distributed among three orders (Procellariformes, Coraciiformes, Piciformes) and one passerine family (Hirundinidae), comprising in all about 250 species—about 3 percent of the total number of living bird species.

The Procellariformes, including the petrels and shearwaters, have perhaps the greatest potential geologic importance.

Some pelagic islands, especially in the Southern Ocean, are reported to be literally riddled with burrows (as much as 3 m or so in length) made by various sorts of petrels (Oliver, 1930). The amount of terrain suitable for burrow construction seems to act in many cases as a limiting factor in the population density of the birds. The making of the burrow—often in stony and (or) frozen ground—may involve several months' effort and is obviously of considerable biological importance to the birds (see Murphy, 1936, p. 476–477). I may mention here that penguins (Order Sphenisciformes) of several kinds also excavate burrows, using their feet to do so. Intense bioturbation, in sediments not likely to be affected by other animals, must certainly result from this activity of oceanic birds; but to my knowledge, no structures attributable to this source have been reported in the geological literature.

In more temperate and tropical latitudes, especially along freshwater streams and lakes, members of the Order Coraciiformes—kingfishers, bee-eaters, and their relatives—are noisily familiar. Most of these birds excavate deep burrows in vertical banks. The American belted kingfisher (*Megaceryle alcyon*), for example, burrows horizontally for 2 m (sometimes as much as 5 m) and lays its eggs in a brood chamber at the end of the tunnel. Both sexes, using their beaks as tools, cooperate in the excavation. They are highly selective about their homesites, and according to Sutton (1967): "the proper bank need not be near a stream or pond, sometimes it is literally miles from water." An interesting feature

[3] Spheniscidae—penguins
Apterygidae—kiwis
Procellariidae—shearwaters
Hydrobatidae—storm petrels
Pelecanoididae—diving petrels
Rallidae—rails
Dromadidae—crabplovers
Laridae—gulls
Alcidae—auks
Columbidae—pigeons
Psittacidae—parrots
Strigidae—owls

Aegothelidae—owlet-
    frogmouths
Halcyonidae—kingfishers
Todidae—todies
Momotidae—motmots
Meropidae—bee-eaters
Coraciidae—rollers
Upupidae—hoopoes
Galbulidae—jacamars
Bucconidae—puffbirds
Capitonidae—barbets
Picidae—woodpeckers

Jyngidae—wrynecks
Furnariidae—ovenbirds
Rhinocryptidae—tapaculos
Hirundinidae—swallows
Corvidae—crows
Paridae—titmice
Timaliidae—babblers
Turdidae—thrushes
Sturnidae—starlings
Dicaeidae—flowerpeckers

of the kingfisher burrow is the semilithified "cup" of fish bones often found in the brood chamber (Eastman, 1969, p. 62–63). Kingfishers (Family Halcyonidae) are mostly solitary, so their burrows are mostly isolated, whereas the tropical family Meropidae (bee-eaters) are colonial, and their burrows therefore occur in multiples.

The woodpeckers and their relatives (Order Piciformes) are, of course, wood-borers *par excellence*; less well known is that a great many of them, particularly in the South American and African grasslands, are earth burrowers—making holes in termite nests and adobe walls, as well as in erosional scarps (Short, 1971a and 1971b). In contrast with the kingfishers and most other burrowing birds, some of the terrestrial woodpeckers excavate in level ground as well as on steep banks.

Although isolated genera and species among that most diversified of avian orders —the Passeriformes—are burrowers,[4] the principal fossorial passerines are various swallows (Family Hirundinidae), including such well-known birds as the bank swallow or sand martin (*Riparia riparia*). As its vernacular and Linnean names imply, this species burrows into banks, usually above water, often in a sandy substrate. The birds are highly colonial creatures, and river bluffs pockmarked with dozens or hundreds of their burrow entrances are a familiar sight on all continents except Australia and Antarctica (see photograph in Wetmore, 1964, p. 128). The wings are used in shaping and enlarging the burrows. Swallow burrows are usually shallower (0.5 to 1 m) than those of kingfishers, and very commonly include a right-angle bend.

Fossil Examples

Although their rarity as fossils has often been exaggerated, birds do have an unsatis-

[4] The tiny (8 cm) Australian flowerpecker *Pardalotus punctatus*, for example, digs a deep burrow —as much as 1 m (Chapman, 1970).

factory paleontological record, particularly before the Pleistocene. As is true with other vertebrate classes, the development of burrowing habits among various groups of birds has no apparent relationship to fossilization potential. This conclusion is suggested by a comparison of the ratios of extinct species to living species in various burrowing and nonburrowing families (Table 15.1).

So far as I know, no fossil bird burrow has previously been reported; therefore, a brief account of some structures that I believe to be Pleistocene swallow burrows may be of interest. The structures (Fig. 15.2) are exposed in a cut bank of the East Branch of Verdigre Creek in northern Antelope County, Nebraska. Modern erosion is rapidly exhuming an outcrop of Tertiary sandstone from beneath a Pleistocene channel fill. The walls of the Pleistocene channel are nearly vertical and display a dozen or more gravel-filled horizontal tubes about 5 cm in diameter, which I interpret as ancient swallow nests.

Two features of these burrows immediately suggest that they are in fact fossil and not modern swallow holes, which are also present at the same exposure but several meters higher on the bank: (1) the structures are completely filled with gravel identical to that in the adjacent Pleistocene terrace fill, and (2) they are only about 1 m above the modern stream—easily within reach of snakes or mammalian predators and therefore in a position most unlikely for a modern burrow.

To test the assumption of ancient origin, I dug away the Pleistocene gravel to expose several square meters of the Miocene sandstone forming the original valley wall, against which the gravel was deposited. Six more gravel-filled tubes were thus exposed, proving that the burrows are indeed older than the gravels. How much older cannot be definitely determined, but because the Miocene sandstone is relatively soft (it can be crushed with the fingers) and therefore

TABLE 15.1 Ratios of Extinct Species to Living Species in Three Groups of Bird Families Showing Differences in Fossorial Activity.* †

Many or Most Species in Family Burrow		Few Species Burrow		No Species Burrow	
Procellariidae	16/56	Columbiformes	9/289	Ardeidae (herons)	18/58
Hydrobatidae	1/18	Psittaciformes	7/315	Podicepedidae (grebes)	8/20
Coraciidae	1/17	Laridae	14/82	Cracidae (curassows)	28/38
Todidae	0/5	Strigidae	10/123	Phasianidae (pheasants)	57/165
Halcyonidae	2/87	Rallidae	64/132	Accipitridae (hawks)	68/205
Bucconidae	1/30	Alcidae	17/32	Anatidae (ducks)	99/145
Capitonidae	1/72	Dromadidae	0/1		

* Numbers of extinct species taken from Brodkorb's catalog (1963–1971); numbers of modern species from van Tyne and Berger (1959).
† Fossorial habits seem to have little effect on chances of fossilization; other factors, such as size, geographic distribution, etc., are probably more important.

easily eroded, the burrows were probably made no more than a few tens of years before being filled with gravel. The age of the gravels is established as late Pleistocene by the presence within them at nearby exposures of *Bison* sp. and mammoth remains.

That the burrows were probably made by birds and not by rodents or other

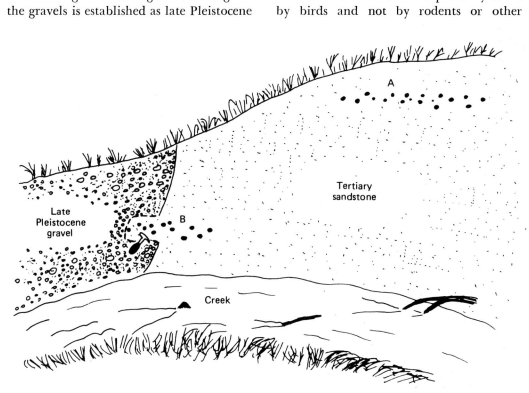

Fig. 15.2 Pleistocene and recent swallow burrows. A, modern bank swallow (*Riparia riparia*) burrows in semiconsolidated sandstone; lower part of Valentine Formation (late Miocene). B, similar structures filled with upper Pleistocene gravel; part of gravel bed removed to show that burrows pre-date gravel. Sketch of outcrop on right bank of East Branch, Verdigre Creek, NE¼, NW¼, NE¼, sec. 9, T 28 N, R 7 W, Antelope County, Nebraska. 30-cm geology hammer for scale.

terrestrial animals is indicated by the following observations: (1) the burrows are entirely horizontal, uniformly shallow culs-de-sac, not connected vertically with the surface; this configuration was determined by digging them out, and (2) their long axes are parallel with each other and normal to the vertical scarp surface. No fossils were found within the burrows, so the identity of the excavators cannot be proved. However, the only burrowing birds now present in northeastern Nebraska (and therefore presumably present in the Pleistocene as well) are kingfishers (*Megaceryle alcyon*) and bank swallows (*Riparia riparia*). The small diameter (4 to 5 cm), shallow depth (less than 1 m), and presence of a right-angle turn in most of the burrows, combined with their occurrence in multiples, point to the latter as the species most probably responsible.

One geological fact not otherwise easily ascertained is provided by the fossil swallow burrows—namely that the terrace gravels must be quite thick, extending several meters below the present streambed. This thickness is suggested by the low position of the burrows, which if constructed like typical modern *Riparia* nests, originally were at least 3 m above water level when they were formed. Because they were filled with gravel before being destroyed by erosion, a rather rapid rate of deposition for the gravel—2 or 3 m in a few tens of years—is indicated. To see the modern swallows still burrowing into the same bank after an interval of at least 10,000 years is somehow reassuring.

CLASS MAMMALIA

"The burrow is the most widely used type of seasonal or permanent shelter among terrestrial mammals," according to Bourliere (1964, p. 72), who offered an excellent, short account of the natural history of fossorial mammals. Table 15.2 shows that the burrowing habit occurs over a large part of the spectrum of mammalian families. Of the 117 living families whose habits are known, 55 (47 percent) include at least one burrowing species. If bats are excluded from the tabulation, the figures become 55 of 91 (60 percent); and if only nonflying and nonaquatic families are considered (bats, whales, sea cows, and seals being omitted), the ratio reaches 55/67 (82 percent).

This tabulation, of course, does not mean that 82 percent of all terrestrial mammal species are burrowers, but that 82 percent of the families contain burrowers. However, in most families containing one burrowing species—especially among the rodents, insectivores, and carnivores—the majority of the remaining species also burrow to some extent. Some of the "largest" mammalian families, both in numbers of included species and in numbers of individuals, are listed in the table. Detailed information about the habits of each of the several thousand species of terrestrial mammals is not available, but I estimate that at least half of the living species and perhaps three-fourths of the individual terrestrial mammals (excluding man) excavate burrows.

Mammals as Sedimentologic Agents

Before surveying the digging habits of various mammalian groups individually, assessing the possible geologic consequences of mammalian burrowing in general may be of interest. Darwin's (1881) well known inquiry into the role of earthworms as geological agents and Grinnell's (1923) similar study of burrowing rodents in California will serve as models for the following crude attempt to quantitatively estimate the potential effect of burrowing mammals on the stratigraphic record.

How much sediment has been moved by burrowing mammals in North America during the Cenozoic? If the following assumptions are made:

TABLE 15.2 Families of Mammals Known to Include at Least One Species that Digs its Own Burrow.*

Order Monotremata 2/2	Order Pholidota 1/1	Hystricidae—Old World
Tachyglossidae—	Manidae—pangolins	porcupines
spiny anteaters	Order Tubulidentata 1/1	Caviidae—cavies
Ornithorhynchidae—	Orycteropodidae—	Dasyproctidae—agoutis
platypus	aardvarks	Chinchillidae—chinchillas
Order Marsupialia 6/7	Order Lagomorpha 2/2	Capromyidae—hutias
Didelphidae—opossums	Leporidae—rabbits	Myocastoridae—coypus
Dasyuridae—marsupial	Ochotonidae—pikas	Octodontidae—hedge rats
"mice"	Order Rodentia 26/32	†Ctenomyidae—tuco tucos
†Notoryctidae—marsupial	Aplodontidae—sewellels	Abracomidae—chinchilla
"moles"	Sciuridae—squirrels	rats
Peramelidae—bandicoots	†Geomyidae—gophers	Thryonomyidae—cane rats
Phascolomidae—wombats	Heteromyidae—	†Bathyergidae—African
Macropodidae—kangaroos	pocket mice	mole rats
Order Insectivora 7/7	Castoridae—beavers	Order Carnivora 6/7
†Tenrecidae—tenrecs	Pedetidae—springhaas	Canidae—dogs
Potamogalidae—shrew-	†Cricetidae—New World	‡Ursidae—bears
otters	mice	‡Procyonidae—raccoons
†Chrysochloridae—golden	†Spalacidae—mole rats	Mustelidae—weasels
"moles"	†Rhyzomyidae—bamboo	Viverridae—civets
Erinaceidae—hedgehogs	rats	Hyaenidae—hyaenas
Soricidae—shrews	Muridae—Old World mice	Order Hyracoidea 1/1
†Talpidae—moles	Gliridae—dormice	Procaviidae—conies
Macroscelididae—	Seleviniidae—dzalmans	Order Artiodactyla 2/9
elephant shrews	Zapodidae—jumping mice	‡Suidae—pigs
Order Edentata 1/3	Dipodidae—jerboas	‡Tayassuidae—peccaries
Dasypodidae—armadillos		

* Numbers following name of order = number of burrowing families/number of families in order.
† Indicates family including at least one species highly specialized for fossorial life.
‡ Indicates family whose members burrow only occasionally or in only part of their geographic range.
The following families were omitted from the tabulation because of lack of information about their habits: Caenolestidae, Solenodontidae, Dinomyidae, Heptaxodontidae. Information on habits mostly from Walker (1968), Matthews (1971), and Anderson and Jones (1967).

1. 10^{19} individual terrestrial mammals have inhabited the land area of North America during the Cenozoic.[5]
2. One animal in ten was a burrower. (This estimate seems conservative by modern standards, but as mentioned above, fossorial habits probably have become progressively more widespread among mammals during the Cenozoic and may have reached an all-time high during the Quaternary, especially after the differential extinction of large mammals in the late Pleistocene.)

[5] Calculated as follows: A, average emergent land area of North America = 2×10^7 km². (The present area is 2.4×10^7 km², but Cenozoic submergence of marginal areas shrank the habitable land area somewhat.) B, average population density = 5,000 individuals/km². [The absolute number of mammals living under natural conditions in any area is almost impossible to determine; Hall's (1946, p. 56) estimate of the population density of mammals in Nevada (5,000/km²) and Mohr's (1940) compilations were utilized in making the above guess. More recent work on populations of deserts (MacMillen, 1964; Jorgensen and Hayward, 1965), forests (Aulak, 1967; Golley et al., 1965) and prairies (Sanderson, 1950) were also consulted. Birds, which are much more easily censused than mammals, average 1,000 individuals/km² in North America (Peterson, 1948) and are probably outnumbered by mammals by a factor of at least 5 in most areas (see, for example, Williams, 1936).] C, average rate of population turnover = twice annually. (Possibly an underestimate because the most numerous mammals have many litters per year.) D, length of the Cenozoic = 65×10^6 years.

3. Each burrower excavated one liter of sediment. [This figure also seems conservative, considering that a mole can move this much material in a few minutes (Godfrey and Crowcroft, 1960, p. 27–28), and individual prairie dog burrows may be hundreds of liters in volume (Sheets et al., 1971). In contrast, a small mouse or shrew might move only a few thimblesful of sediment in its lifetime.]

then mammals should have moved approximately 10^{18} liters (10^6 km^3) of sediment during the Cenozoic in North America.

Both Darwin and Grinnell, in the works cited above, arrived at considerably higher estimates of the sediment moved by burrowers. Converted to liters per square kilometer per year, our estimates are as follows:

Earthworms in Great Britain (Darwin):
$$5 \times 10^6 \text{ liters/km}^2/\text{yr}$$
Rodents in California (Grinnell):
$$2.5 \times 10^6 \text{ liters/km}^2/\text{yr}$$
Burrowing mammals in North America (this chapter):
$$10^3 \text{ liters/km}^2/\text{yr}$$

Perhaps the above differences reflect the particularly favorable habitats studied by the earlier workers. In any case, I suspect that the estimate given here is a minimum. Keeping in mind the uncertainties involved in arriving at this figure, to compare it with the volume of nonmarine Cenozoic sediments in North America is instructive, in order to estimate the percentage of such strata that have been burrowed. No estimate of the total volume of nonmarine Cenozoic strata in North America is known to me, but I have calculated this quantity to be about 2,000,000 cubic kilometers.[6] This quantity is only about twice the estimated volume of sediment burrowed by Cenozoic mammals. Of course, only about 30 percent of the mammals were living in areas now covered by nonmarine Cenozoic strata (assuming that population densities averaged about the same in Cenozoic and non-Cenozoic bedrock areas). Therefore, if no biases were operating against the burrowers or the burrows, about one-sixth

of the total volume of nonmarine Cenozoic strata should show evidence of burrowing. Even workers who, like myself, are convinced that fossil burrows are often overlooked would not suggest that the burrows approach such a degree of abundance. The record must be strongly biased, but what are the sources of the bias?

Distribution and Preservation of Mammalian Burrows

Most workers agree that nonmarine Cenozoic strata are primarily fluvial and lacustrine in origin. The comparative rarity of mammal burrows in waterlaid sediments is not unexpected. Most burrowing mammals avoid wet areas (exceptions include beavers, nutrias, and desman moles), probably because they are particularly vulnerable to drowning. Also, the cut-and-fill mode of deposition characteristic of stream channels also tends to remove any evidence of burrowing that occurs before the channel eventually becomes inactive. Floodplains, particularly in their more distal parts, may seem to combine a more suitable habitat for burrowers, i.e., a less destructive mode of sedimentation, and therefore to be more favorable for burrow preservation; this indeed seems to be the case. But even the infrequently inundated floodplains seem to support smaller populations of burrowers than do adjacent uplands. Many observers

[6] Based on the following assumptions: A, 10.3×10^6 km^2 of North America is covered with Cenozoic strata [Gilluly's (1949) estimate for the U.S. (43 percent), extrapolated to include the entire continent.] B, 70 percent of this area (7×10^6 km^2) is blanketed with sediment of nonmarine origin (Olson, 1965, p. 34). C, the average thickness of sediment over this area is 250 m [based on Kuenen's (1941) figure of 1,200 m as the average thickness of post-Proterozoic strata in North America, and assuming that even though the Cenozoic comprises only 10 percent of post-Proterozoic time, it may be represented by 20 percent or more of the preserved sediments for reasons discussed by Gilluly (1949)]. D, the calculated volume of nonmarine Cenozoic strata is thus roughly 1.75×10^{18} liters, or 1.75×10^6 km^2.

(e.g., Grinnell, 1939) have noted the very destructive effects of high water on populations of such burrowing rodents as pocket gophers, pocket mice, and kangaroo rats living on floodplains. Furthermore, disruptions of sediment textures or fabrics by heavy root growth of floodplain forest trees and shrubs may obscure any original burrows.

The immense blankets of Pleistocene loess in the Holarctic region attests to the magnitude of aeolian deposition within relatively recent geologic time, but the importance of wind as a depositional agent in pre-Quaternary time has not been conclusively established. Pleistocene loess (see below) has a relatively high incidence of fossil burrows, and more ancient loesses, if they exist, should also contain abundant evidence of fossorial activity. Positive identification of Tertiary loesses, using modern sedimentological techniques (see also Chapter 23), is needed before we can answer the question: Is widespread burrowing among mammals a relatively recent evolutionary phenomenon?

In summary, fossorial activity is important in the lives of most mammals, particularly in harsh environments; 72 of the 90 species of mammals that inhabit the steppes of Central Asia are burrowers (Formozov, 1966). Underground rodents often provide homes for less efficient diggers. The microclimate inside a burrow is usually much milder and more uniform than that outside; thus, by using burrows, many organisms are allowed to extend their ranges into steppe, tundra, or desert areas not otherwise habitable by them. For instance, Vaughn (1961) found 22 species of terrestrial vertebrates in occupied or abandoned pocket gopher burrows in a 0.5-km² grassland area in northeastern Colorado. Included were 3 species of amphibians, 5 reptiles, 1 bird, and 11 mammals; 15 of these species were found in gopher holes on a regular basis, and probably could not have successfully occupied the area without the gophers' pioneering work.

Notes on Selected Modern Mammalian Burrowers

Monotremata

The platypus and the echidna, whether or not one regards them as mammals, are both burrowers. Burrell (1927) described the distinctive, backfilled burrow of the platypus. Its riverbank tunnels (as much as 13 m long) have a flat floor, the roof arching over it. The end is plugged with earth and vegetation in order to maintain constant humidity for the eggs. Whether the burrowing habit of monotremes is a carryover from therapsid predecessors, or evolved anew in Australia, cannot be determined on the basis of the almost nonexistent fossil record of the former. None of the known therapsids seems particularly adapted to digging, but the postcranial anatomy of the smaller forms is largely unknown. Did burrowing play any part in the origin of the mammals?

Marsupialia

The "marsupial mole" *Notoryctes*, of Australia, is the only completely fossorial marsupial, but a few members of nearly all the metatherian families dig extensive burrows—even one of the kangaroos (*Bettongia lesueur*) constructs burrows about 3 m deep (Troughton, 1947, p. 158). Distinctive spiral burrows are dug by the "rabbit-bandicoots" of the genus *Thylacomys* (Troughton, 1947, p. 69). The yapok, or water opossum (*Chironectes*), of South America, the only aquatic marsupial, burrows in riverbanks.

Insectivora

The insectivores as a whole are the most diversified group of mammalian burrowers. At least some members of all the living families are fossorial, and the Talpidae and Chrysochloridae include some of the most highly specialized of all subterranean animals. Shrews (Family Soricidae) are not so highly specialized for digging as moles, but according to Matthews (1971, p. 59), all the approximately 300 species dig burrows. Thanks to the studies of Eisenberg and Gould (1970) on the life histories of tenrecs in Madagascar, we now know that most members of this family of "living fossils" are also burrowers. One genus, *Oryzorictes*, is very mole-like in adaptations.

The most successful of fossorial insectivores in the Holarctic region are unquestionably the moles (Family Talpidae). Extended discussions of talpid burrows and burrowing techniques were given by Godfrey and Crowcroft (1960) for the European mole, and by Hisaw (1923a, 1923b) for the American *Scalopus aquaticus*. Moles excavate three types of passages: (1) deep tunnels, which are permanent runways as much as 3 m below the surface and often tens of meters long, the making of which entails building the famous "molehills"; (2) shallow surface burrows, which pose no sediment-disposal problem; and (3) surface grooves (*traces d'amour*), said to be dug during the rutting season. An example of the modification of burrowing behavior to meet local environmental conditions is provided by Brown's (1972) study of *Scalopus aquaticus* in Florida. In warm, moist, sandy, subtropical soils, this mole digs only very shallow (average: 20 cm) burrows, in contrast with the large burrows of the species in temperate climates, often many meters deep.

In Africa the "golden moles" (Chrysochloridae) occupy the ecological niche of talpids, although the two are only distantly related. Like true moles, the chrysochlorids construct elaborate underground passageways and move enormous amounts of sediment (Genelly, 1965).

Edentata, Pholidota, Tubulidentata

Although perhaps not closely related phylogenetically, the Orders Edentata, Pholidota, and Tubulidentata include fairly large terrestrial mammals in which the teeth have been reduced in connection with an insectivorous diet; many are termite feeders. The armadillos (Edentata: Dasypodidae) are, next to rodents, the most important burrowing mammals in South America; the digging ability of such forms as *Dasypus novemcinctus* is legendary (Taber, 1945). *Manis*, the sole genus of scaly anteaters (Order Pholidota), includes both arboreal and terrestrial species, the latter living in burrows. The aardvark is the only living representative of the Order Tubulidentata. Having a length of 2 m and a weight of 65 kg, it is probably the largest animal to spend a large part of its life burrowing.

Lagomorpha

Most rabbits and hares dig only shallow depressions ("forms") for temporary shelter and for rearing their young. The European rabbit *Oryctolagus cuniculus*, however, digs extensive communal burrow systems called "warrens." In America, only 1 of the 24 living species, *Sylvilagus* (*Brachylagus*) *idahoensis*, digs its own burrows, although other species of *Sylvilagus* and *Lepus* may resort to burrows made by other animals. Pikas (Family Ochotonidae), which live in mountainous regions, usually nest in rocky crevices; but members of lowland colonies in Asia are known to dig their own burrows, and captive American pikas readily burrow (Markham and Whicker, 1972).

Rodentia

Gnawing mammals far surpass all other terrestrial mammals in terms of both diversity and population. Most are burrowers. They construct holes ranging in size and complexity from temporary underground shelters to permanent burrows in which a single animal spends a whole season, or even its entire life. In some cases, males and females excavate distinguishable burrows; in others, communal burrow systems are constructed, which may be inhabited continuously for many generations. Many species dig two types of burrows, one for summer and one for winter [e.g., *Dipus sagitta*, the rough-legged jerboa (Ognev, 1948, p. 272)].

Two categories of burrowing rodents are of special interest: (1) those that dig extensive burrow systems but which use them mostly for sleeping, hiding, and reproducing, and (2) those that are perpetually subterranean and come to the surface only accidentally. In the first category are such groups as the ground squirrels (*Citellus, Spermatophilus*), marmots (*Marmota*), prairie dogs (*Cynomys*), kangaroo rats (*Dipodomys*), and various voles (Microtinae) in the Holarctic region; degus (*Octodon*) and viscachas (*Lagostomus*) in South America; and desert jerboas (*Jaculus*) in northern Africa. Figure 15.3 shows some of the variety of architecture found in burrow systems of surface-feeding rodents. Some structures are very elaborate, having multiple exits and entrances, nest chambers, food storage rooms, latrines, etc.

The second category—completely subterranean groups—includes members of the six rodent families marked by a "dagger" in Table 15.2. The pocket gophers (*Geomys, Thomomys*) are endemic to North America. Their enormous incisor teeth are used in digging their extensive tunnels, which they rarely leave. Individual burrows of *Geomys bursarius* up to 0.14 km long have been reported (Scheffer, 1940). Although their range overlaps that of various moles, the two groups tend to favor different habitats; gophers inhabit relatively dry open areas, and moles moister forested terrain. In South America the superficially very

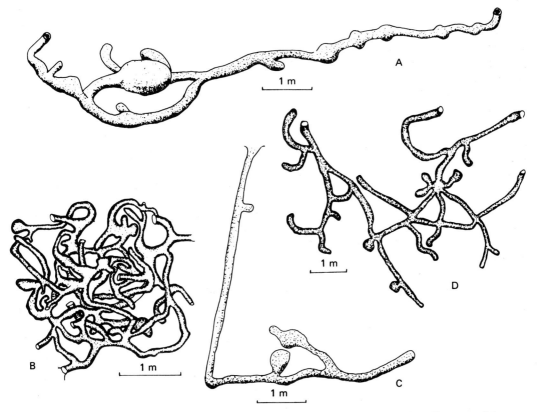

Fig. 15.3 Burrow systems of recent surface-feeding rodents. A, perspective diagram of burrow system of chipmunk (*Tamias striatus*); Wisconsin (from Panuska and Wade, 1956.) B, plan view of kangaroo rat (*Dipodomys spectabilis*) burrow system; Arizona (from Vorhies and Taylor, 1922). C, lateral view of prairie dog (*Cynomys ludovicianus*) burrow; Nebraska (from Merriam, 1902). D, plan view of prairie dog (*C. gunnisoni*) burrow system; Arizona (from Foster, 1924).

gopher-like tuco-tucos are the principal underground foragers. In earlier times their burrows completely undermined and denuded the soil over large areas (Walker, 1968, p. 1048). The mole mice (Cricetidae: *Notiomys*) of Patagonia are so completely fossorial that they emerge from their burrows only when the ground becomes saturated.

The African mole-rats, like most burrowers, have minute eyes and ears, enlarged incisors, and short limbs and tail. One genus, *Heterocephalus*, is also practically hairless, making it look like "a little pink sausage with a pair of legs at each end and four enormous protruding incisor teeth at the front" (Matthews, 1971, p. 94). *Heterocephalus* burrows in a unique manner, described by Jarvis and Sale (1971) as follows:

> A chain of mole-rats operates rather like a chain saw with each animal in the lower string removing the soil with its incisors, pushing it back to the exit and then clambering over the retreating individuals' backs to the working surface of the tunnel. Alternatively, and less frequently, the chain may remain still and the soil passed from one individual to the next.

H. glaber burrow systems may cover 10,000 m² and accommodate as many as 20 individuals.

The only rodents having no external trace of eyes are the Mediterranean mole rats in the genus *Spalax*. Like the pocket gophers, they build large mounds with the sediment excavated from their deep tunnels. The mounds may be 2.5 m in diameter and 1 m high; in poorly drained areas they tend to be larger, offering more protection from floods.

Carnivora

Most burrowing carnivores dig straight, simple tunnels, contrasting sharply with the elaborate labyrinths favored by rodents. Cats (Felidae) seem to be the only carnivore family showing no fossorial tendencies. Canids, both wild and domestic, are efficient but opportunistic diggers; foxes and coyotes, for example, do not hesitate to excavate their own burrows, although they also willingly occupy burrows vacated by other animals. In contrast, bears seldom burrow, preferring to use caves, hollow trees, etc., for winter dens. Where the terrain does not offer such natural cavities, however—for example, in parts of Minnesota (Johnson, 1930) or Siberia (Stroganov, 1952, p. 119—*Ursus americanus* and *U. arctos* will dig their own burrows. Perhaps the largest burrow ever described (1.5 m in diameter; 4 m long) was dug by a polar bear, *U. maritimus* (Doutt, 1967). Raccoons also prefer to nest in hollow trees, but utilize ground burrows in some parts of their range (Dorney, 1954; Gysel, 1961); whether they dig the burrows themselves is not known. Giant pandas are said to dig earthen dens (Pen, 1962).

Most species in the Old World family Viverridae are not particularly specialized for burrowing, but the South African mierkats (*Cynictis*) dig extensive warrens (Matthews, 1971, p. 276). All four species of Hyaenidae sometimes dig their own burrows, but are said usually to enlarge those made by aardvarks or other animals (Matthews, 1971, p. 279). The carnivores most highly adapted for digging are the badgers (*Meles* and *Arctonyx* in Eurasia, *Taxidea* in North America), although most other members of the Family Mustelidae are also fossorial to some extent. Many mustelids prey upon fossorial rodents by digging them out of their holes. Contrary to the rule with most carnivores, badgers excavate complicated, multichambered, permanent burrows.

Artiodactyla

Hoofed animals are obviously poorly suited for burrowing, but a few are known to make use of underground cavities. Old World pigs (*Sus*, *Babyrousa*) disturb considerable volumes of sediment by rooting, but do not burrow. The African wart hog, however, like many other open-savannah animals, finds shelter in burrows—sometimes those made by aardvarks, sometimes their own (Bradley, 1971). New World pigs, or peccaries, usually find shelter in overhanging banks or caves; but where natural shelters are not available, they occasionally excavate their own [(Miller, 1930) found collared peccaries in burrows probably of their own making, and the digging ability of these animals is attested to by Neal (1959)].

Fossil Examples

Mammalian limb bones having peculiarities interpreted as adaptations for digging are known from rocks as old as Paleocene (Reed, 1954), but the oldest mammal burrows described in the literature are Miocene in age. M. F. Skinner (1972, personal communication) found burrows in the Brule Formation (Oligocene) at two localities in western Nebraska.[7] At both sites the burrows are excavated in pink clay and filled with strongly contrasting blue-gray volcanic ash. The identity of the burrowers is not known; several dozen species of Oligocene rodents and insectivores are in the appropriate size range.

Casts of large (as much as 3 m long) helical rodent burrows weathering out of the Harrison Formation (lower Miocene) in western Nebraska first attracted scientific attention in 1892, when E. H. Barbour described them as "a new order of gigantic fossils" and named them *Daimonelix* (Fig. 15.4). These peculiar "devil's corkscrews" were at various times interpreted as freshwater sponges and as lianas, but Schultz (1942), in his extensive review of the problem, confirmed earlier suggestions that they are indeed burrows. Skeletons of the primitive beaver *Paleocastor* are often found within them, and the remains of a carnivore

Fig. 15.4 Three examples of *Daimonelix*, burrow of the primitive beaver *Paleocastor*. Burrows exhibit size variation and both dextral and sinistral coils; numerous root casts visible. Harrison Formation (lower Miocene); Sioux County, Nebraska. (From photograph in Barbour, 1892.)

—perhaps trapped while on a hunting expedition—were found in one burrow. Although occasional individual modern sciurids dig spiral burrows (Schultz, 1942) and, as noted above, the marsupial *Thylacomys* is reported to do so also, no other known beavers developed the habit. The behavior pattern may, in fact, have been confined only to certain populations of *Paleocastor*, because the beaver is found in other areas and horizons yet *Daimonelix* is restricted to the upper part of the Harrison Formation (*sensu stricto*) in westernmost Nebraska and immediately adjacent parts of Wyoming and South Dakota.

Much remains to be learned about *Daimonelix*. Why the helical shape? Possibly it solves the problem of digging closely spaced but noninterconnecting burrows without subjecting the moderately large occupants to climbing a grade of more than about 40°. (See also Toots, 1963.) Was *Paleocaster* a hibernator? The answer might be found by studying the age distribution of individual beavers found in burrows, using the techniques employed by Kurtén (1958) in his study of cave bears.

Three Pliocene examples of mammal burrows have been published. Lugn (1941) presented photographs of burrows filled with volcanic ash in the Ash Hollow Formation in Sheridan County, Nebraska, but did not further locate them geographically or discuss them in any detail. Voorhies and

[7] (1) Badlands 2.5 mi NW of Chadron, Nebraska, in W 1/2, NE 1/4, sec. 6, T 33 N, R 48 W, Dawes County, Nebraska, and (2) Brule exposures northeast of Scott's Bluff, between Country Club and Platte River, in Scottsbluff County, Nebraska.

Toots (1970) described a large burrow at the unconformity between the Ash Hollow Formation and underlying Cretaceous chalk in south-central Nebraska. We interpreted it as possibly the work of a badger-like carnivore, on the basis of size and shape alone; no skeleton was found in the burrow. Voorhies (1974) reported fossil pocket mouse burrows in the Ash Hollow Formation.

Two other Pliocene burrows not previously reported may be discussed briefly here; both serve to emphasize the requirement that a lining or filling of contrasting material be present, else burrows are likely to be overlooked. The first (Fig. 15.5) occurs in northeastern Nebraska, in poorly consolidated fine sands of probable floodplain origin (upper part of Ash Hollow formation). Gray volcanic ash fills the burrows, to depths of about 1 m. Unfortunately, no bones were found that would identify the excavators.

The other occurrence is considerably more informative. I discovered this burrow (Fig. 15.6) a few miles from the previous ones, at approximately the same stratigraphic level. The most interesting feature is the "food storage chamber" partially filled with hackberry seeds. While preparing this structure for study, I found in it an upper molar of a ground squirrel (*Citellus* sp.); unfortunately, the remainder of the skeleton, if it had ever been present, was destroyed during the road construction that exposed the burrow. [Similar food caches were reported in burrows of modern *Citellus tridecimlineatus* by Rongstad (1965), who also noted that, in the population studied by him, mortality usually occurred in the burrows.] Much of the fossil ground squirrel burrow has unfortunately been bulldozed away, but enough remains to show that it had been a rather elaborate structure having several "escape tunnels" and numerous side branches. The typical burrowing pattern of ground squirrels thus seems to have been developed by late Clarendonian time (late early Pliocene, on the North American land-mammal time scale).

In the Pleistocene, burrows appear to become much commoner. All published occurrences known to me are from upper

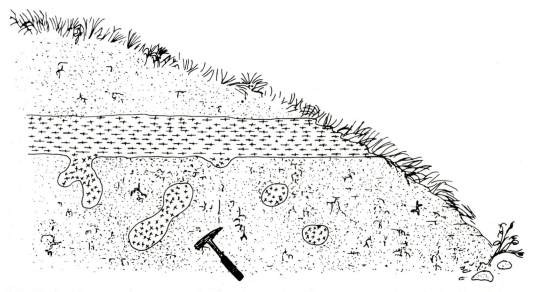

Fig. 15.5   (?)Rodent burrows filled with volcanic ash. Burrows do not penetrate the ash and thus must have been present when the ash fell. Upper part of Ash Hollow Formation (lower Pliocene), exposed in ravine in NW¼, SE¼, NE¼, sec. 21, T 28 N, R 7 W, Antelope County, Nebraska. 30-cm geology hammer for scale.

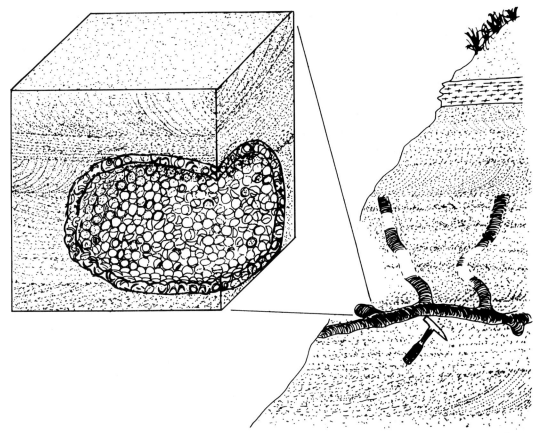

Fig. 15.6 Part of burrow system of an early Pliocene ground squirrel (*Citellus* sp.). Inset shows detail of apparent food-storage chamber lined with hackberry (*Celtis*) seeds. The use of cross-bedded channel sands as a substrate for burrow construction is unusual. The sand may have been partially lithified when burrowing occurred, as indicated by the sharp contact between unbedded sand filling the burrows and well-bedded sand forming the wall. Ash Hollow Formation; road-cut in NW¼, SW¼, NW¼, sec. 22, T 28 N, R 8 W, Antelope County, Nebraska. 30-cm geology hammer for scale.

Pleistocene deposits, but M. F. Skinner (1972, personal communication) reported that rodent burrows occur in the lower Pleistocene beds of north-central Nebraska.[8] Wood and Wood (1933) described and illustrated large burrow casts at the upper Pleistocene Rock Creek locality in the Texas

---

[8] On the Harold Johnson ranch, NW corner, NE ¼, sec. 29, T 31 N, R 21 W, Brown County, Nebraska, on west end of large outcrop ¼ mi below ranch buildings, on north side of Sand Draw; burrows extend from massive reddish tan sand into a gray sand. Although no bones were found in the burrows, the fact that the most abundant fossil mammal collected in the Sand Draw beds is a large pocket gopher, *Geomys quinni* (see Skinner and Hibbard, 1972), is probably not coincidental.

Panhandle; although they referred the burrows to *Daimonelix*, the lack of coiling in the structures makes this designation inappropriate. Burrows and nests of ground squirrels and lemmings occur in the Pleistocene of Alaska (Porsild et al., 1967, and references cited). Pleistocene burrows attributed to weasels were discovered by Brown (1908) in the Conard fissure in Arkansas; quantities of rodent and rabbit remains, some of them bearing tooth marks, were found in the burrows. Vertebrate burrows in Pleistocene loess have often been mentioned but seldom described. Fossil burrows, frequently called "crotovinas," are often associated with buried soils in European loess deposits. In an extremely interesting study, Dinesman (1971) showed that detailed records of climatic and biotic changes over several thousands of years may be preserved in soil profiles built by the activities of deep burrowers such as foxes and marmots. By analyzing the bone and pollen content in sequential, radiocarbon-dated levels in an arctic fox lair in the U.S.S.R., he documented the transition from taiga to tundra over a period of 5,400 years.

Burrows containing skeletons of ground squirrels are said to be so common in certain upper Pleistocene buried soils in Nebraska that their presence defines the "*Citellus* zone"; Condra et al. (1950) stated, however, that this is not a true stratigraphic zone because it crosses formation boundaries, and the burrows may extend through several biostratigraphic units. Unfortunately, descriptions of these burrows and their inhabitants have not been published. Thorpe et al. (1951) gave the only more or less specific location for an outcrop of the "*Citellus* zone" known to me: "Yankee Hill Brick Plant, Lincoln, Nebraska." These burrows, if as abundant as they are alleged to be, would be well worth studying. The environmental requirements of different species of modern rodents differ enough that their presence should be useful in defining with some precision the Pleistocene climate, groundwater levels, etc., wherever they are found.

CONCLUSIONS

Fossil vertebrate burrows are scarcer in publications than they are in sedimentary rocks. Their potential importance as clues in paleoecological reconstruction has barely been exploited, except in the case of Permian lungfish and, to a lesser extent, Miocene beavers. The structure of the sediment immediately surrounding vertebrate fossils is almost never described in sufficient detail to allow an interested worker to determine whether or not a particular animal may have died in a burrow—an item of information at least as significant as the number of bumps on the *processus anconaeus* and which, unlike the latter, cannot be determined once the enclosing matrix is gone. The possibility exists that burrowing on a large scale (by mammals) did not begin before Pleistocene glaciation; but so few critical observations have been made that this possible "behavioral evolution" remains, for the present, an open question.

ACKNOWLEDGMENTS

Donald Baird, E. H. Colbert, R. W. Frey, A. S. Romer, and W. A. S. Sarjeant read the manuscript and made many helpful suggestions. One could hardly ask for a more illustrious panel of reviewers. All remaining errors are, of course, my own.

REFERENCES

Ager, D. V. 1963. Principles of paleoecology. McGraw-Hill, 371 p.

Allen, J. R. L. 1961. Sandstone-plugged pipes in the lower Old Red Sandstone of Shropshire, England. Jour. Sed. Petrol., 31:325–335.

Anderson, S. and J. K. Jones, Jr. 1967. Recent mammals of the world: a synopsis of families. New York, Ronald, 453 p.

Auffenberg, W. 1959. A Pleistocene *Terrapene* hibernaculum, with remarks on a second complete box turtle skull from Florida. Florida Acad. Sci., Quart. Jour., 22:49–53.

———— and W. G. Weaver, Jr. 1969. *Gopherus berlanderi* in southeastern Texas. Florida State Mus., Bull. 13:141–203.

Aulak, W. 1967. Estimation of small mammal density in three forest biotopes. Ekologia Polska, Ser. A, 15:755–778.

Barbour, E. H. 1892. Nature, structure, and phylogeny of *Daimonelix*. Geol. Soc. America, Bull., 8:305–314.

Bellairs, A. 1957. Reptiles: life history, evolution and structure. London, Hutchinson, 192 p.

————. 1970. The life of reptiles (2 vols.). New York, Universe, 590 p.

Bigelow, H. B. and W. C. Schroeder. 1948. Fishes of the western North Atlantic, pt. 1. Sears Found. Marine Res., Mem. 1(1), 576 p.

———— and W. C. Schroeder. 1953. Fishes of the western North Atlantic, pt. 2. Sears Found. Marine Res., Mem. 1(2), 588 p.

Boekschoten, G. J. 1964. Tadpole structures again. Jour. Sed. Petrol., 34:422–423.

Bogert, C. M. and J. A. Oliver. 1945. A preliminary study of the herpetofauna of Sonora. Amer. Mus. Nat. Hist., Bull., 83:297–426.

Bourliere, F. 1964. The natural history of mammals. New York, Knopf, 387 p.

Bradley, R. M. 1971. Warthog (*Phacochoerus aethiopicus*) burrows in Nairobi National Park. East African Wildlife Jour., 9:149–152.

Bragg, A. N. 1944, 1945. The spadefoot toads in Oklahoma with a summary of our knowledge of the group. Amer. Naturalist, 78:517–533; 79:52–72.

Brodkorb, P. 1963–1971. Catalogue of fossil birds, Pts. 1–4. Florida State Mus., Bull., 7:179–293; 8:195–335; 11:99–220; 15:163–266.

Bromley, R. G. 1970. Borings as trace fossils and *Entobia cretacea* Portlock, as an example. In T. P. Crimes and J. C. Harper (eds.), Trace fossils. Geol. Jour., Spec Issue 3:49–90.

Brown, B. 1908. The Conard fissure, a Pleistocene bone deposit in northern Arkansas: with descriptions of two new genera and twenty new species of mammals. Amer. Mus. Nat. Hist., Mem. 9:155–208.

Brown, L. N. 1972. Unique features of the tunnel systems of the eastern mole in Florida. Jour. Mammal., 53:394–395.

Burrell, H. 1927. The platypus. Sydney, Angus and Robertson, 227 p.

Cameron, B. and R. Estes. 1971. Fossil and recent "tadpole nests": a discussion. Jour. Sed. Petrol., 41:171–178.

Carlson, K. J. 1968. The skull morphology and estivation burrows of the Permian lungfish *Gnathorhiza serrata*. Jour. Geol., 76:641–663.

Carr, A. 1963. The reptiles. New York, Time-Life, 183 p.

Carroll, R. L. 1965. Lungfish burrows from the Michigan coal basin. Science, 148:963.

Chapman, B. V. 1970. Common Australian birds. Melbourne, Landsdowne, 143 p.

Colbert, E. H. 1969. Evolution of the vertebrates. New York, John Wiley, 535 p.

Condra, G. E. et al. 1950. Correlation of the Pleistocene deposits of Nebraska (revised version). Nebraska Geol. Survey, Bull. 15A:1–74.

Cook, D. O. 1971. Depressions in shallow marine sediment made by benthic fish. Jour. Sed. Petrol., 41:577–602.

Cox, C. B. 1972. A new digging dicynodont from the Upper Permian of Tanzania. In K. A. Joysey and T. S. Kemp (eds.), Studies in vertebrate evolution. Edinburgh, Oliver & Boyd, p. 173–189.

Darwin, C. 1881. The formation of vegetable mould through the action of worms. London, Murray, 326 p.

Dinesman, L. G. 1971. Mammalian lairs in paleoecological studies and palynology. Jour. Palynol., 7:48–55.

Dorney, R. S. 1954. Ecology of marsh raccoons. Jour. Wildlife Manage., 18:217–225.

Doutt, J. K. 1967. Polar bear dens on the Twin Islands, James Bay, Canada. Jour. Mammal., 48:468–470.

Eastman, R. 1969. The kingfisher. London, Collins, 159 p.

Eisenberg, J. F. and E. Gould. 1970. The tenrecs: a study in mamalian behavior and evolution. Smithsonian Contrib. Zool., 27:1–137.

Estes, R. and M. H. Wake. 1972. The first fossil record of caecilian amphibians. Nature, 239:228–231.

Formozov, A. N. 1966. Adaptative modifications of behavior in mammals of the Eurasian steppes. Jour. Mammal., 47:208–222.

Foster, B. E. 1924. Provision of prairie dogs to escape drowning when town is submerged. Jour. Mammal., 5:266–268.

Freeman, J. R. 1958. Burrowing in the salamanders *Pseudobranchus striatus* and *Siren lacertina*. Herpetologica, 14:130.

Frey, R. W. and J. D. Howard. 1969. A profile of biogenic sedimentary structures in a Holocene barrier island–salt marsh complex, Georgia. Gulf Coast Assoc. Geol. Socs., Trans., 19:427–444.

Gans, C. 1969. Amphisbaenians: reptiles specialized for a burrowing existence. Endeavor, 28:146–151.

Genelly, R. E. 1965. Ecology of the common mole-rat (*Cryptomys hottentotus*) in Rhodesia. Jour. Mammal., 46:647–665.

Gilluly, J. 1949. Distribution of mountain building in geologic time. Geol. Soc. America, Bull., 60:561–590.

Godfrey, G. and P. Crowcroft. 1960. The life of the mole. London, Museum Press, 152 p.

Golley, F. B. et al. 1965. Number and variety of small mammals on the AEC Savannah River Plant. Jour. Mammal., 46:1–18.

Grinnell, J. 1923. The burrowing rodents of California as agents in soil formation. Jour. Mammal., 4:137–149.

————. 1939. Effects of a wet year on mammalian populations. Jour. Mammal., 20:62–64.

Guggisberg, C. A. W. 1972. Crocodiles. Harrisburg, Pa., Stackpole, 195 p.

Gysel, L. W. 1961. An ecological study of tree cavities and ground burrows in forest stands. Jour. Wildlife Manage., 25:12–20.

Hall, E. R. 1946. Mammals of Nevada. Berkeley, Calif., Univ. California Press, 710 p.

Hallinan, T. 1923. Observations made in Duval County, northern Florida, on the gopher tortoise (*Gopherus polyphemus*). Copeia, 1923:11–20.

Halstead, L. B. 1968. The pattern of vertebrate evolution. San Francisco, Freeman, 209 p.

Hisaw, F. L. 1923a. Feeding habits of moles. Jour. Mammal., 4:9–20.

————. 1923b. Observations on the burrowing habits of moles (*Scalopus aquaticus machrinoides*). Jour. Mammal., 4:79–88.

Hoogerwerf, A. 1970. Udjung Kulon: the land of the last Javan rhinoceros. Leiden, Brill, 512 p.

Hotton, N. 1963. Dinosaurs. New York, Pyramid, 192 p.

Hutchison, J. H. 1968. Fossil Talpidae (Insectivora, Mammalia) from the later Tertiary of Oregon. Univ. Oregon, Mus. Nat. History, Bull. 11:1–117.

Jarvis, J. V. M. and J. B. Sale. 1971. Burrowing and burrow patterns in East African mole-rats, *Tachyoryctes*, *Heliophobius*, and *Heterocephalus*. Jour. Zool. Soc. London, 163:451–479.

Johnels, A. G. and G. S. O. Svensson. 1955. On the biology of *Protopterus annectens* (Owen). Archiv Zool., 7:131–164.

Johnson, C. E. 1930. Recollections of the mammals of northwestern Minnesota. Jour. Mammal., 11:435–452.

Jorgensen, C. D. and C. L. Hayward. 1965. Mammals of the Nevada test site. Brigham Young Univ., Sci. Bull., Biol. Ser., 6(3):1–81.

Kevan, D. K. M. 1962. Soil animals. London, Witherby, 237 p.

Kuenen, P. H. 1941. Geochemical calculations concerning the the total mass of sediments in the earth. Amer. Jour. Sci., 239:161–190.

Kühnelt, W. 1950. Bodenbiologie. Vienna, Herold, 368 p.

Kurtén, B. 1958. Life and death of the Pleistocene cave bear: a study in paleoecology. Acta Zool. Fennica, 95:1–59.

Lagler, K. F. et al. 1962. Ichthyology. New York, John Wiley, 545 p.

Langston, W. 1963. Fossil vertebrates and the late Paleozoic red beds of Prince Edward Island. Nat. Mus. Canada, Bull. 187:1–36.

Lugn, A. L. 1941. The origin of *Daimonelix*. Jour. Geol., 49:673–696.

MacMillen, R. E. 1964. Southern California semidesert rodent fauna. Univ. California Publ. Zool., 71:1–66.

McNab, B. K. 1966. The metabolism of fossorial rodents: a study of convergence. Ecology, 47:712–733.

Markham, O. D. and F. W. Whicker. 1972. Burrowing in the pika (*Ochotona princeps*). Jour. Mammal., 53:387–389.

Matthews, L. H. 1971. The life of mammals, Vol. II. New York, Universe, 440 p.

Merriam, C. H. 1902. The prairie dog of the Great Plains. In U.S. Dept. Agriculture Yearbook, 1901. Washington, D.C., Govt. Printing Office, p. 257–270.

Miller, F. W. 1930. Notes on some mammals of southern Matto Grosso, Brazil. Jour. Mammal., 11:10–22.

Mohr, C. O. 1940. Comparative populations of game and other mammals. Amer. Midland Natur., 24:581–584.

Murphy, R. C. 1936. Oceanic birds of South America. New York, Macmillan, 1245 p.

Neal, B. J. 1959. Techniques of trapping and tagging the collared peccary. Jour. Wildlife Manage., 23:11–16.

Neill, W. T. 1971. The last of the ruling reptiles: alligators, crocodiles, and their kin. New York, Columbia Univ. Press, 486 p.

Obruchev, D. V. (ed.) 1964. Agnatha, Pisces. In Y. A. Orlov (ed.), Fundamentals of paleontology. Translated 1967, Israel Prog. Sci. Transl., 11:825 p.

Ognev, S. I. 1948. Rodents. In Mammals of the USSR and adjacent countries. Translated 1963, Israel Prog. Sci. Transl., 6:508 p.

Oliver, W. R. B. 1930. New Zealand birds. Wellington, Fine Arts, 541 p.

Olson, E. C. 1956. Faunas of the Vale and Choza: 11, *Lysorophus*, Vale and Choza; *Diplocaulus* and Eryopidae, Choza. Fieldiana, Geol., 10:313–322.

————. 1965. The evolution of life. New York, Mentor, 302 p.

————. 1967. Early Permian vertebrates from Oklahoma. Oklahoma Geol. Survey, Circ., 74:1–111.

————. 1970. New and little known vertebrates from the Early Permian of Oklahoma. Fieldiana, Geol., 18:395–444.

————. 1971. A skeleton of *Lysorophus tricarinatus* (Amphibia: Lepospondyli) from the Hennesey Formation (Permian) of Oklahoma. Jour. Paleont., 45:443–449.

Panuska, J. A. and N. J. Wade. 1956. The burrow of *Tamias striatus*. Jour. Mammal., 37:23–31.

Pen, H.-S. 1962. Animals of western Szechuan. Nature, 196:14–16.

Peterson, R. T. 1948. Birds over America. New York, Dodd, Mead, 342 p.

Pooley, A. C. 1969. The burrowing behavior of crocodiles. Lammergeyer, 10:60–63.

Pope, C. H. 1939. Turtles of the United States and Canada. New York, Knopf, 343 p.

Porsild, A. E. et al. 1967. *Lupus arcticus* Wats. grown from seeds of Pleistocene age. Science, 158:113–114.

Reed, C. A. 1954. Some fossorial mammals from the Tertiary of western North America. Jour. Paleont., 28:102–111.

Romer, A. S. 1966. Vertebrate paleontology (3rd ed.). Chicago, Univ. Chicago Press, 468 p.

———— and E. C. Olson. 1954. Aestivation in a Permian lungfish. Breviora, 30:1–8.

Rongstad, O. J. 1965. A life history study of thirteen-lined ground squirrel in southern Wisconsin. Jour. Mammal., 46:76–87.

Sanderson, G. C. 1950. Small-mammal population of a prairie grove. Jour. Mammal., 31:17–25.

Scheffer, T. H. 1940. Excavation of a runway of the pocket gopher (*Geomys bursarius*). Kansas Acad. Sci., Trans, 43:473–478.

Schultz, C. B. 1942. A review of the *Daimonelix* problem. Nebraska Univ., Stud. Sci. Technol., 2:1–30.

Seilacher, A. 1964. Biogenic sedimentary structures. In J. Imbrie and N. D. Newell (eds.), Approaches to paleoecology. New York, John Wiley, 432 p.

Sheets, R. G. et al. 1971. Burrow systems of prairie dogs in South Dakota. Jour. Mammal., 52:451–453.

Short, L. L. 1971a. The evolution of terrestrial woodpeckers. Amer. Mus. Nat. Hist., Novitates, 2467:1–23.

————. 1971b. Woodpeckers without woods. Nat. Hist., 80:66–74.

Skinner, M. F. and C. W. Hibbard. 1972. Early Pleistocene preglacial and glacial rocks and faunas of north-central Nebraska. Amer. Mus. Nat. Hist., Bull., 148:1–148.

Smith, H. M. 1956. Handbook of amphibians and reptiles of Kansas (2nd ed.). Univ. Kansas, Mus. Nat. Hist., Misc. Publ. 9, 356 p.

Smith, H. W. 1931. Observations on the African lungfish, *Protopterus aethiopicus*, and on evolution from water to land environments. Ecology, 12:164–181.

Stroganov, S. V. 1952. Carnivorous mammals of Siberia. Translated 1969, Israel Prog. Sci. Transl., 522 p.

Sutton, G. M. 1967. Oklahoma birds. Norman, Okla., Univ. Oklahoma Press, 674 p.

Taber, F. W. 1945. Contributions on the life history and ecology of the nine–banded armadillo. Jour. Mammal., 26:211–226.

Taylor, E. H. 1968. The caecilians of the world. Lawrence, Kan., Univ. Kansas Press, 848 p.

Thorpe, J. et al. 1951. Some post-Pliocene buried soils of central United States. Jour. Soil Sci., 2:1–19.

Toots, H. 1963. Helical burrows as fossil movement patterns. Contr. to Geol., 2:129–134.

Troughton, E. 1947. Furred animals of Australia. New York, Scribner's, 374 p.

van Tyne, J. and A. J. Berger. 1959. Fundamentals of ornithology. New York, John Wiley, 624 p.

Vaughn, P. P. 1964. Evidence of aestivating lungfish from the Sangre de Cristo Formation, Lower Permian of northern New Mexico. Los Angeles County Mus., Contr. Sci., 80:1–8.

Vaughn, T. A. 1961. Vertebrates inhabiting pocket gopher burrows in Colorado. Jour. Mammal., 42:171–174.

Voorhies, M. R. 1974. Fossil pocket mouse burrows in Nebraska. Amer. Midland Natur., 91:492–498.

———— and H. Toots. 1970. An unusual burrow of a Tertiary vertebrate. Contr. to Geol., 9:7–8.

Vorhies, C. T. and W. R. Taylor. 1922. Life history of the kangaroo rat. U.S. Dept. Agriculture, Bull. 1091, 41 p.

Walker, E. P. 1968. Mammals of the world (2nd ed.). Baltimore, Johns Hopkins Press, 1500 p.

Walls, G. L. 1942. The vertebrate eye and its adaptive radiation. Cranbrook Inst. Sci., Bull. 19, 785 p.

Westoll, T. S. 1949. On the evolution of the Dipnoi. In G. L. Jepsen et al. (eds.), Genetics, paleontology, and evolution. Princeton, Princeton Univ. Press, 474 p.

Wetmore, A. 1964. Song and garden birds of North America. Washington, D.C., Nat. Geogr. Soc., 400 p.

Williams, A. B. 1936. The composition and dynamics of a beech-maple climax community. Ecol. Monogr., 6:317–408.

Wood, H. E. and A. E. Wood. 1933. *Daimonelix* in the Pleistocene of Texas. Jour. Geol., 41:824–833.

CHAPTER 16

PROBLEMS IN INTERPRETING UNUSUALLY LARGE BURROWS

RICHARD G. BROMLEY
Institut for historisk Geologi og Palæontologi
Københavns Universitet, Øster Voldgade 10
København, Denmark

H. ALLEN CURRAN
Department of Geology, Smith College
Northampton, Massachusetts, U.S.A.

ROBERT W. FREY
Department of Geology, University of Georgia
Athens, Georgia, U.S.A.

RAYMOND C. GUTSCHICK
Department of Earth Sciences, University of Notre Dame
Notre Dame, Indiana, U.S.A.

LEE J. SUTTNER
Department of Geology, Indiana University
Bloomington, Indiana, U.S.A.

SYNOPSIS

Although marine burrows of unusually large dimensions have long been known in certain areas, they are probably much more widespread in the rock record than is generally recognized. Such burrows constitute a heterogeneous group, having little in common other than "exceptional" size. Yet their size alone unites them in difficulty of interpretation: e.g., densely spaced ?dwelling burrows or combined dwelling-escape burrows as much as 12 cm in diameter and 5 m long; vertical dwelling burrows only 0.5 cm in diameter but up to 9 m long; possible escape structures as much as 24 cm in diameter and 3 m long, subsequently penetrated in some cases by secondary burrow-like structures.

Numerous special problems are encountered in the study and interpretation of burrows of these extreme dimensions: (1) field exposure and accessibility, so that the full extent, or a large part, of the structures can be studied; (2) preservation of the burrows in continuity, not merely in places where they pass through certain beds or within concretion horizons; (3) the "fossilization barrier"; our knowledge of comparable modern structures of similar dimensions or of the animals responsible for them is negligible; and (4) the possibility that certain of these unusual structures were formed by

The authors are here listed alphabetically; respective parts written by each are indicated in the text, except that all collaborated on the synopsis, introduction, and conclusions.

physical rather than organic processes; again, our criteria for comparisons are limited.

The examples selected by us—from the Permian of Montana, Idaho, and Wyoming, the Cretaceous and Paleocene of northwestern Europe, and the Pleistocene of North Carolina *—are intended primarily (1) to call additional attention to such intriguing structures, and (2) to illustrate some of the problems involved in interpreting their origin and function. Hopefully, future work will solve many of these problems.*

INTRODUCTION

Cross-cutting tubular sedimentary structures are in many ways difficult to interpret in the geological record, especially when their dimensions exceed those of structures usually found in modern environments. The first question that arises is: are they of organic or inorganic origin? Of course, both types are possible. Because this book is about trace fossils, we focus here on what we feel to be structures of organic origin, or physical structures that exhibit strikingly burrow-like features.

When the geologist in fact encounters a possible burrow so large that its diameter may be measured in decimeters or its length in many meters, a first thought is apt to be either "this must be artificial or inorganic!," or "this was made by a fairly large vertebrate animal." The popularly known "corkscrew" burrow *Daemonelix* is a classic example of the latter (see Fig. 15.4), and other vertebrate burrows of very large size have been documented. But these occur in continental sediments. Where the structure is found in marine sediments, and bears attributes of smaller burrows made commonly by invertebrate animals, one's awe increases. And then the problems in interpretation begin to unfold.

Presence of the actual trace-making organism buried and preserved within the burrow is conclusive; however, field experience in the search for such evidence usually reveals that the organism was effectively elusive. In the absence of body fossils representing animals that may have been responsible for the structures, what is the evidence that links the structures to burrowing organisms?

The principle of uniformity adds very little in resolving the problem. Our knowledge of comparable modern structures of similar dimensions, or of the animals responsible for such structures, is negligible. In the rock record, we may even have problems in determining the size of the tracemaker. Difficulties inherent in observing such structures in situ in modern marine environments currently preclude our making comparisons of the modern analog with possible fossil examples (see Chapter 2). Furthermore, circumstances of preservation and exposure on the outcrop obscure structures of this kind; rocks are too soft or too hard, fractured, jointed, and erodable, so that complete exposure of single elongate burrows or burrow systems is unlikely. Also, only rarely may we totally discount the possibility that the large structures are physical in origin, masquerading as biogenic sedimentary structures. (See Chapter 5.)

We of course are not the first to observe structures of this kind. The "paramoudras" of Northern Ireland have been known for a century and a half; Seilacher (1964) interpreted some burrows to be 4 m long, in turbidites. But each of us has been awed by such structures, and have spent considerable time pondering their meaning. We still have many questions; but we decided to collaborate here in the hope of drawing increased interest from other quarters, thus accelerating further work on such enigmatic burrows. Our framework for discussing the structures is mainly stratigraphic, oldest specimens first.

PERMIAN BURROWS

Raymond C. Gutschick and Lee J. Suttner

Peculiar cylindrical sedimentary structures, as much as 6 m long, are common in sandstone, chert, and shale-mudstone facies of Permian rocks in southwest Montana, southeast Idaho, and northwest Wyoming. These enigmatic structures have been regarded both as organic and inorganic in origin. Similar cylindrical structures in (?)Permian siltstones of western Colorado "were probably formed by the movement of fresh or salt water or silt through the bed, because other agencies fail to explain the structures satisfactorily" (Gableman, 1955, p. 223). Based on morphology and distribution, such cylindrical columns were regarded as burrows of bottom-dwelling animals by Cressman and Swanson (1964), primarily because they could not envision a suitable inorganic process. Conversely, Yochelson (1968) did not consider the large size and regularity of the columns to be indicative of burrows, primarily because he could not find a suitable organism. Gutschick and Suttner (1972) concluded that the structures are burrows, and Peterson (1972, p. 70) referred to the large-scale, vertical, "boiler-pipe concretions" as evidence of burrowing.

Burrows or burrow-like forms such as those mentioned above are most abundant in the shelf facies of the Shedhorn Sandstone, and they occur in the platform facies of the Tosi Chert and Retort Phosphatic Shale Members of the Phosphoria Formation. All these units intertongue westward —in complicated fashion—with slope-basinal chert, phosphorite, mudstone, and carbonate sediments (Figs. 16.1 and 16.2). The Shedhorn was deposited in water less than 50 m deep, most probably less than 9 m, within 160 km (100 mi) offshore (Cressman and Swanson, 1964, p. 366). Fossil evidence suggests deposition on a firm bottom in clear water, in beach or

near-beach environments (Yochelson, 1968, p. 617). Certain Shedhorn sand bodies have also been interpreted variously as sheet sands, longshore sands, and barrier island complexes (Peterson, 1972; Sheldon, 1972). Among the body fossils present, clams, snails, scaphopods, and brachiopods are the most abundant ones reported from the Shedhorn and Tosi; sponge spicules, bryozoans, and fish remains are also found (Yochelson, 1968).

A wealth of detailed stratigraphical, geochemical, and paleontological information is available for in-depth study of the relations of Permian strata in the northern Rocky Mountains: McKelvey et al. (1959), Sheldon (1963), Cressman and Swanson (1964), McKee et al. (1967a, 1967b), Yochelson (1968), Peterson (1972), and references cited by them.

BURROW CHARACTERISTICS

Observations on the Permian structures include size and shape, configuration (within and between beds), fill, and host rock composition, textures, and structures, and body fossil and trace fossil associations. The array of large structures may represent the work of more than one kind of trace-making organism. Our observations supplement those of Cressman and Swanson (1964, p. 351–354, Figs. 141A–H); however, the maximum length that they reported is only 1.5 m. Total vertical extent of a single long burrow is obscured by partial erosional destruction, so that only in exceptional, fortuitous cases does the rock exposure parallel the structure's entire length.

The large burrows are elongate, cylindrical, columnar structures having an oval, elliptical, or circular cross section. Maximum diameter ranges from 2.0 to 12.5 cm; the mean value lies between 5 and 8.5 cm.

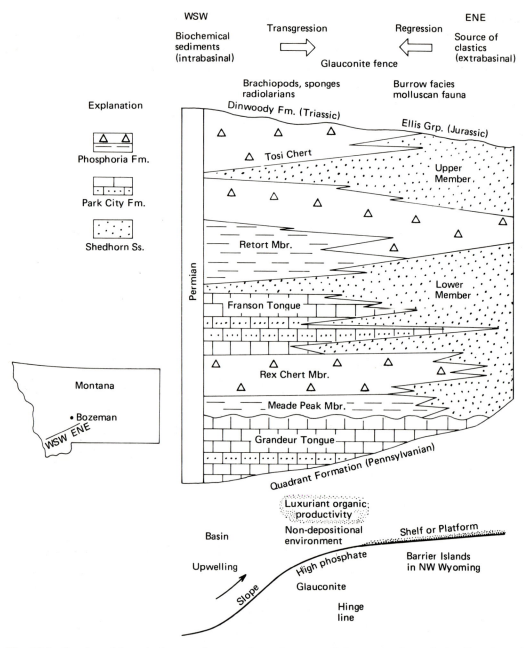

Fig. 16.1 Stratigraphic relations and general environmental interpretation of the Permian of southwest Montana.

Largest values are for burrows in chert matrix (Table 16.1A). Sample measurements made at several localities are given in Table 16.1B.

The greatest density of cylindrical burrows occurs in the upper Shedhorn of the West Fork of the Gallatin River sec-

tion (Fig. 16.3A). A bed 2.5 m thick has more than 75 percent of its volume occupied by burrows. These structures are 4 m long, and some cut through other lithologies above and below. Three beds in the upper Shedhorn at Cinnabar Mountain, totaling more than 5 m in thickness, have columnar chert-filled burrows comprising 50 to 60 percent of the rock (Fig. 16.3B) (Peterson, in Cressman and Swanson, 1964). In this same section, 195 discrete burrows were counted on 1 m^2 of the bedding surface! The unusually high density of burrows was seen in cross section, on such surfaces (Fig. 16.4).

The burrow structures vary in orientation and shape within beds, relative to the lithology of enclosing strata. Within cherts and fine, dark, organic shales, the burrows are typically inclined at low angles to the beds; in sandstones, they tend to be normal to bedding (Table 16.1C). All structures in any one bed generally have approximately

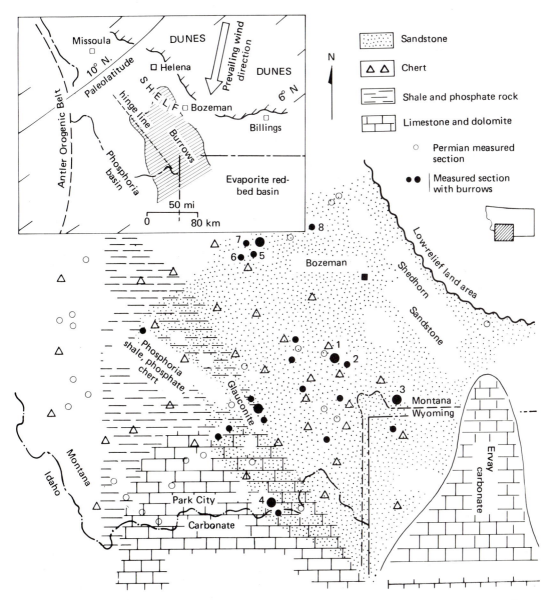

Fig. 16.2 Distribution of burrows relative to the Permian sedimentary framework in southwest Montana. (1), West Fork, Gallatin River. (2), Porcupine Mtn. (3), Cinnabar Mtn. (4), Centennial Mtns. (5), London Hills. (6), South Boulder. (7), Jefferson Canyon. (8), Logan. [After Cressman and Swanson (1964), Sheldon (1964), and Peterson (1972).]

TABLE 16.1   Major Characteristics of Permian Burrows.

A—Mean long diameter of burrows relative to matrix lithology.

Matrix Lithology	Number of Measurements	Mean Long Diameter (cm)	Standard Deviation
Dolomitic sandstone	176	3.1	1.6
Sandstone	855	3.7	2.4
Interbedded sandstone and chert	117	5.4	2.1
Chert	105	6.4	1.6

B—Representative measurements of burrow size, southwest Montana.

Number of Measurements	Cross-sectional Axes (mm)			Average Length (m)	Stratigraphic Section
	Minimum	Mean	Maximum		
100	16 x 11	40 x 31	65 x 50	5	Cinnabar Mountain
184	15 x 11	48 x 37	105 x 80	0.5	Porcupine Mountain
100	43 x 38	68 x 57	100 x 94	0.5	West Fork Gallatin River
101	45 x 40	85 x 71	135 x 105	4	West Fork Gallatin River
100	25 x 15	55 x 45	100 x 92	1 to 1.5	West Fork Gallatin River

C—Shape and orientation of burrows with respect to bedding.

Matrix Lithology (Number of Beds Containing Burrows)	Straight; Normal to Bedding	Straight; Inclined ca. 15° to Bedding	Irregularly Curved; Bulbous
Sandstone (11)	11	—	—
Mudstone (10)	4	4	2
Interbedded chert and mudstone (10)	3	3	4
Sandy dolomite (1)	1	—	—

the same orientation. Cylindrical burrows, where these can be observed, are usually singular and normal to the beds, and issue from an enlarged base (Fig. 16.5). A few such bulbous bases give rise to several parallel burrows, which extend vertically through the sediments. Branching is very uncommon, although a few, very short, stubby, lateral projections do occur. Single vertical burrows in sandstone may be traced upward into shale, where their orientation changes. Some become irregular, laterally anastomosing chambers, but continue across the bedding into the overlying sandstone unit.

Externally, the burrow is smooth in sandstone, and rugose, or wrinkled, in mudstones or cherts (Fig. 16.6A). The latter situation is undoubtedly due to differential compaction between burrow fill and host matrix. [Wrinkled, U-shaped burrows—*Diplocraterion*, several meters long—in Lower Carboniferous graywackes of the Rheinisches Schiefergebirge, Germany were used by Plessmann (1965) as evidence for 30 to 50 percent lateral compaction of sediments, while still in a horizontal position, before folding of the rocks.] Other, similar types of large Permian burrows have nodular, (?)pelleted type walls resembling *Ophiomorpha* and other crustacean burrows (Fig. 16.6B). Internally, the knobby walled burrows have intestine-like meander structures (Fig. 16.6C). Whether or not the meander structures are primary or secondary is unknown. They suggest a foraging pattern made by a smaller sediment-feeding animal than the one that formed the larger overall structure.

The burrow fill is commonly quartz sandstone or chert, but some fills are calcareous. Chert and quartz clasts, spicules, and apatite pellets may be tightly packed or floating in a chert, quartz, or carbonate cement. This fabric does not suggest any systematic pattern of rearrangement of sediment by the burrowing organism. Overgrowths occur on some quartz grains.

Internally, the burrow fills are generally structureless; however, some are vertically laminated, and in cross section thus have a discontinuous, concentric-banded pattern of annular rings. (Cf. Gableman, 1955.) The banding (Fig. 16.7A) is accentuated by alternations of (1) dark, spicular chert layers containing phosphate peloids and quartz sand grains, and (2) light-colored layers of chalcedonic chert. Spreiten-like patterns of active-fill meniscus structures (Fig. 16.7B) were observed in a few burrow fills having convex-upward laminae ("Stopfgefüge" of Valeton, 1971, Fig. 9). Another specimen has menisci convex downward. These patterns suggest that the organisms used the burrows for infaunal protection as well as paths of upward escape from sediment influx. X-radiographs of other burrows reveal a faint lamination, as

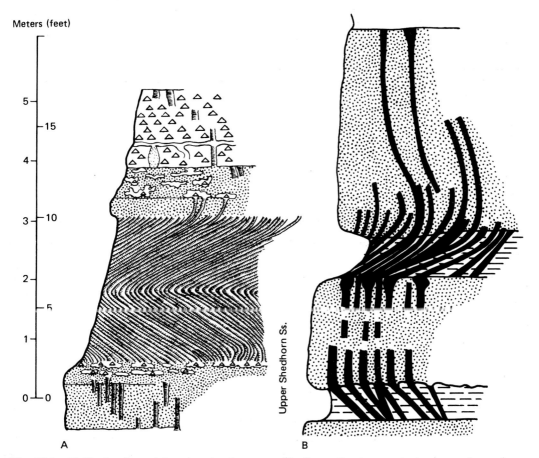

Fig. 16.3 Field sketches of burrow development, Shedhorn Sandstone. A, interpreted maximum burrow development. Chevron pattern = postulated pre-lithification collapse of unit. West Fork, Gallatin River section. B, Cinnabar Mtn. section, north of Yellowstone Park.

Fig. 16.4 Burrow cross sections on bottom of bed. Shedhorn Sandstone; South Boulder section. Scale in 0.1 ft (ca. 3 cm).

defined by oriented sponge spicules, inclined at a high angle to bedding. The laminae do not resemble minisci in that they are not curved, but they probably represent passive infilling from above. No well-preserved fossil organism, thought to be capable of producing the burrows, has been found preserved in any of them.

Fig. 16.5 Burrows having slightly enlarged bases, outlined in ink. A, single burrow. Logan section. B, two burrows; partially deformed bedding adjacent to burrow bases. Jefferson Canyon section.

Other trace fossils found in association with the long burrow structures include *Chondrites*, snails found within their burrows, herringbone-patterned *Cruziana*, questionable lungfish burrows, and other smaller traces. Bioturbate structures similar to those made by echinoids, and spiral, three-dimensional *Zoophycos* also occur in these rocks, but not in association with the long structures.

PROBLEMS IN INTERPRETATION

Analysis of Shedhorn burrows is made partly by comparing sandstone grain-size parameters with the dynamic properties of erosion-transport and associated feeding habits of burrowing organisms (Fig. 16.8). Deposit- and suspension-feeding organisms rely on currents for a continuous supply of organic material at the sediment–water interface; therefore, an optimum relationship exists between current velocity and sediment size. Infaunal suspension feeders thrive at velocities of about 2 cm/sec and thus occur most commonly in fine sand. Deposit feeders live in lower energy environments, where they can rework finer sediments for food (Driscoll, 1969).

Were the Permian burrowers deposit

Fig. 16.6 Small-scale features of burrows. A, rugose or wrinkled burrow. Top unknown. B, nodular, *Ophiomopha*-type burrow; nodules are pelletal white silica. C, internal, intestine-like meander structure of burrow in B.

feeders, living in the fine sediments (clays, muds, and siliceous oozes), escaping upward with rapid sand influx? This is suggested by the concave-upward meniscus fill and by the close correlation between burrow shape and orientation with matrix lithology. Also possible is that they were suspension feeders that made their homes initially in the Shedhorn sands, keeping pace with subsequent mud deposition. Overall length and regularity of burrows, and the fact that a single burrow penetrates more than one lithology, would seem to preclude the possibility that the organisms descended far below the sediment–water interface in order to reach extranutritious sediment, and at the same time, to avoid a high-energy environment at the interface.

If one consults the published fossil record of the Shedhorn and Tosi, few—if any—known organisms were capable of producing these burrows (Yochelson, 1968). Perhaps clams are the most likely possibility among the organisms reported. Infaunal, siphon-feeding clams are among the deepest burrowers (Stanley, 1968) and have the greatest escape potential (Kranz, 1970). Unfortunately, not much is known about their late Paleozoic history (according to Stanley, siphonate clams having fused mantles had not evolved yet), and no clam body fossils were found within the burrow structures.

Sponge spicules are the immediate source of silica in Phosphoria chert, constituting perhaps as much as two-thirds of the silica in this large volume of rock. The spicules probably underwent some transportation, as suggested by the fact that most are broken; yet a large population of benthic animals somewhere furnished this large volume of spicules. The single sponge

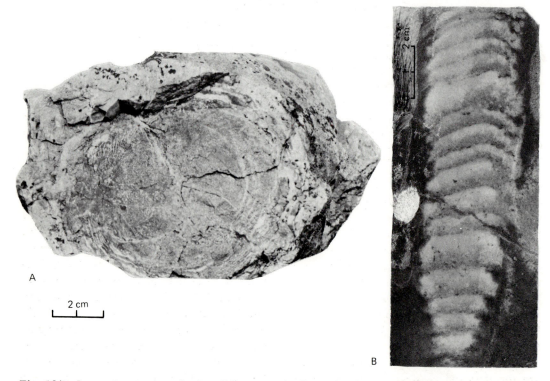

Fig. 16.7 Internal structure of selected burrows. A, discontinuous, concentric-banded cross section of burrow. B, longitudinal section of burrow exhibiting well-developed spreite-like meniscus fill. Shedhorn Sandstone; Centennial Mountain, Montana.

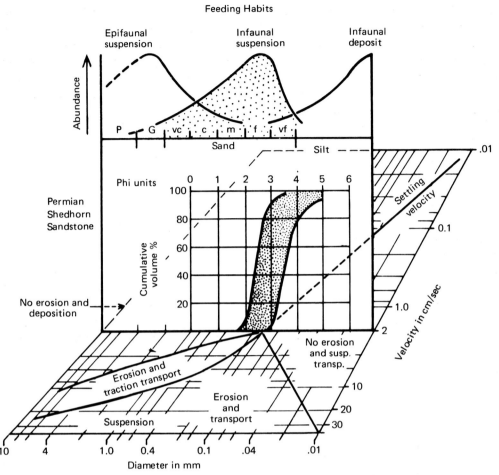

Fig. 16.8 Diagram showing relationship between infaunal feeders, water velocity, and sediment size for the Shedhorn Sandstone. [Interpretation from Driscoll (1969). Data for Shedhorn from Cressman and Swanson (1964).]

Actinocoelia maeandrina occurs in carbonates of the Franson in Wyoming and is common in chert nodules of the Permian Kaibab Limestone of Arizona. R. P. Sheldon and R. M. Finks (1972, personal communication) reported that the sponges are responsible for the long columnar "burrows," but no information has been published. How such organisms would have fared in the environment represented by the Shedhorn is not clear.

Other prime suspects of burrow making are the decapod crustaceans. No fossil remains of arthropods, other than trilobites and ostracods, have been reported from the Phosphoria and Shedhorn of Montana, and none were found by us. Evidence for *Ophiomorpha* from Permian strata of Utah is known (Chamberlain and Baer, 1973), and a pelleted wall structure is indicated —by geopetal matrix and supported quartz-grain fabric—for some burrow walls in Montana (W. A. Pryor, 1972, personal communication). The characteristic long, fairly straight, smooth, columnar, unbranched burrows of the Permian of Montana are unlike modern callianassid and related burrows; at this time, however, the evidence is inconclusive as to the organism(s) responsible for these unusual burrows.

CRETACEOUS AND PALEOCENE BURROWS

Richard G. Bromley

At most horizons, the chalks and fine-grained calcarenites of the Upper Cretaceous and Danian of northwestern Europe are thoroughly bioturbated. Development of flint nodules in these sediments provided a well-preserved record of burrow types involved in the bioturbation. Genera thus preserved include *Thalassinoides*, *Zoophycos*, *Chondrites*, rare *Gyrolithes*, and others (Bromley, 1967; Kennedy, 1970). (See also Chapter 17.) Spectacularly large flint nodules draw attention to the presence of unusually long burrows, which are otherwise inconspicuous where unsilicified. Felder (1971, Fig. 6), for example, recorded a horizontal burrow in the Senonian chalk at Fécamp, Normandie, France, that—because of an enveloping flint concretion—was traceable for 11 m, although having a diameter of only a few centimeters.

One type of giant flint nodule, having a particularly characteristic orientation and shape, is widespread at several horizons within the flint-bearing Upper Cretaceous chalk of northwest Europe. These nodules were first described under the name "paramoudra" by Buckland (1817) from the Campanian White Limestone of Northern Ireland, but have since been reported from chalk of the same age in Norfolk, England (Peake and Hancock, 1961, p. 318; see references); they are also common in Maastrichtian chalk of Denmark and Germany.

The environmental setting for these paramoudras is generally that of a restricted energy, marine shelf seafloor. Water depth probably ranged between 100 and 300 m, and the sedimentation rate was slow to moderately slow. Minor omission surfaces are visible in some regions (Fig. 18.2) and indicate a cyclicity of sedimentation. Additional details may be found among references cited herein.

BURROW CHARACTERISTICS

Paramoudras are vertically extended ring-, barrel-, or pear-shaped to cylindrical flints, variably massive, having a central core of unsilicified chalk that is typically open at each end. The concretions are characteristically stacked in columns (Fig. 16.9), although they may occur alone. Related phenomena include vertical cylinders of flint having wider cores, thinner walls, and greater length than do typical paramoudras (Fig. 16.10; see Felder, 1971, Fig. 5).

Within the central core a burrow shaft can be detected, having a diameter of about 0.4 cm. The shaft is generally vertical but sinuous in detail (Fig. 16.11). In some cases, as it passes through a particular bed, the shaft describes a broad spiral 1 m or more in diameter.

No individual burrow has been traced uninterrupted over its full length. However, associated diagenetic phenomena, particularly columns of cylindrical flints, indicate that the burrows commonly have a length of 5 m and that examples more than 8 m long do occur. Paramoudra columns described in the literature suggest burrow lengths of 8 to 9 m (Lyell, 1865, Fig. 286; Steinich, 1972, "Anlage" 6). The diameter of the shaft is thus characteristically 1/1,000 of its length, and in some cases may reach 1/2,000!

The upper end of the shaft has no special aperture structure, and at the lower end the burrow also ends simply, without a swelling or special chamber. Side branches, having the same diameter as the central shaft, emerge irregularly at most levels and radiate from the shaft more or less horizontally and straight (Fig. 16.11).

The shaft is rendered visible by complex mineralization of its wall and sur-

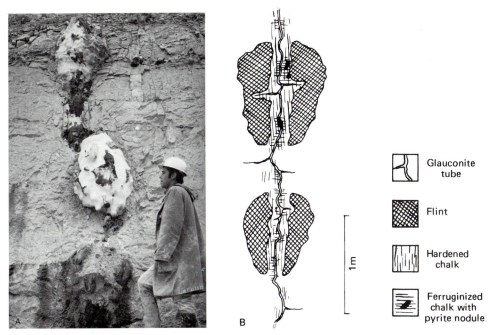

Glauconite
tube

Flint

Hardened
chalk

Ferruginized
chalk with
pyrite nodule

Fig. 16.9 Paramoudras. A, part of column of paramoudras in chalk. Paramoudra at knee level is broken, but its central core is not visible. The cylinder of flint connecting successive paramoudras is unusual. Upper Maastrichtian; Hemmoor, Niedersachsen, Germany. B, composite sketch of section through two paramoudras in a column, showing relationship of burrow to associated diagenetic phenomena.

rounding chalk. Because of incomplete mineralization, one cannot follow the shaft through its complete length (maximum distance, ca. 3 m). Another problem is logistics—tracing a faint, pencil-thick tube in jointed, dirty, often soft and sludgy chalk! Although the shaft is more or less vertical, faces of exposures usually slope irregularly, truncating the burrows at various levels.

The side branches are mineralized only near the shaft; the tinting of the chalk fades progressively as the branches are traced away from the shaft (Fig. 16.11), thus they are rarely visible for more than 20 cm. The side branches therefore appear to have been culs-de-sac and not to have interconnected with neighboring paramoudra burrows.

No structure can be detected in the fill, which lithologically resembles the surrounding sediment.

The shaft is usually extensively lined with a film of glauconite or pyrite, or both; in other cases, the chalk immediately surrounding the burrow wall is slightly glauconitized for a distance of a few millimeters from the burrow. In rare cases the green tube has a complex wall structure (Fig. 16.11), which possibly represents fracture crumpling of the brittle glauconite skin due to compaction. The chalk surrounding the burrow, at a distance of 5 to 10 cm from its wall, is typically iron stained; in some examples it is darkly discolored (Fig. 16.11). Chalk in the vicinity of the burrow is patchily cemented with calcite (Fig. 16.9B).

Beyond this zone of hardened and stained chalk, a zone of flint replacement resulted—at stratigraphic horizons rich in flint layers—in the production of monstrous flints. In their minimum development, the flints are massive rings 20 to 30 cm high, but at many localities they are extended

Fig. 16.10 Flint cylinder (ca. 0.6 m diameter) sectioned in chalk. Upper Campanian; Weybourne, Norfolk, England.

vertically along the shaft as pear-shaped or barrel-like concretions as much as 2 m high (Fig. 16.9A). A central core of unsilicified chalk 10 to 20 cm in diameter invariably runs up the flint and corresponds to the mineralized chalk around the burrow shaft. Paramoudras occur wherever the burrow shaft passes down through the horizon of a normal flint-nodule bed, and also between these horizons, so that a vertical chain or column of paramoudras marks the position of each burrow. In flint-poor chalk, the paramoudras are separated by several meters of chalk, or are completely isolated and occur only at flint-nodule horizons. Paramoudras are absent where all other flint is lacking. Thus, at several localities, only the staining and hardening help one to find the slender green tubes.

PROBLEMS IN INTERPRETATION

The chief difficulties in interpretation of paramoudra burrows lie in the unusual dimensions of the burrows and in their apparent (although not proved) isolation from neighboring shafts. Discussion of these difficulties amounts to little more than speculation until similar structures are described in recent seas.

The burrow was presumably produced either by an animal that had dimensions similar to those of the shaft, or by shorter animals that moved up and down within the shaft. In the first case, the shaft would have been largely filled by a long, worm-like animal that was capable of living in reducing conditions, its respiratory organs exposed at the sea floor.

Fig. 16.11 Section of burrow in chalk. Shaft curves in upper part of picture; complexly "broken" glauconite wall is clearly visible. Basal section includes two side branches. Lower Maastrichtian; Hemmoor, Niedersachsen, Germany.

Few living animal groups have dimensions suitable to fulfill this role. Pogonophores, nemertines, and polychaetes have long, thin, endofaunal representatives. Certain pogonophores are remarkably slender and long (Ivanov, 1960, p. 1541), the most extreme showing a width 1/600 of their length. But the absolute length of these forms is too small, a matter of decimeters. Possibly some species, known only from their upper end as truncated by the dredge, may eventually prove to be much longer. Pogonophores are largely deep-sea forms today, and are immobile, dwelling in unbranched tubes. Attribution of the paramoudra burrow to this group produces problems of interpretation of the side branches, which do not correspond to any structures in known pogonophore tubes.

The more active nemertines include several extremely elongated burrowing forms, some of which reach a length of 2 m (Barnes, 1968, p. 157). The problem here is that most of these worms are active carnivores, unlikely to construct permanent, vertical shafts.

The other possibility, a short animal that shuttled actively up and down the shaft, widens the field of possible trace-makers. But the reason for the descent of a "shorter worm" to such a distance beneath the sea floor remains unexplained. The tendency to spiral, seen in a few examples, may suggest a crustacean architect, but again the diameter of the shaft restricts the great task of excavation to very small crustaceans.

The shaft is almost certainly a dwelling burrow. But no special evidence indicates the particular function of the side branches, whether as feeding structures, egg incubators, burrows of commensal or other cohabitors, etc. Clearly, much more information is needed before these problems can be solved. A more detailed examination, documentation, and discussion of these burrows is given in Bromley et al. (in press).

PLEISTOCENE STRUCTURES

H. Allen Curran and Robert W. Frey

Strip-mining operations near Aurora, North Carolina, on the Atlantic Coastal Plain, reveal a stratigraphic section of beds ranging from middle Miocene to Holocene in age. The upper part of this section contains poorly consolidated marine Pleistocene (?Sangamon) sediments that bear distinctive associations of biogenic and physical sedimentary structures (Welch et al., 1972). Trace fossils commonly found in these units include *Ophiomorpha*, *Planolites*, *Skolithos*, and *Thalassinoides* (Curran and Frey, 1973; Curran et al., 1973). One of the Pleistocene units contains possible escape burrows of very distinctive form and unusually large size, as much as 3 m in length and 24 cm in diameter. This unit also contains some large, vertical, cylindrical structures of uncertain origin; at least one of these cylindrical structures resembles a reburrowed escape structure.

A current-bedded unit in the Pleistocene sequence (Belt et al., in press) has a maximum thickness of 4.7 m and consists of a planar, tabular cross-bedded member overlain by a wavy- and flaser-bedded[1] member. The tabular cross beds are composed of fine to very coarse quartz sand. Coarse and very coarse sands are irregularly interbedded with finer sands of the foreset beds. Ripples outlined by heavy-mineral and silt-clay laminae are common in fine sands. The tabular cross-bed sets are separated by thin, horizontal-bedded, rippled sequences of fine to medium sand. This member contains

---

[1] Terms as defined by Reineck and Wunderlich (1968).

the burrows *Ophiomorpha nodosa* and *Skolithos*, and also the large structures described below. Burrows are sparse in the medium to very-coarse-grained basal cross-bedded sands of the member. Density of burrows, particularly *Skolithos*, greatly increases in the fine to medium sands of sets in the upper part of the member.

Physical and biogenic structures found in this member suggest conditions of rapid deposition in a shallow sublittoral, shoaling environment having strong currents. Welch et al. (1972) interpreted the beds as a possible tidal-delta deposit, or a laterally accreting large bar or spit.

The wavy- and flaser-bedded upper member of the current unit is composed of fine to medium quartz sand and clay. This member is moderately bioturbated, containing *Ophiomorpha nodosa*, *Skolithos*, and *Planolites*. The increased clay content and the flaser and wavy bedding indicate lower energy depositional conditions, and suggest a protected, shallow subtidal or intertidal environment (Welch et al., 1972), such as a tidal flat, shoal, or large bar.

BURROW–LIKE CHARACTERISTICS

"Escape" Structures

In longitudinal section, the large "escape burrows" are marked by V- to U-shaped, in echelon laminae of coarse to very coarse quartz sand layers interbedded with fine to medium quartz sands (Fig. 16.12A). Coarse

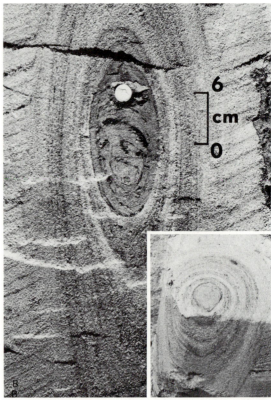

Fig. 16.12 "Escape burrows" in tabular cross-bedded quartz sands. A, sediment core at top of structure, weathered out naturally from mine wall; remainder of surface has been scraped smooth. Core is conical in shape and composed of unlaminated sand. B, close up of oblique section through "escape burrow." Core of structure filled with unlaminated fine sand, rich in organic detritus, from overlying flaser-bedded member. *Inset*: three-dimensional block view of structure. (Minimum overall length of structure, 2 m.)

sand layers in the tabular cross-bedded sets containing these structures can be traced to the margin of the overall structure, where they dip sharply downward. Numerous laminae can then be followed continuously across the structure, in distinctive V- to U-shaped patterns, and then traced back into the cross-bedded set on the opposite side. In transverse sections, a pattern of concentric circles is revealed, also formed by alternating layers of coarse and finer sand. A distinct central core of unlayered fine to medium sand, containing clay blebs and woody organic fragments, is well defined toward the top of some of the structures (Fig. 16.12B). In three dimensions, these structures consist of a set of nested, cone-in-cone laminations, many having an unlaminated core at the top.

Six such structures were found, all of unusually large size, ranging from 1.2 to 3 m in length and 10 to 24 cm in diameter. All lengths reported here are minimal because the structures were recognized by the protrusion of their weathered, truncated tops out of the friable, poorly consolidated sediments of the mine walls (Fig. 16.12A). The structures maintain a constant diameter through much of their length, e.g., the one shown in Figure 16.13 has a diameter of 9 to 10 cm through a length of 1.2 m. The structures taper inward abruptly toward their base, but the actual characteristics of the base could not be determined with certainty because all terminated at the contact of the current unit with an underlying, wet, sandy clay unit.

Laminae that could be traced continuously across the structures were deflected downward as much as 50 cm from their original position in the cross-bedded sets. The laminae were separated across the structures by a distance averaging 10 to 14 cm, reaching a maximum of 24 cm.

Vertical Cylindrical Structures

In general form and size, the large cylindrical structures resemble those de-

Fig. 16.13 Continuous section (1.2 m) of an "escape burrow." Laminae of dark sand, many of which can be traced continuously across the structure, form U- to V-shaped patterns.

scribed previously. They differ in that the pattern of V- to U-shaped, in echelon laminae is largely obliterated. All occur in the tabular cross-bedded member of the current unit, and they are filled with fine quartz sand containing numerous clay blebs and organic fragments from the overlying flaser- and wavy-bedded member.

In one case, features that appear to be cone-in-cone laminations of a large "escape structure" have a secondary structure superimposed on them (Fig. 16.14A). This structure is evocative of reburrowing activity; the path of this "reburrowing" is marked by a thin but continuous dark gray clay layer that lines the inner surface of the shaft, an irregular vertical path having a minimum length of 1.65 m. The diameter of the clay-lined shaft is 7.5 cm in its central part, and gradually narrows toward the top.

At its base, the clay-lined shaft opens up abruptly into an irregularly shaped, thinly clay-lined chamber having a maximum diameter of 15 cm and variable height of 15 to 20 cm.

Figures 16.14B and 16.15 illustrate a similar vertical structure of even larger size, having a minimum length of 2.5 m and maximum diameter of 37 cm. The thin (2 to 4 mm) clayey sand lining is less continuous than that of the structure shown in Figure 16.14A, but this lining can be traced for the length of the structure and shows up clearly in transverse sections (Fig. 16.14B). At its base, the structure gradually

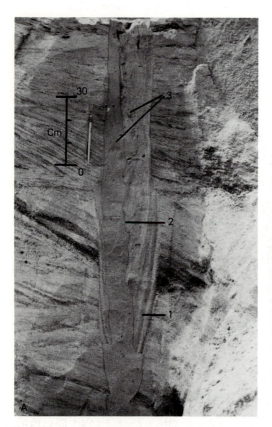

Fig. 16.14 Problematical "burrow" structures. A, ?reburrowed "escape structure" in tabular cross-bedded sand. Laminations of the original "escape burrow" (1) are distinct toward base of structure. Thin clay lining (2) marks the bounds of the "reburrowed" structure. Much of the structure is filled with unlaminated fine sand rich in organic detritus (3), from overlying flaser-bedded member. B, large vertical cylindrical structure. Lining (1) is a thin, continuous, sandy clay layer. Sand filling the structure is from overlying flaser-bedded member. Laminations (2) tend to be deflected downward at base. Small-displacement normal faults (3) adjacent to the structure probably originated during its formation. Machete 68 cm long.

tapers inward to a blunt terminus having a U-shaped profile. A well-defined chamber, such as that in the "reburrowed" structure described previously, is lacking. Numerous normal faults involving small displacement (as much as 25 cm) disrupt layers adjacent to the margins of this structure (Figs. 16.14B and 16.16). In general the fault planes dip toward the structure's margins; most likely, this faulting occurred during formation of the structure.

PROBLEMS IN INTERPRETATION

"Escape" Structures

The Pleistocene structures superficially resemble burrows characterized by downward deflected laminae described by Boyd (1966, p. 45–46) from the Middle Cambrian Flathead Sandstone of Wyoming. Hallam and Swett (1966, p. 103–106) described similar trace fossils from the Lower Cambrian Pipe Rock of Scotland, which they assigned to *Monocraterion*. Features of this type were classified by Hanor and Marshall (1971, p. 128, Fig. 1b) as structures formed by shearing. Experiments conducted by Boyd (1966, p. 46–50) indicated that the downwarping and truncation of laminae in the Flathead burrows probably resulted from subsurface removal of sediment, although the mechanism of subsurface sediment removal was not satisfactorily explained. Schäfer (1972, p. 288, Fig. 165) noted that the modern cerianthid anemone *Cerianthus* deflects sediment downward and truncates laminae during both upward and downward burrowing activity. (See Chapter 22.)

The North Carolina "burrows" differ from those mentioned previously in that numerous deflected laminae can be traced continuously across longitudinal sections of the structure (Figs. 16.12 and 16.13) and were not truncated by downward "burrow-

Fig. 16.15 Cross-sectional view of structure shown in Figure 16.14B. The structure is circular in section and has a thin, sandy clay lining. Scale 15 cm long.

Fig. 16.16 Two vertical cylindrical structures in close proximity, showing characteristic fills. Offsets in thin dark beds reveal small faults.

ing" activity. This configuration of laminae seemingly argues for an inorganic origin for the structures, as by collapse of sediment, e.g., structures strikingly resembling escape burrows are produced by the collapse of sediment into open vertical burrows of *Callianassa major* (Howard, 1971, p. 200, Fig. 9). Some of these collapse structures attain very large sizes. However, we have found no evidence of a cavity into which the Pleistocene sediments might have collapsed. We thus note the similarity of these to dwelling structures formed by an organism forced to move upward periodically in order to escape burial and maintain a constant position relative to the sediment–water interface during rapidly shoaling conditions.

The depth of deflection of laminae (as much as 50 cm) and width of separation of

laminae across the structures (as much as 24 cm) would provide absolute maximum-size dimensions for the "burrower." However, the "burrower" probably would have considerably smaller body dimensions. Penetration experiments by Boyd (1966, p. 48) produced deformed sediment zones having twice the diameter of the penetrating object. (See also Schäfer, 1956.) The unlaminated, cone-shaped, central cores found in the upper part of some of the Pleistocene structures would more accurately reflect the diameter of a burrower's body, because these cores probably would represent fillings formed after the burrower evacuated the burrow site. The core zone shown in Figure 16.12B has a diameter of 8 cm, but laminae are deflected downward across a diameter of 20 cm.

A possible ethological analog for these "escape burrows" are the structures formed by burrowing actinian sea anemones. Shinn (1968) studied burrows of *Phyllactis conguilegia*, an anemone living on the Bahama Banks in cross-bedded oolitic sands, where large migrating megaripples are common. When covered by oolitic and skeletal sand in an aquarium, specimens of this anemone burrowed upward by peristaltic action and resumed their normal position at the sediment–water interface. The resulting burrow recorded by Shinn (1968, Pl. 112, figs. 1 and 2) consisted of cone-in-cone laminae formed by sand trickling down from above as the animal moved upward. Shinn (p. 889, Pl. 112, figs. 3 and 4) reported similar structures from the cross-bedded Pleistocene Miami Oolite and Pleistocene oolites in the Bahamas. Frey (1970a, p. 308–309) studied burrows of the actinian *Paranthus rapiformis* near Beaufort, North Carolina, and found that these burrows are very similar to those of *Phyllactis conguilegia*. The North Carolina Pleistocene structures are strikingly similar in form to those described by Shinn and Frey, including deflected laminae that may be traced across the structure. They differ primarily in their much larger size, having a maximum diameter of as much as 24 cm, as compared to approximately 3.5 cm for the modern anemone burrows.

A matter more problematical than mere size itself, however, is the respective thicknesses of sediment involved. Recorded dimensions for escape structures created by these recent actinians involve only a few centimeters of sediment, whereas the Pleistocene structures involve as much as 3 m of sediment in a given, continuous trace. Such traces, if indeed biogenic, suggest remarkable prowess by the tracemakers in combatting prolonged, unstable conditions in the depositional environment. Little is known about the possible tenacity of anemones under such conditions, but certain of them are at least sufficiently long-lived (as discussed subsequently), and actinians exhibit the requisite behavior on a small scale. Anemones thus remain our best biogenic analog for the Pleistocene tracemakers. But could any ancient anemones (or indeed *any* infaunal invertebrates) have attained the size necessary to produce these huge structures? [One group of animals that warrants further attention in this regard are the marine eels (see Clark, 1972).]

Several modern species of burrowing actinians and cerianthids are known from shallow-marine, sandy-substrate environments along the middle and north Atlantic coast of the United States (Gosner, 1971, p. 150–151, 153, 160; Kirby-Smith and Gray, 1971, p. 8), and others may be present. Individuals of one or more of these latter species conceivably may have formed the Pleistocene escape burrows. Unfortunately, the burrow characteristics of most modern burrowing sea anemones have not been studied. A possible candidate for construction of the Pleistocene burrows might be the large, tube-dwelling cerianthid *Cerian-theopsis brasiliensis*, living offshore near Beaufort (Frey, 1970a, p. 311). However, according to Arai (1972, p. 314), the musculature of cerianthids is such that they may not be capable of the strenuous

burrowing represented by long escape structures. The observations of Schäfer (1972, p. 288, Fig. 165) indicate that cerianthids tend to truncate laminae during burrowing rather than to form the cone-in-cone structure that characterizes the Pleistocene "escape burrows."

If these are escape structures, their exceptional length (at least 3 m) indicates that the organism continuously occupied a single geographic position on the sea floor and moved progressively upward with sediment accretion. The unlaminated, conical sediment core that fills the upper part of some of the structures would represent filling that occurred after the organism moved from the burrow site or died and decayed in place. Individuals of certain species of actinian and cerianthid anemones are known to have life-spans greater than 50 years (Annandale, 1912, p. 607; Cutress, in Frey, 1970a, p. 311). If the "escape" structures were produced by anemones, and assuming a possible life-span for the anemones of as much as approximately 50 years, this is a maximum time for accumulation of the 4-m thickness of tabular cross-bedded Pleistocene sands at this locality; of course, the sediments could have been deposited much more rapidly.

Vertical Cylindrical Structures

The structures shown in Figures 16.14–16.16 occur in sands of the current unit and are filled with unlaminated sand, clay blebs, and organic fragments from the overlying wavy- and flaser-bedded member. Figure 16.14A illustrates a structure that apparently is superimposed on a previously formed "escape burrow"; the cylindrical structure seems to have been excavated from above and to have followed the general path of the preexisting "burrow." The structure is characterized by a nearly continuous, thin clay lining, which is notable evidence for a biogenic origin; conceiving of a physical mechanism that could form this lining is difficult. The lining was traced to a large, irregular chamber at the base.

Could this distinctive structure represent burrowing from above by an animal seeking to form a dwelling chamber? The phenomenon of selective burrowing on the site of previously formed burrows is not uncommon (e.g., Frey, 1970b, p. 26), although the large size of this Pleistocene structure is exceptional. The clay lining of the structure is too thin to have functioned as a strong, protective wall, but it might mark the path followed by the animal. This lining, if biogenic, could have formed during initial penetration of the sediment, after occupancy of the burrow, or after partial filling of the burrow while the animal burrowed upward through sediment fill in the process of exiting the dwelling chamber. The last event seems to be most plausible for the upper part of the structure, where fill material is found both inside and outside of the clay lining. Inorganic mechanisms also must be considered for the origin of this structure, of course, as discussed below.

Figure 16.14B illustrates a structure that has a thin, discontinuous, sandy clay lining, somewhat similar to that of the structure shown in Figure 16.14A; but this structure does not terminate with a chamber at its base. Its form is essentially that of a large inverted cone, filled with unlaminated sediment from above (Fig. 16.15). The process of formation apparently triggered small displacements through normal faulting in the sediments adjacent to the structure's margin (cf. Fig. 16.16). Similar faults were found in other places not in direct association with the large structures, however, thus the two phenomena are not uniquely interdependent.

Two structures of this last type were found in the North Carolina deposits. In size and overall appearance, they closely resemble a large structure described by Dionne and Laverdière (1972) from a Quaternary sand and silt deposit in Quebec Province, Canada.

A variety of origins—physical and biogenic, summarized by Dionne and Laverdière (1972, p. 532–533)—have been postulated for vertical structures of this general type. Dionne and Laverdière suggested that the Canadian structure was formed by the action of a whirlpool eroding a cylindrical hole in unconsolidated sediments, or by the action of spring waters rising through the sediments. Of these two suggestions, the whirlpool mechanism seems most plausible for the North Carolina structures, because they definitely are filled with sediment from the overlying wavy- and flaser-bedded member. Yet the sediments of this member are interpreted as having been deposited on an intertidal sand flat or in a shallow subtidal environment, and to assess the probability of whirlpools or springs being active in either environment is difficult.

The possibility that these structures were formed by a vertebrate animal cannot be ruled out. When startled, some marine snake eels (Ophichthidae) and worm eels (Echelidae) burrow very quickly to depths of 1 m, the approximate length of the animal (Lagler et al., 1962, p. 198). The morays (Muraenidae) are even larger; modern ones are predominantly crevice-nestlers in reefs or rubble, but some ancient ones or little-known modern ones conceivably may burrow in sediments. However, no modern analog for these Pleistocene structures is known, and if biogenic, the identity of the tracemaker(s) remains a mystery—made all the more perplexing by the close correspondence between associated trace fossils and their recent counterparts: e.g., burrows of *Callianassa major* and *Upogebia affinis* for the *Ophiomorpha* and *Thalassinoides*, respectively, and dwelling tubes of *Onuphis microcephala* for the *Skolithos* (Curran and Frey, 1973; Curran et al., 1973).

Other structures of similar form, occurring in generally similar environmental settings, have been reported to us: in Pleistocene oolitic sands from the Bahamas (A. C. Neumann, 1973, personal communication) and Upper Cretaceous quartz sands of Utah (J. D. Howard, 1973, personal communication). We suspect that these kinds of huge structures are more common than is now generally known, and their origins—whether physical or biogenic—certainly warrant further investigation.

DISCUSSION AND CONCLUSIONS

The foregoing examples should illustrate some of the variations, peculiarities, and complexities of unusually large burrows or burrow-like structures. Of course, the interpretation of large burrows in general must be tempered somewhat by the actual function of the structure, as well as by its overall size and probable maker. The length of an escape structure, for example, is more a matter of sedimentation rate and survivorship or longevity of the tracemaker than of body length. The rate of lithification of the substrate is also very important in determining burrow characteristics (e.g., Chapter 18). Similarly, well-integrated burrow systems are theoretically endless; the interconnected shafts and tunnels are in effect a giant commune maintained by many individual animals. The shrimp *Callianassa major* constructs such burrow networks in Georgia beaches (Frey and Mayou, 1971), and certain ancient examples are equally striking; one of us (RGB) has traced *Thalassinoides* systems continuously along the Chalk Rock hardground (see Bromley, 1967), in one enormous quarry, for the entire 2-km extent of the exposure, and suggests that the system is, in fact, continuous through the 200-km extent of the hardground itself!

Nevertheless, the diameter of exceptionally large escape structures may be impressive, important, and problematical, as is the length of the structure when considered in terms of the actual tenacity and adaptations of the tracemaker. And the paramoudras are striking in their singular occurrences: individual burrows that are extremely small in diameter relative to their length.

One of the main problems in interpreting the latter, as well as the Permian forms, is whether the animal(s)—individually or through successive generations—occupied the entire structure at any given time, or whether basal parts were abandoned gradually as the tracemakers kept pace with sediment accumulation. In neither case were the entire burrows likely to have been inhabited simultaneously, throughout their length: the Permian ones because little or no evidence indicates that the burrow walls were structurally reinforced—as is generally seen in durable dwelling structures; and the Cretaceous–Paleocene ones because of the sheer length of time involved—the walls may well have been reinforced, but the rate of deposition of the sediments containing the burrows was so slow that vertical extension with time, as the animals kept pace with sediment accumulation, would require that the burrows be inhabited continuously for millions of years. "Cut-offs" and abandoned burrow components are common among shrimp domiciles (e.g., Weimer and Hoyt, 1964); but such evidence is lacking for the Permian and Cretaceous–Paleocene burrows studied by us. If the animals inhabited the total structure, how did they aerate the lower part? Irrigation is not difficult in burrows having more than one aperture (e.g., Vogel and Bretz, 1972), or even in small vertical burrows (Mangum, 1964); but how so in a single large burrow that is tremendously longer than wide?

The density of large burrows is also important, especially in cases such as those illustrated in Figures 16.3 and 16.4. The ultimate density of trace fossils is as much a matter of depositional rates and function and "preservability" of the original traces as it is of initial animal density (see Chapters 8, 9, and 20). In fact, the dense Permian burrows argue against high animal density; considering the engineering properties of the host sediments, how could the substrate retain its integrity if 80 percent of its volume consisted of open burrows several meters long? Nevertheless, we cannot imagine a very small number of organisms building such large numbers of extensive, geometrically regular structures.

Overall, the problem of interpreting ancient burrows of such size is a matter of understanding the tracemaker's food requirements, need for domicile or protection, and manner of escape from pressures imposed by predation or physiochemical conditions of the environment. In the abscence of body fossils attributable to the tracemaker preserved within the burrow, however, the evidence for the tracemaker remains circumstantial and speculative. Evolutionary changes must also be considered; for example, although the general behavioral patterns of tracemakers have remained more or less constant for hundreds of millions of years (Seilacher, 1964), we have no direct assurance that the Permian burrower has a close counterpart today. The uniformitarian principle is hampered further by a dearth of information on recent tracemakers that construct unusually large burrows, or on their limits of escape through rapidly accumulating sediments.

Finally, in some cases at least, we must also consider the possibility that the "burrows" are, in fact, inorganic in origin. Indeed, the situation here can be paradoxical: we may concede that the structures are inorganic simply because we cannot envision a suitable biogenic process, or that they are biogenic because we cannot conceive of a suitable physical process! Such problematical structures require meticulous study (e.g., Allen, 1961; Walter, 1972; Asgaard and Bromley, 1974); the evidence can be much more subtle than we might expect, and in many cases we probably fail to recognize the significance of the evidence itself (see Chapter 5).

The numerous difficulties notwithstanding, we suspect that unusually large burrows and burrow-like structures are much more abundant in recent and ancient sediments than the literature would suggest. Future work will hopefully reveal more of them and more criteria for their recognition and interpretation.

ACKNOWLEDGMENTS

We are grateful to D. W. Boyd, University of Wyoming, R. Goldring, University of Reading, and J. D. Howard, Skidaway Institute of Oceanography, for their helpful review of the manuscript. Responsibility for these interpretations is, of course, our own.

REFERENCES

Allen, J. R. L. 1961. Sandstone-plugged pipes in the lower Old Red Sandstone of Shropshire, England. Jour. Sed. Petrol., 31:325–335.

Annandale, N. 1912. Aged sea anemones. Nature, 89:607.

Arai, M. N. 1972. The muscular system of *Pachycerianthus fimbriatus*. Canadian Jour. Zool., 50:311–317.

Asgaard, U. and R. G. Bromley. 1974. Sporfossiler fra den mellemmiocæne transgression i Søby–Fasterholt området. Dansk Geol. Foren., Årsskrift 1973:11–19.

Barnes, R. D. 1968. Invertebrate zoology (2nd ed.). Philadelphia, Saunders, 743 p.

Belt, E. S. et al. in press. Pleistocene coastal marine sequences, Lee Creek (Texas Gulf) phosphate mine, eastern North Carolina. In C. E. Ray (ed.), The geology and paleontology of the Lee Creek Mine. Smithsonian Contr. Paleobiol., R. Kellogg Mem. Vol.

Boyd, D. W. 1966. Lamination deformed by burrowers in Flathead Sandstone (Middle Cambrian) of central Wyoming. Contr. to Geol., 5:45–53.

Bromley, R. G. 1967. Some observations on burrows of thalassinidean Crustacea in chalk hardgrounds. Geol. Soc. London, Quart. Jour., 123:157–182.

———— et al. in press. Paramoudras: giant flints, long burrows and the early diagenesis of chalks. Kgl. Dansk Vidensk. Selsk., Biol. Skr.

Buckland, W. 1817. Description of the paramoudra, a singular fossil body that is found in the chalk of the north of Ireland. Geol. Soc. London, Trans., (1)4:413–423.

Chamberlain, C. K. and J. L. Baer. 1973. *Ophiomorpha* and a new thalassinid burrow from the Permian of Utah. Brigham Young Univ. Geol. Stud., 20(1):79–94.

Clark, E. 1972. The Red Sea's gardens of eels. Natl. Geogr. Magazine, 142:724–735.

Cressman, E. R. and R. W. Swanson. 1964. Stratigraphy and petrology of the Permian rocks of southwestern Montana. U.S. Geol. Survey, Prof. Paper 313–C:C275–C569.

Curran, H. A. and R. W. Frey. 1973. Pleistocene and recent biogenic sedimentary structures as paleoenvironmental indicators (abs.). Geol. Soc. America, Abs. Prog., 5(7):588.

———— et al. 1973. Pleistocene trace fossils and recent analogues as paleoenvironmental indicators (abs.). Geol. Soc. America, Abs. Prog., 5(5):391–392.

Dionne, J.-C. and C. Laverdière. 1972. Structure cylindrique verticale dans un dépôt meuble Quaternaire, au nord de Montréal, Québec. Canadian Jour. Earth Sci., 9:528–543.

Driscoll, E. G. 1969. Animal–sediment relationships of the Coldwater and Marshall Formations of Michigan. In K. S. W. Campbell (ed.), Stratigraphy and paleontology. Canberra, Australian Nat. Univ. Press, p. 337–352.

Felder, W. M. 1971. Een bijzondere vuursteenknol. Grondboor en Hamer, 1971:30–38.

Frey, R. W. 1970a. The lebensspuren of some common marine invertebrates near Beaufort, North Carolina. II. Anemone burrows. Jour. Paleont., 44:308–311.

————. 1970b. Trace fossils of Fort Hays Limestone Member of Niobrara Chalk (Upper Cretaceous), west-central Kansas. Univ. Kansas Paleont. Contr., Art. 53, 41 p.

———— and T. V. Mayou. 1971. Decapod burrows in Holocene barrier island beaches and washover fans, Georgia. Senckenbergiana Marit., 3:53–77.

Gableman, J. W. 1955. Cylindrical structures in Permian(?) siltstone, Eagle County, Colorado. Jour. Geol., 63:214–227.

Gosner, K. L. 1971. Guide to identification of marine and estuarine invertebrates. New York, Wiley-Interscience, 693 p.

Gutschick, R. C. and L. J. Suttner. 1972. Sandstone and chert columns in Permian rocks of southwest Montana: biogenic or inorganic? (abs.). Amer. Assoc. Petrol. Geol., Bull., 56:621.

Hallam, A. and K. Swett. 1966. Trace fossils from the Lower Cambrian Pipe Rock of the north-west Highlands. Scottish Jour. Geol., 2:101–106.

Hanor, J. S. and N. F. Marshall. 1971. Mixing of sediment by organisms. In B. F. Perkins (ed.), Trace fossils, a field guide. Louisiana State Univ., School Geosci., Misc. Publ. 71–1:127–135.

Howard, J. D. 1971. Trace fossils as paleoecological tools. In J. D. Howard et al., Recent advances in paleoecology and ichnology. Amer. Geol. Inst., Short Course Lect. Notes, p. 184–212.

Ivanov, A. V. 1960. Embranchement des Pogonophores. In P.–P. Grassé (ed.), Traité de Zoologie, 5:1521–1622.

Kennedy, W. J. 1970. Trace fossils in the chalk environment. In T. P. Crimes and J. C. Harper (eds.), Trace fossils. Geol. Jour., Spec. Issue 3:263–282.

Kirby-Smith, W. W. and I. E. Gray. 1971. A checklist of common marine animals of Beaufort, North Carolina. Duke Univ. Marine Lab. Mus., 31 p.

Kranz, P. M. 1970. Bivalve escape behavior as an indication of sedimentary rates and environments (abs.). Geol. Soc. America, Abs. Prog., 2(7):599.

Lagler, K. F. et al. 1962. Ichthyology. New York, John Wiley, 545 p.

Lyell, C. 1865. Elements of geology (6th ed.). London, John Murray, 794 p.

Mangum, C. P. 1964. Activity patterns in metabolism and ecology of polychaetes. Comp. Biochem. Physiol., 11:239–256.

McKee, E. D. et al. 1967a. Paleotectonic investigations of the Permian System in the United States. U.S. Geol. Survey, Prof. Paper 515, 271 p.

————. 1967b. Paleotectonic maps of the Permian system. U.S. Geol. Survey, Misc. Geol. Invest., Map I–450.

McKelvey, V. E. et al. 1959. The Phosphoria, Park City, and Shedhorn Formations in the western phosphate field. U.S. Geol. Survey, Prof. Paper 313-A:A1–A47.

Peake, N. B. and J. M. Hancock. 1961. The Upper Cretaceous of Norfolk. Norfolk Norwich Natural. Soc., Trans., 19:293–339.

Peterson, J. A. 1972. Permian sedimentary facies, southwestern Montana. Montana Geol. Soc., 21 Ann. Field Conf., p. 69–74.

Plessmann, W. 1965. Laterale Gesteinsverformung vor Faltungsbeginn im Unterkarbon des Edersees (Rheinisches Schiefergebirge). Geol. Mitt., 5:271–284.

Reineck, H.-E. and F. Wunderlich. 1968. Classification and origin of flaser and lenticular bedding. Sedimentology, 11:99–104.

Schäfer, W. 1956. Wirkungen der Benthos-Organismen auf den jungen Schichtverband. Senckenbergiana Leth., 37:183–263.

————. 1972. Ecology and palaeoecology of marine environments. Edinburgh and Chicago, Oliver & Boyd and Univ. Chicago Press, 568 p.

Seilacher, A. 1964. Biogenic sedimentary structures. In J. Imbrie and N. D. Newell (eds.), Approaches to paleoecology. New York, John Wiley, p. 296–316.

Sheldon, R. P. 1963. Physical stratigraphy and mineral resources of Permian rocks in western Wyoming. U.S. Geol. Survey, Prof. Paper 313-B:B47–B271.

————. 1964. Paleolatitudinal and paleogeographic distribution of phosphorite. U.S. Geol. Survey, Prof. Paper 501-C:C106–C113.

————. 1972. Phosphate deposition seaward of barrier islands at edge of Phosphoria sea in northwest Wyoming (abs.). Amer. Assoc. Petrol. Geol., Bull., 56:653.

Shinn, E. A. 1968. Burrowing in recent lime sediments of Florida and the Bahamas. Jour. Paleont., 42:879–894.

Stanley, S. M. 1968. Post-Paleozoic adaptive radiation of infaunal bivalve molluscs—a consequence of mantle fusion and siphon formation. Jour. Paleont., 42:214–229.

Steinich, G. 1972. Endogene Tecktonik in den Unter-Maastricht—Vorkommen auf Jasmund (Rügen). Geologie, 20 (Beiheft 71/72):1–207.

Valeton, I. 1971. Tubular fossils in the bauxites and the underlying sediments of Surinam and Guyana. Geol. en Mijnbouw, 50:733–741.

Vogel, S. and W. L. Bretz. 1972. Interfacial organisms: passive ventilation in the velocity gradients near surfaces. Science, 175:210–211.

Walter, M. R. 1972. Tectonically deformed sand volcanoes in a Precambrian greywacke, Northern Territory of Australia. Jour. Geol. Soc. Australia, 18:395–399.

Weimer, R. J. and J. H. Hoyt. 1964. Burrows of *Callianassa major* Say, geologic indicators of littoral and shallow neritic environments. Jour. Paleont., 38:761–767.

Welch, J. S. et al. 1972. Physical and biogenic sedimentary structures as depositional indicators in the Pleistocene of North Carolina (abs.). Geol. Soc. America, Abs. Prog., 4(2):113.

Yochelson, E. L. 1968. Biostratigraphy of the Phosphoria, Park City, and Shedhorn Formations. U.S. Geol. Survey, Prof. Paper 313-D:D571–D660.

CHAPTER 17

TRACE FOSSILS IN CARBONATE ROCKS

W. J. KENNEDY

Department of Geology and Mineralogy, University of Oxford
Oxford, England

SYNOPSIS

Most terrigenous clastic sequences have calcareous equivalents, and trace fossil suites well known from terrigenous clastic sediments have their counterparts in carbonates. Well-studied examples include trace fossils from situations in which carbonate facies predominate: the Bahaman-type shallow-water environment—as represented by European Mesozoic and tropical Pleistocene limestones; and pelagic oozes—as represented by shelf-sea and deep-sea chalks, the latter now available for study as a result of the Deep Sea Drilling Project. In shallow-water carbonates, trace fossil associations range from beach-shoal assemblages having Ophiomorpha, *to* Thalassinoides-Rhizocorallium-Chondrites-*dominated intertidal and sub-tidal assemblages, modified by the development of hard-substrate (hardground) associations. Among pelagic limestones, shallow-water chalks are typically dominated by* Thalassinoides-Planolites-Chondrites *suites, also modified by associated hardgrounds; but in deep-sea chalks, a* Zoophycos-Teichichnus-Chondrites *association dominates.*

Special problems associated with carbonates are (1) the influence of penecontemporaneous cementation on trace-producing organisms, (2) hardgrounds, and (3) the role of organisms in the diagenesis of sediment: differential cementation, concretion formation, and aggregation into pelletal limestones.

INTRODUCTION

Carbonate sediments are linked together only by their composition. These sediments can occur in virtually all marine, quasimarine, terrestrial, and freshwater environments, and most terrigenous clastic sequences have calcareous equivalents.

Ichnologists generally have placed less emphasis upon carbonates than upon most other sedimentary rocks (trace fossils are ordinarily less conspicuous and more difficult to study in carbonates than in siliceous sediments, and the typically greater abundance of body fossils in carbonates diverts even more attention away from trace fossils in these rocks). Yet trace fossil suites well known in terrigenous clastic successions (as ably documented elsewhere in this volume), and representing environments that range from abyssal to aeolian, all occur in carbonates and have equally significant environmental implications.

One can thus cite *Glossifungites* ichnofacies from the Jurassic and Cretaceous hardgrounds of western Europe (Bromley, 1967, 1968; Voigt, 1959; Hölder and Hollmann, 1969; Fürsich, 1971), whereas *Cruziana* ichnofacies dominate Mesozoic carbonates of Europe and North America (Rieth,

1932; Fiege, 1944; Farrow, 1966; Kennedy, 1967a, 1967b; Frey, 1970) and also Caribbean and Indo-Pacific Pleistocene occurrences. *Zoophycos* ichnofacies occur widely in southern Europe—the Lower to Middle Jurassic *Cancellophycus* Limestone (Lessertisseur, 1955; Middlemiss et al., 1970; Gouvernet et al., 1971; Hallam, 1967, 1971), and the *Nereites* ichnofacies is represented in the southern European calcareous flysch (Crimes, 1973).

However, carbonate sedimentation predominates in certain environments, and these are the subject of this chapter—rather than the more unusual occurrences, such as trails on the soles of crinoidal turbidites and insect burrows in calcareous aeolianites. I have therefore concentrated here on (1) intertidal and shallow marine carbonates, drawing my examples chiefly from European Jurassic and tropical Pleistocene deposits, and (2) pelagic limestones, comparing the trace fossils of Cretaceous shallow-water, shelf-sea chalks with those of the deep-sea Cretaceous and Tertiary chalks now available as a result of the Deep Sea Drilling Project.

Trace fossils in the occurrences just mentioned are distinctive and rather well known. In general, however, trace fossils in Paleozoic carbonates have been studied less than those in Mesozoic and Cenozoic carbonates. Work such as that by Walker and Laporte (1970), and a few other examples cited here and in Chapter 18, are notable exceptions; but additional work is especially needed on the ichnology of Paleozoic limestones.

PROBLEMS OF CARBONATE SUBSTRATES

Two features separate carbonate substrates from other kinds. First, carbonates are especially susceptible to synsedimentary and early diagenetic cementation. These lithified horizons have long been recognized by geologists, and the term "hardground" is widely used by European workers to describe these horizons of submarine cementation (Voigt, 1959; Bromley, 1965, 1967, 1968; Bathurst, 1971).

Hardgrounds vary greatly in their morphology. In some, the lithified surface is planar, and substantial erosion has occurred—as in some western European Jurassic examples (e.g., Purser, 1969). Hardgrounds of this type show a predominance of borings, which post-date lithification. In others, for example the highly irregular chalk hardgrounds, evidence of borers is often subsidiary to that of burrowing organisms, which were contemporaneous with cementation in many cases (Bromley, 1968).

Trace fossils in carbonate sequences show several peculiar associations, which can be regarded as a result either of the availability of hard substrates—as in the case of boring organisms (see Chapters 11 and 12), or the restriction placed upon burrowing organisms by the presence of hard, impenetrable surfaces at or just below the sediment–water interface (Chapter 18).

The second feature peculiar to carbonates is that, when cemented, their chemistry makes them susceptible to attack by a wide variety of organisms that bore not by physical but by chemical means, as do some of the mytilid bivalves (e.g., *Lithophaga*). Cemented carbonates thus have the potential of supporting a more diverse association of boring organisms than their noncalcareous counterparts.

TRACE FOSSILS IN CHALKS AND RELATED SEDIMENTS

Chalks are predominantly coccolith limestones having varying proportions of clay minerals, coarser terrigenous material, authigenic synsedimentary minerals, calcispheres (*Oligostegina*), and other coarse organic debris. Nannofossil oozes of this type are an important category of Holocene oceanic sediment, which the Deep Sea Drilling Program shows to have been equally widespread in the Tertiary and

Mesozoic. During Cretaceous time, a combination of climatic, biotic, and geographic factors generated a situation in which this essentially oceanic sediment was widely deposited in shelf seas. Chalks typically ranging in age from Cenomanian to Maastrichtian (but also extending to the Paleocene) are thus widespread in western Europe: the British Isles, France, Belgium, Holland, Germany, Denmark, and eastward into Poland and the U.S.S.R. Similar deposits of Late Cretaceous age are known in the United States and Australia.

These shelf-sea chalks can be interpreted, on the basis of general faunal and sedimentological grounds, as accumulating in depths of 50 to 300 m (Reid, 1968; Kennedy, 1967–1970), and thus are relatively shallow-water deposits. Their oceanic counterparts can equally be shown to have accumulated at depths on the order of 3,000 m. A comparison of these similar facies, deposited at widely different depths, provides a basis for examining the bathymetric value of trace fossil assemblages.

Shallow-Water Chalks

The chief references on chalk trace fossils are Bather (1911), Voigt and Häntzschel (1956), Bromley (1965, 1967, 1968) and Kennedy (1967a, 1967b, 1970)—dealing with European chalks; and Frey (1970), Frey and Howard (1970), and Hattin (1971) —dealing with North American chalks. In both areas, a striking feature is that the sediment has been intensely churned, and preexisting inorganic sedimentary structures have been largely destroyed (Fig. 17.1). Burrows are seen chiefly in vertical section (bedding-plane exposures are rare in these rocks), and most of the disturbance appears as undiagnostic mottling, the general trace fossil assemblage being rather restricted. Forms typically occurring are summarized in Figure 17.2.

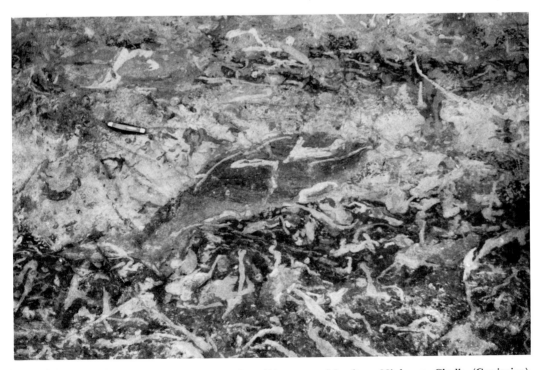

Fig. 17.1 Bioturbate textures in Fort Hays Limestone Member, Niobrara Chalk (Coniacian), Kansas. (From Frey and Howard, 1970.)

Trace Fossils	Morphology	Shelf-Sea Chalks	Deep-Sea Chalks
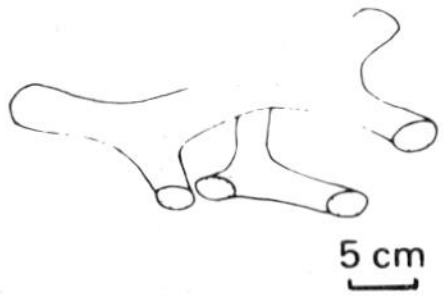 *Thalassinoides* (5 cm)	Burrow systems consisting of vertical shafts connecting largely horizontal tunnel networks. Burrows 2 to 20 cm in diameter, typically showing Y-shaped branching patterns, some joined to from polygons. Burrow surfaces may bear ridges (scratch marks); fills may be laminated, packed with fecal pellets, or consist of concavo-convex laminae (backfills). Vertical elements are common in hardground-associated systems. Produced by a variety of decapods (and other?) crustaceans.	Ubiquitous, often associated with hardgrounds.	Not known to occur.
 *Chondrites* (1 cm)	Plant-like ramifying burrow systems having a few vertical elements and abundant horizontal or gently inclined tunnels. Burrows are circular in cross section, of constant diameter, and have smooth walls. Branching is fairly regular, tending to be pinnate, especially at the periphery of the system. Branching is lateral, never equal, with many orders of branching. No symmetry is apparent, other than a radial tendency. Phobotaxis is obvious. Dwelling, feeding burrows, producer unknown (worm-like).	Widespread; the dominant constituent of burrow mottling at some levels.	Widespread.
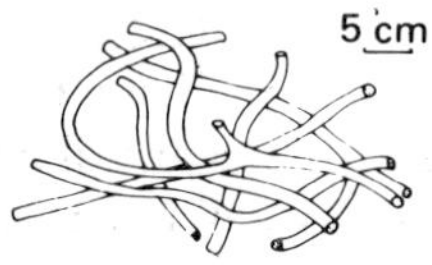 *Planolites* (5 cm)	Irregular, horizontal or inclined burrows, generally on the order of 1 cm in diameter, and filled by sediment presumably passed through the gut of some vermiform animal. Burrows meander and undulate more or less randomly, and may branch.	Ubiquitous; often dominant.	Widespread.
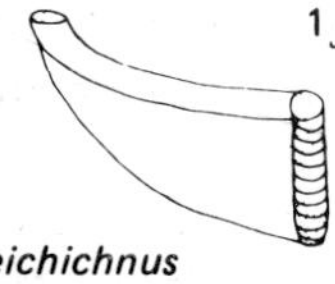 *Teichichnus* (1 cm)	Largely vertical, internally laminated sheet-like structures (spreiten) produced by upward or downward migration of horizontal burrows, which are often preserved at the upper termination of the structure. Rarely branched; feeding burrows.	Widespread, but stratigraphically restricted in western Europe.	Widespread.
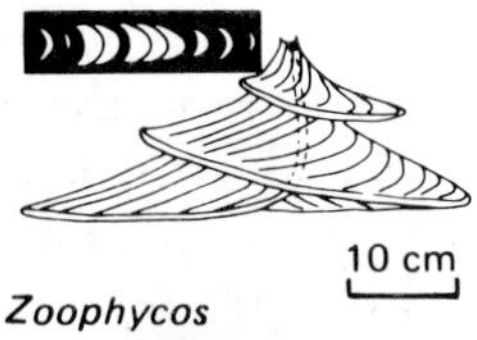 *Zoophycos* (10 cm)	Spreiten burrows produced by reworking of sediment in a series of J-shaped motions, closely spaced. Complete system variable in shape, from tabulate to helicoid (as shown here), with or without a marginal tubular structure recognizable. In section, a spreite consists of alternating reworked and undisturbed lunae of sediment (inset).	Occurs only at very restricted levels.	Widespread.
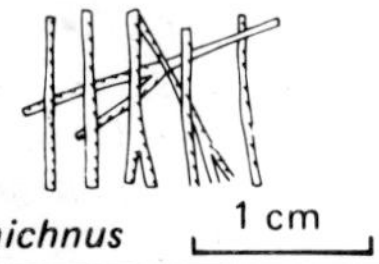 *Trichichnus* (1 cm)	Branched or unbranched, hair-like cylindrical to sinuous burrows, less than 1 mm in diameter, oriented at various angles to bedding. Burrow walls more or less distinct, commonly lined with diagenetic minerals.	Widespread.	Not known in chalk; occurs in foraminiferal oozes.
Pseudobilobites Side view / Base  (1 cm)	Rounded to oval masses of sand-grade microfossils and shell fragments 3-7 cm long, having a convex lower surface and flat or slightly concave upper surface. Lower surface (shown here) bears groups of short, parallel ridges, which reflect scratch marks produced by arthropods. May represent lag floor deposits at the base of burrows.	Widespread.	Not known.

Fig. 17.2 Comparison of trace fossils from shelf-sea and deep-sea chalks.

In both Europe and America, the most common trace is the cylindrical, irregularly branched burrow that Frey (1970) termed *Planolites*. The name embraces a multitude of traces which, if better preserved, could probably be shown to represent several taxa. Also common is *Chondrites* and tiny thread-like burrows called *Trichichnus*. All these traces are of unknown origin. Structures produced by arthropods are particularly important in both regions, especially the genus *Thalassinoides;* it has several distinctive species, some of which contain floor deposits like those described from recent arthropod tunnel systems by Shinn (1968a). Additional types of arthropod-produced structures include *Rusophycus*-like traces, spiral *Gyrolithes* systems, and claw-scratched burrows termed *Spongeliomorpha*. In North American chalks, two spreiten burrows are widespread: *Zoophycos* and *Teichichnus*. These two also occur, less widely, in Europe. To these can be added a few curiosities: *"Asterosoma*-like" structures, *Laevicyclus*, lined tubes—commonly referred to as *Terebella* —and several unnamed forms.

Trace Fossil Associations

In North America, Frey (1970; see also Frey and Howard, 1970) demonstrated a vertical succession in the Fort Hays Limestone—from an association having *Thalassinoides* prominent, at the base of the sequence, up into chalks having *Zoophycos* prominent. Taken with other paleontological and sedimentological evidence (Frey, 1972), this succession can be related to progressive deepening and transgression: the trace fossil spectra thus represent a change from the shallow, inner part of Seilacher's (1967) *Cruziana* "facies" to the deeper, outer part of that "facies," possibly into the upper part of the *Zoophycos* "facies."

In European chalks, trace fossil associations can be subdivided similarly. The bulk of the sequence is dominated by a suite that can be regarded as a reduced *Cruziana* assemblage: *Thalassinoides, Pseudobilobites,* "Planolites," *Chondrites,* "Terebella," *?Spongeliomorpha,* and scarce *Teichichnus.* Forms such as *Rhizocorallium* and *Diplocraterion* are striking by their absence. This reduced assemblage suggests the deeper part of the *Cruziana* "facies." Chalks of this type have long been recognized in Europe; marly varieties are termed "Kreidemergel" (chalk marl), and clay-free varieties are known as "Schreibekreide" (writing chalk). Two important variations on this association can be recognized:

1. Glauconitic chalks: widespread horizons of condensed, winnowed chalks having abundant pelletal glauconites occur either as basal transgressive deposits (Glauconitic Marl, Chalk Basement Beds) or condensed sequences on local swell areas (e.g., Craie de Rouen). The two are invariably characterized by a trace fossil association dominated by *Thalassinoides* and the curious burrow system *"Spongeliomorpha" annulatum* (Kennedy, 1967a): branching, glauconite-free, clay cylinders covered with scratch marks and set in a glauconitic matrix.

2. Hardgrounds: the second variation is the hardground-nodular chalk trace fossil association, admirably documented by Voigt (1959, 1968) and Bromley (1965, 1967, 1968). This association is so closely linked to the actual formation of the hardground that I digress here in order to outline the process (see also Chapter 18):

> With local shallowing, reduced sedimentation, and stabilization of the sea floor on restricted swell areas, colonies of burrowing arthropods become established in a soft bottom. A short distance beneath the sea floor, contemporaneous with the burrowers, calcite precipitation begins around local foci; HCO_3 and Ca^{2+} concentration gradients are set up, and early diagenetic concretions begin to appear. These initially subspherical to lensoid nodules grow progressively into irregular, interlocking lumps, and eventually fuse into a solid lithified layer. Growth of the nodules inhibits the activities of burrow-

ers, and the Thalassinoides systems associated with nodular chalks become highly irregular and distorted, as the animals avoid contact with the growing nodules. Eventually, cementation results in "fossilization" of the burrow system into the classic, irregular, hardground Thalassinoides system (Figs. 17.3–17.5). The reality of this process can be demonstrated by the frequent presence of hard nodules protruding into burrow systems and by the avoidance of inoceramid (bivalve) debris and echinoid tests by the burrowing arthropods, which could not cope with such harder objects.

Formation of the chalk hardground may be divided into a series of stages; an interruption of the process—by renewed sedimentation or other factors—can result in preservation at any stage. The nodular chalks of the Lower and Middle Turonian of the Anglo-Paris Basin, alternating rhythmically with smooth, burrowed chalk, clearly represent arrested "incipient" hardgrounds. Every variation,

from chalk having nodules to truly lithified, rocky substrates, can be recognized in the Upper Turonian to Maastrichtian chalks of western Europe. Curiously, hardgrounds have not been reported from American chalks.[1]

Once generated, the chalk hardground may have a complex history. Diagenetic alteration gives rise to widespread phosphatization and glauconitization, whereas exhumation can give rise to a highly irregular, convoluted surface, which can be shown to be composite: local patches of later sediment are cemented into depressions. Bromley's Chalk Rock hardgrounds (Fig. 17.3) are of this last type (see references cited above). A further type, having more in common with shallow-water examples from the European Jurassic, and

[1] According to R. G. Bromley and R. W. Frey (1973, personal communications), the Arcola Limestone Member of the Mooreville Chalk in Alabama qualifies as a hardground, albeit a poorly developed one (see Monroe, 1941).

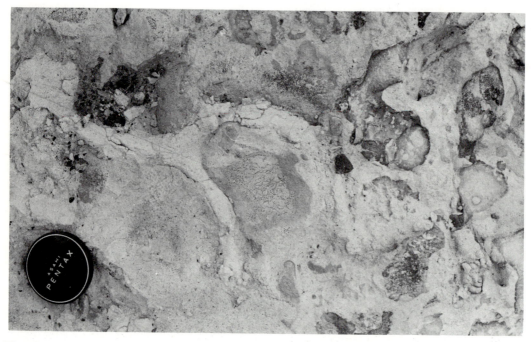

Fig. 17.3 A classic hardground association. View of subhorizontal section, showing highly irregular, white, chalk-filled Thalassinoides (burrow) among glauconitized and phosphatized pebbles and bosses of the hardground surface (darker areas). Where the hardground surface is broken, borings of clionid sponges and other organisms appear as white patches against glauconitized chalk. Upper Turonian (Subprionocyclus neptuni Zone) Chalk Rock; Kensworth, Bedfordshire, England. (Photo courtesy R. G. Bromley.)

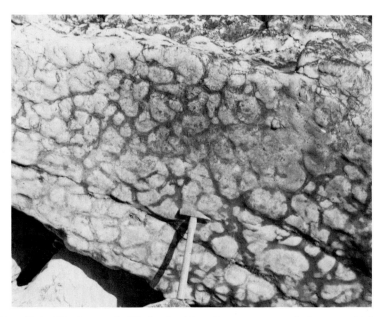

Fig. 17.4 Hardground *Thalassinoides*. Burrows extend from a higher horizon to meet this surface; the animals, being unable to penetrate the hardground, spread out their burrow system as a dense horizontal net, now noticed as darker burrow fills against a lighter, chalky matrix. Summit of Division B, Cenomanian limestone at Beer Head, Devon, England. (From Kennedy, 1970.)

showing obvious signs of mechanical and biological erosion, has a planar surface that often is traceable over many square kilometers. Examples of this latter type occur in the Upper Cenomanian, Turonian, and Coniacian of Normandy, and the Campanian of southern England.

The chalk hardground substrate provided a base for a variety of epifaunal and boring organisms. Distinctive trace fossil associations arose, including borings of thallophytes, clionid sponges, polychaetes, ctenostome bryozoans, bivalves, cirripeds, and other forms (see Bromley, 1968, 1970; Chapter 18). Burrowing organisms were less fortunate; they could penetrate downward only by following earlier burrow systems that remained unfilled, or the fillings of which had not been cemented. As a result, the fillings of chalk hardground burrows were frequently reburrowed, and even provided homes for secondary inhabitants (Bromley, 1967). Where hardgrounds were closely stacked, further problems arose, as in the Cenomanian hardground sequences at Hunstanton, Norfolk. Here, Thalassinoides systems spread out along the surface of earlier hardgrounds (Fig. 17.4), joining earlier burrows and

passing down through these into softer chalks below. Interestingly, Farrow (1971) demonstrated a comparable phenomenon where Pleistocene rock surfaces are covered by a thin veneer of unconsolidated sediment: arthropod burrow systems almost invariably vanish down crevices.

One difference between hardground and "soft ground" Thalassinoides systems is the common preservation of the vertical elements in the former (Fig. 17.5), whereas the latter show only the horizontal basal network. Soft-chalk trace fossil assemblages apparently represent only the deeper elements of systems, their upper part being destroyed by intense churning of sediment by shallow burrowers. With the development of hardgrounds, this later destruction is prevented.

Bänderkreide

At many levels in the western European chalk successions, *Zoophycos* appears in abundance, and this facies has long been known in Germany as "Bänderkreide" (Voigt, 1929; Voigt and Häntzschel, 1956). The chalk is, as the name suggests, banded;

Fig. 17.5 Hardground (top of photo) showing ramifying, three-dimensional *Thalassinoides*, either empty or having silicified fills. In Cerithium Limestone (Lower Danian), Rødvig, Stevns Klint, Denmark; the hardground rests on Fish Clay (basal Danian), in turn resting on Maastrichtian chalk. (Photo courtesy R. G. Bromley.)

the bands are sections of *Zoophycos* of various types, including forms identified as *Zoophycos brianteus* (Häntzschel, 1962). This facies ranges in age from Coniacian to Maastrichtian, chiefly as intercalations in white chalk sequences; it is absent from chalk marls. Broad relationships thus seem superficially rather similar to those described by Frey (1970) from America—reduced *Cruziana* assemblages being replaced by possible *Zoophycos* assemblages. Here the picture is not so simple, however; in the English Coniacian, *Zoophycos* appears in association with hardgrounds, the animal having reworked the soft chalk of burrow fillings in the hardground and between chalk nodules! Any simplistic view of depth-related occurrence must clearly be viewed with some doubt in this case.

Deep-Sea Chalks

The shallow-water chalks described above are remarkably similar in some ways to deep-sea Cretaceous and Tertiary chalks.

Through the courtesy of J. E. Warme (Rice Univ.) and N. Schneidermann (Univ. Illinois), I was able to study sections of Tertiary and Cretaceous chalks from Leg 15 of the Deep Sea Drilling Program in the Caribbean (Warme et al., 1973). All the material studied to date is in the form of vertical sections, but even so, comparisons with shelf-sea chalks are striking (Figs. 17.2, 17.6, and 17.7). In general, bioturbation was intense, and closely resembled that of the European Kreidemergel or the American Fort Hays Limestone (Fig. 17.1). The trace fossils present are also similar. *Chondrites, Zoophycos, "Planolites,"* and *Teichichnus* identical with those described by Kennedy (1967a) and Frey (1970) are present, amid comparable background burrow mottling.

The most obvious difference between traces is the striking absence in deeper chalks of the arthropod burrows that dominate shallower chalks; an overall difference in trace fossil assemblages is also clear. Depth as such is now "unfashionable" as a

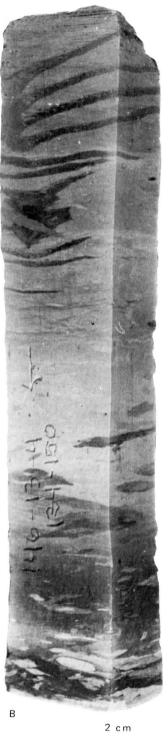

A

B

2 cm

Fig. 17.6 Bioturbate textures in deep-sea chalks. A, general burrow mottling, chiefly *Planolites* and *Chondrites*. B, section through complete *Zoophycos* system. Both in Maastrichtian chalks; Caribbean (site 145 of Deep Sea Drill. Proj. Leg 15). Depths = 537 and 501 m, respectively. (Photos courtesy J. E. Warme.)

Fig. 17.7 *Zoophycos* in deep-sea chalk. Maastrichtian; Caribbean (site 146 of Deep Sea Drill. Proj. Leg 15). Depth = 537 m. (Photo courtesy J. E. Warme.)

governing factor for trace fossil assemblages, but some depth-related parameter seems to be responsible for these differences, as discussed subsequently.

Jurassic and Lower Cretaceous Pelagic Carbonates

Finding pre-Upper Cretaceous deposits to compare with chalks is fraught with difficulty, because the special combination of biologic and geographic factors generating this facies does not seem to have existed previously. However, several nonchalk facies show certain common features that can be discussed meaningfully here.

Much of the Cretaceous in the Vocontian region of southern France contains lithologies broadly comparable to the Chalk Marl: the partly depositional and partly diagenetic rhythmic alternations of more or less calcareous marly limestones. The lithologies show extensive *Thalassinoides - Chondrites - Planolites* - dominated bioturbate textures in the more basinal facies; shoreward, or onto swell regions, they are interrupted by nonsequences. The latter are associated with hardgrounds: burrows and borings like those of their Upper Cretaceous counterparts, even glau-

conitic beds having a *"Spongeliomorpha" annulatum-Thalassinoides*-dominated association. In some of these swell or nearer shore areas, slightly different trace fossil assemblages appear; *Spongeliomorpha iberica* is in many places abundant, and *Rhizocorallium* may be present—paralleling the original Spanish Miocene association noted by Saporta (1887), interpreted by Darder (1945) as beds of sexually dimorphic algae.

This type of limestone–marl rhythmic sequence was studied in detail in the western European Lias by Sellwood (1970a, 1971). With an elegant and comprehensive interpretation of cycles—using trace fossils, body fossils, and lithological criteria—he extended Hallam's (1964) observation that the presence of burrows indicates a part-primary origin for the rhythmicity. Sellwood recognized three main kinds of cycles, interpreted as coarsening-upward units. Two are predominantly clastic, and need not concern us here; the third—the limestone–shale rhythm of previous authors—was summarized by Sellwood (1971, Fig. 4; see also Fig. 18.3B). The cycles are interpreted as minor regressive, shallowing-upward, low- to high-energy sequences. Lithological changes are accompanied by the replacement of essentially deposit-feeding organisms (represented by body fossils and trace fossils) in the clay part of the sequence, by predominantly suspension-feeding forms in the limestone unit.

Other analogs of chalk sedimentation are found in the pelagic Calcaire Rouge-Ammonitico Rosso facies (Jurassic) of southern Europe (Hallam, 1967). Many lithological similarities exist between these and nodular chalks; *Thalassinoides* tunnel systems are conspicuous locally, and in the Calcaire Rouge, well-developed hardground sequences are also present.

TRACE FOSSILS IN SHALLOW-WATER CARBONATES

Recent shallow-water carbonates have received a great deal of attention in the past decade, and burrows in these sediments have been studied in two main areas. Shinn (1968a), working in Florida and the Bahamas, recognized five distinctive burrow types: four produced by crustaceans (*Alpheus, Callianassa, Cardisoma,* and *Uca*), and one by a coelenterate (*Phyllactis*). Farrow (1971), working on Aldabra atoll, and Braithwaite and Talbot (1972), working in the Seychelles of the western Indian Ocean, also studied chiefly arthropod burrows (*Ocypode, Neaxius, Albunea, Callianassa, Alpheus, Macropthalmus,* and *Uca*). These three papers form an excellent basis for discussing comparable traces in ancient carbonates, although problems arise where no recent analogs to ancient burrow types are available, as with *Rhizocorallium* and similar traces. A wealth of information may also be gleaned from Multer's (1971) *Field Guide to Some Carbonate Rock Environments: Florida Keys and Western Bahamas,* and various earlier guidebooks [see Multer (p. 3, and extensive bibliography)] or via the bibliography in Bathurst's (1971) extensive review.

Supratidal Deposits

As examples of supratidal carbonates, consider beach-berm or natural-levee areas and supratidal mud flats, separated from the open sea or flooded only periodically by wind tides. Such environments present problems to the palichnologist in that minor changes of geography may destroy or inundate these areas, and older sediments are colonized by later faunas very different from those typical of the actual supratidal situation.

Supratidal deposits are best recognized by features other than trace fossils, such as laminated algal mats, desiccation cracks, and more importantly, birdseye structures. Development of anhydrite and dolomite are also important clues, especially where the classic "sabkha cycle" (Shearman, 1963, 1966) can be recognized [see Shinn (1968b) and Shinn et al. (1969) for other features].

The most obvious biogenic structures to seek in ancient examples of such sediments are of two kinds: (1) simple helical dwelling burrows comparable to those of *Cardisoma, Uca,* and *Ocypode* (Verway, 1930; Shinn, 1968a; Farrow, 1971), together with fecal pellet deposits or pellets produced during burrow excavation (Shinn, 1968b, Fig. 11B), and (2) rhizoliths—root casts or external molds produced as a result of the decay of woody matter. (See Chapter 10.) In particular, recognition of mangrove root systems in Pleistocene and Tertiary intertidal to supratidal carbonates has potential as an environmental tool.

Ancient examples of supratidal carbonates are numerous, but few accounts of trace fossils within them have been given. Multer and Hoffmeister (1968) pointed out the common occurrence of rootlet casts in the Pleistocene laminated crusts of the Florida Keys. Braithwaite et al. (1973) recorded similar crusts in the western Indian Ocean, and observed rootlet molds associated with land snails and fossil tortoises in the mid-Pleistocene calcarenites of Aldabra atoll. A truly impressive example of rhizoliths comes from the Key Biscayne area of Florida, where Hoffmeister and Multer (1965) figured and described mangrove "reefs" consisting of carbonate tubes; these developed around *Avicennia* roots as the woody material decomposed, releasing carbon dioxide and ultimately leading to carbonate precipitation.

The bulk of ancient "examples" of rootlet beds of this type are, in fact, references to arthropod burrows, as in the early interpretations of "tubulures" in chalk hardgrounds (Bromley, 1967, p. 158). Care is needed in identifying root molds; the presence of lignite is the best clue. Nevertheless, rhizoliths and rootlet beds seem to be very widespread in the Great Oolite (Bathonian) White Limestones of the English Midlands (T. Palmer, unpublished), and indeed seem to indicate temporarily emergent carbonate areas. These structures even occur with desiccation features and birdseyes; their potential is obvious.

Seemingly significant is that the simple arthropod burrows—which characterize this environment today—and comparable signs of bioturbation are absent in ancient supratidal deposits.

Littoral and Sublittoral Deposits

Work on recent carbonates indicates the great difficulties in separating the trace fossil associations of intertidal and subtidal deposits based on traces alone; thus, to point out some specific examples is perhaps safer than attempting generalities.

One of the now-classic littoral and sublittoral sand-body trace fossils is *Ophiomorpha* (Häntzschel, 1952; Toots, 1961; Weimer and Hoyt, 1964; Kennedy and MacDougall, 1969; Chapter 2). *Ophiomorpha* consists of medium-size, three-dimensional burrow systems that branch dichotomously at acute angles—having swollen branching points—and that are lined with ovoid pellets. In recent sediments, such structures are produced by a variety of crustaceans, the best known being the ghost shrimp *Callianassa major* (Weimer and Hoyt, 1964). Most such crustaceans are described as occurring primarily in the lower littoral zone, between mean sea level and low tide on beaches and in sounds and tidal flats having moderate currents and near-normal salinity. They also extend into the shallow sublittoral zone.

Fossil examples of these burrows are known as far back as the Permian. They commonly occur in well-sorted, massive-bedded sandstones, and here indicate littoral or shallow sublittoral conditions with high energy. Where they occur in poorly sorted, well-bedded silty sandstones, less current energy or deeper sublittoral conditions are suggested. (See Table 2.1.)

The tunnel-lining process is presumably a response to prevent burrow collapse in unstable substrates, and the obvious carbonate environment in which to seek such structures are beaches and oolite shoals. *Ophiomorpha* is common locally in the Pleistocene Miami Oolite of Florida. F. Fürsich (1972, personal communication) also found an equivalent association in the Oxfordian (Upper Jurassic) oolite shoal facies of the Dorset Corallian. Predictably, the other trace fossils in these unstable environments are deep burrows such as *Skolithos*.

Most traces described from recent littoral and sublittoral carbonates fall within the classic *Cruziana* "facies" of Seilacher (1967); *Thalassinoides,* the ancient analog of certain callianassid and alpheid burrows, is predominant. For several years, workers puzzled over the absence of likely burrow producers preserved in these sediments, the few fossil associations described being of seemingly nonburrowing crustaceans such as *Meyeria* and *Glyphea*. Recently, however, Rice and Chapman (1971) showed that *Nephrops* and *Goneplax,* a lobster and a crab exhibiting no structural adaptations for burrowing, produce *Thalassinoides*-like burrows in recent sediments, and *Glyphea* has been found within *Thalassinoides* (Fig. 2.6); thus, this problem can probably be dismissed.

Thalassinoides is common—and often dominant—in European Bajocian, Bathonian, Oxfordian, and Portlandian limestones, in facies ranging from muddy calcarenites to oolites. A variety of species are present; most resemble the alpheid traces described by Farrow (1971). *Gyrolithes*-like structures also occur, although rarely. Many burrows are packed with fecal pellets (e.g., Kennedy et al., 1969; Sellwood, 1970b), and the quantity of arthropod feces may reach rock-forming proportions. In Mesozoic shallow-water carbonate environments, as in the Holocene, arthropods seem to have been the dominant burrowers [see also American examples figured by Shinn et al. (1969), and European occurrences figured by Weigelt (1929), Rieth (1932), Seilacher (1955), and Fürsich (1971)].

A common trace fossil in Mesozoic shallow-water carbonates is the horizontal U burrow *Rhizocorallium*. This is probably also a crustacean trace, and the ethology of the *Rhizocorallium* animals seemingly varied, depending on the local environment. Certain short traces were probably produced by suspension feeders, whereas longer, meandering traces represent deposit-feeding activities. Farrow (1966) analyzed these relationships in the mixed carbonate-clastic Middle Jurassic Scarborough Beds of the Yorkshire coast; similar patterns emerge in other European Mesozoic limestones.

In older carbonates, the absence of abundant arthropod burrows gives a slightly different picture. Reis (1910) described *Thalassinoides* from Triassic limestones, and Shinn (1968a, Fig. 16) figured signs of activity from Permian dolomites identical with recent callianassid traces. Kaźmierczak and Pszczółkowski (1969) suggested that burrowing in the mid-Triassic shallow-water Muschelkalk limestones of the Holy Cross Mountains is due primarily to enteropneusts (Fig. 17.8). Soergel (1923) also suggested that burrows in the Thuringian Muschelkalk were produced by *Balanoglossus;* Mägdefrau (1932) described similar structures, also from Thuringian Triassic limestones. Seilacher (1955, 1964)

noted, however, the presence of *Rhizocorallium, Thalassinoides, Teichichnus,* and *Scolicia* from the south German Muschelkalk, so that the Triassic *Cruziana* facies can be linked, via Farrow's (1966) Jurassic associations, to recent assemblages.

Hardgrounds

Penecontemporaneous cementation is now known to be widespread in recent shallow-water and supratidal carbonates [see Bathurst (1971) for references], and rocky substrates consisting of Pleistocene limestones are very widespread in Holocene carbonate environments.

As with the chalk hardgrounds discussed previously, the evidence indicates that lithification can proceed while burrowing animals are actually living in the sediment, or where burrows are vacated but unfilled by sediment. In recent examples, Shinn (1968a, p. 890) suggested that open burrows in the Caribbean, now buried approximately 2 m below present sea level, are on the order of 2,000 years old; he concluded that lithification might even take place before burrow infilling. Fürsich (1971) recently documented precisely this phenomenon in a closely comparable Jurassic example: the multiple hardgrounds in the Bajocian limestones of Normandy.

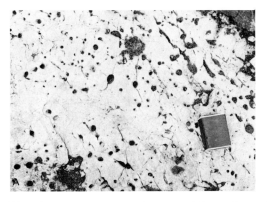 

Fig. 17.8 Enteropneust burrows. Views of weathered surface of bed (A), and section (B). On the surface, apertures of main burrows were scoured and accentuated prior to infilling; smaller anal outlets also visible. Middle Triassic Lukowa Beds (Muschelkalk) limestones; Holy Cross Mountains, Poland. (Photos courtesy J. Kaźmierczak.)

He noted the same limitation on burrowers by superficial cementation as has been recorded in chalk hardgrounds (Fig. 17.9).

Only one burrow system, *Thalassinoides,* is consistently associated with hardgrounds. Borings, in contrast, are far more diverse. In recent shallow-water lithified carbonate substrates, a large variety of organisms bore (Bromley, 1970; Chapter 11). Fossil analogs have been described from the Miocene of Poland (Radwański, 1964, 1965, 1968, 1970); here the borings are sufficiently close in form to recent ones that an interpretation of affinities to living genera of borers could be attempted.

Comparable Eocene examples of hardgrounds, interpreted as rocky shores, were described by Hecker et al. (1963). Cretaceous and Jurassic bored surfaces have also been documented in detail: Pianovsky and Hecker (1966), Hölder and Hollmann (1969), Purser (1969), Głazek et al. (1971), Fürsich (1971), Kennedy and Klinger (1972), and Bromley (Chapter 18).

Hardgrounds are also widespread in older limestone sequences, but their trace fossil assemblages differ in that borings of sponges and bivalves are much rarer. Paleozoic examples cited by Oriviku (1940) and Hecker (1960, 1970) are dominated by *Trypanites* and show little variety. To what extent the trace systems in Ordovician hardgrounds recorded by Jaanusson (1961, and references therein) and Lindström (1963) are burrows, as opposed to borings, remains unclear.

DIAGENESIS

Preceding discussions of hardgrounds show the complex interrelationships between the activities of burrowing organisms and sea-floor diagenesis. Some further points are worth noting. The role of sediment-eating organisms in the alteration of Eh/pH conditions in sediments is clear from data provided by Davidson as early as 1891; he suggested that *Arenicola* dwelling on sand flats off the Northumberland coast ingested 2,000 tons (18 x 10^5 kg) of sand per acre (4,000 m^2) per annum. The effect of this sort of activity on carbonate sediments would be even more severe, as evidenced by Taylor's (1964) observation that 80 to 90 percent of the carbonate sands in the Bahamas had passed through echinoderm (chiefly holothurian) intestinal tracts. In bioturbated ancient carbonates, we may thus expect evidence of similar effects.

A further important result of organisms is an aggregation of fecal-pellet sediments (e.g., Folk and Robles, 1964). Vast quantities of calcilutite are aggregated into pelletal form by mollusks, echinoderms, and arthropods. The hydrodynamic properties of the sediment are altered, and microenvironments for diagenetic change are created. Aggregation thus seems to be prerequisite for such diagenetic processes as glauconitization and phosphatization. Microcoprolites are widespread in ancient carbonate sequences, and should be treated as trace fossils. Häntzschel et al. (1968) provided an annotated bibliography of coprolites from a biological viewpoint; the extensive literature on pelletal limestones was reviewed by Bathurst (1971) and various other authors (in Pray and Murray, 1965).

One of the more curious aspects of these pellets is that many are highly diagnostic of their producers (Kennedy et al., 1969, Pl. 99). Progressive morphological changes in some arthropod intestines, as manifested in their fecal pellets, has given these objects a biostratigraphic value (e.g., Brönnimann and Norton, 1960; Elliot, 1962; Kornicker, 1962; Sartoni and Crescenti, 1962; Masse, 1966; Brönnimann and Masse, 1968).

In the later stages of diagenesis, trace fossils have an important role to play. Many burrows initially have organic or mucus-rich linings, and are thus potential sites for the development of reducing microenvironments. Burrow fill and adjacent sediment may have different porosity, Eh,

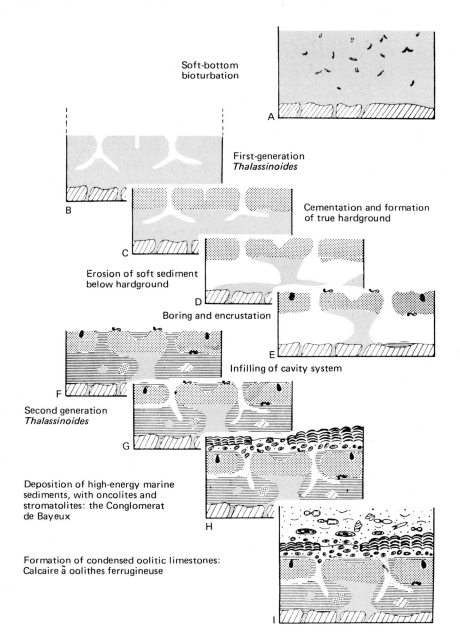

Soft-bottom
bioturbation

First-generation
Thalassinoides

Cementation and formation
of true hardground

Erosion of soft sediment
below hardground

Boring and encrustation

Infilling of cavity system

Second generation
Thalassinoides

Deposition of high-energy marine
sediments, with oncolites and
stromatolites: the Conglomerat
de Bayeux

Formation of condensed oolitic limestones:
Calcaire à oolithes ferrugineuse

Fig. 17.9 Diagrammatic representation of sequence of sedimentary events (from top to bottom) associated with formation of a complex hardground sequence in the Bajocian of the Normandy coast. A, on top of slightly indurated burrow horizon, lime mud is deposited, then bioturbated by sediment eaters. B, after initial compaction and erosion, the same mud forms a firm ground in which crustaceans can make more permanent burrows of *Thalassinoides* type. C, cementation from the surface downward produces a true hardground. D, erosion undermines the hardground by washing out burrows. E, encrusting and boring organisms settle on the exposed hardground surfaces. F, sedimentary infills of cavities in hardground are burrowed by second generation of *Thalassinoides*-animals, during period of nonsedimentation. G–I, stromatolites and ferruginous ooliths form during long period of nonsedimentation, giving added complexity to environmental history. (Data courtesy of F. Fürsich.)

pH, and organic content, thus acting as foci for cementation and replacement.

Examples of the above processes are widespread in fine-grained sediments of many types. Pyritic films develop along burrow fill-burrow wall interfaces, especially in chalks and chalky shales, and may subsequently develop into pyrite cylinders and nodules. These are the "pyritisierte Gangsysteme" of German workers (e.g., Schloz, 1968). Various other examples of pyritic burrows were noted by Kennedy (1967b) and Frey (1970). Differences in sediment consistency can result in precipitation of calcium carbonate in burrow fillings, and is one of the reasons why burrows are in fact visible in many limestones. In the English Corallian (Oxfordian), for instance, burrows of this type are well known. Not widely realized, however, is that overprecipitation of carbonate in burrow fills can give rise to interlocking systems that, in section, seem to be merely nodular limestones (Fürsich, 1973). Many concretions owe their origin to deposition of $CaCO_3$ in burrows, with or without subsequent overgrowth. Such nodules and concretions are widespread in both limestones and calcareous clays. (See Chapter 4.)

The origin of siliceous nodules in many carbonates provides one of the most spectacular examples of the role of physical discontinuities associated with burrow horizons as foci for diagenesis. Most (but not all) of the flints in European chalks, and cherts in some Mesozoic limestones, are in fact silicified burrows; a transition from partly silicified burrow fill to perfect burrow pseudomorph can often be demonstrated. Voigt and Häntzschel (1956) and Häntzschel (1960) figured many flint *Zoophycos* from the north German Bänderkreide. Bromley (1967) demonstrated horned and finger-like flints to be fragments of silicified or oversilicified *Thalassinoides* (e.g., Figs. 17.10 and 17.11). Similar silicified burrows can be recognized in the Tuffeau facies of northern France and Holland; they also occur in British Volgian limestones (Portland Formation, southern England) and North American Carboniferous limestones (Chamberlain, 1971, p. 223).

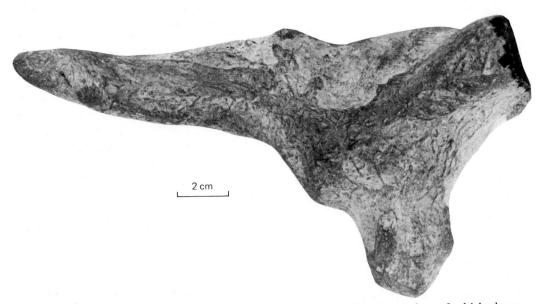

2 cm

Fig. 17.10 Flint representing silicified *Thalassinoides* (burrow fill), the surface of which shows a reticulum of *Chondrites* and other burrows. Silicification accentuates detailed bioturbate textures invisible in original white, homogenous chalk. Upper Turonian (*Subprionocyclus neptuni* Zone) Upper Chalk; Dover, Kent, England. (From Kennedy, 1970.)

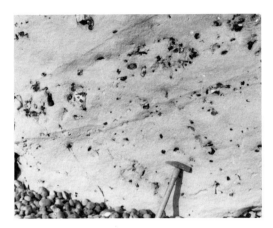

Fig. 17.11 Finger-like flint nodules. Silicification proceeded along otherwise invisible, three-dimensional *Thalassinoides* (burrow systems) in homogeneous white chalk. Middle Turonian (*Collignoniceras woolgari* Zone) Middle Chalk; near Beer Head, Devon, England. (From Kennedy, 1970.)

Oversilicification leads to the development of courses of horned cherts. "Bedwell's Columnar Band," an important marker in the monotonous Santonian chalks of southeastern England, is a prime example of a silicified-burrow horizon that is used in correlation. This course of cavernous horned flints, having regularly spaced, finger-like, upward extensions 50 cm high, represents a widespread, synchronous halt in sedimentation associated with extensive development of thalassinid burrow systems.

The spectacular "paramoudras"—huge barrel-shaped flints, figured and described from Norfolk and Ireland by Buckland (1817), Lyell (1838), Peake and Hancock (1961), and others (Chapter 16)—are also related to the presence of burrow systems. Some "fairy ring"-type flints are associated with huge *Asterosoma*-like structures.

SUMMARY AND DISCUSSION

Carbonate equivalents exist for the bulk of standard terrigenous clastic sequences, and comparable ranges of environments (and hence trace fossil assemblages) are present in each. Carbonate sedimentation is largely confined to two broad environments:

1. The shallow-water Bahamas–Persian Gulf situation, represented by many Indo-Pacific and Caribbean Pleistocene–Tertiary successions, and much of the northwest European Jurassic. Recent–ancient analogs are numerous; bioturbate textures are widespread, and arthropod structures are the most conspicuous trace fossils. Particularly important in both ancient and modern carbonates of this type is the widespread occurrence of synsedimentary and early-diagenetic cementation. Hardgrounds are common in both situations; these provide suitable substrate conditions for boring organisms of many types, but place severe restraints on many burrowers.

    In many carbonate sediments, bioturbate textures reflect the wholesale fragmentation of sediment grains by deposit-feeding organisms; an overall alteration of pH–Eh conditions also resulted from their activities. The latter is of great significance in subsequent diagenesis. The widespread occurrence of recent pelletal lime muds and ancient fecal-pellet limestones (trace-fossil sediments, no less), is another important aspect of shallow-water carbonate sedimentation.

2. Pelagic carbonates, exemplified by chalks, both ancient and modern, and their pre-Cretaceous analogs. Trace fossils in these rocks provide a great deal of information about substrate conditions; also, burrowing animals played an important part in early diagenesis (hardgrounds and nodular limestones) as well as providing foci for later diagenesis.

Finally, the preceding discussions provide a base for reviewing the concept of trace fossil paleocommunities—Seilacher's now-classic "universal marine ichnofacies." No doubt these exist, and scores of publications attest to the validity of this approach [see the publications by Seilacher cited here, the papers in Crimes and Harper (1970), Chamberlain (1971), and Chapters 2, 7, and 9)]. Problems arise only when the precise factors limiting their occurrence are suggested.

Controlling factors can be proposed at a variety of levels. At a low level, factors such as food supply, oxygen availability, and substrate are clearly important. These parameters can be linked by the concept of turbulence and environmental energy and, in turn, to a depth control, as Seilacher originally proposed (1964, 1967, etc.).

Depth as such does not greatly affect the distribution of organisms. Rather, those factors just mentioned generally change progressively with depth—in addition to overriding factors such as environmental stability and predictability. This depth relationship, although often true, is clearly not invariable (cf. Fig. 2.8); quiet-water "predictable" environments, accumulating fine-grained sediment, can occur at all depths, although sediments deposited in high-energy conditions are less likely to occur in deep water. In this way, the occurrence of *Zoophycos* in siltstones alternating with coals and rootlet beds in the Yoredale facies (Upper Carboniferous, northern England) or the Albian Upper Greensand Malmstone facies (Albian, southern England) is explicable.

Trace fossils may thus appear in shallower water deposits than predicted by Seilacher's model; but the reverse—the extension of, say, traces from *Cruziana* "facies" into the abyss—is less likely. The foregoing account of shallow and deep-water chalks well illustrates this generality. Bänderkreide *Zoophycos* chalks represent the temporary spread of deeper water organisms into the "deep" end of the *Cruziana* "facies," but traces such as *Thalassinoides*, *Pseudobilobites*, and *Rhizocorallium* do not appear in true deep-water chalks.

ACKNOWLEDGMENTS

I thank R. G. Bromley, J. E. Warme, J. Kaźmierczak, F. Fürsich, and R. W. Frey for supplying various illustrations used in this chapter, and am grateful to these workers, and also to B. W. Sellwood, T. J. Palmer, and A. Hallam, for their help and constructive criticisms.

REFERENCES

Bather, F. A. 1911. Upper Cretaceous terebelloids from England. Geol. Magazine, 8:481–487, 549–556.

Bathurst, R. G. C. 1971. Carbonate sediments and their diagenesis. Developments in Sedimentology, 12, 620 p.

Braithwaite, C. J. R. and M. Talbot. 1972. Crustacean burrows in the Seychelles, Indian Ocean. Palaeogeogr., Palaeoclimatol., Palaeoecol., 11:265–285.

———— et al. 1973. The evolution of an atoll: the depositional and erosional history of Aldabra. Royal Soc., Philos. Trans. (B), 266:307–340.

Bromley, R. G. 1965. Studies in the lithology and conditions of sedimentation of the Chalk Rock and comparable horizons. Unpubl. Ph.D. Dissert., London Univ., 355 p.

————. 1967. Some observations on burrows of thalassinidean Crustacea in chalk hardgrounds. Geol. Soc. London, Quart. Jour., 123:157–182.

————. 1968. Burrows and borings in hardgrounds. Dansk Geol. Foren. Meddr., 18:248–250.

————. 1970. Borings as trace fossils and *Entobia cretacea* Portlock, as an example. In T. P. Crimes and J. C. Harper (eds.), Trace fossils. Geol. Jour., Spec. Issue 3:49–90.

Brönnimann, P. and J. P. Masse. 1968. Thalassinid (anomura) coprolites from Barremian-Aptian passage beds, Basse-Provence, France. Rev. Micropaleont., 11:153–160.

———— and P. Norton. 1960. On the classification of fossil fecal pellets and description of new forms from Cuba, Guatemala and Libya. Eclogae Geol. Helvetiae, 53: 832–842.

Buckland, W. 1817. Description of the Paramoudra, a singular fossil body that is found in the chalk of the north of Ireland, with some general observations on flints in chalk. Geol. Soc. London, Trans., 4:413–423.

Chamberlain, C. K. 1971. Morphology and ethology of trace fossils from the Ouachita mountains, southeastern Oklahoma. Jour. Paleont., 45:212–246.

Crimes, T. P. 1973. From limestones to distal turbidites: a facies and trace fossil analysis in the Zumaya flysch (Paleocene–Eocene), north Spain. Sedimentology, 20:105–131.

———— and J. C. Harper (eds.) 1970. Trace fossils. Geol. Jour., Spec. Issue 3, 547 p.

Darder, B. 1945. Estudio geologico del Sur de la Provinca de Valencia y Norte de la Alicante. Inst. Min. Geol. Espana, Bol. 57:59–362.

Davidson, C. 1891. On the amount of sand brought up by lobworms to the surface. Geol. Magazine, 8:489–493.

Elliot, G. F. 1962. More micro-problematica from the Middle East. Micropaleontology, 8:29–44.

Farrow, G. E. 1966. Bathymetric zonation of Jurassic trace fossils from the coast of Yorkshire, England. Palaeogeogr., Palaeoclimatol., Palaeoecol., 2:103–151.

————. 1971. Back-reef and lagoonal environments of Aldabra atoll distinguished by their crustacean burrows. Zool. Soc. London, Symp., 28:455–500.

Fiege, K. 1944. Lebensspuren aus dem Muschelkalk Nordwestdeutschlands. Neues Jahrb. Mineral. Geol. Paläont., Abh. B, 88:401–426.

Folk, R. L. and R. Robles. 1964. Carbonate sands of Isla Perez, Alacran reef complex, Yucatan. Jour. Geol., 72:255–292.

Frey, R. W. 1970. Trace fossils of Fort Hays Limestone Member of Niobrara Chalk (Upper Cretaceous), west-central Kansas. Univ. Kansas Paleont. Contr., Art. 53, 41 p.

————. 1972. Paleoecology and depositional environment of Fort Hays Limestone Member, Niobrara Chalk (Upper Cretaceous), west-central Kansas. Univ. Kansas Paleont. Contr., Art. 58, 72 p.

———— and J. D. Howard. 1970. Comparison of Upper Cretaceous ichnofaunas from siliceous sandstone and chalk, western interior region, U.S.A. In T. P. Crimes and J. C. Harper (eds.), Trace fossils. Geol. Jour., Spec. Issue 3:141–166.

Fürsich, F. 1971. Hartgründe und Kondensation im Dogger von Calvados. Neues Jahrb. Geol. Paläont., Abh., 138:313–342.

————. 1973. *Thalassinoides* and the origin of nodular limestone in the Corallian Beds (Upper Jurassic) of southern England. Neues Jahrb. Geol. Paläont., Mh., 1973:136–156.

Głazek, J. et al. 1971. Lower Cenomanian trace fossils and transgressive deposits in the Cracow Upland. Acta Geol. Polonica, 21:433–448.

Gouvernet, C. et al. 1971. Provence. Paris, Masson, 229 p.

Hallam, A. 1964. Origin of the limestone–shale rhythm in the Blue Lias of England: a composite theory. Jour. Geol., 72:157–169.

————. 1967. Sedimentology and palaeogeographic significance of certain red limestones and associated beds in the Lias of the Alpine region. Scottish Jour. Geol., 3:195–220.

————. 1971. Facies analysis of the Lias in west central Portugal. Neues Jahrb. Geol. Paläont., Abh., 139:226–265.

Häntzschel, W. 1952. Die Lebensspuren von *Ophiomorpha* Lundgren im Miozän bei Hamburg, ihre weltweite Verbreitung und Synonymie. Mitt. Geol. Staatsinst. Hamburg., 21:142–153.

————. 1960. Spreitenbauten (*Zoophycos* Massalongo) im Septarienton Nordwest-Deutschlands. Mitt. Geol. Staatsinst. Hamburg., 29:95–100.

————. 1962. Trace fossils and problematica. In R. C. Moore (ed.), Treatise on invertebrate paleontology, Pt. W, Miscellanea. Lawrence, Kan., Geol. Soc. America and Univ. Kansas Press, p. W177–W245.

———— et al. 1968. Coprolites: an annotated bibliography. Geol. Soc. America, Mem. 108, 132 p.

Hattin, D. E. 1971. Widespread, synchronously deposited, burrow-mottled limestone beds in Greenhorn Limestone (Upper Cretaceous) of Kansas and southeastern Colorado. Amer. Assoc. Petrol. Geol., Bull., 55:412–431.

Hecker, R. T. 1960. Fossil facies of smooth rocky sea floor. Akad. Nauk. Est. S.S.R., Trudy Geol. Inst., 5:199–227.

————. 1970. Palaeoichnological research in the Palaeontological Institute of the Academy of Sciences of the U.S.S.R. In T. P. Crimes and J. C. Harper (eds.), Trace fossils. Geol. Jour., Spec. Issue 3:215–226.

———— et al. 1963. Fergana Gulf of Paleogene sea of central Asia, its history, sediments, fauna and flora, their environment and evaluation. Amer. Assoc. Petrol. Geol., Bull., 47:617–631.

Hoffmeister, J. E. and J. G. Multer. 1965. Fossil mangrove reef of Key Biscayne. Geol. Soc. America, Bull., 76:845–852.

Hölder, H. and R. Hollmann. 1969. Bohrgänge mariner Organismen in jurassischen Hart- und Felsboden. Neues Jahrb. Geol. Paläont., Abh., 113:79–88.

Jaanusson, V. 1961. Discontinuity surfaces in limestones. Geol. Inst. Univ. Uppsala, Bull., 40:221–241.

Kaźmierczak, J. and A. Pszczółkowski. 1969. Burrows of Enteropneusta in Muschelkalk (Middle Triassic) of the Holy Cross Mountains, Poland. Acta Palaeont. Polonica., 14:299–324.

Kennedy, W. J. 1967a. Burrows and surface traces from the Lower Chalk of southern England. British Mus. (Nat. Hist.), Geol., Bull., 15:127–167.

——. 1967b. Field meeting at Eastbourne, Sussex. Lower Chalk sedimentation. Geologists' Assoc., Proc., 77:365–370.

——. 1969. The correlation of the Lower Chalk of south-east England. Geologists' Assoc., Proc., 77:459–560.

——. 1970. Trace-fossils in the chalk environment. In T. P. Crimes and J. C. Harper (eds.), Trace fossils. Geol. Jour., Spec. Issue 3:263–282.

—— and H. C. Klinger. 1972. Hiatus concretions and hardground horizons in the Cretaceous of Zululand (South Africa). Palaeontology, 15:539–549.

—— and J. D. S. MacDougall. 1969. Crustacean burrows in the Weald Clay (Lower Cretaceous) of south-eastern England and their environmental significance. Palaeontology, 12:459–471.

—— et al. 1969. A *Favreina-Thalassinoides* association from the Great Oolite of Oxfordshire. Palaeontology, 12:549–554.

Kornicker, L. S. 1962. Evolutionary trends among molluscan fecal pellets. Jour. Paleont., 36:829–834.

Lessertisseur, J. 1955. Trace fossiles d'activite animale et leur significance paléobiologique. Soc. Géol. France, Mém. 74, 150 p.

Lindström, M. 1963. Sedimentary folds and the development of limestone in an Early Ordovician sea. Sedimentology, 2:243–292.

Lyell, C. 1838. Elements of geology. London, John Murray, 794 p.

Mägdefrau, K. 1932. Über einige Bohrgänge aus dem unteren Muschelkalk von Jena. Paläont. Zeitschr., 14:150–160.

Masse, J. P. 1966. Sur la présence en Basse-Provence d'un niveau a *Favreina* aff. *salvenis* (Parejas) a la limite Barremien-Aptien. Soc. Géol. France, Compt. Rend., 8:298–300.

Middlemiss, F. A. et al. 1970. Summer field meeting in Provence. Geologists' Assoc., Proc., 81:363–396.

Monroe, W. H. 1941. Notes on deposits of Selma and Ripley age in Alabama. Alabama Geol. Survey, Bull. 48, 150 p.

Multer, H. G. 1971. Field guide to some carbonate rock environments: Florida keys and western Bahamas. Rutherford, N.J., Fairleigh Dickinson Univ., 159 p.

—— and J. E. Hoffmeister. 1968. Subaerial laminated crusts of the Florida keys. Geol. Soc. America, Bull., 79:183–192.

Oriviku, K. 1940. Lithologie der Tallinna-Serie (Ordovizium, Estland). I. Tartu Ulik. Geol. Inst. Taim., 5.

Peake, N. B. and J. M. Hancock. 1961. The Upper Cretaceous of Norfolk. Norfolk Norwich Nat. Soc., Trans., 19:239–399.

Pianovsky, I. A. and R. T. Hecker. 1966. Rocky shores and hardground of the Cretaceous and Palaeogene seas in central Kizil-Kum and their inhabitants. In Organism and environment in the geological past, a symposium. "Nauka."

Pray, L. C. and R. C. Murray (eds.) 1965. Dolomitization and limestone diagenesis: a symposium. Soc. Econ. Paleont. Mineral., Spec. Publ. 13, 180 p.

Purser, B. H. 1969. Syn-sedimentary marine lithification of Middle Jurassic limestones in the Paris Basin. Sedimentology, 12:205–230.

Radwański, A. 1964. Boring animals in Miocene littoral environments of southern Poland. Acad. Pol. Sci. (Sér. Sci. Géol. Géogr.), Bull., 12:57–62.

——. 1965. Additional remarks on Miocene littoral structures of southern Poland. Acad. Pol. Sci. (Sér. Sci. Géol. Géogr.), Bull., 13:167–173.

——. 1968. Tortonian cliff deposits at Zahorska Bystrica near Bratislava (southern

Slovakia). Acad. Pol. Sci. (Sér. Sci. Géol. Géogr.), Bull., 16:97–102.

————. 1970. Dependence of rock-borers and burrowers on the environmental conditions within the Tortonian littoral zone, southern Poland. In T. P. Crimes and J. C. Harper (eds.), Trace fossils. Geol. Jour., Spec. Issue 3:371–390.

Reid, R. E. H. 1968. Bathymetric distribution of Calcareata and Hexactinellida in the present and the past. Geol. Magazine, 105:546–559.

Reis, O. M. 1910. Beobachtungen über Schichtenfolge und Gesteinsausbildungen in der fränkischen Trias. I. Muschelkalk und untere Lettenkohle. Geognost. Jahresh., 22:1–258.

Rice, A. L. and C. J. Chapman. 1971. Observations on the burrows and burrowing behaviour of two mud-dwelling decapod crustaceans, Nephrops norvegicus and Goneplax rhomboides. Marine Biol., 10:330–342.

Rieth, A. 1932. Neue Funde spongeliomorpher Fucoiden aus dem Jura Schwabens. Geol. Paläont., Abh., 19:257–294.

Saporta, G. de. 1887. Nouveaux documents relatifs aux organismes problématiques des anciennes mers. Soc. Géol. France, Bull., 15:285–302.

Sartoni, S. and U. Crescenti. 1962. Richerche biostratigrafiche nel mesozoico dell' Appenino Meridionale. Giorn. Geol., 29.

Schloz, W. 1968. Über Beobachtungen zur Ichnofacies und über umgelagerte Rhizocorallien im Lias α Dogger Schwabens. Neues Jahrb. Geol. Paläont., Mh., 11:691–698.

Seilacher, A. 1955. Spuren und Lebensweise der Trilobiten: Spuren und Fazies im Unterkambrium. In O. H. Schindewolf and A. Seilacher, Beiträge zur Kenntnis des Kambriums in der Salt Range (Pakistan). Akad. Wiss. u. Lit. Mainz, math-naturw. Kl., Abh, 10:86–143.

————. 1964. Biogenic sedimentary structures. In J. Imbrie and N. D. Newell (eds.), Approaches to paleoecology. New York, John Wiley, p. 296–316.

————. 1967. Bathymetry of trace fossils. Marine Geol., 5:413–428.

Sellwood, B. W. 1970a. The relation of trace fossils to small-scale sedimentary cycles in the British Lias. In T. P. Crimes and J. C. Harper (eds.), Trace fossils. Geol. Jour., Spec. Issue 3:489–504.

————. 1970b. A Thalassinoides burrow containing the crustacean Glyphaea udressiei (Meyer) from the Bathonian of Oxfordshire. Palaeontology, 14:589–591.

————. 1971. Regional environmental changes across a Lower Jurassic stageboundary in Britain. Palaeontology, 15:125–157.

Shearman, D. J. 1963. Recent anhydrite, gypsum, dolomite and halite from the coastal flats of the Arabian shore of the Persian Gulf. Geol. Soc. London, Proc., 1607:63–65.

————. 1966. Origin of marine evaporites by diagenesis. Inst. Mining Met., Trans. (B), 75:208–215.

Shinn, E. A. 1968a. Burrowing in recent lime sediments of Florida and the Bahamas. Jour. Paleont., 42:879–894.

————. 1968b. Practical significance of birdseye structures in carbonate rocks. Jour. Sed. Petrol., 38:212–223.

———— et al. 1969. Anatomy of a modern carbonate tidal flat, Andros Island, Bahamas. Jour. Sed. Petrol., 39:1202–1228.

Soergel, W. 1923. Beiträge zur Geologie von Thuringen II. Spuren mariner Würmer im mittleren Buntsandstein (Bausandstein) und im unterem Muschelkalk Thuringens. Neues Jahrb. Mineral, 49:510–549.

Taylor, J. H. 1964. Some aspects of diagenesis. Advmt. Sci., London, 18:417–436.

Toots, H. 1961. Beach indicators in the Mesaverde Formation. Wyoming Geol. Assoc., 16 Ann. Field Conf. Guidebook, p. 165–170.

Verwey, J. 1930. Einiges über die Biologie ostindischer Mangrovekrabben. Treubia, 12:167–261.

Voigt, E. 1929. Die Lithogenese der Flach- und Tiefwassersedimente des jüngeren Oberkreidemeeres. Jahrb. Hall. Verb. Erforschg. Mitteldeutsch. Bodensch., 8:1–136.

————. 1959. Die ökologische Bedeutung der Hartgründe ("Hardgrounds") in der oberen Kreide. Paläont. Zeitschr., 33:129–147.

————. 1968. Über Hiatus-Konkreteionen (dargestellt an Beispielen aus dem Lias). Geol. Rundschau, 58:281–296.

———— and W. Häntzschel. 1956. Die grauen Bander in der Schreibkreide Nordwest-Deutschlands und ihre Deutung als Lebensspuren. Mitt. Geol. Staatsinst. Hamburg, 25:104–122.

Walker, K. R. and L. F. Laporte. 1970. Congruent fossil communities from Ordovician and Devonian carbonates of New York. Jour. Paleont., 44:928–944.

Warme, J. E. et al. 1973. Trace fossils in Leg 15 cores. Deep Sea Drill. Proj., Initial Rept., 15:813–831.

Weigelt, J. 1929. Fossile Grabschächte brachyurer Decapoden als Lokalgeschiebe in Pommern und das *Rhizocorallium*-Problem. Zeitschr. Geschiebeforsch., 5:1–42.

Weimer, R. J. and J. H. Hoyt. 1964. Burrows of *Callianassa major* Say, geologic indicators of littoral and shallow neritic environments. Jour. Paleont., 38:761–767.

CHAPTER 18

TRACE FOSSILS AT OMISSION SURFACES

RICHARD G. BROMLEY

Institut for historisk Geologi og Palæontologi, Københavns Universitet

Østervoldgade 10, København, Denmark

SYNOPSIS

Trace fossils at discontinuity surfaces may be classified as three distinct assemblages, according to their time relationship with the depositional hiatus. These assemblages may be called the "preomission," "omission," and "postomission" suites. Discrimination of these suites is essential for the stratinomic interpretation of the discontinuity surface. The three suites may represent the same ichnocoenose in different modes of preservation, or the appearance of a new ichnocoenose may suggest the environmental conditions responsible for the omission of sediments and the resumption of deposition.

Synsedimentary lithification of the discontinuity surface to produce a hardground brings about profound changes in the trace fossil assemblage. The omission suite is then sub-divided into two distinct ichnocoenoses: pre-lithification burrows and postlithification borings. In certain carbonate sequences, the processes of lithification and bioturbation show mutual intereference, causing irregular distribution of the cement and stenomorphism of the burrows. Mature hardgrounds commonly contained a "fossilized" system of empty burrows, the hardened walls of which provided substrates for postlithification borings, so that the two distinct ichnocoenoses form an intimately inter-mingled assemblage. In many cases the omission suite reveals temporary periods of sedimentary cover on the hardground, in the form of multiple excavation of the fill of the burrows and super-imposition of several suites of borings.

INTRODUCTION

At one time, geologists generally assumed that sedimentation was widespread and continuous in the marine environment. Consequently, any break in an ancient sedimentary sequence was liable to be interpreted as a period of emersion. In the last 100 years, however, the advance of oceanography revealed the existence of numerous areas of nondeposition on the present sea floor, and workers have come to realize that most minor breaks in the sedimentary column are submarine in origin. Nevertheless, the terms associated with these hiati —particularly "erosion"—have received the connotation of subaerial processes and still retain this sense today. I must, therefore, necessarily begin with a definition of the principal terms as used in this chapter; other terms are discussed at their place in the text.

Discontinuity surfaces: minor breaks in the sedimentary column, chiefly intraformational but including interformational junctions that have not involved large-scale erosion. More minor in rank than "disconformity." The following types of discontinuity surfaces are considered:

Minor erosion surfaces: discontinuity surfaces involving secondary omission of sediment, i.e., sediments were deposited but subsequently removed by denudation. As already stated, many authors (e.g., Termier and Termier, 1963) restrict this term to continental levels of denudation. Heim (1924) was aware of the lack of a term for submarine denudation by physical processes and suggested "dereption" to fill the role. This term has hardly been used since, however, and is not reemployed here. Instead, "erosion" and "scour" are used to denote physical denudation of hard and soft sea floors, respectively.

Subsolution surfaces [Heim's (1958) term "subsolution" for submarine dissolution of carbonates is now widely used]: a cemented, carbonate sea floor that has been corroded by subsolution processes.

Omission surfaces [this term was introduced by Heim (1924)]: discontinuity surfaces of the most minor nature, which mark temporary halts in deposition but involve little or no erosion (primary omission). They correspond to Hadding's (1958) "hidden hiatuses" and fall genetically between "condensed beds" and "erosion surfaces."

Hardgrounds: synsedimentarily lithified sea floors. Owing to the profound influence that lithification has on the ichnocoenoses, hardgrounds occupy a major part of this chapter; the definition of the term is discussed subsequently.

OCCURRENCE OF TRACE FOSSILS AT DISCONTINUITY SURFACES

Discontinuities in the sedimentary sequence represent changes in the depositional environment that, in many cases, are more clearly registered in the ichnological record than by other features of the sediment. The mode of preservation and distribution of trace fossils at the discontinuity surface is largely controlled by the characteristics of the hiatus. An extreme example is offered by turbidite sequences, where individual turbidites are separated by omission surfaces or by thin beds of pelagic sediments. The infauna produces burrows from these surfaces only, before being blanketed by the

next flow (Seilacher, 1962; Laughton et al., 1972, p. 764).

Omission surfaces also occur repeatedly in many sequences of rhythmic sedimentation; an omission surface terminates each rhythmic unit. Heim (1924, p. 17) called such sequences "omission bedding" (Omissionsschichtung) and characterized the rhythm as 12301230, where 0=omission. In bedding such as this, the terminal omission surface of each unit is characteristically overlain by a contrasting type of sediment, with the onset of the next cyclothem. This change in character of sediment emphasizes the trace fossils connected with the omission surface. For the same reason, isolated intraformational discontinuity surfaces commonly have noticeable trace fossils because of the contrasting fill from superjacent sediment.

During the nondeposition at the omission surface, the sea floor was commonly mineralized, e.g., by a ferruginous impregnation, which further emphasizes the associated trace fossils. Synsedimentary lithification of the sea floor as a hardground during the depositional hiatus produces complex relationships between trace fossils originating before, during, and after hardening of the substrate.

Trace fossils in sediment subjacent to the discontinuity surface belong to three categories, according to their time relationship with the hiatus: the preomission, omission, and postomission suites of trace fossils.

Preomission Suite

The earliest suite of trace fossils in such sequences belongs to the parent sediment of the discontinuity surface, and represents activity of organisms in sea floors before the hiatus (Fig. 18.1A). The discontinuity surface may be an erosion surface that was cut down into this sediment; preomission trace fossils that are in physical contact with the erosion surface may thus predate the hiatus by a considerable period of time.

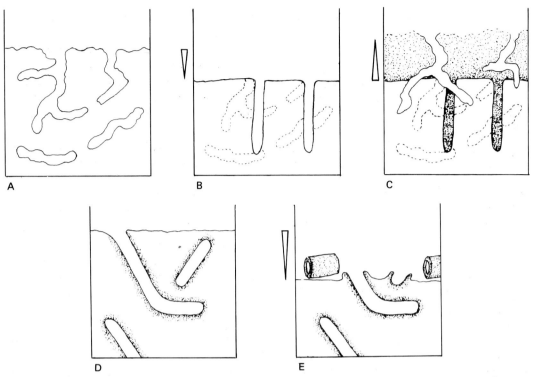

Fig. 18.1 Interpenetrating and eroded burrows. A–C, superimposition of three suites of trace fossils at omission surface. A, preomission suit burrows in parent sediment. B, slight scour followed by nondeposition, and development of omission suite burrows. C, renewed sedimentation fills omission suite; burrows constructed in new sediment cut across earlier structures as a postomission suite of traces. D, E, development of reworked trace fossils at omission surface, where wall of a burrow has been more or less cemented by the inhabitant (modified after Goldring, 1964).

Omission Suite

A suite of trace fossils is commonly present that penetrates the sediment from the discontinuity surface and that dates from the period of sedimentary omission (Fig. 18.1B). This suite may differ from the preomission suite merely in quantity, because omission surfaces represent the superimposition of time lines, with the consequent crowding of burrows.

However, the ichnocoenose of the omission suite may also differ in kind from that which preceded the interruption. This difference may be due to changes in the behavior patterns of the infauna, owing to an increase in energy of the environment or in response to the stabilization of the sea floor. This modified trace fossil assemblage was termed the *Glossifungites* ichnofacies by Seilacher (1967; see Table 2.1).

The preomission and omission suites usually differ in mode of preservation. The fill of omission suite trace fossils is very significant because it commonly comprises sediment and fossils that elsewhere are unrepresented at the omission surface. However, if the burrows remained open in the sea floor through the duration of the hiatus, they contain the sediment that directly overlies the omission surface. In either case, the fill usually differs in constitution from that of the surrounding sediment ("bed-junction preservation" of Simpson, 1957, p. 479; Chapter 4); the burrow walls in many cases have been more or less mineralized during their exposure, so that the omission suite trace fossils are generally conspicuous. This enhancement is particularly true in hardgrounds.

The special case of "reworked trace fossils" may be included under this heading. Although generally true that trace fossils are absolutely autochthonous, in some cases—e.g., *Chondrites* and *Siphonites* —the walls of the burrows are sufficiently strong to withstand gentle exhumation and resedimentation at minor erosion surfaces (Simpson, 1957, p. 480—"burial preservation"; Goldring, 1962, p. 245; Chapter 4) (Fig. 18.1E). Such cases are not common. In contrast, rapid diagenesis may "fossilize" traces and render them suitable for exhumation. Schloz (1968) recorded reworked *Rhizocorallium* at a Lower Jurassic minor erosion surface.

Postomission Suite

With the return of deposition, a third suite of traces may be cut into the sediment beneath the omission surface and superimposed onto the earlier suites (Fig. 18.1C). Postomission burrows are filled with postomission sediment, and the ichnocoenose reflects the environmental conditions that allowed the return of sedimentation.

TRACE FOSSILS IN UNLITHIFIED DISCONTINUITY SURFACES

The complex relationships among the three suites of trace fossils described above are best illustrated by some examples. The two that follow contrast a minor omission surface and a local erosion surface, neither involving synsedimentary lithification.

Nonerosional Omission Surfaces in Chalk

In Upper Cretaceous chalk of northwest Europe, omission surfaces are abundant through much of the succession. Omission bedding is particularly apparent in Cenomanian gray chalks of southern England; burrows at the omission surfaces were described by Kennedy (1967a; 1967b, p. 128) and Destombes and Shephard-Thorn (1971).

In white chalk of north Germany, the more noticeable omission surfaces have been termed "Grabganglagen" (Ernst, 1963; = "Pseudo-Hardgrounds" of Steinich, 1972). At these horizons, bioturbation of the chalk is rendered more clearly visible by pale gray, clay-mineral-enriched chalk overlying the omission surface and by slight ferruginous staining of the pure white chalk immediately beneath the interface (Fig. 18.2). This change of sediment and faint mineralization allows one to distinguish the three trace fossil suites:

1. The preomission suite (Bänderkreide facies) comprises chiefly *Thalassinoides*, *Chondrites*, and *Zoophycos* having pale blue-gray fill, the color apparently due to disseminated pyrite.
2. The omission suite consists of *Thalassinoides* alone. The fill is gray, from the relatively clayey overlying chalk, and the walls are faintly ferruginized from exposure to sea water during the period of nondeposition.
3. The postomission suite of the overlying sediment comprises *Thalassinoides*, *Chondrites*, and *Zoophycos*, as in the preomission suite, but these penetrate below the omission surface only through reexcavation of the fill of *Thalassinoides* of the omission suite. Elsewhere, postomission *Thalassinoides* run along the surface of omission but do not cross it.

The failure of animals to penetrate the omission surface was not due to hardness of the older sediment; their avoidance of it may have been due to its low nutrient content. In many chalk omission surfaces, e.g., those beneath pelletal phosphatic horizons, postomission burrows cross the omission surface in such numbers as to completely break up the junction. In the resulting tangle of burrows, to distinguish omission suite from postomission suite is usually impossible (Fig. 18.3A).

Lack of new elements in the omission suite ichnocoenose in this example testifies to the negligible change in the depositional environment that brought about the sedimentary omission. This situation is in sharp contrast to certain discontinuity surfaces, e.g., those described by Sellwood (1970) in the Lower Pliensbachian Belemnite Marl

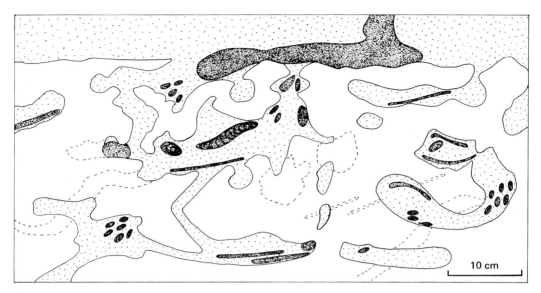

Fig. 18.2 Vertical section through unlithified omission surface in white chalk overlain by slightly gray chalk. Preomission suite hardly visible, shown by broken lines; omission suite has pale gray fill; postomission suite has dark gray fill, confined to fill of omission suite. Features combined from three surfaces, at and closely beneath Campanian–Maastrichtian boundary (Upper Cretaceous); Saturn quarry, Kronsmoor, Schleswig-Holstein, Germany.

of England (Fig. 18.3B). In this case, the omission suite is represented by a new ichnocoenose, characterized by the appearance of *Diplocraterion*. (See Chapter 17.)

An Erosional Clay–Chalk Omission Surface

A completely different form of discontinuity surface, involving local scouring, is illustrated by the following example (Fig. 18.4). The Lower Campanian "smectite," a blue-gray clay, is overlain directly by the Upper Campanian "craie glauconifére" together with the secondary omission of the otherwise intervening "Hervian greensand"— also Upper Campanian (Calembert, 1956; Schmid, 1959).

The smectite displays a very characteristic preomission suite of trace fossils, comprising *Gyrolithes* and *Chondrites*; the latter is restricted within the thick, glauconitized wall material of the former (Saporta, 1884, Pls. 5 and 6; Bromley and Frey, 1974). The fill of *Gyrolithes* consists of clay identical to that in the surrounding matrix. Truncation of the tops of *Gyro-*

lithes testifies to scouring, which caused the omission of the greensand.

The uppermost 25 cm of the smectite is penetrated by a postomission suite of trace fossils that consists of oblique, cylindrical shafts filled with glauconitic chalk and cutting through the *Gyrolithes* indiscriminately. The origin of many of these burrows can be traced to a level well above the erosion surface, and the existence of a true omission suite originating at the erosion surface itself is doubtful.

Turbidites

Turbidites represent a very special form of omission bedding. Seilacher (1962) confirmed by a study of trace fossils that certain flysch sediments were laid down as turbidites. Two distinct groups of trace fossils are preserved on sole surfaces of flysch psammites. The first group was termed "predepositional" in relation to the overlying, casting sandstone; these burrows correspond to the preomission suite. Their state of preservation is due to their exposure

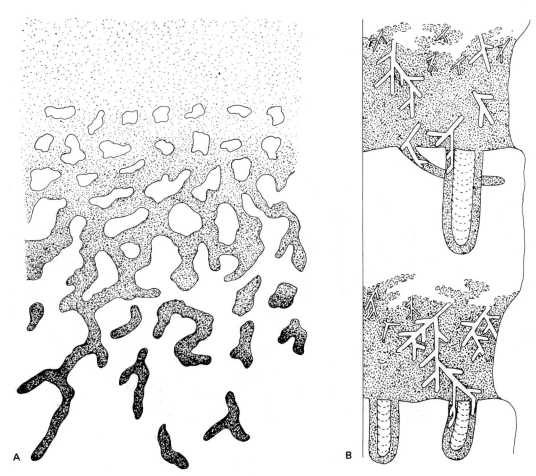

Fig. 18.3 Trace fossils in nonerosional omission surfaces. A, a pseudobreccia. Omission and post-omission *Thalassinoides* of 2 cm diameter "pipe" the phosphatic chalk down into almost un-lithified white chalk, so that little remains of the original omission surface. Base of Upper Brown Chalk (Santonian, Upper Cretaceous); Taplow, Buckinghamshire, England. B, diagrammatic representation of omission surfaces and their associated trace fossils in a rhythmic limestone-marl succession. Preomission suite may include *Thalassinoides*, as indicated by sediment mixing at base of the limestones. Omission suite at top of the limestones includes *Diplocraterion* and *Rhizocorallium*, and their fills are entered by large postomission *Chondrites* that descend from the base of the next limestone above. Pliensbachian (Lower Jurassic), southwest England. (Modified after Sellwood, 1970.)

at the sea floor, affected by the special process of scouring that immediately precedes sedimentation with each turbidity current, and their details are consequently somewhat degraded. The other group, post-depositional burrows, were produced by the infauna reworking the pelite–psammite interface after deposition of the turbidite. These burrows correspond to the postomission suite, relative to the buried discon-tinuity surface at which they are preserved (Seilacher, 1964, p. 304, Fig. 4; see also Crimes, 1973, p. 122–125).

Further Examples from the Literature

The following publications were selected as further reading, from the large literature on omission surfaces, because they record the details of trace fossils and also represent

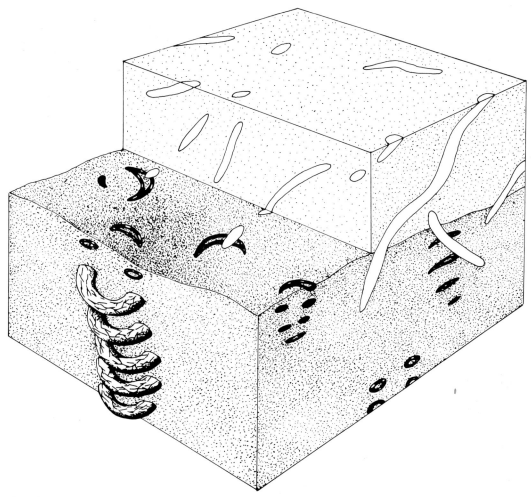

Fig. 18.4 Erosional junction between glauconitic chalk and underlying marly clay (smectite). The preomission *Gyrolithes* have thick, glauconitic walls riddled with *Chondrites*. Campanian; exposed in the loop-line cutting at Bon Espérance quarry near Visé, Belgium. Height of block, 25 cm.

some of the variety of trace fossil relationships that can occur: Cayeux (1939, p. 240; 1941b, p. 536, Pl. 31; 1950, p. 891), Hallam (1960, 1964), van Straaten (1967), Kaźmierczak and Pszczółkowski (1968, 1969), Sellwood (1970, 1971), and Schloz (1972, p. 191).

LITHIFIED DISCONTINUITY SURFACES

Hardground: A Definition

The term "hardground" has been defined or used recently by some authors (e.g., Termier and Termier, 1963; Fabricius, 1968; Purser, 1969) as a synonym of Twen-

hofel's (1950, p. 275) general term "hard bottom." The origin of "hardground" in oceanographical literature (Murray and Renard, 1891) and its literal sense carry such connotation. Since the time of its early introduction into geological literature (Cayeux, 1897, p. 553), however, most authors have used the term in the narrower sense of intraformational, synsedimentary lithification surfaces (e.g., Calembert, 1953; Lombard, 1956; Calembert and Meyer, 1956; Hofker, 1958; Schmid, 1959; Voigt, 1959; Rosenkrantz, 1966; Pożaryska, 1965; Bromley, 1967b; Shinn, 1970; Bathurst, 1971; Schloz, 1972).

The restricted sense of the word is used in this chapter, because the early-diagenetic lithification of sea floors is (1) a special phenomenon constituting an end member in the evolutionary series of omission surfaces and (2) is of the highest significance in their ichnological development.

The long and involved history of the study of Cretaceous chalk hardgrounds, which intimately concerns the interpretation of their trace fossils, may be compiled from papers by Cayeux (1939, p. 34), Ellenberger (1946, 1948) and Voigt (1959); that of Jurassic oolitic and micritic hardgrounds from Cayeux (1922), Richardson (1933, p. 48), and Bigot (1941); and of Paleozoic hardgrounds generally from Jaanusson (1961).

Trace Fossil Occurrences in Hardgrounds

Synsedimentary lithification fundamentally changes the sedimentological and ecological history of the sea floor. The omission suite of trace fossils then becomes subdivided into two entirely separate ichnocoenoses: prelithification burrows and postlithification borings. [Radwański (1964, p. 57) introduced the term "lithophocoenosis" for the postlithification ichnocoenose.] This subdivision of the omission suite is displayed dramatically in certain Cretaceous hardgrounds in Texas, where dinosaur footprints impressed into the omission surface were bored subsequently by pholads (Moore, 1964, p. 11; Perkins and Stewart, 1971b, p. 58) (Fig. 18.5A).

In many cases, burrows of the prelithification omission suite remained open and more or less empty in the hardground during long periods of nondeposition, and their hardened walls became strongly mineralized as an extension of the sea floor (Bromley, 1967b). Many authors recorded the characteristic association of glauconite and phosphate with omission surfaces (e.g., Goldman, 1922; Brückner, 1951; Bromley, 1967a); in some cases, pyrite (Hallam, 1969) or manganese crusts (Jenkyns, 1971) occur. The hardened burrow walls offered an extensive and protected substrate to an infauna that produced the postlithification omission suite borings (Fig. 18.5B).

Hardening stabilizes the sea floor and largely protects it against retrogressive

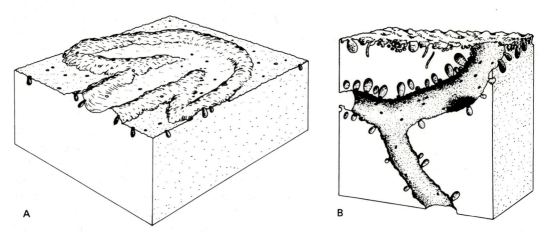

Fig. 18.5 The omission suite: postlithification traces superimposed on prelithification traces. A, dinosaur footprint and pholad borings at summit of Glen Rose Limestone (Lower Cretaceous) near Leander, Winson County, Texas. Height of block, 15 cm. B, hardground terminating a Cretaceous section. Floor, walls, and roof of prelithification *Thalassinoides* bored chiefly by bivalves; these borings escaped subsequent erosion, unlike those on upper surface of the hardground. Curfs quarry, east of Maastricht, Holland. Height of block, 18 cm.

processes. Yet in certain hardgrounds, sub-solution removed several centimeters of limestone (Hollmann, 1964; Jurgan, 1969). In two such cases, Wendt (1969, 1970) recorded boring algae as possibly abetting the dissolution process. Overall, however, subsolution surfaces show few trace fossils.

Trace Fossils and the Sculpture of Hardgrounds

The sculpture of hardgrounds is produced and modified by a variety of processes. Wendt (1970) showed that the sculpture of subsolution surfaces varies in relief according to the amount of subsolution that takes place and the inhomogeneity of the sediment under attack. But these are largely inorganic processes and do not concern us here.

In most hardgrounds, the form of sculpture is closely dependent on the burrowing or boring activities of organisms within it. Three primitive types of sculpture provide a base upon which biogenic and erosive modifications are superimposed; these may be called "hummocky", "flat", and "convoluted," respectively.

Hummocky Hardgrounds

In simple examples of hummocky hardgrounds, i.e., where the sculpture has not been modified by boring organisms, the entrances to prelithification omission suite burrows lie in valleys between hummocks (e.g., Figs. 18.7, 18.10). This configuration suggests that lithification took place sufficiently rapidly to preserve the topography of the bioturbated sea floor, i.e., more or less as a lithified equivalent of the omission surface shown in Figure 18.2.

Flat Hardgrounds

Most hardgrounds have flat or nearly flat surfaces. In many cases (Fig. 18.6) the pre-omission suite is insignificant, so the sea floor may have been flat originally. In other cases (Fig. 18.7), however, the preomission suite includes *Thalassinoides* and is no different from that of hummocky hardgrounds in the same sediment. In these cases the flatness may be attributed to pre-lithification scouring followed by rapid lithification; other features of the surface, such as truncated burrow systems, also indicate gentle scouring.

Convoluted Hardgrounds

In the chalk environment, particularly, the intromission of scouring at a stage of incomplete lithification of the sea floor (see below, stage 2) resulted in high-relief, convoluted hardgrounds (Fig. 18.8).

Burrows in Hardgrounds

At first sight, the presence of soft-substrate burrows in hardgrounds seems to be paradoxical. However, more often than not, hardgrounds contain networks of large, branching burrows of the prelithification omission suite which, because of their generally excellent preservation, have received much attention in the literature.

Hardgrounds in a variety of carbonate sediments in the lower Tertiary and Upper Cretaceous of north Africa and northwest Europe (Figs. 18.7 to 18.10) contain well-preserved systems of *Thalassinoides*, ascribed to the work of crustaceans (Cayeux, 1939, p. 241, and 1941b; Voigt, 1959; Bromley, 1967b and 1968; Rasmussen, 1971). Identical burrow systems also occur in hardgrounds in the Upper Jurassic of Poland (Kaźmierczak and Pszczółkowski, 1968) and Middle Jurassic of England (e.g., Richardson, 1933, p. 48). In the Triassic Muschelkalk of Germany and Poland, similar burrow systems were ascribed not to crustaceans but to enteropneusts—and named *Balanoglossites* (Mägdefrau, 1932; Müller, 1956; Kaźmierczak and Pszczółkowski, 1969). This enteropneust interpretation has been extended to burrows in Devonian hardgrounds of the northern Russian plat-

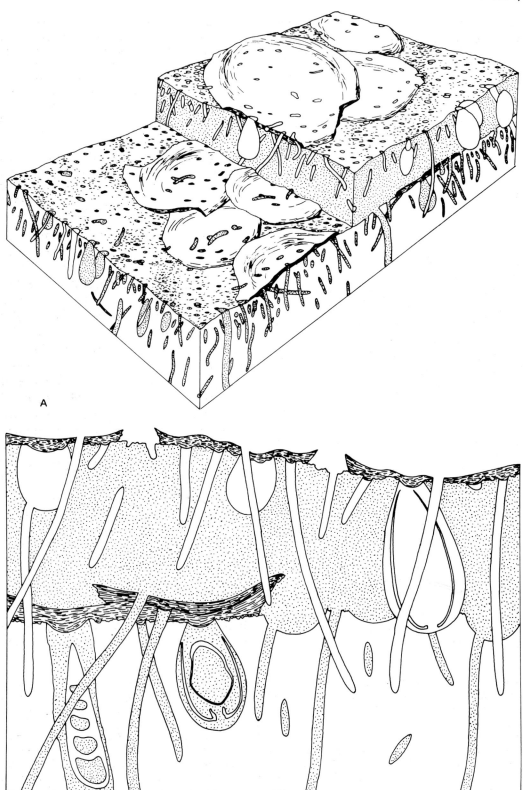

A

B

form (Hecker, 1970, p. 219, Fig. 2). The Ordovician limestones of Öland, Sweden, contain closely repeated hardgrounds, the burrows in which were accurately depicted by Lindström (1963, Fig. 9).

Hardground Burrows and Lithification

The omission suite burrows in some hardgrounds provide evidence that the sediment around them was lithified while the burrowing infauna remained active (Bromley, 1967b; Zankl, 1969, p. 244). In the case of European Cretaceous chalk hardgrounds, the activity of the infauna and concurrent processes of lithification apparently had a strong mutual influence on each other, causing irregularity of both the burrows and the distribution of cement.

Incipient or embryonic hardgrounds arrested in partial development by burial through the return of deposition, reveal the process of cementation of the chalk at all of its stages (Fig. 18.11):

Stage 1. In the sediment immediately beneath the discontinuity surface, many concretionary centers of lithification developed (Fig. 18.12). The nodules are more or less rounded or somewhat irregular in shape, 2 to 6 cm in diameter, and well-lithified but having diffuse boundaries. The surrounding sediment remained soft and was bioturbated.

Stage 2. Further cementation increased the size, number, and irregular shape of the concretions, which now joined to form an irregular, continuous network of indurated chalk (cf. Kennedy, 1970a, p. 651). This stage in nodular chalk was well described by Cayeux (1941a, p. 47), in terms of a hard stony "armature" entangling the remaining unhardened chalk. Removal of the soft chalk by present-day weathering processes leaves an intricate "limestone skeleton." In contrast, contemporaneous removal of unhardened chalk by the intervention of scouring at this stage would lead to the development of a highly convoluted hardground.

Stage 3. With continued nondeposition, the hard chalk encroached farther upon the soft; diffuse boundaries separated the two, and a point was passed at which equal amounts of the sediment were lithified and unlithified. After this point, the continuous network of bioturbated soft chalk became increasingly restricted until, in the fully developed hardground, the remaining chalk was lithified and an empty system of *Thalassinoides* was left as a cavity network within it (Bromley, 1967b, p. 168).

Although nodularity of limestones is in many cases a late feature of diagenesis (e.g., Jurgan, 1969; Dvořák, 1972), several workers recorded synsedimentary nodular cement in hardground development (e.g., Aubouin, 1965, p. 127; van Straaten, 1967). Taylor and Illing (1969) recorded a case of lithification of sediments in the intertidal environment today in which continuous hardgrounds develop from coalescence of nodules; they also attributed the initial nodularity to burrowing activity of crustaceans, which produced branching tunnels between and within the lumps.

Taylor and Illing (1969) suggested that cementation may be triggered by chemical influence of the infauna. Worth noting in this connection is that hardgrounds having preserved omission burrow systems are characteristically lithified to depths of nearly 30 cm beneath the surface—still more deeply in coarse–grained sediments— whereas those lacking burrows are almost invariably thinner. Thus, the burrowers likely have a further influence on cementation, probably through maintaining open

◄ **Fig. 18.6** The Subvesulian unconformity at Nunney quarry, near Shepton Mallet, Somerset, England. A, erosion surface cut in Carboniferous limestone is plastered with oysters and contains three types of borings. This surface overlain by Upper Inferior Oolite (Middle Jurassic, stippled), within which a hardground is developed just above the base, forming top of block. Height of block, 4 cm. B, polished vertical section of same rock. Two bivalve borings contain the preserved shells of the borer; one of them contains in addition the shell of a nestling bivalve, which subsequently occupied the boring. A third category of associated body fossil is seen to the lower left, where a gastropod shell accidentally fell into a boring. Superimposition of borings shows that the fill of the earlier suite lithified before introduction of the next suite. The bivalve borings show various degrees of truncation by erosion. Height of section, 4 cm.

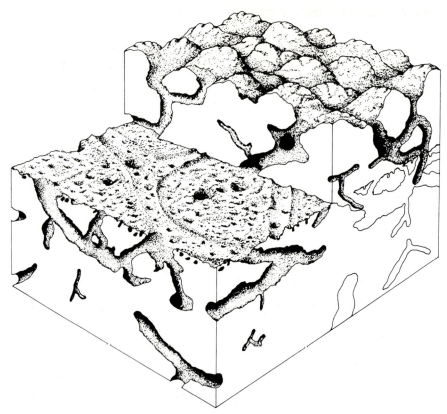

Fig. 18.7 Two hardgrounds from the Chalk Rock (Turonian, Upper Cretaceous), Charnage Down, near Mere, Wiltshire, England. Upper hardground has
hummocky relief; the lower is flat. Omission *Thalassinoides* of lower hardground
forms horizontal network just below the surface. A preomission horizontal network of *Thalassinoides* was exposed by scouring, and is preserved on the hardground surface as paired ridges. Omission *Thalassinoides* of upper hardground
is xenomorphic in that horizontality was imposed upon it by the flat hardground.
Two sizes of *Thalassinoides* are seen in each hardground. (Borings in upper
hardground not shown.) Height of block, 26 cm.

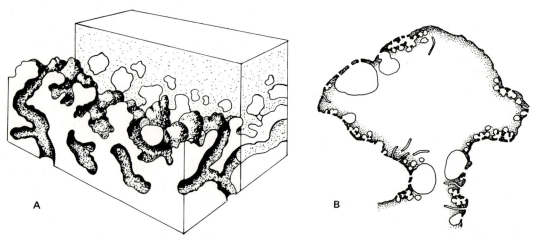

Fig. 18.8 Hardground having convoluted surface, terminating the Chalk Rock at High Wycombe,
Buckinghamshire, England. A, the sculpture consists of irregular, exaggerated bosses between
Thalassinoides entrances. Height of block, 30 cm. B, vertical section through a boss from same
locality, showing attack by boring organisms (chiefly sponges), weakening the stalk of the boss.

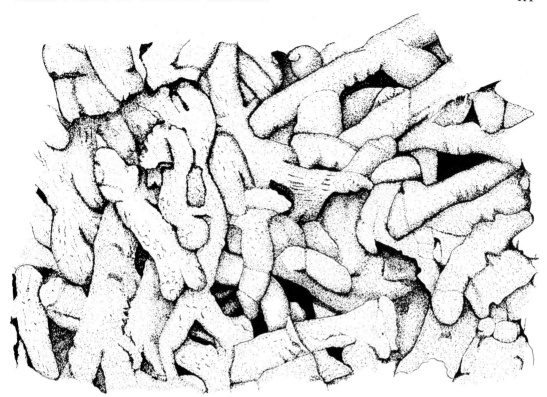

Fig. 18.9 Idiomorphic but crowded *Thalassinoides*, weathered out on underside of an omission surface and viewed from below. Phosphatic chalk (lower Eocene) near Sétif, Constantine, Algeria. (Drawn after Cayeux, 1941.)

passages in the sediment and pumping sea water through them, thereby allowing the interface reaction to extend to a greater depth beneath the sea floor.

Fischer and Garrison (1967, p. 490) noted the downward extension of cement along the walls of burrows beneath a modern hardground. In hardgrounds in the topmost Cretaceous calcarenites of Holland, Voigt (1959, p. 142) described intraclasts having a somewhat tubular shape, suggesting break-up of case-hardened burrow walls. Certainly the burrows are uncontorted (idiomorphic) in these hardgrounds, and the sediment is not noticeably nodular, which together might indicate that rapid case-hardening from burrow walls arrested the activity of the infauna at an early stage.

Although bioturbation influenced the progress of lithification, progressive lithification clearly had a reciprocal effect on the form of the burrows. In several chalk hardgrounds, systems of *Thalassinoides* can be distinguished that hardly differ from those in unhardened sediment. For example, Kennedy (1967b, Pls. 3 and 4) dissected out of a hardground a well-proportioned system comprising parts of two horizontal networks. This burrow was unquestionably constructed in largely soft sediment. In the same hardground, however, are smaller, constricted networks of burrows (stenomorphic *Thalassinoides*) interconnected with the larger, idiomorphic *Thalassinoides* systems. The stenomorphic burrows were clearly constructed at a later stage, when the sediment contained hard nodules (Bromley, 1968) (see also Fig. 18.10).

In later publications, Kennedy (in Bromley, 1967b, p. 181; Kennedy, 1970a) usefully restricted the name *Thalassinoides paradoxicus* [=*Spongia paradoxica*] to the stenomorphic hardground systems.

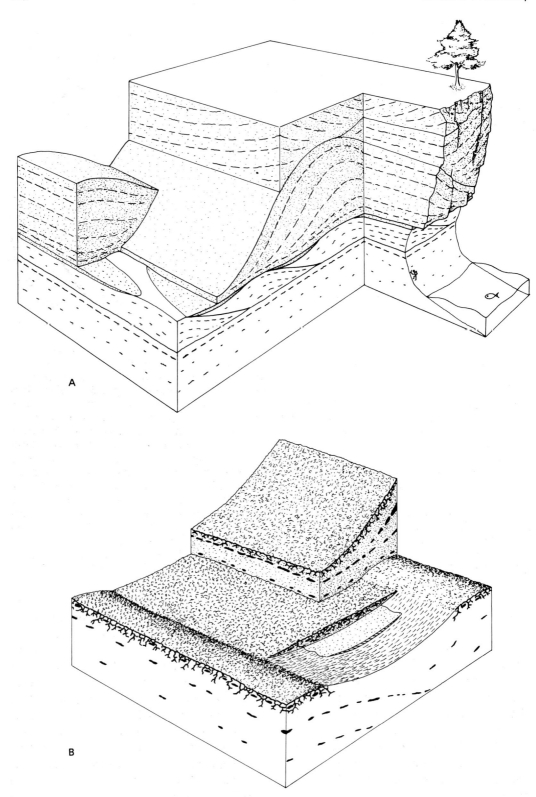

Fig. 18.10 *See facing page, for caption.*

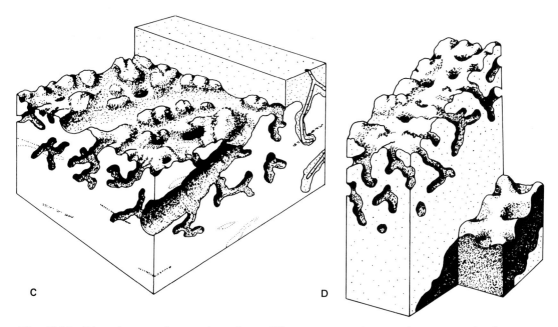

Fig. 18.10 Dissection, on three scales, of sea cliff exposures at Stevns Klint, Denmark (data compiled from several localities). A, at right, the sculpture of the exposure. Remainder of block is dissected along omission surfaces. The complex junction is shown between uppermost Cretaceous chalk (Maastrichtian) and lowermost Tertiary bryozoan bioherms (Danian; stippled). Broken lines represent beds of nodular flint; solid lines are omission surfaces. Height of block, 25 m. (Some details after Håkansson, 1971.) B, details of Maastrichtian–Danian junction at the nearest corner of block A. A hardground cuts across the undulating Mesozoic–Tertiary junction so that its parent sediment is part Cretaceous ("gray chalk") and part Danian ("Cerithium limestone") in age. Hardgrounds ("crab layers") are also developed locally on the sloping terminal surfaces of overlying bioherms. C, detail of hardground at nearest corner of block B. Prelithification scour exposed systems of giant *Thalassinoides* at the omission surface, still recognizable as wide grooves in the sculpture of the hardground. Omission suite within the hardground comprises large idiomorphic and small stenomorphic *Thalassinoides*, which cut the preomission suite *Thalassinoides* and *Zoophycos* (broken lines). Postomission burrows extend into the fill of omission *Thalassinoides* from overlying sediment. Scale as in D. D, "crab layer" hardground from far corner of block B. Flint preserves idiomorphic preomission *Thalassinoides* networks, which contrast with the smaller, somewhat stenomorphic omission *Thalassinoides* that descend from the hummocky hardground. Height of block, 70 cm.

Lag Intraclasts and Hiatus Concretions

A slight increase in energy at the sea floor when lithification had reached Stage 1 (above) caused gentle scouring and the exhumation of discrete concretions as intraclasts. These intraclasts are distinguishable from undisturbed concretions by their sharp boundaries, which are commonly emphasized by impregnations of glauconite, phosphate, or pyrite, and are bored by organisms.

The interplay of processes of deposition, scouring, and lithification can produce compound intraclasts or, as Voigt (1968) called them, "hiatus concretions." On the basis of Jurassic examples, Voigt showed that hiatus concretions record the following sequence of events. The lag intraclast, bored and encrusted by organisms, was reburied and grew by further accretion. This growth was interrupted by reexhumation through scouring, and the new surface was encrusted and bored by organisms. The cycle of events may be repeated several times, each instance producing a micro-

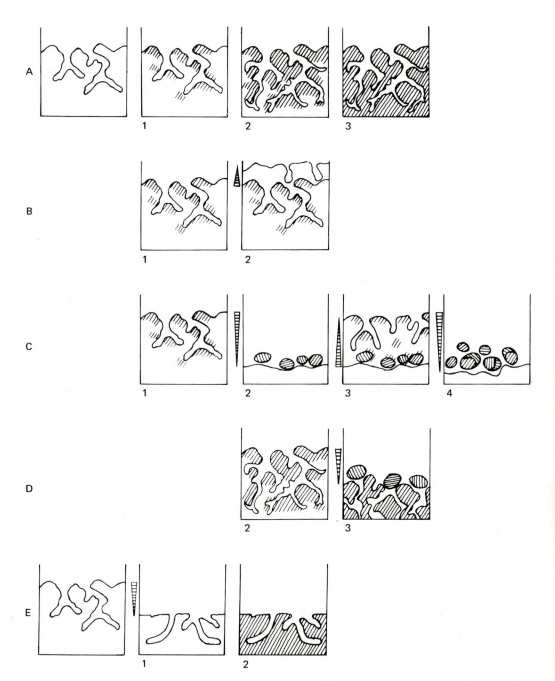

Fig. 18.11 Relationships between burrowing, lithification, scouring, and sedimentation at omission surfaces, based on data from the chalk facies. Shading indicates extent of lithification. A, development of hummocky hardground from unlithified omission surface, showing lithification stages 1, 2, 3 (see text); new burrows are noticeably stenomorphic after stage 2. B, lithification interrupted by return of sedimentation; an incipient hardground is preserved (2). C, scouring at stage 1 leaves lag intraclasts (2); renewed sedimentation, lithification to stage 1 (3), and re-exhumation by scouring (4) produces hiatus concretions (compound intraclasts). D, scouring at lithification stage 2 produces convoluted hardground (3). E, flat seafloor topography produced by scouring (1) may be lithified rapidly to produce a flat hardground (2).

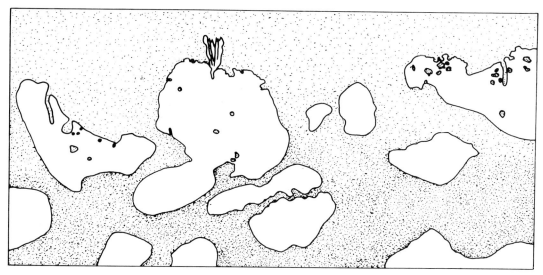

Fig. 18.12 Vertical section of incipient hardground in chalk, showing stage-1 lithification. Only topmost chalk at omission surface was lithified, forming a minimally indurated substrate for a few boring organisms and the attachment of a solitary coral (its base extends into a boring). The surface was then buried under slightly gray chalk. At or near the Turonian–Coniacian junction, east of Dover, England. From a photograph; natural size.

omission surface within the intraclast (Figs. 18.11 and 18.13).

The Fill of Hardground Burrows

Characteristics of the fill of hardground burrows are very variable, and depend largely on the time at which the burrows were filled. Burrows that remained empty until final burial of the hardground are filled with superjacent sediment. The earliest deposits of renewed sedimentation possess some of the qualities of a basal conglomerate and are often rich in "im-

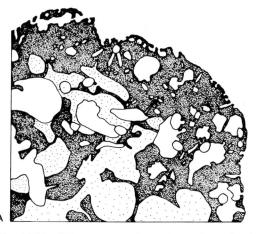
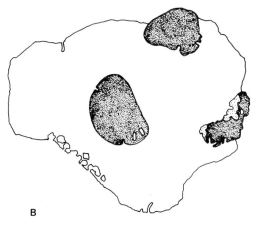

Fig. 18.13 Hiatus concretions. A, section of a lag intraclast in which much original sediment was replaced by fill of sponge borings. Three suites of borings are superimposed, the oldest fill lightly stippled. Chalk Rock, top hardground, Turonian; High Wycombe, England. Slightly enlarged. B, chalk hiatus concretion incorporating three discrete sediments: the oldest darkly stippled, the youngest white. Totternhoe Stone, Cenomanian (Upper Cretaceous); Blackdyke Farm quarry, Norfolk, England. Natural size.

purities," which decrease progressively upward (e.g., Voigt, 1959, Fig. 1). Thus, the hardground burrows in cyclic, phosphatic chalk sequences contain the richest phosphates (Cayeux, 1941b, p. 395); those in the gray–white rhythms of the white chalk facies contain the grayest chalk (Figs. 18.2 and 18.12), and hardground burrow fills in the English Middle Jurassic Great Oolite Series show exaggerated graded bedding.

Some of the material in the fill may date from the period of sedimentary omission represented by the hardground. For example, in the hardground that terminates the Campanian chalk in the Belgian and Dutch Limburg district, overlain by Upper Maastrichtian sediments, Hofker (1959) extracted Lower Maastrichtian foraminifers from the burrow fills.

Where late-diagenetic lithification has not affected the sediments, the fill of hardground burrows normally remains unconsolidated because it is protected from compaction. In many cases these soft fills contain exceedingly finely preserved, delicate fossils that were spared reworking, transport, and exposure at the sea floor (Bromley, 1967b, p. 165). Unbroken, fragile bryozoans were described by Voigt (1959) from burrow fills in a hardground, including species found exclusively in this microenvironment of high fossilization potential.

Reexcavation of the Fill

In hardgrounds that show evidence of many periods of temporary sedimentation, periodic filling and reexcavation of the burrows often can be detected (Fig. 18.14). The new burrows were confined to the old, because of the hardness of the surrounding sediment; in most cases, owing to incomplete reexcavation, some earlier sediment remained in the old system. In purer limestones, detection of these remnants of fills in burrows is greatly assisted by use of luminescence by ultraviolet radiation (Bromley, 1965, p. 212; 1967b, p. 166) (see Fig. 23.2).

Borings in Hardgrounds

As is true of hardground burrows, the postlithification omission trace fossils (borings) are also conspicuous in many hardgrounds, owing to mineralization and contrasting fill material. Unlike burrows, however, the walls of borings cut sharply through the fabric of the hardground—both grains and cement (Hölder and Hollmann, 1969; Purser, 1969; Shinn, 1969; Rose, 1970; Perkins, 1971). This feature imparts a sharp outline to the borings which, together with their crowded occurrence at the discontinuity surface, further emphasizes them. Not surprisingly, therefore, bored hardgrounds have been used as local stratigraphic marker horizons: e.g., by Staring (1860, p. 317, 330—"boormossel-lagen") in the uppermost Cretaceous of Holland; Richardson (1907) in the Middle Jurassic of southwest England; Klüpfel (1916a, b) in the Middle and Lower Jurassic of Lorraine, France; and Moore (1964—"bored surfaces") and Perkins (1966) in the Lower Cretaceous of Texas.

Identification of a bored (as opposed to a burrowed) sedimentary surface is conclusive proof of sedimentary omission and of the presence of a hard sea floor. In completely lithified sequences, evidence for hardgrounds and omission may nevertheless be difficult to detect, and may rest largely on the identification of true borings (e.g., Zankl, 1969; Warme and Olson, 1971, p. 41; Boyd and Newell, 1972, p. 12).

Whether the quantity of borings in a hardground provides an indication of the duration of sedimentary omission, as suggested by Hofker (1959, p. 324), is doubtful. In a favorable environment, sponges and bivalves can completely break down a surface of limestone to a depth of several centimeters within a period of a few years (e.g., Ryder, 1879). According to Neumann (1966, p. 107), erosion by a clionid sponge in the limestone coasts of Bermuda proceeds at a rate of 1 m per 70 years. Warme and Marshall (1969) made a similar estimate

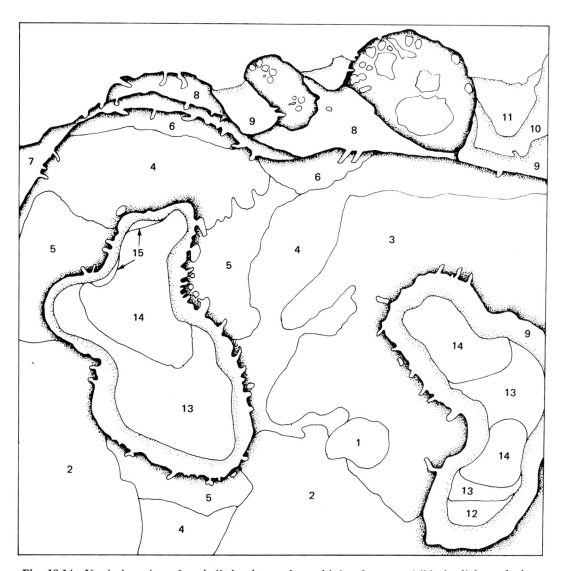

Fig. 18.14 Vertical section of a chalk hardground, combining features visible by light and ultra-violet luminescence. At least 14 successive sediments comprise the specimen (numbered chronologically). Progressive induration of the seafloor is shown by the gradual advance of lithified sediment on soft, bioturbated sediment, until a mature hardground was reached (sediment 6). The burrows then stood empty, and their walls and the seafloor were bored and glauconitized (dense stippling). Sediments 7 and 8 were added, and each was glauconitized. Subsequent sedimentary additions (9–11) were phosphatized (light stippling), and one of these sediments, perhaps 9, was plastered evenly on all burrow walls as active fill. The hardground was then finally buried, yet the burrows contain three further filling stages (12–14), representing postomission reexcavation. Small spaces in burrow roofs were never filled with sediment (15). Chalk Rock, Turonian; Medmenham, Buckinghamshire, England. Natural size.

for a whole community of borers. In contrast, a few weeks of burial by a wandering patch of sediment on the sea floor is sufficient to suffocate the entire community boring in the hardground. Many fossil hardgrounds apparently represent the omission of hundreds or thousands of years of sediment in shallow, oxygenated water

(Hofker, 1958, p. 148; Bathurst, 1971, p. 399), and yet they are bored to a surprisingly slight degree.

The chief inhibiting factors of boring in hardgrounds are (1) insufficient hardness of the sediment and (2) burial by new sediment. Several minor factors may also play an important role in some hardgrounds. For example, (3) an extremely rich overgrowth of encrusting organisms threatens to cover over the entrances of the borings (e.g., Voigt, 1959). One may also envision that (4) the strong superficial mineralization of $CaCO_3$ to glauconite or phosphate may inhibit the entry of boring organisms that are dependent on carbonate substrates.

Borings as Indicators of Hardness

Most groups of boring organisms are restricted to fully hard substrates, particularly those groups employing chemical means of boring (Bromley, 1970; see also Chapter 11). However, one group of borers in particular —the mechanically boring bivalves—crosses the boundary between burrowing and boring habits of life. Several genera of bivalves living today can penetrate sediments of all grades of hardness. The shape both of the shell of these borers and of the boring itself show characteristic differences in substrates of different hardness (Evans, 1968, 1970). Thus, borings in harder substrates tend to be shorter and wider than those in softer substrates.

Borings as Indicators of Erosion

Several groups of organisms produce borings of such characteristic shape that they can be recognized even when severely damaged by erosion. The damaged remains indicate the amount of erosion that has taken place since the borings were constructed (Voigt, 1959; Evans, 1968; Perkins, 1971). This relation was recognized at an early date; Buckland and De la Beche (1830, p. 31) made a remarkably detailed

reconstruction of the paleoenvironment and succession of events leading to the formation of a bed of bored pebbles at a Jurassic intraformational omission surface, on the basis of the degree of erosion of bivalve borings.

Perkins (1971) showed that the form of borings by acrothoracic cirripeds, bivalves, and sponges is considerably altered by erosion (Fig. 18.15), which breaks open the restricted surface pores and exposes the main cavity of the borings. The shape of borings by acrothoracicans, in particular, renders these structures sensitive indicators of very minor abrasion, and the presence of intact borings shows that absolutely no erosion took place after the death of the borers (Bromley, 1970, p. 70).

Perkins (1971) also pointed out that the shallower types of borings are destroyed in succession, according to their depth of penetration into the substrate, until in a deeply eroded hardground, only the distal ends of the deepest bivalve and "worm" borings remain. In many hardgrounds, however, periods of erosion alternated with quieter periods, when the abraded surface was recolonized by boring organisms. Thus, at final burial the hardground contains borings showing all degrees and stages of erosion (Figs. 18.5, 18.6, and 18.16) (Umbgrove, 1925, Pl. 9).

Boring Organisms as Agents of Erosion

As stated previously, the erosional rate of organisms boring in rock is remarkably high. In many hardgrounds, the sculpture of the omission surface was completely modified by the activity of boring organisms (Figs. 18.16 and 18.17). The result of this activity is the production of clastic material of all grades. Clionid sponges and the mechanical borers produce large quantities of fine-sand to silt-grade material, as a result of their boring processes (Bromley, 1970). Fines are usually removed by currents, whereas small to large, angular intraclasts remain; the latter are products

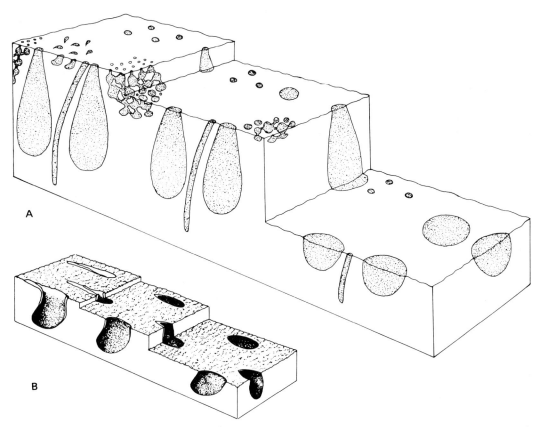

Fig. 18.15 Effect of abrasion on borings. A, generalized borings of sponges, acrothoracican cirripeds, bivalves, and "worms." (Modified after Perkins, 1971.) B, details of fresh, slightly abraded, and deeply abraded borings of acrothoracican cirripeds, based on *Rogerella* in Cretaceous echinoid tests. Height of block, 4 mm.

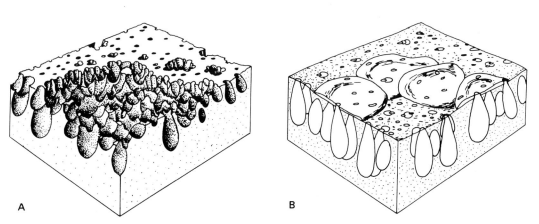

Fig. 18.16 Bivalve erosion of hardgrounds. A, advanced attack by bivalves on a sandstone hardground. In places, thin rock partitions between borings have broken down, and a new generation of borers entered the substrate from the base of older, vacated borings. Erosion almost entirely due to the agency of the bivalves. Based on specimens from Cretaceous–Paleocene junction at Tunorqo, Nûgssuaq peninsula, West Greenland. B, hardground terminating Edwards Limestone (Lower Cretaceous), south-central Texas. Although at least as many borings are present here as in block A, intermittent sedimentation "repaired" the surface by filling the borings. After lithification of the fill, a new suite of borings was superimposed on the earlier. Three successive suites are represented, with contemporaneous encrustation by oysters. Height of blocks, 4 cm.

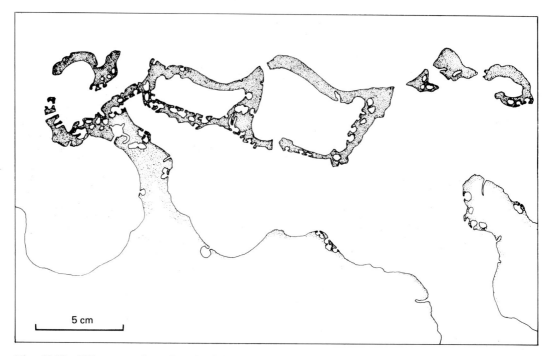

5 cm

Fig. 18.17 Filigree erosion of a chalk hardground, largely the work of sponges, which produced large cavities and riddled the remaining walls of these cavities with small chambers. Physical breakdown of the brittle surface was beginning when the hardground was buried. Stippling indicates extent of phosphatization, which may have protected the sediment from further boring. Chalk Rock (Turonian); Medmenham, England.

of mechanical breakage of the substrate through the weakening effect wrought by boring activity. Unsorted samples of these intraclasts are normally preserved in the fill of the surviving borings (Fig. 18.18).

In highly convoluted hardgrounds, the upstanding bosses between entrances to the hardground burrows commonly have a mushroom or bridge-like form (Fig. 18.8). Boring organisms preferentially attacked these structures on the undersides of overhangs, weakening their bases and causing them to break off as large, irregular intraclasts. These pieces are not easily distinguishable from lag intraclasts, which may accompany the dislodged bosses on the hardground as a result of periods of temporary deposition.

Inverted Borings

The sheltered environment offered by the walls of empty hardground burrows just below the sea floor was a favorable substrate for borers. Their crypts enter the rock at all angles, from the sides, floors, and particularly the roofs of these burrows. In some hardgrounds, extensive horizontal galleries were excavated during the period of lithification (Fürsich, 1971) (Fig. 18.18). The floors of such galleries were intermittently covered with sediment, which was disturbed continually by the infauna, whereas the roofs were permanently free of sediment. This relation resulted in poor preservation of the floors, yet many of the roofs are sharp, well mineralized, and richly bored; they have the appearance of "inverted hardgrounds." Such "inverted borings" were illustrated by Voigt (1959, Pl. 15), Kaźmierczak and Pszczółkowski (1968, Fig. 4), Shinn (1969, Fig. 10), and Purser (1969, Fig. 6).

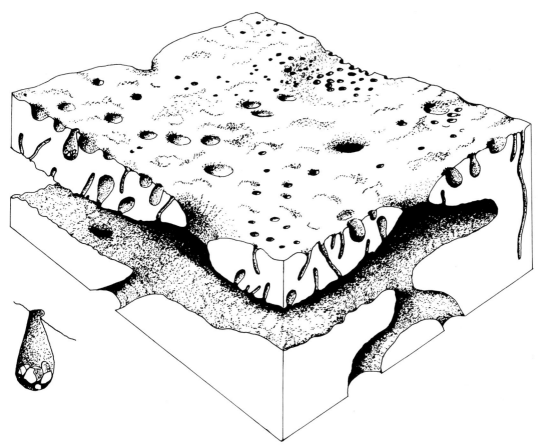

Fig. 18.18 Extensive horizontal gallery within a chalk hardground. Inverted borings in roof are unabraded. Borings in upper surface of the hardground trapped intraclasts, which are otherwise unpreserved (see enlarged boring). Campanian–Maastrichtian junction; Hallembaye quarry, near Visé, Belgium. Height of block, 8 cm.

The Fill of Borings

As is true of hardground burrows, the fill of borings sometimes preserves sediment that is otherwise omitted at the surface. Several examples have been described in which pebbles, showing no signs of transportation over significant distances, contain borings filled with sediment differing from that surrounding and constituting the pebble (e.g., Lutze, 1967; Schloz, 1972, p. 176). In some cases, different borings within the same pebble contain different fills, and thereby provide evidence of several periods of boring separated by periods of deposition and scouring (Cayeux, 1939, p. 225).

Erosional intraclasts, where they have been swept away from the hardground surface by currents, are commonly trapped and preserved within the fill of hardground borings and burrows (Fig. 18.18).

Also in analogy with burrow fills, those of borings provide a protected environment of high fossilization potential. The spicules of boring sponges are preserved within their borings in some cases (Bromley, 1970, p. 80); boring bivalves are commonly so well preserved that they may be identified to species. Ubaghs (1865, p. 35) identified six thin-shelled boring bivalves within their borings in the hardground terminating the Cretaceous in Holland, and Abel (1935, Fig. 414b) illustrated a Miocene bivalve boring in which not only the hard shell is well preserved but also the form of the siphons.

Nestling organisms commonly invade vacated borings in rock surfaces. These nestlers include bivalves and sponges, the skeletal remains of which are liable to be mistaken for those of borers (Kühnelt, 1951; Evans, 1967; Warme and Marshall, 1969; Bromley, 1970, p. 64, 74). As Evans (1970) pointed out, the special ecological niche and preservational environment of bivalve borings in some cases results in preservation of a fauna that is otherwise unrepresented at the omission surface (Fig. 18.6).

Superimposed Borings

Alternation between periods of deposition and nondeposition led in some instances to a vertical series of hardgrounds, which follow so closely upon one another that borings descending from one extend into the sediment of the next one below, cutting through boring structures in the older hardground (Hecker, 1965, Pl. 8; 1970, Fig. 1b) (Fig. 18.6).

Greater condensation superimposes the hardgrounds upon one another so that they cannot be distinguished individually ("compound hardgrounds"). Sediment deposited on the hardground was subsequently removed by scouring before it could be lithified, except for that detritus which clogged the borings in the reexhumed surface. This fill of the borings was then lithified. A new assemblage of borers invaded the reexhumed hardground, their new borings cutting randomly through the hardened fill and walls of the older suite of borings. Multiple repetition of this cycle of events brings about the superimposition of many suites of borings (Purser, 1969, Fig. 5) (Figs. 18.13A and 18.16B). In the ultimate stage of this process, described by Zankl and Schroeder (1972, p. 532), the lithification, internal sedimentation, and boring processes follow so closely upon each other that the substance of the original substrate is entirely replaced by the fill of superimposed suites of borings. Superimposition of borings is therefore evidence for the following sequence of events: sedimentation–scouring–lithification–boring–sedimentation.

Postomission Suite in Hardgrounds

Downward progress of organisms burrowing in the postomission sediment is inhibited by the buried hardground. The new burrows are deflected along this impenetrable surface, and enter the unhardened fill of hardground burrows within it (Fig. 18.7). Postomission burrows are thus compelled to assume a foreign form, controlled by the sculpture of the hardground or the restricting shape of the burrow fills within it, so that these new burrows may be considered xenomorphic. This feature has been observed in Holocene (Shinn, 1969, p. 123), Mesozoic (Bromley, 1967b—"imposed horizontality"), and Paleozoic hardgrounds (Hessland, 1949, p. 452). In some places in the European Cretaceous chalk, at levels of closely spaced hardgrounds, reexcavation of the fill of hardground *Thalassinoides* by postomission *Thalassinoides*–animals produced the confusing effect of a continuous burrow system crossing several omission surfaces without interruption (Bromley, 1967b, p. 162).

Further Reading

Within the copious literature on hardgrounds, only a few clear descriptions of trace fossils in these beds have been given, and these are almost entirely limited to the conspicuous omission suite: De la Beche (1846, p. 289, Figs. 43–44), Rutot and Van den Broeck (1886, p. 226), Richardson (1907), Klüpfel (1916b), Dacqué (1921, p. 449), Mägdefrau (1932), Marlière (1933), Cayeux (1939, p. 240; 1950, p. 891), Bigot (1941), Ellenberger (1948), Lombard (1956, p. 374), Müller (1956), Voigt (1959), Lindström (1963), Radwański (1964, 1965), Moore (1964), Moore and Martin (1966), Klein (1965), Hecker (1965, 1970), Jahnke

(1966), Bromley (1967b), Kennedy (1967b, p. 142; 1970a; 1970b, p. 266), Kaźmierczak and Pszczółkowski (1968, 1969), Purser (1969), Hölder and Hollmann (1969, Fig. 1), Rose (1970), Sellwood et al. (1970, p. 729), Bathurst (1971, Fig. 290), Warme and Olson (1971, p. 41), Perkins and Stewart (1971a), and Sellwood (1972, p. 145, Pl. 29).

CONCLUSIONS

Trace fossils at discontinuity surfaces fall into three clearcut groups, according to their time relationship with the hiatus. Trace fossils that predate the hiatus are here called the preomission suite; those produced during the period of nondeposition represent the omission suite; and traces formed after renewal of sedimentation constitute the postomission suite.

Distinction of these three trace fossil categories at discontinuity surfaces serves an essential purpose in the stratinomic interpretation of environmental conditions controlling deposition and nondeposition. The special significance of trace fossils at discontinuity surfaces may be summed in the following points:

1. The omission suite is usually emphasized by bed-junction preservation (i.e., change of sediment at the omission surface).
2. Appearance of new forms in the omission ichnocoenose reflects environmental causes for the omission of sediment.
3. The occurrence, and sometimes the degree, of scouring of the sea floor are indicated by truncation of burrows.
4. Synsedimentary lithification of the sea floor to form a hardground profoundly affects the characteristics of the omission suite, dividing it into two distinct ichnocoenoses: prelithification burrows and postlithification borings.
5. Interaction of bioturbation and the process

of cementation causes malformation of burrow systems (stenomorphism) and nodular distribution of the cement.
6. Postomission burrows that encounter a buried hardground are diverted along its impenetrable surface, or enter the confines of the fills of omission burrows within the hardground; these are said to be xenomorphic.
7. The intervention of scouring at early stages of lithification is indicated by beds of lag intraclasts.
8. Repeated burial and reexhumation of these intraclasts, which accrete during temporary burial, is indicated by bored surfaces within them (hiatus concretions).
9. Open hardground burrows in the sea floor acted as sediment traps, preserving otherwise omitted sediment. Multiple fills in such burrows reveal several temporary periods of sedimentation on the hardground and reexcavation of the burrow systems.
10. The presence of borings indicates a hard substrate. The morphology of some boring bivalves and their borings may be a useful indicator of the degree of hardness of past sea floors.
11. Truncation of borings indicates the intervention of erosion and, in some cases, allows one to estimate the amount of rock removed.
12. Superimposed borings indicate a definite sequence of events: boring–sedimentation–scour–lithification–boring.
13. Inverted borings are sometimes preserved in the roofs of hardground burrows and galleries. (Thus, borings clearly cannot be used indiscriminately as geopetals.)
14. The fill of borings offers a microenvironment of high fossilization potential for the remains of borers and nestlers.

ACKNOWLEDGMENT

It is a pleasure to record my gratitude to E. Nordmann, without whose close teamwork in the preparation of the illustrations, this chapter would not have met the editor's deadline.

REFERENCES

Abel, O. 1935. Vorzeitliche Lebensspuren. Jena, G. Fischer, 644 p.

Aubouin, J. 1965. Geosynclines. Developments in Geotectonics, 1, 335 p.

Bathurst, R. G. C. 1971. Carbonate sediments and their diagenesis. Developments in Sedimentology, 12, 620 p.

Bigot, A. 1941. Les surfaces d'usure et les remaniements dans le Jurassique de Basse-Normandie. Soc. Géol. France, Bull., (10)5:165–176.

Boyd, D. W. and N. D. Newell. 1972. Taphonomy and diagenesis of a Permian fossil assemblage from Wyoming. Jour. Paleont., 46:1–14.

Bromley, R. G. 1965. Studies in the lithology and conditions of sedimentation of the Chalk Rock and comparable horizons. Unpubl. Ph. D. Dissert., Univ. London, 355 p.

———. 1967a. Marine phosphorites as depth indicators. Marine Geol., 5:503–509.

———. 1967b. Some observations on burrows of thalassinidean Crustacea in chalk hardgrounds. Geol. Soc. London, Quart. Jour., 123:157–182.

———. 1968. Burrows and borings in hardgrounds. Dansk Geol. Foren. Meddr., 18:247–250.

———. 1970. Borings as trace fossils and *Entobia cretacea* Portlock, as an example. In T. P. Crimes and J. C. Harper (eds.), Trace fossils. Geol. Jour., Spec. Issue 3:49–90.

——— and R. W. Frey. 1974. Redescription of the trace fossil *Gyrolithes* and taxonomic evaluation of *Thalassinoides*, *Ophiomorpha* and *Spongeliomorpha*. Geol. Soc. Denmark, Bull., 23:311–335.

Brückner, W. 1951. Lithologische Studien und zyklische Sedimentation in der helvetischen Zone der Schweizeralpen. Geol. Rundschau, 39:196–212.

Buckland, W. and H. T. De la Beche. 1830. On the geology of the neighbourhood of Weymouth and the adjacent parts of the coast of Dorset. Geol. Soc. London, Trans., (2)4:1–46.

Calembert, L. 1953. Sur l'extension régionale d'un hard ground et d'une lacune stratigrafique dans le Crétacé supérieur du Nord-Est de la Belgique. Acad. Roy. Belgique, Bull., Cl. Sci., (5)39:724–733.

———. 1956. Le Crétacé supérieur de la Hesbaye et du Brabant. Excursion du 19 sept. 1955. Soc. Géol. Belgique, Ann., 80:B129–B156.

——— and M. Meyer. 1956. Sur l'extension d'une lacune stratigraphique dans le Crétacé supérieur du Pays de Herve et du Limbourg hollandais. Soc. Géol. Belgique, Ann., 79:B413–B423.

Cayeux, L. 1897. Contribution à l'étude microscopique des terrains sédimentaires. Lille, Bigot Frères, 589 p.

———. 1922. Les minerais de fer oolithique de France. II. Minerais de fer secondaires. Études des gîtes mineraux de la France. Paris, Impr. National, 1051 p.

———. 1939. Les phosphates de chaux sédimentaires de France. I. France metropolitaine. Études des gîtes mineraux de la France. Paris, Impr. National, 349 p.

———. 1941a. Causes anciennes et causes actuelles en géologie. Paris, Masson, 81 p.

———. 1941b. Les phosphates de chaux sédimentaires de France. II. Égypte, Tunesie, Algérie. Études des gîtes mineraux de la France. Paris, Impr. Nationale, p. 351–659.

———. 1950. Les phosphates de chaux sédimentaires de France. III. Maroc et conclusions générales. Études des gîtes mineraux de la France. Paris, Impr. Nationale, p. 661–1019.

Crimes, T. P. 1973. From limestones to distal turbidites: a facies and trace fossil analysis in the Zumaya flysch (Paleocene–Eocene), north Spain. Sedimentology, 20:105–131.

Dacqué, E. 1921. Vergleichende biologische Formenkunde der fossilen niederen Tiere. Berlin, Borntraeger, 777 p.

De la Beche, H. T. 1846. On the formation of the rocks of south Wales and south western England. Geol. Surv. Great Britain, Mem. 1:1–296.

Destombes, J. P. and E. R. Shephard-Thorn. 1971. Geological results of the Channel Tunnel site investigation 1964-65. Inst. Geol. Sci. London, Rept. 71/11, 12 p.

Dvořák, J. 1972. Shallow-water character of the nodular limestones and their paleogeographic interpretation (Upper Devonian–Lower Carboniferous of the Rhenoherzynicum and Sudeticum). Neues Jahrb. Geol. Paläont., Mh., 1972:509–511.

Ellenberger, F. 1946. Sur la signification de la craie à tubulures de Meudon. Soc. Géol. France, Bull., (5)15:497–507.

———. 1948. Le problème lithologique de la craie durcie de Meudon. Bancs-limites et "contacts par racines": lacune sous–marine ou émersion? Soc. Géol. France, Bull., (5)17:255–274.

Ernst, G. 1963. Stratigraphische und gesteinschemische Untersuchungen im Santon und

Campan von Lägerdorf (SW-Holstein). Mitt. Geol. Staatsinst. Hamburg, 32:71–127.

Evans, J. W. 1967. Relationship between *Penitella penita* (Conrad, 1837) and other organisms of the rocky shore. Veliger, 10:148–151.

————. 1968. The effect of rock hardness and other factors on the shape of the burrow of the rock-boring clam, *Penitella penita*. Palaeogeogr., Palaeoclimatol., Palaeoecol., 4:271–278.

————. 1970. Palaeontological implications of a biological study of rock-boring clams (Family Pholadidae). In T. P. Crimes and J. C. Harper (eds.), Trace fossils. Geol. Jour., Spec. Issue 3:127–140.

Fabricius, F. 1968. Calcareous sea bottoms of the Raetian and Lower Jurassic from the west part of the Northern Calcareous Alps. In G. Müller and G. M. Friedman (eds.), Recent developments in carbonate sedimentology in central Europe. Berlin, Springer-Verlag, p. 240–249.

Fischer, A. G. and R. E. Garrison. 1967. Carbonate lithification on the sea floor. Jour. Geol., 75:488–496.

Fürsich, F. 1971. Hartgründe und Kondensation im Dogger von Calvados. Neues Jahrb. Geol. Paläont., Abh., 138:313–342.

Goldman, M. I. 1922. Basal glauconite and phosphate beds. Science, 56:171.

Goldring, R. 1962. The trace fossils of `the Baggy Beds (Upper Devonian) of North Devon, England. Paläont. Zeitschr., 36:232–251.

————. 1964. Trace-fossils and the sedimentary surface in shallow-water marine sediments. In L. M. J. U. van Straaten (ed.), Deltaic and shallow marine deposits. Developments in Sedimentology, 1:136–143.

Hadding, A. 1958. Hidden hiatuses and related phenomena. Some lithological problems. Kgl. Fysiogr. Sällsk. Lund, Förhand., 28:159–171.

Håkansson, E. 1971. Stevns Klint. In S. Floris et al., Geologi på Øerne 1, Sydøstsjælland og Møn. Varv Ekskursionsfører, 2:25–36. København.

Hallam, A. 1960. The White Lias of the Devon coast. Geologists' Assoc., Proc., 71:47–60.

————. 1964. Origin of the limestone-shale rhythm in the Blue Lias of England: a composite theory. Jour. Geol., 72:157–169.

————. 1969. A pyritized limestone hardground in the Lower Jurassic of Dorset (England). Sedimentology, 12:231–240.

Hecker, R. T. 1965. Introduction to paleoecology. New York, American Elsevier, 166 p.

————. 1970. Palaeoichnological research in the Palaeontological Institute of the Academy of Sciences of the USSR. In T. P. Crimes and J. C. Harper (eds.), Trace fossils. Geol. Jour., Spec. Issue 3:215–226.

Heim, A. 1924. Über submarine Denudation und chemische Sedimente. Geol. Rundschau, 15:1–47.

————. 1958. Oceanic sedimentation and submarine discontinuities. Eclogae Geol. Helvetiae, 51:642–649.

Hessland, I. 1949. Investigations of the Lower Ordovician of the Siljan District, Sweden. IV. Geol. Inst. Univ. Uppsala, Bull., 33:437–510.

Hofker, J. 1958. Foraminifera from the Cretaceous of Limburg, Netherlands. XXXVIII. The gliding change in *Bolivinoides* during time. Natuurhist. Maandblad, 47:145–159.

————. 1959. Les foraminiferes du Crétacé supérieur de Harmignies, Bassin de Mons. Soc. Géol. Belgique, Ann., 82:B319–B333.

Hölder, H. and R. Hollmann. 1969. Bohrgänge mariner Organismen in jurassischen Hart- und Felsböden. Neues Jahrb. Geol. Paläont., Abh., 113:79–88.

Hollmann, R. 1964. Subsolutions-Fragmente (Zur Biostratinomie der Ammonoidea im Malm des Monte Baldo/Norditalien). Neues Jahrb. Geol. Paläont., Abh., 119:22–82.

Jaanusson, V. 1961. Discontinuity surfaces in limestones. Geol. Inst. Univ. Uppsala, Bull., 40:221–241.

Jahnke, H. 1966. Beobachtungen an einem Hartgrund (Oberkante Terebratelbank mu^{r2} bei Göttingen). Der Aufschluss, 17(1):2–5.

Jenkyns, H. C. 1971. The genesis of condensed sequences in the Tethyan Jurassic. Lethaia, 4:327–352.

Jurgan, H. 1969. Sedimentologie des Lias der Berchtesgadener Kalkalpen. Geol. Rundschau, 58:464–501.

Kaźmierczak, J. and A. Pszczółkowski. 1968. Sedimentary discontinuities in the Lower Kimmeridgian of the Holy Cross Mts. Acta

Geol. Polonica, 18:587–612. [In Polish, with English summary.]

———— and A. Pszczółkowski. 1969. Burrows of Enteropneusta in Muschelkalk (Middle Triassic) of the Holy Cross Mountains, Poland. Acta Palaeont. Polonica, 14:299–324.

Kennedy, W. J. 1967a. Field meeting at Eastbourne, Sussex. Lower Chalk sedimentation. Geologists' Assoc., Proc., 77:365–370.

————. 1967b. Burrows and surface traces from the Lower Chalk of Southern England. British Mus. (Nat. Hist.), Geol., Bull., 15:127–167.

————. 1970a. A correlation of the uppermost Albian and the Cenomanian of south-west England. Geologists' Assoc., Proc., 81:613–677.

————. 1970b. Trace fossils in the chalk environment. In T. P. Crimes and J. C. Harper (eds.), Trace fossils. Geol. Jour., Spec. Issue 3:263–282.

Klein, G. de V. 1965. Dynamic significance of primary structures in the Middle Jurassic Great Oolite Series, southern England. Soc. Econ. Paleont. Mineral., Spec. Publ. 12:173–191.

Klüpfel, W. 1916a. Zur Kenntnis des Lothringer Bathonien. Geol. Rundschau, 7:1–29.

————. 1916b. Über die Sedimente der Flachsee im Lothringer Jura. Geol. Rundschau, 7:97–109.

Kühnelt, W. 1951. Contribution à la connaissance de l'endofaune des sols marins durs. Année Biol., (3)27:513–523.

Laughton, A. S. et al. 1972. Site 119. In A. S. Laughton et al., Initial reports of the Deep Sea Drilling Project, 12:753–901.

Lindström, M. 1963. Sedimentary folds and the development of limestone in an early Ordovician sea. Sedimentology, 2:243–292.

Lombard, A. 1956. Géologie sédimentaire. Les séries marines. Paris, Masson, 727 p.

Lutze, G. F. 1967. Ein Emersions-Horizont im Bathonium von Hildesheim. Senckenbergiana Leth., 48:535–548.

Mägdefrau, K. 1932. Über einige Bohrgänge aus den unteren Muschelkalk von Jena. Paläont. Zeitschr., 14:150–160.

Marlière, R. 1933. De nombreux bancs phosphatés dans la craie à Actinocamax quadratus du Bassin de Mons. Soc. Géol. Belgique, Bull., 56:289–301.

Moore, C. H. 1964. Stratigraphy of the Fredericksburg Division, south-central Texas. Univ. Texas, Bureau Econ. Geol., Rept. Invest., 52:1–48.

———— and K. G. Martin. 1966. Comparison of quartz and carbonate shallow marine sandstones, Fredericksburg Cretaceous, central Texas. Amer. Assoc. Petrol. Geol., Bull., 50:981–1000.

Murray, J. and A. F. Renard. 1891. Deep sea deposits. Report of the scientific results of the exploring voyage of H.M.S. Challenger, 1873-76. London, H.M.S.O., 525 p.

Müller, A. H. 1956. Weitere Beiträge zur Icnologie, Stratinomie und Ökologie der germanischen Trias. I. Geologie, 5:405–414.

Neumann, A. C. 1966. Observations on coastal erosion in Bermuda and measurements of the boring rate of the sponge, Cliona lampa. Limnol. Oceanogr., 11:92–108.

Perkins, B. F. 1966. Rock-boring organisms as markers of stratigraphic breaks (abs.). Amer. Assoc. Petrol. Geol., Bull., 50:631.

————. 1971. Traces of rock-boring organisms in the Comanche Cretaceous of Texas. In B. F. Perkins (ed.), Trace fossils, a field guide. Louisiana State Univ., School Geosci., Misc. Publ. 71-1:137–148.

———— and C. L. Stewart. 1971a. Stop 3: Whitestone Quarry. In B. F. Perkins (ed.), Trace fossils, a field guide. Louisiana State Univ., School Geosci., Misc. Publ. 71-1:17–22.

———— and C. L. Stewart. 1971b. Stop 7: Dinosaur Valley State Park. In B. F. Perkins (ed.), Trace fossils, a field guide. Louisiana State Univ., School Geosci., Misc. Publ. 71-1:56–59.

Pożaryska, K. 1965. Foraminifera and biostratigraphy of the Danian and Montian in Poland. Palaeont. Polonica, 14:1–156.

Purser, B. H. 1969. Syn-sedimentary marine lithification of Middle Jurassic limestones in the Paris Basin. Sedimentology, 12:205–230.

Radwański, A. 1964. Boring animals in Miocene littoral environments of southern Poland. Acad. Pol. Sci., (Sér. Sci. Géol. Geogr.) Bull., 12:57–62.

————. 1965. Additional notes on Miocene littoral structures of southern Poland. Acad. Pol. Sci. (Sér. Sci. Géol. Geogr.) Bull., 13:167–173.

Rasmussen, H. W. 1971. Echinoid and crustacean burrows and their diagenetic significance in the Maastrichtian–Danian of Stevns Klint, Denmark. Lethaia, 4:191–216.

Richardson, L. 1907. The Inferior Oolite and contiguous deposits of the Bath–Doulting district. Geol. Soc. London, Quart. Jour., 63:383–436.

————. 1933. The country around Cirencester. Geol. Surv. England Wales, Mem., 1–119.

Rose, P. R 1970. Stratigraphic interpretation of submarine versus subaerial discontinuity surfaces: an example from the Cretaceous of Texas. Geol. Soc. America, Bull., 81:2787–2798.

Rosenkrantz, A. 1966. Die Senon/Dan–Grenze in Dänemark. Deutsch. Gesell. Geol. Wiss., Ber. A, Geol. Paläont., 11:721–727.

Rutot, A. and E. Van den Broeck. 1886. La géologie de Mesvin–Ciply. Soc. Géol. Belgique, Ann., 13:197–260.

Ryder, J. A. 1879. On the destructive nature of the boring sponge, with observations on its gemmules or eggs. Amer. Naturalist, 13:279–283.

Saporta, G. de. 1884. Les organismes problématiques des anciennes mers. Paris, Masson, 102 p.

Schloz, W. 1968. Über Beobachtungen zur Ichnofazies und über umgelagerte Rhizocorallien im Lias α Schwabens. Neues Jahrb. Geol. Paläont., Mh., 1968:691–698.

————. 1972. Zur Bildungsgeschichte der Oolithenbank (Hettangium) in Baden-Württemberg. Inst. Geol. Paläont. Univ. Stuttgart, Arb., N.F., 67:101–212.

Schmid, F. 1959. Biostratigraphie du Campanien-Maastrichtien du NE de la Belgique sur la base des Bélemnites. Soc. Géol. Belgique, Ann., 82:B235–B256.

Seilacher, A. 1962. Paleontological studies on turbidite sedimentation and erosion. Jour. Geol., 70:227–234.

————. 1964. Biogenic sedimentary structures. In J. Imbrie and N. D. Newell (eds.), Approaches to paleoecology. New York, John Wiley, p. 296–316.

————. 1967. Bathymetry of trace fossils. Marine Geol., 5:413–428.

Sellwood, B. W. 1970. The relation of trace fossils to small scale sedimentary cycles in the British Lias. In T. P. Crimes and J. C.

Harper (eds.), Trace fossils. Geol. Jour., Spec. Issue 3:489–504.

————. 1971. The genesis of some sideritic beds in the Yorkshire Lias (England). Jour. Sed. Petrol., 41:854–858.

————. 1972. Regional environmental changes across a Lower Jurassic stage-boundary in Britain. Palaeontology, 15:125–157.

———— et al. 1970. Field meeting on the Jurassic and Cretaceous rocks of Wessex. Geologists' Assoc., Proc., 81:715–732.

Shinn, E. A. 1969. Submarine lithification of Holocene carbonate sediments in the Persian Gulf. Sedimentology, 12:109–144.

————. 1970. Submarine formation of bored surfaces (hardgrounds) and possible misinterpretation in stratigraphic applications (abs.). Amer. Assoc. Petrol. Geol., Bull., 54:870.

Simpson, S. 1957. On the trace-fossil Chondrites. Geol. Soc. London, Quart. Jour., 107:475–499.

Staring, W. C. H. 1860. De bodem van Nederland. II. Haarlem, Kruseman, 480 p.

Steinich, G. 1972. Pseudo-Hardgrounds in der Unter-Maastricht–Schreibkreide der Insel Rügen. Wiss. Zeitschr. Ernst–Moritz-Arndt-Univ. Greifswald, Math.-Naturwiss. Jahrg., 21:213–223.

Taylor, J. C. M. and L. V. Illing. 1969. Holocene intertidal calcium carbonate cementation, Qatar, Persian Gulf. Sedimentology, 12:69–107.

Termier, H. and G. Termier. 1963. Erosion and sedimentation. London, D. van Nostrand, 433 p.

Twenhofel, W. H. 1950. Principles of sedimentation. New York, McGraw-Hill, 673 p.

Ubaghs, J. C. 1865. Die Bryozoen-Schichten der Maastrichter Kreidebildung. Naturhist. Verein. Preuss. Rheinl. Westph., Verhhand., 22:31–62.

Umbgrove, J. H. F. 1925. De Anthozoa uit het Maastrichtsche Tufkrijt. Leidse Geol. Med., 1:83–126.

van Straaten, L. M. J. U. 1967. Solution of aragonite in a core from the SEn. Adriatic Sea. Marine Geol., 5:241–248.

Voigt, E. 1959. Die ökologische Bedeutung der Hartgründe ("Hardgrounds") in der oberen Kreide. Paläont. Zeitschr., 33:129–147.

————. 1968. Über Hiatus-Konkretionen (dargestellt an Beispielen aus dem Lias). Geol. Rundschau, 58:281–296.

Warme, J. E. and N. F. Marshall. 1969. Marine borers in calcareous terrigenous rocks of the Pacific coast. Amer. Zool., 9:765–774.

———— and R. W. Olson. 1971. Stop 5: Lake Brownwood Spillway. In B. F. Perkins (ed.), Trace fossils, a field guide. Louisiana State Univ., School Geosci., Misc. Publ. 71–1:27–43.

Wendt, J. 1969. Foraminiferen-"Riffe" im karnischen Halstätter Kalk des Feuerkogels (Steiermark, Österreich). Paläont. Zeitschr., 43:177–193.

————. 1970. Stratigraphische Kondensation in triadischen und jurassischen Cephalopodenkalken der Tethys. Neues Jahrb. Geol. Paläont., Mh., 1970:433–448.

Zankl, H. 1969. Structural and textural evidence of early lithification in fine-grained carbonate rocks. Sedimentology, 12:241–256.

———— and J. H. Schroeder. 1972. Interaction of genetic processes in Holocene reefs off North Eleuthera Island, Bahamas. Geol. Rundschau, 61:520–541.

PART FOUR

Recent Aquatic Lebensspuren

CHAPTER 19

RECENT LEBENSSPUREN IN NONMARINE AQUATIC ENVIRONMENTS

C. KENT CHAMBERLAIN
Department of Geology, Ohio University
Athens, Ohio, U.S.A.

SYNOPSIS

Diverse faunas exist in nonmarine aquatic environments, and the animals make distinctive tracks, trails, tubes, and burrows. For example, certain beetles make dwellings or feeding burrows and pupal chambers. Midgefly larvae and aquatic earthworms extensively rework lake bottoms and lentic parts of rivers. Caddisfly larvae use clastic grains and plant material to construct unique, mobile dwelling cases. Snails and clams make abundant surface traces and resting burrows. Distinctive shore tracks and trails, dwelling burrows and similar structures, hibernation burrows, feeding traces, and nesting structures are made by aquatic and semiaquatic, freshwater vertebrates of diverse types.

The principles of ecology and ichnology that apply to nonmarine aquatic animals and environments are the same as marine ones; only the parameters are different. Consequently, ichnological studies made on local streams or lakes can yield equally interesting and instructive results; and much work remains to be done.

INTRODUCTION

Traces of nonmarine aquatic animals are both abundant and diverse. Freshwater and other nonmarine aquatic animals, like marine animals, make dwelling structures, resting and crawling traces, and feeding burrows. Invertebrates make feeding traces and also pupal, brood, hibernation, and aestivation chambers, and vertebrates make hibernation burrows and nesting structures. As in the marine realm, the types of behavior are varied in shallow-water environments; but in deep waters, deposit feeding by invertebrates is most common.

The faunas responsible for the lebensspuren are also diverse. For example, shore beetles and crickets make extensive feeding burrows. Tiger-beetle larvae make deep dwelling burrows, and caddisfly larvae make dwelling cases by agglutinating organic debris or sediment grains together. The shape of many dwelling cases is diagnostic of a particular habitat and a particular species. Burrows in the substrate of deep lakes are made mainly by aquatic earthworms (oligochaetes) and amphipods. Some animals, such as the eubranchiopods, are diagnostic of ephemeral ponds, where they crawl upon or plough through the substrate and where only one or two species are found associated together. Even some dipterans (flies) make traces; for example, larvae of the midgefly make extensive dwelling burrows in lake sediments.

Vertebrates that make dwelling burrows in banks of rivers and lakes include muskrats, nutria, and the duck-billed platypus.

Beavers make dwelling structures (lodges) from wood and mud, and leave gnawed wood as evidence of their building and feeding activities. Shore tracks and trails are made in abundance by many vertebrates, particularly the numerous mammals. Among the lower vertebrates, sea lampreys build dwelling burrows and nesting structures.

The study of recent freshwater lebensspuren has some distinct advantages. The traces can be studied in local streams, rivers, lakes, or ponds, and the same principles can be applied there as in the generally more distant or otherwise inaccessible marine environments. Another advantage is that many new, original observations can be made, without the burden of too much previous information or possible dogmatic misconception.

At present, nonmarine aquatic environments have been studied so little by ichnologists that few reliable criteria are known for recognizing ancient analogs. Neither has the transition from marine to nonmarine environments been studied sufficiently.

My purpose in this chapter is to summarize the character of the lebensspuren of nonmarine aquatic animals as they are known at present, neglecting coprolites and fecal pellets, cysts and borings, and rasping traces (cf. Chapters 10 to 13).

Quantitative classifications (e.g., first- and second-order streams) and formal definitions are not necessary in describing nonmarine aquatic environments cited in this chapter, because the zoogeographic extents of trace-making animals are not precisely known. (Indeed, the animals seem to have a particular disdain for such artificial classifications.) Usually, "stream" refers to any body of flowing water. "Brook," "creek," and "river" imply successive increases in the size of a stream; "bayou" refers to a small, secondary, sluggish stream. "Lake" refers to any standing body of inland water, whether fresh, alkaline, or saline. A "pond" is a small, shallow lake, and a "pool" is a small, deep

lake. "Lotic" refers to running water, and "lentic" to standing water. Streams are mostly lotic but have some lentic parts, and lakes are mostly lentic but have some lotic parts (e.g., wave-swept beaches).

NONMARINE AQUATIC INVERTEBRATES

Invertebrates living in freshwater environments are both numerous and diverse, although less so than in marine environments. Approximately 30 groups (phyla, classes, or orders) of invertebrates have representatives living in freshwater environments (Table 19.1). A few of these animals—eubranchiopods, nematomorphs, and hydracarins—are mainly or entirely freshwater denizens. The remaining groups are variably represented by a few to many species in freshwater environments, and generally have more species in marine waters (see Pennak, 1953, Table 2).

Insects and mollusks represent most of the species in the world. Only approximately 4 percent of the insects are aquatic or have aquatic stages, however, and only a small percentage of mollusks are freshwater forms. Nevertheless, insects and mollusks are among the major macroscopic invertebrates of benthic communities in lakes and streams, although amphipods and oligochaetes are also very abundant.

LEBENSSPUREN OF NONMARINE AQUATIC INVERTEBRATES

Although less than half of the freshwater invertebrates listed in Table 19.1 make lebensspuren, some groups include many species that are very active on or in the substrate; thus, abundant and interesting lebensspuren do exist in freshwater environments. Crayfish, gastropods, bivalves, aquatic earthworms, and certain insects are responsible for most of the lebensspuren. Nematodes, ostracods, amphipods, and eubranchiopods also make traces, although the extent of these is less well known.

TABLE 19.1 Invertebrates of Freshwater Environments.

Phylum Protozoa (protozoans)
  Class Mastigophora (flagellates)
  Class Sarcodina (rhizopods)
  Class Sporozoa (sporozoans)
  Class Ciliata (ciliates)
  Class Suctoria (suctoriales)
Phylum Porifera (sponges)
Phylum Coelenterata (jellyfish, sea anemones, corals)
  Class Hydrozoa (hydrozoans)
Phylum Platyhelmintha (flatworms)
  Class Turbellaria (flatworms)
Phylum Nemertea (proboscis worms)
Phylum Gastrotricha (gastrotrichs)
Phylum Rotatoria (rotifers)
Phylum Nematoda (round worms)
Phylum Nematomorpha (horsehair worms)
Phylum Tardigrada (water bears)
Phylum Ectoprocta (bryozoans)
Phylum Endoprocta (endoprocts)

Phylum Annelida (segmented worms)
  Class Oligochaeta (aquatic earthworms)
  Class Polychaeta (polychaetes)
  Class Hirudinea (leeches)
Phylum Arthropoda (arthropods)
  Class Crustacea (crustaceans)
    Order Eubranchiopoda (fairy, tadpole, and clam shrimps)
    Order Cladocera (water fleas)
    Order Ostracoda (seed shrimp)
    Order Copepoda (copepods)
    Order Mysidacea (opossum shrimp)
    Order Isopoda (aquatic sow bugs)
    Order Amphipoda (scuds)
    Order Decapoda (crayfish, crabs, shrimp)
  Class Hydracarina (water mites)
  Class Arachnida (spiders)
  Class Insecta (insects)
Phylum Mollusca (mollusks)
  Class Gastropoda (snails)
  Class Bivalvia (pelecypods—clams, mussels)

Sponges, bryozoans, and most hydrozoans are sessile and consequently do not make traces. Certain animals, such as the opossum shrimp and water mites, are mainly nektonic and have little deliberate contact with the substrate. Other animals, such as the nemerteans and many turbellarians and insects, prefer a plant substrate rather than a clastic one. Protozoans, gastrotrichs, water bears, and rotifers are small enough that they probably do not leave an obvious trace when they move through or across the substrate; or at least, special techniques would be required to observe them. Among the intermediate-small animals, such as water fleas, rotifers, and copepods, whatever lebensspuren they might make have not been reported. Some of the rotifers do make small dwelling cases and tubes, by agglutinating clastic and woody grains together; but these inconspicuous structures range in size from less than 1 to about 5 mm across—even in colonial ones—and may easily be overlooked. (Such lebensspuren are nevertheless significant, as emphasized in Chapter 9, and warrant further study.) Water mites are very common in many freshwater bodies; the animals have been observed to make brush marks on the substrate of an aquarium, and plough marks on a drying substrate, but this sort of small feature probably would go unrecognized in natural situations.

The types of lebensspuren made by certain freshwater animals are summarized in Table 19.2. The table is deceptive in that not much is known about several of the groups of lebensspuren; the characteristics of some are inferred from the behavior of the animals and not from actual reports on observed traces. The type of lebensspur—resting trace, dwelling structure, etc.—corresponds generally to the classical behavioral types of trace fossils (see Chapter 3). Pupal, brood, hibernation, and aestivation chambers, however, are another category of lebensspuren. Unfortunately, not much information is available concerning these structures. Most of them are simple oval chambers corresponding to the size of the pupae or adults. The exact size, shape, depth, and habitat location of the trace, as well as the nature of the access tube, may

TABLE 19.2 Types of Lebensspuren Made by Nonmarine Aquatic Invertebrates.

Key *—the animals probably make this structure x—the animals do make this structure a—aestivation or hibernation chamber b—brood chamber p—pupal chamber	Resting Trace	Dwelling Structure	Crawling Trail	Feeding Trail	Feeding Burrow	Chamber
Platyhelmintha						
Turbellaria (flatworms)			*			
Nematoda (roundworms)			x			
Nematomorpha (horsehair worms)			*			
Annelida						
Oligochaeta (aquatic earthworms)					x	
Polychaeta (polychaetes)		x				
Hirudinea (leeches)			x			a
Arthropoda						
Crustacea						
Eubranchiopoda						
Conchostraca (clam shrimps)	x	*	x	*		
Notostraca (tadpole shrimps)	*	*	*	*		
Anostraca (fairy shrimps)	*		*			
Isopoda (aquatic sow bugs)	*					a
Ostracoda (seed shrimps)				x	x	
Amphipoda (scuds)	*	x	x		*	
Decapoda						
Astacidae (crayfish)	*	x	*			a
Potamidae (crabs)	*	x	*			
Atyidae and Paleomonidae (shrimps)	*		*			
Insecta						
Plectoptera (stoneflies)	*		*			
Ephemeroptera (mayflies)	*	x	*			
Odonata (dragonflies)	*	x	*			
Megaloptera (Alder and Dobson flies)	*		*			p
Hemiptera (bugs)		?				
Trichoptera (caddisflies)		x	x			
Coleoptera (beetles)						
Carabidae (carabs)			?			
Dytiscidae (predaceous diving beetle)	x					
Georýssidae (minute mud-loving beetles)	x					
Heteroceridae (variegated mud-loving beetles)					x	p
Hydraenidae (hydraenids)					?	
Gyrinidae (whirligig)						a

TABLE 19.2—Continued.

Key *—the animals probably make this structure x—the animals do make this structure a—aestivation or hibernation chamber b—brood chamber p—pupal chamber	Resting Trace	Dwelling Structure	Crawling Trail	Feeding Trail	Feeding Burrow	Chamber
Hydrophilidae (water scavenger)	?		?			
Noteridae (burrowing water beetle)					x	
Ptilodactylidae (ptilodactylids)					x	
Staphylinidae (rove beetles)			x		?	
Cicindelidae (tiger beetles)			x			a,b,p
Diptera (flies)						
Syrphidae (hoverflies)	x					
Tabanidae (horseflies)					?	
Chironomidae (blood-worm)			x			
Hymenoptera						
Formicidae (ants)			x			
Sphecidae (mud daubers)				(excavation)		
Orthoptera (crickets)						
Gryllotalpinae (mole crickets)					x	
Tridactylidae (sand cricket)					x	
Dermaptera (earwigs)			x			
Mollusca						
Gastropoda (snails)	x		x			a
Bivalvia (clams and mussels)	x		x			a

be important in ultimately recognizing different traces when the animal is no longer present.

Pupal, brood, hibernation, and aestivation chambers are similar to agglutinated tests of the polychaete *Pectinaria* and arenaceous foraminiferans, but the chambers are true traces whereas the two tests are potential body fossils. (See Chapter 3.) Dwelling cases of caddisfly larvae are analogous to the latter and thus technically are "body parts" also.

An undetermined number of animals, particularly insects, commonly walk or creep along the shores of rivers or lakes, or through ephemeral puddles, but are not regular inhabitants of these environments. The crisscrossing furrows made through puddles by terrestrial earthworms, following a storm, are familiar to almost everyone. A reasonable attempt cannot be made at enumerating all of the lebensspuren possible under such fortuitous and random situations, particularly considering the thousands of active insects prone to tread hither and thither without regard to the plight of the neoichnologist. Perhaps the subdivision of these animals into major groups of treaders could be attempted after completion of additional careful field observation and some experimental neoichnology.

Experimental neoichnology has not been employed nearly often enough! (cf. Chapter 22). Based on the surprising complexity of insect locomotion traces reported by Graber (1884, 1886) and Demoor (1880, 1890), and salamander tracks by Evans (1946), further experimental neoichnology would be both interesting and replete with instructive surprises (Fig. 19.1A–C).

The burrow of a wasp and the trail of an earwig, each made in ephemeral storm puddles, are included in my descriptions. They are not aquatic in the sense of the other animals discussed here, but they typify forms that visit fresh muds. Some species, such as the mud-dauber wasp, come to gather building materials; but certain other forms, such as the earwig, come fortuitously. These particular ones were selected because they are distinctive, resemble other important lebensspuren, and information was already available on them.

Platyhelmintha: Turbellaria (Flatworms)

Freshwater turbellarians are widespread on many substrates. They range from a few to 30 mm in length, and are narrow. Some move by smooth, gliding movements of cilia on a thin coat of mucus; others crawl on the substrate by peristaltic waves of muscular contraction. These movements result in shallow, rounded furrows; further details are not known.

Nematoda (Roundworms)

Nematodes are cosmopolitan in almost all waters and substrates. Under proper conditions, their small sinuous trails may be preserved in sediments or other substrates (e.g., Sandstedt et al., 1961; Gray and Lissmann, 1964; Wallace, 1968; Moussa, 1970). The trails range from 0.5 to 1 mm across, and have a sine-curve regularity (Figs. 19.4N, 19.6A). The trails are known from the Pleistocene (Tarr, 1935 = *Chironomous* larvae) and Tertiary (Moussa, 1970), but might well occur in much older rocks.

Nematomorpha (Horsehair or Gordian Worms)

Although horsehair worms are nowhere very abundant, they are fairly cosmopolitan —particularly in shallow, roadside water bodies—and are commonly found writhing about in damp shore sediments. They range from 0.3 to 2.5 mm across, and 10 to 70 cm in length. The swimming movements are feeble, slow undulations or writhings, and result in erratic brush and writhe marks against the substrate (Fig. 19.4O).

Annelida

Oligochaeta (Aquatic Earthworms)

Aquatic earthworms extensively rework sediments in many freshwater environments. They occur, sometimes in profusion, at all depths; but mainly they occur in shallow, quiet waters (less than 1 m deep). Some are most abundant in polluted waters. Oligochaetes construct small tubes, 1 to 2 mm across and projecting 2 to 5 mm above the substrate surface (Figs. 19.4M, 19.6C, 19.7I). The tubes are made of agglutinated mud, silt, or sand, depending on the substrate. Elongate fecal strings of ingested

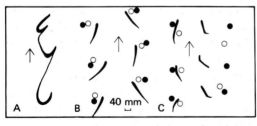

Fig. 19.1 Results of experimental ichnology, in which beetles were allowed to walk after treating each leg with different colors of paint. A, tibial spines of right hind leg of *Dyticus*, the swimming beetle. B, walking pattern of *Blaps mortisaga*; dots are tracks of the foreleg, circles of the middle leg, and slashes of the hind leg. C, *Trichodes*, using same codes as in B; beetle subjected to a 30° slope. (After Graber, 1884.)

sediment are extruded to form mounds or circular ridges ranging from 5 to 20 mm across. Full details on the irregular, branching, feeding burrows are not known. They radiate from the surface tube, and most of them branch horizontally and continue a few centimeters laterally; others continue downward and simply fork at a wide angle. They have a fairly constant diameter; depending on the size or species, it ranges from less than 1 to more than 2 mm. Certain burrows are similar to the trace fossil *Chondrites*, except that they do not display phobotaxis (see Chapter 6). Irregular chambers are developed in some, apparently from complete mining of particularly nutritious areas.

Kozhov (1963) reported some large oligochaetes in Lake Baikal, Russia, that attain lengths of 12 to 20 cm and are 2 to 5 cm in diameter. One of these large forms occurs at the greatest depth of the lake; to know something of its burrowing behavior would be extremely interesting.

Polychaeta (Polychaetes)

Polychaetes have not become generally established in freshwater environments. Most of them are found in lakes or streams presently or recently joined with the ocean. A few have been found in high-altitude streams. Consequently, polychaetes are considered to have adapted only recently to the freshwater environment. Certainly their limited occurrence represents a low level of success at invading fresh waters through past eons.

Most freshwater polychaetes are small, ranging from 3 to 15 mm in length. They build small chitinous tubes containing agglutinated mud or silt, in silty bottoms, or moveable tubes on shore stones (Fig. 19.4L).

Hirudinea (Leeches)

Leeches generally do not live on a substrate suitable to form or retain lebensspuren, but some make creeping, looping, or inchworm movements on clastic substrates. These lebensspuren are not known in detail but probably are worth further study. Leeches are generally cosmopolitan, and range in length from 5 to 50 mm (Pennak, 1953).

Arthropoda: Crustacea: Eubranchiopoda

Conchostraca (Clam Shrimp)

Clam shrimp prefer vernal ponds and puddles, and seldom appear in lakes or ponds containing carnivorous animals. They range from Devonian to Holocene. Tasch (1964) illustrated and described crawling and resting lebensspuren of some extant species in an artificial environment. He observed a serpentine configuration approximately 2.6 mm wide. It had a median ridge at a lower elevation than the sides of the trails (Fig. 19.2F). At the bow of the looped area, the clam shrimp apparently crossed its previous trail. Another shorter, hairpin trail seemed to have been the older of the two trails. The parallel depressions of the trail, at a constant distance apart, indicated that they were excavated by the animal's paired appendages and that the sediment was moved posteriorly, as in burrowing. The width of the trail corresponded closely to the width of the animal's body.

Notostraca (Tadpole Shrimp)

Tadpole shrimp creep or burrow superficially in soft substrates much of the time (Pennak, 1953) and also spend much time swimming gracefully, by wave-like beating movements of the legs. Tadpole shrimp prefer vernal, alkaline, muddy waters in temperate-arctic regions, and range in age from Cambrian to Holocene.

For lack of specific data, I only conjecture here as to the nature of the lebensspuren left by tadpole shrimp (Fig. 19.2G–K; also see Bromley and Asgaard, 1972, Fig. 3). Resting, walking, and ploughing traces of tadpole shrimp would compare favorably with trace fossils made by trilobites, in

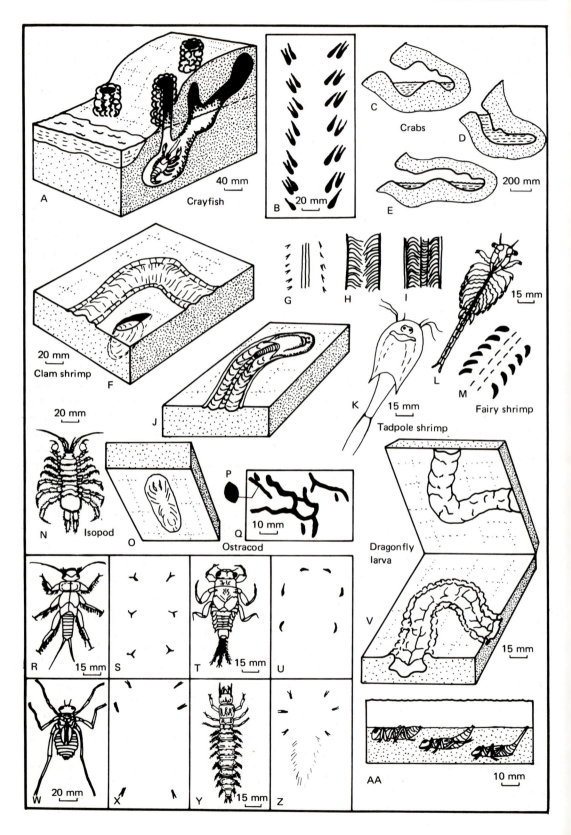

A — Crayfish — 40 mm

B — 20 mm

C — Crabs

D — 200 mm

E

F — Clam shrimp — 20 mm

G   H   I

J

K — Tadpole shrimp — 15 mm

L — 15 mm

M — Fairy shrimp

N — Isopod — 20 mm

O

P

Q — Ostracod — 10 mm

R — 15 mm

S

T — 15 mm

U

V — Dragonfly larva — 15 mm

W — 20 mm

X

Y — 15 mm

Z

AA — 10 mm

which the body morphology is shown in various degrees of fidelity, depending mainly on the activities of the animal. Ploughing by means of the numerous paired legs would leave a striated, bilobate furrow. If the shield-like carapace is brought against the substrate, then lateral grooves would parallel the furrow. If the abdomen-telson is brought in contact with the substrate, a central concave groove would result; and with the paired caudal rami in contact, paired grooves would be produced. Savage (1971) described a resting trace and several crawling trails that compare very favorably with traces that would be left by tadpole shrimp.

Anostraca (Fairy Shrimp)

Fairy shrimp occupy ephemeral puddles and ponds, and include the brine shrimp *Artemia salina* of the Great Salt Lake. They range from Oligocene to Holocene. Fairy shrimp are swimmers that make graceful movements by wave-like beating of the legs; they normally swim on their backs.

Again, no specific information is available on their lebensspuren; my speculation is based on body morphology, in order to suggest something of possible traces. If these forms brushed the bottom, with the ventral side down rather than the dorsal, they would leave a series of paired appendage marks, and perhaps a median paired groove as the abdomen-telson

dragged the substrate (Fig. 19.2L, M). These traces would be distinguishable from tadpole shrimp trails by having shorter and stouter appendage marks. The trace fossil *Umfolozia* (Savage, 1971, Figs. 5, 6) shows the type of trail one might expect from fairy shrimp.

Crustacea: Isopoda (Aquatic Sow Bugs)

Allee (1929) studied isopod aggregations in a stream. The animals oriented themselves up–current, making resting traces before they made another attempt against the current. The details of this lebensspur are not known; it would conform generally to the morphology of the isopod, and would have a bilobate, oval shape (Fig. 19.2N, O).

Crustacea: Ostracoda (Seed Shrimps)

Ostracods inhabit all types of waters and substrates, and are very cosmopolitan. Most species do prefer quiet mud bottoms, less than 1 m deep. Typical animals are less than 1 mm long and seldom more than 3 mm. One South African freshwater species attains a length of nearly 8 mm (Pennak, 1953).

Most species of ostracods are nektonic animals, but some creep or scurry along the substrate. Species of the Candoninae burrow as deeply as 5 cm in soft substrates, but most remain within 2 cm of the surface. The burrows are ramifying tubes having an

◀ **Fig. 19.2** Selected arthropods and their lebensspuren. A, crayfish dwelling burrows and stacks, the latter being incidental to construction of burrows. B, theoretical crayfish tracks; single slashes made by 4th and 5th pereiopods and double ones by the 1st and 3rd pereiopods. C–E, crab burrows made in river banks (after Peters and Panning, 1933). F, trail of clam shrimp and burrow (lower right corner) (after Tasch, 1964). G–J, possible trails of tadpole shrimp: G, assuming very light contact of appendages and telson with substrate; H, interpretation assumes ploughing with appendages in the substrate; I–J, assume deep ploughing, the edge of the carapace and the telson leaving furrows. K, dorsal view of *Apus*, a tadpole shrimp. L, *Branchinecta paludosa*, a fairy shrimp. M, possible trail made by fairy shrimp. N, *Asellus communis*, an isopod. O, probable resting trace of aquatic isopod. P, cross section of ostracod burrow, enlarged. Q, cross section of ostracod burrow system. R, stonefly. S, claw and tarsal traces of a stonefly. T, mayfly larva. U, claw traces of mayfly larva. V, upper, cast of burrow of dragonfly naiad; lower, burrow of dragonfly naiad in substrate surface. W, dragonfly naiad. X, claw and tarsal traces of dragonfly naiad. Y, *Cory-dalus* larva. Z, resting trace of hellgrammite, showing claw and tarsal traces of appendages. AA, burrowing depth of different genera of dragonfly naiads (after Needham and Heywood, 1929).

oval cross-section only slightly larger than the burrower (Fig. 19.2P, Q). Systematic movement is not evident, and a boxwork of small burrows is quickly developed in suitable substrates.

Voigt and Hartmann (1970) described zigzag-like traces made by ostracods on the bottom of a desiccating pool in a limestone quarry in northern Germany. The ostracods moved forward with the dorsal rim inclined in the direction of movement.

Crustacea: Amphipoda (Scuds)

Amphipods are cosmopolitan in clear, unpolluted waters. They make a variety of lebensspuren, most of which have not been observed sufficiently to permit detailed descriptions. Kozhov (1963) reported that several forms burrow in Lake Baikal, Russia, both in shallow and deep water. He did not provide any further details. Amphipods make two types of crawling traces. Ones made when the body remains vertical consist of a criss-cross of slash marks, as the appendages scrape the substrate. The other form consists of irregular concave plough marks made when the scud chose to crawl across the substrate with its side resting on the substrate. The dwelling tubes described by Mills (1967) and the feeding burrows described by Howard and Elders (1970) from shallow marine environments may be considered as models for similar amphipod structures made in fresh water, until further studies are made. (See Fig. 22.11.)

Crustacea: Decapoda

Astacidae (Crayfish); Atyidae and Paleomonidae (Shrimp)

Crayfish (also called crawfish, crawdads, or "crabs") are common in temperate and tropical zones, and prefer shallow streams and lakes. Dwelling burrows are common (Figs. 19.2A, 19.7A). Some species habitually build burrows only when streams or ponds dry up or temperatures are lowered,

whereas others build them only in wet pastures and marshy areas, and still others do not make burrows but remain in permanent waters (Pennak, 1953). The burrows differ widely in construction, depending on the species, substrate, and depth of the water table. Usually only one entrance exists, although as many as three have been observed. The tube leading from the entrance may proceed vertically, at an angle, or almost laterally in a sloping bank. In some the galleries are branched or irregular, but a chamber is always present at the lower end, where the crayfish remains during the hours of daylight. Certain burrows have a lateral chamber. The depth of a burrow ranges from a few centimeters to as much as 2 to 3 m, and is partially determined by the level of the water table; the chamber must contain water in order to keep the animal's gills wet. Burrows near the edge of a pond or stream are shallow; those farther away are deeper. Except during the animal's breeding season, each burrow houses a single crayfish. Burrows are constructed only at night, and the crayfish brings up pellets of mud and deposits them at the entrance to form a chimney. Such chimneys are approximately 15 cm high, but a few as high as 45 cm have been reported; they do not serve any particular purpose, but simply represent the safest and most convenient method of disposing of the mud pellets (Pennak, 1953, p. 456). [Also see Tack (1941) and Ortmann (1906).] The central tube in these burrows ranges from 1 to 5 cm, and the chimney from 4 to 15 cm, in diameter. The inside of the chimney is fairly smooth but the outside is very knobby, superficially resembling the burrows (*Ophiomorpha*) of *Callianassa major*, the marine ghost shrimp (cf. Fig. 2.2A).

Resting and walking traces have not been reported for crayfish or shrimp but can be inferred from the behavior and morphology of the animal, and would be comparable to those reported for fossil shrimp (Glaessner, 1969). Figure 19.2B is a speculative sketch of a walking trace for a shrimp or crayfish, the pereiopods placed as they might occur in a series.

Potamidae (Crabs)

Freshwater crabs are probably recently adapted to this habitat and have a limited distribution up-river, a few miles from the ocean. They build dwelling burrows several centimeters above water level, in the banks of rivers. The burrows are several centimeters in diameter and tens of centimeters deep (Fig. 19.2C–E). As in crayfish burrows, secondary galleries and large, open chambers are typical. The burrow may be completely filled with water, or only parts of the open chambers may be filled. [See Peters and Panning (1933) and Chace et al. (1959).]

Insecta

Plectoptera (Stoneflies)

Stonefly nymphs are abundant in well-oxygenated waters, particularly lotic environments. Stonefly nymphs generally remain on debris, aquatic plants, or under stones. Stonefly, dragonfly, dobsonfly, and mayfly nymphs or larvae are very similar, and make similar lebensspuren when they occasionally sprawl, creep, or scurry on clastic substrates (Fig. 19.2R–Z). Differentiation between mayfly traces and other lebensspuren is possible, based on preservation of markings made by a single claw on the tarsus of the mayfly leg. Dragonfly, dobsonfly, and stonefly larvae have paired claws (unguis) on the tip of the appendages.

Ephemeroptera (Mayflies)

The dwelling burrows made by some mayflies are much more significant than their sprawling, resting, or creeping traces (Fig. 19.2T, U). The dwelling burrows are horizontal or inclined U forms, and very regularly shaped (Figs. 19.3A, B; 19.6D, F; and 19.7G). [See Carpenter (1928), Ide (1935), Needham et al. (1935), Wesenberg-Lund (1943), Seilacher (1967, Fig. 1).] They occur in fine sand, silt, firm mud, and even in fine sediments between conglomerate clasts. Some are lined with a layer of finer particles. The Us range from 1 to 5 mm in tube diameter and are 1 to 2 cm across; they range from 5 to 15 cm in length. Although the animals occur worldwide in well-oxygenated shallow waters, particular species prefer particular environments. Almost all ephemerids burrow; baetids clamber, swim, and sprawl; and heptageniids are sprawlers in streams.

Odonata (Dragonflies)

In addition to sprawling and creeping traces (Fig. 19.2W, X), some dragonfly naiads burrow or plough shallowly in the substrate (Fig. 19.2V). Among the petalurids and gomphids, this burrowing is a search for aquatic insects, annelids, mollusks, and small crustaceans, on which they feed. The cordulegasterids construct a resting lebensspur, where the naiad awaits its prey. The traces range—with the size of the naiad —from 0.5 to 15 mm across and 10 mm to an indefinite length. The exact nature of the trace is not known, but in general it is a deep, irregular-bottomed furrow. Needham and Heywood (1929) compared the depth of burrowing of different genera, as shown here in Figure 19.2AA.

Dragonflies are widespread in all fresh waters, and are commonly found on the bottoms of ponds, streams, marshes, and shallows of lakes, in unpolluted waters. [See Wesenberg-Lund (1943), Pennak (1953), Smith and Pritchard (1956).]

Megaloptera (Hellgrammites)

The larvae of dobsonflies—called "hellgrammites"—are customarily found along the margins of ponds and lakes and under or between stones; but sometimes they lie buried, or crawl around on muddy substrates (Pennak, 1953). These crawling trails would be superficially similar to those of dragonfly and stonefly larvae (Fig. 19.2Y,

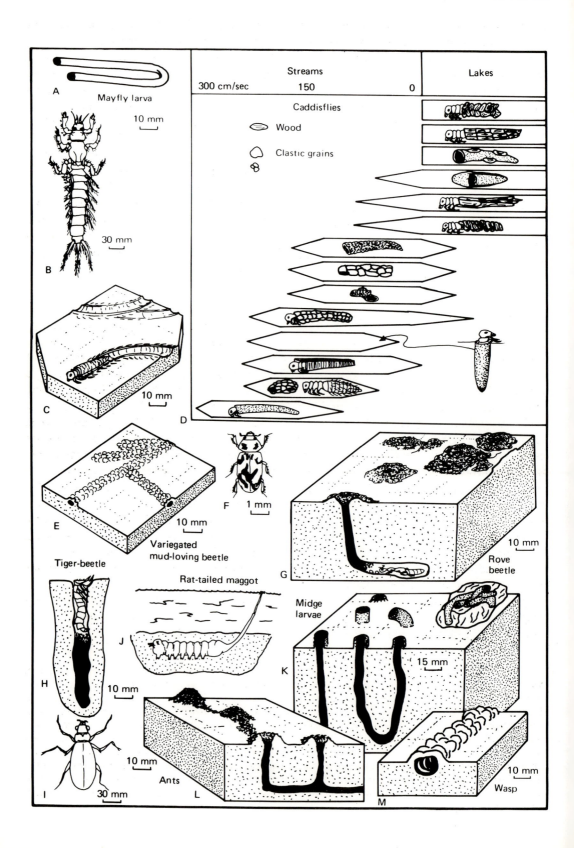

A Mayfly larva

B

C

D

Streams

300 cm/sec   150   0

Lakes

Caddisflies

Wood

Clastic grains

E Tiger-beetle

F Variegated mud-loving beetle

G Rove beetle

H

I

J Rat-tailed maggot

K Midge larvae

L Ants

M Wasp

Z). Resting traces might be distinguished by impressions of the abdominal spiracles, but speculation rather than observation is the basis for my remarks. Pupal chambers are constructed as much as 50 m landward from the water, and may be situated under various objects or 5 to 10 cm within the earth (Pennak, 1953).

Hemiptera: Salidadae (Shore-Bugs)

Comstock (1966) reported that shore-bugs burrow but did not provide details of their burrows. They abound along the shore of streams and lakes, especially in damp soils of marshes near coasts.

Trichoptera (Caddisflies)

Caddisfly larvae sometimes build dwelling cases very meticulously, choosing the materials and fashioning the exact form of the case. Some use only particular types of plants, mineral grains, small abandoned shells, or woody fragments. The animals may agglutinate these particles in (1) helical cases, (2) elongate, curving, or straight tapering tubes having round or square cross-sections, (3) "turtle shell" cases, or (4) structures agglutinated to rocks. Dobbs and Hisaw (1925) studied the relationship between case form and their distribution in lotic and lentic environments; that information is modified here as Figure 19.3D. The same forms may occur in lentic parts of both lakes or streams, and others in the lotic parts; but an intriguing zoogeography is nevertheless present. In general, heavy cases occur in swift lotic environments and plant cases in more lentic environments. Some forms build dwelling tubes by burrowing into the sand on the bottom of streams and cementing the walls of the burrow (Denning, 1956). Crawling trails occur very regularly when the larvae drag their dwelling cases across the substrate (Fig. 19.3C). This lebensspur may be distinguished from other rounded furrows where leg marks are preserved on both sides of the drag furrow.

Caddisfly larvae are widespread in almost all suitably oxygenated streams and lakes. They inhabit all types of substrates but are more prone to occupy those at shallower depths in lakes.

Coleoptera (Beetles)

Carabidae. Silvey (1936) studied the burrows of four species of *Dyschirius*, one of *Bembidion*, one of *Agonoderus*, and two of *Omophron*, all found on the shore of Douglas Lake, Cheboygan County, Michigan. The detailed description of those species are available in Silvey's paper. Some traces are simple inclined tubes; others are irregular or branched, and others are complex, having multiple branches.

Dytiscidae. The little predaceous diving beetle *Hydroporous mellitus* was described by Shelford (1937, p. 102) as burying itself in sand on the bottom of streams. Presumably it creates a resting trace, while it awaits prey; but details of the lebesspur were not given.

Georyssidae. The minute mud-loving beetle *Georyssus pusillus* is approximately 1.7 mm long. It is a shore species that dwells

◄ **Fig. 19.3** Selected arthropods and their lebensspuren. A, burrow of mayfly nymph. B, *Pentagenia*, mayfly nymph responsible for burrows found near Bryan, Texas. C, surface trails made by caddisfly larvae. D, environmental range of certain types of cases of caddisfly larvae, showing overlap from lakes to streams of larvae adapted to lotic or lentic conditions. Second example from bottom represents naked forms and turtle-shell forms; fourth from bottom represents tube-making form (modified after Dobbs and Hisaw, 1925). E, burrows made along damp shores by variegated mud-loving beetle, *Heterocerus*. F, adult *Heterocerus flexuosus*. G, rove beetle burrows and mounds (after Smith and Hein, 1971). H, larva of tiger-beetle in burrow, found in shore zone. I, adult tiger-beetle. J, *Tubifera*, rat-tailed maggot, in (?)burrow in mud. K, burrows and surface tubes of midge larvae. L, ant mounds and tunnels, found along river and lake shores. M, burrow of wasp, made in temporary pond.

in mud along banks of rivers and lakes (Borror and Delong, 1955). The nature of the lebensspur has not been reported. According to Comstock (1966), *Georyssus* covers itself with a coating of mud or fine sand.

Heteroceridae. The variegated mudloving beetle *Heterocerus* is a small insect, seldom longer than 2 mm. It makes distinctive burrows in mud and silt along the shores of streams and lakes (Figs. 19.3E, F; 19.5C–D; 19.7B). The burrow is largely superficial—just beneath the surface—and is made by pushing sediment upward bit-by-bit while forging forward. The result is a small tunnel, having striated walls and a hummocky ridge overhead, tracing its course. The animal seems to have no systematic plan in making this feeding burrow. Many perpendicular and angular branches are seen, and these may cross other burrows. In addition to the dwelling burrows, pupal chambers have also been reported. Williams and Hungerford (1911) illustrated urnshaped mud cases of *Heterocerus* sp. that they thought were made by a larval stage. [See Claycomb (1919), Larsen (1936), Silvey (1936), and Wesenberg-Lund (1939).]

Hydraenidae. Leech and Chandler (1956) and Leech and Sanderson (1959) reported that some hydraeniids tunnel in damp sand near streams and that the larvae are predaceous, occurring also in the damp sand and mud at the edge of the body of water.

Hydrophilidae. Laccobius, a water scavenger beetle, crawls or dabbles for concealment in mud at the water's edge (Pennak, 1953).

Noteridae (or Noterinae). The burrowing water beetles have fossorial larvae that burrow and dig through mud around the roots of aquatic plants (Leech and Chandler, 1956).

Ptilodactylidae. Like the noteriids, some of the ptilodactylids burrow into the substrate in order to feed upon roots of water plants (Leech and Chandler, 1956).

Staphylinidae. Dwelling burrows of

Bledius, one of the rove beetles, were described by Smith and Hein (1971) and mentioned by Leech and Chandler (1956). Smith and Hein observed the burrows in sandy areas after receding flood waters, along the Platte and Loup Rivers in eastern Nebraska. Small sandy mounds were heaped up by *Bledius* as spoil from the excavations. Smith and Hein showed the galleries to be first inclined, in shallow parts, and then horizontal, in deeper parts, but did not provide any further details. If continued burrowing was to be maintained in search for food, then the burrows would probably be more complex than that shown in Figure 19.3G.

Cicindelidae. Larvae of *Cicindela hirticollis*, the beach tiger-beetle, have been observed to build simple vertical dwelling tubes in moist areas on the shore of streams and lakes (Shelford, 1937). The tubes are as much as 15 cm deep and a few millimeters across (Fig. 19.3H, I). Wallis (1961) reported that *C. hirticollis* prefers dry sand. *C. repanda* and *C. duodecimgutta* burrow in heavy, moist soil along river banks—especially mud flats and sandy bars—as much as 30 cm or so from the water's edge. *C. oregona* prefers margins of lakes and streams, in clay or sandy soil. *C. nevadica* was found on wet mud along the margins of saline or alkaline lakes and streams. Wallis did not describe the burrows, except to write that *C. repanda* burrows are approximately 15 cm long.

Adults make shallow burrows at night or in the heat of day. The aestivation burrows of adults and larvae are about equally deep. When building hibernation chambers, the adult initially burrows several centimeters, throwing the soil out behind; then the soil is packed in the burrow behind the beetle as it burrows deeper. The burrow is kept large enough for the animal to turn around. In the chamber at the bottom, enough room remains both for turning and for backfilling when the beetle ends its hibernation. (See also Fig. 2.1.)

Diptera (Flies)

Syrphidae. Rat-tailed maggots are the larvae of flower or hover flies. *Tubifera* burrows into mud or silt, but details of the burrow are not known. They are limited to shallow water (a few centimeters) because they extend their elongate caudal respiratory tube to the surface (Fig. 19.3J) [see Wigglesworth (1964, Fig. 21)].

Tabanidae. Horsefly larvae burrow in the substrate in order to feed on organic matter or on snails, oligochaetes, and insect larvae. Details of the burrow are not known, but it probably reflects a random search pattern. The larvae are cosmopolitan, some even occupying swift waters; but most are found in shallow muddy waters of ponds and swamps (Pennak, 1953).

Chironomidae. Blood worms, or midge larvae, are widely distributed in sluggish streams, ponds, and lakes, and even in fine sediments dispursed among gravels in swift streams. They construct dwelling tubes 0.5 to 3 mm across and as much as 15 cm deep (see Figs. 19.3K—left side, 19.5F, 19.7I). Most tubes are irregular U forms having two entrances, but some seem to be blind ends or juxtaposed, vertical tubes lacking another outlet. Tubes are extended above or onto the surface by the agglutination of organic detritus, algae, or fine-sand and silt grains. Under certain conditions (e.g., low water or oxygen-poor habitats) the tubes are constructed irregularly for several centimeters across the substrate surface. Many tubes are fixed to stones or plant debris (Fig. 19.3K—right side).

Curry (1954) studied midge larvae in Hunt Creek, Michigan, and observed definite substrate preferences exhibited by different genera. Three made tunnel-like chambers in sand masses fixed to mossy stones, in lentic waters, and twelve genera preferred lotic waters. Larvae in lotic waters built tubes in mud and sand substrates.

Hymenoptera: Formicidae (Ants) and Sphecidae (Wasps)

Ants occur down to the saturated edge of many streams and lakes, where they build small mounds, shafts, and shallow galleries (Figs. 19.3L, 19.7E). These galleries are irregular, extend indefinitely, and range to depths of 5 to 10 cm. They are linked to the surface mounds by irregular vertical shafts spaced 3 to 10 cm apart. The galleries average approximately 3 mm across. Although some food is probably encountered during construction of the galleries, extensive above-ground foraging is probably the main means of food gathering, and the galleries are mainly for dwelling.

In an ephemeral puddle, I observed an unidentified wasp tunneling into the cohesive mud, by scraping the mud and removing it, apparently to build pupal chambers (Fig. 19.3M). During the process, the tunnel was constructed more than 60 cm long; it had scrape marks on the inside and a hummocky ridge tracing its course on the surface. The structure was almost identical to that made by mole crickets, discussed below.

Orthoptera (Crickets)

Gryllotalpinae. The mole crickets (Fig. 19.4D) are widespread and build dwelling tunnels in moist sand or mud, particularly along moist margins of streams and ponds (although many gardeners will declare that they are most common elsewhere). Comstock (1966) stated that their burrows extend 20 to 30 cm below the surface, but those observed by me near waterways generally have galleries just beneath the surface. Some do dip down several centimeters under obstacles, or to terminate in a resting chamber. Near the water's edge, they have been observed to branch repeatedly—the animal apparently processing the sediment adjacent to water level (Figs. 19.4A, B; 19.6B). The gallery is constructed by pushing the sediment forward and lift-

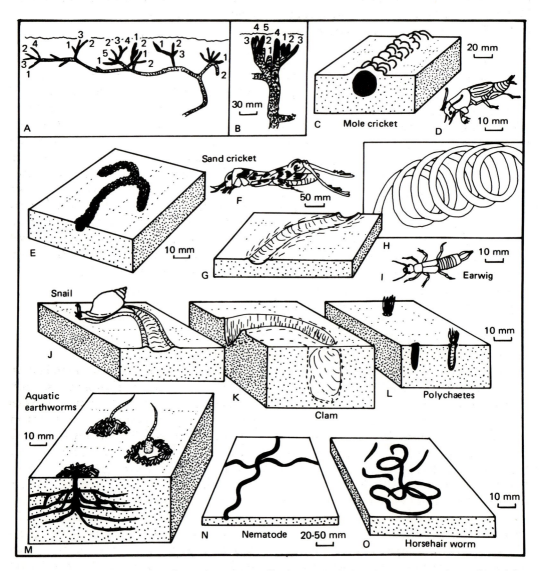

Fig. 19.4 Selected worms, arthropods, and mollusks, and their lebensspuren. A, mole cricket burrows (plan view) along a bayou near Houston, Texas; numbers indicate order of uncovered burrows at edge of stream. B, mole cricket burrows, as in A. C, mole cricket burrow. D, mole cricket *Gryllotalpa*. E, burrows of sand cricket *Tridactylis*, as found in lakes, stream shores, and wet gardens. F, *Tridactylis apicalis*, the sand cricket (or pygmy mole cricket). G, H, trails of earwigs in ephemeral puddles, after a storm; H, a pattern made by an injured earwig. I, adult earwig (Dermaptera). J, snail trail, as in rivers and lakes, showing only simple surface expression. K, trail of bivalve, typical of large rivers and lakes. L, small agglutinated tubes of polychaetes found in lakes or streams. M, *Tubifex*, an aquatic earthworm, extending from small agglutinated surface tube (on right) and surrounded by circular ridge of clastic fecal rods. N, nematode trails. O, possible bottom traces left by horsehair worm (Nematomorpha).

ing it bit-by-bit, so that a hummocky ridge traces the course of the burrow (Figs. 19.4C, 19.6B). The galleries range from approximately 0.5 to 1 cm across and may comprise a continuous system traceable for a few meters. Burrows shown by Hanley et al. (1971) from Seminoe Reservoir, Wyoming, and those by Frey and Howard (1969, Pl. 4, fig. 4) are mole cricket burrows.

Tridactylidae. The sand or pygmy mole cricket *Tridactylus* (Fig. 19.4F) builds distinctive dwelling burrows in moist sand or mud, particularly on shores or bars of streams and lakes. The burrows described by Blatchley (1920) occurred in sandy margins of ponds; the upper parts were vertical, and lower parts ran horizontally —not more than 3 cm below the substrate surface. The latter were approximately 1 mm in diameter.

Urquhart (1937) observed *Tridactylis apicalis* in northeastern Toronto throughout the year. Hibernation chambers were found 45 to 60 cm below the surface, in a soft sand underlying a sandy clay. Brood chambers were observed at another time; a tunnel extended to depths of 2 to 4 cm below the surface, and widened out into a small chamber at the far end. Solitary females were found inside, guarding 10 to 27 eggs in each batch.

In Texas, the lebensspuren of sand crickets seldom went below the surface when I observed them in July and August. Sand crickets there built a superficial burrow by working a few grains of sand into small clusters, using their maxillary palpi, and adeptly sticking the balls together in an arch over the excavation (Figs. 19.4E, 19.5A, B). The burrows were laid out irregularly, branching perpendicularly or angularly, and curving or looping. Most of them were less than 10 cm long, and the width of an individual tunnel was less than 5 mm. I have seen regions adjacent to the Brazos River, Texas, that were extensively worked in a zone a meter or more wide paralleling the shore. In wet muds, sand crickets observed along the Brazos and the Hocking River of Ohio built wider burrows —as much as a centimeter across and ramifying through an area of several centimeters. The surface was hummocky, suggesting that it had been lifted bit-by-bit

rather than being built by piecing separate mud balls together.

Dermaptera (Earwigs)

Among the fortuitous trails made in ephemeral rills and puddles are those of earwigs (Fig. 19.4G–I). After a severe storm, I observed several crawling trails that consisted of a round furrow approximately 4 mm across, having appendage marks along the side. Another, more intriguing pattern was made by injured earwigs; they made erratic circular traces that were superficially reminiscent of many of the complex grazing patterns of marine animals noted in the geologic record [e.g., *Cruziana semiplicata* (Seilacher, 1970, Pl. 1)].

Mollusca

Gastropoda (Snails)

Snails are cosmopolitan and abundant in almost all freshwater environments. Most of their trails seem to be simple furrows corresponding to the width of the foot of the animal (Figs. 19.4J, 19.8B—left). Details of their burrows and trails are not known. Based on the complex structure of the trace fossil *Scolicia*, attributed mainly to snails, a great deal of variability is probably present in freshwater trails. Commonly, the trace involves more than a ploughing of the substrate as the foot moves peristaltically across it. Often, layers of sediment are indiscriminately pushed aside or behind the foot. In some forms, the shell may be carried in such a way that it also makes a print in the substrate; in others, a continuous fecal string may be part of the trail. Whether the snail is on top of the substrate, or partially or totally concealed within it, also makes a difference. Obviously, a great deal of study is needed in order to determine the exact nature of the burrows and trails. Aestivation and hibernation chambers several centimeters deep are made by some forms; again, the characteristics

Fig. 19.5 Arthropod lebensspuren. A, B, burrows of *Tridactylis* in shore of bayou, near Houston, Texas. Small rod-like burrow at arrow, in A. Many such burrows present in wet and dry sediments along the stream have survived several rains. (10 mm scales.) C–E, burrows of *Heterocerus*, the variegated mud-loving beetle. D, from shore of Brazos River, Bryan, Texas; bird tracks also present. (Quarter-dollar coin for scale.) C, E, from bayou: Houston, Texas. F, midge larvae tubes from Dow Lake, Athens, Ohio; larva in lower left, at arrow.

and variability of these structures are unknown.

Bivalvia (Clams and Mussels)

Clams and mussels are widespread in unpolluted fresh water. They prefer stable gravel, sand, and mixed sand-silt substrates in the shallows of large rivers. The lebensspuren are essentially grooves ranging from a few millimeters to several centimeters in width, depending on the age and species of the clam (Figs. 19.4K, 19.6E, 19.7H, 19.8B).

Pryor (1967) studied clam burrows on point bars in the Whitewater River of western Ohio and Wabash River in western Indiana, and found distinctive patterns. His observations are summarized in a subsequent part of this chapter.

ASSEMBLAGES OF NONMARINE AQUATIC LEBENSSPUREN

Information available on the zoogeography of freshwater trace-making animals is both limited and widely dispersed in texts and journals. No general synthesis has been made (cf. Chapter 2, Table 2.1), and consequently, only a few generalizations and speculations can presently be made about these assemblages. Northern Hemisphere, temperate lakes and streams were selected —mainly because of availability of data at the time of writing—and great limitations are inherent in the particular selections.

Ephemeral Ponds and Lakes

Branchiopods are highly selective as to the ponds they inhabit, and seldom occur where carnivorous animals persist. Consequently, one or only a few species seem to characterize particular ephemeral, saline or alkaline, muddy ponds, puddles, or lakes. A distinctive and easily described assemblage of freshwater lebensspuren should occur in such lakes, and should include the scrape, crawl, or plough marks of tadpole shrimp, clam shrimp, and (or) fairy shrimp. The trace fossil assemblage described by Savage (1971) from a late Paleozoic varvite in Natal is not too different from those both known and to be expected in similar Holocene lakes and ponds.

Shores

River and lake shores seem to be dominated by a beetle-trace assemblage that includes larval tubes of the tiger-beetle *Cicindela*, dwelling burrows of the variegated mud-loving beetle *Heterocerus*, and dwelling or feeding burrows of species of the rove beetle *Bledius*. The orthopterid mole crickets and sand crickets seem to be almost equally common, and locally the crayfish burrows may form a significant part of the stream-shore assemblage of lebensspuren.

Shelford (1937) recorded the occurrence of beach tiger-beetle larvae (*Cicindela hirticollis*) in moist parts of the shore of Lake Michigan. The tubes were simple vertical forms. No other burrows were mentioned. Wallis (1961) reported *C. hirticollis* from dry sands, and *C. rapanda*, *C. duodecimguttata*, *C. oregona*, and *C. nevadi* from wet shores of lakes or streams.

The rove beetles *Bledius pallipennis* and *B. bellicosus* were found by Smith and Hein (1971) burrowing in sandy bars and anabranches along the Platte and Loup Rivers of Nebraska, following recession of high waters.

Silvey (1936) found eight species of carab beetles, one species of heterocerid, and one rove beetle along lake shores. Both adults and larvae were found to make distinctive burrows in the inner beach—from the water's edge up to the dry-sand line.

Along the shore of the Brazos River, west of Bryan and at Richmond, Texas, and along the shore of various bayous in and around Houston, I observed a distinctive freshwater lebensspuren assemblage that contains the burrows of *Heterocerus*, mole crickets, sand crickets, and ants.

Fig. 19.6 Worm, arthropod, and mollusk lebensspuren. A, nematode trail from bayou near Houston, Texas. B, mole cricket burrows from shore of bayou near Houston, Texas; water edge at top. C, *Tubifex* in tubes; worm at arrow, and fecal rods around tubes. Small dark holes are ostracod burrows, and larger one at upper right is midge larva burrow. D, F, burrows of mayfly larva, *Pentagenia*, from near Bryan, Texas. Burrows are Holocene, made in Tertiary shales. D, paired tube openings; quarter-dollar coin for scale. F, plan view of paired tubes (horizontal), the basal U missing. E, burrow of clam *Anodontoides*, from Richmond, Texas, along Brazos River; refuse for scale.

Classical U-shaped burrows of mayfly nymphs are very abundant west of Bryan. Although burrows are made below water level, they can be collected above the water line where Holocene burrows in Tertiary shales are exposed at low water. Bivalves burrowing near the shore and just below the water line also may be considered as part of this shore assemblage. Various birds and mammal tracks are present, and along the smaller, less permanent waterways, numerous crayfish burrows are found. The traces made by crayfish, ants, mayfly nymphs, and bivalves are essentially dwelling burrows whereas the coleopteran and orthopteran traces are horizontal feeding burrows. I observed a distinct zonation within this overall shore assemblage (Fig. 19.7). Sand crickets are most abundant in the algae-rich, water-saturated muds and silts (but do not range exactly to the waters edge), sparse in the transition zone to damp sediments, and abundant again in the damp shore zone. Mole crickets are abundant in the saturated muds and silts all the way to the water's edge, and sparse in the damp shore. Variegated mud-loving beetles and

ants occur more or less evenly through the saturated and damp shore, but not down to the water's edge.

Streams

Lateral zonations related to bathymetry may be established in lakes relatively easily; but in streams, few definitive criteria seem to be available for describing assemblages. The work by Dobbs and Hisaw (1925) is an exception, and suggests the kind of results that might be possible with proper study. Through high- and low-energy streams and in lotic and lentic parts of lakes, they observed a distinctive distribution of caddisfly larvae (Fig. 19.3D).

The assemblage of freshwater molluskan lebensspuren described by Pryor (1967) occurred in the bar-foreshores and back-bar sloughs of the Wabash River in western Indiana and the Whitewater River of western Ohio. The larger species and individuals were found on the upstream parts of the bars, where the sediment size is coarser and the flow regime higher. The size of individual mollusks gradually de-

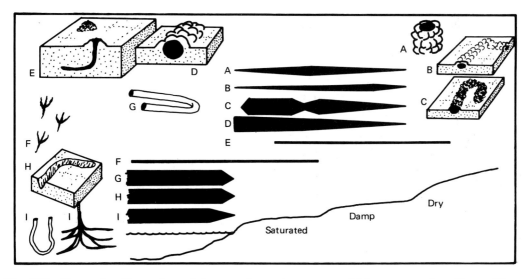

Fig. 19.7 Composite illustration of shore and nearshore range of lebensspuren from Brazos River, near Bryan and Richmond, Texas, and a bayou near Houston, Texas. A, crayfish. B, *Heterocerus,* the variegated mud-loving beetle. C, *Tridactylis,* a sand cricket. D, mole cricket. E, ants. F, vertebrates. G, mayfly larvae. H, bivalves. I, midge larvae (U tube) and aquatic earthworms (dichotomous burrow).

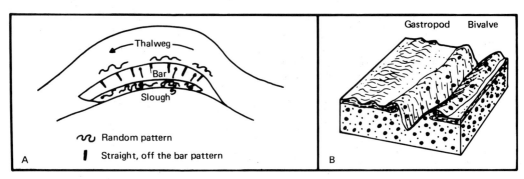

Fig. 19.8 Distribution of mollusk lebensspuren on point bars of the Wabash River, Indiana, and Whitewater River, Ohio. A, straight, off-the-bar traces made as flood waters recede. Irregular and crisscross patterns in the slough occur after water recedes and slough begins to dry. Random patterns nearshore, along the bar, are normal, daily patterns. B, plough-depth patterns observed: snails disturbed top, muddy layer; clams ploughed up coarse material. (After Pryor, 1967.)

creases downstream, toward the quieter parts of the bars. During periods of stable water level, the mollusks burrow randomly below water level (Fig. 19.8A). When the water rises, they climb higher in order to attain their optimum water depth. As the water level drops, the clams plough directly down the length of the bar, toward the water. If they are left behind by the receding water, they burrow. Mollusks caught in a back-bar slough initially burrow randomly, without much crisscrossing of trails; but as the water level drops farther, they make numerous meandering, crisscrossing trails. As the water level drops below the substrate surface, the bivalves burrow in and the snails die. Bivalves in the back-bar slough ploughed deep enough to bring coarser sediment up, in lateral ridges, whereas snails ploughed through only the finer, muddy sediment, leaving a central furrow and two lateral ridges (Fig. 19.8B—left).

Along lentic and lotic parts of the Hocking River of Ohio, Ludwig (1932) observed the distribution of freshwater animals. In the lentic areas, midge larvae and aquatic earthworms were abundant. Snails and bivalves were also present, but less abundantly. In the lotic parts, caddisfly larvae, isopods, and amphipods were more abundant. In a small, abandoned part of the Hocking River—where the stream is approximately 2 m across and normally 20 cm deep—aquatic earthworms are common along the edges, where mud is trapped by cattails and grasses. Snail burrows course randomly through mud, sand, or fine gravel on the bottom. Dragonfly naiads plough randomly through the mud, remaining just below the substrate surface.

Lakes

The distribution of animals that might, or do, make lebensspuren in lakes is similar in Lake Simcoe, Canada; Esrom Lake, Denmark; and Lake Baikal, Russia (Fig. 19.9). Most forms are restricted to shallow depths (a few tens of meters), and include leeches, flatworms, ostracods, isopods, mayfly nymphs, caddisfly larvae, and gastropods. Bivalves and midges extend into intermediate depths (several tens of meters). Amphipods and oligochaetes occur at all depths—from the shore to hundreds of meters (where they are the dominant benthic forms).

As in marine environments, many deposit feeders live in the deeper waters of lakes. Resting, dwelling, crawling, and suspension-feeding behaviors are typical of the animals in shallower waters.

Esrom Lake in Denmark is 8 km long, 2 to 3 km wide, and 22 m deep. It is situated in glaciated terrain, amid forests and farms. At depths greater than 14 m, the substrate is mud; at shallower depths, the substrate

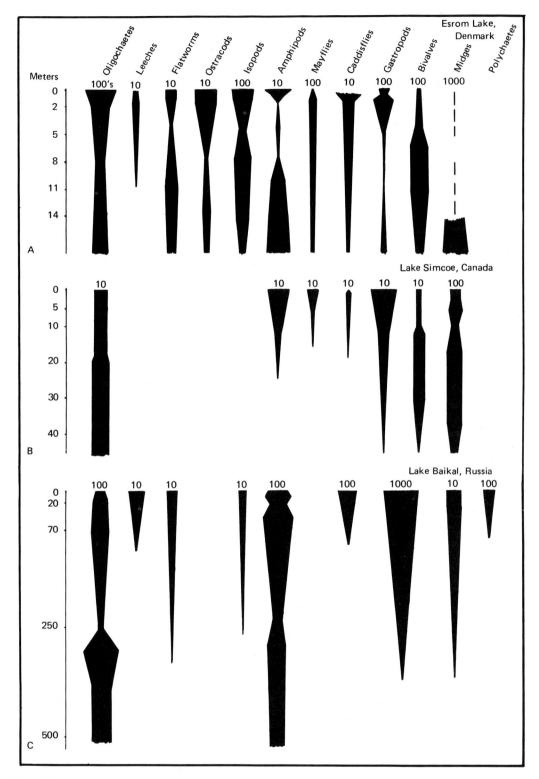

Fig. 19.9 Bathymetric distribution of some possible trace-making invertebrates of certain shallow (A), intermediate (B), and deep-water (C) lakes of northern temperate zones. Available data sketchy, and expressed in units not easily standardized; distributions expressed in relative numbers of specimens.

is sand or sand and gravel. Berg studied the fauna extensively, and the data from his 1938 book, as summarized by Macan (1966), are the basis for Figure 19.9A.

Rawson (1930) studied Simcoe Lake and provided rather good data on this glacial lake, situated just northwest of Lake Ontario. Lake Simcoe is more or less of equal dimensions and covers approximately 200 km². Figure 19.9B summarizes Rawson's data, but only includes those forms that might have left traces.

Lake Baikal is the world's deepest lake, and one of the oldest. It is located in central Siberia; thus, like Lake Simcoe and Esrom Lake, it is in the northern temperate zone. The data in Figure 19.9C are taken from Kozhov (1963).

FRESHWATER VERTEBRATE LEBENSSPUREN

Insufficient information and limited space allow only a cursory report on vertebrate lebensspuren in nonmarine aquatic environments. Consequently, the purpose of this section is to provide—in the context of overall assemblages—at least some idea of the types of tracks, trails, burrows, or other structures made by fish, amphibians, reptiles, birds, and mammals. This topic is covered further in Chapters 14 and 15.

Dwelling structures, feeding traces, hibernation burrows, and nesting structures are made by aquatic or semiaquatic freshwater vertebrates, but shore tracks and trails are probably the most common traces.

Dwelling Structures

The sea lamprey is an extant representative of the first vertebrates, the jawless fish (Agnatha). The young individuals mature inside dwelling burrows in the bottom of streams (Fig. 19.10A; see Applegate and Moffett, 1971). Most dwelling burrows made by other vertebrates are dug into the banks of rivers or lakes. The duck-billed platypus (*Ornithorhynchus*)—a monotreme mammal

—makes a long burrow in the bank, just above water line; the structure has a leaf-lined chamber at the end, and other auxillary galleries branching off (Fig. 19.10B; see Bergaminii, 1967). The water opossum (*Chironectes minimus*) of Central and South America is a marsupial mammal that makes burrows similar to those of the platypus; it is the only marsupial adapted to an aquatic life, having webbed hind feet. Walker et al. (1968) reported at least eight insectivores that are either fully aquatic or semiaquatic, most of which maintain dwelling burrows in the banks of streams or lakes: for example, the rice tenrec (*Oryzorictes hova*) of Madagascar, the otter shrew (giant African water shrew, *Potamogale velox*), the web-footed water shrew (Tibetan water shrew, *Nectogale elegans*), and the Russian desmans (*Desmana moschata*). Beaver, muskrats, and nutria are some of the more familiar aquatic rodents. Beaver and muskrats build lodges, using sticks and mud, and burrow extensively into river banks (Fig. 19.10C). Muskrats use smaller material than do beavers. Nutria burrows, built in banks, consist of a main tunnel having a slightly enlarged chamber at the back (Collins, 1959). The flat-tailed otter (*Pteronura brasiliensis*) is not only a carnivore that is adapted to an aquatic life style, but it also maintains a den in the bank of the river (Walker et al., 1968).

Tracks and Trails

Tracks and trails made on the shores of lakes and rivers are very extensive and diverse, and certainly are not limited to the strictly aquatic or semiaquatic vertebrates; included are the fortuitous tracks of vertebrates merely passing through the area, or more commonly those made on the shore while animals are drinking. Such tracks, mainly mammal, are treated in several texts, such as those by Mason (1943), Collins (1959), Carrington (1963, p. 186–187), and Ormond (1965). Some of the characteristic tracks of aquatic or semiaquatic vertebrates

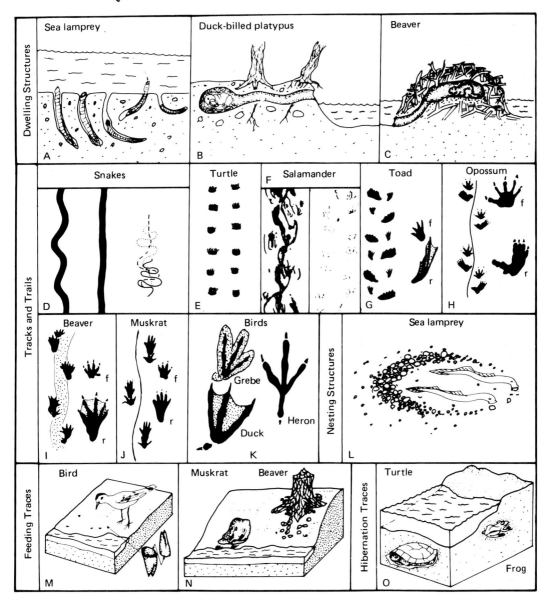

Fig. 19.10 Selected vertebrates and their lebensspuren. A, sea lamprey larvae living in dwelling burrows in bottom of stream, as they mature (after Applegate and Moffett, 1971). B, duck-billed platypus in dwelling burrow, in bank of stream (after Bergaminii, 1967). C, beaver lodge made of sticks and mud, in pond. Muskrats also construct such dwellings, using smaller sticks, mud, and pieces of water plants. D, snake crawling traces, as found along lake or river shores, might include the classic serpentine pattern (left), inchworm crawl (middle), or concertina (right)—solid line= initial state, dashed=forward advance of head, and dotted=forward advance of tail. E, turtle trail through a bayou near Houston, Texas. F, copy of salamander locomotion traces, made on a carbon drum (Evans, 1946). Left trace shows rapid movement, the body on the ground; right trace shows slow movement, the body raised. G, toad walking trace.* H, opossum tracks and trail, commonly found near streams.* I, beaver tracks and trail.* J, muskrat tracks and trail.* K, tracks of aquatic and semiaquatic birds. L, nesting trace of sea lampreys; similar structures are made by sunfish. M, bird feeding traces, made on shores of lakes and streams, consist of conical pits that may be bifurcated. N, muskrat feeding traces, in the form of small burrows or scraped-out areas, and beaver-gnawed stumps and limbs, and wood chips. O, turtle and frogs hibernating, as in substrate of stream or pond. (* f=front, r=rear foot.)

include those of walking catfish; amphibians such as toads, frogs, or salamanders (Fig. 19.10F, G); reptiles such as certain turtles (Fig. 19.10E), alligators and crocodiles, and certain snakes (Fig. 19.10D); numerous birds (Figs. 19.5D, 19.10K); and mammals such as opossums (Fig. 19.10H), beavers (Fig. 19.10I), muskrats (Fig. 19.10J), hippopotamuses, and probably the Baikal seals. Alligators, crocodiles, and hippopotamuses all make extensive wallow holes.

Resting Traces

Hiding or resting traces made within the substrate are probably most characteristic of the lower vertebrates; the traces conform partly to body morphology but mostly to animal movements within the sediment. Frogs, fish, and turtles are commonly seen darting into soft substrates in order to evade predators; some species conceal themselves while awaiting prey.

Feeding Traces

Feeding structures comparable to those made by invertebrates are lacking for vertebrates. Birds commonly make peck marks in the beach as they seek infaunal invertebrates (Fig. 19.10M). Many aquatic animals are herbivores, and while feeding or during construction of lairs, may leave gnaw marks on pieces of wood or plant material (e.g., beaver; Fig. 19.10N); others, such as the muskrat, make irregular burrows in shores and banks as they gather small plants (Fig. 19.10N).

Nesting and Hibernation Structures

Several fish, such as the sea lamprey (Fig. 19.10L), sunfish, tilapia, pumpkinseed, and port, make nesting structures by (1) gathering pebbles into a circle or hemicircle, and scouping out a central depression or (2) by merely brushing out a depression. The stickleback constructs a tubular nest by agglutinating plant material together; some structures include a little sand around the base (Ommanney, 1964).

Excluding the more or less permanent dwellings of the mammals, hibernation or aestivation structures are made mainly by turtles, frogs (Fig. 19.10O), and lungfish.

REFERENCES

Allee, W. C. 1929. Studies in animal aggregations: natural aggregations of the isopod, *Asellus communis*. Ecology, 10:14–36.

Applegate, V. C. and J. W. Moffett. 1971. The sea lamprey. In J. R. Moore (ed.), Oceanography. Readings from Scientific American, W. H. Freeman, San Francisco, p. 391–396.

Bergaminii, D. 1967. The land and wildlife of Australia. New York, Time Inc., 198 p.

Blatchley, W. S. 1920. Orthoptera of northeastern America. Nature Publ. Co., 748 p.

Borror, D. J. and D. M. Delong. 1955. An introduction to the study of insects. Rinehart, 1030 p.

Bromley, R. G. and U. Asgaard. 1972. Notes on Greenland trace fossils. I. Freshwater *Cruziana* from the Upper Triassic of Jameson land, East Greenland. Geol. Surv. Greenland, Rept. 49:7–13.

Carpenter, K. E. 1928. Life in inland waters. New York, MacMillan, 267 p.

Carrington, R. 1963. The mammals. New York, Time Inc., 192 p.

Chace, F. A. et al. 1959. Malacostraca. In W. T. Edmondson (ed.), Fresh-water biology (2nd ed.). John Wiley, p. 869–901.

Claycomb, G. B. 1919. Popular and practical entomology: notes on the habits of *Heterocerus* beetles. Canadian Entomol., 51:25.

Collins, H. H., Jr. 1959. Complete field guide to American wildlife: east, central, and north. New York, Harper & Row, 683 p.

Comstock, J. H. 1966. An introduction to entomology (9th ed.). Ithaca, N.Y., Comstock Publ. Assoc., 1064 p.

Curry, L. L. 1954. Notes on the ecology of the midge fauna (Dipera:Tendipedidae) of Hunt Creek, Montmorenay Co., Mich. Ecology, 35:541–550.

Demoor, J. 1880. Recherches sur la marche des insectes et des arachnides. Étude experimentale d'Anatomie et de Physiologie comparées. Archiv Biologie, 42 p.

————. 1890. Über das Gehen der Arthropoden mit Berücksichtigung der Schwankungen des Körpers. Compt. Rend. Acad. Sci. Paris, p. 839–840.

Denning, D. G. 1956. Trichoptera. In R. L. Usinger (ed.), Aquatic insects of California. Berkeley, Calif., Univ. California Press, p. 237–270.

Dobbs, G. S. and F. L. Hisaw. 1925. Ecological studies on aquatic insects. III. Adaptations of caddisfly larvae to swift streams. Ecology, 6:123–137.

Evans, F. G. 1946. Anatomy and function of the foreleg in salamander locomotion. Anat. Rec., 95:257–281.

Frey, R. W. and J. D. Howard. 1969. A profile of biogenic sedimentary structures in a Holocene barrier island-salt marsh complex, Georgia. Gulf Coast Assoc. Geol. Socs., Trans., 19:427–444.

Glaessner, M. F. 1969. Decapoda, burrows and trails. In R. C. Moore (ed.), Treatise on invertebrate paleontology, Pt. R, Arthropoda 4; V. 2, Decapoda. Geol. Soc. America and Univ. Kansas Press, p. R429–R431.

Graber, B. 1884. Über die Mechanik des Insektenkorpers. Biol. Centralbl., p. 560–570.

————. 1886. Die ausseren mechnischen Werkzeuge der Tiere. II Teil. Wirbellose Tiere, p. 175–182, 208–210.

Gray, J. and H. W. Lissmann. 1964. The locomotion of nematodes. Jour. Experiment. Biol., 41:135–154.

Hanley, J. H. et al. 1971. Trace fossils from the Casper Sandstone (Permian), southern Laramie basin, Wyoming and Colorado. Jour. Sed. Petrol., 41:1065–1068.

Howard, J. D. and C. A. Elders. 1970. Burrowing patterns of haustoriid amphipods from Sapelo Island, Georgia. In T. P. Crimes and J. C. Harper (eds.), Trace fossils. Geol. Jour., Spec. Issue 3:243–262.

Ide, F. P. 1935. Life history notes on *Ephoron, Potamanthus, Leptophlebia* and *Blasturus* with descriptions (Ephemeroptera). Canadian Entomol., 67:113–125.

Kozhov, M. 1963. Lake Baikal and its life. W. Junk, Publ., 344 p.

Larsen, E. B. 1936. Biologische Studien über die tunnelgrabenden Käger auf Skallingen. Skalling-Lab., Medd., 3:1–231.

Leech, H. B. and H. P. Chandler. 1956. Aquatic Coleoptera. In R. L. Usinger (ed.), Aquatic insects of California. Berkeley, Calif., Univ. California Press, p. 293–371.

———— and M. W. Sanderson. 1959. Coleoptera. In W. T. Edmondson (ed.), Freshwater biology (2nd ed.). New York, John Wiley, p. 981–1023.

Ludwig, W. B. 1932. The bottom invertebrates of the Hocking River. Ohio State Univ., Bull., 36:222–249.

Macan, T. T. 1966. Freshwater ecology. New York, John Wiley, 338 p.

Mason, G. F. 1943. Animal tracks. New York, Wm. Morrow, 95 p.

Mills, E. L. 1967. The biology of an ampeliscid amphipod crustacean sibling species pair. Jour. Fish. Res. Board Canada, 24:305–334.

Moussa, M. T. 1970. Nematode fossil trails from the Green River Formation (Eocene) in the Uinta Basin, Utah. Jour. Paleont., 44:304–307.

Needham, J. G. and H. B. Heywood. 1929. A handbook of the dragonflies of North America. C. C. Thomas Publ., 378 p.

———— et al. 1935. The biology of mayflies. Comstock Publ. Co., 759 p.

Ommanney, F. D. 1964. The fishes. New York, Time Inc., 192 p.

Ormond, C. 1965. Complete book of outdoor life. New York, Harper & Row, 498 p.

Ortmann, A. E. 1906. The crawfishes of the state of Pennsylvania. Mem. Carnegie Mus., 2:343–521.

Pennak, R. W. 1953. Fresh-water invertebrates of the United States. New York, Ronald Press, 769 p.

Peters, N. and A. Panning. 1933. Die chinesische Wollhandkrabbe (*Eriocheir sinensis* H. Milne-Edwards) in Deutschland. Zool. Anz., Suppl., 104 p.

Pryor, W. A. 1967. Biogenic directional features on several recent point-bars. Sediment. Geol., 1:235–245.

Rawson, D. S. 1930. The bottom fauna of Lake Simcoe and its role in the ecology of the lake. Univ. Toronto Studies, Biol. Ser., Publ. Ontario Fish Res. Lab., 40:1–183.

Sandstedt, R. et al. 1961. Nematode tracks in the study of movement of *Meloidogyre in-*

cognita incognita. Nematologica, 6:261–265.

Savage, N. M. 1971. A varvite ichnocoenosis from the Dwyka Series of Natal. Lethaia, 4:217–233.

Seilacher, A. 1967. Bathymetry of trace fossils. Marine Geol., 5:413–426.

————. 1970. *Cruziana* stratigraphy of "nonfossiliferous" Palaeozoic sandstones. In T. P. Crimes and J. C. Harper (eds.), Trace fossils. Geol. Jour., Spec. Issue 3:447–476.

Shelford, V. E. 1937. Animal communities in temperate America, as illustrated in the Chicago region—a study in animal ecology. Geog. Soc. Chicago, Bull. 5, 368 p.

Silvey, J. K. G. 1936. An investigation of the burrowing inner-beach insects of some freshwater lakes. Mich. Acad. Sci. Arts, Letters, Pap., 21:655–696.

Smith, N. D. and F. J. Hein. 1971. Biogenic reworking of fluvial sediments by staphylinid beetles. Jour Sed. Petrol., 41:598–602.

Smith, R. F. and A. E. Pritchard. 1956. Odonata. In R. L. Usinger (ed.), Aquatic insects of California. Berkeley, Calif., Univ. California Press, p. 106–153.

Tack, P. I. 1941. The life history and ecology of the crayfish *Cambarus immunis* Hagen. Amer. Midland Natur., 25:420–446.

Tarr, W. A. 1935. Concretions in the Champlain Formation of the Connecticut River Valley. Geol. Soc. America, Bull., 46:1493–1534.

Tasch, P. 1964. Conchostracan trails in bottom clays, muds and on turbid water surfaces. Kansas Acad. Sci., Trans., 67:126–128.

Urquhart, F. A. 1937. Some notes on the sand cricket *Tridactylus apicalis* Say. Canadian Field Nat., 51:28–29.

Voigt, E. and G. Hartmann. 1970. Über rezente vergipste Ostracodenfährten. Senckenbergiana Marit., 2:103–118.

Walker, E. P. et al. 1968. Mammals of the world (2nd ed.). Baltimore, Johns Hopkins Press, 1500 p.

Wallace, H. R. 1968. The dynamics of nematode movement. Ann. Rev. Phytopathol., 6:91–114.

Wallis, J. B. 1961. The Cicindelidae of Canada. Toronto, Ont., Univ. Toronto Press., 74 p.

Wigglesworth, V. B. 1964. The life of insects. New York, World Publ. Co., 360 p.

Williams, F. X. and H. B. Hungerford. 1911. Notes on Coleoptera from western Kansas. Ent. News, 25:1–9.

Wesenberg-Lund, C. 1939. Biologie der Süsswassertiere; wirbellose Tiere. Wien, Julius Springer, 817 p.

————. 1943. Biologie der Süsswasserinsekten. Copenhagen, Glydendalske Boghandel. Nordisk Forlay., 682 p.

CHAPTER 20

RECENT BIOCOENOSES AND ICHNOCOENOSES IN SHALLOW-WATER MARINE ENVIRONMENTS

JÜRGEN DÖRJES AND GÜNTHER HERTWECK

Institut für Meeresgeologie und Meeresbiologie "Senckenberg"
Wilhelmshaven, Germany

SYNOPSIS

Interrelated biocoenoses, ichnocoenoses, and ichnofacies in different shallow-marine environments are discernible, based on the relations between environmental conditions and macrobenthic organisms; these are very important ecologically and ichnologically.

The first part of this chapter (Biological Aspects) is devoted to the most important ecological parameters influencing animal life in these environments, and to the main trends in faunal distribution. Comparisons of the distribution, abundance, and zonation of macrobenthic animals between the area above the high-water line and shallow shelf of the North Sea (German Bight), Mediterranean Sea (Gulf of Gaeta), and Georgia coastal region (Sapelo Island), reveal the following general trends: (1) supralittoral: very few terrestrial and marine species, relatively few individuals; (2) eulittoral: few marine species, although some are represented by numerous individuals; (3) upper sublittoral: slightly more marine species, very few individuals; (4) middle sublittoral: numerous species and numerous individuals; and (5) lower sublittoral: fewer numbers of species and individuals. Certain "parallel" communities (iso-communities) were also observed in the three areas investigated.

The second part of this chapter (Ichnological Aspects) is concerned with the distribution, abundance, and zonation of lebensspuren. The total number of lebensspuren depends mainly on the numbers of trace-making species and individuals occurring in each environment. The greatest abundances of animals and traces are observed in the subtropical coastal region of Georgia, the lowest abundances in the temperate-boreal North Sea.

Rates of sedimentation and sediment reworking are important in determining the actual preservability of the lebensspuren. For example, traces are much less likely to be preserved in the high-energy conditions of the upper sublittoral zone than in the quieter conditions of the middle or lower sublittoral.

In most of the environments studied, characteristic lebensspuren provide a conspicuous ichnofacies. Such ichnofacies are mainly dependent on the preservability of certain traces as characteristic lebensspuren, under proper conditions. Therefore, to compare or correlate ichnofacies occurring in corresponding environmental zones of different geographical regions is difficult. A comparison of similarly constructed lebensspuren from different regions, however, leads to the distinction of types of lebensspuren

---

The authors are here listed alphabetically; respective parts written by each are indicated in the text.

that reflect ecological conditions or animal adaptations, mainly the trophic relationships of the trace-producing organisms. Such lebens- *spuren show a distinct zonation in a generalized beach-offshore profile.*

BIOLOGICAL ASPECTS

Jürgen Dörjes

INTRODUCTION

In addition to systematical or morphological problems, marine zoologists have traditionally tried to investigate the various relations between organisms and the different parameters of their environments (e.g., Allen, 1899; Petersen, 1913, 1914, 1918; Davis, 1923, 1925; Ford, 1923; Hagmeier, 1923; Linke, 1930; Thorson, 1957). (See also Howard et al., 1971.)

With the inception of marine ecology, the sea bottom and the water column (benthic and pelagic environments, respectively) were established as the two main realms of the oceans (Hesse et al., 1951). In ichnology, only the sea bottom—whether rocky, or covered by soft sandy, muddy, or mixed sediments—is primarily important.

In this part of the chapter I wish to discuss the distribution and abundance of recent macrobenthic animals from different soft-bottom littoral environments, and to point out the main relationships between such local faunas and their different ecological conditions, as we know them.

CLASSIFICATION OF BOTTOM ENVIRONMENTS

First, I must outline a distinct division of the main shallow-water marine bottom environments located between the supratidal backshore and the edge of the continental shelf (Fig. 20.1). In contrast to the classifications given by Zenkevich (1951, 1959), Pérés (1957), Hedgpeth (1957), and Ager (1963), the total shelf area influenced by salt water is here termed the littoral zone and

can be conveniently divided into three major environments: (1) the supralittoral, above mean high water, is more or less always emerged but is influenced by saline ground water and wind-driven salt water, or is sporadically submerged during storm tides; (2) the eulittoral, located between mean high and mean low-water line, is characterized predominantly by tidal processes and surf action; this zone is rhythmically emerged during low tides and submerged during high tides; and (3) the sublittoral, which is always submerged, includes all environments between the mean low-water line and the continental edge.

Faunistic investigations in the sublittoral show that mean wave base, and the border between recent and relict sediments (storm wave base?), are distinct boundaries between animal communities. In this chapter both borders are used to divide the sublittoral into three subenvironmental zones: (1) the upper sublittoral, located between mean low-water line and mean wave base, is mainly influenced by surf, breakers, and strong wave oscillations; (2) the middle sublittoral, located between mean wave base and the boundary between recent and relict sediments, is mainly influenced by currents; and (3) the lower sublittoral is located between the recent-relict sediments border and the continental shelf edge, where high-energy physical processes are less important. (See Table 20.1.)

SOFT-BOTTOM ENVIRONMENTS AND ENVIRONMENTAL PARAMETERS

The supralittoral zone includes all environments above mean high-water line that are

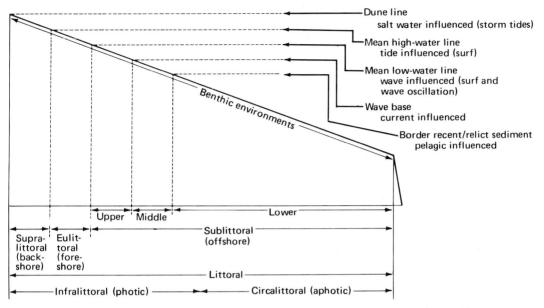

Fig. 20.1 Classification of marine environments. (Compare with Table 20.1.)

somehow influenced by salt water. These sediments, ranging between clean sand and clay, are rarely moistened or submerged by salt water. Generally, the sediments are irrigated by rain or are completely dry. In addition to variations in interstitial water content and salinity, the daily and annual temperature changes are high—even extreme. Therefore, the infauna consists mainly of terrestrial insects (including their larvae); only a few marine species occur here, such as highly mobile amphipods, isopods, and several crabs, which are more or less well adapted to a subaerial life (Pérés, 1966). Virtually all of these crustaceans combat unfavorable environmental conditions by burrowing down to sediment layers that are permanently moist, which insulates them against environmental changes (Dörjes, 1972). (See Chapter 9.)

The eulittoral zone, which is characterized predominantly by tidal processes, includes all environments that are regularly emerged during low tide and submerged during high tide, as well as tidal channels, runnels, and creeks. The changes in environmental conditions are great, but they generally follow a very rhythmic pattern.

No other marine zone embraces such diversity of subenvironments, with such intense changes in environmental conditions, as does the eulittoral. Geomorphologically, the eulittoral includes such features as coastal lagoons, estuaries, tidal flats, salt marshes, barrier beaches, shoals, and other associated shallow-water or intertidal habitats. Each of these subenvironments may be subdivided into smaller units, such as runnels, ridges, bars, banks, troughs, inlets, channels, creeks, and sheltered and unsheltered tidal flats. All of these units may be delimited by abrupt gradients in ecological conditions, such as water content of sediments, types of sediments, salinity, temperature, oxygen, rate of sedimentation, erosion, current velocities, substrate consistency, sediment reworking, availability of food, wave action, and many others.

Many of these environmental units are characterized by dense populations of very few benthic species, which have wide environmental tolerances and are thus highly adapted to the rigorous environmental conditions here. Generally, the number of species and individuals decreases with the increase of wave action and current velocities, and the decrease of salinity and period of submersion.

The sublittoral includes all marine subenvironments between low-water line and the continental edge, and is characterized mainly by a continuous covering of salt water. In its shallow parts, wave action, sediment reworking, and types of sediment mainly influence animal life. Below wave base, the availability of food, current velocities, and different types of sediments are the main factors controlling the numbers of animal species and specimens (Dörjes, 1971).

Along transects investigated in the North Sea, Mediterranean Sea, and Georgia coastal region, running offshore from the low-water line, the number of species and individuals increases with increasing water depth. Below wave base, the location of which depends upon local hydrographic conditions, the numbers of species and specimens reach maxima because sediment reworking and wave action are low and the food supply is more than sufficient. Below this area, the decrease in number of species and individuals per unit sample is probably conditioned by the decrease in food supply (Dörjes, 1972).

Substrate position and animal mobility are also important, e.g., whether a given organism is epifaunal (living upon the sediment surface) or endofaunal (living within the sediment), and whether vagile (free moving), hemisessile (restricted movements), or sessile (fixed to or within the substrate).

In general, Friedrich (1965) and others stressed the importance of "sediment quality" to benthic inhabitants. The epifauna of mud (epipelos) is characterized by vagile polychaetes, isopods, mysidaceans, and certain sea stars. Sessile species are rare, those present being mostly coelenterates. On the surface of sandy sediments (epipsammon), only a very few sessile and vagile species are typically present; worth mentioning are isopods, crangonids, true crabs, and some gastropods. Many more species are endobenthic. In muddy sediments occur many

vagile, sessile, and hemisessile polychaetes, pelecypods, gastropods, sea stars, brittle stars, holothurians, and certain anthozoans. In sandy sediments, vagile faunas are less important. Most of the animals inhabiting these sediments—polychaetes, enteropneusts, crustaceans, pelecypods, and holothurians—are hemisessile or sessile. Many authors mention that with increasing mud content in sediments, the numbers of suspension feeders decrease whereas the numbers of deposit feeders increase.

EXAMPLES OF ZONATION OF MARINE BENTHOS

The occurrence of each species in an environment is conditioned by its demands and tolerances for environmental conditions, i.e., each benthic species and each bottom community is more or less related to a specific environment having specific conditions for life.[1] The following investigations are cited as examples.

Faunistic investigations carried out in the North Sea (Dörjes, in preparation), the Mediterranean Sea (Dörjes, 1971), and on the Atlantic coast of the United States (Dörjes, 1972) show comparable as well as different trends in the distribution of species, specimens, and communities. However, in all three areas, distinct relations between faunal distribution or abundance and environmental conditions were found. Transects were run at right angles to the shoreline, between high-water line and the offshore area; samples of comparable size (0.2 m²) were taken by ship and by SCUBA diving. Important differences among the areas investigated stem mainly from (1) hydrographic conditions, which are responsible for nearshore geomorphological features, distribution of sediments, rates of

[1] Some biologists (Mills, 1969) argue whether a community is really a structural unit or merely a "collection" of individual animals that happen to live together because of their individual responses to given conditions, not because they are inherently interrelated in a functional sense.

sediment reworking by physical processes, and other tidal and current effects, and (2) the mean water temperature, which considerably influences the total number of species (Friedrich, 1965).

The Gulf of Gaeta, located on the west coast of Italy in the Mediterranean Sea, is a long, slender bight having a high-energy coast, longshore bars scattered among nearshore shallow-water areas, and a tide range of only 0.3 m. The width of the shelf, between low-water line and the continental edge, is only 10 km. Therefore, the bottom slope is relatively steep, and the environments, running parallel to the shoreline, are very narrow (Fig. 20.2). The sediments consist of (1) clean fine sand in the supralittoral (backshore), eulittoral (foreshore), and upper sublittoral (including the shoreface); (2) muddy sediments in the middle sublittoral; and (3) pure mud in the lower sublittoral. Runnels and ridges are not

developed in the extremely narrow (1.0 m) eulittoral zone. Nearshore water temperatures fluctuate between 13°C in winter and 24°C in summer.

Sapelo Island, a barrier island located on the coast of Georgia (southeastern United States) is characterized by low energy and large tidal range (2.1 m). Subtidal longshore bars are absent in nearshore shallow-water areas. Instead, migrating ridges and runnels are well developed on the wide intertidal beach slope of the eulittoral (foreshore) (Figs. 20.3, 20.4). The shelf, having a maximum width of more than 100 km, is extremely wide; therefore, the bottom surface dips gently toward the continental edge. The sublittoral (offshore) environments, running approximately parallel to the shoreline and the main axis of the Gulf Stream, are also relatively wide. The sediments consist of (1) clean fine sand in the supralittoral, eulittoral,

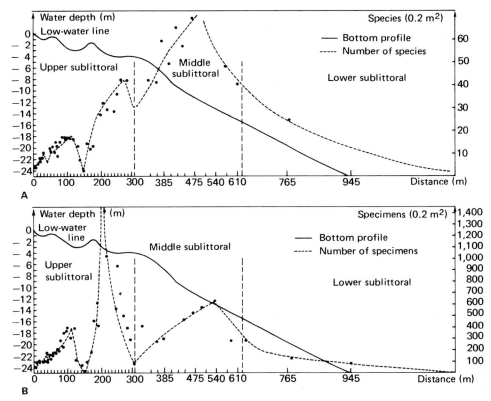

Fig. 20.2 Gulf of Gaeta (Mediterranean Sea): a transect between low-water line and the 25-m-depth line. A, distribution and abundance of species. B, distribution and abundance of individuals.

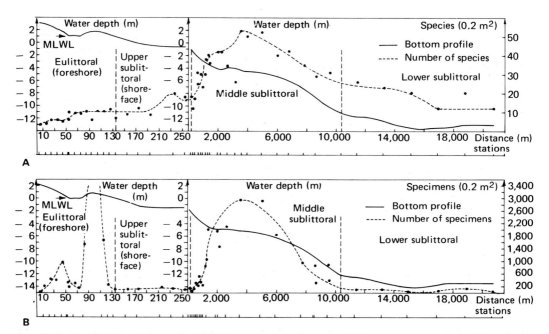

Fig. 20.3 Sapelo Island (Georgia, U.S.A.): a transect between high-water line and the 15-m-depth line. A, distribution and abundance of species. B, distribution and abundance of individuals.

and upper sublittoral; (2) muddy sediments in the middle sublittoral; and (3) coarse relict sands in the lower sublittoral. Water temperatures vary seasonally, between approximately 8°C in winter and 30°C in midsummer.

Norderney, another barrier island, is located in the German Bight, which is part of the North Sea. This high-energy coast exhibits a series of nearshore, subtidal, longshore bars and troughs in the shallow-water area, and migrating ridges and runnels in the eulittoral (Fig. 20.5). The tidal range averages 2.4 m; consequently, the intertidal eulittoral zone is extremely wide. The shelf, extending between the European mainland and the continental edge north of the Dogger Bank, includes the entire North Sea. Because of the exceptional width of the shelf, steep slopes on the sea bottom are more or less developed only between the shoreline and the 20-m-depth line. Sediments in this area consist mainly of clean fine sand or coarse relict sand. Muddy sediments are very rare because of the low supply of fine organic or inorganic detritus

in this part of the North Sea. Water temperatures vary between approximately 3°C in winter and 19°C in summer.

In spite of obvious differences in the hydrographic, topographic, physical, and chemical conditions of the different areas of investigation, the distribution, abundance, and zonation of macrobenthic animals reveal many comparable trends along the transects (Figs. 20.2–20.5). Only a very few terrestrial and marine migrant species live in the supralittoral, or landward of the high-water line. All of these migrants are either crustaceans or insects, which are highly adapted to a quasi-terrestrial life. In the eulittoral, very few marine species (sometimes represented by high numbers of individuals) are able to live; the conditions for life are very rigorous here as a result of tide and surf influences. Below these zones the numbers of individuals and species increase almost linearly with increasing water depth and decreasing surf influence, until they reach maxima a short distance offshore. Interruptions in this otherwise regular increase are

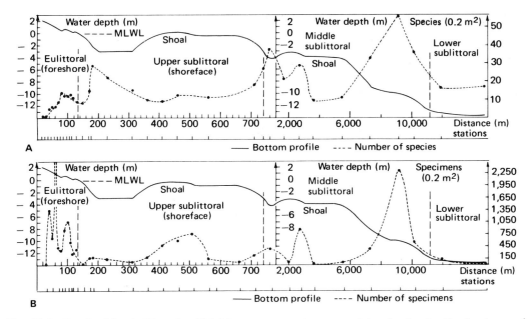

Fig. 20.4 Sapelo Island (Georgia, U.S.A.): a transect interrupted by shoals. A, distribution and abundance of species. B, distribution and abundance of individuals.

caused by shoals, ridges, runnels, and long-shore bars, where environmental conditions differ from those of the adjacent areas. Off-shore the numbers of species and individuals decrease again if the supply of food decreases (Figs. 20.2–20.5).

Further conspicuous similarities among the three areas studied include: (1) the dominance of polychaete species, followed in importance by mollusks and crustaceans; (2) the exceptionally high proportion of individual polychaetes, mollusks, and crustaceans relative to the total number of individuals (Norderney 99 percent; Sapelo 98 percent; Gaeta 95 percent); the next biggest taxonomic group is comprised by echinoderms (which are represented by only 0.3 percent each); (3) the dominance of only a very few species; the 10 most common species include 75 percent or more of the total number of individuals (Norderney 88 percent; Sapelo 88 percent; Gaeta 75 percent); and (4) the respective maxima of individuals and species are located at the same water depth, although a second conspicuous maximum of individuals, comprised by very few species, may be located

in the intertidal area or near low-water line (Figs. 20.2–20.4).

Distinct faunistic differences among the areas of investigation are also apparent. Influenced by the different water temperatures, only 68 species were found along the Norderney transect between high-water line and the 20-m-depth line, whereas 152 species were found along the Gaeta transect and 179 along the Sapelo transect. This relation between increasing water temperature and increasing number of benthic species, or the increase of species from arctic seas to the tropics, is a well-known phenomenon (Friedrich, 1965).

In the different areas reported here, the maxima of species and individuals, developed along all transects investigated, are located at different water depths due to different hydrographic conditions. As became apparent from the three studies, the maxima of species and individuals are located just below wave base, where the rate of sediment reworking by waves is low and the supply of detrital food high. In contrast to Sapelo, the coasts of Gaeta and Norderney are characterized by high energy;

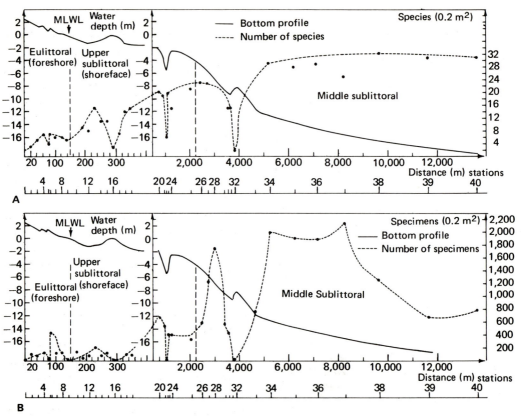

Fig. 20.5 Norderney Island (North Sea): a transect between high-water line and the 19-m-depth line. A, distribution and abundance of species. B, distribution and abundance of individuals.

the influence of surf and wave activities is correspondingly greater. In Norderney, tidal currents reinforce wave activities so that the net wave energy is higher here than in the Gulf of Gaeta. Therefore, the maxima along Sapelo transects are located at a water depth of only 5 to 6 m (Figs. 20.3 and 20.4), whereas they are located in the Gulf of Gaeta at a water depth of 11 to 12 m (Fig. 20.2) and off Norderney at a water depth of 16 to 17 m (Fig. 20.5).

Smaller differences exist in the total number of individuals from the three areas, based upon an equal number of samples. In Gaeta and off Norderney, approximately 22,000 individuals were found; in contrast, 48,000 individuals were recovered from the Sapelo samples. Because of the extremely narrow foreshore and very steep bottom slope in the Gulf of Gaeta, the increase in number of species is abrupt (Fig. 20.2).

Also, equal numbers of polychaetes, mollusks, and crustaceans are included in the total number of individuals here, whereas on Sapelo and Norderney the polychaetes are highly dominant.

Generally, however, the following patterns may be discerned among the different environments of these sand-barrier beaches and shallow-shelf areas:

1. *Supralittoral and supralittoral–eulittoral transition:* very few terrestrial and marine migrant species; relatively few individuals, which are more or less well adapted to subaerial life.
2. *Eulittoral beaches and eulittoral shoals:* few marine species, some represented by numerous individuals, belonging mostly to the crustaceans. The limited number of species living in these environments results mainly from tide influences and wave activities. Only a very few species are adapted to these rigorous environmental conditions. With increasing protection by sand ridges or other

geomorphological structures, which reduce surf action and sediment reworking, the number of species increases.

3. *Upper sublittoral* (including the shoreface): slight increase in the number of species, due to permanent submersion. Very few individuals, due to the high rate of sediment reworking by waves and surf.

4. *Middle sublittoral:* numerous species and numerous individuals; these reach maxima below wave base, where surf and wave activities are comparatively insignificant and the supply of food is high. Fine-grained organic and inorganic material, brought by rivers, sounds, and inlets, are deposited here if a sufficient source is available.

5. *Lower sublittoral:* decreasing numbers of species and individuals, due to the decreasing availability of food; this depends upon water depth and agitation as well as upon distance from the mainland. Generally, a decrease in the number of species and individuals is known from the littoral (0 to 200 m) to the bathyal (200 to 4,000 m), abyssal (4,000 to 5,000 m), and hadal (5,000 to 11,000 m) zones of the oceans [Thorson (1957), Hedgpeth (1957), Ager (1963, Fig. 2.3)], although Hessler and Sanders (1967) obtained different results.

ANIMAL COMMUNITIES

Benthic organisms are grouped into various ecological units (see Remane, 1940; Thorson, 1957). The composition of these units is conditioned by the tolerance, pretension, and adaptation of each species to the existing physiochemical environmental factors and also by the relations existing between different species. Therefore, marine environments having comparable or similar ecological conditions are inhabited by the same groups of animal species, forming an animal community.

The significance of given species within the community is highly varied, due to such things as complex life cycles, average duration of life, and degrees of tolerance for environmental conditions and other species inhabiting the same environment. Abundant and widely distributed species that occur throughout the year in an environment and are more or less limited to the

environment are characteristic species of the community. Each community ideally should have several characteristic species.

Thorson (1957), as well as Fischer (1960), mentioned that only the Arctic, Antarctic, and deep-sea communities are exceptionally stable. In contrast, tropical communities are rather unstable due to the long pelagic life of most inhabitants. Communities of different geographical regions, inhabiting environments that are controlled by many similar ecological factors, are characterized mostly by different species of the same or closely related genera. Groups of such closely related communities are known as "isocommunities" or "parallel communities" (Thorson, 1957).

Results of the investigations on Norderney and Sapelo Island and in the Gulf of Gaeta (Dörjes, 1971, 1972, in preparation) show several conspicuous similarities with regard to bottom communities. The supralittoral of the barrier beaches investigated are generally characterized by crustaceans. As presently known, true supralittoral crabs occur only in tropical and subtropical environments; they are absent at higher latitudes (Pérés, 1966). Therefore, among the areas investigated, a true crab community occurs only in the supralittoral of Sapelo; in the same environment of Norderney and Gaeta, amphipod and isopod species are dominant.

Distinct eulittoral beaches are developed only on Sapelo and Norderney, due to the high tidal range. Both areas are characterized mainly by haustoriid amphipods, after which the communities are named. These communities are isocommunities with respect to the well known *Bathyporeia-Haustorius* community (see Remane, 1940). On sandy sediments, these are succeeded sublittorally by communities that are named after tellinid bivalves (pelecypods) of the genera *Macoma*, *Angulus*, and *Tellina*. On Norderney, the *Macoma baltica* community, described by Petersen (1913), was found. In the Gulf of Gaeta an isocommunity, named after *Macoma (Angulus) exigua*, occurs. Other comparable *Macoma* or tellinid isocommunities were described by Thorson (1933, 1934) from the East Greenland fjords; by Kirsop (1922) from Puget Sound; and by Miyadi (1941) from the Pacific. On Sapelo a comparable iso-community is absent, possibly in conse-quence of the low-energy conditions in that area.

The different *Macoma* communities of Norderney and Gaeta are followed by *Venus gallina* communities, described by Petersen (1913), which also occur in sandy sediments. As with *Macoma*, the *Venus* community is also absent on Sapelo. Well known isocommunities exist in Greenland, described by Thorson (1934); the Medi-terranean Sea, described by Vatova (1934); and the New England coast, described by Lee (1944).

In all three areas investigated (Dörjes, 1971, 1972), the above communities are followed offshore—on mixed sediments—by brittle star communities that are spread worldwide (Thorson, 1957). Off the coast of Gaeta, the *Amphiura chiajei* community occurs. On similar sediments off the coast of Sapelo, an isocommunity named after *Hemipholis elongata* was observed. Along the Norderney transect, no brittle stars were found due to the absence of muddy sedi-ments; farther offshore, where the content of mud in the sediments increases, an *Amphiura filiformis* community was de-scribed by Hagmeier (1923). With regard to sediment types, the lower sublittoral of the three areas investigated are very differ-ent. Therefore, the communities occurring in these areas cannot be compared with each other.

In summary, with increasing water depth and changing environmental con-ditions, the composition of communities generally changes due to the different tolerances, pretensions, and adaptations of each species to the respective environmental conditions. Such communities are very im-portant in paleoecology generally, and should also be considered in ichnological investigations.

ICHNOLOGICAL ASPECTS

Günther Hertweck

GENERAL TERMINOLOGY

Ichnocoenoses and Ichnofacies

An ichnocoenose is an association of lebensspuren reflecting life activity of the individuals of a biocoenose, i.e., a com-munity of organisms, on and (or) in the substratum. Most of the lebensspuren found in marine environments originate from benthic organisms. Epibenthic organisms mainly produce surface traces; endobenthic organisms mainly produce internal traces (if vagile) or dwelling structures (if sessile, hemisessile, etc.), some of which are con-nected with surface traces (Hertweck, 1970a).

Shallow marine environments character-ized by mobile substrates are generally populated by endobenthic communities. The ichnocoenose of these substrates con-sists mostly of dwelling structures and in-ternal traces. Such lebensspuren may be enhanced by preservation after death or emigration of the trace-producing organ-isms. Surface traces, however, are more ephemeral and tend to be preserved only exceptionally. The preserved record of an ichnocoenose provides the ichnofacies of a sediment body. The ichnofacies is one aspect of the biofacies.

The terms "ichnocoenose" and "ichno-facies" reflect two different aspects of ichnology. An ichnocoenose is an associa-tion of lebensspuren that can be related to one definite biocoenose. The lebensspuren may be surface traces, dwelling structures, or other internal traces of a contemporary animal community, as well as preserved lebensspuren originating from a previous or emigrated community. The preserved traces may be open, or filled with sediment. (See Chapter 4.) Also, preserved lebensspuren themselves may originate from one or more population periods (generations, or phases) of a biocoenose. However, an important distinction is that the association of lebensspuren being considered proves to be a "coenose," i.e., an assemblage of ecologically coherent lebensspuren produced in one definite environment by members of a given biocoenose.

"Ichnocoenose" is a paleontological term meaning a "community" (coenose) of traces (ichnia) that can be recognized and described as individuals. In accordance with its etymology, use of the term "ichnocoenose" should be conformable with that of the term "biocoenose," i.e., in the sense outlined above. (See also Radwański and Roniewicz, 1970.)

"Ichnofacies" is a sedimentologic term meaning a certain appearance ("face" = facies) imparted to sediment by lebensspuren (ichnia). The term "ichnofacies" should be used in harmony with the term "biocoenose." The original definition of "facies" by Gressly (1838) distinctly specified that the characteristic fossils of a facies represent organisms that had found their most favorable conditions for life in the depositional environment reflected by the lithofacies. (See also Moore, 1957.) Thus, a biofacies leads to the reconstruction of a paleobiocoenose, and the ichnofacies to the recognition of an ichnocoenose. The individual lebensspuren of an ichnocoenose reflect ethological and ecological responses of the trace-making organisms of a paleobiocoenose to the substratum. Sometimes the functional morphology of trace-making organisms may be reconstructed, even in cases of uncertain systematic and taxonomic position of the organism.

Characteristic Lebensspuren

The close relationship between biocoenoses on the one hand, and ichnocoenoses and ichnofacies on the other, suggests the profitable use of terms and criteria in ichnology that indicate the environmental significance of trace-making organisms within a community. The most successful approach to marine benthic animal communities is the concept of Petersen (1913, 1924), who distinguished bottom communities by their relationship to different sediments. Particularly, he recognized as characteristic species of the 1st order those species that are restricted to a definite sedimentary environment. Characteristic species of the 2nd order are found only in one part of the same overall environment, thus characterizing a subcommunity. Characteristic species of the 3rd order are found not only in this environment but also in adjoining ones, yet they occur in especially large numbers in the environment being considered. Other common species, but ones having lesser environmental significance, Petersen defined as "associated species." The terms of Petersen (1924) were redefined by Thorson (1957), who added some quantitative criteria by Spärk (1937) and correlated these terms with those provided in other concepts.

Petersen (1913, p. 27) also discussed the significance of his concept in paleontology. He considered the characteristic species of animal communities to be the "leading fossils" of the present. In fact, experience with skeletal remains in modern marine environments shows that characteristic species of animal communities are potentially the characteristic fossils of the biofacies being formed by a given community. This relationship holds true if no major sediment reworking occurs (Cadee, 1968; Hertweck, in Reineck et al., 1968; Hertweck, 1971; Warme, 1971). Lebensspuren may also become characteristic fossils if the trace-making organism is the characteristic species of a community (Hertweck, in

Reineck et al., 1968; Hertweck, 1970b, 1972).

Inasmuch as the definitions by Petersen (1924) and Thorson (1957) were concerned with characteristic and associated species of animal communities, similar definitions can be made for ichnology (Hertweck, 1972, p. 153–155). Characteristic lebensspuren thus are conspicuous, preservable lebensspuren occurring principally in one definite environment and having relatively high frequency and abundance, i.e., those found at most sampling sites in considerable quantities. This definition corresponds generally to trace-making organisms that are characteristic species of the 1st order.

Ideally, the above definition can be differentiated to indicate lebensspuren of 1st and 2nd order. In this case, a characteristic lebensspur of the 2nd order should be restricted to a subenvironment consisting of somewhat different grain size, fabric, etc. However, in the preserved record, either modern or fossil, fewer individual details are usually recognizable. Thus, to evaluate such an order of ranks among fossil or recent traces may be difficult. Probably, finding lebensspuren that correspond to characteristic species of the 3rd order would be easier. Such lebensspuren should occur in large quantities (i.e., high frequency and abundance) in one environment, but should also be found, in lower quantities, in adjoining environments. In these adjoining environments, they would appear as "associated lebensspuren."

Associated lebensspuren are conspicuous, preservable lebensspuren occurring with significant frequency and abundance in an environment, i.e., they are found at several sampling sites in appreciable quantities (Hertweck, 1972, p. 155). Associated lebensspuren provide additional characterization of an ichnofacies.

The distribution of characteristic and associated lebensspuren is not necessarily identical with the distribution of the trace-producing animals, however. Certain animals produce different lebensspuren in different environments, i.e., in or on different substrates. Such specific structures may appear as characteristic lebensspuren of different ichnofacies if the above conditions concerning frequency and abundance are fulfilled.

Extremely poor conditions for preservation of a lebensspur in one environment, opposed to exceedingly favorable conditions for its preservation in another, should also be considered. In such case, preservability is the main factor determining whether a lebensspur will be characteristic. However, the above definition does not claim that a characteristic lebensspur must necessarily be produced by a characteristic animal species; it claims only that a characteristic lebensspur be preservable, restricted mainly to one definite environment, occurring in considerable quantities.

Also, the environmental significance of a lebensspur may be exaggerated when compared to that of the trace-producing animal: certain complex burrow systems or crawling traces may occupy a major part of a given sediment volume whereas the producing animal species itself may not be present in comparable abundance (MacGinitie and MacGinitie, 1968, p. 77–81; Frey, 1971, p. 105; Hertweck, 1972, p. 150–151). (See Chapter 9.)

The main way to determine characteristic lebensspuren is to investigate profiles (or other *quantitative* sampling arrangements) consisting of a considerable number of representative samples from different environments. By comparing the ichnocoenoses related to different substrates, one can evaluate bioturbation features that are common to several or to all environments of the profile, and discern which lebensspuren are restricted to one specific environment.

In addition to the above, which constitutes an investigation of "frequency"— the number of sampling sites at which the lebensspuren of one species are found within a definite area—evaluating their abundance is also important. "Abundance" means the number of specimens of the lebensspuren of one species occurring in a standard sampling unit.

Also, the physical environment of an ichnocoenose or a characteristic lebensspur must be defined by its particular sedimentological characters, such as grain size, content of organic matter, aeration, and depositional sedimentary structures—including indicators of currents, wave action, and rates of deposition, reworking, or erosion.

The final goal of a quantitative ichnological investigation is to obtain a zonation of ichnofacies, each characterized by one or several characteristic lebensspuren (fossil or recent). However, lebensspuren of this quality are not necessarily present in all environmental zones of every profile investigated. This disparity may be due to (1) the total absence of trace-making organisms within a zone, (2) the lack of organisms producing conspicuous (characteristic) traces, or (3) poor conditions for preservation of lebensspuren, e.g., high rates of erosion or reworking. In these cases, body fossils alone may characterize the biofacies of this zone. Where neither trace fossils nor body fossils are preserved, only inorganic features will be available for the characterization of facies. (Ideally, of course, we would hope to find both body fossils and trace fossils present, in addition to distinctive physical features.)

EXAMPLES OF ENVIRONMENTAL ZONATION OF LEBENSSPUREN

Shallow-water marine environments are particularly well suited for the study of modern ichnocoenoses. A variety of benthic animal communities, represented by numerous species and individuals, occurs in the shelf-bottom region. The nearshore area especially shows a typical zonation of environments, depending on such things as tidal range and energy conditions, as outlined in the first part of this chapter.

A general classification of nearshore environments, i.e., the littoral system, is represented by the terms "supralittoral, eulittoral," and "sublittoral" (Fig. 20.1). In coastal areas of the beach type, these areas are known as "backshore, foreshore, shoreface (inshore)," and "offshore." The terminology of the beach-offshore classification is subordinated to the littoral classification, which—in addition to beaches—is also valid for rocky coasts (where it was developed by Lorenz, in 1863), reef environments, tidal flat-tidal channel systems, etc. The beach-offshore classification pays more regard to geomorphology and sedimentology. In the definitions outlined by Hunt and Groves (1965), wave action and sediment movements provide important aspects; this reflects the fact that in coastal regions of the beach type, a particular regime of sedimentary processes, especially reworking, can be discerned. Correlation of the two terminologies is shown in Table 20.1.

The distribution of lebensspuren in three shallow, nearshore marine regions are outlined as follows: (1) the German Bight between the coast and Helgoland, (2) the Gulf of Gaeta, Italy, and (3) Sapelo Island, Georgia (see Fig. 20.6).

The German Bight

The coastal area of the German Bight is the classical region of ichnological investigations in modern environments. Rudolf Richter, Walter Häntzschel, and Wilhelm Schäfer were the pioneers of this research (see Schäfer, 1972). The preponderant part of the ichnological information was obtained from tidal flats and beaches, however. Only a few of the lebensspuren found in intertidal environments play an important role in the beach-offshore profile of the southern North Sea. (In the entrances of the Baltic Sea, many of the "tidal flat species" do occur in submerged areas.)

Seilacher (1953a) described feeding behavior and surface traces of the polychaete *Scolecolepis squamata*. Tube construction and feeding by the polychaete *Lanice conchilega* were studied by Seilacher (1951,

TABLE 20.1. Zonation, Terminology, and Environmental Characteristics in a Generalized Beach-Offshore Profile.

| *Boundaries between environmental zones* | | *Mean high-water line* | | | |
| | | | *Mean low-water line* | | |
					Wave base
General (littoral) classification	Supralittoral	Eulittoral	Upper sublittoral	Middle sublittoral	Lower sublittoral
Beach-offshore classification	Dunes and backshore	Foreshore	Shoreface	Upper offshore	Lower offshore
Energy sources and sedimentary processes	Eolian reworking; currents and wave action during spring tides and storms	Bidiurnal submersion; mainly wave action; periodically high reworking rate; low rate of bioturbation	Strong wave action; breakers, surf; high reworking rate; low rate of bioturbation	Low wave influence to sea bottom; sedimentation prevails over reworking; high rate of bioturbation	Exceptional storm wave influence to sea bottom; currents are main energy source; scarce or no sedimentation or reworking; considerable bioturbation
Sediments	Fine sand	Fine sand with medium sand	Fine sand with medium sand	Silty fine sand	Silt to clay, or coarse relict sediments

1953a) and Ziegelmeier (1952, 1969). Röder (1971) investigated the ecology, behavior, and burrow patterns of the polychaete *Paraonis fulgens*. Schäfer (1939) described burrows and burrowing behavior of talitrid amphipods in beaches. Sedimentological and ichnological research in the deeper part of the German Bight was initiated by Reineck et al. (1967).

The beach, shoreface, and upper offshore areas of Norderney (East Frisian Islands) were investigated in 1970 and 1971. Results of this study are not yet published, but a brief report is sketched as follows.

In the backshore area, no lebensspuren of marine animals have been found. The foreshore area is a broad (150 m) zone, reflecting the tidal range of 2.4 m. Several beach runnels and ridges occur. The main lebensspuren found within this area are

burrows of the polychaete *Scolecolepis squamata*. They were found, in slightly varying abundance, in nearly all can cores investigated. Röder (1971) suggested that *S. squamata* prefers the upper foreshore and the beach ridges, whereas the lower foreshore and the beach runnels are populated by the polychaete *Paraonis fulgens*. Burrows of *P. fulgens*, exhibiting characteristic spiral and meandering parts, are the second most important lebensspuren of the foreshore.

In the submerged area, the boundary between shoreface and offshore (*sensu stricto*) is not yet determined. Presumably, the tidal currents represent an energy level below wave base similar to that of waves in the lower part of the shoreface (H.–E. Reineck, 1973, oral communication). Thus, a similar rate of reworking occurs in both areas. Zoologically, the nearshore area is bounded at a depth of 15 m, 6 km away from mean low-water line (Fig. 20.5). Up to this depth, the characteristic species of the intertidal and nearshore community, *Macoma baltica*, occurs. The most important lebensspuren of this area are produced by two polychaete species; tubes of *Lanice conchilega* and burrows of *Capitella capitata* are found in nearly all box cores taken at water depths less than 21 m, 12 km offshore. Tubes of the polychaete *Spiophanes bombyx* were also found at several stations.

In none of the box cores investigated were old, preserved lebensspuren observed. Apparently, energy conditions of the East Frisian nearshore zone are not favorable for the preservation of lebensspuren.

More evidence on preservability of lebensspuren in the German Bight was obtained in a profile from the area south of Helgoland, toward Büsum, in the eastern coastal region of the Bight (Reineck et al., 1968). The eastern part of this profile is an area consisting of muddy fine sand that extends from 10 to 15 m water depth. The only common lebensspuren of this area are tubes of *Lanice conchilega*. This distribution resembles the ichnological situation in the nearshore area off Norderney; tubes preserved after death of the animal are not found, apparently due to a high rate of sediment reworking.

In contrast, the adjoining area of mud and fine sand, in the west (15 to 30 m depth), exhibits intense bioturbation and many preserved features (Reineck et al., 1967, 1968). Characteristic lebensspuren are the burrows of *Echiurus echiurus*. Burrows of the polychaete *Notomastus latericeus* appear as associated lebensspuren. Preserved in places are callianassid burrows and *Echinocardium cordatum* crawling traces, both found in deeper parts of the sediment (Reineck et al., 1967). Dörjes (in Reineck et al., 1968) designated *E. echiurus* as the characteristic species of the 1st order of the animal community occurring in this environment. The biofacies was named by Hertweck (in Reineck et al., 1968) as the *Echiurus echiurus* biofacies (cf. Hertweck, 1970b). In this case, the ichnofacies is the dominant biogenic feature of the biofacies in general.

The western part of the profile consists of an area of mud containing medium sand (ca. 30 m water depth). The sediment is totally mottled by *Echinocardium cordatum*; at places, distinct crawling traces of this heart urchin may be observed (Reineck et al., 1967; Reineck, 1968). Other distinct lebensspuren are the burrows of *Notomastus latericeus*. Although *Echinocardium cordatum* affects total bioturbation of the sediment, its crawling traces should not be designated as characteristic lebensspuren; heart urchin traces are found in the adjoining muddy area to the east, and *E. cordatum* may occur also in exposed sand flats of the intertidal zone (Dörjes et al., 1969). Thus, in the offshore region of the North Sea, the lebensspuren of *E. cordatum* may correspond to the characteristic species of the 3rd order.

In general, distinct lebensspuren of nine species are found in the beach and offshore areas of the German Bight; three of these seem to be usually preservable (Fig. 20.6). The number of different lebensspuren imparted to the sediment thus contrasts strikingly with the total number of animals present (Fig. 20.5).

Only in one of the environments does an ichnofacies restricted to this area occur: the *Echiurus echiurus* ichnofacies of the muddy area south of Helgoland. The adjoining central area of the German Bight, consisting of mud and some medium sand, is dominated by *Echinocardium cordatum* crawling traces. In the nearshore belt, mainly in the upper offshore area, the long tubes of *Lanice conchilega*, extending several decimeters below the sediment surface, suggest that they have some chance of being preserved under favorable conditions. In this case, the tubes may occur as characteristic lebensspuren of a nearshore ichnofacies zone.

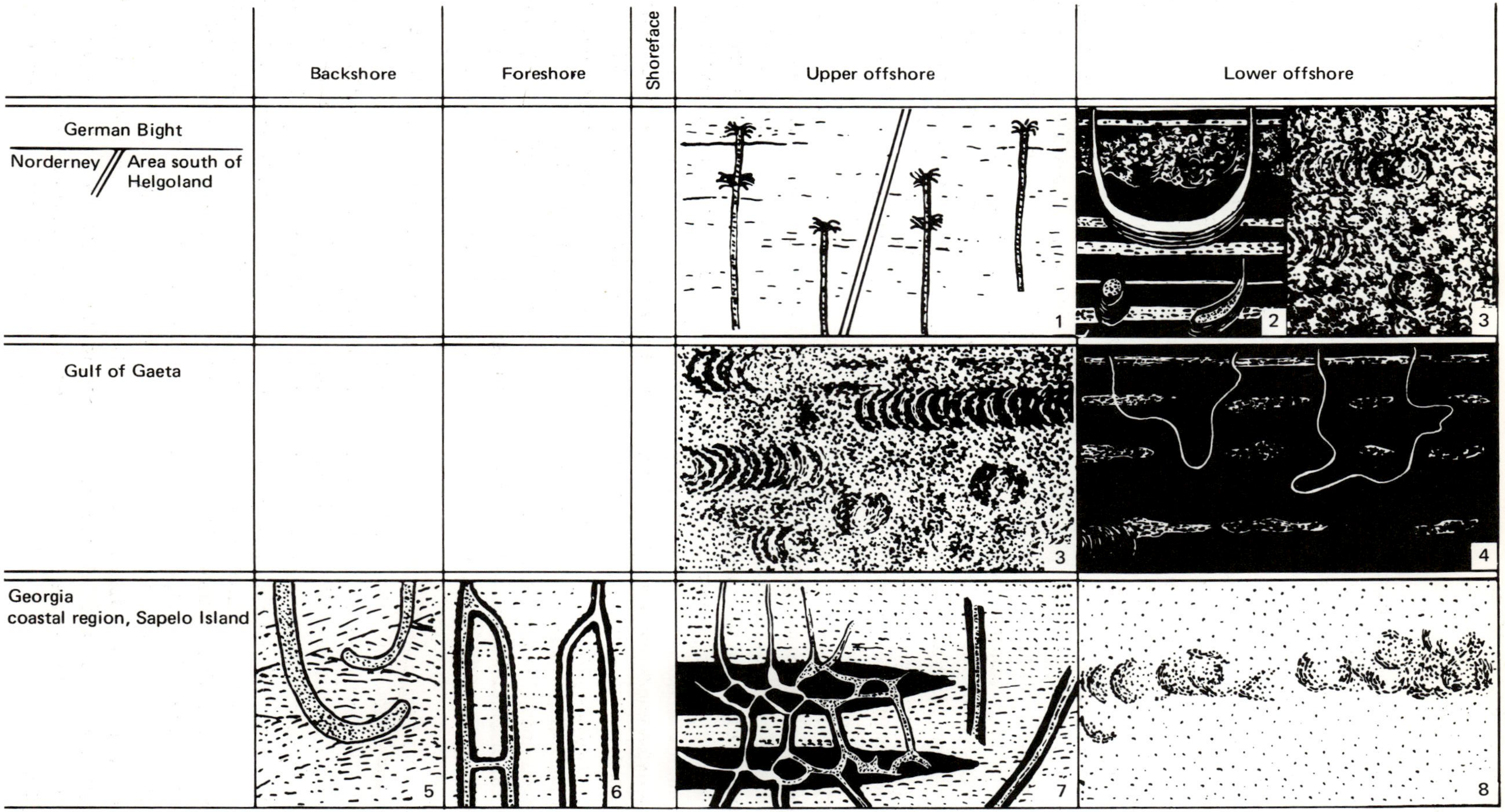

Fig. 20.6 Zonation of ichnofacies in shallow nearshore environments of the German Bight, Gulf of Gaeta, and Georgia coastal region. 1, *Lanice conchilega* ichnofacies. 2, *Echiurus echiurus* ichnofacies. 3, areas dominated by *Echinocardium cordatum* traces. 4, *Nephthys hombergi* ichnofacies. 5, *Ocypode quadrata* ichnofacies. 6, *Callianassa major* ichnofacies. 7, *Callianassa biformis* ichnofacies. 8, *Moira atropos* ichnofacies.

Gulf of Gaeta, Italy

A detailed study on the distribution of macrobenthic animals and communities in the Gulf of Gaeta was conducted by Dörjes (1971). (See Fig. 20.2.) Based on his results, Hertweck (1971) outlined the zonation of biofacies with respect to the distribution of mollusk skeletal remains. Bioturbation of the shoreface and shelf sediments was discussed by Reineck and Singh (1971). Detailed descriptions of some selected lebensspuren of the beach, shoreface, and offshore areas were provided by Hertweck (1973).

In the backshore area the isopod *Tylos europaeus* produces vertical, straight burrows. The foreshore area is a very narrow zone, due to the extremely low tidal range (30 cm). In this area a dense population of the spionid *Scolecolepis squamata* occurs, the worms having closely spaced vertical burrows.

The shoreface area is a zone of medium to fine sand, 350 m wide, characterized by several longshore bars. The water depth increases to 5 m. In the upper shoreface (0 to 2 m depth), the area populated by *S. squamata* continues, but with decreasing animal density. Other important lebensspuren are burrows of the enteròpneust *Balanoglossus clavigerus* and the polychaete *Lumbrinereis impatiens*. Conditions for preservation of lebensspuren are unfavorable in the upper shoreface, due to both a high rate of reworking and the continuous shifting of longshore bars.

In the lower shoreface (2 to 5 m depth), the burrows of *Capitomastus minimus* are found in high frequency and abundance. However, chances for preservation are also poor in the lower shoreface. Because of heavy storms during the winter season, a high rate of reworking occurs seasonally (Reineck and Singh, 1971). Important tubes found mainly in the lower shoreface are those of the polychaetes *Diopatra neapolitana*, *Onuphis eremita*, and *Owenia fusiformis*; the last species occurs also in the upper offshore.

The upper offshore area (6 to 14 m depth), consisting of silty fine sand, is the zone of the most intense bioturbation found in the region; it is mainly due to the heart urchin *Echinocardium cordatum*. Although the sediment is totally bioturbated, crawling traces of heart urchins remain recognizable in all relief peels examined by Reineck and Singh (1971). However, these traces are not found exclusively in this environmental zone; they occur also in the muddy sediment of the lower offshore.

Other important lebensspuren of the upper offshore are tubes of the polychaetes *Owenia fusiformis*, *Lanice conchilega*, and *Spiophanes bombyx*. Trumpet tubes of *Pectinaria koreni* are also worth mentioning; although they are not lebensspuren in a strict sense (see Chapter 3).

The lower offshore area (below 15 m water depth), varying from silt to clay, is also an area of high bioturbation, but remnants of many silty layers occur. The indistinct bioturbation features may be related to the snail *Turritella communis*, which is the characteristic species of the 1st order of the lower offshore animal community. In contrast, distinct open burrows produced by the polychaete *Nephthys hombergi* occur with high frequency and abundance in the lower offshore. The animal itself is found in considerable quantities throughout the shoreface-offshore profile of the Gulf of Gaeta; this species has the highest frequency of all animals in the overall region (Dörjes, 1971). However, we have no evidence for the occurrence of burrows of this polychaete in the shoreface and upper offshore areas; obviously, this vagile, carnivorous animal produces ephemeral crawling traces in coarser grained sediments. Crawling traces of *Echinocardium cordatum* also occur in the lower offshore, as associated lebensspuren.

A general view of the ichnocoenoses of the Gulf of Gaeta is presently limited by incomplete sampling; no comprehensive census of all lebensspuren occurring in this region exists. The most conspicuous traces observed are produced by 10 animal species; the lebensspuren of 3 species are usually found preserved (cf. Fig. 20.2).

The determination of characteristic lebensspuren is difficult here. However, two ichnofacial zones are recognized in the sediments (Hertweck, 1973): the upper offshore, dominated by crawling traces of *Echinocardium cordatum*, and the lower offshore, dominated by *Nephthys hombergi* burrows, which are characteristic lebensspuren due to their exclusive preservability in the muddy sediments of this zone (Fig. 20.6).

Sapelo Island, Georgia

A first approach to determining the diversity of lebensspuren of the Sapelo Island region was given by Frey and Howard (1969). The environmental significance of the conspicuous burrows of *Callianassa major* was discussed by Hoyt and Weimer (1963), Weimer and Hoyt (1964), and Hoyt et al. (1964). Frey and Mayou (1971) provided an ichnological zonation of the beach areas, based on decapod burrows. Zoological investigations concerning the animal communities of the sea bottom off Sapelo Island were conducted by Smith (1971) and Dörjes (1972). (See Figs. 20.3 and 20.4.) Based on these faunistic results and previous literature, Hertweck (1972) provided a detailed study of the morphology, as well as the distribution and environmental significance, of lebensspuren in the offshore areas.

In the backshore and dunes, burrows of the ghost crab *Ocypode quadrata* are the characteristic lebensspuren. Seaward, the occurrence of this quasiterrestrial crab is limited approximately by mean high-water level; the burrows also may be found several hundred meters inland. The greatest abundance of burrows, mainly related to adult individuals, is observed in the upper backshore. Young specimens prefer the lower backshore and the backshore-foreshore transition because they apparently need

more moisture for respiration than do the adults. Thus, an ichnological subzonation of the backshore area, based on abundance of ghost crab burrows, is possible (Frey and Mayou, 1971).

The foreshore area is characterized by burrow systems of the ghost shrimp *Callianassa major*. The highest density of burrow openings is observed in the lower foreshore, i.e., between mean sea level and mean low-water level; in the upper foreshore, burrow openings are relatively rare (Frey and Mayou, 1971). *C. major* burrows are also characteristic of other intertidal (i.e., eulittoral) environments of relatively high energy, such as tidal flats, shoals, and point bars in tidal streams. Other important lebensspuren of the foreshore area are burrows of the polychaete *Nerenides agilis* and siphon passages of the clam *Donax variabilis*, both of which occur with high density (Frey and Howard, 1969; Dörjes, 1972). (See Chapter 2, Fig. 2.10.) These burrows are usually destroyed, after death of the animals, by sediment reworking— especially by the gradual shifting of beach ridges and runnels. The same is usually true for the distinctive bioturbation features caused by haustoriid amphipods, such as *Neohaustorius schmitzi* and *Haustorius* sp.; Howard and Elders (1970) studied bioturbation by foreshore and shoreface haustoriids in aquaria. However, where ancient foreshore sediments are preserved, bioturbation structures not usually preservable may indeed be found (e.g., Frey and Mayou, 1971, Pl. 4, fig. 3a).

The most abundant bioturbation structures of the foreshore area are produced by the haustoriid amphipods *Parahaustorius longimerus* and *Acanthohaustorius* sp. (cf. Dörjes, 1972). Other important lebensspuren are tubes of the polychaetes *Onuphis microcephala* and *Diopatra cuprea*. [A detailed description of *Diopatra* tube construction was given by Myers (1970, 1972).] However, chances for preservation of lebensspuren are relatively poor in the shoreface; a considerable rate of reworking is the main sedimentologic parameter of this zone (Howard and Reineck, 1972).

In the upper offshore area, a great variety of lebensspuren are found (Hertweck, 1972). Although polychaetes are dominant among the species producing lebensspuren, the characteristic lebensspuren of the upper offshore area are produced by a crustacean—the decapod *Callianassa biformis*. Preserved abandoned burrow systems of these animals, as well as inhabited ones, are found throughout the area. Two modifications of this lebensspur occur in the two subenvironments of the upper offshore (Hertweck, 1972, p. 136–137, Figs. 6, 7). In muddy intercalations of the nearshore area, more complex burrow systems are developed; these have irregularly branched interlacing patterns (Fig. 20.6). In sandy sediments of the deeper part of the upper offshore, the burrows are simple, having few branchings but more prominent, knobby, silty linings. Associated lebensspuren of the upper offshore are burrows of the stomatopod crustacean *Squilla* sp. and the polychaete *Glycera* sp.—probably *G. americana*.

Only the characteristic lebensspuren were found to be preserved, unoccupied, with diagnostic frequencies and abundances. Numerous other lebensspuren occur in the upper offshore area, but with few exceptions, they were found to be "active" or inhabited. The highest frequency and abundance are exhibited by tubes of *Spiophanes bombyx* and burrows of *Capitomastus* cf. *C. aciculatus*, both polychaetes.

The following lebensspuren, occurring to considerable depths in sediment, are usually preservable: tubes of the polychaetes *Diopatra cuprea*, *Mesochaetopterus taylori*, *Spiochaetopterus oculatus*, *Onuphis eremita*, *O. microcephala*, and *Owenia fusiformis*, and of the anthozoan *Cerianthus* sp. Also, burrows of the polychaetes *Drilonereis longa* and *Notomastus latericeus* may usually be preservable. Although the burrows of *Capitomastus* cf. *C. aciculatus* occur with high abundance, to a sediment depth of 50 cm, only inhabited burrows having aerated linings were found. Apparently, the burrows collapse after death of the animals. [Seilacher (1964) suggested that such tiny structures may reappear in the fossil record, as enhanced by diagenesis. In this case, *Capitomastus* burrows may become another characteristic trace fossil of this environment.]

Poorly suited for preservation are the following lebensspuren, developed only a few centimeters below the sediment surface: burrows of the anthozoans *Edwardsia* sp. and *Haliactus* sp., the isopod *Cirolana polita*, the brittle star *Hemipholis elongata*, and the polychaete *Magelone* sp., biogenic mud accumulations by the polychaete *Pectinaria gouldi*, and tubes of *Spiophanes bombyx*.

In the lower offshore area (below 10 m water depth) the number of lebensspuren decreases abruptly, corresponding to the decrease in number of animal species and individuals (Fig. 20.3; see Dörjes, 1972). Characteristic lebensspuren of the lower offshore are crawling traces of the heart urchin *Moira atropos*. Other important and preservable lebensspuren are tubes of the polychaetes *Mesochaetopterus taylori*, *Spiochaetopterus oculatus*, *Onuphis nebulosa*, and *Petaloproctus socialis*.

In general, lebensspuren of 34 species were observed in the beach-offshore system of Sapelo Island. [The 40 species reported by Hertweck (1972) relate also to skeletal remains in living position.] The lebensspuren of 11 species were found as preserved in sediment, after the animal had died or abandoned the structure. The most dense ichnocoenose (Hertweck, 1972), as well as the highest degree of bioturbation (Howard and Reineck, 1972), are observed in the upper offshore area, corresponding to the peak density of the biocoenose (Figs. 20.3, 20.4; see Dörjes, 1972); lebensspuren of 19 species were observed, 5 as preserved in the sediment.

Four ichnofacies are developed in the sediments of the five environmental zones

(Hertweck, 1972); the *Ocypode quadrata* ichnofacies in the backshore; the *Callianassa major* ichnofacies in the foreshore; the *Callianassa biformis* ichnofacies in the upper offshore; and the *Moira atropos* ichnofacies in the lower offshore (Fig. 20.6).

COMPARISON OF FACIES AND TYPES OF LEBENSSPUREN

Correlation of Environments

Based on the results of ichnological work in the North Sea, Gulf of Gaeta, and Georgia coastal region, a general synthesis of common features in shallow, nearshore marine ichnocoenoses may be attempted.

As a first thought, the contrast between shallow sea and deep sea (see Chapter 21) may suggest a more detailed bathymetric zonation within shallow marine environments. Seilacher (1967a) provided a representative ichnofacies zonation, from continental and shallow marine to turbidite and deep-sea environments; this zonation represents "relative" bathymetry. (See Chapter 2, Table 2.1.) In contrast, some average "absolute" values of bathymetric levels are known, which represent the shallow sea (shelf), 0 to 200 m; continental edge (shelf break), ca. 200 m; deep-sea floor, ca. 5,000 m; and deep-sea trenches, ca. 10,000 m (Ager, 1963, Fig. 2.3). However, to reproduce the absolute bathymetry of the oceans, as well as the general ichnofacies zonation by Seilacher (1967a), within such a narrow-spaced area as the shallow sea, is difficult. Experience with shallow marine environments suggests that numerical bathymetric values are largely of regional meaning only. This situation is obvious if we consider the different distances from the shoreline, and depths of the continental edge, in the three shallow marine regions discussed here—North Sea, Gulf of Gaeta, and Georgia shelf.

However, another kind of "absolute" bathymetric zonation of environments,

based on different energy levels or energy steps, is possible in the nearshore, shallow marine area. Sedimentologically, the most important hydrographic parameters of nearshore environments are tidal range and average wave height (Reineck and Singh, 1971). Thus, the levels of mean high water, mean low water, and average wave base provide a bathymetric zonation that is independent of numerical depth values. Accordingly, such environmental zones as backshore, foreshore, shoreface (inshore), and upper offshore may be correlated in different regions. A correlation of lower offshore areas in different regions is difficult, however. Different sedimentary processes occur, due to decreases of energy at different rates. Currents, whether tidal or nontidal, may play a major role. Thus, other criteria must be found for the comparison of lower offshore regions.

The main criteria for recognizing energy conditions of a depositional environment are characteristic depositional sedimentary structures (Reineck, 1963; Reineck and Singh, 1971; Howard and Reineck, 1972). Another diagnostic feature is the quantitative relationship between depositional and bioturbation structures. (See Chapter 8.) Reineck (1963) distinguished six degrees of bioturbation intensity, based on percentages of the total sediment volume involved (cf. Howard and Reineck, 1972, Pl. 2). In beach-offshore sequences, a definite boundary exists between the shoreface and offshore areas, i.e., a level of relatively high energy and a level of relatively low energy. This boundary is constituted by the average wave base. Above this energy boundary, a high rate of reworking occurs, taphocoenoses are mainly allochthonous, conditions for preservation of lebensspuren are limited, and inorganic sedimentary structures are predominant. Below wave base, the rate of sediment reworking is low, taphocoenoses are generally autochthonous, conditions for preservation of lebensspuren are favorable, and bioturbation structures are predominant. Thus, in nearshore environments above wave base, the distribution of both skeletal remains and lebensspuren is controlled mainly by hydrographic-energy conditions. In nearshore environments below wave base, the distribution of both skeletal remains and lebensspuren is controlled mainly by the productivity of benthos (cf. Hertweck 1971, 1972, 1973). Therefore, the greatest variety of lebensspuren is realized in environments where autochthonous skeletal remains provide a high degree of environmental significance. In contrast, exclusive environmental significance is provided by preserved lebensspuren occurring in environments where allochthonous skeletal remains do not have definite diagnostic value.

Comparison of Ichnofacies

In Figure 20.6 the different ichnofacies recognized in the beach-offshore profiles of the North Sea, Gulf of Gaeta, and Georgia coastal region are compared. Obviously, the ichnofacies are not similar in corresponding environmental zones of the different regions. Especially, the bioturbation structures by heart urchins occurring in all profiles cannot be correlated, or considered as diagnostic of a particular environmental zone. In the two European marine environments, heart urchin bioturbation is affected by the same species—*Echinocardium cordatum*. However, in the Gulf of Gaeta their crawling traces are found mainly in the upper offshore, whereas in the North Sea they occur mainly in the lower offshore, in coarser relict sediments. In the Georgia coastal region, the crawling traces of another heart urchin species, *Moira atropos*, are a characteristic lebensspur restricted to the lower offshore relict sediments.

Other characteristic lebensspuren shown in Figure 20.6 are specific for the environmental zones in which they occur. This specificity is also true for burrow systems of the two callianassid species in the Georgia profile. Each system represents a different type of geometric configuration of burrows: *Callianassa major* builds more vertical and straight parts (Hoyt and Weimer, 1965; Frey and Mayou, 1971), whereas *C. biformis* constructs irregular patterns (Frey and Howard, 1969; Hertweck, 1972). Ghost crabs of the genus *Ocypode* are typical backshore inhabitants of tropical and subtropical regions. Cowles (1908), Frey and Mayou (1971), and Hill and Hunter (1973) described burrows of *O. quadrata* from American beaches; burrows of other species occurring in China and Japan were reported by Hayasaka (1935), Krejci-Graf (1935, 1937), and Utashiro and Horii (1965). Compared with the beach environments of Georgia, the backshore and foreshore areas of the North Sea and Gulf of Gaeta are "anichnial facies" (lacking traces) in terms of the preservability of characteristic lebensspuren. The same is true for shoreface sediments of all three regions.

The comparison made in Figure 20.6 does not depreciate the concept of ichnofacies based on characteristic lebensspuren, however. Due to their size, preservability, frequency, and abundance, characteristic lebensspuren in the sense outlined above (cf. Hertweck, 1972) are the most conspicuous, diagnostic patterns within a sediment containing lebensspuren; they are restricted closest to one definite environment, whether by their origin from a characteristic animal species, or by their preservability as related exclusively to this environment. In the case of complex burrow systems or densely spaced crawling traces, their environmental significance may be exaggerated when compared to abundances of the trace-producing animals. Thus, many biological, ichnological, and sedimentological factors determine whether a given bioturbation structure becomes a characteristic lebensspur or not. Therefore, in corresponding environmental zones of different regions, characteristic lebensspuren do not necessarily belong to the same ichnial pattern or originate from related taxonomic groups.

Comparison of Individual Lebensspuren

Characteristic lebensspuren are usually not the only bioturbation structures in an ichnocoenose. Associated lebensspuren also provide representative frequencies and abundances. In addition, other lebensspuren may be preserved. In modern ichnocoenoses, a great number of lebensspuren may also be observed that are inhabited but that are usually not preservable. The larger the number of samples taken in an area, and the more thoroughly they are examined, the greater the quantity of lebensspuren to emerge and provide comprehensive information about an ichnocoenose. In search of features common to ichnocoenoses of corresponding environmental zones, the complete variety of lebensspuren should be compared.

In Table 20.2, important lebensspuren from the German Bight, Gulf of Gaeta, and Georgia coastal region are presented in terms of the environmental zones where they occur. In the backshore area, no comparable lebensspuren are found. The foreshore area, however, is favorable for comparison. In the North Sea and Gulf of Gaeta, densely spaced burrows of the spionid *Scolecolepis squamata* are the main feature. They are very comparable with burrows of another polychaete belonging to the same family, *Nerenides agilis*, in the Georgia foreshore.[2]

In the shoreface area, tubes of several polychaete species represent comparable features. The tubes of *Lanice conchilega* (North Sea), *Diopatra neapolitana* (Gulf of Gaeta), and *D. cuprea* (Georgia coast) occur throughout the shoreface; those of *Onuphis eremita* (Gulf of Gaeta, Georgia coast) and *O. microcephala* (Georgia) are found in the lower shoreface. In the upper shoreface of the North Sea and Gulf of Gaeta, burrows of *Scolecolepis squamata* occur. Capitellid burrows are another comparable feature:

[2] In fact, this species was reported previously as *Scolecolepis agilis* (Frey and Howard, 1969).

in the North Sea, *Capitella capitata* occurs throughout the shoreface; in the Gulf of Gaeta, *Capitomastus minimus* is restricted to the lower part of the shoreface.

The upper offshore is the environment containing the most lebensspuren and the highest degree of bioturbation of sediments. This density corresponds to the peak numbers of both animal species and individuals in this zone (Figs. 20.2–20.5; see Dörjes, 1971, 1972). Most of the comparable lebensspuren found in the upper offshore are polychaete tubes, such as those of *Lanice conchilega* (North Sea, Gulf of Gaeta), *Spiophanes bombyx* (North Sea, Gulf of Gaeta, Georgia coastal region), *Owenia fusiformis* (Gulf of Gaeta, Georgia coastal region), and *Onuphis microcephala*, *O. eremita*, *Mesochaetopterus taylori*, and *Spiochaetopterus oculatus* (Georgia coastal region). Comparable burrows in the upper offshore are those of *Capitella capitata* (North Sea) and *Capitomastus* cf. *C. aciculatus* (Georgia coastal region).

In the lower offshore area, a decrease in abundance of comparable features is obvious. Especially, the equation of lebensspuren from different regions presents some difficulties. In addition to the characteristic lebensspuren discussed above, only those features comparable to lebensspuren of the upper offshore—e.g., burrows of *Callianassa* sp. in the North Sea and tubes of *Spiochaetopterus oculatus*, *Petaloproctus socialis*, and *Onuphis nebulosa* in the Georgia coastal region—occur in the lower offshore area.

Relationships Between Food Supply and Lebensspuren Types

Trophic Adaptations by Animals

The interrelationship between types of lebensspuren, food supply, and bathymetry was outlined by Seilacher (1964, 1967a). Rhoads (1966) and Frey (1971) contributed some important ideas about this relation. (See Chapters 8 and 9.) General results of these studies provide a predictable sequence of lebensspuren configurations, ranging from straight, vertical dwelling structures of suspension feeders in shallow marine environments, through horizontally oriented, spreiten-bearing grazing traces and burrows of deposit feeders in deeper bottom regions.

Walker (1972) discussed trophic relationships in ancient animal communities, based on the method of trophic-group analysis by Turpaeva (1957). In this method, a stratification of four feeding levels was demonstrated: (1) high-level suspension feeders, (2) low-level suspension feeders, (3) animals collecting food from the sediment surface, and (4) animals extracting food from within the sediment. Walker recognized all of these trophic groups among the fossils of eight Paleozoic animal communities. However, his approach was entirely successful only among body fossils. For trace fossils, only two groups could be distinguished: (1) straight vertical burrows and U-shaped burrows, both indicating low-level suspension feeding, and (2) horizontal burrows indicating intrasedimentary feeding. (Cf. Walker and Bombach, 1974.)

Experience with ichnocoenoses in modern environments suggests that this narrower result may be mainly a consequence of limited preservation of the trace fossils considered by Walker (1972). Microscopic analysis of the walls of straight, vertical structures among the trace fossils should reveal whether they were originally unlined burrows or dwelling tubes. (See also Rhoads, 1970.)

Tubes are built so as to have the opening of the dwelling structure independent of the substratum surface and its changes. Tubes are usually of the straight, vertical, basic type. A few exceptions are known, such as the U-shaped tubes of *Chaetopterus variopedatus* (see Fig. 2.5). In most tubes the upper parts, having the openings, project above the substratum surface. Thus, tubes generally suggest high-level feeding by the inhabitants. The animals may be sus-

pension feeders as well as carnivores. However, certain tube-constructing animals build their tube openings at the level of the substratum surface, e.g., *Clymenella torquata* and *Cerianthus* sp. In the fossil record, tube parts sticking up above the contemporary sediment surface may be not preserved, or may be embedded by sediment. Thus, the original relation between tube opening and substratum surface is usually not recognizable.

Burrow openings are more closely related to the sediment surface and its changes—sedimentation, reworking, and erosion. In contrast to tubes, considerable variety is evident in burrow morphology. Accordingly, burrowing animals comprise various different feeding types. Yet, to recognize the feeding type of an animal by the morphology of its burrow is difficult, especially for burrows having distinct openings at the sediment surface. Such burrows, whether vertical and straight or basically U-shaped, are typically dwelling structures. Their inhabitants may be either low-level suspension feeders or collectors of food from the sediment surface. Usually the ultimate food source is the same in the first three of Walker's trophic groups; suspended food particles are obtained at different levels above the substratum, or retrieved as freshly deposited on the sediment surface. Certain other collectors feed on organic matter that is produced at the bottom, e.g., diatoms and other algae. Thus, the collectors are a trophic group at the borderline between suspension feeding and substratum feeding. Ichnologically, they should be treated together with suspension feeders, according to the basic orientation of their dwelling structures. (Some tubes are also inhabited by collectors, e.g., *Pygospio elegans* in North Sea tidal flats).

Other dwelling burrows having straight, vertical main parts but also various branchings, especially in the lower part, suggest intrasedimentary feeding. The chief burrow types of this trophic group, however, are ones showing a distinct horizontal orienta-

TABLE 20.2. Distribution and Zonation of Lebensspuren in Shallow Nearshore Environments of the German Bight, Gulf of Gaeta, and Georgia Coastal Region.

Area studied	Backshore	Foreshore	Shoreface (Upper → Lower)		Upper offshore	Lower offshore
German Bight Norderney area south of Helgoland		Burrows of *Scolecolepis squamata* *Paraonis fulgens*	Burrows of *Scolecolepis squamata*	Burrows of *Capitella capitata*	Burrows of *Capitella capitata*	Burrows of *Notomastus latericeus* *Echiurus echiurus* *Callianassa* sp.
		Bioturbation by *Haustorius arenarius*	← Tubes of *Lanice conchilega* →		Tubes of *Lanice conchilega* *Spiophanes bombyx*	Bioturbation by *Echinocardium cordatum*
Gulf of Gaeta	Burrows of *Talitrus saltator* *Tylos europaeus*	Burrows of *Scolecolepis squamata*	Burrows of *Scolecolepis squamata* Burrows of ← *Balanoglossus clavigerus* → ← *Lumbrinereis impatiens* → Tubes of ← *Diopatra neapolitana* →	Burrows of *Capitomastus minimus* Tubes of *Owenia fusiformis* *Onuphis eremita*	Tubes of *Lanice conchilega* *Spiophanes bombyx* *Owenia fusiformis* Bioturbation by *Echinocardium cordatum*	Burrows of *Nephthys hombergi* Bioturbation by *Echinocardium cordatum*

TABLE 20.2.—Continued.

Georgia coastal region (Sapelo Island)	Burrows of Ocypode quadrata	Burrows of Callianassa major Nerenides agilis		Burrows of Callianassa biformis Squilla sp. Glycera sp. Capitomastus cf. C. aciculatus Notomastus latericeus Hemipholis elongata	Tubes of Onuphis eremita Onuphis nebulosa Spiochaetopterus oculatus Petaloproctus socialis
			↕ Tubes of Diopatra cuprea Tubes of Onuphis eremita Onuphis microcephala ↕	Tubes of Diopatra cuprea Tubes of Onuphis eremita Onuphis microcephala Spiochaetopterus oculatus Mesochaetopterus taylori Owenia fusiformis Spiophanes bombyx Cerianthus sp.	
		Bioturbation by Haustoriidae	↕ Bioturbation by Haustoriidae ↕		Bioturbation by Moira atropos

tion. Their shapes may range from irregular burrows or networks to regular, geometrical patterns. In contrast to burrows of filterers and collectors, the burrows of intrasedimentary feeders are easily recognized in the fossil record. Intrasedimentary (substratum) feeders are obliged to an active search for food within the sediment. This ingestion process may be represented by different stages of systematization. Obviously, irregular feeding burrows in sediment represent the lowest stage. The lower the quantity of food in sediment, the more systematic the search for food becomes. Episedimentary grazing animals, showing a highly systematized search for food, are characteristic in deep-sea environments; therefore, spiraling and meandering grazing traces are found especially on deep-sea sediment surfaces (Seilacher, 1967a, b; Raup and Seilacher, 1969; Heezen and Hollister, 1971; Chapter 21).

Examples from Shallow Marine Environments

Among the individual lebensspuren listed in Table 20.2 are various dwelling tubes. The greatest variety of tubes is found in the upper offshore, i.e., the environmental zone immediately below wave base. This area is characterized by a relatively low rate of reworking and a considerable abundance of food, especially suspended matter (Dörjes, 1971, 1972). High-level feeding is reported for many tube-constructing animals, such as the omnivorous polychaete *Diopatra cuprea* (Mangum et al., 1968), and the suspension-feeding polychaetes *Lanice conchilega* (Watson, 1890; Seilacher, 1951, 1953a; Ziegelmeier, 1952, 1969), *Spiochaetopterus oculatus* (Barnes, 1964), and *Mesochaetopterus taylori* and other chaetopterids (Barnes, 1965). Straight vertical tubes represent one of the main types of lebensspuren that reflect ecological conditions in the upper offshore area (Fig. 20.7).

In the turbid waters above wave base,

the availability of suspended food may be greater than below wave base, but the higher rate of reworking is less favorable to development of biocoenoses having high species diversity than in the upper offshore (Dörjes, 1971, 1972). In the lower offshore, almost no sediment reworking occurs, but the amount of suspended food presumably decreases, affecting a decrease in the number of both species and individuals (Dörjes, 1971, 1972). Thus, the total number of dwelling tubes originating from one or more suspension-feeding animal species shows a marked decline in the shoreface and in the lower offshore (Fig. 20.7).

In the foreshore and upper shoreface, straight vertical burrows such as those of the spionids *Scolecolepis* and *Nerenides* are the main lebensspuren types reflecting ecological conditions. In these areas the sediment reworking rate is so high that the animals must continuously repair their dwelling structures (cf. Hertweck, 1972, p. 148). Unlined burrows are generally easier to repair than are tubes, and burrowing animals can thus respond quicker to reworking than can tube-building animals. Under moderate conditions of reworking, tube-constructing spionids, such as *Spiophanes bombyx*, also occur in the upper offshore. Seilacher (1953a) described the feeding of *Scolecolepis squamata*; in agitated water the worm catches suspended food particles by means of tentacles, whereas in calm water the worm moves its tentacles along the bottom, thus agitating and resuspending the deposited food particles.

Dwelling burrows having two openings suggest low-level suspension feeding, because the respiratory current moved by the animal through the burrow may also be used for obtaining food. Basically, such burrows are U-shaped, but diverse variations are known. *Lumbrinereis impatiens* (Gulf of Gaeta) and *Echiurus echiurus* (North Sea) have simple U-shaped burrows. Gislen (1940) reported that *E. echiurus* uses its proboscis to collect food particles deposited on the sediment surface. Gut-

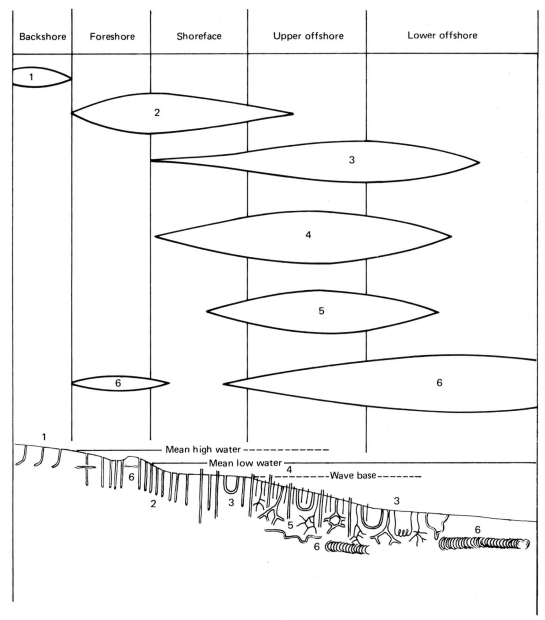

Fig. 20.7 Distribution and zonation of ecologically indicative lebensspuren in a generalized beach-offshore profile. 1, burrows of terrestrial crustaceaceans (mostly scavengers). 2, straight vertical burrows of low-level suspension feeders. 3, burrows having two openings (basically U-shaped) of low-level suspension feeders or collectors. 4, tubes of high-level feeders. 5, dwelling burrows of intrasedimentary feeding animals. 6, crawling traces of intrasedimentary feeding animals.

mann (in Reineck et al., 1967) suggested that *E. echiurus* gains most of its food from suspended matter in the respiratory current. *Balanoglossus clavigerus* (Mediterranean Sea) employs a J-shaped modification of the U-type burrow (Hesse et al., 1951, Fig. 24), resembling the burrow of *Arenicola marina* (see Chapter 22, Fig. 22.5). Feeding may be similar in both species, i.e., accumulation of food particles in the sediment at the lower end of the J by respiratory currents, and then feeding upon this sediment (see Krüger, 1959; Jacobsen, 1967). J-shaped burrows were also reported for *Balanoglossus* sp. from the Georgia coastal region (Frey and Howard, 1969, 1972).

Burrows of the brittle star *Hemipholis elongata* (Georgia coastal region) are irregularly star- or X-shaped; but a V-shaped part is used for the respiratory and feeding function. Two or three arms stick up above the sediment surface, catching suspended food in the water or collecting deposited food particles from the bottom. This burrow also shows that orientations other than horizontal are possible in brittle star traces [cf. the normal *Asteriacites* type, in Seilacher (1953b, Figs. 1–3); Fig. 4.5].

U-shaped dwelling burrows, including various modifications, are the main type of lebensspuren indicative of ecological conditions among low-level suspension feeders in shallow, nearshore marine environments. They occur under decreased wave-energy conditions, from the shoreface through the lower offshore, having their highest abundance in the transition area between upper and lower offshore (Fig. 20.7).

Burrows of the capitellid *Notomastus latericeus*, occurring in the lower offshore of the North Sea and the upper offshore of the Georgia coastal region, also show a basic U shape: two vertical shafts are connected by a middle part that spirals around a horizontal axis (Reineck et al., 1967; Hertweck, 1972). This burrow suggests low-level suspension feeding from the respiratory current. However, short, blind branches observed in some *Notomastus* burrows also indicate a search for food within the sediment (Reineck et al., 1967, Fig. 9).

Other capitellids, such as *Capitella capitata* (North Sea), *Capitomastus minimus* (Gulf of Gaeta), and *Capitomastus* cf. *C. aciculatus* (Georgia coastal region) have tiny, irregular, multi-branched burrow systems spreading considerably below the sediment surface. Capitellids are known as

intrasedimentary feeders (Linke, 1939; Schäfer, 1972). Burrows of *Callianassa biformis* (Georgia coastal region) and *Callianassa* sp. (North Sea) show a similar, although much enlarged, general pattern. D. Meischner (1973, oral communication) reported that *C. subterranea* (Mediterranean Sea) feeds upon sediment within its burrows (cf. Paul, 1970).

Thus, such multibranched burrow systems, occurring with high abundances, may be considered as the main, ecologically indicative lebensspuren of intrasedimentary feeders in shallow, nearshore marine environments. They are especially well developed in the upper offshore, where the increased content of mud and the processes of organic decomposition in the sediment provide considerable food within the substratum. Thus, burrows of intrasedimentary feeding animals have about the same environmental significance in the upper offshore as do the tubes of high-level suspension feeders (Fig. 20.7).

A similar noncompetitive relation exists between the intrasedimentary feeding polychaete *Paraonis fulgens* and the low-level filterer *Scolecolepis squamata* in the foreshore area of the North Sea coast. Sediment reworking at every flood tide completely redistributes the diatoms within the sandy sediment. However, this amount of food is relatively low if compared with that in the dark, reduced, silty fine sands in the upper offshore area. Therefore, a highly systematic search for food is necessary, as is demonstrated by the spiraling and meandering burrows of *P. fulgens* (Röder, 1971).

Crawling traces of heart urchins may be considered as the first step in the systematization of searches for food within sediment, if compared with the irregular burrow patterns of some intrasedimentary feeders. Heart urchin traces are found in different environmental zones of the beach-offshore sequence, in sediments ranging from mud to medium sand. Probably, competitive relationships to other animals have an important influence upon their distribution.

Burrows of *Nephthys hombergi*, found in lower offshore muds of the Gulf of Gaeta, are irregularly shaped and have two openings, i.e., a basic U. *N. hombergi* is a carnivore, thus its burrows obviously have a dwelling and respiratory function in the muddy substratum (Hertweck, 1973).

As indicated previously, the approximate environmental significance of various ecologically indicative lebensspuren types are summarized in Figure 20.7. Results of this comparison conform to the abundance relations of different trophic groups outlined for modern animal communities by Turpaeva (1957) and for ancient communities by Walker (1972). In the preserved record of ichnocoenoses, however, these original relationships may be considerably changed through differences in preservability of the lebensspuren. As mentioned above, preservation of lebensspuren is influenced mainly by energy conditions and sediment types in a depositional environment, and by the size, strength of construction, and subsurface extent of the lebensspuren themselves. Some of the lebensspuren in an ichnocoenose usually appear as characteristic or associated lebensspuren, due either to their original

environmental significance or to their particular preservability. The result of this process of selective preservation is the ichnofacies of a sediment.

One of the main goals of paleoecological studies in ichnology is to determine the full spectrum of ecologically indicative lebensspuren and their environmental significance. The reconstructed ichnocoenose that emerges from the study may be considerably different from the first impression provided by the ichnofacies.

ACKNOWLEDGMENTS

We wish to express our gratitude to the initiator and editor of this book, R. W. Frey, for inviting us to write this chapter. With many discussions, he contributed to the essential concept of the paper. And last, he helpfully reviewed the text.

We are also indebted to H. Böger, Kiel, J. D. Howard, Skidaway Island, Georgia, C. A. Elders, Athens, Georgia, and D. C. Rhoads, New Haven, Connecticut, who critically read the manuscript. Many important comments and suggestions given by S. Gadow and F. Wunderlich, Wilhelmshaven, I. B. Singh, Lucknow, and W. Katzmann, Vienna, are also gratefully acknowledged.

The chapter is based on studies that were supported by Deutsche Forschungsgemeinschaft, Bad Godesberg, and the National Science Foundation, Washington, D. C.

REFERENCES

Ager, D. V. 1963. Principles of paleoecology. New York, McGraw-Hill, 371 p.

Allen, E. J. 1899. On the fauna and bottom-deposits near the thirty-fathom line from Eddystone Grounds to Start Point. Jour. Marine. Biol. Assoc. United Kingdom, 5:365–542.

Barnes, R. D. 1964. Tube-building and feeding in the chaetopterid polychaete, *Spiochaetopterus oculatus*. Biol. Bull., 127:397–412.

———. 1965. Tube-building and feeding in chaetopterid polychaetes. Biol. Bull., 129:217–233.

Cadee, G. C. 1968. Molluscan biocoenoses and thanatocoenoses in the Ria de Arosa, Galicia, Spain. Zool. Verhand., 95:1–121.

Cowles, R. P. 1908. Habits, reactions, and associations in *Ocypoda arenaria*. Papers Tortugas Lab., Carnegie Inst. Washington, 2:1–41.

Davis, F. M. 1923. Quantitative studies on the fauna of the sea bottom. 1. Preliminary investigation of the Dogger Bank. Great Britain Fish. Invest. II, 6:1–54.

———. 1925. Quantitative studies on the fauna of the sea bottom. 2. Southern North Sea. Great Britain Fish. Invest. II, 8:1–50.

Dörjes, J. 1971. Der Golf von Gaeta (Tyrrhenisches Meer). IV. Das Makrobenthos und seine küstenparallele Zonierung. Senckenbergiana Marit., 3:203–246.

———. 1972. Georgia coastal region, Sapelo

Island, U.S.A.: sedimentology and biology. VII. Distribution and zonation of macro-benthic animals. Senckenbergiana Marit., 4:183–216.

———. (in preparation). Sedimentology and biology off the Isle of Norderney (North Sea). IV. Distribution and zonation of macrobenthic animals.

——— et al. 1969. Die Rinnen der Jade (Südliche Nordsee). Sedimente und Makro-benthos. Senckenbergiana Marit., 1:5–62.

Fischer, A. J. 1960. Latitudinal variations in organic diversity. Evolution, 14:64–81.

Ford, E. 1923. Animal communities of the level sea-bottom in the waters adjacent to Plymouth. Jour. Marine Biol. Assoc. United Kingdom, 13:164–224.

Frey, R. W. 1971. Ichnology—the study of fossil and recent lebensspuren. In B. F. Perkins (ed.), Trace fossils, a field guide. Louisiana State Univ., School Geosci., Misc. Publ. 71–1:91–125.

——— and J. D. Howard. 1969. A profile of biogenic sedimentary structures in a Holocene barrier island-salt marsh complex, Georgia. Gulf Coast Assoc. Geol. Socs., Trans., 19:427–444.

——— and J. D. Howard. 1972. Georgia coastal region, Sapelo Island, U.S.A.: sedimentology and biology. VI. Radiographic study of sedimentary structures made by beach and offshore animals in aquaria. Senckenbergiana Marit., 4:169–182.

——— and T. V. Mayou. 1971. Decapod burrows in Holocene barrier island beaches and washover fans, Georgia. Senckenbergiana Marit., 3:53–77.

Friedrich, H. 1965. Meeresbiologie, Gebr. Borntraeger, 436 p.

Gislen, T. 1940. Investigations on the ecology of Echiurus. Lunds Univ. Ärsskr., N. F., Avd. 2, 36:1–39.

Gressly, A. 1838. Observations géologiques sur le Jura soleurois. Soc. Helvetiae Sci. Nat., N. Mém., 2:7–26.

Hagmeier, A. 1923. Vorläufiger Bericht über die vorbereitenden Untersuchungen der Bodenfauna der deutschen Bucht mit dem Petersen-Bodengreifer. Deutsch. Wiss. Komm. Meeresforsch., Ber., N. F., 1:247–272.

Hayasaka, I. 1935. The burrowing activities of certain crabs and their geologic significance. Amer. Midland Natur., 16:99–103.

Hedgpeth, J. W. (ed.) 1957. Treatise on marine ecology and paleoecology. Vol. 1. Ecology. Geol. Soc. America, Mem. 67, 1296 p.

Heezen, B. C. and C. D. Hollister. 1971. The face of the deep. Oxford Univ. Press, 659 p.

Hertweck, G. 1970a. Die Bewohner des Wattenmeeres in ihren Auswirkungen auf das Sediment. In H.-E. Reineck (ed.), Das Watt, Ablagerungs- und Lebensraum. Frankfurt, W. Kramer, p. 106–130.

———. 1970b. The animal community of a muddy environment and the development of biofacies as effected by the life cycle of the characteristic species. In T. P. Crimes and J. C. Harper (eds.), Trace fossils. Geol. Jour., Spec. Issue 3:235–242.

———. 1971. Der Golf von Gaeta (Tyrrhenisches Meer). V. Abfolge der Biofaziesbereiche in den Vorstrand- und Schelfsedimenten. Senckenbergiana Marit., 3:247–276.

———. 1972. Georgia coastal region, Sapelo Island, U.S.A.: sedimentology and biology. V. Distribution and environmental significance of lebensspuren and in-situ skeletal remains. Senckenbergiana Marit., 4:125–167.

———. 1973. Der Golf von Gaeta (Tyrrhenisches Meer). VI. Lebensspuren einiger Bodenbewohner und Ichnofaziesbereiche. Senckenbergiana Marit., 5:179–197.

Hesse, R. et al. 1951. Ecological animal geography (2nd ed.). New York, John Wiley, 715 p.

Hessler, R. R. and H. L. Sanders. 1967. Faunal diversity in the deep-sea. Deep-Sea Res., 14:65–78.

Hill, G. W. and R. E. Hunter. 1973. Burrows of the ghost crab Ocypode quadrata (Fabricius) on the barrier islands, south-central Texas coast. Jour. Sed. Petrol., 43:24–30.

Howard, J. D. and C. A. Elders. 1970. Burrowing patterns of haustoriid amphipods from Sapelo Island, Georgia. In T. P. Crimes and J. C. Harper (eds.), Trace fossils. Geol. Jour., Spec. Issue 3:243–261.

——— and H.-E. Reineck. 1972. Georgia coastal region, Sapelo Island, U.S.A.: sedimentology and biology. IV. Physical and biogenic sedimentary structures of the nearshore shelf. Senckenbergiana Marit., 4:81–123.

——— et al. 1971. Recent advances in paleoecology and ichnology. Amer. Geol. Inst., Short Course Lect. Notes, 268 p.

Hoyt, J. H. and R. J. Weimer. 1963. Comparison of modern and ancient beaches, central Georgia coast. Amer. Assoc. Petrol. Geol., Bull., 47:529–531.

———— and R. J. Weimer. 1965. The origin and significance of *Ophiomorpha* (*Halymenites*) in the Cretaceous of the Western Interior. Wyoming Geol. Assoc., 19 Field Conf. Guidebook, p. 203–207.

———— et al. 1964. Late Pleistocene and recent sedimentation, central Georgia coast, U.S.A. In L.M.J.U. van Straaten (ed.), Deltaic and shallow marine deposits. Developments in Sedimentology, 1:170–176.

Hunt, L. M. and D. G. Groves (eds.) 1965. A glossary of ocean science and undersea technology terms. Compass Publications, 173 p.

Jacobsen, V. H. 1967. The feeding of the lugworm, *Arenicola marina* (L.). Quantitative studies. Ophelia, 4:91–109.

Kirsop, F. M. 1922. Preliminary study of methods of examining the life of the sea-bottom. Puget Sound Mar. Biol. Station, Publ., 3:129–139.

Krejci-Graf, K. 1935. Beobachtungen am Tropenstrand. I. Bauten und Fährten von Krabben. Senckenbergiana, 17:21–32.

————. 1937. Über Fährten und Bauten tropischer Krabben. Geol. Meere Binnengewässer, 1:177–182.

Krüger, F. 1959. Zur Ernährungsphysiologie von *Arenicola marina* L. Deutsch. Zool. Gesell., Verhandl., Frankfurt 1959:115–120.

Lee, R. F. 1944. A quantitative survey of the invertebrate bottom fauna in Menemsha Bight. Biol. Bull., 86:83–97.

Linke, O. 1939. Die Biota des Jadebusenwattes. Helgoländer Wiss. Meeresuntersuch., 1:201–348.

Lorenz, J. 1863. Physikalische Verhältnisse und Vertheilung der Organismen im Quarnerischen Golfe. Akad. Wiss. Wien, 379 p.

MacGinitie, G. E. and N. MacGinitie. 1968. Natural history of marine animals. McGraw-Hill, 523 p.

Mangum, C. P. et al. 1968. Distribution and feeding in the onuphid polychaete *Diopàtra cuprea* (Bosc). Marine Biol., 2:33–40.

Mills, E. L. 1969. The community concept in marine zoology, with comments on continua

and instability in some marine communities: a review. Jour. Fish Res. Board Canada, 26:1415–1428.

Miyadi, D. 1941. Ecological survey of the benthos of the Ago-wan. Annot. Zool. Japan, 20:169–180.

Moore, R. C. 1957. Modern methods in paleoecology. Amer. Assoc. Petrol. Geol., Bull., 41:1775–1801.

Myers, A. C. 1970. Some palaeoichnological observations on the tube of *Diopatra cuprea* (Bosc): Polychaeta, Onuphidae. In T. P. Crimes and J. C. Harper (eds.), Trace fossils. Geol. Jour., Spec. Issue 3:331–334.

————. 1972. Tube-worm-sediment relationships of *Diopatra cuprea* (Polychaeta: Onuphidae). Marine Biol., 17:350–356.

Paul, J. 1970. Sedimentgeologische Untersuchungen im Limski kanal und vor der istrischen Küste (nördliche Adria). Göttinger Arbeit. Geol. Paläont., 7, 75 p.

Pérés, J. 1957. Problème de l'étagement des formations benthiques. Station Mar. Endoume, Rec. Trav., 21:4–21.

————. 1966. Benthonic zonation. In R. W. Fairbridge (ed.), Encyclopedia of oceanography. Reinhold Publ. Corp., 1021 p.

Petersen, C. G. J. 1913. Valuation of the sea. II. The animal communities of the sea bottom and their importance for marine zoogeography. Danish Biol. Station, Rept., 21:1–44.

————. 1914. On the distribution of the animal communities of the sea bottom. Danish Biol. Station, Rept., 22:1–7.

————. 1918. The sea bottom and its production of fish food. A survey of the work done in connection with the valuation of the Danish waters from 1883–1917. Danish Biol. Station, Rept., 25:1–62.

————. 1924. A brief survey of the animal communities in Danish waters. Amer. Jour. Sci., 7:343–354.

Radwański, A. and P. Roniewicz. 1970. General remarks on the ichnocoenose concept. Acad. Pol. Sci. (Sér. Sci. Geol. Geogr.), Bull., 18:51–56.

Raup, D. M. and A. Seilacher. 1969. Fossil foraging behavior: computer simulation. Science, 166:994–995.

Reineck, H.-E. 1963. Sedimentgefüge im Bereich der südlichen Nordsee. Senckenberg. Naturforsch. Gesell., Abh., 505, 138 p.

————. 1968. Lebensspuren von Herzigeln. Senckenbergiana Leth., 49:311–319.

———— and I. B. Singh. 1971. Der Golf von Gaeta (Tyrrhenisches Meer). III. Die Gefüge von Vorstrand- und Schelfsedimenten. Senckenbergiana Marit., 3:185–201.

———— et al. 1967. Das Schlickgebiet südlich Helgoland als Beispiel rezenter Schelfablagerungen. Senckenbergiana Leth., 48:219–275.

———— et al. 1968. Sedimentologie, Faunenzonierung und Faziesabfolge vor der Ostküste der inneren Deutschen Bucht. Senckenbergiana Leth., 49:261–309.

Remane, A. 1940. Einführung in die zoologische Ökologie der Nord- und Ostsee. In G. Grimpe and E. Wagler (eds.), Die Tierwelt der Nord- und Ostsee. Leipzig, Becker and Erler, 1:1–238.

Rhoads, D. C. 1966. Missing fossils and paleoecology. Discovery, Yale Peabody Mus., 2:19–22.

————. 1970. Mass properties, stability, and ecology of marine muds related to burrowing activity. In T. P. Crimes and J. C. Harper (eds.), Trace fossils. Geol. Jour., Spec. Issue 3:391–406.

Röder, H. 1971. Gangsysteme von *Paraonis fulgens* Levinsen 1883 (Polychaeta) in ökologischer, ethologischer und aktuopaläontologischer Sicht. Senckenbergiana Marit., 3:3–51.

Schäfer, W. 1939. Beobachtungen an sandwühlenden Flohkrebsen der Nordsee–Küste. Natur u. Volk, 69:512–518.

————. 1972. Ecology and paleoecology of marine environments. Oliver & Boyd and Univ. Chicago Press, 568 p.

Seilacher, A. 1951. Der Röhrenbau von *Lanice conchilega* (Polychaeta). Senckenbergiana, 32:267–280.

————. 1953a. Studien zur Palichnologie. I. Über die Methoden der Palichnologie. Neues Jahrb. Geol. Paläont., Abh., 96:421–452.

————. 1953b. Studien zur Palichnologie. II. Die fossilen Ruhespuren (Cubichnia). Neues Jahrb. Geol. Paläont., Abh., 98:87–124.

————. 1964. Biogenic sedimentary structures. In J. Imbrie and N. D. Newell (eds.), Approaches to paleoecology. New York, John Wiley, p. 296–316.

————. 1967a. Bathymetry of trace fossils. Marine Geol., 5:413–428.

————. 1967b. Fossil behavior. Scientific Amer., 217:72–80.

Smith, K. L., Jr. 1971. Structural and functional aspects of a sublittoral community. Unpubl. Ph. D. Dissert., Univ. Georgia, 170 p.

Spärk, R. 1937. The benthic animal communities of the coastal waters. Zool. Iceland, 1(6):1–45.

Thorson, G. 1933. Investigations on shallow water animal communities in the Franz Joseph Fjord (East Greenland) and adjacent waters. Medd. Om Grønland, 100:1–68.

————. 1934. Contributions to the animal ecology of the Scoresby Sound fjord complex (East Greenland). Medd. Om Grønland, 100:69–135.

————. 1957. Bottom communities (sublittoral and shallow shelf). In J. W. Hedgpeth (ed.), Treatise on Marine Ecology and Paleoecology. Vol. 1. Ecology. Geol. Soc. America, Mem. 67:461–534.

Turpaeva, E. P. 1957. Food interrelationships of dominant species in marine benthic biocoenoses. Acad. Sci. U.S.S.R., Inst. Oceanol. Marine Biol., Trans., 20:137–148.

Utashiro, T. and Y. Horii. 1965. Some knowledge of *Ocypoda stimpsoni* Ortmann and on its burrows. Biological study of "Lebensspuren." Pt. VI. Niigata Univ., Takada Branch, Faculty Res. Rept., 9:121–141.

Vatova, A. 1934. Ricerche quantitative sul Benthos del Golfo di Rovigno. Inst. Biol. Rovigno, Not., 12:1–12.

Walker, K. R. 1972. Trophic analysis: a method for studying the function of ancient communities. Jour. Paleont., 46:82–93.

———— and R. K. Bombach. 1974. Feeding by benthic invertebrates: classification and terminology for paleoecological analysis. Lethaia, 7:67–78.

Warme, J. E. 1971. Paleoecological aspects of a modern coastal lagoon. Univ. California, Publ. Geol. Sci., 87, 110 p.

Watson, A. T. 1890. The tube-building habits of *Terebella littoralis*. Jour. Royal Micros. Soc., 1890:685–689.

Weimer, R. J. and J. H. Hoyt. 1964. Burrows of *Callianassa major* Say, geologic indicators of littoral and shallow neritic environments. Jour. Paleont., 38:761–767.

Zenkevich, L. A. 1951. Fauna and biological productivity of the sea. Leningrad, Sov. Nauk., 506 p. (In Russian)

————. 1959. Certain zoological problems connected with the study of the abyssal and ultra-abyssal zones in the ocean. 15 Internat. Zool. Congr., London, Proc., 215–218.

Ziegelmeier, E. 1952. Beobachtungen über den Röhrenbau von *Lanice conchilega* (Pallas) im Experiment und am natürlichen Standort. Helgoländer Wiss. Meeresuntersuch., 4:107–129.

————. 1969. Neue Untersuchungen über die Wohnröhren-Bauweise von *Lanice conchilega* (Polychaeta, Sedentaria). Helgoländer Wiss. Meeresuntersuch., 19:216–229.

CHAPTER 21

ANIMAL TRACES ON
THE DEEP-SEA FLOOR

C. D. HOLLISTER

Woods Hole Oceanographic Institution
Woods Hole, Massachusetts, U.S.A.

B. C. HEEZEN

Lamont–Doherty Geological Observatory
Palisades, New York, U.S.A.

K. E. NAFE

Department of Geology, Princeton University
Princeton, New Jersey, U.S.A.

SYNOPSIS

The deep-sea floor is a cold, dark, forbidding place, yet it harbors a significant number of trace-making organisms. These animals include enteropneusts, polychaetes, arthropods, holothurians, echinoids, and stelleroids, especially, and scattered representatives of other groups. Most tracemakers are mobile deposit feeders specifically adapted for gathering food in the abyss, and they leave behind a characteristic array of tracks, trails, shallow burrows, and fecal castings.

Most deep-sea animal traces are made on the substrate surface or shallowly within it; but because very slow depositional rates generally permit extensive sediment reworking—even by very sparsely populated animals—the entire length of cores may exhibit bioturbate textures. Structures observed in cores and in photographs of the modern abyssal seafloor are valuable in the interpretation of suspected deep-water trace fossils contained in the geologic record.

INTRODUCTION

This chapter deals not with fossils but rather with features that we surmise can, with time and lithification, eventually become fossils.[1]

To date, the main parallels drawn by geologists between recent and ancient deep-water deposits have centered around bathymetric zonations (Seilacher, 1967) and the salient characteristics of turbidites or molasse-flysch transitions (Heezen et al., 1955; Seilacher, 1958, 1962, 1974; Kuenen, 1968, Ksiazkiewicz, 1970; Chamberlain, 1971; Scholle, 1971; Crimes, 1973, 1974). Suspected deep-water trace fossils are now

[1] A point in semantics is raised here: the traces being made today certainly cannot be considered as trace fossils, yet the bioturbations seen in cores may grade with depth into those of totally unconsolidated Mesozoic and Cenozoic sediments; biogenic structures in these ancient deposits must be considered as fossils (Warme et al., 1973), on the basis of age alone.

Contribution number 3126, Woods Hole Oceanographic Institution. Contribution number 2168, Lamont-Doherty Geological Observatory.

493

rather well known, giving increased impetus to the search for modern analogs. Spectacularly close analogs have indeed been found (e.g., Bourne and Heezen, 1965), and many of the early speculations based upon trace fossils have been confirmed in the deep sea. Modern oceanographic work is now gradually piecing together a more complete picture (e.g., Ewing and Davis, 1967; Heezen and Hollister, 1971), and the job continues—as evidenced by the numerous new photographs included in this chapter.

Because few studies have yet been made of trace fossils in strata cored by the Deep Sea Drilling Project [the paper by Warme et al. (1973) being a notable exception], we confine our discussion here to traces that can be seen on the modern deep-sea floor through direct observation and photography, and to those observed in cores of the unconsolidated upper layer of sediments. We hope that this description will be of additional assistance to those geologists attempting to interpret trace fossils of possible deep-sea origin observed in the lithified sedimentary sequence.

Space does not permit a detailed account of the abyss, but the following brief comments on environmental conditions 'will hopefully place in better perspective the deep-sea tracemakers discussed and illustrated in this chapter.

THE ABYSSAL ENVIRONMENT

The irregularities of the deep-sea floor are being rounded by the gradual accumulation of eroded detrital particles and "raining" tests of microorganisms. This slowly growing blanket of sediment provides both a home and the food needed by organisms residing on or just beneath its surface. Compared to the continental shelf and slopes (sediments at less than 3 km water depth), the abyss appears stark and barren. The density of life is very low, and the seascape bears little evidence of habitation

beyond an occasional track, trail, burrow, or fecal coil left in the wake of some passing animal, going about its search for sustenance and shelter (Heezen and Hollister, 1971).

The deep-sea sediment that comprises this soft, easily marked substrate is loosely assembled. Water content in the uppermost few centimeters of this abyssal blanket ranges from 60 to 90 percent, and compaction is nearly imperceptible in the upper meter of sediment. The bearing capacity of the substrate ranges from 5 to 25 g/cm², making it resistant enough to bear the weight of the majority of animals walking or plowing across it (Richards, 1966; Hamilton et al., 1956). Abyssal brown clays accumulate at depths below 5 km, at a rate of from less than 1 to 2 mm per 1,000 years. The creamy, calcareous, foraminiferal ooze of moderate, midocean depths (2 to 4 km) is laid down at a rate of from 1 to 3 cm per 1,000 years, whereas the green and gray silts and clays at places on the continental margin may accumulate in thicknesses of as much as 60 cm during the same period (Heezen and Hollister, 1971).

Abyssal sediments are subject to substantial current reworking and transport locally (Hollister and Heezen, 1972). Turbidity currents and their characteristic deposits are of course significant features of the bathyal and abyssal zones, and much has been written about them (e.g., Heezen et al., 1955; Emery, 1964; Stanley, 1971; and numerous others).

Abyssal faunas are not fundamentally different from their shallow-water counterparts; representatives of many of the same phyla and classes are present in both places, and all reflect nature's scheme of reproduction, evolution, and utilization of available resources. But deep-sea faunas do have strikingly different population densities and community structures (Sokolova, 1959; Carey, 1965; Sanders et al., 1965; Sanders and Hessler, 1969; Griggs et al., 1969), and members of these faunas exhibit several distinct adaptations related to life on or in a dark, cold, mud bottom (e.g., Allen and Sanders, 1966). The meshing of form and function is everywhere apparent, the morphology of the various animal groups seemingly well suited to their particular modes of existence and methods of feeding. For instance, many crustaceans and fishes have overcome blindness by developing long tactile antennae and fins. By and large, the animals are delicately built and notably lacking in carbonate skeletons and other hard parts that are subject to dissolution by cold, bottom water, e.g., scores of small, fragile animals, in addition to the more familiar enteropneusts or holothurians (Figs. 21.1, 21.20).[2] Sessile filter-feeding animals have developed stalks in order to raise themselves up off the mud and into the paths of nutrient-rich currents (Fig. 21.2), and with increasing depth, the scavengers and carnivores of shallower waters are replaced by deposit feeders, the principal producers of traces on the abyssal sea floor.

Food is scarce in the abyss, and survival depends on the ability of the animals to detect and utilize every available food item. The quantity of organic matter and bacteria in the sediment is volumetrically small, thus the detritus feeders must ingest enormous volumes in order to derive a sufficient supply of food. The drift of organic particles into the deep sea, from shallower areas or from the overlying water, is the ultimate source of food. The remains of bottom dwellers and planktonic and nektonic organisms are decomposed by bacteria and consumed by the roving deposit feeders that "scavenge" the floor of the abyss.

[2] Of course, numerous species of deep-sea animals are in fact capable of producing skeletons of calcium carbonate or calcium phosphate, including pelecypods, urchins, stelleroids, crinoids, and fishes; but producing them is a thermodynamic struggle, and the skeletons generally disappear quickly once the animals die. The result is another example of a situation in which ichnology (traces) is potentially more usable than paleontology (body remains).

Fig. 21.1 Holothurian (upper left) egesting feces. These jelly-like organisms are indistinct in many deep-sea photographs; however, their feces and surface trails are easily recognizable. 3,420 m; southwestern Pacific. (NSF ELTANIN photo.)

Bacteria—having densities of at least 1,000,000 individuals per ml of recently deposited sediment—may be one of the primary food items of deposit feeders. The concentration of bacteria decreases rapidly with depth beneath the sea floor; several studies have shown the bacterial count to drop as much as a hundredfold between the surface and a depth of 1 cm (e.g., Morita and Zobell, 1955). Small crustaceans

Fig. 21.2 Attached pennatulid (horny coral) swaying in gentle current. (Shadow of stalk extends toward upper left.) Large mass of holothurian feces to left of coral. 4,450 m; southeastern Indian Ocean. (NSF ELTANIN photo.)

living within the upper few centimeters, such as isopods (Hessler, 1970) and cumaceans (Jones and Sanders, 1972), may also be a significant source of food.

Animals grazing over the surface film of mud encounter the richest crop of organic matter. Detritus-feeding enteropneusts and elasipod holothurians skim off this material. Echinoids plow through the somewhat less nutritious, upper 5 to 10 cm of sediment. Other organisms, such as worms, probe to slightly greater depths. The search for food rarely requires penetration beyond the upper 10 to 15 cm of mud, although the burrowers may go much deeper (e.g., Griggs et al., 1969).

ANIMAL TRACES IN THE ABYSS

As in shallow-water deposits, burrows are potentially more apt to be preserved, and later observed by geologists, than are surface traces; tracks and trails tend to be more ephemeral than burrows, and at the outcrop or in a core, one usually views sediments in a vertical rather than a horizontal sense. Because of extremely slow rates of deposition, even the very sparse animals here have adequate time to rework the accumulating sediments. Descriptions or illustrations of burrows and bioturbate textures in deep-sea cores are thus very important, and have been presented by several workers, e.g., Bramlette and Bradley (1942), Arrhenius (1952), Ericson et al. (1961), Berger and Heath (1968), Clarke (1968), Glass (1969), Piper and Marshall (1969), Griggs et al. (1969), Donahue (1971), Hanor and Marshall (1971), Warme et al. (1973), and Reineck (1973). (See also Chapter 2, Fig. 2.8.)

Nevertheless, experience in the rock record shows that surficial traces are indeed preserved and "recovered" in significant quantities among various deep-water deposits (see references cited in the introduction), and in this chapter we choose to concentrate mainly upon these features.

Our scheme of classification is a simple one, based mainly upon convenience in discussion. A useful descriptive classification of the various shapes and sizes of deep-sea traces was published by Ewing and Davis (1967, p. 263–267).

Tracks and Trails

Tracks and trails left on the sea floor are natural consequences of the endless search for food (Fig. 21.3). Variations in the traces arise from the different modes of feeding, and different structural forms used by the animals in carrying out this vital process.

Echinoderms are responsible for the majority of large, discrete tracks and trackways seen on the abyssal floor; within this group, most animals walk on multiple feet, their bodies raised off the sediment surface. The deep-sea elasipod holothurians possess few, although large and long, podia. Two such giants, *Psychropotes* and *Benthodytes*, have a double row of podia on the central part of their ventral creeping soles; and both have a fused mantle and modified podia (Fig. 21.4). When walking, they produce a wide trail (10 to 20 cm) consisting of four rows of tiny, shallow holes; the outer two rows are less perfectly aligned than the inner two (Fig. 21.5), having been produced by the lateral podia of the mantel. Less common are traces consisting of multiple rows of tracks, eight, ten, or twelve holes wide (Fig. 21.6); they are recognized only under optimum photographic conditions. Both types of trails occasionally begin or end abruptly; some of the deep-sea holothurians are almost neutrally buoyant, and may swim or drift just off the bottom.

Asteroids may move entirely with their tube feet, or walk with their arms alone, producing long, blunt tracks in mud. The Porcellanasteridae are a group of large, mud-eating sea stars found generally at depths greater than 4,000 m. Having no intestine or anus, they take mud into their

Fig. 21.3 A multitude of tracks, trails, and grooves of unknown origin. The double row of footprints traversing left side of illustration may have been produced by holothurians. 2,787 m; western Arctic Ocean. (Photo courtesy K. Hunkins.)

oral cavities, where all digestible materials are utilized (Hyman, 1955). These sea stars are probably responsible for the large, distinct, multihole "tractor tread" traces seen occasionally on the deep-sea floor (Fig. 21.7).

Brittle stars (Ophiuroidea) produce feathery, pinnate traces consisting of rows of partial body impressions, created as they pull themselves over the bottom (Fig. 21.8).

Abyssal bathypteroid fishes should leave in their wake a delicate pattern of "fin-prints" produced by the slender fins as they touch and probe the sediment lightly in search of potential food (Heezen and Hollister, 1971, Figs. 3.5, 3.6).

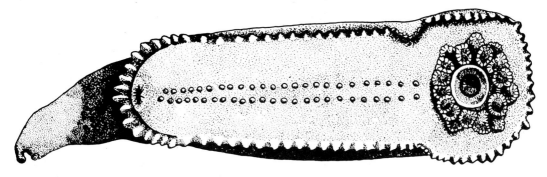

Fig. 21.4 Sketch of ventral surface of the holothurian *Psychropotes*. This organism should leave a wide trackway consisting of four parallel rows of holes (see Fig. 21.5). (From Heezen and Hollister, 1971.)

Fig. 21.5 Holothurian path. On left end of trace, the two centrally located rows of holes indicate that the tracemaker may have been the holothurian *Psychropotes*. Long shadow crossing from bottom to top produced by camera-tripping wire. 4,426 m; southeastern Indian Ocean. (NSF ELTANIN photo.)

Grooves and Furrows

Plowmarks and plowers are the principal elements in bioturbation on the deep-sea floor; and of the plowers, sea urchins are responsible for the majority of meandering furrows left in mud (Fig. 21.9). In order to adapt to this mode of life, irregular urchins gradually lost their radial symmetry several million years ago. Their spines became shorter and more specialized, to serve a locomotory function. Their mouths migrated to an anterior position, their anuses to a posterior one, and they began to depend entirely on digestible matter derived from the sediment.

Fig. 21.6 An elasipod holothurian (*Peniagone*) producing a path consisting of numerous rows of holes. Movement was from left to right; animal photographed in the process of changing directions. 4,445 m; southeastern Indian Ocean. (NSF ELTANIN photo.)

Fig. 21.7 Large, ribbed, multihole path probably produced by a mud-eating abyssal sea star. Abrupt termination of the trail on the right, and tendency of the trace to become more obscure toward the left, suggest that the animal moved from left to right and then was either eaten or swam away. Fecal string, lower center, probably produced by a deep-sea holothurian. 3,155 m; southwestern Pacific Ocean. (NSF ELTANIN photo.)

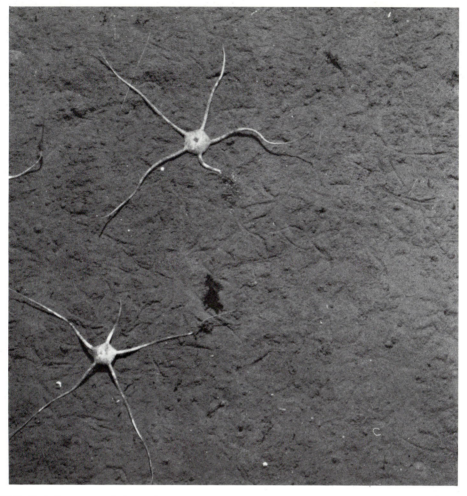

Fig. 21.8 Brittle stars and their characteristic pinnate, partial body impressions. 1,600 m; Hudson Submarine Canyon, off New York.

Spines on the ventral side of irregular urchins are divided into three rows. A central longitudinal row of paddle-like spines, extending the full length of the animal, is used to lever the animal forward. The lateral margins of the underside bear rows of longer, curved spines; and due to this arrangement the urchins produce three rows of transversely crenulated ridges as they lurch along in search of food. The trace is divided longitudinally by two parallel, narrow furrows that are the impressions of the spineless areas between the three locomotory bands on the animal's underside (Heezen and Hollister, 1971, Fig. 4.24). Occasionally, the crenulate trail becomes a broader, simpler groove, W or U shaped in cross-section, when the urchin plows more deeply through the bottom (Fig. 21.10). These grooves may terminate in small sediment-covered mounds having dark, central holes that mark the respiratory shafts of the buried echinoids.

Pseudostichopus, unlike the majority of large holothurians, plows a broad U-shaped path through sediment, somewhat resembling the furrow of an irregular echinoid, because it is propelled along the sea floor by innumerable small podia on its ventral surface.

Bivalves plow randomly over the soft bottom sediments, leaving narrow V-shaped

Fig. 21.9 Abyssal irregular sea urchin (upper left), producing a meandering ridge having delicate lateral crenulations. These echinoids frequently recross their own trails as they search for food, a trait not normally seen in the trails of deep-sea holothurians and enteropneusts. Animal moving toward upper center when picture was taken. 4,694 m; southeastern Indian Ocean. (NSF ELTANIN photo.)

Fig. 21.10 Irregular sea urchin plowing a furrow in sand-size material. Feathery decorations on the furrow are produced by locomotory podia. This kind of sea urchin can move across the surface without making a furrow; or, when searching for food within the upper few centimeters of sediment, can make a furrow as wide as its body. 146 m; continental shelf off Nova Scotia. (Photo courtesy L. H. King.)

grooves in their wake. Although small and fragile, their frequent recovery in deep-sea trawls indicates that they may be responsible for many of the small meandering furrows seen in exceptionally clear, bottom photographs (Fig. 21.11).

Burrows

Abyssal infaunal and "epifaunal" animals burrow in search of bacteria, other microorganisms, and the decomposed remains of larger organisms. Burrowing is one of the principal means of avoiding predation; consequently, many mollusks, crustaceans, holothurians, and fishes (Fig. 21.12) hide in holes—reaching out or swimming out at the prospect of a passing meal, only to retreat again at the first sign of danger.

Animals burrow or otherwise rework sediment in a variety of ways. Elasipod holothurians live off the surface film of sea-floor sediment, ingesting and remolding it into compact fecal coils and strings. They

hardly qualify as burrowers. Irregular echinoids "burrow" somewhat deeper, mixing sediment with feces in their furrowed wake. Other burrowers penetrate the ooze and leave behind a subsurface trace filled with reworked sediment. Burrows recovered in deep-sea cores vary from less than 1 to more than 50 mm in diameter; they stand out from the surrounding sediments by virtue of their contrasting colors and textures (Fig. 21.13). Well-illustrated accounts of such textures in deep-sea cores are given in the references cited previously.

Both asteroids and ophiuroids create distinctive star-shaped "burrows" (resting or feeding traces) in the soft-bottom muds (Fig. 21.14), making identification of the burrowers an easy task. Ophiuroids are known to bury their central discs, leaving only their ray tips visible above the surface.

Echiurids are unsegmented worms that

Fig. 21.11 Narrow, jagged, V-shaped groove made by an unidentified organism (?bivalve). 4,218 m; southwestern Pacific Ocean (NSF ELTANIN photo.)

Fig. 21.12 Bottom-dwelling fish emerging from its hiding place. Observations from submersibles along continental margin of the eastern United States indicate that many of the larger holes in the substrate are produced by bottom-dwelling fishes seeking concealment and protection. Holes produced inorganically, or by other organisms, often are soon occupied by bottom-dwelling fishes. As the fishes move in and out of the holes during their feeding activities, they continually enlarge them and reexcavate sediment, which is transported away from the hole by bottom currents. In this case, absence of a surrounding crater indicates that the excavated sediment was transported elsewhere. 1,300 m; Gulf of Mexico. (Photo courtesy R. Church.)

reside permanently in U-shaped burrows (e.g., Hertweck, 1970; Chapter 23, Fig. 23.10). Openings to the surface, 10 to 100 cm apart, allow continuous circulation of bottom water. The burrows are cleared of debris intermittently by powerful jets of water from the anus of the animal (see Chapter 22). This process gradually creates a small cone around one of the two openings. While feeding, the echiurid slowly extends its proboscis out of the burrow and onto the sediment surface, moving in one direction. Food is entangled in the mucus surrounding the proboscis, and drawn into the mouth in a series of slow stages. Repeated extension and withdrawal in this manner produces a circular, spoke-like pattern on the sea floor (cf. Fig. 21.15). Similar spoke-like patterns may be created by the siphons of pelecypods, or by other animals residing in simple vertical burrows. Such patterns have been seen in both recent and ancient sediments.

Most annelids feed on detritus or plankton, and many of them—especially in shallow water—build permanent, leathery burrow linings. They feed by extending their tentacles or proboscis out onto the surface muds, probing the area for possible food. This activity also produces a fine radial pattern surrounding the tube opening on the sea floor (cf. Frey, 1971, Fig. 3). Other annelids produce mounds of re-

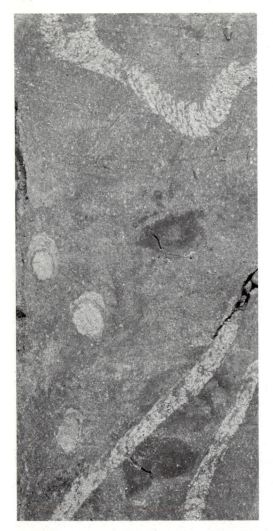

Fig. 21.14 Large sea star impression in moderately shallow water. These kinds of impressions are perhaps the easiest of all to identify with the tracemaker. 1,723 m; Australian continental slope. (NSF ELTANIN photo.)

been seen in eruption in shallower waters (Cousteau and Dugan, 1953; Shinn, 1968). As sediment gradually covers the cone, the apical vent is filled in and obliterated, making the aging feature merely another insignificant mound on the sea floor. Cones containing craters appear singly, in pock-like clusters randomly scattered on the

Fig. 21.13 Bioturbations in deep-sea cores. These traces probably result from burrowing activity by worms or crustaceans in the uppermost accumulation of sediments (cf. Chapter 17, Figs. 17.6, 17.7). Longitudinal sections seen at lower right and top, transverse sections on left. Core 6 cm wide. 4,485 m; North Atlantic. (From Heezen and Hollister, 1971.)

worked sediment (cf. Heezen and Hollister, 1971, Fig. 6.25; Gordon, 1966).

   A continuous succession of cone-shaped accumulations—ranging from ones triangular in cross section having no apical vent, through cones having large apical vents, to large craters lacking an apparent cone—have been observed in deep-sea photographs (Figs. 21.16 and 21.17). Such cones have

Fig. 21.15 Spoke-like impressions radiating from a central hole. These types of traces may be produced by the extensible siphons of pelecypods, or by polychaete or echiurid worms as they extend themselves from a central hiding place in search of food (see Risk, 1973). 4,225 m; southeastern Indian Ocean. (NSF ELTANIN photo.)

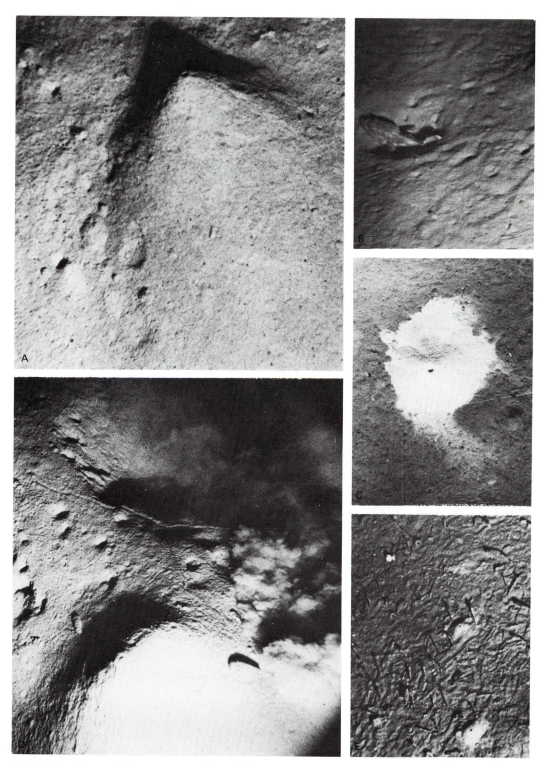

Fig. 21.16 Assorted deep-sea mounds and burrow openings. Very few deep-sea photographs show large symmetrical cones having apical vents (A, C). Most are much less regular (B, D, E). Fish are associated with craters shown in B, D. (From Heezen and Hollister, 1971.)

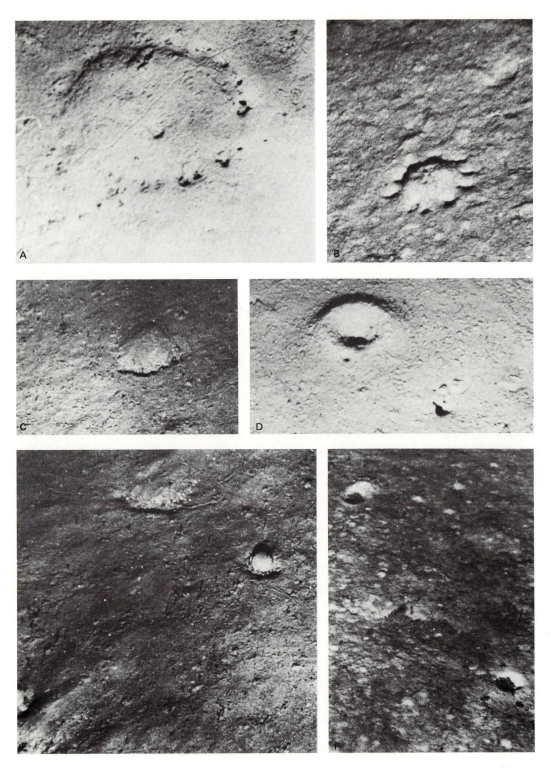

Fig. 21.17 Selected photos of holes, mounds, and craters produced by various bottom-dwelling organisms. Those in A and B may have been produced by sea stars. Makers of the other mounds and holes (C to F) are unknown. (From Heezen and Hollister, 1971.)

sea bottom, or in straight rows; the most mysterious of these craters bear no evidence of excavated sediments, which typically form the cone rims of most other miniature volcanoes.

Mounds, bumps of uncertain origin, and vague impressions, which cause sea-floor irregularities on the scale of a few centimeters, are the predominant features of most sea-floor photographs. Some of these structures may be fecal masses, others are inconspicuous animals, and some are the results of the efforts of burrowers observed in various stages of construction or destruction of sedimentary features (Fig. 21.18). But whatever the configuration, most are the result of animal activity. The lifeless floors of such stagnant basins as the Cariaco Trench, Walvis Bay, some fjords, and the Black Sea (Fig. 21.19) exhibit no tracks, trails, cones, or burrows—the signposts of the inhabited deep.

Excrement

Excrement is the natural by-product of the abyssal creatures' successful searchings for food. Carnivorous animals produce feces

Fig. 21.18 Raised ridge-like "furrow" produced by an irregular echinoid (upper left). An old abandoned trail of the same kind traverses from lower left to upper right; abundant holes and mounds surround the trails. 3,451 m; southeastern Indian Ocean. (NSF ELTANIN photo.)

Fig. 21.19 The featureless, azoic bottom of the Black Sea. The only biogenic textures seen on such anaerobic bottoms are the occasional feces and other organic remains of pelagic or purely surface-dwelling organisms. No mounds, holes, grooves, or furrows of infaunal organisms are seen originating from within these lifeless substrates. 2,104 m; eastern margin of Black Sea. (Photo courtesy A. C. Vine.)

of loose consistency; vegetarians produce firmer ones; and deposit feeders produce the most resistant of all (Moore, 1939). Those of the last category are the predominant ones seen on the deep-sea floor. But regardless of consistency or maker, all feces consist largely of remolded sediment, because it is the primary source of nutrients for the vagile benthos.

Large, striking, planispiral coils are seen frequently in photographs from high, southern latitudes (Fig. 21.20). At one camera station, the animal responsible for the coils was photographed in the act of producing them. It was a large enteropneust, or acorn worm, a hemichordate about 1 m long and 5 cm thick (see Bourne and Heezen, 1965). As this animal feeds upon the ooze, particles trapped in a strand of mucus secreted by its proboscis are passed into the mouth by ciliary action. A strip of mucus, mixed with bits of sediment, can be seen running back from the collar to the fecal coil.

Less spectacular, but by far the most common excrement found throughout the abyss, resembles a piece of coiled clothesline. Circular in cross-section (0.25 to 2 cm diameter) and evenly segmented, the strings

may be piled, unevenly coiled, or lie in open loops on the sea floor (Fig. 21.21). The tracks of elasipod holothurians are often associated with these knots and loops; and these animals have been photographed in the process of egesting this form of re-molded sediment (Heezen and Hollister, 1971, Fig. 5.17).

Leptosynapta inhaerens ejects its feces with great force, shrinking to one-third its natural body length during the process of emission (Fenton and Fenton, 1934). The feces is thrown a distance of two to eight times the body length, and changes in form from spirals to loosely looped coils (cf. Fig. 21.22). This species has not been found in the deep sea, and does not always defecate in the manner just described (cf. Frey, 1971, Fig. 18); but it demonstrates the force employed by certain holothurians in ejecting their excrement, and probably has its analog in the abyss.

Holothurians are the most common mud-eating benthic animals of the abyss. Giant elasipods skim only the uppermost few millimeters off the bottom, leaving little evidence of their passage—other than their tube-feet impressions. The tentacles surrounding the mouth are used as fingers; particles of digestible food are trapped in their mucus coatings and then drawn into the mouth. Their habit of concentrating sediment in their intestines and ejecting it at various intervals has been observed in photographs; only rarely does one see a continuous fecal string. The feces are generally tightly coiled, freely looped, in random piles, or in discrete masses (Heezen and Hollister, 1971, Fig. 5.10).

Although minute ovoid or rod-shaped fecal pellets, less than 1 mm in diameter, have been described in scientific literature (e.g., Moore, 1939; Kraeuter and Haven, 1970), the larger feces of the abyss are rarely mentioned. The principal reason for this lack of information is the difficulty in obtaining good samples. Fragile fecal strings are destroyed in the process of dredging or trawling at sea; and the few

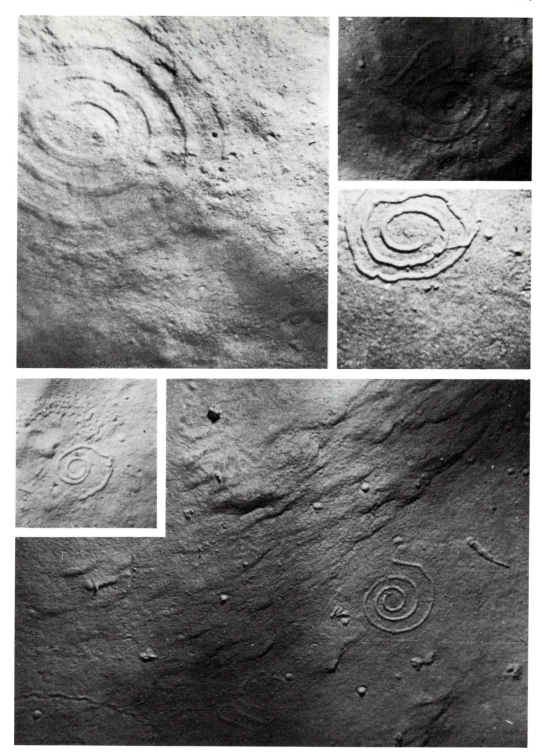

Fig. 21.20 Various large, planispiral traces produced by detritus-feeding acorn worms (hemichordates). These characteristic spirals are found over much of the deepest Pacific and Atlantic floors. Many animals have been photographed while producing these unique traces. (From Heezen and Hollister, 1971.)

Fig. 21.21 Large "clothesline-type" fecal "knots" left by an abyssal holothurian. 4,221 m; southeastern Indian Ocean. (NSF ELTANIN photo.)

specimens recovered are usually washed away with the fine material when animals are sieved from the sediment. Finally, the generally nondescript form of feces in bottom samples and photographs make their recognition difficult. (Cf. Chapter 2, Fig. 2.8D.)

DISCUSSION AND CONCLUSIONS

Searching through more than 100,000 photographs taken at more than 2,000 locations on the deep-sea floor has yielded only about 100 examples of animals recorded in the act of producing tracks, trails, and feces. A preliminary study of the morphology and arrangement of the central surfaces of bottom-dwelling animals nevertheless shows that the makers of certain traces may be ascertained with some degree of certainty, without their actually being caught "in the act." Many of the organisms in these photographs are unidentifiable, due to their small size, but the larger animals, responsible for the majority of these markings, are members of the Phylum Echinodermata; lesser contributions are made by the larger arthropods, worms, and occasional fishes. Innumerable organisms probably produce the mounds and holes seen in abyssal sediments, but their specific identification must await the probings of deep-diving vehicles equipped with an external sampling apparatus.

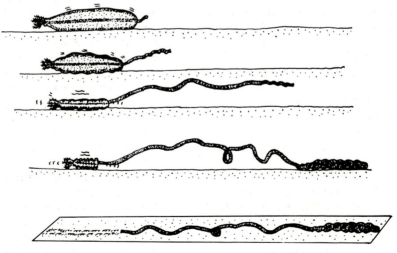

Fig. 21.22 Sketches illustrating the style of defecation by a holothurian from shallower water. Many of the deep-sea fecal coils were probably produced in similar fashion. (From Heezen and Hollister, 1971.)

Once the largely descriptive stage of deep-sea ichnology has been surpassed, we will have a much better basis for documenting ancient analogs—especially the broad transitions from freshwater or shallow marine environments to the abyss, as stressed in Chapters 7 to 9, 19 and 20.

ACKNOWLEDGMENTS

We wish to express our sincere thanks to the many sea-going oceanographers who painstakingly obtained bottom photographs: the necessary raw data for studies of modern deep-sea traces. Special thanks are due G. T. Rowe and R. H. Backus for their helpful suggestions while preparing this manuscript. We also thank the Oxford University Press for permission to publish Figures 21.4, 21.13, 21.16, 21.17, 21.20, and 21.22, originally published in *The Face of the Deep* by Heezen and Hollister, 1971. This research was partially supported at the Woods Hole Oceanographic Institution by the National Science Foundation, NSF grant 16098, and at Columbia University by the Office of Naval Research, contract N00014–67–A–0108–0036.

REFERENCES

Allen, J. A. and H. L. Sanders. 1966. Adaptions to abyssal life as shown by the bivalve *Abra profundorum* (Smith). Deep-Sea Res., 13:1175–1184.

Arrhenius, G. O. S. 1952. Sediment cores from the east Pacific. Swedish Deep-Sea Exped., Rept., 5:1–227.

Berger, W. H. and G. R. Heath. 1968. Vertical mixing in pelagic sediments. Jour. Marine Res., 26:134–143.

Bourne, D. W. and B. C. Heezen. 1965. A wandering enteropneust from the abyssal Pacific and the distribution of "spiral" tracks on the sea floor. Science, 150:60–63.

Bramlette, M. N. and W. H. Bradley. 1942. Geology and biology of North Atlantic deep-sea cores. U.S. Geol. Survey, Prof. Paper 196:1–34.

Carey, A. G., Jr. 1965. Preliminary studies on animal-sediment interrelationships off the central Oregon coast. Ocean Sci., Ocean Engineer., 1:100–110.

Chamberlain, C. K. 1971. Bathymetry and paleoecology of Ouachita geosyncline of southeastern Oklahoma as determined from trace fossils. Amer. Assoc. Petrol. Geol., Bull., 55:34–50.

Clarke, R. H. 1968. Burrow frequency in abyssal sediments. Deep-Sea Res., 15:397–400.

Cousteau, J. Y. and J. Dugan. 1953. The living sea. New York, Harper & Row, 228 p.

Crimes, T. P. 1973. From limestones to distal turbidites: a facies and trace fossil analysis in the Zumaya flysch (Paleocene-Eocene), north Spain. Sedimentology, 20:105–131.

————. 1974. Colonisation of the early ocean floor. Nature, 248:328–330.

Donahue, J. 1971. Burrow morphologies in north-central Pacific sediments. Marine Geol., 11:M1–M7.

Emery, K. O. 1964. Turbidites—Precambrian to present. Stud. Oceanogr., 1964:486–495.

Ericson, D. B. et al. 1961. Atlantic deep-sea sediment cores. Geol. Soc. America., Bull., 72:193–285.

Ewing, M. and R. A. Davis. 1967. Lebensspuren photographed on the ocean floor. In H. B. Hersey (ed.), Deep-sea photography. Johns Hopkins Oceanogr. Stud., 3:259–294.

Fenton, C. L. and M. A. Fenton. 1934. *Lumbricaria;* a holothuroid casting? Pan-American Geol., 61:291–292.

Frey, R. W. 1971. Ichnology—the study of fossil and recent lebensspuren. In B. F. Perkins (ed.), Trace fossils, a field guide. Louisiana State Univ., School Geosci., Misc. Publ. 71–1:91–125.

Glass, B. P. 1969. Reworking of deep-sea sediments as indicated by the vertical dispersion of the Austalasian and Ivory Coast microtectite horizons. Earth Planet. Sci. Letters, 6:409–415.

Gordon, D. C., Jr. 1966. The effects of the deposit feeding polychaete *Pectinaria gouldii* on the intertidal sediments of Barnstable Harbor. Limnol, Oceanogr., 11:327–332.

Griggs, G. B. et al. 1969. Deep-sea sedimentation and sediment-fauna interaction in Cascadia channel and on Cascadia abyssal plain. Deep–Sea Res., 16:157–170.

Hamilton, E. L. et al. 1956. Acoustic and other physical properties of shallow water sediments off San Diego. Jour. Acoustical Soc. America, 28:1.

Hanor, J. S. and N. F. Marshall. 1971. Mixing of sediment by organisms. In B. F. Perkins (ed.), Trace fossils, a field guide. Louisiana State Univ., School Geosci., Misc. Publ. 71-1:127–135.

Heezen, B. C. and C. D. Hollister. 1971. The face of the deep. New York, Oxford Univ. Press, 659 p.

——— et al. 1955. The influence of submarine turbidity currents on abyssal productivity. Oikos, 6:170–182.

Hertweck, G. 1970. The animal community of a muddy environment and the development of biofacies as effected by the life cycle of the characteristic species. In T. P. Crimes and J. C. Harper (eds.), Trace fossils. Geol. Jour., Spec. Issue 3:235–242.

Hessler, R. R. 1970. The Desmosomatidae (Isopoda, Asellota) of the Gay Head-Bermuda transect. Scripps Inst. Oceanogr., Bull., 15:1–185.

Hollister, C. D. and B. C. Heezen. 1972. Geologic effects of ocean bottom currents. In A. L. Gordon (ed.), Studies in physical oceanography—a tribute to George Wust on his 80th birthday. New York, Gordon & Breach, 2:37–66.

Hyman, L. H. 1955. The invertebrates: Echinodermata. New York, McGraw–Hill, 761 p.

Jones, N. S. and H. L. Sanders. 1972. Distribution of Cumacea in the deep Atlantic. Deep-Sea Res., 19:737–745.

Kraeuter, J. and D. S. Haven. 1970. Fecal pellets of common invertebrates of lower York River and lower Chesapeake Bay, Virginia. Chesapeake Sci., 11:159–173.

Ksiażkiewicz, M. 1970. Observations on the ichnofauna of the Polish Carpathians. In T. P. Crimes and J. C. Harper (eds.), Trace fossils. Geol. Jour., Spec. Issue 3:283–322.

Kuenen, P. H. 1968. Turbidity currents and organisms. Eclogae Geol. Helvetiae, 61:525–544.

Moore, H. B. 1939. Fecal pellets in relation to marine deposits. In P. D. Trask (ed.), Recent marine sediments. Amer. Assoc. Petrol. Geol., p. 516–524.

Morita, R. Y. and C. E. Zobell. 1955. Occurrence of bacteria in pelagic sediments collected during the mid-Pacific expedition. Deep-Sea Res., 3:66–73.

Piper, D. J. W. and N. F. Marshall. 1969. Bioturbation of Holocene sediments on La Jolla Deep–Sea Fan, California. Jour. Sed. Petrol., 39:601–606.

Reineck, H.-E. 1973. Schichtung und Wühlgefüge in grundproben vor der Ostafriknischen Küste. "Meteor" Forsch. Ergebnisse, 16:67–81.

Richards, A. (ed.) 1967. Marine geotechnique. Chicago, Univ. Illinois Press, 327 p.

Risk, M. J. 1973. Silurian echiuroids: possible feeding traces in the Thorold Sandstone. Science, 180:1285–1287.

Sanders, H. L. and R. R. Hessler. 1969. Ecology of deep-sea benthos. Science, 163:1419–1424.

——— et al. 1965. An introduction to the study of deep-sea benthic faunal assemblages along the Gay Head-Bermuda transect. Deep–Sea Res., 12:845–867.

Scholle, P. A. 1971. Sedimentology of fine-grained deep-water carbonate turbidites, Monte Antola flysch (Upper Cretaceous), northern Apennines, Italy. Geol. Soc. America, Bull., 82:629–658.

Seilacher, A. 1958. Zur ökologischen Charakteristik von Flysch und Molasse. Eclogae Geol. Helvetiae, 51:1062–1078.

———. 1962. Paleontological studies on turbidite sedimentation and erosion. Jour. Geol., 70:227–234.

———. 1967. Bathymetry of trace fossils. Marine Geol., 5:413–428.

———. 1974. Flysch trace fossils: evolution of behavioral diversity in the deep-sea. Neues Jahrb. Geol. Paläont., Mh., 1974:233–245.

Shinn, E. A. 1968. Burrowing in recent lime sediments of Florida and the Bahamas. Jour. Paleont., 42:879–894.

Sokolova, M. N. 1959. On the distribution of deep-water bottom animals in relation to their feeding habits and the character of the sediment. Deep-Sea Res., 6:1–4.

Stanley, D. J. 1971. Bioturbation and sediment failure in some submarine canyons. Vie et Milieu, 3rd Symp. Europé. Biol. Marine, Supp. 22:541–555.

Warme, J. E. et al. 1973. Biogenic sedimentary structures (trace fossils) in Leg 15 cores. Deep-Sea Drill. Proj., Initial Repts., 15:813–831.

PART V

Techniques in the Study of Lebensspuren

CHAPTER 22

EXPERIMENTAL APPROACHES IN NEOICHNOLOGY

CHRISTOPHER A. ELDERS

New Hampshire Vocational–Technical College
Claremont, New Hampshire, U.S.A.

SYNOPSIS

A variety of experimental approaches have been taken in behavioral studies of mobile invertebrates, ranging from simple observations in aquaria to the use of sophisticated equipment. Studies involving these approaches are of greatest benefit to ichnology when destruction of primary sedimentary structures and formation of biogenic structures are also noted, but behavioral studies themselves are of practical value in reconstructing paleoethologies.

Invertebrates may be split into two artificial groups, according to their respective burrowing mechanisms. "Soft-bodied" animals rely on the generation of high internal-fluid pressures within specific body parts to produce anchors in the sediment. The musculature can then work against these anchors to pull the organism forward. "Hard-bodied" invertebrates rely on the action of appendages or spines to effect

locomotion. The similarity of mechanisms within each group is due to limitations placed on the organism by the substrate in which it must move and by the required life functions of the animal itself.

In general, however, no such clear, parallel division is seen between sedimentary traces produced by members of the two groups. This disparity is due to the large variety of specific adaptations shown by these animals, and the somewhat greater independence of these individual adaptations from the sedimentary environment, relative to possible modes of locomotion. Herein lies the main justification for experimental neoichnology: to study these adaptations and their role in the formation of specific traces by specific organisms, thereby gaining information that is invaluable in interpretations of trace fossils.

INTRODUCTION

This chapter deals with experimental laboratory studies in neoichnology; such studies are valuable because of the importance of neoichnology in the interpretation of trace fossils. Neoichnology lays a foundation for consideration of fossil lebensspuren, as well as making paleontological information gained from present-day traces useful without comparable fossil examples necessarily being considered (e.g., Seilacher,

1951). Ladd (1959), Schäfer (1972), and Frey (Chapter 2) also stressed the importance of the study of present-day organisms and their habits in the interpretation of body fossils and trace fossils. Certain groups of animals (e.g., bivalves) have been studied more than others, of course, and therefore are better represented in this chapter.

A series of approaches that asymptotically approximates natural conditions has been taken in experimental neoichnology. The simplest is a natural history approach, involving only the observation of burrowing among given organisms placed in aquaria. Description centers on these observations.

A second approach, similar to the first, relates the functional morphology of the animal to its burrowing, as observed in aquaria. This work often involves extensive anatomical description, emphasizing the adaptations of various organs for burrowing.

Unfortunately, little or no quantification is involved in these first two approaches. In addition, observations are limited to features or behavior that can be seen on or near the substrate surface; this restriction is significant, because differences may exist between initial burrowing at the surface and subsequent burrowing once a firm anchorage is achieved within the substrate. Recently, A. D. Ansell and E. R. Trueman, in a series of studies (see respective references), partly overcame these problems by using pressure moment and force transducers connected to pen recorders, and photographic analysis. Another group of workers (Howard, 1968; Howard and Elders, 1970; Stanley, 1970; Frey and Howard, 1972) has used x-ray radiographic techniques to study burrowing within sediments, thereby minimizing disturbances to the organism.[1]

The use of such sophisticated equipment is of limited benefit to ichnology, however, if the researcher describes only burrowing behavior and not its effect upon the substrate. Few studies take this aspect into account, yet only in such work do we see specific discussion of traces that may later be preserved. Much of the neoichnological work done in Germany conformed to this "sedimentological" approach (e.g., Schäfer, 1972, and references cited).

The final approach is a dynamic one, in which such sedimentary processes as erosion and sedimentation are simulated, and the resulting biogenic sedimentary structures investigated. This procedure most nearly approximates natural conditions.

Inherent in all of these approaches is the artificial restriction of organisms within aquaria or other experimental apparatus; the result may be a modification of animal behavior. Therefore, field verification of experimental results is desirable. (Numerous examples of field observations could indeed be cited, but I have confined my review exclusively to laboratory observations.)

Evaluation of work done within the above approaches also reflects the viewpoint of the investigator. The simple and the more sophisticated observational approaches have been taken largely by zoologists, whereas geologists have usually studied the sedimentary structures produced under both static and dynamic conditions. Workers are now needed to bridge this gap, by relating functional morphology directly to the structures produced in sediment during burrowing.

In this review I have concentrated strictly upon burrowing marine animals. Boring and "rooting" organisms are equally intriguing, as are nonmarine burrowers, and references to these may be found in Chapters 10 to 19.

The burrowing animals discussed in this review may be separated into two major groups: one that either lacks hard skeletal parts (worms) or does not rely exclusively upon hard parts for burrowing (mollusks), and one that uses spines or appendages in burrowing (decapods, echinoids). "Soft-bodied" animals are treated first.

"SOFT-BODIED" INVERTEBRATES

Mollusca

Bivalvia

One of the most obvious groups of burrowing organisms is the bivalve mollusks. Early

---

[1] The effects of the x-rays themselves must of course be considered.

workers studying the primitive protobranch mollusks (Drew, 1900; Vlès, 1906) recognized the importance of the foot as an organ of penetration and anchorage in burrowing. Indeed, this organ is responsible for locomotion in almost all bivalves; however, specific adaptations of the foot for burrowing differ in various bivalve groups. In protobranchs, such as *Nucula, Solemya,* and *Yoldia,* the foot consists of two ventral muscular flaps, which develop side by side. This foot is thrust into the sediment and the flaps are spread, forming a terminal anchor against which pedal retractor muscles work to pull the animal down. The protobranch foot is generally found in the anterior half of the mantle cavity; burrowing is thus directed diagonally downward, the anterior end foremost (Yonge, 1939; Owen, 1961). Among razor clams (Superfamily Solenacea), the foot is usually cylindrically shaped and very extensible at the tip. This organ "bores" into the substrate until fully extended, then swells at the tip to a size twice the diameter of the remainder of the foot (Drew, 1907). The terminal swelling forms a pedal anchor, against which retraction occurs. Pedal emergence generally is at the anterior end, causing burrowing to be vertical. Drew also noted the use of the valves as a second (penetration) type of anchor, in which the valves open and hold the organism in position as the foot extends into the substrate.

Another characteristic trait of bivalve burrowing is the expulsion of a water jet from the pedal gape. First seen in *Ensis directus* (Drew, 1907), the jet loosens and washes away sand from beneath the shell, aiding in its downward movement. Water expulsion occurs just before final pedal retraction and is produced by shell adduction after the siphons and ventral mantle opening are closed.

Trueman (1966a) emphasized the fluid dynamics involved in protraction and dilation of the foot, and in production of the water jet. The bivalves' musculature works against a hydrostatic skeleton consisting of blood in the hemocoele and water in the mantle cavity. As a result, high internal pressures are produced during adduction and pedal retraction. Adduction pressures produce water ejection and terminal anchorage of the foot; this swollen foot is maintained during downward movement by pressures generated through retraction. Some deep-burrowing forms, such as *Mya* and *Ensis,* show extensive fusion of the marginal mantle folds—except at the siphonal and pedal (and sometimes the fourth pallial) apertures—which is used to regulate these pressures.

A third characteristic of burrowing seen in certain bivalves is a rocking around an imaginary axis joining the valves; the exact position of the axis depends on the arrangement and strength of pedal retractors. Vlès (1906) attributed this rocking, observed in *Glycymeris (Pectunculus) glycymeris,* to successive contraction of the anterior and posterior pedal retractor muscles. *Mactra glauca* and *M. corallina* exhibit secondary pedal retractor contractions, producing a double rocking motion with each downward movement (Trueman, 1968c). The need for rocking during burrowing is apparently related to the degree of shell elongation (Stanley, 1970). The result is that circular bivalves employ a large angle of rotation and go vertically downward. Moderately elongate species utilize some rotation, which often produces a forward component in downward movement due to the eccentric axis of rotation. Very elongate species utilize little or no rocking; they penetrate vertically or nearly vertically, if pedal emergence is terminal (*Solemya, Ensis*), or obliquely, if pedal emergence is oblique (*Yoldia, Macrocallista*).

Integration of these characteristics into a general scheme of bivalve burrowing was attempted by Ansell (1962). More recently, Trueman et al. (1966a) described the digging of *Tellina tenuis, Macoma balthica, Donax vittatus,* and *Cardium edule* as occurring in two phases. The first phase involves foot probing and anchorage, followed by movement raising the shell to a vertical position after it has been lying on the substrate surface. This is followed by the second phase, in which cycles of digging movements lead to burial (Fig. 22.1). Each cycle consists of six stages: (1) foot probing, accompanied by slight lifting of the shell, which acts as a penetration anchor; (2) siphon closure and continued probing; (3) adduction, producing increased internal pressure that leads to dilation and terminal anchoring of the foot, and water ejection from the mantle cavity; (4) pedal retractor contraction, causing the shell to be pulled into the sand, sometimes accompanied by a rocking motion caused by sequential contraction of the anterior and posterior retractors; (5) adductor relaxation, followed by valve gaping and loss of pedal anchorage; and (6) a static period, during which foot probing recommences, while the shell gapes to act as an anchor.

These two phases of burrowing are

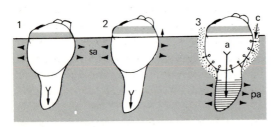

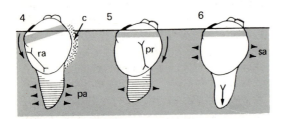

Fig. 22.1 Positions of a bivalve during six stages in the burrowing cycle. Emphasized are the terminal pedal (pa) and shell (sa) anchors, adduction (a), water ejection (   ), sand loosened around the shell (c), foot probing (   ) and contraction of the anterior (ra) and posterior (pr) retractor muscles. (Modified from Trueman, 1968b.)

generally characteristic of all bivalve mollusks. However, some notable exceptions are known in terms of the details of digging cycles. In *Lyonsia norvegica* (Ansell, 1967), movement is slow because stage 3, where terminal pedal anchorage and water ejection normally occur, is poorly developed; in fact, no water is ejected through the pedal gape. In *Glycymeris glycymeris* (Ansell and Trueman, 1967a), the mantle cavity extends so far dorsally that the visceral mass is not in contact with the valves. For this reason, adduction only indirectly effects the hemocoele, and transverse muscles (*G. glycymeris* has more of these than most other bivalves) assume the role of maintaining pressure in the foot during digging. Swelling of the foot therefore occurs in stage 1 of the digging cycle, rather than stage 3, and anchorage is increased in stage 4 by spreading of the two pedal flaps.

Mercenaria mercenaria (Ansell and Trueman, 1967b) differs slightly from most bivalves in that siphonal closure, characteristic of stage 2, may precede final foot probing (stage 1). Also, later digging cycles exhibit a secondary phase of siphonal movement, between stages 5 and 6, in which the siphons close and withdraw for about five seconds; reopening of the siphons follows their slow reextension to the surface. Movements of the siphons are accompanied by similar movements of the foot. These secondary movements may increase pressure in the hemocoele and mantle cavity, to aid in relaxation of the adductors and in opening the valves. Such aid is necessary because of the presence in *Mercenaria* of a weak hinge ligament. A secondary opening movement is also found in *Margaritifera margaritifera* (Trueman, 1968a) and in *Petricola pholadiformis* (Ansell, 1970). In all other respects, however, the burrowing of the above bivalves conforms to the six-stage cycle of other bivalves.

As a result of the repetition of this burrowing cycle, bivalves move into the safety and stability of the sediment, usually having their valves oriented vertically. However, certain deposit-feeding tellinid bivalves are known to occupy a horizontal position in the substrate (Holme, 1961). In order to assume this position, various species of *Tellina* turn themselves within the substrate. *Arcopagia crassa* burrows obliquely when placed on its left side, but turns itself over before burrowing when placed on its right side. Because these species are surface deposit feeders, they must be able to move easily on the horizontal plane, and maintaining a horizontal position facilitates such movement.

Consideration of some of the adaptations to burrowing in bivalves, relative to the mechanical aspects of sediment penetration, is pertinent here. Studies of both *Ensis* (Trueman, 1967) and *Glycymeris* (Ansell and Trueman, 1967a) show that, as sediment grain size is increased experimentally, weaker anchorage in retraction tends to be balanced by easier penetration of the valves. However, this relationship is modified in nature by functional adaptations, including water ejection from the mantle cavity, rocking of the shell, and gaping of the shell during burrowing (Nair and Ansell, 1968). Stanley (1970) evaluated the relation of shell morphology to such aspects as burrowing rate and depth, and shell orientation after burrowing. Some of the shell characteristics considered were obesity, elongation, thickness, position of pedal and siphonal gapes and maximum shell width, and shell ornamentation. Stanley found that most rapid burrowers are cylindrical, blade-like, or disk-like. Nearly all are also thin shelled. Thick shells are characteristic of shallow burrowers, which sacrifice burrowing speed for stability (Fig. 22.2).

Once buried, bivalves must create a water current flowing into the mantle cavity, to satisfy their respiratory and nutritional needs. Exhalent currents are also required, to rid the organism of gaseous and fecal wastes. In fact, although bivalves show various specific adaptations in shape and construction, a basic scheme of endobenthic existence is imposed by these necessary life functions (Schäfer, 1956). In *Solemya parkinsoni*, an inhalent current is maintained at the anterior end, around the pedal gape, and an exhalent current flows out a posterior aperture (Owen, 1961); in *S. velum*, the limbs of its U- or Y-shaped burrow function as siphonal passages leading to the substrate surface (Stanley, 1970). However, most bivalves maintain a connection with the surface by means of true inhalent and exhalent siphons, which may be either separate or fused. Where the siphons are short, as in *Corbula (Aloidis) gibba* (Yonge, 1946a), the organism lies flush with, or just beneath, the sediment surface. Longer siphons permit greater depths of burial (Fig. 22.2).

Among certain bivalves, deep siphons actually lie in specially constructed siphonal tubes. In the superfamily Lucinacea (Allen, 1958a), the slender, vermiform foot is extensible to four or five times the body length, and can swell at the tip even when extended. By emerging at the anterior end of the anterior adductor and forcing its way to the substrate surface after burial of the organism, the foot forms a mucus-lined, anterior, inhalent siphonal tube leading to the surface. In the family Lucinidae, the foot also has a well-developed heel, which can be expanded by itself. In this family, the heel is used almost exclusively as a plow in burrowing, whereas the vermiform part of the foot is used almost exclusively for tube building. Siphonal tube construction is also exhibited by *Cochlodesma praetenue* (family Laternulidae), which lives horizontally, often on its right valve, at a sediment depth of about 7 cm (Allen, 1958b). An inhalent siphonal tube opens to the surface, and an exhalent tube extends horizontally into the substrate for about 7 cm. The siphons lie in mucus-lined tubes, which in the lab are rebuilt by the siphons every 12 to 72 hours.

Because of the general similarity in life-styles and adaptations shown by in-

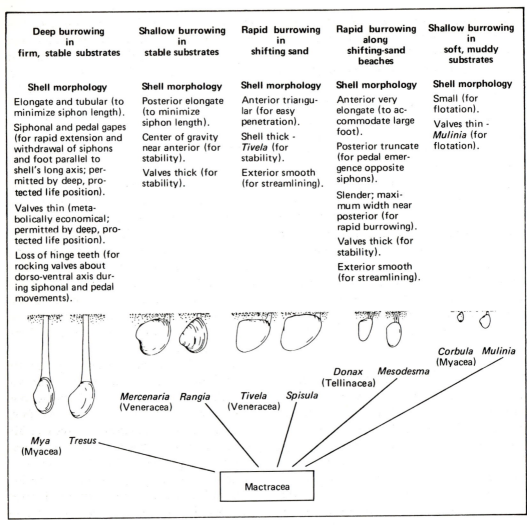

Deep burrowing in firm, stable substrates	Shallow burrowing in stable substrates	Rapid burrowing in shifting sand	Rapid burrowing along shifting-sand beaches	Shallow burrowing in soft, muddy substrates
Shell morphology	**Shell morphology**	**Shell morphology**	**Shell morphology**	**Shell morphology**
Elongate and tubular (to minimize siphon length).	Posterior elongate (to minimize siphon length).	Anterior triangular (for easy penetration).	Anterior very elongate (to accommodate large foot).	Small (for flotation).
Siphonal and pedal gapes (for rapid extension and withdrawal of siphons and foot parallel to shell's long axis; permitted by deep, protected life position).	Center of gravity near anterior (for stability). Valves thick (for stability).	Shell thick - *Tivela* (for stability). Exterior smooth (for streamlining).	Posterior truncate (for pedal emergence opposite siphons).	Valves thin - *Mulinia* (for flotation).
Valves thin (metabolically economical; permitted by deep, protected life position).			Slender; maximum width near posterior (for rapid burrowing).	
Loss of hinge teeth (for rocking valves about dorso-ventral axis during siphonal and pedal movements).			Valves thick (for stability). Exterior smooth (for streamlining).	

Fig. 22.2 Relation of various bivalve morphological characteristics to the life style of the organisms. Diagram stresses adaptive radiation of the Mactracea. (Modified from Stanley, 1970.)

faunal bivalves, these organisms leave relatively few characteristic types of traces in the sediment. Bivalves having movements involving large horizontal components leave furrows in the surface layers (Yonge, 1939). Shallow vertical burrowing often results in funnels that are V shaped in cross-section, as with *Donax variabilis* and *Mulinia lateralis* (Frey and Howard, 1972). Deep vertical burrowing ordinarily leaves a cylinder of bioturbated sediment above the animal, approximately the width of the shell, except when this burrowing is the direct result of growth. In this latter case

(Fig. 8.5A1), the cylinder of reworked sediment increases in width with increasing depth in the substrate (Reineck, 1958). Also, where sedimentary laminae are thick relative to the size of the bivalve, pocket-like bending of the laminae may occur (Schäfer, 1956). Distinct, elongate, smooth-walled burrows are produced only by bivalves that habitually move about within their domiciles.

During conditions of slight sedimentation or erosion, any sediment falling into the burrow is usually incorporated into the burrow walls; the result is a secondary lining, which Reineck (1957) termed "raumauskleidung" (as opposed to a lining arising from actual construction of the burrow itself, or "bauauskleidung"). An example is the behavior of *Mya arenaria*; sediment falling down its siphonal shaft is pressed into the shaft walls and held in place by mucus. In cross-section, the shaft appears as a series of concentric layers. In contrast, marked erosion often leads to attempts by the animal to dig down, leaving a new structure above the bivalve that is as wide as the shell (Fig. 8.5A3). More often, however, the animal is washed free of the sediment, whereupon it reburrows quickly, is swept away, or dies on the surface. With rapid sedimentation, bivalves produce "climbing" (escape) structures below themselves, the width of the shell; these structures often display cup-in-cup layering and downward bending of adjacent laminae.

Whereas these individual traces are local in extent, populations of bivalves can have more widespread effects on the sediment. Rhoads (1963) found that *Yoldia limatula* sits more than half-covered with sediment, feeding by means of specialized palps. Fecal pellets are ejected from siphons that project at an angle from the substrate; the result is a mound of sediment in which effective grain size is larger than that of the surrounding sediment. This removal of sediment from below the depositional interface, and its reintroduction into the water, also distorts or obliterates primary sedimentary structures. Rhoads calculated that the population of *Yoldia* in each square meter of Buzzards Bay reworks 5,664 ml of sediment per year in this way.

Rhoads and Young (1970) found that populations of *Nucula proxima* placed in aquaria create a granular sediment surface of sand-size pellets and mud clasts. The pellets arise from reworked silt- and clay-size particles, and the clasts from displacement during lateral movement of *Nucula*. Oscillating currents ranging from 1.3 to 13 cm/sec were generated above this

granular surface; these experiments demonstrated that greater resuspension and turbidity occur over the reworked sediment when current speeds exceed 4 cm/sec. Water content of the sediment is also increased by the burrowing of *Nucula*.

Gastropoda

The gastropod foot is a flat sole adapted for creeping over various substrates. Most of the early work on gastropod locomotion (Gersch, 1934; Lissman, 1945; Vlès, 1913) dealt with the production and role of muscular waves passing along the foot, either in an anterior (direct) or a posterior (retrograde) direction; these aspects of behavior are not discussed in detail herein. The various workers agreed that these waves represent the main locomotor force in most gastropods.

Burrowing in gastropods is apparently a continuation of normal surface locomotion. Among naticid gastropods, monotaxic locomotor waves passing along the foot, from posterior to anterior, are responsible for movement (Trueman, 1968d). These waves are produced by the columellar muscle and other pedal muscles, operating antagonistically on a fluid skeleton formed of blood in the pedal hemocoele and, in some species, water taken up by the animal.

Burrowing in *Bullia melanoides* is also a continuation of surface locomotion (Ansell and Trevallion, 1969), involving retrograde monotaxic waves passing along the foot. These waves pass sand up and back, over lateral parts of the foot, as the propodium is extended. The shell is then drawn down and forward over the foot, by columellar muscle contraction. However, propodial anchorage is not as important in naticid gastropods.

Gastropods exhibit a four-stage burrowing cycle (Trueman, 1968d), comparable to the six-stage cycle of bivalves. The gastropod stages are: (1) cessation of propodial extension and onset of its dilation, through blood forced forward by dorsoventral muscle contraction; (2) maximum propodial dilation and pedal retraction, which draws the shell and posterior part of the foot into the sediment; (3) a decrease in propodial dilation and a relaxation of the columellar muscles; and (4) propodial extension and mesopodial swelling, the shell acting as a penetration anchor. Water may supplement the hydrodynamic role of blood among some species (Brown, 1964).

Burrowing gastropods must create water currents to oxygenate the gills, and these currents are often passed through connections maintained with the substrate surface. Once *Turritella communis* is buried, for example, its foot creates an inhalent depression by pushing mud to the right; the sediment, partly consolidated by mucus from pedal glands, forms a low mound in front of the head (Yonge, 1946b). An exhalent siphon projects out of a smaller depression found to the right of the mound, in front of the head. The gastropod *Aporrhais pes-pelecani* reportedly uses its proboscis to form incurrent and excurrent openings to the surface, once buried (Yonge, 1937).

Little work has been done on the traces left by gastropods. However, due to the horizontal mode of burrowing of gastropods, and their lack of extensible siphons, gastropod traces should be expected only on or near the surface. Howard (1968) found that movement in *Polinices duplicatus* creates sand ridges ahead and alongside the organism, and a furrow behind. Comparable features have also been observed among other species (Frey and Howard, 1972). Similar traces by *Littorina littorea* consist of a median band and two sidewalls (Gräf, 1956). The median band may be either smooth or segmented, and the sidewalls smooth, segmented, or absent; six kinds of trail formations are thus possible. Segmentation of the median band consists of forward-convex cross-grooves and elevated intergroove areas. Sometimes a furrow or ridge is seen in the middle of the median band; this is caused by the median

line, which divides the sole of the snail foot into two halves. Relief along the trail is slight in finer size sediments; an increase in water content of the sediment produces a deeper trail, even though less mucus is secreted.

Schäfer (1956) described the trace of *Lunatia nitida* as being similar to the feature produced by a ball moving through the substrate. A completely bioturbated zone is formed, at the edge of which horizontal sediment layers are deformed.

Scaphopoda

As with other mollusks, scaphopods depend on the foot in burrowing, first to penetrate the substrate and then to act as a terminal anchor (Morton, 1959). The foot is long and tubular, and is encircled distally by an erectile fold of tissue (the epipodium), into which the pedal hemocoele extends (Trueman, 1968e). The digging cycle occurs in four stages: (1) foot dilation begins terminally and spreads to the epipodium, which erects to form a wide, circular, flange anchor, (2) contraction of the paired, longitudinal retractor muscles pulls the shell down, after which (3) pedal dilation is lost and the epipodial lobes flatten against the foot; pedal protraction also occurs, followed by (4) probing caused by contraction of circular pedal muscles. During stage 4, the shell acts as a penetration anchor; the shell's effectiveness is enhanced by its posteriorly tapering shape. The first digging cycle always involves a twisting of the shell to bring the concave dorsal side uppermost. In comparing this four-stage digging cycle to the six-stage cycle common in bivalves, Trueman (1968c) stressed that both scaphopods and bivalves use essentially the same two types of anchors in burrowing —a terminal pedal anchor and a penetration (shell) anchor.

One trace reportedly left by *Dentalium conspicuum* is produced during the righting reflex; the shell twists at the onset of burrowing, to bring the dorsal concave side upward (Dinamani, 1964). On mud, a mark is left by the shell apex, indicating the direction and extent of the twist. Also, after entry into the sediment, a saucer-shaped depression is generally seen above the front end of the animal (Fig. 22.3). The trace is distinct in fine sand and mud but not in coarser sediments. This structure is due to action by the foot, creating a cavity in front of the organism in which the feeding organs, or captacula, can work. As the cavity is cleared, some particles trickle down into it from above, creating the surface depression.

Cephalopoda

Dorso-ventral elongation, shell reduction, and body fins are adaptations to a pelagic way of life among some of the highly developed cephalopods. However, many cephalopods bury themselves in sediment during the day and come out to feed at night.

Von Boletzky and von Boletzky (1970) described the burrowing of several species of *Sepiola* and *Sepietta*. The process consists of two phases: (1) the siphon washes away sand from under the body, and (2) the dorso-lateral arms rake sand over the body. Apparently, the fins are used for

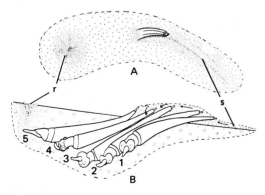

Fig. 22.3 Lebensspuren produced by scaphopod *Dentalium conspicuum*. A, surface view and B, a cross-section, showing markings left by the righting reflex (s) and in the feeding cavity (r). Numbers 1 to 5 indicate successive positions of organism during the righting reflex. (After Dinamani, 1964.)

burial only where the sediment is very coarse relative to the body size of the animal.

Only the traces of *Heterosepiola atlantica* have been described (Shäfer, 1956); these are pockets of bioturbated sediment. No special traces are produced during active sedimentation, because cephalopods normally leave the sediment under these conditions.

Annelida

Polychaete annelids have also been obvious subjects for investigation, because they are abundant, diverse, and well adapted to an infaunal existence. One of the most intensively studied polychaetes is *Arenicola marina*. In this organism, body fluids and body-wall muscles act to produce a "resting" coelomic pressure, the chief role of which is to maintain muscle tone. Maximum pressure during burrowing (as much as 40 times the resting pressure) lasts only 1 to 3 seconds, but is associated with definite muscular configurations. However, Trueman (1966b) found that a gradual increase in the magnitude of major pressure maxima occurs only after the second annulus (fleshy parapodial ridge) has penetrated. For this reason, he felt that the pressure peaks are related to dilation of the anterior end, serving as a terminal anchor while the posterior end of the worm is pulled into the substrate. A penetration anchor, formed by raising the fleshy parapodial ridges on the first three segments, is utilized during proboscis eversion (Fig. 22.4). Thus, burrowing starts by proboscis eversion, without high coelomic pressures, to affect initial penetration. The terminal anchor, accompanied by high coelomic pressure and the pulling of the posterior part of the worm into the burrow, then alternates with the penetration anchor, accompanied by body lengthening and proboscis eversion. Recently, Seymour (1971), using the techniques pioneered by Ansell and Trueman, dis-

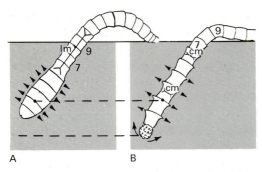

Fig. 22.4 Burrowing of polychaete *Arenicola marina*, using the terminal (A) and penetration (B) anchors, produced by action (▶) of the circular (cm) and longitudinal (lm) muscles. (Segments 7, 9 for reference.) (Modified from Trueman, 1966b.)

cussed in even greater detail the muscular configurations and resulting coelomic pressures in *Arenicola*.

The polychaete *Nephthys* lacks the circular muscle that, in *Arenicola*, usually acts counter to longitudinal muscles (Clark and Clark, 1960). In *Nephthys*, dorsoventral muscles are used; contractions in the longitudinal plane are compensated by changes in segment height rather than width. The proboscis, everted by coelomic pressure generated by longitudinal muscle contraction, penetrates the sediment while the worm is anchored by the widest segments (15 through 45). Forward movement is then by undulatory waves, similar to those seen in swimming. The backward power stroke is performed by parapodia of the first fifteen segments, as they make contact with the burrow wall. Thus, because *Nephthys* tapers in shape at both ends and lacks circular muscles, anchorage can be affected only by a short part of the body. Therefore, the peristaltic burrowing movements seen in *Arenicola* would be ineffective in *Nephthys*.

Intermediate between *Arenicola* and *Nephthys* is the polychaete *Glycera alba* (Ockelmann and Vahl, 1970). In this organism, the proboscis is initially thrust into the sediment; here it swells distally to provide an anchor, against which the body can be pulled into the sediment. However,

proboscis eversion terminates after burial is complete; subsequent burrowing is a combination of parapodial creeping and peristaltic body movements.

The role of the proboscis in burrowing by these polychaetes is intimately related to thixotropic qualities of the sediment. In *Arenicola marina* (Chapman and Newell, 1947, 1948), the force that the worm can exert for its initial thrust, by itself, is not sufficient for penetration; this force cannot exceed the weight of the worm. Therefore, prodding of the sand by extrusion of the proboscis decreases the resistance of the sand and converts it to a semifluid, into which the anterior part of the worm is thrust. The sand then drains and becomes firmer, allowing a grip to be obtained by dilation of the anterior chaetigerous segments of the worm. Finally, longitudinal muscle contractions draw the worm forward.

The variety of adaptations by polychaetes results in numerous types of lebensspuren left in sediment. Significantly, the shape and structure of these traces depend on the morphology and necessary life functions of the organisms. For example, the burrowing activities of *Arenicola* can result in a U-shaped burrow consisting of a sediment-filled "head shaft," a "gallery," and an open "tail shaft" (Fig. 22.5). Wells (1945) indicated that feeding, burrow irrigation, and upward excursion of the worm are all involved in setting up the head shaft. (See also Jacobsen, 1967.) Contrary to earlier thinking, the tube of *Glycera alba* (Ockelmann and Vahl, 1970) is fairly permanent. It consists of an initial vertical tube, from which a system of galleries extend. The deepest parts of the system are "retreat tubes" and the shorter ones reaching the surface are "waiting tubes," according to their function in the life of the animal. Finally, Nicol (1930) demonstrated that the morphology of the tube of *Sabella pavonina* is intimately related to the functional anatomy of the worm. Mud particles are stored and coated

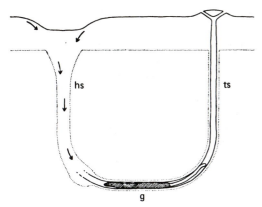

Fig. 22.5 Diagrammatic representation of *Arenicola* burrow, showing the head shaft (hs), gallery (g), and tail shaft (ts). Arrows indicate movement of sand down the head shaft, allowing *Arenicola* to feed inside its burrow. (Modified from Wells, 1945.)

with mucus in ventral sacs lying just under the mouth, and are then formed into a mucus string by the parallel folds below the ventral sacs. The string is passed on to the collar folds, which lay the string along the tube edge, and press and cement it in

this position. This mode of construction often results in striated tube walls.

Many vagile polychaetes penetrate the sediment fairly cleanly, leaving a simple cylindrical burrow which may or may not collapse and which may be surrounded by a zone of flexed sedimentary laminae. In many species, however, fixed tubes are constructed. The dwelling tube of *Lanice conchilega* is U- or W-shaped (Fig. 22.6), the latter arising through growth of the worm; *Lanice* uses one limb of its old tube, onto which it adds a new bend and second limb (Seilacher, 1951). The unused part of the old tube is then either partly filled with sediment and feces, or is sealed off. The tube mouth is characteristically surrounded by sand fringes (Fig. 22.6B), and is often split into two fan-like lips that are oriented perpendicular to the predominant current. Using artificially produced currents, Ziegelmeier (1969) found that the tube crown is built so that the fringed, fan-like lobes are transverse or perpendicular to the current direction. Best results were achieved with

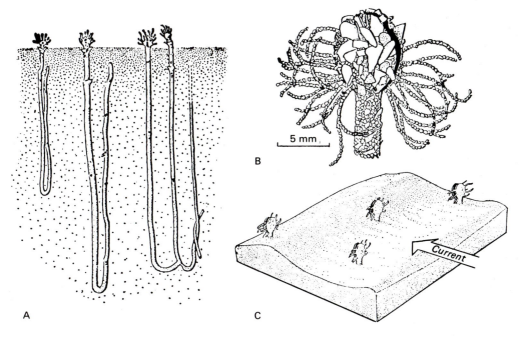

Fig. 22.6 Dwelling tube of polychaete *Lanice conchilega*, illustrating U and W forms (A), and tube aperture (B) having two lips and sand fringes, oriented perpendicular to currents (C). (Modified from Seilacher, 1951.)

the fastest current used (60 mm/sec). In addition, fringes always border those parts of the tube directed toward or away from the current; however, fringes are added indiscriminately in still water. The tube has an inner skin-like layer and an outer layer of transient bottom materials (Ziegelmeier, 1952), and projects 20 to 30 mm above the substrate surface. Ziegelmeier found that no selection of either size or composition of particles is evident in the tube, and that the shaft is produced simply by sticking bottom material to the mucus produced by the animal, rather than by active insertion.

The tube of *Chaetopterus variopedatus* (Enders, 1909) is also U- or W-shaped. It consists initially of mucus, which hardens into a parchment-like material, giving the tube a laminated appearance. However, the tube of *Chaetopterus* is atypical of the Chaetopteridae (Barnes, 1965), which generally build straight vertical tubes having one opening that projects above the substrate; the tube has an organic, secreted layer, which may be parchment-like or cornified, and usually has a series of transverse partitions.

Durable vertical tubes are also characteristic of the Onuphidae. The position or configuration of these structures is ordinarily constant, but *Diopatra cuprea* is able to adjust the upper part of its dwelling tube by moving it through the sediment. Such movements may produce conspicuous funnel-like areas of deformed sediment laminae surrounding the tube (Frey and Howard, 1972).

The reactions of polychaetes to various sedimentary conditions, and the resulting traces, are fairly standard. Under conditions of slight sedimentation or erosion, polychaetes either extend or close off the burrow opening (*Heteromastus filiformis, Magelona papillicornis, Nephthys hombergii*), or press the incoming sediment into the burrow walls, as does *Nereis diversicolor* (Reineck, 1958). In the latter case, upward displacement of the burrow may result from repeated sedimentation, resulting in thicker burrow linings on the floor than on the roof (Reineck, 1957).

Rapid accumulation of sediment elicits the formation of escape structures to the surface; these differ from living and feeding structures by being vertical and straight, and by lacking mucus linings or multiple walls. These structures are surrounded by downward-flexed sedimentary laminae. Also, the number or density of escape tunnels is less than that of "permanent" burrows in the new horizon of settlement. Ziegelmeier (1952) showed that *Lanice conchilega*, already established in an aquarium, is forced to make such escape structures when buried by new sediment. In this case, a new fringed crown, characteristic of *Lanice* tubes, is made at the tube's upper end. In areas of pronounced sedimentation, therefore, the result is a long tube having a series of crowns at various levels indicating previous substrate surfaces.

At the population level, Schäfer (1952) showed that benthic animals can produce biogenic layering in areas of irregular sedimentation. Fecal pellets arising from such organisms as the polychaetes *Nereis* and *Heteromastus* accumulate on the surface during a period of little or no sedimentation. When sedimentation resumes, the sedimentary layer of biogenic origin is buried.

Biogenic grading of sediments can also be produced by populations of certain polychaetes, such as *Clymenella torquata* (Rhoads and Stanley, 1965). This sorting is generally due to the worms' tendency to selectively ingest and move medium-size sand to the surface. As a result, coarser grains—too large for ingestion—are concentrated around the base of the tubes. Because of the removal of finer sediment grades, slumping and other disruption of laminae occur, and sediment permeability increases. Based on a natural density of 318 worms per m^2, 87 liters of sediment are reworked per m^2 per year in this way.

Similar measurements by Gordon (1966) on *Pectinaria gouldii* indicate that in every square meter, a population of this worm works the sediment to a depth of 6 cm in 15 years; each worm reworks 400 cm³ per year. *Pectinaria* sits obliquely just below the surface, digging with special setae. Sand is transported either through the gut or between the body and tube; either way, sediment is deposited as a small mound near the posterior end of the animal. Caverns excavated by the worm in this way continually collapse. Although Gordon found no sorting of sediment by the field population in Barnstable, Massachusetts, lab experiments showed sorting to occur where the substrate contains grains larger than 1 mm, which *Pectinaria* is unable to rework.

Coelenterata

Actiniaria

Recent work on *Peachia hastata* (Ansell and Trueman, 1968) and *Phyllactis* sp. (Mangum, 1970), indicates that these coelenterates utilize peristaltic waves of circular-muscle contraction to burrow. The waves pass from the anterior (scaphus) region to the base (physa), forcing fluid toward the base to produce physa eversion. This action displaces sand, forming a substrate depression into which the column can move. The eversion is immediately preceded by base introversion, produced by retractor muscles in the column. Once initial penetration occurs, further burial is accomplished by the use of two anchors: terminal physa dilation and anchorage, accompanied by pulling of the column into the sand, alternate with penetration anchorage by the column, during which physa eversion occurs, until burial is complete. These physa and column anchors are analogous to the dilation and flange anchors of *Arenicola* (cf. Trueman, 1966b), and to the pedal and shell anchors of bivalves.

Actinians build no tube or permanent burrow, and therefore leave only pockets of bioturbated sediment. However, upward peristaltic movements elicited by slow sedimentation often leave vertical sac-shaped zones in the sediment (Schäfer, 1956). Shinn (1968) noted such "cone-in-cone" structures left by *Phyllactis conquilega* as it moves to the surface in response to sedimentation. The surface creeping that occurs in some actinians can also result in traces left on the substrate (Willem, 1927).

Ceriantharia

In contrast to actinians, cerianthid coelenterates build felted tubes of cnidae (shed nematocysts) and mucus, containing various amounts of sand and shell fragments (Schäfer, 1956; Frey, 1970). With slow sedimentation, the tube of *Cerianthus lloydii* is built higher, and the lower, unused end of the tube fills with layers of detritus and sediment (Schäfer, 1956). In this case, surrounding sedimentary laminae are not bent (Fig. 8.5C). However, with rapid sedimentation, the cerianthid moves to the surface without constructing a tube. As a result, sedimentary laminae are bent down, surrounding a cylinder of bioturbated sediment.

Priapulida

The use of anchors is evident in the burrowing of the cylindrical priapulid worms *Priapulus caudatus* and *Halicryptus spinulosus* (Friedrich and Langeloh, 1936). In *Priapulus*, the skin musculature and comparatively large proboscis work together to produce forward locomotion. When the proboscis is expanded, longitudinal muscle contraction pulls the animal forward while the proboscis acts as a terminal anchor. Following proboscis retraction, the circular and longitudinal muscles contract simultaneously, forcing the proboscis out into the sediment again and starting a wave of swelling that runs toward the tail. This swelling anchors the body during further proboscis extension.

A more detailed study on *Priapulus* (Hammond, 1970) explains the coordinated muscle movement used in locomotion (Fig. 22.7). Such coordinated action is less evident in *Halicryptus* (Friedrich and Langeloh, 1936), which has a smaller proboscis. In this priapulid, peristaltic waves arise at the end of the anterior third of the body, caused by circular muscle contraction followed immediately by relaxation. These waves spread in both directions from the point of origin, and are a necessary precondition for proboscis protraction (however, not every wave results in such expulsion). The proboscis is then quickly pulled in while the body moves forward over it, as a result of longitudinal muscle relaxation. The proboscis is finally ex-

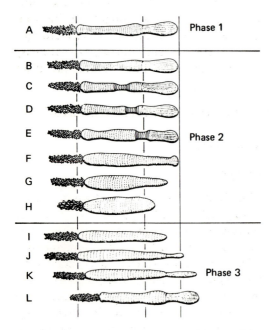

Fig. 22.7 Mechanism of burrowing in *Priapulus caudatus*. Anterior end of animal to right. *Phase 1*: animal is stationary and the proboscis is extended (A). *Phase 2*: proboscis contracts (B-F) and is then retracted into body (G-H). *Phase 3*: trunk elongates and is thrust into the substrate (I), after which the proboscis is shot out (J) and swells to form an anchor (K). Finally, the trunk shortens to pull the body forward (K, L). (From Hammond, 1970.)

tended again, backward movement being prevented by swelling of the anterior part of the body.

No traces have been reported from experimental lab work on these organisms.

Sipunculida

Peebles and Fox (1933) described thrusting of the introvert (=head and anterior part of body) into sand by *Dendrostomum zostericolum* (=*Dendrostoma zostericola*). After swelling at the anterior end, the introvert is used as a terminal anchor, enabling the remainder of the body to be pulled forward by the introvert muscles. At least in *Sipunculus nudus* (Zuckerkandl, 1950), however, no strict coordination exists between introvert expulsion and coelomic pressure variations. Under conditions of little external resistance, low pressures suffice to cause introvert protraction. However, protraction does not occur below a certain coelomic pressure. The characteristic humping of *Sipunculus* during burrowing anchors the animal against backward movement during eversion of the introvert; the latter reduces substrate resistance, as in the polychaete *Arenicola* (cf. Chapman and Newell, 1947).

The trace left by the burrowing of *Sipunculus nudus* (Fig. 22.8) is a more or less straight cylinder of totally bioturbated sand, surrounded by downward-flexed sediment laminae (Schäfer, 1956).

Hemichordata

In the worm-like enteropneusts *Balanoglossus occidentalis* and *Saccoglossus pusillus* (=*Deilichoglossus pusillis*), burrowing involves a combination of ciliary and muscular action (Ritter, 1902). The proboscis tip is driven forward by cilia and by waves of circular-muscle contraction. The waves, which move from the tip to the base of the proboscis, often remain stationary; they act as anchors, against which longitudinal proboscis and collar muscles

Fig. 22.8 Burrowing of the worm *Sipunculus nudus*, and the trace produced in sediment. (From Schäfer, 1956.)

contract to draw the body forward. Similarly, *Saccoglossus cambrensis* uses peristaltic contractions of the proboscis to effect locomotion (Knight-Jones, 1952). Apparently, three distinct contractile waves may appear simultaneously on the proboscis.

Most shallow-water hemichordates seem to inhabit well-defined, branched, U-shaped burrows, as reported for *Balanoglossus biminensis* and *Saccoglossus kowalevskii* (Frey and Howard, 1972). (Deep-sea enteropneusts commonly leave combined crawling-grazing traces; see Chapter 21.) Laminae penetration is neater in the posterior limb than in the anterior limb of U-shaped burrows, and the walls of the former are smooth and distinct.

Echiurida

Although the mechanics of burrowing in echiurids have not been reported, *Echiurus echiurus* is known to produce irregular

U-shaped burrows (Reineck et al., 1967). Many of these burrows exhibit multiple walls, and a spreite under or over the base of the U; both features were originally thought to be due to sediment filtering into the burrow. Slight sedimentation results in the worm crawling back and forth in the burrow, pressing detrital sediment into the floor and forming a spreite below the U. When sedimentation is sufficient to plug the burrow, the worm "reburrows" with its anterior end, which tends to press downward because of the dorsally held proboscis. As a result, downward displacement of the burrow occurs, and the spreite is formed above the U. However, aquaria observations have shown that spreiten are produced even when no passive sedimentation occurs, and that respiratory currents only occasionally transport material out of the burrow. Therefore, the multiple walls and spreiten are now thought to arise from fecal pellets and sediment actively brought into the burrow of *Echiurus*, as well as by passive sedimentation. (See also Risk, 1973.)

"HARD-BODIED" INVERTEBRATES

Arthropoda

Decapoda

The similar burrowing mechanisms noted among soft-bodied organisms is in direct contrast to the methods employed by other invertebrates, such as the crustaceans. These organisms depend almost totally on their appendages, which are used as levers in burrowing, as illustrated by the mole crab *Emerita portoricensis* (Trueman, 1970). This decapod burrows posterior-end-first, burying itself until only the second antennae are above the substrate surface (Fig. 22.9). After initial contact with the sediment, the first three pairs of legs move back and forth, pushing the body backward and the sand antero-laterally. At the same time, the fourth pair of legs and the uropods move in unison at about twice the speed of the other appendages, forcing sand anteriorly as they excavate the burrow and pull the crab backward. The role of the fourth limbs and uropods is greatest in the initiation of digging and in initial substrate penetration, whereas the first three pairs of limbs become more important thereafter.

The burrowing thalassinidean shrimps use their legs to loosen the substrate. The sediment is then carried out of the burrow in a basket formed by the third maxillipeds and the first and (or) second pairs of legs (Sankolli, 1965; Thompson, 1972). The first antennae (=antennules) may also play a part in basket formation (Devine, 1966). "Turn-around" cavities formed in the burrow, e.g., *Callianassa californiensis* (MacGinitie, 1934), allow the shrimp to turn around after excavating sediment at the bottom of the burrow, in order to carry the load to the surface, front-first. Once buried, the shrimp may shove sand under the abdomen, from where the pleopods

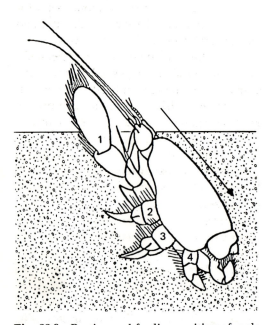

Fig. 22.9 Resting and feeding position of mole crab *Emerita talpoida* after burrowing. Arrow indicates direction of burrowing, effected by means of first four thoracic legs (1–4). (Modified from Trueman and Ansell, 1969.)

move it up and out of the burrow (Lunz, 1937). Often, some of the sand is plastered into the burrow walls after being coated with mucus by the third maxillipeds; the remaining sand is dumped outside the burrow. Thompson (1972) reported that mud gathered into the basket of *Upogebia pugettensis* is mixed with mucus secreted by the hind-gut gland, and is then pressed into the sediment to form the burrow wall. The pelletoidal texture of such burrow walls is well known (e.g., Fig. 2.2A), although not all thalassinideans make these walls.

The kinds of traces left by decapods depend on whether or not a true burrow is constructed. The mole crabs *Lepidopa websteri* (Howard, 1968) and *Emerita talpoida* (Frey and Howard, 1972) leave a path consisting of nested U-shaped laminae in the sediment; the open end of the U indicates the direction of movement. The complex burrow of *Alpheus floridanus* (Shinn, 1968) has a rough roof but a smooth floor. The latter is laminated due to size sorting, by water currents, of sediment falling from the roof. The tunnels of the lobster *Nephrops norvegicus* (Fig. 22.10) have a front entrance opening into a crater-like depression outside the tunnel, and a smaller rear opening (Dybern and Höisaeter, 1965). The rear part of the tunnel is narrower than the front, barely allowing the animal to back out, and the tunnel length and depth are in fairly direct proportion to the length of the organism. Branching tunnels are produced sometimes, as when a homeless lobster establishes itself in a tunnel already occupied.

A little investigated but evidently very important aspect of decapod behavior concerns larvae that remain inside the burrow of parents during early ontogeny. Most decapods have planktonic larvae; individuals settle to the sea floor to establish burrows only in the early post-larval stage (e.g., Thompson, 1972). But Forbes (1973) showed that the larvae of *Callianassa*

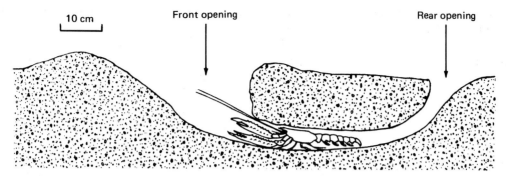

Fig. 22.10 Diagrammatic representation of burrow of *Nephrops norvegicus*, showing the lobster in life position. (Cf. Chapter 2, Fig. 2.6A.) (After Dybern and Höisaeter, 1965.)

kraussi forego a planktonic stage; the post-larvae establish their burrows by tunneling through the walls of the parent burrow, into the adjacent sediments. As the young continue to grow, their burrows are enlarged and eventually connect with the substrate surface. The burrows of the juveniles are then cut off from the parent burrow. In the fossil record this adaptation would probably be represented by a maze of large and small burrow systems.

Amphipoda

The action of appendages is of utmost importance in amphipod burrowing. In some cases, the shovelling and sweeping activities of the perieopods are aided by currents produced by the pleopods. An example is the amphipod *Haustorius arenarius*, which forms a tunnel under its body during burrowing; the coxal plates and the expanded lobes of the third perieopods form the walls (Dennell, 1933). Sand loosened by the first and second perieopods is then swept posteriorly through the tunnel by a current created by beating of the pleopods. This action also tends to drive the animal forward. Similar burrowing is seen in the amphipod *Urothoë marina* (Watkin, 1940), except that more appendages are involved.

In contrast to *Haustorius* and *Urothoë*, Nicolaisen and Kanneworff (1969) found

the typical position of several *Bathyporeia* species to be upside-down in a cavity, the roof of which is supported by the second gnathopods and the first and second uropods. Walls of the cavity are formed by these limbs and their coxal and epimeral plates. The first and second antennae loosen sand and pass it to the mandibular palps, which throw the sand back to the first gnathopods. These appendages hold the sand grains against the mouthparts for feeding, then pass the sand to the first and second perieopods. The latter convey the grains to the remainder of the perieopods, until the sand is deposited behind the animal. The fourth and fifth perieopods also aid in pushing the organism forward. Friction against the sand raises movable spines on the appendages when the limbs move backward, and presses the spines against the limbs when the latter move forward. Muscles play no part in spine movement.

Traces left by haustoriid amphipods consist of distinct, backfilled cylinders; the width of these approximates that of the animal (Howard and Elders, 1970). With time, primary lamination or initial individual burrows are gradually destroyed by the activities of these organisms (Fig. 22.11).

In contrast to haustoriid amphipods, *Corophium volutator* builds a U-shaped tube, which is lined with mucus at the base of the U in order to prevent collapse of the tube (Reineck, 1957). This amphipod also makes star-like traces on the surface of mud flats. In the lab, Trusheim (1930) found that *Corophium* constructs these traces by sitting at one end of its U-shaped burrow, stretching out its long, second antennae while holding them together, and sinking the antennae into the sediment. The antennae are then pulled back as the organism backs into its burrow, pulling some sediment into the burrow and leaving a small furrow on the surface leading to the burrow opening. Repetition of this activity produces the star-like traces, which differ from similar ones made by certain polychaetes in having straight rays that vary in length. Although some of these traces are as much as 28 mm in diameter, they average 10 to 20 mm in diameter, and exhibit no more than 20 rays. The behavior responsible for the traces arises from nutritional needs; detritus in the sediment thus pulled into the burrow is sifted out and eaten.

Slight sedimentation or erosion causes *Corophium* to extend the burrow opening above the surface; any foreign material entering the tube is washed out by currents produced by the pleopods. With increased sediment accumulation, escape structures are made to the new substrate surface (Reineck, 1958).

Echinodermata

Echinoidea

Echinoids rely on spines covering the test to effect locomotion. Integumentary cilia play no role in burrowing, and the ambulacral feet are of minor significance, as shown in *Echinarachnius parma* (Parker and van Alstyne, 1932). In this echinoid, any individual spine moves anterior-to-posterior in a vigorous vertical stroke, but returns to its original position by a sideways swing near the surface of the test. Waves of such movement begin at the anterior

2 cm

Fig. 22.11 Bioturbation of sediment by amphipod *Acanthohaustorius*, radiographed after several hours in aquarium. Continuation of the process would result in total derangement of sediment fabric (cf. Howard and Elders, 1970).

margin of the test and then sweep back
over the aboral surface and the lateral
edges.

Metachronal movement of long spines
in three areas are responsible for the burrowing of the echinoid *Mellita lata* (Kenk,
1944): the marginal spines, the ambulacral
spines, and the spines on each side of the
posterior lunule. Spines are also most important in the burrowing of *Spatangus
purpureus* (Nichols, 1959) and *Echinocardium cordatum* (Buchanan, 1966;
Nichols, 1959). Nichols attributed the differences between the two echinoids in
burrowing depth, respiratory tubes, and
sanitation tubes partly to substrate particle
size. *Spatangus* lives in gravel near the
surface, and needs no respiratory funnel.
Its subanal fasciole, from which the sanitation tube arises by the action of subanal
spines and tube feet, is bilobed; the sanitation device thus consists of two separate
tubes extending horizontally. The walls of
the tubes are coated with mucus to prevent
their immediate collapse. *Echinocardium*
lives deeper in sand (Fig. 22.12), and creates
respiratory funnels every few centimeters
along its path; the funnels are built and
maintained by dorsal tube feet. A series of
depressions on the surface mark the old
funnels. The subanal fasciole is not bilobed;
a single sanitation tube extends as far as
the subanal tube feet can stretch to maintain it (often 8 to 12 cm). In both echinoids,
the plastron spines contribute the actual
forward component in burrowing. The
spines are spatulate to work against sand
in *Echinocardium*, but are only slightly
flattened in *Spatangus*.

Traces left by echinoids depend greatly
on the shape of the organism, but all traces
result from combined shoveling motions of
the spines. In the approximately spherical
Echinocardium cordatum, disturbed sand is
pressed into the cavity left as the animal
moves forward (Reineck, 1968). As a result,
this sand is more compressed than the
surrounding sand, and in cross-section

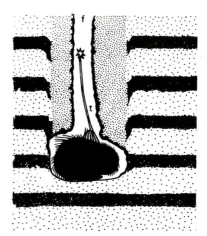

Fig. 22.12 *Echinocardium* in its burrow, showing tube foot (t) being extended up the respiratory funnel (f). (Modified from Schäfer, 1956.)

resembles a flat shell having a sunken upper
border. The sunken border is due to the
dorsal spines dividing the sand current
flowing back along the body. In longitudinal section, erect half-moon-like layers
are seen in the trace; in the bottom third of
these, a plug of feces is visible. In contrast
to *Echinocardium*, the dorsoventrally
flattened *Mellita quinquiesperforata* tends
to glide between sedimentary layers, without appreciably altering the horizontal
position of the sand (Bell and Frey, 1969).
However, vertical displacement disrupts the
sediment fabric, leaving a vestigial trail.

Holothuroidea and Ophiuroidea

Relatively few members of the other classes
of echinoderms are totally infaunal. Among
these are some holothurians, which burrow
either with their tentacles (Gerould, 1897)
or by waves of muscle contraction (Pearse,
1908). In *Thyone*, the burrow is distinctly
U-shaped (Howard, 1968).

The ophiuroid *Amphioplus* sp. constructs sand-and-mucus tubes running down
from the surface to a 10 to 20 mm hole, in
which the disk and two or three of the
arms sit (Fricke, 1970). Sand from the hole
and tube is conducted outside by laying the

remaining arms in the tube next to each other; the podia then push the sand out in long cylinders, which pile up on the surface around the tube mouth. With increased sediment accumulation, *Hemipholis* produces escape structures that appear as vertical streaks of bioturbated sediment (Howard, 1968).

SUMMARY AND DISCUSSION

Essentially, only two general types of burrowing or locomotor mechanisms are seen among marine invertebrates, although each organism has its own specific adaptations. "Hard-bodied" invertebrates rely on the action of appendages or spines, acting as levers, to effect locomotion. In contrast, the use of a hydrostatic skeleton, consisting of blood in the hemocoele or coelomic fluid in the coelom (and sometimes also of water taken into body spaces), is of prime importance in the burrowing of "soft-bodied" invertebrates (Trueman and Ansell, 1969). Action of body musculature on this hydrostatic skeleton results in the production of anchors, through elevated internal coelomic or hemocoelomic pressures. As best illustrated by bivalves, these anchors are generally of two types: penetration anchors and terminal anchors. However, bivalves have an additional, unique fluid muscle system (the pallial system) by which water jets are produced to aid penetration.

The paucity of truly different burrowing mechanisms that have evolved is due to limitations set by the environment and by various life functions that the organism must perform. Although each taxon has at its disposal specific functional and structural means for a benthic existence, such limitations impose a basic plan on the body and its functions. Therefore, a certain gradation in the ordering of these functions exists, regardless of taxonomic rank (Schäfer, 1956). As a result, similar locomotory mechanisms have tended to evolve within different taxonomic groups.

However, not even a general differentiation may be made between traces produced by the "hard-" and "soft-bodied" invertebrates. Both groups include animals that simply plough through the sediment, and also animals that construct fairly elaborate tubes or burrows. Both include shallow-burrowing and deep-burrowing organisms, and no apparent correlation exists between members of the two groups and orientation of their traces relative to the sediment surface. For example, bivalve traces often conform to shell shape, are vertically oriented, and may penetrate fairly deeply. However, traces made by members of other mollusk classes are generally more horizontally oriented and are not related as directly to the organisms' size. Finally, in all taxonomic groups, appreciable sedimentation elicits characteristic reactions, often resulting in escape structures (either vertical "escape ways" or cup-in-cup "climbing" structures).

Trace morphology and orientation are therefore more dependent on the specific functional anatomy of the organism than on its mechanism of locomotion, which is severely limited by the environment in which it occurs. However, use of the same locomotor mechanism to produce structures having different functional roles in the life of the organism must be considered in this context, because these structures may differ biostratonomically. Clearly, then, each trace must be considered within its own specific context, before conclusions may be made concerning the production and functional role of the trace. This principle reveals the greatest potential value of experimental neoichnology to ichnology as a whole.

ACKNOWLEDGMENTS

I am grateful to the following persons, who critically reviewed the manuscript: D. V. Ager, University College of Swansea; R. W. Frey, University of Georgia; H.–E. Reineck, Senckenberg Institut; E. R. Trueman, University of Manchester; and Marc and Ann Zimmerman, University of Georgia.

REFERENCES

Allen, J. A. 1958a. On the basic form and adaptations to habitat in the Lucinacea (Eulamellibranchia). Royal Soc. London, Philos. Trans., Ser. B, 241:421–484.

————. 1958b. Observations on *Cochlodesma praetenue* (Pulteney) (Eulamellibranchia). Jour. Marine Biol. Assoc. United Kingdom, 37:97–112.

Ansell, A. D. 1962. Observations on burrowing in the Veneridae (Eulamellibranchia). Biol. Bull., 123:521–530.

————. 1967. Burrowing in *Lyonsia norvegica* (Gmelin) (Bivalvia: Lyonsiidae). Malacol. Soc. London, Proc., 37:387–393.

————. 1970. Boring and burrowing mechanisms in *Petricola pholadiformis* Lamarck. Jour. Experiment. Marine Biol. Ecol., 4:211–220.

———— and A. Trevallion. 1969. Behavioral adaptations of intertidal molluscs from a tropical sandy beach. Jour. Experiment. Marine Biol. Ecol., 4:9–35.

———— and E. R. Trueman. 1967a. Observations on burrowing in *Glycymeris glycymeris* (L.) (Bivalvia, Arcacea). Jour. Experiment. Marine Biol. Ecol., 1:65–75.

———— and E. R. Trueman. 1967b. Burrowing in *Mercenaria mercenaria* (L.) (Bivalvia, Veneridae). Jour. Experiment. Biol., 46:105–115.

———— and E. R. Trueman. 1968. The mechanism of burrowing in the anemone, *Peachia hastata* Gosse. Jour. Experiment. Marine Biol. Ecol., 2:124–134.

Barnes, R. D. 1965. Tube-building and feeding in chaetopterid polychaetes. Biol. Bull., 129:217–233.

Bell, B. M. and R. W. Frey. 1969. Observations on ecology and the feeding and burrowing mechanisms of *Mellita quinquiesperforata* (Leske). Jour. Paleont., 43:533–560.

Brown, A. C. 1964. Blood volume, blood distribution and sea-water spaces in relation to expansion and retraction of the foot in *Bullia* (Gastropoda). Jour. Experiment. Biol., 41:837–854.

Buchanan, J. B. 1966. The biology of *Echinocardium cordatum* (Echinodermata: Spatangoida) from different habitats. Jour. Marine Biol. Assoc. United Kingdom, 46:97–114.

Carlson, A. J. 1905. The physiology of locomotion in gasteropods. Biol. Bull., 8:85–92.

Chapman, G. and G. E. Newell. 1947. The role of the body fluid in relation to movement in soft-bodied invertebrates. I. The burrowing of *Arenicola*. Royal Soc. London, Proc., Ser. B, 134:431–455.

———— and G. E. Newell. 1948. Burrowing of the lugworm. Nature, 162:894–895.

Clark, R. B. and M. E. Clark. 1960. The ligamentary system and the segmental musculature of *Nepthys*. Microscopic. Sci., Quart. Jour., 101:149–176.

Dennell, R. 1933. The habits and feeding mechanism of the amphipod *Haustorius arenarius* Slabber. Linnean Soc. London, Zool. Jour., 38: 363–388.

Devine, C. E. 1966. Ecology of *Callianassa filholi* Milne-Edwards 1878 (Crustacea, Thalassinidae). Royal Soc. New Zealand, Trans., 8:93–110.

Dinamani, P. 1964. Burrowing behavior of *Dentalium*. Biol. Bull., 126:28–32.

Drew, G. A. 1900. Locomotion in *Solenomya* and its relatives. Anatom. Anzeiger, 17:257–266.

————. 1907. The habits and movements of the razor-shell clam, *Ensis directus*, Con. Biol. Bull., 12:127–140.

Dybern, B. I. and T. Höisaeter. 1965. The burrows of *Nephrops norvegicus* (L.). Sarsia, 21:49–55.

Enders, H. E. 1909. A study of the life-history and habits of *Chaetopterus variopedatus*, Renier and Claparède. Jour. Morphol., 20:479–532.

Forbes, A. T. 1973. An unusual abbreviated larval life in the estuarine burrowing prawn *Callianassa kraussi* (Crustacea: Decapoda: Thalassinidea). Marine Biol., 22:361–365.

Frey, R. W. 1970. The lebensspuren of some common marine invertebrates near Beaufort, North Carolina. II. Anemone burrows. Jour. Paleont., 44:308–311.

———— and J. D. Howard. 1972. Georgia coastal region, Sapelo Island, U.S.A.: sedimentology and biology. VI. Radiographic study of sedimentary structures made by beach and offshore animals in aquaria. Senckenbergiana Marit., 4:169–182.

Fricke, H. W. 1970. Beobachtungen über Verhalten und Lebensweise des im Sand lebenden Schlangensternes *Amphioplus* sp. Helgoländer Wiss. Meeresuntersuch., 21:124–133.

Friedrich, H. and L. P. Langeloh. 1936. Untersuchungen zur Physiologie der Bewegung und des Hautmuskelschlauches bei *Halicryptus spinulosus* und *Priapulus caudatus*. Biol. Zentralblatt, 56:249–260.

Gerould, J. H. 1897. The anatomy and histology of *Caudina arenata* Gould. Boston Soc. Nat. Hist., Proc., 27:7–75.

Gersch, M. 1934. Zur experimentellen Veränderung der Richtung der Wellenbewegung auf der Kriechsohle von Schnecken und zur Rückwärtsbewegung von Schnecken. Biol. Zentralblatt, 54:511–518.

Gordon, D. C., Jr. 1966. The effects of the deposit feeding polychaete *Pectinaria gouldii* on the intertidal sediments of Barnstable Harbor. Limnol. Oceanogr., 11:327–332.

Gräf, I. E. 1956. Die Fährten von *Littorina littorea* Linné (Gastropoda) in verschiedenen Sedimenten. Senckenbergiana Leth., 37:305–317.

Hammond, R. A. 1970. The burrowing of *Priapulus caudatus*. Jour. Zool., 162:469–480.

Holme, N. A. 1961. Notes on the mode of life of the Tellinidae (Lamellibranchia). Jour. Marine Biol. Assoc. United Kingdom, 41:699–703.

Howard, J. D. 1968. X-ray radiography for examination of burrowing in sediments by marine invertebrate organisms. Sedimentology, 11:249–258.

———— and C. A. Elders. 1970. Burrowing patterns of haustoriid amphipods from Sapelo Island, Georgia. In T. P. Crimes and J. C. Harper (eds.), Trace fossils. Geol. Jour., Spec. Issue 3:243–262.

Jacobsen, V. H. 1967. The feeding of the lugworm, *Arenicola marina* (L.). Quantitative studies. Ophelia, 4:91–109.

Kenk, R. 1944. Ecological observations on two Puerto Rican echinoderms, *Mellita lata* and *Astropecten marginatus*. Biol. Bull., 87:177–187.

Knight-Jones, E. W. 1952. On the nervous system of *Saccoglossus cambrensis* (Enteropneusta). Royal Soc. London, Philos. Trans., Ser. B, 236:315–354.

Ladd, H. S. 1959. Ecology, paleontology, and stratigraphy. Science, 129:69–78.

Lissman, H. W. 1945. The mechanism of locomotion in gastropod molluscs. II. Kinetics. Jour. Experiment. Biol., 22:37–50.

Lunz, G. R. 1937. Notes on *Callianassa major* Say. Charleston (South Carolina) Museum, Leaflet 10:1–15.

MacGinitie, G. E. 1934. The natural history of *Callianassa californiensis* Dana. Amer. Midland Natur., 15:166–177.

Mangum, D. C. 1970. Burrowing behavior of the sea anemone *Phyllactis*. Biol. Bull., 138:316–325.

Morton, J. E. 1959. The habits and feeding organs of *Dentalium entalis*. Jour. Marine Biol. Assoc. United Kingdom, 38:225–238.

Nair, N. B. and A. D. Ansell. 1968. Characteristics of penetration of the substratum by some marine bivalve molluscs. Malacol. Soc. London, Proc., 38:179–197.

Nichols, D. 1959. Changes in the chalk heart-urchin *Micraster* interpreted in relation to living forms. Royal Soc. London, Philos. Trans., Ser. B, 242:347–437.

Nicol, E. A. T. 1930. The feeding mechanism, formation of the tube, and physiology of digestion in *Sabella pavonina*. Royal Soc. Edinburgh, Trans., 56:537–598.

Nicolaisen, W. and E. Kanneworff. 1969. On the burrowing and feeding habits of the amphipods *Bathyporeia pilosa* Lindstrom and *B. sarsi* Watkin. Ophelia, 6:231–250.

Ockelmann, K. W. and O. Vahl. 1970. On the biology of the polychaete *Glycera alba*, especially its burrowing and feeding. Ophelia, 8:275–294.

Owen, G. 1961. A note on the habits and nutrition of *Solemya parkinsoni* (Protobranchia: Bivalvia). Microscopic. Sci., Quart. Jour., 102:15–21.

Parker, G. H. and M. A. van Alstyne. 1932. Locomotor organs of *Echinarachnius parma*. Biol. Bull., 62:195–200.

Pearse, A. S. 1908. Observations on the behavior of the holothurian, *Thyone briareus* (LeSeur). Biol. Bull., 15:259–288.

Peebles, F. and D. L. Fox. 1933. The structure, functions and general reactions of the marine sipunculid worm, *Dendrostoma zostericola*. Scripps Inst. Oceanogr., Bull., Tech. Ser., 3:201–234.

Reineck, H.-E. 1957. Über Wühlgänge im Watt und deren Abänderung durch ihre Bewohner. Paläont. Zeitschr., 31:32–34.

———. 1958. Wühlbau–Gefüge in Abhängigkeit von Sediment-Umlagerungen. Senckenbergiana Leth., 39:1–23, 54–56.

———. 1968. Lebensspuren von Herzigeln. Senckenbergiana Leth., 49:311–319.

——— et al. 1967. Das Schlickgebiet südlich Helgoland als Beispiel rezenter Schelfablagerungen. Senckenbergiana Leth., 48:219–275.

Rhoads, D. C. 1963. Rates of sediment reworking by *Yoldia limatula* in Buzzards Bay, Massachusetts, and Long Island Sound. Jour. Sed. Petrol., 33:723–727.

——— and D. J. Stanley. 1965. Biogenic graded bedding. Jour. Sed. Petrol., 35:956–963.

——— and D. K. Young. 1970. The influence of deposit-feeding organisms on sediment stability and community trophic structure. Jour. Marine Res., 28:150–178.

Risk, M. J. 1973. Silurian echiuroids: possible feeding traces in the Thorold Sandstone. Science, 180:1285–1287.

Ritter, W. E. 1902. The movements of the Enteropneusta and the mechanism by which they are accomplished. Biol. Bull., 3:255–261.

Sankolli, K. N. 1965. On the occurrence of *Thalassina anomala* (Herbst.), a burrowing crustacean in Bombay waters, and its burrowing methods. Jour. Bombay Nat. Hist. Soc., 60:600–605.

Schäfer, W. 1952. Biogene Sedimentation im Gefolge von Bioturbation. Senckenbergiana, 33:1–12.

———. 1956. Wirkungen der Benthos-Organismen auf den jungen Schichtverband. Senckenbergiana Leth., 37:183–263.

———. 1972. Ecology and palaeoecology of marine environments. Edinburgh, Oliver & Boyd, and Chicago, Ill., Univ. Chicago Press, 568 p.

Seilacher, A. 1951. Der Röhrenbau von *Lanice conchilega* (Polychaeta). Senckenbergiana, 32:267–280.

Seymour, M. K. 1971. Burrowing behavior in the European lugworm *Arenicola marina* (Polychaeta: Arenicolidae). Jour. Zool., 164:93–132.

Shinn, E. A. 1968. Burrowing in recent lime sediments of Florida and the Bahamas. Jour. Paleont., 42:879–894.

Stanley, S. M. 1970. Relation of shell form to life habits in the Bivalvia. Geol. Soc. America, Mem. 125, 296 p.

Thompson, R. K. 1972. Functional morphology of the hind-gut gland of *Upogebia pugettensis* (Crustacea, Thalassinidea) and its role in burrow construction. Unpubl. Ph.D. Dissert., Univ. California, Berkeley, 202 p.

Trueman, E. R. 1966a. The fluid dynamics of the bivalve molluscs, *Mya* and *Margaritifera*. Jour. Experiment. Biol., 45:369–382.

———. 1966b. The mechanism of burrowing in the polychaete worm, *Arenicola marina* (L.). Biol. Bull., 131:369–377.

———. 1967. The dynamics of burrowing in *Ensis* (Bivalvia). Royal Soc. London, Proc., Ser. B, 166:459–476.

———. 1968a. The locomotion of the freshwater clam *Margaritifera margaritifera* (Unionacea: Margaritanidae). Malacologia, 6:401–410.

———. 1968b. The burrowing activities of bivalves. Zool. Soc. London, Symp., 22:167–186.

———. 1968c. A comparative account of the burrowing process of species of *Mactra* and of other bivalves. Malacol. Soc. London, Proc., 38:139–151.

———. 1968d. The mechanism of burrowing of some naticid gastropods in comparison with that of other molluscs. Jour. Experiment. Biol., 48:663–678.

———. 1968e. The burrowing process of *Dentalium* (Scaphopoda). Jour. Zool., 154:19–27.

———. 1970. The mechanism of burrowing of the mole crab, *Emerita*. Jour. Experiment. Biol., 53:701–710.

——— and A. D. Ansell. 1969. The mechanisms of burrowing into soft substrata by marine animals. Oceanogr. Marine Biol., Ann. Rev., 7:315–366.

——— et al. 1966. The dynamics of burrowing of some common littoral bivalves. Jour. Experiment. Biol., 44:469–492.

Trusheim, F. 1930. Sternförmige Fährten von *Corophium*. Senckenbergiana, 12:254–260.

Vlès, F. 1906. Notes sur la locomotion du *Pectunculus glycymeris* Lk. Soc. Zool. France, Bull., 31:114–117.

————. 1913. Observations sur la locomotion d'*Otina otis* Turt. Remarques sur la progression des Gastéropodes. Soc. Zool. France, Bull., 38:242–250.

von Boletzky, S. and M. V. von Boletzky. 1970. Das Eingraben in Sand bei *Sepiola* und *Sepietta* (Mollusca, Cephalopoda). Rev. Suisse Zool., 77:536–548.

Watkin, E. E. 1940. The swimming and burrowing habits of the amphipod *Urothoë marina* (Bate). Royal Soc. Edinburgh, Proc., 60:271–280.

Wells, G. P. 1945. The mode of life of *Arenicola marina* L. Jour. Marine Biol. Assoc. United Kingdom, 26:170–207.

Willem, V. 1927. Observations sur la locomotion des Actinies. Acad. Royale Sci., Lettres, Beaux-Arts Belgique, Bull., 13:630–650.

Yonge, C. M. 1937. The biology of *Aporrhais pes-pelecani* (L.) and *A. serresiana*. Jour. Marine Biol. Assoc. United Kingdom, 21:687–704.

————. 1939. The protobranchiate mollusca; a functional interpretation of their structure and evolution. Royal Soc. London, Philos. Trans., Ser. B, 230:79–147.

————. 1946a. On the habits and adaptations of *Aloidis (Corbula) gibba*. Jour. Marine Biol. Assoc. United Kingdom, 26:358–376.

————. 1946b. On the habits of *Turritella communis* Risso. Jour. Marine Biol. Assoc. United Kingdom, 26:377–380.

Ziegelmeier, E. 1952. Beobachtungen über den Röhrenbau von *Lanice conchilega* (Pallas) im Experiment und am natürlichen Standort. Helgoländer Wiss. Meeresuntersuch., 4:107–129.

————. 1969. Neue Untersuchungen über die Wohnröhren-Bauweise von *Lanice conchilega* (Polychaeta, Sedentaria). Helgoländer Wiss. Meeresuntersuch., 19:216–229.

Zuckerkandl, E. 1950. Coelomic pressures in *Sipunculus nudus*. Biol. Bull., 98:161–173.

CHAPTER 23

TECHNIQUES FOR THE STUDY OF FOSSIL AND RECENT TRACES

GEORGE E. FARROW

Department of Geology, Glasgow University
Glasgow, Scotland

SYNOPSIS

The study of trace fossils demands an acute observational technique because the structures must be visualized in three dimensions. In lithified rocks, a massive "structureless" bed should be regarded with suspicion; the homogenization created by intense burrowing can sometimes be revealed only in the laboratory. Techniques for enhancing obscure bioturbation structures differ but little from those used for physically formed sedimentary structures: initial sandblasting or staining may be followed by x-ray radiography and infrared photography. Thin sections of burrowed sediments should be slightly thicker than the normal 0.03 mm, to produce greater contrast; if of suitable areal dimensions (5 × 5 cm), they may be projected directly onto a screen by a slide projector.

Great advances have been made in the replication of burrows and trails from unconsolidated sediments through the use of resins. Whereas lacquer peeling is limited by lithology and wetness, resins harden under water and can be used both to cast open burrow systems (divers have used polyesters at depths of 40 m) and to peel a wide range of bioturbated sediments. Epoxy relief peels produce outstanding results, and when taken in block form, they enable the shape, configuration, and density of burrows to be comprehended fully.

The Senckenberg box and the can corer are the most satisfactory devices for sampling both intertidal environments and—in modified form—offshore regions. These corers collect rectangular chunks of undisturbed sediment that are ideal for serial slicing, peeling, and impregnation; vacuum impregnation with resin permits casts to be taken of burrows less than 1 mm in diameter. Thus, highly detailed block diagrams may be constructed for a variety of sedimentary facies, regardless of depth, and then compared in the closest detail with their ancient counterparts.

INTRODUCTION

In this chapter I attempted to analyze the total assortment of techniques that have been used to describe, preserve, and enhance traces, whether from ancient or modern environments. For many years such biogenic structures tended to lurk in the background of otherwise extensive sedimentological investigations; only recently have ichnologists come to develop their own methods of studying them.

With the following account, I have tried to give references both to the original description of each technique (in whatever field) and to the most pertinent example of its application to ichnology, the more important ones being illustrated. The standard works by Kummel and Raup (1965) and Carver (1971) are indispensable;

detailed procedural instructions, together with lists of suppliers, can be found in the book by Bouma (1969)—the essential manual for anyone intending to apply these techniques.

Many of the methods outlined here are relatively inexpensive yet highly instructive, providing stimulating material for display in student laboratories. By their effective "fossilizing," these replications permit the closest comparison between the present and the past.

Experimental techniques for studying the traces and ethology of live animals are discussed in Chapter 22, and microborings in Chapter 12.

TRACE FOSSILS IN LITHIFIED SEDIMENTS

Observational Method

The indiscriminate assignment of trace fossils to the "fucoids," which persisted for many years after their true origin had been demonstrated (e.g., Hancock, 1858), is a sign of the lack of rigor that long attended their study. (See Chapter 1.) An acute observational technique is imperative; appreciation of structures in three dimensions is more important here than in other branches of paleontology because correct identification often depends upon it. Attempts to introduce such rigor have included the use of quadrats (Farrow, 1966, p. 109) and statistics (Frey and Cowles, 1969; Crimes, 1970, p. 62). Quadrats or line transects ensure that the observer is not biased toward the more spectacular traces at the expense of less conspicuous although possibly more significant ones (cf. Ager, 1963, p. 228–229). In addition, quadrats and transects facilitate the study of mutual relations between taxa and comparisons between different trace fossil populations; again, statistics also help (Frey, 1970, Table 4). A natural extension of the accurate recording of burrowing patterns is the analysis of their possible ethological significance through modeling (Hanor and

Marshall, 1971) and computer simulation (Raup and Seilacher, 1969). (For techniques in vertebrate ichnology, see Chapter 14.)

Field Enhancement, Photography, and Replication of Biogenic Structures

The photography of trace fossils requires a style slightly different from that adopted with body fossils. Burrows are generally accentuated by wetting the rock surface (Frey, 1970, p. 11) or by smearing ink over it and then washing it off (Farrow, 1966, p. 110). In other instances, traces having delicate claw scratches or other fine details should be whitened and photographed with strong side lighting (Farrow, 1966, Pl. 7) (Fig. 23.2A). Drawing field sketches may be helped by first delimiting inconspicuous trace fossils with a felt-tipped pen. If neither the specimen nor a good photograph can be obtained in the field, a latex mold could be taken of critical areas, such as a burrow packed with fecal pellets, and photographed back in the laboratory.

Many of the staining techniques normally performed indoors can also be applied outdoors—a fact that few workers seem to realize! Spraying carbonate-cemented rocks with dilute HCl, or siliceous rocks with KOH, may increase the relief on fresh or sawed quarry faces; from these, good peels can sometimes be obtained by either spraying or painting lacquer directly onto the face. Many different lacquers are available, one of the most straightforward being celluloid film dissolved in acetone. The Overlau method (in Bouma, 1969, p. 69–74) enables peels 30 × 30 cm to be taken from actively worked limestone faces and was developed for the study of Carboniferous stromatolites. In general, however, lithified rocks cannot be peeled in the field as successfully as can unconsolidated sediments.

Staining can be a very useful tool. It has recently affected paleogeographic interpretations of the Upper Chalk of England, where "featureless" strata previously considered to be bathyal in origin were shown to exhibit bioturbate textures typical of much shallower depth (Goldring and Crichton, 1966; see Chapter 17). Ink smearing produces a stain controlled by differences in porosity, whereas organic dyes such as Alizarin Red and Methylene Blue (see Chapter 18) are preferentially adsorbed by clay minerals; thus, fine-grained sediments generally accept stains more readily than do sandstones.

Laboratory Techniques

X-Ray Radiography (see Hamblin, in Carver, 1971, Chapter 11)

The truly structureless bed is probably something of a rarity. Most strata usually reveal some degree of internal form when subjected to either x-ray radiography or infrared photography. Recent discussions about the possible existence of massive beds in fluvial sandstones from the English Carboniferous (Allen, 1971, in discussion of Collison, 1970) demonstrated the wisdom of regarding all such beds with suspicion. Hamblin (1962a, 1965) showed that x-ray radiography can unmask many original structures, including intense bioturbation, that had been adversely affected by diagenesis. This observation is especially important to ichnologists; with maximum activity, animals tend to modify or destroy their own characteristic burrowing patterns and to produce almost thoroughly churned sediments (Fig. 23.1) which, after diagenesis, might appear homogeneous in unprepared samples. (Cf. Hamblin, Figs. 1, 2, in Carver, 1971.)

Bouma (1969) provided an extensive passage on the x-ray radiography of sediments, in which he showed the importance of using only a thin slab of sediment (Bouma, 1969, Fig. 3.23); no burrow is visible when the slab is 7 cm thick, yet one appears clearly when the thickness is reduced to 1 cm. Superimposed images produce a fuzzy picture of a thick slab, and increased exposure time and higher kilovoltage are necessary for adequate penetration. The likelihood of success with this method therefore depends on one's ability to cut sufficiently thin slabs of material. Well-jointed rocks, which tend to fracture during sawing, can be impregnated or embedded; but any slice of consolidated sediment must necessarily be thinner than an unconsolidated one, because of its greater density (see Bouma, 1969, for examples).

A fine illustration of the technique applied to ichnology is the recent work of Hester and Pryor (1972), who took x-ray radiographs of serial sections of burrows in order to elucidate the development of Eocene *Ophiomorpha*.

X-radiography is also valuable in studying bioturbation by animals in aquaria (see Chapter 22).

Infrared Photography

X-ray and infrared radiation differ in their sensitivity to organic matter, thus they cannot be regarded as alternative techniques. Relatively costly equipment is needed for x-ray work whereas taking infrared photographs merely requires a special film and filter. Rhoads and Stanley (1966) pioneered application of the latter to sediments. A thin slab, cut to less than 5 mm (rather than a thin section), is photographed by transmission; exposure time is proportional to the organic matter in the sample, sandstones being more transparent to infrared than are finer lithologies. Considering the ease and inexpense involved, surprisingly little use has been made of the technique.

Ultraviolet Photography

Among certain lithologies, ultraviolet photography can be used to good advantage in studying traces that otherwise would remain obscure or unnoticed (Chapter 18, Fig. 18.14). Best results are obtained with relatively pure carbonates, having little iron content. For the rock shown in Figure 23.2, the light source is a mercury arc having a Wood's glass filter—a UV positive filter that

Fig. 23.1 X-radiography. Epoxy relief peel (A) exhibits little detail in lower two-thirds of box core, yet a radiograph of same sediments (B) reveals thorough biogenic reworking by heart urchins. Peel and radiograph are in mirror image of one another. Modern continental shelf sediments (shallow-water shelly sands), Georgia.

passes only radiation of 3650 Å (95 percent). The camera has a Wratten (Kodak) 2E filter, which is a total UV negative filter that passes visible radiation but not UV.

Sandblasting

A fresh rock that appears homogeneous simply because it has not been allowed to weather can be "weathered" artificially in the laboratory within 5 minutes, by sandblasting (Hamblin, 1962b). Inexpensive abrasive units can be obtained that adapt to standard laboratory air jets. The greatest detail is achieved by using an abrasive of unsorted sand in which maximum grain size is slightly less than that of the sample.

Sectioning

In trace fossil petrography, one should use thin sections that are slightly thicker than the normal 0.03 mm section (e.g., Bouma, 1969, Fig. 2.16b). Reineck (1970), working with impregnated modern sediments, makes his thin sections 5 × 5 cm in area and projects them onto a screen by means of a lantern slide projector. For studying large burrows or extended burrowing sequences, one can produce very long sections; instead of the usual small glass slide sizes, plate glass is cut to the appropriate size and the rock slice fixed with Eastman 910, an adhesive that dries in seconds and requires no heating. Chisholm (1970a, b) and Goldring (1962) made considerable use of thin sections in their studies of *Teichichnus* and *Diplocraterion*.

Serial sectioning (cf. Frey and Cowles, 1969, Pl. 3, fig. 2) does not seem to have been used widely in ichnology, although Orme and Brown (1963) demonstrated its value in their study of the problematical "stromatactis" cavities in Carboniferous reefs.

Fig. 23.2 Chalk block (horizontal plane) photographed in ordinary light (A) and ultraviolet radiation (B). A, oblique lighting, showing surface relief features (a slightly hardened network of fractures and tiny fragments of calcitic fossils). B, fluorescent lighting due to radiation (UV invisible). Burrow fills are dark, the different generations of burrows having different color intensities; white flecks are skeletal calcite grains, set in matrix of coccolithic micrite. Trace fossils visible include *Thalassinoides* and *Zoophycos* (spreiten burrows). Rock surface is 8 cm wide. White Chalk, Upper Maastrichtian; Stevns Klint, Denmark.

Three stains have been employed in studying sedimentary structures: Hamblin (1962b) used Alizarin Red; West (1965) mixed India Ink with carborundum powder during the grinding process; and Goldring and Crichton (1966) used Methylene Blue. Dense rocks generally yield poor results, although some limestones can be stained very successfully, pellets and burrow linings thus being accentuated (Scoffin, 1973). Unconsolidated sediments, particularly dried out core samples, are amenable to Methylene Blue (Pantin, 1960). (See also Chapter 18.)

BIOGENIC STRUCTURES IN UNCONSOLIDATED SEDIMENTS

The expertise required to replicate bioturbation structures from soft sediment increases sharply from onshore to offshore, as does the cost of the equipment. Land-based lacquer peeling techniques for dealing with moist sands are straightforward and inexpensive. Waterlogged sediments require resins, which are expensive but which produce outstanding results. Offshore, however, the recovery and preservation of burrows and other traces has been perfected by only a small number of marine stations, most notably the Senckenberg Institut, where the accomplishments of Reineck et al. (1967) rank alongside the most elegant technical achievements in the history of paleontology and sedimentology.

Observational Method

The observation of trace-making activities under water presents certain difficulties that cannot often be solved solely by diving operations, which may interfere with typical animal behavior patterns. In shallow water, as much as 4 m deep, an underwater "hide" having an access hatch on top can be employed for obtaining continuous day and night in situ records; Fuss and Ogren (1965) designed such a chamber for observing penaeid shrimp burrowing. In deeper water, underwater television is frequently used by marine biologists (e.g., Chapman and Rice, 1971—at a depth of 30 m). For use at similar depths, Rhoads (1970) developed a sediment-water interface camera for the time-lapse photography of bioturbation created by the shallow infauna of fluid subtidal muds.

Problems of accurate sampling and observation are not, however, confined to subtidal regions. Mobile elements of the benthos (whether epifaunal or infaunal) create special difficulties, as do organisms intimately associated with dense root systems; the latter makes salt marsh and eel grass environments particularly challenging. Frey et al. (1973) used special high-walled quadrats, emplaced at high tide, and have developed a combined sieving, staining, and flotation procedure for the consistent separation of burrowing animals from dense grass roots.

Peeling Soft Sediments in the Field (see Klein, in Carver, 1971, Chapter 10)

Lacquer Peeling

Lacquer peeling is a well-established technique (Voigt, 1936) that applies as readily to a horizontal surface as to a vertical one, provided that the sand is neither too wet nor too dry (Fig. 23.3). Many fine museum displays have been prepared, using lacquer that is sprayed onto a smoothed face of sand (e.g., Hähnel, 1962). The key to obtaining good peels is to set fire to the face with acetone, if too wet, or to spray it with water, if too dry. Even where no textural contrast is apparent in the field after a face has been prepared, a peel should be taken because the lacquer will infiltrate to varying degrees, and some relief should appear. Spraying is better than brushing because the fabric is not disturbed (Fig. 23.3A), although, using the Shell method, one avoids

Fig. 23.3 Lacquer peeling. A, spraying smoothed face for initial stabilization (trench in inter-tidal sand spit). B, brushing lacquer onto linen cloth pegged over stabilized sand (aeolian dune). Modern carbonate sands, Persian Gulf.

this by laying linen over the face before painting on the lacquer (Fig. 23.3B).

The use of three peels, taken mutually at right angles, is of the greatest value in replicating traces, because they enable the shape, density, and dimensions of burrows and other structures to be depicted fully in three dimensions. A slow-drying mixture gives maximum penetration and secures high relief (Fig. 23.4).

Fig. 23.4 Lacquer peel, taken as in Figure 23.3B. Conspicuous relief produced by porosity variation around burrows (*Ophiomorpha*). Less conspicuous bioturbate textures also are recorded (lighter areas of low relief). Miocene sands; South Frimmersdorf Quarry, near Cologne.

Because several problems exist in this method, it is being superseded by multi-purpose polyester resins. Direct sun and strong wind, for example, may adversely affect the early stages of the lacquer process, although the use of toluene in place of acetone as the solvent retards the rate of evaporation of the mixture. However, too much water loosens the film because the lacquer cannot penetrate pores. Similarly, clay laminae cannot be peeled.

Polyester Resin Peels

Resins will harden under water and therefore are ideal for intertidal and subtidal environments. McMullen and Allen (1964) replicated both vertical sections of burrows and their surface morphology (Fig. 23.5), using a spraying technique; styrene monomers reduce the viscosity of the polyester so that it can be sprayed. The method is quicker than using lacquer and avoids the latter's pitfalls.

Epoxy Relief Peels

Although originally described in terms of laboratory conditions by Bouma (1969, p. 58–62), the method for obtaining epoxy relief peels can also be used in the field (e.g., Frey and Howard, 1969). Such peels merit separate designation because two-dimensional peels appear with three-dimensional effect, as a result of the extreme penetration of the resin (Fig. 23.6). The technique described by Barr et al. (1970) for use in trenches employs a backing board of masonite, which is normal procedure. Thomson (in Bouma, 1969, p. 51) used

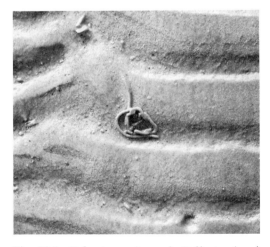

Fig. 23.5 Polyester resin peel. Delicate *Arenicola* casting replicated by spraying resin directly onto seawater-saturated muddy surface at low water, the resin hardening before rise of next tide. Modern intertidal flat; Poole Harbour, Dorset.

Fig. 23.6 Epoxy relief-peel. Extreme penetration of epoxy into the prepared face of box-core sample effectively replicated the shell-lined burrow (dwelling tube) of the polychaete *Owenia fusiformis* in three dimensions. Cracks in peel due to desiccation of clay in sediments. Modern shallow-water offshore sands, Georgia.

Fig. 23.7 Burrow casting underwater: the Shinn resin method. Cast of small *Callianassa* burrow being retrieved at 7 m water depth. Resin was poured from a plastic bag into a fruit-juice can having both ends removed, which acted as a funnel and provided a hydraulic head, forcing resin into the burrow. Modern carbonate sands, Bahama Banks.

plexiglas instead, which enables the resulting peel to be viewed in transmitted light. His method is quickest of all and has an obvious advantage, i.e., when several peels have to be taken in a short time, as at low tide.

The time required for resins to cure is inversely proportional to temperature;

curing is thus rapid in the tropics but may be slow in temperate latitudes, especially in the presence of water, which further increases the hardening time. Resins are currently being developed, however, that will harden within a few hours at temperatures of only a few degrees Centigrade, even in the presence of salt water.

Resin Casting of Burrows in the Field: the Shinn Method

Plaster of Paris was used as long ago as 1928 by Stevens, to cast burrows of the intertidal crustacean *Upogebia*, and more recently silicone rubber was tried by Crichton (1960, p. 10); but neither substance will enter burrows filled with water. Shinn (1968) pioneered the adoption of polyester resin to combat this difficulty. The method involves pouring a mixture of resin, hardener, and catalyst down burrow openings, which may be above or below water (Fig. 23.7). Only open burrows are suitable, especially those like the lined callianassid dwellings shown in Figure 23.8A; the resin cannot readily be applied to the burrows of *Arenicola* and certain other sediment-ingesting worms (various phyla) living in sandy substrates. Nevertheless, the range of burrows that have been cast is impressive (Frey and Howard, 1969; Farrow, 1971; Frey et al., 1973). Not only the producer[1] of a burrow may be trapped but also its commensals, often in life position (Fig. 23.8B). For very small burrows, where surface tension retards the flow of the resin, expendable syringes may be used to inject the material, or a resin of very low viscosity may be used; either way, one can successfully cast burrows as small as 1 mm in diameter (Fig. 23.9).

The resin casting method of "fossilizing" burrows is analogous to the bed-junc-

tion type of preservation seen in many trace fossils [e.g., *Chondrites* (Simpson, 1957); see Chapters 4, 18], and when combined with impregnation of the sediments surrounding the burrow, the method enables very detailed comparisons with ancient examples to be undertaken.

The feasibility of using resins under

Fig. 23.8 Resin burrow casts. A, two simple, lined burrows of the callianassid shrimp *Neaxius*; basal chambers incompletely filled because of airlock (this problem is less serious in systems having several openings). Arrow indicates position of enlargement shown in B. B, commensal mollusks from the burrow lining of one chamber, trapped in life position by the resin. Modern eel grass bed, Aldabra Atoll.

[1] An extremely mobile burrower can hinder the casting process. Such animals can generally be flushed out of their burrow system with formalin, before introducing the resin.

Fig. 23.9 Small resin casts. U-shaped burrow (inverted view) of tiny amphipod *Corophium volutator*, cast in the field by pouring low-viscosity epoxy resin directly onto sediment surface. Modern tidal flat; Solway Firth, Scotland. (Diameter of coin, 17 mm.)

and Rice and Chapman (1971) in their study of the Norway lobster *Nephrops norvegicus*. In an excellently organized investigation of crustacean burrowing in muds 30 m deep in Loch Torridon, Scotland, they initially used underwater television and direct observations to map the distribution of burrow openings and to study burrowing behavior. Connections between burrows were then discerned by squirting potassium permanganate into the system; internal configuration of the system was later determined by resin casting.

Sampling Devices

Hand-Operated Corers

Three well-tried samplers that are straightforward to construct and to operate are the van Straaten tube, the Senckenberg box, and the can corer; none produces serious compaction effects.

The van Straaten tube enables cores as much as 2 m long to be obtained, either on land or in water of wading depth. The device is very simple, consisting of a brass pipe having a valve-plate on top to facilitate removal of the core; yet van Straaten (1954) obtained outstanding results, the key to which is his use of drying tins to enhance the contained structures. By photographing cores as soon as they were dry enough to reveal their burrows, more detail was recorded than could actually be seen by eye, owing to the greater sensitivity of photographic paper (see Bouma, 1969, Fig. 4.7).

The Senckenberg box involves two stages (Reineck, 1957; or in English,

water was demonstrated many years before they came to be applied specifically to burrows. Brown and Patnode (1953) developed a resin-charged corer to impregnate loose sand actually on the sea floor, although little advantage seems to have been taken of their device, possibly because cleaning the equipment is such a problem with resins—an alternative is to have mostly expendable items. Hertweck and Reineck (1966) impregnated their cores after retrieval, using a vacuum chamber to facilitate the casting of extremely thin burrows (less than 1 mm in diameter).

A striking example of the success of the Shinn method in comparatively deep water was provided by Chapman and Rice (1971)

Fig. 23.10 Offshore sampling of traces: the Senckenberg technique. A, box corer being lowered ▶ into 30 m of water off the Elbe estuary; the frame support, weighted box corer, and pivotted baseplate are visible. B, box core removed from corer and prepared with wire cheese-cutter; the open U-burrow of *Echiurus* penetrates recent light-colored storm layers, whereas earlier burrows are sediment filled. C, block diagram prepared from box-core samples, showing the appearance of burrows in vertical section (cf. cliff section, thin section). D, block diagram prepared by serial slicing of box-core samples, showing the appearance of burrows in three dimensions (cf. weathered outcrop). In diagrams, Ce = *Cerianthus*; Ec = *Echiurus*; En = *Echinocardium*; No = *Notomastus*; P–V = *Pectinaria* (test and bioturbation); Sc = *Scalibregma*; T = Thalassinidea.

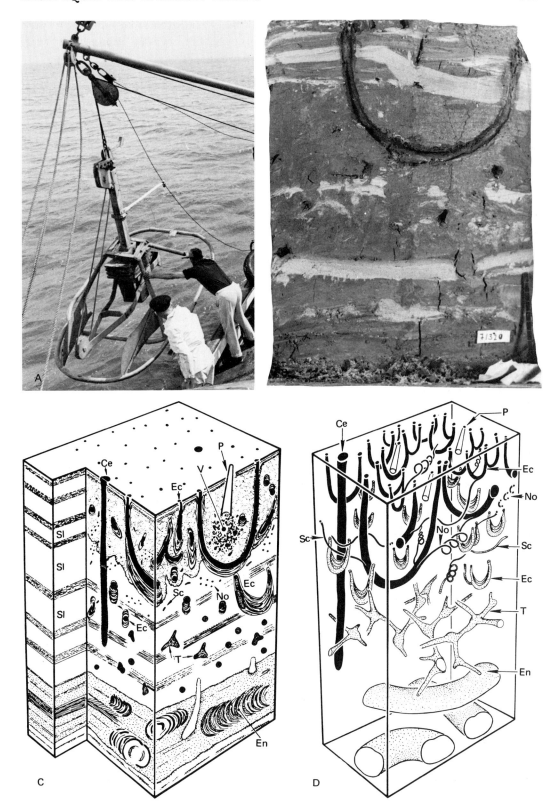

Bouma, 1964). First, a steel device, shaped like a drawer without its back, is forced into the sediment; then an inverted, L-shaped, flanged cover slides over the open side. This closure secures a rectangular sediment sample that is ideally suited to the three-dimensional study of burrows, by serial slicing with a cheese-cutter. Retrieval from waterlogged intertidal substrates often can be achieved only by digging around the box. In water, a modification is necessary to prevent the sample from falling out of the open bottom during recovery; this is accomplished by a curved baseplate, which pivots into position before the box is lifted from the sediment (Fig. 23.10A).

The can corer is perhaps the easiest device for a wader or diver to use when sampling burrowed sediment. With its base cut off and the screw-cap removed, a 1-gal oil can is forced into the seafloor; the cap is then screwed on and the corer withdrawn. [Where sediments are bound by root masses, these must first be cut through with a knife (Frey et al., 1973).] A plate secured to the base of the can, by rope or bands, prevents subsequent loss of the sample during transport. In the laboratory, the core sample is extruded by gentle air pressure from a compressor nozzle plugged into the can opening (Reineck and Rosenboom, 1969, p. 213, Fig. 4; Howard and Reineck, 1972, p. 83).

Recently a freeze-coring technique has been developed; this involves a diver pumping liquid nitrogen or solid carbon dioxide + acetone into a previously inserted corer, and then recovering the solid core (D. C. Rhoads, 1972, personal communication).

Ship-Operated Corers

Cylindrical corers have been used in many studies involving bioturbation. Moore and Scruton (1957), for instance, tested the lateral continuity of burrowed horizons in the pro-delta deposits of the Mississippi, using a double-barreled corer that had a constant separation of 80 cm. In many cases, however, an offshore version of the Sencken-

berg box has proved to be superior, recovering cores where piston and gravity corers had failed (Reineck, 1963). Rectangular box samples $20 \times 30 \times 45$ cm deep have been the mainstay of recent advances made in the study of burrowing in relation to storm sedimentation off the Elbe Estuary. By careful serial slicing, relief peeling, and resin impregnation, Reineck et al. (1967, 1968) documented with meticulous block diagrams the burrow forms diagnostic of a wide range of epicontinental environments. Their technique (Fig. 23.10) is a model for all workers. It has been adopted by Howard (1969), Howard and Reineck (1972), Smith and Howard (1972), and Howard and Frey (1973) along the Georgia coast, and an enlarged form is also being used at Scripps Institute for the investigation of bioturbation on the deep-sea floor (e.g., Piper and Marshall, 1969; Hanor and Marshall, 1971, p. 132–136).

SUMMARY AND SUGGESTED READING

The techniques outlined above are summarized in Table 23.1, which shows the relative merits of each and the sort of problems likely to be encountered. In the references below I have indicated with an asterisk eight papers that constitute a good guide to the way in which modern techniques are being applied to ichnology. For those who wish to go further and attempt some of the techniques themselves, the books by Bouma (1969) and Carver (1971) are the essential starting points. (See also Chapters 12, 22.)

ACKNOWLEDGMENTS

Most of this review concerns work that is not my own. Critical reading of the manuscript by H.–E. Reineck, G. Hertweck, D. C. Rhoads, J. D. Howard, and R. W. Frey has ensured a more thorough coverage than might have occurred otherwise. In attempting to strike a desirable balance among the illustrations, I enlisted the help of many colleagues, whose cooperation is much appreciated: J. D. Howard

TABLE 23.1. Techniques for Studying Fossil and Recent Traces.

Problem	Technique	Application	Advantages	Disadvantages
Biogenic structures poorly developed, or sediment appears massive	Acid etching	Enhancing relief	Field use	Calcareous sediments only
	Base etching	Enhancing relief	Field use	Siliceous sediments only
	Sandblasting	Artificial weathering	Quick	Sandstones only
	Staining	Heightening contrast	Inexpensive	Porous strata only; dense rocks unsatisfactory
	Serial sectioning	Grain, fabric, sorting, fecal pellets	3D reconstruction	Destroys specimen
	X-ray radiography	Unmasking diagenetic effects	Suitable for lithified and unlithified sediments	Basic equipment costly; thin slab must be cut (sample may need impregnating)
	Infrared photography	Variation in texture of organic-rich sediment	Straightforward, inexpensive	Poor on well-sorted sediments
Peeling bioturbated sediment	Lacquer	Large unwaterlogged areas	Inexpensive	Moist porous sand only; clay not replicated
	Polyester resin	Waterlogged surfaces	Unaffected by salt water; suitable for all grades of sediment	Lengthy curing time at low temperature
	Epoxy Relief	Waterlogged vertical faces	Good penetration gives 3D effect	Expensive for large areas
Burrow casting	Plaster of Paris	Simple open burrows above water	Inexpensive	Fragile; unsuitable for salt-water filled systems
	Silicone rubber	Simple open burrows above water	Flexible	Sagging of rubber after excavation can distort true burrow configuration
	Polyester resin	Underwater systems	Strong; heavier than seawater	Expensive for large complex systems

(Fig. 23.1); R. G. Bromley and N. Svendsen (Fig. 23.2); E. A. Shinn (Figs. 23.3, 23.7); H. R. Grunau (Fig. 23.4); J. R. L. Allen (Fig. 23.5); R. W. Frey (Fig. 23.6); and G. Hertweck (Fig. 23.10). I thank the Zoological Society of London and the Senckenbergische Naturforschende Gesellschaft for permission to reproduce Figure 23.8 and part of Figure 23.10, respectively.

My own work on resin casting was carried out during the 1968 Aldabra Expedition and financed by the Royal Society, for which I am duly thankful.

REFERENCES [2]

Ager, D. V. 1963. Principles of paleoecology. New York, McGraw-Hill, 371 p.

Allen, J. R. L. 1971. Massive beds in the central Pennine basin: a discussion. Yorkshire Geol. Soc., Proc., 38:293–294.

Barr, J. L. et al. 1970. Large epoxy peels. Jour. Sed. Petrol., 40:445–449.

Brown, W. E. and H. W. Patnode. 1953. Plastic lithification of sands *in situ*. Amer. Assoc. Petrol. Geol., Bull., 37:152–157.

Bouma, A. H. 1964. Sampling and treatment of unconsolidated sediments for study of internal structures. Jour. Sed. Petrol., 34:349–354.

*————. 1969. Methods for the study of sedimentary structures. New York, Wiley-Interscience, 458 p.

*Carver, R. E. 1971. Procedures in sedimentary petrology. New York, Wiley-Interscience, 653 p.

Chapman, C. J. and A. L. Rice. 1971. Some direct observations on the ecology and behaviour of the Norway lobster *Nephrops norvegicus*. Marine Biol., 10:321–329.

Chisholm, J. I. 1970a. Lower Carboniferous trace-fossils from the Geological Survey boreholes in west Fife (1965-6). Geol. Survey Great Britain, Bull. 31:19–35.

————. 1970b. *Teichichnus* and related trace-fossils in the Lower Carboniferous at St. Monance, Scotland. Geol. Survey Great Britain, Bull. 32:21–51.

Collinson, J. D. 1970. Deep channels, massive beds and turbidity current genesis in the central Pennine basin. Yorkshire Geol. Soc., Proc., 37:495–515.

Crichton, O. W. 1960. Marsh crab, intertidal tunnel-maker and grass-eater. Estuarine Bull., 5:3–10.

Crimes, T. P. 1970. Trilobite tracks and other trace fossils from the Upper Cambrian of north Wales. Geol. Jour., 7:47–68.

Farrow, G. E. 1966. Bathymetric zonation of Jurassic trace fossils from the coast of Yorkshire, England. Palaeogeogr., Palaeoclimatol., Palaeoecol., 2:103–151.

*————. 1971. Back-reef and lagoonal environments of Aldabra Atoll distinguished by their crustacean burrows. Zool. Soc. London, Symp., 28:455–500.

Frey, R. W. 1970. Trace fossils of Fort Hays Limestone Member of Niobrara Chalk (Upper Cretaceous), west-central Kansas. Univ. Kansas Paleont. Contr., Art. 53, 41 p.

———— and J. Cowles. 1969. New observations on *Tisoa*, a trace fossil from the Lincoln Creek Formation (mid-Tertiary) of Washington. The Compass, 47:10–22.

*———— and J. D. Howard. 1969. A profile of biogenic sedimentary structures in a Holocene barrier island–salt marsh complex, Georgia. Gulf Coast Assoc. Geol. Socs., Trans., 19:427–444.

———— et al. 1973. Techniques for sampling salt marsh benthos and burrows. Amer. Midland Natur., 89:228–234.

Fuss, C. M. and L. H. Ogren. 1965. A shallow water observation chamber. Limnol. Oceanogr., 10:290.

Goldring, R. 1962. The trace fossils of the Baggy Beds (Upper Devonian) of North Devon, England. Paläont. Zeitschr., 36:232–251.

———— and W. Crichton. 1966. Bioturbation structures in the Upper Chalk revealed by staining with Methylene Blue. Unpubl. Rept. (Demonstration Mtg.), Palaeont. Assoc.

Hähnel, W. 1962. The lacquer-film method of conserving geological objects. Curator, 5:353–368.

Hamblin, W. K. 1962a. X-ray radiography in the study of structures in homogeneous sediments. Jour. Sed. Petrol., 32:201–210.

————. 1962b. Staining and etching techniques for studying obscure structures in clastic rocks. Jour. Sed. Petrol., 32:530–533.

[2] For explanation of asterisks, see the text section "Summary and Suggested Reading."

————. 1965. Internal structures of "homogeneous" sandstones. Geol. Surv. Kansas, Bull. 175(1):1–37.

Hancock, A. 1858. Remarks on certain vermiform fossils found in the Mountain Limestone districts of the North of England. Ann. Mag. Nat. Hist., 3:443–457.

Hanor, J. S. and N. F. Marshall. 1971. Mixing of sediment by organisms. In B. F. Perkins (ed.), Trace fossils, a field guide. Louisiana State Univ., School Geosci., Misc. Publ. 71-1:127–135.

*Hertweck, G. and H.-E. Reineck. 1966. Untersuchungsmethoden von Gangbauten und anderen Wühlgefügen mariner Bodentiere. Natur u. Museum, 96:429–438.

Hester, N. C. and W. A. Pryor, 1972. Blade-shaped crustacean burrows of Eocene age: a composite form of *Ophiomorpha*. Geol. Soc. America, Bull., 83:677–688.

Howard, J. D. 1969. Radiographic examination of variations in barrier island facies; Sapelo Island, Georgia. Gulf Coast Assoc. Geol. Socs., Trans., 19:217–232.

———— and R. W. Frey. 1973. Characteristic physical and biogenic sedimentary structures in Georgia estuaries. Amer. Assoc. Petrol. Geol., Bull., 57:1169–1184.

———— and H.-E. Reineck. 1972. Georgia coastal region, Sapelo Island, U.S.A.: sedimentology and biology. IV. Physical and biogenic sedimentary structures of the nearshore shelf. Senckenbergiana Marit., 4:81–123.

Kummel, B. and D. Raup. 1965. Handbook of paleontological techniques. W. H. Freeman, 582 p.

McMullen, R. M. and J. R. L. Allen. 1964. Preservation of sedimentary structures in wet unconsolidated sands using polyester resins. Marine Geol., 1:88–97.

Moore, D. G. and P. C. Scruton. 1957. Minor internal structures of some recent unconsolidated sediments. Amer. Assoc. Petrol. Geol., Bull., 41:2723–2751.

Orme, G. R. and W. W. M. Brown. 1963. Diagenetic fabrics in the Avonian limestones of Derbyshire and north Wales. Yorkshire Geol. Soc., Proc., 34:51–66.

Pantin, H. M. 1960. Dye-staining technique for examination of sedimentary microstructures in cores. Jour. Sed. Petrol., 30:314–316.

Piper, D. J. W. and N. F. Marshall. 1969. Bioturbation of Holocene sediments on La Jolla deep sea fan, California. Jour. Sed. Petrol., 39:601–606.

Raup, D. and A. Seilacher. 1969. Fossil foraging behavior: computer simulation. Science, 166:994–995.

Reineck, H.-E. 1957. Stechkästen und Deckweisz, Hilfsmittel des Meeresgeologen. Natur u. Volk, 87:132–134.

————. 1963. Der Kastengreifer. Natur u. Museum, 93:102–108.

————. 1970. Reliefguss und projizierbarer Dickschliff. Senckenbergiana Marit., 2:61–66.

———— and W. Rosenboom. 1969. Stechkasten zur Entnahme von Watten– und Unterwasserproben. Natur u. Museum, 99:45–55.

*———— et al. 1967. Das schlickgebiet südlich Helgoland als Beispiel rezenter Schelfablagerungen. Senckenbergiana Leth., 48:219–275.

———— et al. 1968. Sedimentologie, Faunenzonierung und Faziesabfolge vor der Ostkuste der inneren Deutschen Bucht. Senckenbergiana Leth., 49:261–309.

Rhoads, D. C. 1970. Mass properties, stability, and ecology of marine muds related to burrowing activity. In T. P. Crimes and J. C. Harper (eds.), Trace fossils. Geol. Jour., Spec. Issue 3:391–406.

———— and D. J. Stanley. 1966. Transmitted infra-red radiation: a simple method for studying sedimentary structures. Jour. Sed. Petrol., 36:1144–1149.

*Rice, A. L. and C. J. Chapman. 1971. Observations on the burrows and burrowing behaviour of two mud-dwelling decapod crustaceans, *Nephrops norvegicus* and *Goneplax rhomboides*. Marine Biol., 10:330–342.

Scoffin, T. P. 1973. Crustacean faecal pellets, *Favreina*, from the Middle Jurassic of Eigg, Inner Hebrides. Scottish Jour. Geol., 9:145.

*Shinn, E. A. 1968. Burrowing in recent lime sediments of Florida and the Bahamas. Jour. Paleont., 42:878–894.

Simpson, S. 1957. On the trace-fossil *Chondrites*. Geol. Soc. London, Quart. Jour., 112:475–496.

Smith, K. L., Jr. and J. D. Howard. 1972. Comparison of a grab sampler and large volume corer. Limnol. Oceanogr., 17:142–145.

Stevens, B. A. 1928. Callianassidae from the west coast of North America. Puget Sound Mar. Biol. Station, Publ., 6:315–369.

van Straaten, L. M. J. U. 1954. Composition and structure of recent marine sediments in the Netherlands. Leidse Geol. Med., 19:1–110.

Voigt, E. 1936. Die lackfilmmethode, ihre Bedeutung und Anwendung in der Paläontologie, Sedimentpetrographie und Bodenkunde. Zeitschr. Deutsch. Geol. Gesell., 88:272–292.

West, I. M. 1965. A new method of displaying microstructures in porous limestone. Jour. Sed. Petrol., 35:250–251.

INDEX

Because this book is concerned mainly with "principles, problems, and procedures" in ichnology, an author index is not included; also excluded are lithologies, stratigraphic units, divisions of geologic time (with the single exception of the Precambrian), and geographic names. Terms pertaining to materials, conditions, and processes are indexed in a way that stresses the basic concepts involved and then gives pertinent examples of their application. Numerous cross-references help show interrelationships among the topics. A generic name followed by a "B" indicates that it is a biological rather than an ichnological taxon. Bold-faced page numbers refer to illustrations or tabulations.